Higher Spin Gauge Theories

Higher Spin Gauge Theories

Special Issue Editors

Nicolas Boulanger

Andrea Campoleoni

MDPI • Basel • Beijing • Wuhan • Barcelona • Belgrade

Special Issue Editors
Nicolas Boulanger
Groupe de Mécanique et Gravitation
Université de Mons–UMONS
Belgium

Andrea Campoleoni
Institut für Theoretische Physik
ETH Zürich
Switzerland

Editorial Office
MDPI
St. Alban-Anlage 66
Basel, Switzerland

This is a reprint of articles from the Special Issue published online in the open access journal *Universe* (ISSN 2218-1997) from 2017 to 2018 (available at: http://www.mdpi.com/journal/universe/special_issues/HS)

For citation purposes, cite each article independently as indicated on the article page online and as indicated below:

LastName, A.A.; LastName, B.B.; LastName, C.C. Article Title. *Journal Name* **Year**, *Article Number*, Page Range.

ISBN 978-3-03842-997-5 (Pbk)
ISBN 978-3-03842-998-2 (PDF)

Contents

About the Special Issue Editors

Nicolas Boulanger was born in Namur, Belgium, in 1977. He obtained his Ph.D. Degree in Physics from ULB Brussels (Belgium) in 2003. After post-doctoral stays at the DAMTP (Cambridge, U.K.), University of Mons (Mons, Belgium) and Scuola Normale Superiore (Pisa, Italy), he got a permanent F.R.S.-FNRS research position at the University of Mons (UMONS) in 2009. In October 2015, he took the direction of the group Mécanique et Gravitation inside the unit of Theoretical and Mathematical Physics of UMONS. The same year, he received the Théophile De Donder prize for Mathematical Physics from the Royal Academy of Belgium. In 2017, he was promoted Senior Research Associate of the F.R.S.-FNRS. The research activities of his group are focused on the quantization of gauge theories, duality, supergravity and higher-spin gauge theory, the main expertise of the group. He has been Visiting Professor at the Universities of Tours and Aix-Marseille, France.

Andrea Campoleoni was born in Varese, Italy, in 1981. He studied physics at the University of Pisa and at Scuola Normale Superiore di Pisa (Italy). He obtained his Ph.D. in Physics from Scuola Normale Superiore in 2009. He then held postdoctoral fellowships at the Max Planck Institute for Gravitational Physics of Potsdam (Germany) and at the Université libre de Bruxelles (Belgium). He is currently a postdoctoral fellow at ETH Zurich (Switzerland). In 2015 he received the Renato Musto Prize for Theoretical Physics, awarded under the patronage of the University Federico II of Naples, Italy. His research focuses on higher-spin gauge theories; in particular, on the study of mixed-symmetry fields and of three-dimensional models for higher-spin interactions.

Preface to "Higher Spin Gauge Theories"

Higher-spin gauge theory has been a fascinating field of research since the dawn of quantum field theory and remains a topic of active fundamental investigations. After the 1984 String Theory revolution, it also triggered important researches from string theorists. It was indeed argued that higher-spin gauge theory should describe a maximally symmetric phase of string theory, emerging in the tensionless limit of the string.

Many years of efforts aiming at building consistent interactions among massless fields of arbitrary spin, along the lines of the program formalized by Fronsdal in the late seventies, led to Vasiliev's equations. The latter are, to this day, the only known set of non-linear equations that describe, after linearisation around four dimensional (anti-)de Sitter spacetime, the free propagation of an infinite set of higher-spin gauge fields featuring the graviton. These last few years have seen a surge of activities in the study of these equations, their perturbative expansion, their generalizations as well as their links with string theory and conformal field theory via the AdS/CFT correspondence. The new results partly confirmed some expectations on the subject and they also showed that more work is required for a better understanding of the weak-field expansion of Vasiliev's equations around (A)dS_4. On the one hand, the holographic reconstruction from the free O(N) model in three dimensions seems to reproduce, as expected from the AdS/CFT conjecture, a non-Abelian gauge theory of higher spin fields in (A)dS_4, at least up to cubic order in the weak fields. On the other hand, the Vasiliev equations display some seemingly unavoidable non-localities in their perturbative regime around (A)dS_4, which opens a possible new window on classical field theory. In parallel, recent works performed in the light-cone gauge brought back to the scene some results obtained by Metsaev in the early nineties, pointing to the existence of a consistent and interacting higher-spin theory in flat background. Other recent exciting developments include, for instance, a quantitative connection between string theory and higher-spin gauge theory via topological open string theories of the Cattaneo-Felder type and the discovery of a remarkable web of dualities within CFT_3, leading to 3D bosonisation, triggered by the higher-spin/CFT duality.

The present collection of reprints is gathering together original contributions focusing on exciting and timely questions at the forefront of research in higher-spin gauge theories. The topics covered include conformal higher-spin theory, AdS/CFT duality in various dimensions, infinite-spin theories, matrix models and higher-spin extension of BMS symmetries, to cite but a few. In addition, four review papers are provided, that will prove very useful to both experts and newcomers in the active field of higher-spin gauge theories.

Nicolas Boulanger, Andrea Campoleoni

Special Issue Editors

universe

MDPI

Review

On Exact Solutions and Perturbative Schemes in Higher Spin Theory

Carlo Iazeolla [1], Ergin Sezgin [2,*] and Per Sundell [3]

1 NSR Physics Department, G. Marconi University, via Plinio 44, 00193 Rome, Italy; c.iazeolla@gmail.com
2 Mitchell Institute for Fundamental Physics and Astronomy, Texas A&M University, College Station, TX 77843, USA
3 Departamento de Ciencias Físicas, Universidad Andres Bello, Sazie 2212, Santiago, Chile; per.anders.sundell@gmail.com
* Correspondence: sezgin@tamu.edu

Received: 9 November 2017; Accepted: 12 December 2017; Published: 1 January 2018

Abstract: We review various methods for finding exact solutions of higher spin theory in four dimensions, and survey the known exact solutions of (non)minimal Vasiliev's equations. These include instanton-like and black hole-like solutions in (A)dS and Kleinian spacetimes. A perturbative construction of solutions with the symmetries of a domain wall is also described. Furthermore, we review two proposed perturbative schemes: one based on perturbative treatment of the twistor space field equations followed by inverting Fronsdal kinetic terms using standard Green's functions; and an alternative scheme based on solving the twistor space field equations exactly followed by introducing the spacetime dependence using perturbatively defined gauge functions. Motivated by the need to provide a higher spin invariant characterization of the exact solutions, aspects of a proposal for a geometric description of Vasiliev's equation involving an infinite dimensional generalization of anti de Sitter space are revisited and improved.

Keywords: higher-spin gravity; exact solutions; higher-spin geometry

---

1. Introduction

Higher spin (HS) theory in four dimensions, in its simplest form and when expanded about its (anti-)de Sitter vacuum solution, describes a self-interacting infinite tower of massless particles of spin $s = 0, 2, 4 \ldots$. The full field equations, proposed long ago by Vasiliev [1–3] (for reviews, see [4,5]), are a set of Cartan integrable curvature constraints on master zero-, one- and two-forms living on an extension of spacetime by a non-commutative eight-dimensional twistor space. The latter is fibered over a four-dimensional base, coordinatized by a Grassmann-even $SL(2,\mathbb{C})$-spinor oscillator $Z^A = (z^\alpha, \bar{z}^{\dot{\alpha}})$, and the fiber is coordinatized by another oscillator $Y^A = (y^\alpha, \bar{y}^{\dot{\alpha}})$; the master fields are horizontal forms on the resulting twelve-dimensional total space, valued an infinite-dimensional associative algebra generated by Y^A, that we shall denote by $\mathcal{A}$, and subject to boundary conditions on the base manifold.

A key feature of Vasiliev's equations is that they admit asymptotically (anti-)de Sitter solution spaces, obtained by taking the HS algebra $\mathcal{A}$ to be an extension of the Weyl algebra, with its Moyal star product, by involutory chiral delta functions [6,7], referred to as inner Klein operators, relying on a realization of the star product using auxiliary integration variables [4]. Introducing a related class of forms in Z-space, that facilitates a special vacuum two-form in twistor space, the resulting linearized master fields can be brought to a special gauge, referred to as the Vasiliev gauge, in which their symbols defined in a certain normal order are real analytic in twistor space, and the master zero- and one-forms admit Taylor expansions in Y at $Z = 0$ in terms of Fronsdal fields on the mass shell and subject to physical boundary conditions.

Although the Vasiliev equations take a compact and elegant form in the extended space, their analysis in spacetime proceeds in a weak field expansion which takes an increasingly complicated form beyond the leading order. Indeed, they have been determined so far only up to quadratic order. In performing the weak field expansion, a number of challenges emerge. Firstly, obtaining these equations requires boundary conditions in twistor space, referring to the topology of Z space and the classes of functions making up $\mathcal{A}$ [2,8,9]. The proper way of pinning down these aspects remains to be determined. Second, the cosmological constant, Λ, which is necessarily nonvanishing in Vasiliev's theory (as the transvection operators of the isometry algebra are realized in $\mathcal{A}$ as bilinears in Y), appears in the effective equations to its first power via critical mass terms, but also to arbitrary negative powers via non-local interactions [4,10]. Thus, letting ϕ denote a generic Fronsdal field, it follows that $\partial\phi \sim \sqrt{|\Lambda|}\phi$ on-shell, and hence interactions with any number of derivatives are of equal relevance (at a fixed order in weak field amplitudes). This raises the question of just how badly nonlocal are the HS field equations, the attendant problem of divergences arising even at the level of the amplitudes [11,12], and what kind of field redefinitions are admissible. One guide available is the holographic construction of the bulk vertices [13–15]. Clearly, it would be desirable to find the principles that govern the nonlocal interactions, based on the combined boundary conditions in twistor space as well as spacetime, such that an order by order construction of the bulk vertices can proceed from the analysis of Vasiliev equations. The simple and geometrical form of Vasiliev equations, in turn, may pave the way for the construction of an off-shell action that will facilitate the computation of the quantum effects.

In an alternative approach to the construction of HS equations in spacetime, it has been proposed to view Vasiliev's equations as describing stationary points of a topological field theory with a path integral measure based on a Frobenius-Chern-Simons bulk action in nine dimensions augmented by topological boundary terms, which are permitted by the Batalin-Vilkovisky formalism, of which only the latter contribute to the on-shell action [16,17].

This approach combines the virtues of the on-shell approach to amplitudes for massless particles flat spacetime with those of having a background independent action, in the sense that the on-shell action is fixed essentially by gauge symmetries and given on closed form, which together with the background independence of Vasiliev's equations provides a machinery for perturbative quantum computations around general backgrounds.

In this context, it is clearly desirable to explore in more detail how the choice of boundary conditions in the extended space influences the classical moduli space of Vasiliev's equations, with the purpose of spelling out the resulting spaces, computing HS invariant functionals on-shell, and examining how the strongly coupled spacetime nonlocalities are converted into physical amplitudes using the aforementioned auxiliary integral representation of star products in twistor space.

The aim of this article is to review three methods that have been used to find exact solutions of the Vasiliev equations, and to describe two schemes for analyzing perturbations around them. In particular we will describe the gauge function method [18,19] for finding exact solutions and summarize the first such solution found in [20], as well as its generalization to de Sitter spacetime studied in [21] together with the solutions of for a chiral version of the theory with Kleinian $(2,2)$ signature. As we shall see, this method uses the fact that the spacetime dependence of the master fields can be absorbed into gauge functions, upon which the problem of finding exact solutions is cast into a relatively manageable deformed oscillator problem in twistor space. The role of different ordering schemes for star product as well as gauge choices to fix local symmetries in twistor space will also be discussed.

Next, we will describe a refined gauge function method proposed in [22], where the twistor space equations are solved by employing separation of twistor variables and holomorphicity in the Z space in a Weyl ordering scheme and enlarging the Weyl algebra in the fiber Y space by inner Kleinian operators. This approach provides exact solution spaces in a particular gauge, that we refer to as the holomorphic gauge, after which the spacetime dependence is introduced by means of a sequence of large gauge transformations, by first switching on a vacuum gauge function, taking the solutions to what we refer to as the L-gauge, where the configurations must be real analytic in Z space, which provides an admissibility condition on the initial data in holomorphic gauge. The solutions can then be mapped further to the Vasiliev gauge, where the linearized, or asymptotic, master fields, are real analytic in the full twistor space and obey a particular gauge condition in Z space which ensures that they consist of decoupled Fronsdal fields in a canonical basis; the required gauge transformation, from the L gauge to the Vasiliev gauge, can be constructed in a perturbation scheme, which has so far been implemented mainly at the leading order. We will describe a black hole-like solution in some detail and mention other known solutions obtained by this method so far, including new solutions with six Killing symmetries [23].

We shall also outline a third method, in which the HS equations are directly tackled without employing gauge functions. In this method, solving the deformed oscillators in twistor space also employs the projector formalism, though the computation of the gauge potentials does not rely on the gauge function method. The black hole-like solution found in this way in [24] will be summarized.

We shall also review two approaches to the perturbative treatment of Vasiliev's equations. One of them, which we refer to as the normal ordered scheme, is based on a weak field expansion around (anti-)de Sitter spacetime [3,4,25]. It entails nested parametric integrals, introduced via a homotopy contraction of the de Rham differential in Z space used to solve the curvature constraints that have at least one form index in Z space, followed by inserting the resulting perturbatively defined master fields into the remaining curvature constraints with all form indices in spacetime. In an alternative scheme, the equations are instead solved exactly in the aforementioned L-gauge, and a perturbatively realized large HS gauge transformation is then performed to achieve interpretation in terms of Fronsdal fields in asymptotically (anti-)de Sitter spacetimes in Vasiliev gauge [7]. The advantages of the latter approach in describing the fluctuations around more general HS backgrounds will be explained.

A word of caution is in order concerning the usage of 'black hole' terminology in describing certain types of exact solutions to HS equations. This terminology is, in fact, misleading in some respects since the notion of a line interval associated with a metric field is not HS invariant. Indeed, the apparent singular behaviour at the origin may in principle be a gauge artifact. This point is discussed in more detail in Section 4.2. Moreover, given the nonlocal nature of the HS interactions, the formulation of causality, which is crucial in describing the horizon of a black hole, is a challenging problem without any proposal for a solution yet in sight; in fact, a more natural physical interpretation of the black hole-like solutions may turn out to be as smooth black-hole microstates [7,26]. Another aspect of the known black hole-like solution in HS theory is that they activate fields of all possible spins, and apparently it is not possible to switch of all spins except one even in the asymptotically AdS region.

So, what is meant by a black hole solution in HS theory? Firstly, the $SO(3) \times SO(2)$ symmetry of the solution (which is part of an infinite dimensional extended symmetry forming a subgroup of the HS symmetry group) is in common with the symmetry group arising in the asymptotically AdS BH solution of ordinary AdS gravity. Second, the solution contains a spin-two Weyl tensor field which takes the standard Petrov type D form, with a singularity at the origin; more generally, the spin-s Weyl tensors are of a generalized Petrov type D form, given essentially by the s-fold direct products of a spin-one curvature of the Petrov type D form. The BH terminology is thus used in the context of HS theory with the understanding that it is meant to convey these properties, albeit they do not constitute a rigorous definition of a black hole in HS theory.

The use of HS invariants for exact solutions to capture their physical characteristics has been considered and in some cases they have been computed successfully. These particular invariants alone do not, however, furnish an answer to the question of whether it makes any sense to think about event horizons in HS theory at all, and if so, how to define them; in fact, their existence rather supports the aforementioned microstate proposal, wherein the HS invariants can be interpreted as extensive charges defining HS ensembles.

Motivated by the quest for giving a physical interpretation of the exact solution in the context of underlying HS symmetries, a geometrical approach to HS equations was proposed in [27]. We shall summarize this proposal in which the HS geometry is based on an identification of an infinite dimensional structure group in a fibre bundle setting, and the related soldering phenomenon that leads to a HS covariant definition of classes of (non-unique) generalized vielbeins and related metrics, and as such an infinite dimensional generalization of AdS geometry. In doing so, we will improve the formulation of [27] by dispensing with the embedding of the relevant infinite dimensional coset space into a larger one that involves the extended HS algebra that includes the twistor space oscillators.

Finally, we are not aware of any exact solutions of HS theories in dimensions $D > 4$ [28,29], while in $D = 3$, assuming that the scalar field is coupled to HS fields, we can refer to [8,30,31] for the known solutions. Purely topological HS theory, which has no dynamical degrees of freedom, and which allows a more rigorous definition of black holes, is known to admit many exact solutions whose description goes beyond the scope of this review.

2. Vasiliev Equations

2.1. Bosonic Model in $(A)dS$

Vasiliev's theory is formulated in terms of horizontal forms on a non-commutative fibered space $\mathcal{C}$ with four-dimensional non-commutative symplectic fibers and eight-dimensional base manifold equipped with a non-commutative differential Poisson structure. On the total space, the differential form algebra $\Omega(\mathcal{C})$ is assumed to be equipped with an associative degree preserving product $\star$, a differential d, and an Hermitian conjugation operation $\dagger$, that are assumed to be mutually compatible. The base manifold is assumed to be the direct product of a commuting real four-manifold $\mathcal{X}_4$ with coordinates x^μ, and a non-commutative real four-manifold $\mathcal{Z}_4$ with coordinates Z^A; the fiber space and its coordinates are denoted by $\mathcal{Y}_4$ and Y^A, respectively. The non-commutative coordinates are assumed to obey

$$[Y^A, Y^B]_\star = 2iC^{AB}\,, \qquad [Z^A, Z^B]_\star = -2iC^{AB}\,, \qquad [Y^A, Z^B]_\star = 0\,, \tag{2.1}$$

where C_{AB} is a real constant antisymmetric matrix. The non-commutative space is furthermore assumed to have a compatible complex structure, such that

$$Y^A = (y^\alpha, \bar{y}^{\dot{\alpha}})\,, \qquad Z^A = (z^\alpha, \bar{z}^{\dot{\alpha}})\,, \tag{2.2}$$

$$(y^\alpha)^\dagger = \bar{y}^{\dot{\alpha}}\,, \qquad (z^\alpha)^\dagger = -\bar{z}^{\dot{\alpha}}\,, \tag{2.3}$$

where the complex doublets obey $[y^\alpha, y^\beta]_\star = 2i\epsilon^{\alpha\beta}$ and $[z^\alpha, z^\beta]_\star = -2i\epsilon^{\alpha\beta}$. The horizontal forms can be represented as sets of locally defined forms on $\mathcal{X}_4 \times \mathcal{Z}_4$ valued in oscillator algebras $\mathcal{A}(\mathcal{Y}_4)$ generated by the fiber coordinates glued together by transition functions, that we shall assume are defined locally on $\mathcal{X}_4$, resulting in a bundle over $\mathcal{X}_4$ with fibers given by $\Omega(\mathcal{Z}_4) \otimes \mathcal{A}(\mathcal{Y}_4)$. The algebra $\mathcal{A}(\mathcal{Y}_4)$ can be given in various bases; we shall use the Weyl ordered basis, and the normal ordered basis consisting of monomials in $a_\pm = Y \pm Z$ with a_+ and a_- oscillators standing to the left and right, respectively. We assume that the elements in $\Omega(\mathcal{Z}_4) \otimes \mathcal{A}(\mathcal{Y}_4)$ have well-defined symbols in both Weyl and normal order. The normal order reduces to Weyl order for elements that are independent of either Y or Z, and in the cases where depend on both Y and Z, they can be composed using the Fourier transformed twisted convolution formula in normal ordered scheme as

$$(f \star g)(Z;Y) = \frac{1}{(2\pi)^4} \int_{\mathbb{R}^4 \times \mathbb{R}^4} d^4U d^4V f(Z+U; Y+U)\, g(Z-V; Y+V)\, e^{iV^A U_A}\,. \tag{2.4}$$

The model is formulated in terms of a zero-form Φ, a one-form

$$A = dx^\mu W_\mu + dz^\alpha V_\alpha + d\bar{z}^{\dot{\alpha}} \bar{V}_{\dot{\alpha}}\,, \tag{2.5}$$

and a non-dynamical holomorphic two-form

$$J := -\frac{ib}{4} dz^{\alpha} \wedge dz_{\alpha}\, \kappa\,, \tag{2.6}$$

with Hermitian conjugate $\bar{J} = (J)^{\dagger}$, where b is a complex parameter and

$$\kappa := \kappa_y \star \kappa_z\,, \qquad \kappa_y := 2\pi\delta^2(y)\,, \qquad \kappa_z := 2\pi\delta^2(z)\,, \tag{2.7}$$

are inner Klein operators obeying $\kappa_y \star f \star \kappa_y = \pi_y(f)$ and $\kappa_z \star f \star \kappa_z = \pi_z(f)$ for any zero-form f, where π_y and π_z are the automorphisms of $\Omega(\mathcal{Z}_4) \otimes \mathcal{A}(\mathcal{Y}_4)$ defined in Weyl order by

$$\pi_y(x; z, \bar{z}; y, \bar{y}) = (x; z, \bar{z}; -y, \bar{y})\,, \qquad \pi_z(x; z, \bar{z}; y, \bar{y}) = (x; -z, \bar{z}; y, \bar{y})\,. \tag{2.8}$$

It follows that $dJ = 0,\ J \star f = \pi(f) \star J$ and $\pi(J) = J$, *idem* $\bar{J}$, with

$$\pi := \pi_y \circ \pi_z\,, \qquad \bar{\pi} := \pi_{\bar{y}} \circ \pi_{\bar{z}}\,. \tag{2.9}$$

It is useful to note that the inner Kleinian takes the following forms in different ordering schemes:

$$\kappa = \begin{cases} e^{iy^{\alpha} z_{\alpha}} & \text{in normal ordering scheme} \\ (2\pi)^2 \delta^2(y) \delta^2(z) & \text{in Weyl ordering scheme} \end{cases} \tag{2.10}$$

The nonminimal and minimal models with all integer spins and only even spins, respectively, are obtained by imposing the conditions

$$\text{Non-minimal model } (s = 0, 1, 2, 3, ...)\ :\ \pi \circ \bar{\pi}(A) = A\,, \qquad \pi \circ \bar{\pi}(B) = B\,, \tag{2.11}$$

$$\text{Minimal model } (s = 0, 2, 4, ...)\ :\ \tau(A) = -A\,, \qquad \tau(B) = \bar{\pi}(B)\,, \tag{2.12}$$

where τ is the anti-automorphism

$$\tau(x^{\mu}; Y^A, Z^A) = f(x^{\mu}; iY^A, -iZ^A)\,, \qquad \tau(f \star g) = \tau(g) \star \tau(f)\,, \tag{2.13}$$

It follows that $\tau(J, \bar{J}) = (-J, -\bar{J})$. Models in Lorentzian spacetimes with cosmological constants Λ are obtained by imposing reality conditions as follows [21]:

$$\rho(B^{\dagger}) = \pi(B)\,, \qquad \rho(A^{\dagger}) = -A\,, \qquad \rho := \begin{cases} \pi\,, & \Lambda > 0 \\ \text{Id}\,, & \Lambda < 0 \end{cases} \tag{2.14}$$

Basic building blocks for Vasiliev equations are the curvature and twisted-adjoint covariant derivative defined by

$$F := dA + A \star A\,, \qquad DB := dB + [A, B]_{\pi}\,, \tag{2.15}$$

respectively, where the π-twisted star commutators is defined as

$$[f, g]_{\pi} := f \star g - (-1)^{|f||g|} g \star \pi(f)\,, \tag{2.16}$$

and

$$d := d_x + d_Z\,, \qquad d_x = dx^{\mu}\partial_{\mu}\,, \qquad d_Z = dz^{\alpha}\partial_{\alpha} + d\bar{z}^{\dot{\alpha}}\bar{\partial}_{\dot{\alpha}}\,. \tag{2.17}$$

Vasiliev equations of motion are given by

$$F + B \star (J - \bar{J}) = 0\,, \qquad DB = 0\,, \tag{2.18}$$

which are compatible with the kinematic conditions and the Bianchi identities, implying that the classical solution space is invariant under the following infinitesimal gauge transformations:

$$\delta A = D\epsilon := d\epsilon + [A,\epsilon]_\star\,, \qquad \delta B = -[\epsilon, B]_\pi\,, \tag{2.19}$$

for parameters obeying the same kinematic conditions as the connection.

It remains a challenging problem to determine if these equations can be derived from a suitable tensionless, or critical tension, limit followed by a consistent truncation of string field theory on a background involving AdS_4. It will be very interesting to also determine if these equations follow from a consistent quantization of a topological string field theory. For a more detailed discussion and progress in this direction, see [32,33]. The component fields of V_A do not transform properly under the Lorentz transformations generated by $(\frac{1}{4i}(y^\alpha y^\beta - z^\alpha z^\beta) - h.c)$. To remedy this problem and achieve manifest Lorentz covariance, one introduces the field-dependent Lorentz generators [4,25]

$$M_{\alpha\beta} = M^{(0)}_{\alpha\beta} + S_{(\alpha} \star S_{\beta)}\,, \qquad M^{(0)}_{\alpha\beta} := y_{(\alpha} \star y_{\beta)} - z_{(\alpha} \star z_{\beta)}\,, \tag{2.20}$$

and their complex conjugates, where

$$S_\alpha = z_\alpha - 2iV_\alpha\,, \qquad \bar{S}_{\dot\alpha} = \bar{z}_{\dot\alpha} - 2i\bar{V}_{\dot\alpha}\,. \tag{2.21}$$

Next one defines

$$W'_\mu = W_\mu - \frac{1}{4i}\left(\omega_\mu^{\alpha\beta} M_{\alpha\beta} + \bar\omega_\mu^{\dot\alpha\dot\beta}\bar{M}_{\dot\alpha\dot\beta}\right)\,, \tag{2.22}$$

where $(\omega_\mu^{\alpha\beta}, \omega_\mu^{\dot\alpha\dot\beta})$ is the canonical Lorentz connection.

It is defined up to tensorial shifts [27] that can be fixed by requiring that the projection of W' onto $M^{(0)}_{\alpha\beta}$ and its complex conjugate, vanish at $Z = 0$, that is

$$\frac{\partial^2}{\partial y^\alpha \partial y^\beta} W'|_{Y=Z=0} = 0\,, \qquad \frac{\partial^2}{\partial \bar{y}^{\dot\alpha}\partial \bar{y}^{\dot\beta}} W'|_{Y=Z=0} = 0\,. \tag{2.23}$$

The above redefinitions ensure that under the Lorentz transformations with parameters

$$\epsilon_L = \frac{1}{4i}\Lambda^{\alpha\beta} M_{\alpha\beta}\,, \qquad \epsilon_L^{(0)} = \frac{1}{4i}\Lambda^{\alpha\beta} M^{(0)}_{\alpha\beta}\,, \tag{2.24}$$

the master fields transform properly under the Lorentz transformations as [25]

$$\delta_L B = [\epsilon_L^{(0)}, B]_\star\,, \tag{2.25}$$

$$\delta_L S_\alpha = [\epsilon_L^{(0)}, S_\alpha]_\star + \Lambda_\alpha{}^\beta S_\beta\,, \qquad \text{idem} \quad \bar{S}_{\dot\alpha}\,, \tag{2.26}$$

$$\delta_L W'_\mu = [\epsilon_L^{(0)}, W'_\mu]_\star + \frac{1}{4i}\left(\partial_\mu \Lambda^{\alpha\beta} M_{\alpha\beta} + h.c.\right)\,. \tag{2.27}$$

Using (2.21), the component form of Vasiliev equations reads

$$d_x W + W \star W = 0\,, \tag{2.28}$$

$$d_x B - [W, B]_\pi = 0\,, \tag{2.29}$$

$$d_x S_\alpha + [W, S_\alpha]_\star = 0\,, \qquad d_x \bar{S}_{\dot\alpha} + [W, \bar{S}_{\dot\alpha}]_\star = 0\,, \tag{2.30}$$

$$[S_\alpha, B]_\pi = 0\,, \qquad [\bar{S}_{\dot\alpha}, B]_\pi = 0\,, \quad [S_\alpha, \bar{S}_{\dot\beta}]_\star = 0\,, \tag{2.31}$$

$$[S^\alpha, S_\alpha]_\star = 4i(1 - bB \star \kappa)\,, \quad [\bar{S}^{\dot\alpha}, \bar{S}_{\dot\alpha}]_\star = 4i(1 - \bar{b}B \star \bar\kappa)\,. \tag{2.32}$$

This is the form of the equations typically used to seek exact solutions as it displays the role of the deformed oscillator algebra in the last two equations. Here one may exploit the technology developed in the study of noncommutative field theories and the construction of projection operators in a suitably defined oscillator space.

2.2. The Nonminimal Chiral Model in Kleinian Space

In this model the spinor oscillators are now representations of $SL(2,\mathbb{R})_L \times SL(2,\mathbb{R})_R$, and as such their hermitian conjugates are now given by

$$(y^\alpha)^\dagger = y^\alpha\,, \quad (z^\alpha)^\dagger = -z^\alpha\,, \quad (\bar{y}^{\dot\alpha})^\dagger = \bar{y}^{\dot\alpha}\,, \quad (\bar{z}^{\dot\alpha})^\dagger = -\bar{z}^{\dot\alpha}\,. \tag{2.33}$$

The field equations are now given by

$$F = J \star B\,, \qquad DB = 0\,, \qquad J := -\frac{i}{4} dz^\alpha \wedge dz_\alpha\, \kappa\,, \tag{2.34}$$

with reality conditions $A^\dagger = -A$ and $B^\dagger = \pi(B)$, and kinematical conditions

$$\text{Minimal model } (s = 0,2,4,...) \;:\; \tau(A) = -A\,, \qquad \tau(B) \;=\; \bar\pi(B)\,, \tag{2.35}$$

$$\text{Non-minimal model } (s = 0,1,2,3,...) \;:\; \pi \circ \bar\pi(A) = A\,, \qquad \pi \circ \bar\pi(B) = B\,. \tag{2.36}$$

The field equations in components take the same form as in (2.28)–(2.32), but now with

$$b = 1\,, \qquad \bar b = 0\,. \tag{2.37}$$

These models are referred to as chiral in view of the half-flatness condition on the twistor space curvature, namely $\bar F_{\dot\alpha\dot\beta} = 0$. These models admit the coset space $H_{3,2} = SO(3,2)/SO(2,2)$ as a vacuum solution, which has the Kleinian signature $(2,2)$. For a detailed description of these spaces, including the curved Kleinian geometries, see [34]. Our motivation for highlighting this case is due to the fact that the first exact solution of the Vasiliev equation in which all HS fields are nonvanishing was found for this model [21], and that the Kleinian geometry is relevant to $N = 2$ superstring as well as to integrable models.

3. Gauge Function Method and Solutions

3.1. The Method

In order to construct solutions to Vasiliev's equations, one may consider the approach [19] in which they are homotopy contracted in simply connected spacetime regions U to deformed oscillator algebras in twistor space at a base point $p \in U$; the constraints

$$F_{\mu\nu} \;=\; 0\,, \qquad F_{\mu\alpha} \;=\; 0\,, \qquad D_\mu B = 0\,, \tag{3.1}$$

are thus integrated in U using a gauge function $g = g(Z,Y|x)$ obeying

$$g|_p = 1\,, \tag{3.2}$$

and initial data

$$B' = B|_p\,, \qquad S'_A \;= S_A|_p\,, \tag{3.3}$$

subject to

$$[S'_\alpha, B']_\pi = 0\,, \qquad [\bar{S}'_{\dot\alpha}, B']_\pi = 0\,, \qquad [S'_\alpha, \bar{S}'_{\dot\beta}]_\star = 0\,, \tag{3.4a}$$

$$[S'^\alpha, S'_\alpha]_\star = 4i(1 - bB' \star \kappa)\,, \quad [\bar{S}'^{\dot\alpha}, \bar{S}'_{\dot\alpha}]_\star = 4i(1 - \bar{b}B' \star \bar\kappa)\,. \tag{3.4b}$$

The fields in U can then be expressed explicitly as

$$W_\mu = g^{-1} \star \partial_\mu g\,, \qquad S_A = g^{-1} \star S'_A \star g\,, \qquad B = g^{-1} \star B' \star \pi(g)\,, \tag{3.5}$$

after which the Lorentz covariant HS gauge fields can be obtained from (2.22) subject to (2.23), which serves to determine the spin connection $\omega_{\mu ab}$. Thus, the deviations in the spacetime HS gauge fields away from the topological vacuum solution, that is the solution with $W_\mu = 0$, thus come from the gauge function g as well as the non-linear shift on the account of achieving manifest Lorentz covariance. The deformed oscillator algebra requires a choice of topology for $\mathcal{Z}_4$, initial data for B' and a flat background connection. In what follows, we shall assume that $\mathcal{Z}_4$ has the topology of $\mathbb{R}^4$ with suitable fall-off conditions at infinity [7,17], and impose

$$C'(Y) \;= B'|_{Z=0}\,, \qquad S'_A|_{C'=0} = Z_A; \tag{3.6}$$

for nontrivial flat connections on $\mathcal{Z}_4$, that are not pure gauge, see [21]. The gauge function represents a gauge transformation that is large in the sense that it affects the asymptotics of gauge fields so as to introduce additional physical degrees of freedom to the system, over and above those contained in the twistor space initial data and flat connection; strictly speaking, in order to define such transformations, one should first introduce a set of classical observables forming a BRST cohomology modulo a set of boundary conditions on ghosts, after which a large gauge transformation is a gauge transformation that does not preserve all the classical observables. In particular, in order to describe asymptotically maximally symmetric, or Weyl flat, solutions, one may take

$$g|_{B'=0} = L\,, \tag{3.7}$$

where $L = L(Y|x)$ is a metric vacuum gauge function, to be described below. In order to obtain exact solutions, we shall choose $g = L$ for all C', that we refer to as the L-gauge. However, in order to extract Fronsdal fields in the asymptotic region, one has to impose a gauge condition in twistor space to the leading order in the weak field expansion in the asymptotic region, which introduces a dressing of the vacuum gauge function by an additional perturbatively determined gauge function; see Section 6.

3.2. Vacuum Solutions

In order to obtain solutions containing locally maximally symmetric asymptotic regions, one may take the gauge function $L(Y|x)$ to be corresponding coset representatives. In what follows, we shall focus on the spaces $AdS_4 = SO(3,2)/SO(3,1)$, $dS_4 = SO(4,1)/SO(3,1)$ and $H_{3,2} := SO(3,2)/SO(2,2)$, which can be realized as the embeddings

$$X^A X^B \eta_{AB} \equiv -(X^0)^2 + (X^1)^2 + (X^2)^2 + (X^3)^2 + \epsilon(X^5)^2 = -\lambda^{-2}\,, \tag{3.8}$$

where

$$\left(\epsilon, \frac{\lambda^2}{|\lambda^2|}\right) = \begin{cases} (-,+) & \text{for } AdS_4 \\ (+,-) & \text{for } dS_4 \\ (-,-) & \text{for } H_{3,2} \end{cases} \tag{3.9}$$

These spaces can be conveniently described in a unified fashion using the stereographic coordinates $x^a_\pm\,(a=0,1,2,3)$ obtained by means of the parametrization

$$X^M|_{U_\pm} \approx \left(\frac{2x^a_\pm}{1-\lambda^2 x^2_\pm}, \pm\ell\frac{1+\lambda^2 x^2_\pm}{1-\lambda^2 x^2_\pm}\right), \qquad -1 \le \lambda^2 x^2_\pm < 1\,, \tag{3.10}$$

$$x^2_\pm := x^a_\pm x^b_\pm \eta_{ab}\,, \qquad \eta_{ab} = \mathrm{diag}(\epsilon, \lambda^2/|\lambda^2|, +, +)\,, \qquad \ell = |\lambda|^{-1}\,, \tag{3.11}$$

where $U_\pm$ denotes the two stereographic coordinates charts, each covering one half of the space (3.9); on the overlap one has $\lambda^2 x^2_\pm = -1$, and the coordinate transition function

$$x^a_\pm = R^a(x_\mp)\,, \qquad \lambda^2 x^2_\pm < 0\,, \tag{3.12}$$

where the reflection map

$$R^a(v) := -\frac{v^a}{\lambda^2 v^2}\,. \tag{3.13}$$

The boundary is given by $\lambda^2 x^2_\pm = 1$, which has the topology of $S^2 \times S^1$ in the case of AdS_4 and $H_{3,2}$, and $S^3 \cup S^3$ in the case of dS_4. Instead of covering the vacuum manifold with two charts, one may extend either one of the charts to $\mathbb{R}^4 \setminus \{x^a : \lambda^2 x^2 = 1\}$, which provides a global cover using a single chart, with the understanding that $\{x^a : \lambda^2 x^2 = 1^-\} \cup \{x^a : \lambda^2 x^2 = 1^+\}$ provides a two-sheeted cover of the boundary. The induced line element $ds^2_0 = dX^A dX^B \eta_{AB}|_{\lambda^2 X^2 = -1}$ is given by

$$ds^2_0 = \frac{4dx^2}{(1-\lambda^2 x^2)^2}\,. \tag{3.14}$$

On $|\lambda^2|x^2 < 1$, the corresponding vacuum gauge function

$$L = \frac{2h}{1+h}\exp(-iy^\alpha a_\alpha{}^{\dot\alpha}\bar y_{\dot\alpha})\,, \tag{3.15}$$

where

$$a_{\alpha\dot\alpha} = \frac{\lambda x_{\alpha\dot\alpha}}{1+h}\,, \qquad x_{\alpha\dot\alpha} = (\sigma^a)_{\alpha\dot\alpha} x_a\,, \qquad h = \sqrt{1-\lambda^2 x^2}\,. \tag{3.16}$$

$$W_0 \equiv e_0 + \omega_0 = L^{-1} \star dL = \frac{1}{4i}\left[\omega_0{}^{\alpha\beta} y_\alpha y_\beta + \bar\omega_0{}^{\dot\alpha\dot\beta}\bar y_{\dot\alpha}\bar y_{\dot\beta} + 2e_0{}^{\alpha\dot\alpha} y_\alpha \bar y_{\dot\alpha}\right]\,, \tag{3.17}$$

where

$$e_0{}^{\alpha\dot\alpha} = -\frac{\lambda(\sigma^a)^{\alpha\dot\alpha}dx_a}{h^2}\,, \qquad \omega_0{}^{\alpha\beta} = -\frac{\lambda^2(\sigma^{ab})^{\alpha\beta}dx_a x_b}{h^2}\,. \tag{3.18}$$

A global description can be obtained using two gauge functions $L_\pm = L(Y|x_\pm)$ defined on $U_\pm$; the Z_2-symmetry implies that if $\Phi_\pm|_{p_\pm} = C'$, where $p_\pm := x_\pm^{-1}(0)$, then the two locally defined solutions on $U_\pm$ can be glued together using the gauge transition function $T^+_- := L_+^{-1} \star L_- = 1$ defined on the overlap region where $\lambda^2 x^2_\pm = -1$.

For later purposes, it is convenient to introduce alternative coordinate systems which are defined by the embeddings (with $|\lambda|^2 = 1$)

$$\begin{aligned} AdS_4:\quad & x^2>0:\quad X^0 = \sinh\tau\sinh\psi\,,\ X^i = n^i\cosh\tau\sinh\psi\,,\ X^5 = \cosh\psi\,,\\ & x^2<0:\quad X^0 = \sin\tau\cosh\psi\,,\ X^i = n^i\sin\tau\sinh\psi\,,\ X^5 = \cos\tau\,, \end{aligned} \tag{3.19}$$

$$\begin{aligned} dS_4:\quad & x^2>0:\quad X^0 = \sinh\tau\sin\psi\,,\ X^i = n^i\cosh\tau\sin\psi\,,\ X^5 = \cos\psi\,,\\ & x^2<0:\quad X^0 = \sinh\tau\cosh\psi\,,\ X^i = n^i\sinh\tau\sinh\psi\,,\ X^5 = \cosh\tau\,, \end{aligned} \tag{3.20}$$

The metrics for (A)dS in these coordinate systems are given in (3.41)–(3.44). In the case of $H_{3,2}$, we will find the following coordinate system to be useful (with $|\lambda^2| = 1$)

$$X^0 = r \sin t \,, \qquad X^i = n^i \sqrt{1+r^2} \,, \qquad X_5 = r \cos t \,, \tag{3.21}$$

where $n^i n^j \delta_{ij} = 1$, $0 \le r < \infty$ and $0 \le t \le 2\pi$.

3.3. *Instanton Solutions of Minimal Model in (anti) de Sitter Space*

Having obtained a vacuum gauge function, the next task is to solve the deformed oscillator problem (3.4) subject to the initial data in twistor space. To this end, it is helpful to constrain the primed configurations further by assuming that they preserve a nontrivial amount of HS symmetries. This can be achieved by imposing symmetry conditions by seeking a subspace $\mathfrak{k}'$ of HS gauge parameters ϵ' that are unbroken, i.e.,

$$\delta_{\epsilon'} B' = 0 \,, \qquad \delta_{\epsilon'} S'_\alpha = 0 \,, \tag{3.22}$$

for all $\epsilon' \in \mathfrak{k}'$; upon switching on the gauge function g, the resulting full solution is invariant under gauge parameters in the space $\mathfrak{k} = g^{-1} \star \mathfrak{k}' \star g$. For example, one may require that an n-dimensional subalgebra $\mathfrak{g}_n$ of the maximal finite dimensional subalgebra $\mathfrak{g}_{10}$ of the HS algebra remains unbroken, which implies that $\mathfrak{k}'$ is given by the intersection of $\mathrm{Env}(\mathfrak{g}_n)$ and the HS algebra. In particular, taking $n = 10$ yields the vacuum solution

$$W' = W_0 \,, \qquad S_A = z_A \,, \qquad B = 0 \,, \tag{3.23}$$

which preserves the HS algebra itself. Taking $n < 10$, the first distinct cases with nontrivial Weyl zero-form arise for $n = 6$; the space $\mathfrak{k}'$ is then given exactly, as we shall describe below for a particular realization of $\mathfrak{g}_6$, or perturbatively. In the latter case the $\mathfrak{g}_6$ will be realized as a subalgebra of the HS algebra in the leading order.

In [20], asymptotically anti-de Sitter solutions with $\mathfrak{g}_6 = \mathfrak{o}(1,3)$ were constructed by taking $\mathfrak{k}$ to be generated by the full Lorentz generators $M_{\alpha\beta}$ from (2.20). Thus, in the primed basis, the corresponding symmetry conditions read

$$[M'_{\alpha\beta}, B']_\pi = 0 \,, \qquad [M'_{\alpha\beta}, S'_\gamma]_\star = 0 \,, \qquad [M'_{\alpha\beta}, \bar{S}'_{\dot\gamma}]_\star = 0 \,, \tag{3.24}$$

and complex conjugates, where $M'_{\alpha\beta} = M^{(0)}_{\alpha\beta} + S'_{(\alpha} \star S'_{\beta)}$, which are given by $y_\alpha y_\beta$ plus perturbative corrections, and whose star product commutators close modulo Lorentz transformations acting on the component fields; thus, consistency of the invariance conditions implies that all canonical Lorentz tensors that are not singlets must vanish. Alternatively, it is possible to use other embeddings of $\mathfrak{o}(1,3)$ into the algebra of primed HS gauge transformations; for example, one can simply take $y_\alpha y_\beta$, as we shall comment on in Section 5, though it remains an open problem whether the resulting solutions are gauge equivalent to those that will be presented below. Taking $\mathfrak{g}_n$ to be generated by unperturbed functions of Y is useful, however, in considering unbroken symmetry algebra involving transvection operators, as we shall spell out in further detail in Section 5 in the case of domain walls and related time dependent solutions.

Turning to (3.24), the simplest possible ansatz for B' is a constant, viz.,

$$B' = \nu \,, \tag{3.25}$$

which leads to a deformed oscillator problem with exact solutions closely related to those of the 3D HS theory constructed by Prokushkin and Vasiliev [8]. Adapting to the 4D Type A model, for which $b = 1$, the following solution for the twistor space connection was found in [20]

$$S'_\alpha = z_\alpha + z_\alpha \int_{-1}^{1} dt\, q(t) \exp\left(\frac{i}{2}(1+t)\, u\right), \qquad u := y^\alpha z_\alpha\,, \tag{3.26}$$

$$q(t) = -\frac{\nu}{4}\left({}_1F_1\left[\frac{1}{2};2;\frac{\nu}{2}\log\frac{1}{t^2}\right] + t\, {}_1F_1\left[\frac{1}{2};2;-\frac{\nu}{2}\log\frac{1}{t^2}\right]\right). \tag{3.27}$$

Expanding $\exp(itu/2)$ results in integrals of the degenerate hypergeometric functions times positive algebraic powers of t, which improve the convergence at $t = 0$. Thus V_α is a power series expansion in u with coefficients that are functions of ν that are well-behaved provided this is the case for the coefficient of u^0. This is the case for ν in some finite region around $\nu = 0$, as discussed in detail in [20,35]. Indeed, as we shall see below after carrying out the integration over t, ν must lie in the interval $-3 \leq \nu \leq 1$.

The solution in spacetime is obtained in the two regions $\lambda^2 x^2 < 1$ and $\lambda^2 \widetilde{x}^2 < 1$ using the stereographic gauge functions $L \equiv L(x|Y)$ and $\widetilde{L} \equiv L(\tilde{x}|Y)$ where x and $\tilde{x}$ are related by the reflection map in the overlap region where $\lambda^2 x^2 < 0$ and $\lambda^2 \widetilde{x}^2 < 0$. From (3.16), one finds [20]

$$B = \nu(1 - \lambda^2 x^2) \exp\left[-i\lambda x^{\alpha\dot{\alpha}} y_\alpha \bar{y}_{\dot{\alpha}}\right]. \tag{3.28}$$

This shows that the physical scalar field is given in the x^a-coordinate chart by

$$\phi(x) = B|_{Y=Z=0} = \nu(1 - \lambda^2 x^2)\,, \qquad \lambda^2 x^2 < 1\,, \tag{3.29}$$

while the Weyl tensors for spin $s = 2, 4, \ldots$ vanish. Using instead $\widetilde{L}$, the physical scalar field in the $\widetilde{x}^a$-coordinate chart is given by

$$\widetilde{\phi} = \nu(1 - \lambda^2 \widetilde{x}^2)\,, \qquad \lambda^2 \widetilde{x}^2 < 1\,. \tag{3.30}$$

As a result, the two scalar fields are related by a duality transformation in the overlap region

$$\widetilde{\phi}(\tilde{x}) = \frac{\nu\phi(x)}{\phi(x) - \nu}\,, \qquad \lambda^2 x^2 = (\lambda^2 \widetilde{x}^2)^{-1} < 0\,. \tag{3.31}$$

Thus, if the transition takes place at $\lambda^2 x^2 = \lambda^2 \widetilde{x}^2 = -1$, then the amplitude of the physical scalar never exceeds 2ν.

The master fields S_A and W'_μ are obtained from (3.5) and (2.22). The generating functional for the spacetime gauge fields is given by [20] [1]

$$W'|_{Z=0} = W_0 - \frac{1}{4i} Q\omega_\mu{}^{\alpha\beta}\left[(1+a^2)^2 y_\alpha y_\beta + 4(1+a^2) a_\alpha{}^{\dot{\alpha}} y_\beta \bar{y}_{\dot{\alpha}} + 4 a_\alpha{}^{\dot{\alpha}} a_\beta{}^{\dot{\beta}} \bar{y}_{\dot{\alpha}} \bar{y}_{\dot{\beta}}\right] - \text{h.c.}\,, \tag{3.32}$$

---

1 An Euclidean version of this solution has been obtained in [21], and as the spin connection plays an eminent role in this solution and assuming that the action, as proposed in [17], is finite on this solution, we use the terminology of "instanton solution".

subject to (2.23), which serve to determine the spin connection, and where $a_{\alpha\dot\alpha}$ is defined in (3.16), and

$$a^2 \;:=\; a^{\alpha\dot\alpha}a_{\alpha\dot\alpha} = \frac{1-\sqrt{1-\lambda^2x^2}}{1+\sqrt{1-\lambda^2x^2}}\,, \qquad -1\le a^2\le 1\,, \tag{3.33}$$

$$Q \;=\; -\frac{1}{4}(1-a^2)^2\int_{-1}^{1}dt\int_{-1}^{1}dt'\,\frac{q(t)q(t')(1+t)(1+t')}{(1-tt'a^2)^4}\,. \tag{3.34}$$

This gives [20]

$$Q = -\frac{(1-a^2)^2}{4}\sum_{p=0}^{\infty}\left[\binom{2p+3}{2p}\left(\sqrt{1-\tfrac{\nu}{2p+1}}-\sqrt{1+\tfrac{\nu}{2p+3}}\right)^2 a^{4p} -\binom{2p+4}{2p+1}\left(\sqrt{1-\tfrac{\nu}{2p+3}}-\sqrt{1+\tfrac{\nu}{2p+3}}\right)^2 a^{4p+2}\right], \tag{3.35}$$

exhibiting branch cuts for $\mathrm{Re}\,\nu\le -3$ and $\mathrm{Re}\,\nu\ge 1$[2]. For $\nu \ll 1$, and in the interval $-1\le a^2\le 1$, this function can be approximated by [20]

$$Q\simeq \frac{\nu^2(1-a^2)^2}{48a^4}\left[1-\frac{2a^2}{(1-a^2)^2}+\frac{(1-a^2)^2}{2a^2}\log\frac{1-a^2}{1+a^2}\right]. \tag{3.36}$$

Using (2.23), one determines $\omega^{\alpha\beta}$ and $e^{\alpha\dot\alpha}$ from (3.32), while the HS Fronsdal potentials $\phi_{\mu a_1\dots a_s}$ vanish for $s>2$:

$$\phi_\mu{}^{a_1\dots a_{s-1}} = \frac{\partial^{2s-2}}{\partial y^{\alpha_1}\cdots\partial y^{\alpha_{s-1}}\bar\partial\bar y^{\dot\alpha_1}\cdots\bar\partial\bar y^{\dot\alpha_{s-1}}}W'|_{Z=Y=0}=0\,, \qquad s>2\,. \tag{3.37}$$

Even though the HS fields vanish, it is to be noted that the solution of the metric and scalar field constitute a solution of a highly nonlinear system of equations in which all higher derivatives play a role. One can reverse engineer a two derivative action describing the coupling of gravity to scalar field that admits the same solution [20] but such an action is clearly of limited use in the context of HS theory.

An advantage of presenting the solutions in stereographic coordinates is that it facilitates their unified description for (A)dS. In these coordinates the solution for the scalar field is given by (3.29) and the metric by

$$ds^2 \;=\; \frac{4\Omega^2(d(g_1x))^2}{(1-\lambda^2g_1^2x^2)^2}\,, \tag{3.38}$$

$$\Omega \;=\; \frac{(1-\lambda^2g_1^2x^2)f_1}{2g_1}\,, \qquad g_1=\exp\left(\frac{1}{2}\int_1^{x^2}\frac{f_2(t)\,dt}{f_1(t)}\right), \tag{3.39}$$

where

$$f_1(x^2)=\frac{2f}{h^2}\left[1+(1-a^2)^2Q\right],\quad f_2(x^2)=\frac{16Qf}{h^2(1+h)^2}\,,$$
$$f(x^2)=\left[1+(1+6a^2+a^4)Q\right]^{-1}, \tag{3.40}$$

2 The unitary representations of Wigner's deformed oscillator algebra can obtained starting from the standard Fock space and factoring out ideals that depend on integer part of $(1+\nu)/2$, that is, the ideal jumps for odd values of ν [36–39]. It would be interesting to examine to what extent it is possible to extend the solution to general ν properly taking into account the branch points in Q at odd ν.

and it is understood that the integration variable in (3.39) is $t = x^2$.

It is also convenient to give the result in the coordinate system defined in (3.19). In these coordinates, the solution takes the form [35]

$$AdS_4: \quad x^2>0 \; : \; ds^2 = d\psi^2 + \eta^2 \sinh^2\psi\left(-d\tau^2 + \cosh^2\tau\, d\Omega_2\right), \qquad \phi = \nu\,\mathrm{sech}^2\frac{\psi}{2}, \tag{3.41}$$

$$x^2<0 \; : \; ds^2 = -d\tau^2 + \eta^2 \sin^2\tau\left(d\psi^2 + \sinh^2\psi\, d\Omega_2\right), \qquad \phi = \nu\sec^2\frac{\tau}{2}, \tag{3.42}$$

$$dS_4: \quad x^2>0 \; : \; ds^2 = d\psi^2 + \eta^2 \sin^2\psi\left(-d\tau^2 + \cosh^2\tau\, d\Omega_2\right), \qquad \phi = \nu\sec^2\frac{\psi}{2}, \tag{3.43}$$

$$x^2<0 \; : \; ds^2 = -d\tau^2 + \eta^2 \sinh^2\tau\left(d\psi^2 + \sinh^2\psi\, d\Omega_2\right), \qquad \phi = \nu\,\mathrm{sech}^2\frac{\tau}{2}, \tag{3.44}$$

where we have set $|\lambda|^2 = 1$ and

$$\eta = \frac{f_1 h^2}{2} = \frac{1+(1-a^2)^2 Q}{1+(1+6a^2+a^4)\,Q}\,. \tag{3.45}$$

$$AdS_4: \; a^2 = \begin{cases} \tanh^2\frac{\psi}{4} & \text{for } x^2>0 \\ -\tan^2\frac{\tau}{4} & \text{for } x^2<0 \end{cases} \qquad dS_4: \; a^2 = \begin{cases} -\tan^2\frac{\psi}{4} & \text{for } x^2>0 \\ \tanh^2\frac{\tau}{4} & \text{for } x^2<0 \end{cases} \tag{3.46}$$

In addition to the $SO(3,1)$ symmetry generated by $M'_{\alpha\beta}$, the solution is also left invariant by additional transformations with rigid HS parameters

$$\epsilon' = \sum_{\ell=0}^{\infty} \epsilon'_\ell\,, \tag{3.47}$$

where the ℓ'th level is given by [20]

$$\epsilon'_\ell = \sum_{m+n=2\ell+1} \Lambda^{\alpha_1\ldots\alpha_{2m},\dot\alpha_1\ldots\dot\alpha_{2n}} M'_{\alpha_1\alpha_2}\star\cdots\star M'_{\alpha_{2m-1}\alpha_{2m}}\star \bar M'_{\dot\alpha_1\dot\alpha_2}\star\cdots\star \bar M'_{\dot\alpha_{2n-1}\dot\alpha_{2n}} - \text{h.c.}\,, \tag{3.48}$$

with constant $\Lambda^{\alpha_1\ldots\alpha_{2m},\dot\alpha_1\ldots\dot\alpha_{2n}}$. The full symmetry algebra is thus a higher-spin extension of $SO(3,1) \simeq SL(2,\mathbb{C})$, that we shall denote by

$$hsl(2,\mathbb{C};\nu) \supset sl(2,\mathbb{C})\,, \tag{3.49}$$

where $sl(2,\mathbb{C})$ is generated by $M'_{\alpha\beta}$ and its hermitian conjugate, and we have indicated that in general the structure coefficients may depend on the deformation parameter ν.

Turning back to the solutions given above, a holographic and cosmological interpretation of (3.41) has been discussed in [35], where a bouncing cosmology scenario was observed, and its comparison with a similar phenomenon occurring supergravities [40,41] was made. In this context, it is useful to examine the behaviour of the solutions, both with AdS and dS asymptotics, near the boundary as well as distant future.

For $|\nu| \ll 1$ and near the boundary, where $\lambda^2 x^2 \to 1$, or equivalently $a^2 \to 1$, the scale factors η and Ω behaves as[3]

$$\lambda^2 x^2 \to 1 \quad \Longrightarrow \quad \eta \to \frac{1}{1-\frac{\nu^2}{3}}\,, \qquad \Omega \to 1\,, \tag{3.50}$$

which means that the solutions are asymptotically maximally symmetric spacetimes with undeformed radius.

Another interesting limit to consider is $\eta \to 0$; for $|\nu| \ll 1$ this takes place for $a^2 + 1 \ll 1$, that is, for a^2 close but not equal to -1, which corresponds to $\tau \to \pm\tau_{\rm crit}$ in the AdS case, and $\psi \to \psi_{\rm crit}$ in the dS case. In the former case, we have[4]

$$AdS_4: \qquad \eta \simeq \frac{\nu^2}{6}\left[\exp\left(\frac{3}{2\nu^2}\right)\right](\tau_{\rm crit}-\tau)\,, \qquad \tau_{\rm crit} \simeq \sin^{-1}\left[2\exp\left(\frac{-3}{2\nu^2}\right)\right]\,. \tag{3.51}$$

In the Einstein frame[5], it takes infinite proper time to reach the critical surface, which means that one may interpret the future region of the solution as a singularity free $SO(3,1)$ invariant cosmology with a finite asymptotic scalar field, as

$$\phi \to \phi_{\rm crit} \simeq 4\nu \exp\left(\frac{3}{\nu^2}\right)\,. \tag{3.52}$$

In the dS case, one may instead interpret the critical limit as a domain wall at an infinite proper space-like distance from the center of the solution.

3.4. Solutions of the Non-Minimal Chiral Model in Kleinian Space

In obtaining the solutions described above, symmetries on the master fields were imposed. In [21] projection operators were used as well. In the case of the non-minimal chiral model[6] in Kleinian space, it is possible to use projectors to build solutions with non-vanishing Weyl zero-form and HS fields. They are

$$B' = (1-P)\star\kappa\,, \qquad S'_\alpha = z_\alpha \star P\,, \qquad \bar{S}'_{\dot\alpha} = \bar{z}_{\dot\alpha}\star(1-2\bar{P})\,, \tag{3.53}$$

where

$$P\star P = P\,, \qquad \bar{P}\star\bar{P} = \bar{P}\,, \qquad [P,\bar{P}]_\star = 0\,, \tag{3.54}$$

and satisfy the conditions $\pi\circ\bar\pi(P,\bar P) = (P,\bar P)$. The projectors P and $\bar P$ are independent. Consider the simplest case in which

$$P = 2e^{-2uv} = 2e^{yby}\,, \qquad \bar{P} = 0\,, \tag{3.55}$$

where, in terms of constant spinors $(\lambda_\alpha, \mu_\alpha)$ we have defined

$$u = \lambda^\alpha y_\alpha\,, \qquad v = \mu^\alpha y_\alpha\,, \qquad b_{\alpha\beta} = 2\lambda_{(\alpha}\mu_{\beta)}\,, \qquad \lambda^\alpha\mu_\alpha = \frac{i}{2}\,. \tag{3.56}$$

---

3 The result for the limits of η and Ω given here correct Equations (4.67) and (4.68) in [20].
4 The expression for η and $\tau_{\rm crit}$ corrects a factor of two in [35].
5 This is the (torsion-free) frame obtained by rescaling the vielbein as $e^a \to \eta^{-1}e^a$.
6 The solution can be constructed for the minimal model as well by working with a convenient integral presentation of the projection operators.

Upon the L-dressing, and expanding the result in Y-oscillators, the following component results are found in [21]:

$$\phi = -1\,, \qquad C_{\alpha_1\ldots\alpha_{2s}} = 0\,, \tag{3.57}$$

while the anti-self-dual Weyl tensors take the form

$$\bar{C}_{\dot\alpha_1\cdots\dot\alpha_{2s}} = -2^{2s+1}(2s-1)!!\left(\frac{h^2-1}{h^2}\right)^s \bar{u}_{(\dot\alpha_1}\cdots\bar{u}_{\dot\alpha_s}\,\bar{v}_{\dot\alpha_{s+1}}\cdots\bar{v}_{\dot\alpha_{2s})}\,, \tag{3.58}$$

where

$$\bar{u}_{\dot\alpha} = \frac{x^a}{\sqrt{x^2}}\,(\bar\sigma_a\lambda)_{\dot\alpha}\,, \qquad \bar{v}_{\dot\alpha} = \frac{x^a}{\sqrt{x^2}}\,(\bar\sigma_a\mu)_{\dot\alpha}\,. \tag{3.59}$$

In stereographic coordinates, the Kleinian space is covered in two charts with $0 \le h^2 \le 2$, and hence the Weyl tensors blow up in the limit $h^2 \to 0$ preventing the solution from approaching $H_{3,2}$ in this limit. In coordinate system introduced in (3.21), the metric reads $ds^2 = -(dr^2 + r^2dt^2) + (1+r^2)d\Omega_2^2$, and the pre-factor in this solution reads

$$\left(\frac{h^2-1}{h^2}\right)^s = 2^{-s}(1 - r\cos t)^s\,. \tag{3.60}$$

which, indeed, diverges at the boundary $r \to \infty$ as noted above. In [21], it was also found that

$$e_\mu^a = \frac{-2}{(h^2+2h^{-2})}\left[(1+2h^{-2})\delta_\mu^a + 2\lambda^2h^{-4}x_\mu x^a + \frac{6\lambda^2}{h^4-4}\,(Jx)_\mu(Jx)^a\right]\,, \tag{3.61}$$

where

$$J_{ab} = (\sigma_{ab})^{\alpha\beta}\,b_{\alpha\beta}\,, \qquad J_a{}^cJ_c{}^b = -\delta_a^b\,. \tag{3.62}$$

For the spin connection, the result is

$$\begin{aligned}
\omega^{\alpha\beta} &= \frac{1}{1-4h^{-4}}\left[\omega_0^{\alpha\beta} - 8h^{-4}(b\,\omega_0\,b)^{\alpha\beta}\right] + \frac{4h^{-4}}{(1-2h^{-4})(1-4h^{-4})}b^{\alpha\beta}\,b_{\gamma\delta}\,\omega_0^{\gamma\delta}\,, \\
\bar\omega^{\dot\alpha\dot\beta} &= \bar\omega_0^{\dot\alpha\dot\beta} + 4(1+h)^2h^{-4}\left[-(\bar{a}ba)^{\dot\alpha\dot\beta}b_{\gamma\delta} + 2(\bar{a}b)^{\dot\alpha\gamma}(\bar{a}b)^{\dot\beta\delta}\right]\omega_{\gamma\delta}\,.
\end{aligned} \tag{3.63}$$

Note that in the last term of the second equation the full spin connection $\omega_{\gamma\delta}$ arises. The metric $g_{\mu\nu} = e_\mu^a e_\nu^b\,\eta_{ab}$ takes the form

$$\begin{aligned}
g_{\mu\nu} &= \frac{4}{(h^2+2h^{-2})^2}\left[(1+2h^{-2})^2\eta_{\mu\nu} + 4h^{-4}\left((1-h^2)h^{-4} + (1+2h^{-2})\right)x_\mu x_\nu\right. \\
&\quad \left. + \frac{12}{1-4h^{-4}}\left(\frac{3(1-h^2)}{1-4h^{-4}} + (1+2h^{-2})\right)(Jx)_\mu(Jx)_\nu\right]\,,
\end{aligned} \tag{3.64}$$

The vierbein has potential singularities at $h^2 = 0$ and $h^2 = 2$. The limit $h^2 \to 0$ is a boundary at which $e_\mu{}^a \sim h^{-2}x_\mu x^a$, i.e., a scale factor times a degenerate vierbein. In the limit $h^2 \to 2$ one approaches the boundary of a coordinate chart. Also in this limit, the vierbein becomes degenerate, viz., $e_\mu{}^a \sim h^{-2}(Jx)_\mu(Jx)^a$.

3.5. Perturbative Construction of Domain-Wall Solution

As mentioned earlier, if we wish to construct a solution of the HS equations that has a symmetry group that includes any translation generators P^a, given that there is no known realization of these generators that would form a closed algebra with the full Lorentz generators $M_{\alpha\beta}$, the symmetry conditions (3.22) need to be imposed in terms of the undeformed generators that are bilinear in the oscillators. The deformation of this symmetry to accommodate nonlinear corrections can then be computed perturbatively in a weak field expansion scheme. This is the framework which was pursued in considerable detail in [20], where the perturbative construction of solutions with 3, 4 and 6 parameter isometry subgroups of the AdS group were considered. Here we shall outline the key aspects of this constructions by describing the example of a domain-wall solution having $ISO(2,1)$ symmetry and its appropriate HS extension [20].

The $ISO(2,1)$ algebra is generated by

$$ISO(2,1): \qquad M_{ij}\,, \qquad P_i = (\alpha M_{ab}L^b + \beta P_a)L_i^a\,, \qquad \alpha^2 - \beta^2 = 0\,, \tag{3.65}$$

where α and β are real parameters and (L_i^a, L^a) is a representative of the coset $SO(3,1)/SO(2,1)$, obeying

$$L^a L_a = \epsilon = \pm 1\,, \qquad L_i^a L_a = 0\,, \qquad L_i^a L_{ja} = \eta_{ij} = \text{diag}(+,+,-)\,, \tag{3.66}$$

and the generators are taken form the oscillator realization of $SO(3,2)$ algebra given by

$$M_{ab} = -\frac{1}{8}\left[(\sigma_{ab})^{\alpha\beta} y_\alpha y_\beta + (\bar{\sigma}_{ab})^{\dot{\alpha}\dot{\beta}} \bar{y}_{\dot{\alpha}} \bar{y}_{\dot{\beta}} \right]\,, \qquad P_a = \frac{1}{4}(\sigma_a)^{\alpha\dot{\beta}} y_\alpha \bar{y}_{\dot{\beta}}\,, \tag{3.67}$$

In particular $[P_i, P_j] = i(\beta^2 - \alpha^2)M_{ij}$ vanishes for $\alpha^2 = \beta^2$, as required for $ISO(2,1)$. Thus, the symmetry conditions to be imposed are

$$[M_{ij}, C'] = 0\,, \qquad [P_i, C']_\pi = 0\,. \tag{3.68}$$

As shown in [20], these conditions are solved by

$$C'(P) = (\mu_1 + \mu_2 P)e^{4iP}\,, \qquad P := \frac{1}{4} L^a (\sigma_a)_{\alpha\dot{\alpha}} y^\alpha \bar{y}^{\dot{\alpha}}\,. \tag{3.69}$$

Denoting the $ISO(2,1)$ transformations discussed above by $\epsilon'_{(0)}$, we can seek its nonlinear deformation by expanding

$$\epsilon' = \epsilon'_{(0)} + \epsilon'_{(1)} + \epsilon'_{(2)} + \cdots\,, \tag{3.70}$$

where $\epsilon'_{(n)}$ is constant in spacetime but may depend on Y and Z. The symmetry condition at first order is satisfied by C' given in (3.69). To establish the symmetry at second order, we need to satisfy

$$\left([\epsilon'_{(1)}, C']_\pi + [\epsilon'_{(0)}, B'_{(2)}]_\pi \right)_{Z=0} = 0\,, \tag{3.71}$$

where $B'_{(2)}$ is obtained from the normal ordered perturbative scheme (see Section 6 below) to be [20]

$$\begin{aligned} B'_{(2)} &= f + \tau\bar{\pi} f + \pi(f + \tau\bar{f})^\dagger\,, \\ f &:= -z^\alpha \int_0^1 dt (V_\alpha'^{(1)} \star C')_{Z\to tZ}\,, \\ V_\alpha'^{(1)} &= -\frac{i}{2} z_\alpha \int_0^1 t dt B'(-tz, \bar{y})\kappa(tz, y)\,. \end{aligned} \tag{3.72}$$

This condition (3.71) is solved in [20] where it is found that

$$\epsilon'_{(1)} = -\int_0^1 dt \int_0^1 t' dt' \left(\frac{itt'}{2} \lambda^{\alpha\beta} z_\alpha z_\beta + \lambda^{\alpha\dot{\beta}} z_\alpha \bar{\partial}_{\dot{\beta}} \right) C'(-tt'z, \bar{y}) e^{itt'yz} - h.c.\,, \tag{3.73}$$

where $\lambda^{\alpha\beta}, \lambda^{\dot\alpha\dot\beta}$ are arbitrary constant parameters. For a more detailed discussion of the procedure outlined above, see [20].

3.6. Other Known Solutions

The solutions described in Section 3.3 were generalized in [21] to find new Lorentz-invariant vacuum solutions, in which, in addition to the continuous parameter ν, an infinite set of independent and discrete parameters $\theta_k = \{0,1\}$, each turning on a Fock-space projector $P_n(u)$, were activated. Should they be proved to be gauge-inequivalent to AdS_4, as they seem to be, they would represent monodromies of flat but non-trivial connections $(V_\alpha, \bar{V}_{\dot\alpha})$ on $\mathcal{Z}$. An interesting limit of this solution arises upon setting $\nu = 0$ and $(\theta_k - \theta_{k+1})^2 = 1$, leading to the degenerate metric

$$ds^2 = \frac{4(x^a dx_a)^2}{\lambda^2 x^2 (1 - \lambda^2 x^2)} \,. \tag{3.74}$$

The methods above can also be extended to the Prokushkin-Vasiliev theory in $D = 3$, giving rise to Lorentz-invariant instanton solutions (with additional twisted sectors of the theory excited, and the characteristic extra deformation parameter λ that allows to vary the mass of the scalar), as well as to the above projector vacua [31].

Finally, in [42] a different class of exact solutions was constructed by means of the gauge function method coupled with a different choice of gauge in twistor space, there referred to as axial gauge. As for the perturbative construction of solutions, it is worth mentioning the plane wave solution of [4,19], whose elevation to an exact solution remains to be investigated, to our best knowledge.

4. Factorization Method and Solutions

4.1. The Method

The method developed in [6,7,22] for finding exact solutions of Vasiliev equations also exploits the gauge function method to solve for (B, V, W) in terms of (B', V') from (3.5) and (2.22), which, in turn, are to be determined by solving (3.4). It is here that the method differs from the method described above, by making the following factorized ansatz for (B', V'):

$$B'(Z,Y) = \nu \Psi(y,\bar{y}) \star \kappa_y \,, \tag{4.1}$$

$$V'_\alpha(Z,Y) = V'_\alpha(z;\Psi) = \sum_{n\geq 1} V^{(n)}_\alpha(z) \star \Psi^{\star n} \,. \tag{4.2}$$

Note that this ansatz for V'_α is holomorphic in z, and in the following we shall refer to the solution in this form as given in *holomorphic gauge* [7]. In this section we shall consider the nonminimal model in which we recall that conditions (2.35) apply. It is shown in [7] that this ansatz solves the fully non-linear Equation (3.4) provided that

$$\pi_z(V^{(n)}_\alpha(z)) = -V^{(n)}_\alpha(z) \,, \tag{4.3}$$

and that

$$s_\alpha := z_\alpha - 2i \sum_{n\geq 1} V^{(n)}_\alpha(z) \nu^n \,, \tag{4.4}$$

obeys the deformed oscillator algebra

$$[s_\alpha, s_\beta]_\star = -2i\epsilon_{\alpha\beta}(1 - \nu\kappa_z) \,, \qquad \kappa_z \star s_\alpha = -s_\alpha \star \kappa_z \,. \tag{4.5}$$

Note that only the first power of ν in (4.4) survives in the commutator (4.5). One class of solutions is given by [6]

$$\sum_{n\geq 1} V_\alpha^{(n)}(z)\nu^n = -\frac{i\nu}{2} z_\alpha \int_{-1}^{+1} \frac{dt}{(t+1)^2} \exp\left(i\frac{t-1}{t+1} z^+_{(0)} z^-_{(0)}\right) {}_1F_1(\tfrac{1}{2};1;b\nu\log t^2)\,, \tag{4.6}$$

where the constant spinors $v^{\pm}_{(0)\alpha}$ are used to defined the projected oscillators

$$v^{+\alpha}_{(0)} v^-_{(0)\alpha} = 1\,, \qquad z^{\pm}_{(0)} = v^{\pm\alpha}_{(0)} z_\alpha\,. \tag{4.7}$$

The presence of z^+z^- term breaks manifest Lorentz covariance, which will be restored when we consider the field dependent gauge transformation in Section 7 that is needed to cast the results into a form that can be interpreted in terms of Fronsdal fields that obey the standard boundary conditions.

At this point, B' and V'_α are determined, with $\Psi(Y)$ representing an arbitrary initial datum. One can proceed to compute (B, V, W') from (3.4), (3.5) and (2.22). However, one needs to ensure that the star products involving Ψ are well defined. The analyticity properties of the resulting (B, V, W') also require special care. The strategy adopted in [6,7,22] is to employ projection operators with well defined group theoretical origin, and easily deducible symmetry properties. We shall illustrate aspects of this procedure below with a relatively simple example, namely the black hole-like solution [7] closely related to that of [24], which we shall also describe in a subsequent section below.

4.2. Black Hole Solution

We are seeking an exact solution of Vasiliev's equations which has the symmetries of 4D static black hole solution that is asymptotically AdS_4, namely spatial rotations and time translations generated by

$$SO(3)\times SO(2): \qquad M_{rs}\,(r,s=1,2,3)\,, \text{ and } M_{05} = E = \frac{1}{4}(\sigma_0)_{\alpha\dot\alpha}\, y^\alpha \bar y^{\dot\alpha}\,. \tag{4.8}$$

The first instance of such a solution was found in [24] in a different approach that will be summarized later. In both approaches the following projector plays an important role:

$$\Psi = \nu P'\,, \qquad P'\star P' = P'\,, \tag{4.9}$$

where P' is given by

$$P' = 4e^{-4E}\,, \tag{4.10}$$

and the reality conditions dictate that

$$\nu = iM\,, \tag{4.11}$$

with $M\in\mathbb{R}$. This projector clearly has the desired symmetry property since $[M_{rs}, P']_\star = 0$ and $[E, P']_\star = 0$. In performing the L-dressing, we need the result

$$P = L^{-1}\star 4e^{-4E}\star L = 4\exp\left(-\frac{1}{2}K_{AB}Y^AY^B\right)\,, \tag{4.12}$$

where K_{AB} are the Killing parameters taking the form

$$K_{AB} = \begin{pmatrix} u_{\alpha\beta} & v_{\alpha\dot\beta} \\ \bar v_{\dot\alpha\beta} & \bar u_{\dot\alpha\dot\beta}\end{pmatrix}\,. \tag{4.13}$$

In terms of the x-dependent eigenspinors of $u_{\alpha\beta}$, these parameters can be expressed as

$$\begin{aligned} u_{\alpha\beta} &= 2r\, u^+_{(\alpha} u^-_{\beta)}\,, \qquad v_{\alpha\dot\beta} = \sqrt{1+r^2}\left(u^+_\alpha \bar u^+_{\dot\beta} + u^-_\alpha \bar u^-_{\dot\beta}\right)\,, \\ u^{-1}_{\alpha\beta} &= \frac{2}{r}\, u^+_{(\alpha} u^-_{\beta)}\,, \qquad 2u^-_{[\alpha} u^+_{\beta]} = \epsilon_{\alpha\beta}\,. \end{aligned} \tag{4.14}$$

The Kerr-Schild vector $k_{\alpha\dot\alpha}$ associated with E is obtained via a projection of $v_{\alpha\dot\alpha}$ with the eigenspinors of $u_{\alpha\beta}$, $\bar u_{\dot\alpha\dot\beta}$, and can be written as[7]

$$k_{\alpha\dot\alpha} = \frac{1}{\sqrt{1+r^2}} u^+_\alpha \bar u^+_{\dot\alpha}\,. \tag{4.15}$$

In obtaining the above results, the computation is first done in stereographic coordinates, and then a coordinate change is made to go over to the spherical (global) coordinate system in which the AdS metric reads $ds^2 = -(1+r^2)dt^2 + (1+r^2)^{-1}dr^2 + r^2 d\Omega_2^2$.

From the point of view of the factorized Ansatz (4.1)–(4.2), the fact that Ψ is proportional to a projector can be seen as a way of enforcing the Kerr-Schild property of a black-hole solution, as it effectively causes a collapse of all non-linear correction in V'_α, $\bar V'_{\dot\alpha}$ down to the linear order, at least from the point of view of the oscillator dependence. Coupled with the gauge freedom on S_α this allows to effectively reach a gauge in which the full solution only contains first order deformations in ν [6]. Another noteworthy fact is that, due to the factorized dependence on Y and Z, one can effectively separately rotate the two oscillators by means of a factorized gauge function

$$g = L(x,Y) \star \widetilde L(x,Z)\,. \tag{4.16}$$

As discussed in [6,7,22], turning on the second factor is useful since by choosing it appropriately one can achieve collinearity between the spin-frame $(v^+, v^-)(x)$ on $\mathcal{Z}$ (which is obtained by pointwise rotation of $v_{(0)}$ given in (4.7) by $\widetilde L$) and the eigenspinors (u^+_α, u^-_β) of $u_{\alpha\beta}$ in order to remove singularities that appear in the solution for the master one-form, that are gauge artifacts. We note that the factor $\widetilde L(x,Z)$, being purely Z-dependent, does not affect B and only acts non-trivially on V_α, $\bar V_{\dot\alpha}$.

Thus, dressing the primed fields given above by the gauge function g defined in (4.16), the following results have been obtained [6]

$$B = \frac{4M}{r}\exp\left[\frac{1}{r^2}\left(\tfrac12 y^\alpha u_{\alpha\beta} y^\beta + \tfrac12 \bar y^{\dot\alpha} \bar u_{\dot\alpha\dot\beta} \bar y^{\dot\beta} + i y^\alpha u_{\alpha\beta} v^\beta{}_{\dot\alpha} \bar y^{\dot\alpha}\right)\right]\,, \tag{4.17}$$

$$S_\alpha = z_\alpha + 8P a_\alpha \int_{-1}^{1} \frac{dt}{(t+1+ir(t-1))^2}\, j(t) \exp\left[\frac{i(t-1)}{t+1+ir(t-1)} a^+ a^-\right]\,, \tag{4.18}$$

$$\begin{aligned} W' = {}& W_0 + \widetilde L^{-1} \star d_x \widetilde L - \Big\{\omega^{--}\left[y^+ y^+ + 8P(f_1 - f_2) a^+ a^+\right] \\ & + \omega^{++}\left[y^- y^- + 8P(f_1 - f_3) a^- a^-\right] - 2\omega^{-+}\left[y^+ y^- - 8P(f_1 + f_4) a^+ a^- - r(f_5 + f_6)\right]\Big\}\,, \end{aligned} \tag{4.19}$$

where $y^\pm = u^{\alpha\pm} y_\alpha$, and similarly$(a^\pm, \omega^{++}, \omega^{--}, \omega^{-+})$ are projections of a_α, $\omega_{\alpha\beta}$ with $u^{\alpha\pm}$, and the function $j(t)$ is defined as

$$j(t) = -\frac{\nu}{4}\,{}_1F_1\left[\frac12; 2; \frac{\nu}{2}\log\frac{1}{t^2}\right]\,. \tag{4.20}$$

---

[7] In stereographic coordinates these read [7] $u_{\alpha\beta} = 2x^i(\sigma_{i0})_{\alpha\beta}/(1-x^2)$ and $v_{\alpha\dot\beta} = (\sigma_0)_{\alpha\dot\beta} - 4x_{[0}x^i(\sigma_{i]})_{\alpha\dot\beta}/(1-x^2)$.

The modified oscillators $(a_\alpha, \bar{a}_{\dot\alpha})$ are defined as

$$\widetilde{Z}_A := (a_\alpha, \bar{a}_{\dot\alpha}) = Z_A + iK_A{}^B Y_B\,, \qquad [\widetilde{Z}_A, \widetilde{Z}_B] = 4i\epsilon_{AB}\,. \tag{4.21}$$

Furthermore, we have defined the functions

$$f_1 = \int_{-1}^{1} dt\, q(t)(t+1)\xi^3 \exp\left[i(t-1)\xi a^+ a^-\right], \tag{4.22a}$$

$$f_2 = \int_{-1}^{1}\int_{-1}^{1} dt\,dt'\, j(t)j(t')\, 2t\, \tilde\xi^3 \exp\left[i(tt'-1)\tilde\xi a^+ a^-\right], \tag{4.22b}$$

$$f_3 = \int_{-1}^{1}\int_{-1}^{1} dt\,dt'\, j(t)j(t')\, 2t' \tilde\xi^3 \exp\left[i(tt'-1)\tilde\xi a^+ a^-\right], \tag{4.22c}$$

$$f_4 = \int_{-1}^{1} dt\, q(t)\xi^2 \exp\left[i(t-1)\xi a^+ a^-\right], \tag{4.22d}$$

$$f_5 = \int_{-1}^{1}\int_{-1}^{1} dt\,dt'\, j(t)j(t')\, (tt'+1)\, \tilde\xi^3 \exp\left[itt'\tilde\xi a^+ a^-\right], \tag{4.22e}$$

$$f_6 = \int_{-1}^{1}\int_{-1}^{1} dt\,dt'\, j(t)j(t')\, \tilde\xi^2 \exp\left[i(tt'-1)\tilde\xi a^+ a^-\right], \tag{4.22f}$$

and

$$\xi := \frac{1}{t+1+ir(t-1)}\,, \qquad \tilde\xi := \frac{1}{tt'+1+ir(tt'-1)}\,. \tag{4.23}$$

The modified oscillators a_α appear naturally in V_α as a consequence of the factorized form of (4.2) with $\Psi^{\star n} \propto P$, and of the fact that $z_\alpha \star P = a_\alpha P$ (analogously for $\bar{a}_{\dot\alpha}$ in $\bar{V}_{\dot\alpha}$). A consequence of the modified oscillator appearing in the solution is that V'_α does not obey the standard Vasiliev gauge $Z^A V_A = 0$.

As noted earlier, this solution is closely related to that of Didenko and Vasiliev [24]. Indeed, the solution for B is the same, while the relationship between the solutions for (S_A, W') is more subtle and it will be discussed in the next section.

Note that B does not depend on Z. Thus, $B|_{Z=0} = B = C(Y)$, and its holomorphic components give the Petrov type -D Weyl tensors

$$C_{\alpha_1\ldots\alpha_{2s}} = \frac{4M}{r^{s+1}} u^+_{(\alpha_1} \cdots u^+_{\alpha_s} u^-_{\alpha_{s+1}} \cdots u^-_{\alpha_{2s})}\,, \tag{4.24}$$

and similarly for the anti-holomorphic components giving $\bar{C}_{\dot\alpha_1\ldots\dot\alpha_{2s}}$. The singularity of individual Weyl tensors does not necessarily imply a physical singularity in HS gravity for the following reasons. At the level of the master fields, $r=0$ also appears as the only point at which V_α, as well as W', acquire a pole on a plane in $\mathcal{Z}\times\mathcal{Y}$ defined by $a_\alpha|_{r=0} = z_\alpha + i(\sigma_0\bar{y})_\alpha = 0$, due to the zero at $t=-1$ that the denominator of the integrand in (4.18) develops at $r=0$. The master-field curvature is however given by $B\star\kappa$ and one can argue that at the master-field level, which is the only sensible way to look at such solution in the strong-field region, B remains, in fact, regular at $r=0$. Qualitatively this can be understood as follows. r appears in (4.17) as the parameter of a delta sequence: away from the origin one has a smooth Gaussian function, approaching a Dirac delta function on $\mathcal{Y}$ as r goes to zero [6]. However, unlike the delta function on a commutative space, the delta function in noncommutative twistor space, thought of as a symbol for an element of a star product algebra, is smooth. Indeed, it is possible to show [6] that by changing ordering prescription one can map the delta function to a regular element, and the smoothness of such change of basis manifests itself in the fact that the solution of the deformed oscillator problem obtained in the new ordering can be mapped back smoothly to the solution above. In this sense, the singularity in $r = 0$ may be an artifact of the ordering choice for the infinite-dimensional symmetry algebra governing the Vasiliev system.

In order to extract the x-dependence of even just the spin-2 component of W', one still needs to evaluate the complicated parametric integrals in (4.22a). However, as the Weyl-tensors take the simple form in x-space given in (4.24), we expect that a suitable gauge transformation exists that will give the metric in the standard Kerr-Schild form, namely $g_{\mu\nu} = g_{\mu\nu}^{AdS} + 2Mk_\mu k_\nu / r$.

Finally, let us note that the basic black hole-like solution reviewed above has a generalization in which infinite sets of projection operators P_n and twisted projection operators $\widetilde{P}_n$ are introduced[8]. The twisted projectors $\widetilde{P}_n$ are invariant under $SO(3) \times SO(2)$ discussed above while the projectors P_n are invariant under $SO(3)$ and the (B', S'_A) sector of the solution takes the form [6,7,22]

$$B' = \sum_{n=\pm1,\pm2,\ldots} \left(\nu_n \widetilde{P}'_n + \widetilde{\nu}_n P'_n \right) , \qquad \widetilde{P}'_n = P'_n \star \kappa_y \,, \tag{4.25}$$

$$S'_A = z_A - 2i \sum_{n=\pm1,\pm2,\ldots} \left(V_{n,A} \star P'_n + \widetilde{V}_{n,A} \star \widetilde{P}'_n \right) , \tag{4.26}$$

with

$$P'_n(E) = 2(-1)^{n-\frac{1+\varepsilon}{2}} \oint_{C(\varepsilon)} \frac{d\eta}{2\pi i} \left(\frac{\eta+1}{\eta-1} \right)^n e^{-4\eta E} , \qquad \varepsilon := n/|n| \,, \tag{4.27}$$

$$\widetilde{P}'_n(E) := P'_n(E) \star \kappa_y = 4\pi(-)^{n-\frac{1+\varepsilon}{2}} \oint_{C(\varepsilon)} \frac{d\eta}{2\pi i} \left(\frac{\eta+1}{\eta-1} \right)^n \delta^2(y - i\eta\sigma_0\bar{y}) \,, \tag{4.28}$$

where the contour integrals are performed around a small contour $C(\varepsilon)$ encircling ε. The expressions for $(V_{n,A}, \widetilde{V}_{n,A})$, and their $g = L \star \widetilde{L}$ dressing can be found in [7]. It turns out that the solutions with only ν_n parameters switched on correspond to black hole-like solutions, of which the case summarized above arises for $n = 1$. Solutions with only nonvanishing $\widetilde{\nu}_n$ parameters correspond to massless particle modes, and surprisingly black hole modes as well entering from second order onwards in a perturbative treatment of the solution; for a detailed description of this phenomenon see [7,26]. Note that the $SO(3)$ invariant projectors are associated with spin-0 modes. To extend this construction to spin-s particle modes, one needs the spin-s generalization of the projectors P'_n discussed above. A particular presentation of such projectors can be found in [43].

4.3. Other Known Solutions

By means of the same factorized Ansatz (4.1)–(4.2) black-hole-like solutions with biaxial symmetry have also been found [6,22,44], some of which being candidate HS generalizations of the Kerr black hole [44]. The separation of variables in holomorphic gauge was also instrumental to finding solutions with $\mathfrak{g}_6$ isometries of cosmological interest, that we shall present in a forthcoming paper [23].

5. Direct Method and the Didenko-Vasiliev Solution

While, as we have seen so far, the gauge function method is in general of great help in constructing exact solutions, it is sometimes possible to attack the equations directly, by virtue of some other simplifying Ansatz or gauge condition. We shall generically refer here to any method which does not rely on the use of gauge function as direct method. One such solution has been found so far in this way by Didenko and Vasiliev [24], which has nonvanishing HS fields, and which contains the Schwarzschild black hole solution in the spin 2 sector.

---

8 Even though $\widetilde{P}'_n \star \widetilde{P}'_n = P'_n$, we refer to $\widetilde{P}'$ as twisted projector to emphasize the fact that it is related to the projector P'_n by the relation $\widetilde{P}'_n = P'_n \star \kappa_y$.

Indeed, motivated by the phenomenon that solutions of Einstein equations that can be put in Kerr-Schild form solve the linearized as well as the nonlinear form of the equations, the Authors of [24] thought of an Ansatz that would generalize some distinctive features of black hole solutions in gravity. First, it is based on an AdS timelike Killing vector, in the sense that the Weyl tensor will be a function of some element K_{AB} as (4.13). More precisely, if $f(K)$ satisfies the Killing vector equation, a proper ansatz for a solution of the linearized twisted adjoint equation will be given by $f(K) \star \kappa_y$. Second, they chose $f(K)$ in such a way that the Ansatz linearize the Vasiliev equations. For the latter purpose it is important that the function $f(K)$ is a projector—in fact, a Fock-space vacuum projector that coincides with (4.12). Such choice in particular reduces Equations (2.31) and (2.32) to two copies of the 3D (anti)holomorphic deformed oscillator problem that arises in Prokushkin-Vasiliev HS theory in 3D [8] in terms of the oscillators in (4.21)[9].

The ansatz [24]

$$B = MP \star \kappa_y \,, \qquad S_\alpha = z_\alpha + P f_\alpha(a|x) \,, \qquad \bar{S}_{\dot\alpha} = \bar{z}_{\dot\beta} + P\bar{f}_{\dot\alpha}(\bar{a}|x) \,, \tag{5.1}$$

where M is a constant and $(f_\alpha, \bar{f}_{\dot\alpha})$ are functions to be determined, indeed reduces the Equations (2.31) and (2.32) to two deformed oscillator problems in terms of the latter functions. A specific gauge choice on the $(f_\alpha, \bar{f}_{\dot\alpha})$, while bringing about a further breaking of the manifest Lorentz covariance, effectively linearizes their equations, which are then solved by means of the standard perturbative methods of Section 6. A further ansatz [24]

$$W = W_0 + P\left[g(a|x) + \bar{g}(\bar{a}|x)\right] \,, \tag{5.2}$$

where g is another function to be determined is then employed to deal with the remaining equations that involve W, namely (2.28), (2.29) and (2.30). The resulting exact solution is given by [24]

$$B = \frac{4M}{r}\exp\left[\frac{1}{r^2}\left(\tfrac{1}{2}y^\alpha u_{\alpha\beta}y^\beta + \tfrac{1}{2}\bar{y}^{\dot\alpha}\bar{u}_{\dot\alpha\dot\beta}\bar{y}^{\dot\beta} + y^\alpha u_{\alpha\beta}v^\beta{}_{\dot\alpha}\bar{y}^{\dot\alpha}\right)\right] \,, \tag{5.3a}$$

$$S^+ = z^+ + MP\frac{a^+}{r}\int_0^1 dt \exp\left(\frac{t}{r}a^+a^-\right) \,, \quad \text{idem } \bar{S}^+ \,, \tag{5.3b}$$

$$S^- = z^- \,, \quad \text{idem } \bar{S}^- \,, \tag{5.3c}$$

$$\begin{aligned} W = {} & W_0 + \frac{1}{2r}MP\left[d\tau^{--}a^+_\alpha a^+_\beta \int_0^1 dt(1-t)\exp\left(\frac{t}{r}a^+a^-\right) + h.c.\right] \\ & + \frac{1}{8r}MP\left[u_{\alpha\beta}\,\omega_0^{\alpha\beta} + h.c. + 2(v_{\alpha\dot\alpha} + \kappa_{\alpha\dot\alpha})\,e_0^{\alpha\dot\alpha}\right] \,, \end{aligned} \tag{5.3d}$$

where

$$\begin{aligned} & \tau_{\alpha\beta} = \frac{u_{\alpha\beta}}{r} \,, \qquad \tau^{--} = u^{\alpha-}u^{\alpha-}\tau_{\alpha\beta} \,, \qquad z^\pm = u^{\alpha\pm}z_\alpha \,, \qquad a^\pm = u^{\alpha\pm}a_\alpha \,, \\ & d\tau_{\alpha\beta} = -\frac{1}{r}\omega_{0(\alpha}{}^\gamma u_{\beta)\gamma} + \frac{1}{r}e_0^{\gamma\dot\gamma}\left(\epsilon_{\gamma(\alpha}v_{\beta)\dot\gamma} + \frac{1}{2r^3}v_{\dot\gamma}{}^\delta u_{\alpha\beta}u_{\gamma\delta}\right) \,. \end{aligned} \tag{5.4}$$

Note also that S_A does not satisfy the Vasiliev gauge $S(Z,Y)|_{Z=0} = 0$, and that W above has not been redefined as in (2.22). Nonetheless, it has been noted in [24] that a HS transformation of the form

9 In this section we shall use the conventions of [24] which differ form ours.

$\delta W = D_0\epsilon^{(1)}$ with $\epsilon^{(1)} = \left(-\frac{1}{2}\int_0^1 dtz^\alpha S_\alpha|_{z\to tz} + h.c. + f(Y|x)\right)$ and arbitrary $f(Y|x)$ maps W to $W^{\rm phys}$ given by

$$W^{\rm phys} = \frac{4M}{r} e_0^{\alpha\dot\alpha} \kappa_{\alpha\dot\alpha} \exp\left(-\frac{1}{2}\kappa_{\beta\dot\beta} y^\beta \bar y^{\dot\beta}\right) , \tag{5.5}$$

whose spin 2 component gives the frame field and the associated metric

$$e_\mu^a = e_\mu{}^{a(AdS)} + \frac{M}{r}k_\mu k^a , \qquad g_{\mu\nu} = g_{\mu\nu}^{\rm AdS} + \frac{2M}{r}k_\mu k_\nu , \tag{5.6}$$

where $k_\mu = e_0^{\alpha\dot\alpha}\,\kappa_{\alpha\dot\alpha}$. This is the metric of a black hole of mass M in AdS_4 in Kerr-Schild form. The terminology of black hole in HS context requires caution as discussed in the introduction. In addition to the $SO(3)\times SO(2)$, the solution summarized above has been shown to also have $1/4$ of the $\mathcal{N}=2$ supersymmetric HS symmetry of the model, and their infinite dimensional extension thereof [24].

The solution (5.3) differs from (4.17)–(4.19) both in the form of the internal connection $(V^\pm, \bar V^\pm)$ and in that of the gauge field generating functions. As for the internal connection, the difference can be ascribed to the two choices employed in [6] and [24] for solving the deformed oscillator problem (referred to as "symmetric" and "most asymmetric", respectively, in [6,22]). One can show that the resulting internal connections can be connected via a gauge transformation (see [6]), although the small or large nature of this transformation is yet to be investigated. The comparison of W' in (4.19) and W in (5.3) is technically more complicated, as the two differ also by the shift of the Lorentz connection (2.22), and it will be postponed to a future work, but we note that having the same B identical in both solutions strongly suggests that the physical gauge fields should be equivalent, and in particular equivalent to (5.5).

It is worth mentioning that even by working without the gauge function method, with a specific choice of gauge the Didenko-Vasiliev solution can be simplified in such a way that the W connection is reduced to the vacuum one W_0. This simplification was studied in [45], along with the embedding of the solution in the $\mathcal{N}=2$ and $\mathcal{N}=4$ supersymmetric extensions of the bosonic Vasiliev equations.

6. Perturbative Expansion of Vasiliev Equations

In this Section, we shall summarize the standard perturbative expansion of the Vasiliev equations, benefiting from [4,6,12,25,27,46] (for more recent treatments, see [47,48]).

In the normal order, defined by the star product formula (2.4), the inner Klein operators become real analytic in Y and Z space, viz.,

$$\kappa = \kappa_y \star \kappa_z = \exp(iy^\alpha z_\alpha) , \qquad \bar\kappa = \kappa_{\bar y}\star\kappa_{\bar z} = \exp(i\bar y^{\dot\alpha}\bar z_{\dot\alpha}) . \tag{6.1}$$

Assuming that the full field configurations are real-analytic on $\mathcal{Z}_4$ for generic points in $\mathcal{X}_4$, one may thus impose initial conditions

$$B|_{Z=0} = C , \qquad W_\mu|_{Z=0} = a_\mu , \tag{6.2}$$

where the x-dependence is understood. In order to compute the Z-dependence of the fields one may choose $V_A|_{C=0}$, which is a trivial flat connection, to vanish, and a homotopy contractor for the de Rham differential on $\mathcal{Z}_4$, which entails imposing a gauge condition on V_A. One may then solve the constraints on $D_A B$, F_{AB} and $F_{A\mu}$ on $\mathcal{Z}_4$ in a perturbative expansion in C. This procedure gets increasingly complex with increasing order in the expansion, which schematically can be written as

$$B = \sum_{n\geqslant 1} B^{(n)}, \qquad B^{(1)}(C) \equiv C, \tag{6.3}$$

$$V = \sum_{n\geqslant 1} V^{(n)}, \tag{6.4}$$

$$W = \sum_{n\geqslant 0} W^{(n)}(a_\mu), \qquad W^{(0)}(a_\mu) \equiv a_\mu, \tag{6.5}$$

where $B^{(n)}$, $V^{(n)}$ and $W^{(n)}(a_\mu)$ are n-linear functionals in C, and $W^{(n)}(a_\mu)$ is a linear functional in a_μ. These quantities, which are constructed using the homotopy contractor on $\mathcal{Z}_4$, depend on Z, and are real-analytic in $\mathcal{Y}_4 \times \mathcal{Z}_4$ provided that C and a_μ are real analytic in Y-space and all star products arising along the perturbative expansion are well-defined. As for the remaining equations, that is, that $F_{\mu\nu} = 0$ and $D_\mu B = 0$, it follows from the Bianchi identities that they are perturbatively equivalent to $F_{\mu\nu}|_{Z=0} = 0$ and $D_\mu B|_{Z=0} = 0$, which form a perturbatively defined Cartan integrable system on $\mathcal{X}_4$ for C and a_μ.

To Lorentz covariantize, one imposes

$$W'|_{Z=0} = w, \tag{6.6}$$

with W' from (2.22), that is

$$a_\mu = w_\mu + \frac{1}{4i}\left(\omega_\mu^{\alpha\beta} M_{\alpha\beta} + \bar{\omega}_\mu^{\dot{\alpha}\dot{\beta}} \bar{M}_{\dot{\alpha}\dot{\beta}}\right)\Big|_{Z=0}, \tag{6.7}$$

where w does not contain any component field proportional to $y_\alpha y_\beta$ and $\bar{y}_{\dot{\alpha}}\bar{y}_{\dot{\beta}}$ in view of (2.23). Upon substituting the above relation into $W^{(n)}(a_\mu; C, \ldots, C)$, it follows from the manifest Lorentz covariance that the dependence of $F_{\mu\nu}|_{Z=0}$ and $D_\mu B|_{Z=0}$ on the Lorentz connection arises only via the Lorentz covariant derivative ∇ and the Riemann two-form $(r^{\alpha\beta}, r^{\dot{\alpha}\dot{\beta}})$. Thus, the resulting equations in spacetime take the form [6]

$$\nabla w + w \star w + \frac{1}{4i}\left(r^{\alpha\beta} y_\alpha y_\beta + h.c.\right) + i \sum_{\substack{n_1+n_2\geqslant 1 \\ n_{1,2}\geqslant 0}} \left(r^{\alpha\beta} V_\alpha^{(n_1)} \star V_\beta^{(n_2)} + h.c.\right)\Big|_{Z=0}$$

$$+ \sum_{\substack{n_1+n_2\geqslant 1 \\ n_{1,2}\geqslant 0}} W^{(n_1)}(w) \star W^{(n_2)}(w)\Big|_{Z=0} = 0, \tag{6.8}$$

$$\nabla C + \sum_{\substack{n_1+n_2\geqslant 1 \\ n_1\geqslant 0,\, n_2\geqslant 1}} [W^{(n_1)}(w_\mu), B^{(n_2)}]_\pi\Big|_{Z=0} = 0, \tag{6.9}$$

where

$$\nabla w := d_x w + [\omega^{(0)}, w]_\star, \quad \nabla C := d_x C + [\omega^{(0)}, C]_\star, \quad \omega^{(0)} = \frac{1}{4i}\omega^{\alpha\beta} M^{(0)}_{\alpha\beta} - \text{h.c.}, \tag{6.10}$$

$$r^{\alpha\beta} := d\omega^{\alpha\beta} - \omega^{\alpha\gamma} \wedge \omega_\gamma{}^\beta, \qquad \bar{r}^{\dot{\alpha}\dot{\beta}} := d\bar{\omega}^{\dot{\alpha}\dot{\beta}} - \bar{\omega}^{\dot{\alpha}\dot{\gamma}} \wedge \bar{\omega}_{\dot{\gamma}}{}^{\dot{\beta}}. \tag{6.11}$$

Alternatively, in order to stress the perturbation expansion around the maximally symmetric background, including the spin-two fluctuations, it is more convenient to work in terms of the original one-form a_μ. The perturbative expansion up to 3rd order in Weyl curvatures reads

$$da = a \star a + \mathcal{V}(a,a,C) + \mathcal{V}^2(a,a,C,C) + \mathcal{O}(C^3)\,, \quad (6.12)$$

$$dC = a \star C - C \star \pi(a) - \mathcal{U}(a,C,C) + \mathcal{O}(C^3)\,, \quad (6.13)$$

where $(\mathcal{V}, \mathcal{V}^2, \mathcal{U})$ are functionals that can be determined from (6.8) and (6.9). One next assumes that the homotopy contraction in Z-space is performed such that

$$z^A V_A = 0\,, \quad (6.14)$$

which we refer to as the Vasiliev gauge, and expands

$$a = W_0 + a_1 + a_2 + \cdots\,, \qquad C = C_1 + C_2 + \cdots\,. \quad (6.15)$$

where W_0 is the maximally symmetric background; a_1 a and C_1 are linearized fields; a_n and C_n are nth order fluctuations. The resulting linearized field equations on X-space provide an unfolded description of a dynamical scalar field

$$\phi = C_1\,|_{Y=0}\,, \quad (6.16)$$

and a tower of spin-s Fronsdal fields

$$\phi_{a(s)} = \left(e_0^{\mu a} \left((\sigma_a)^{\alpha\dot\alpha} \frac{\partial^2}{\partial y^\alpha \partial \bar y^{\dot\alpha}} \right)^{s-1} a_{1,\mu} \right)\Bigg|_{Y=0=Z}\,, \quad (6.17)$$

where we use the convention that repeated indices are symmetrized. Computing the functional $\mathcal{V}(W_0, W_0, C)$, the linearized unfolded system is given by [4]

$$D_0 a_1 = -\frac{i}{4} H^{\alpha\beta}\partial_\alpha\partial_\beta C_1(y,0) - \frac{i}{4}\bar H^{\dot\alpha\dot\beta}\bar\partial_{\dot\alpha}\bar\partial_{\dot\beta} C_1(0,\bar y) \quad (6.18)$$

$$D_0 C_1 = 0\,, \quad (6.19)$$

where

$$D_0 a_1 := d_x a_1 + \{W_0, a_1\}_\star\,, \qquad D_0 C_1 := d_x C + [W_0, C_1]_\star\,, \quad (6.20)$$

$$h^{\alpha\dot\alpha} := e_0{}^{\alpha\dot\alpha}\,, \qquad H^{\alpha\beta} := h^{\alpha\dot\gamma} \wedge h^{\beta}{}_{\dot\gamma}\,, \qquad \bar H^{\dot\alpha\dot\beta} := \bar h^{\dot\alpha\gamma} \wedge \bar h^{\dot\beta}{}_{\gamma}\,. \quad (6.21)$$

The oscillator expansion of (6.18) furnishes the definition of the spin$-s$ Weyl tensors, and gives the field equations for spin-s fields which remarkably do not contain higher than second order derivatives, and indeed they are the well known Fronsdal equations for massless spin s-fields in AdS. As for (6.19), its oscillator expansion gives the AdS massless scalar field equation in unfolded form.

Perturbative expansion around AdS at second order is rather complicated but still manageable. Schematically the equations take the form

$$D_0 a_2 - \mathcal{V}(W_0, W_0, C_2) = a_1 \star a_1 + 2\mathcal{V}(W_0, a_1, C_1) + \mathcal{V}(W_0, W_0, C_1, C_1)\,, \quad (6.22a)$$

$$D_0 C_2 = [a_1, C_1]_\pi + \mathcal{U}(W_0, C_1, C_1)\,. \quad (6.22b)$$

Even though these terms have been known in the form of parametric integrals for some time, their detailed structure and consequences for the three point functions were considered much later in [11,49], where the field equation for the scalar field was examined, and used for computing the three-point amplitude for spins $0 - s_1 - s_2$. If $s_1 \neq s_2$, only the first source term in (6.22b) contributes and gives a finite result, in agreement with the boundary CFT prediction. However, if $s_1 = s_2$, only the second source term in (6.22b) contributes and gives a divergent result [11,49]. This divergence was confirmed later in [12,50] where the divergence in the three point amplitude for spins $s - 0 - 0$, resulting from the last term in (6.22a). Soon after, it was shown that a suitable redefinition of the master zero-form cures this problem [48,51], as has been also confirmed with the computation of relevant three-point amplitudes [52,53]. A similar redefinition in the one-form sector has also been determined so that the divergence problem arising in the last term in the first equation above has also been removed [54].

In determining the higher order terms in the perturbative expansions of Vasiliev equations, it remains to be established in general what field redefinitions are allowed in choosing the appropriate basis for the description of the physical fields. In wrestling with this problem, the remarkable simplicity of the holographic duals of this highly nonlinear and seemingly very complicated interactions may provide a handle by means of their holographic reconstruction. Such reconstruction has been achieved for the three and certain four-point interactions [13–15]. Putting aside the analysis and interpretation of the nonlocalities [14,55,56], which are present, and nonetheless in accordance with holography by construction, the issue of how to extract helpful hints from them with regard to the nature of the allowed field redefinitions in perturbative analysis of Vasiliev equation remains to be seen.

Of course, ultimately it would be desirable to have a direct formulation of the principles that govern the nonlocal interactions, based on the combined boundary conditions in twistor space as well as spacetime, as we shall comment further below.

7. A Proposal for an Alternative Perturbation Scheme

In what follows, we shall show that for physically relevant initial data $\Psi(Y)$ given by particle and black hole-like states, the solutions obtained using the factorization method can be mapped to the Vasiliev gauge used in the normal ordered perturbation scheme at the linearized level. Whether the Vasiliev gauge is compatible with an asymptotic description in terms of Fronsdal fields to all orders in perturbation theory, or if it has to be modified, possibly together with a redefinition of the zero-form initial data, is an open problem. In finding the proper boundary conditions in both spacetime and twistor space it may turn out to be necessary to require finiteness of a set of classical observables involving integration over these spaces.

One can define formally the aforementioned map to all orders of classical perturbation theory by applying a gauge function

$$G = L \star H \,, \tag{7.1}$$

to the holomorphic gauge solution space, where $H = 1 + \sum_{n \geqslant 1} H^{(n)}$ is a field dependent gauge function to be fixed as to impose the Vasiliev gauge in normal ordering. To begin with, let us consider fluctuations around AdS for which $\Psi(Y)$ consists of particle states, focusing, for concreteness, to the case of scalar particle states worked out in [7]. Upon switching on the gauge function L, the field $V_\alpha^{(L)} := L^{-1} \star V'_\alpha \star L$ develops non-analyticities in the form of poles in Y-space in the particle sector [7]. Applying H removes these poles and expresses the results in terms of Fronsdal fields (at the same time restoring the manifest Lorentz covariance, as we shall see), at least in the leading order. This can be see as follows. We want to obtain $V_\alpha^{(G)}$ such that

$$z^\alpha V_\alpha^{(G)} + \text{h.c.} = z^+ V^{(G)-} - z^- V^{(G)+} + \text{h.c.} \;=\; 0 \,. \tag{7.2}$$

The leading order gauge transformation reads

$$V_\alpha^{(G)(1)} = V_\alpha^{(L)(1)} + \partial_\alpha H^{(1)} \,. \tag{7.3}$$

Contracting by z^α and using the fact that by definition $Z^A V_A^{(G)} = 0$, one finds

$$H^{(1)} = -\left(\frac{1}{z^\beta \partial_\beta} z^A V_A^{(L)(1)} + \text{h.c.}\right) . \tag{7.4}$$

In particular, activating only the scalar ground state (and its negative-energy counterpart) via the parameters $\widetilde{\nu}_{\pm 1}$ in the projector ansatz for B (see (4.25)), this was computed in [7] with the result

$$H^{(1)} = -\frac{i}{4}\frac{1-x^2}{1-2ix_0+x^2}\frac{1}{\widetilde{y}^+\widetilde{y}^-}\frac{\widetilde{y}^+z^- + \widetilde{y}^-z^+}{\widetilde{u}}\left(e^{i\widetilde{u}}-1\right)\Big|_{\eta=+1} + \text{idem}|_{\eta=-1} , \tag{7.5}$$

where we have set $b = \widetilde{\nu}_1 = \widetilde{\nu}_{-1} = 1$, and

$$\widetilde{u} := \widetilde{y}^\alpha z_\alpha = \widetilde{y}^+ z^- - \widetilde{y}^- z^+ , \qquad \widetilde{y}_\alpha = y_\alpha + M_\alpha{}^{\dot\beta}(x,\eta)\bar{y}_{\dot\beta} . \tag{7.6}$$

The matrix $M_\alpha{}^{\dot\beta}(x,\eta)$ can be found in Section 5.2.1 of [7].
This result for $H^{(1)}$ is regular in $\mathcal{Z}$ but has a pole in $\mathcal{Y}$.
It follows that

$$V_\alpha^{(G)(1)} = -\frac{1}{2}\frac{1-x^2}{1-2ix_0+x^2}\frac{z_\alpha}{\widetilde{u}}\left[e^{i\widetilde{u}} - \frac{e^{i\widetilde{u}}-1}{i\widetilde{u}}\right] + \text{idem}|_{\eta=-1} , \tag{7.7}$$

$$B(Y|x) = \frac{4(1-x^2)}{1-2ix_0+x^2}\, e^{iy^\alpha M_\alpha{}^{\dot\beta}\bar{y}_{\dot\beta}}\Big|_{\eta=+1} + \text{idem}|_{\eta=-1} . \tag{7.8}$$

We observe that $V_\alpha^{(G)(1)}$ is now real-analytic *everywhere* on $\mathcal{C}$.
Furthermore, it was shown in [7] that the above expressions for (B, V_α) lead to the relation

$$V_\alpha^{(G)(1)} = z_\alpha \int_0^1 dt\, t\, B(-tz,\bar{y})\, e^{ity^\alpha z_\alpha} \tag{7.9}$$

in agreement with the result obtained in the standard perturbative analysis of Vasiliev equations at leading order. The emergence of z_α in (7.7) shows that manifest Lorentz covariance is restored, in comparison with the expression for $V_\alpha^{(L)(1)}$. The prefactor in (7.7) is a consequence of the fact that we are considering the lowest mode alone in the solution for Fronsdal equation the scalar field, as opposed to summing all the full set of modes.

Expressing the exact solutions obtained in holomorphic gauge in terms of Fronsdal fields amounts to setting up an alternative perturbation scheme in which one constructs the higher orders of the gauge function $H^{(n)}$ subject to *dual boundary conditions*, that is, to conditions restricting the both the twistor-space dependence of the master fields and their spacetime asymptotic behaviour. Indeed, one requires that after having switched on H, the master fields have symbols in normal order that are real-analytic at $Y = Z = 0$, and symbols in Weyl order that belong to a (associative) star-product subalgebra with well-defined classical observables defined by traces over twistor space and integrations over cycles in spacetime. The proposal is that this problem admits a non-trivial solutions, and that it fixes $H^{(n)}$ up to residual small HS gauge transformations and the initial data for the master zero-form B to all orders in classical perturbation theory. It would be interesting to see whether these type of field redefinitions are related to those recently proposed by Vasiliev in order to obtain a quasi-local perturbation theory in terms of Fronsdal fields [51,54]. An important related issue is whether the gauge function G is large in the sense that it affects the values of the HS zero-form charges, which are special types of classical observables given by traces over twistor space defining zero-forms in spacetime that are de Rham closed [20,27,57].

A nontrivial test of the factorization approach is first to show that the solution is finite after performing the higher order $H^{(n)}$ gauge transformations, and second, to show that the resulting n-point correlators are in agreement with the result expected from holography. The corrections beyond the leading order remain to be determined, while the computation of correlators has been performed in which the second order solution in standard perturbative scheme has been used. It has been shown that a naive computation of $B^{(2)}$ in the standard perturbative scheme leads to divergences [11,12], and later it was shown that these divergences can be removed by a suitable redefinition of $C(Y|x)$ [48,51]. Whether there exists a principle based on any notion of quasi-locality in spacetime that governs the nature of such redefinitions to all orders in perturbation is not known, to our best knowledge.

An advantage of the factorization method is that here we start from a full solution to Vasiliev's theory, defined as a classical field theory on the product of spacetime and twistor space (not referring a priori to the conventional perturbative approach). This provides a convenient framework for the description of the solutions with particles fluctuating around nontrivial backgrounds. The key principle here is that linear superposition principle holds for the zero-form initial data $\Psi(Y)$. For example, if we want to describe the solution to Vasiliev equations for particles propagating around BH solution of Section 4, one simply takes $\Psi = \nu_n P_n + \widetilde{\nu}_n \widetilde{P}_n$. The exact solution for the combined system is obtained in this way, but in small fluctuations of a particle propagating in a fixed and exact black hole background, one may treat the parameter $\widetilde{\nu}_n$ and ν_n as small and large, respectively. A very interesting open problem is thus to combine this scheme with the aforementioned proposal for dual boundary conditions in order to work out new types of generating functions for HS amplitudes.

8. Aspects of Higher Spin Geometry

HS theory has been mostly studied at the level of the field equations and in terms of locally defined quantities in ordinary spacetimes or twistor space extensions thereof. It is clearly desirable to develop a globally defined framework for the classical theory, in order to provide geometric interpretations of the exact solutions, and shed further light on the important issue of the choice of boundary conditions on the master fields on the total space required for Vasiliev's equations to produce physically meaningful anti-holographic duals of boundary conformal field theories. An attempt at a global description of HS theory was made in [27] where bundle structures based on different choices of structure groups, soldering forms and classical observables were considered[10]. Here, we shall highlight a particularly interesting choice of structure group, and the resulting infinite dimensional coset description involving tensorial coordinates, and associated generalized frame field and resulting metrics.

8.1. Structure Group

Noting that Vasiliev equations are Cartan integrable in arbitrary number of commuting dimensions, yet without changing the local degrees of freedom associated to the zero-forms (more on this below), we consider the formulation of Vasiliev's theory in terms of the master fields $(W, B, S_\alpha, \bar{S}_{\dot{\alpha}})$ thought of as horizontal forms on a noncommutative fibered space $\mathcal{C}$ with eight-dimensional fibers given by $\mathcal{Y}_4 \times \mathcal{Z}_4$ and base given by an infinite-dimensional commuting real manifold $\mathcal{M}$, consisting of charts coordinatized by X^M. In each chart, the one-form field $W = dX^M W_M$ thus takes its values in the HS Lie algebra

$$\widehat{\mathfrak{hs}}(4) := \left\{ \widehat{P}(Y,Z) \ : \ \tau(\widehat{P}) = (\widehat{P})^\dagger \ = \ -\widehat{P} \right\} , \tag{8.1}$$

and the deformed oscillators $(S_\alpha, \bar{S}_{\dot{\alpha}})$ and the deformation field B are thought of as zero-forms on $\mathcal{M}$, valued in representations of the HS Lie algebra as described in Section 2, where the τ-map is also defined. The extended equations of motion are given by Equations (2.28)–(2.32), with the only difference being that the manifold χ_4 is replaced by $\mathcal{M}$.

---

10 We refer the reader to [27] for considerable amount of details albeit using a a considerably different notation.

In order to define the theory globally on $\mathcal{M}$, we glue together the locally defined master fields using transition functions from a structure group, which by its definition is generated by a structure algebra $\mathfrak{h}$ given by a subalgebra of $\mathfrak{hs}(4)$. One interesting choice is [27]

$$\mathfrak{h} = \widehat{\mathfrak{hs}}_+(4) \oplus \mathfrak{sl}(2,\mathbb{C})\,, \qquad \widehat{\mathfrak{hs}}_+(4) := \frac{1}{2}(1+\pi)\widehat{\mathfrak{hs}}(4)\,, \tag{8.2}$$

where $\mathfrak{sl}(2,\mathbb{C})$ is the algebra of canonical Lorentz transformations. The corresponding connection, also referred to as the generalized Lorentz connection, is given by

$$\Omega := W^+ \oplus \omega\,, \qquad W^\pm := \frac{1}{2}(1 \pm \pi)W'\,, \tag{8.3}$$

where ω is the canonical Lorentz connection, while the projection

$$\mathcal{E} := W^-\,, \tag{8.4}$$

is assumed to form a section[11]. The Vasiliev equations and gauge transformations in terms of these master fields are spelled out in [27].

8.2. Soldering Mechanism

The horizontal differential algebra on $\mathcal{C}$, which is quasi-free in the sense that the curvature constraints are cartan integrable modulo zero-form constraints, can be projected to a free horizontal differential algebra on the reduced total space $\mathcal{C}|_{Z=0}$. To this end, one first solves the constraints on $\mathcal{Z}$ given initial data

$$w(Y|X) \equiv dX^M w_M = dX^M\, \widehat{W}'_M\Big|_{Z=0}\,, \qquad C(Y|X) = B\Big|_{Z=0}\,, \tag{8.5}$$

where $w(Y|X)$ belongs to the reduced HS algebra

$$\mathfrak{hs}(4) := \widehat{\mathfrak{hs}}(4)\Big|_{Z=0}\,, \tag{8.6}$$

and $C(Y|X)$ belongs to its twisted adjoint representation. Next, defining

$$\Gamma \oplus \omega := \Omega\Big|_{Z=0} \in \mathfrak{hs}_+(4) \oplus \mathfrak{sl}(2,\mathbb{C})\,, \qquad E := \mathcal{E}\Big|_{Z=0} \in \mathfrak{hs}(4)\,/\,\mathfrak{hs}_+(4)\,, \tag{8.7}$$

where $\mathfrak{hs}_+(4) := \frac{1}{2}(1+\pi)\mathfrak{hs}(4)$, we assume that $\mathcal{M}$ is soldered by E, that is, the tangent space of $\mathcal{M}$ is assumed to be identified with the coset $\mathfrak{hs}(4)\,/\,\mathfrak{hs}_+(4)$ via E. Thus, expanding

$$E = dX^M E^A_M P_A\,, \qquad \pi(P_A) = -P_A\,, \tag{8.8}$$

where P_A is a basis for the π-odd elements of $\mathfrak{hs}(4)$, and denoting the inverse of the frame field $E_M{}^A$ by $E^M{}_A$, the local translations with gauge parameters $\xi = \xi^A P_A$ can be identified by usual means [27] as infinitesimal diffeomorphisms generated by globally defined vector fields $\vec{\xi} = \xi^A \mathcal{E}^M{}_A \vec{\partial}_M$ combined with local generalized Lorentz transformations with parameters $(\imath_{\vec{\xi}}\Gamma, \imath_{\vec{\xi}}\omega)$.

---

11 It is worth noting that in [27], the form $\mathcal{E}$ was considered to be a soldering form on a manifold $\underline{\mathcal{M}}$ with tangent space isomorphic to the coset $\widehat{\mathfrak{hs}}(4)/\widehat{\mathfrak{hs}}_+(4)$ and containing $\mathcal{M}$ as a submanifold. We have simplified the geometrical framework here by formulating the system directly on $\mathcal{M}$.

8.3. Elimination of Tensorial Coordinates

The framework described above can be used to write the zero-form constraint on $\mathcal{M} \times \{Z = 0\}$, up to leading order in C, as

$$D_A(\Gamma, \omega)C + \{P_A, C\}_\star = 0\,, \tag{8.9}$$

where $D(\Gamma, \omega) = \nabla + \mathrm{ad}_\Gamma$ with ∇ representing the Lorentz covariant derivative. This constraint can be analyzed by decomposing $P_A = \left\{P_{A_\ell}\right\}_{\ell=0}^{\infty}$ into levels of increasing tensorial rank, viz.,

$$P_{A_\ell} = \left\{P_{a(2\ell+1),b(2k)}\right\}_{k=0}^{\ell} = \left\{M_{\{a_1 b_1} \star \cdots \star M_{a_{2k} b_{2k}} \star P_{a_{2k+1}} \star \cdots \star P_{a_{2\ell+1}\}}\right\}_{k=0}^{\ell}\,, \tag{8.10}$$

where (M_{ab}, P_a) are the generators of $SO(3,2)$ and $P_{a(2\ell+1),b(2k)}$ is a Lorentz tensor of type $(2\ell+1, 2k)$. The zeroth level of the zero-form constraint reads

$$D_a(\Gamma, \omega)C + \{P_a, C\}_\star = 0\,, \qquad a = 1, ..., 4\,, \tag{8.11}$$

where the translation generator is given by the twistor relation $P_a = (\sigma_a)^{\alpha\dot{\alpha}} y_\alpha \bar{y}_{\dot{\alpha}}/4$, which implies that $C_{\alpha(n+2s),\dot{\alpha}(n)}$ is identifiable as the nth order symmetrized covariant vectorial derivative of the primary spin-s Weyl tensor $C_{\alpha(2s)}$ [25]. On the other hand, the ℓth level of the zero-form constraint implies

$$D_{a(2\ell+1)}(\Gamma, \omega)C + \{P_{a(2\ell+1)}, C\}_\star = 0\,, \tag{8.12}$$

where the higher translation generator is now given by the enveloping formula

$$P_{a(2\ell+1)} = P_{\{a_1} \star \cdots P_{a_{2\ell+1}\}}\,. \tag{8.13}$$

As a result, the tensorial derivatives $\nabla_{a(2\ell+1)}(\Gamma, \omega)$ factorize on-shell into multiple vectorial derivatives; for example, the tensorial derivative of the physical scalar $\phi = C|_{Y=0}$ factorizes into

$$D_{a(2\ell+1)}(\Gamma, \omega)\phi \propto D_{\{a_1}(\Gamma, \omega) \cdots D_{a_{2\ell+1}\}}(\Gamma, \omega)\phi\,. \tag{8.14}$$

It follows that no new strictly local degrees of freedom are introduced due to the presence of the tensorial coordinates of $\mathcal{M}$.

8.4. Generalized Metrics

Taking traces of $\star$-products of generalized vielbeins $\mathcal{E} = dX^M \mathcal{E}_M$ and adjoint operators on twistor space, one can construct structure group invariants that are tensor fields on $\mathcal{M}$ [27]; in particular, we may consider symmetric rank-r tensor fields

$$G_{(r)} := dX^{\underline{M}_1} \cdots dX^{\underline{M}_r} \mathrm{Tr}\left[\mathcal{K} \star \left(\mathcal{E}_{M_1} \star \mathcal{V}^{k_1,\bar{k}_1}_{\lambda_1,\bar{\lambda}_1}\right) \star \cdots \star \left(\mathcal{E}_{M_r} \star \mathcal{V}^{k_r,\bar{k}_r}_{\lambda_r,\bar{\lambda}_r}\right)\right]\,, \tag{8.15}$$

where $\mathcal{K} \in \{1, \kappa\bar{\kappa}\}$ and

$$\mathcal{V}^{k,\bar{k}}_{\lambda,\bar{\lambda}} := \left\{\exp_\star\left[i(\lambda^\alpha S_\alpha + \bar{\lambda}^{\dot{\alpha}} \bar{S}_{\dot{\alpha}})\right]\right\} \star (B \star \kappa)^{\star k} \star (B \star \bar{\kappa})^{\star \bar{k}}\,, \qquad k, \bar{k} \in \mathbb{N}\,, \tag{8.16}$$

where $(\lambda^\alpha, \bar{\lambda}^{\dot{\alpha}})$ are auxiliary twistor variables, and

$$\mathrm{Tr}\left[f(Z, Y)\right] := \int_{\mathcal{Y} \times \mathcal{Z}} \frac{d^4Y d^4Z}{(2\pi)^4} f(Z, Y)\,. \tag{8.17}$$

Given a (compact) p-cycle $[\Sigma]$ in the homology of $\mathcal{M}$, one can consider the formally defined minimum

$$\mathcal{A}_{\min}[\Sigma, G_{(r)}] := \min_{\Sigma' \in [\Sigma]} \mathcal{A}[\Sigma', G_{(r)}]\,, \tag{8.18}$$

of the area functional

$$\mathcal{A}[\Sigma', G_{(r)}] := \int_{\Sigma'} d^p\sigma \left(\epsilon^{m_1[p]} \cdots \epsilon^{m_r[p]} \underbrace{(f^*G)_{m_1 \ldots m_r} \cdots (f^*G)_{m_1 \ldots m_r}}_{p \text{ times}} \right)^{1/r} , \tag{8.19}$$

defined using the totally anti-symmetric tensor density $\epsilon^{m_1 \cdots m_p}$ and induced metric

$$(f^*G)_{m_1 \ldots m_r} = \partial_{m_1} X^{\underline{M}_1} \cdots \partial_{m_r} X^{\underline{M}_s} G_{\underline{M}_1 \ldots \underline{M}_r} , \tag{8.20}$$

on Σ' equipped with local coordinates σ^m. The area functional is structure group invariant, and its minimum, if well-defined, is Diff($\mathcal{M}$) invariant, hence serving as a classical observable for the HS theory. Clearly, the dressings by the vertex operator-like operators result in a large number of inequivalent metrics for each r, but this is not a novelty in HS theory, as it is possible to consider similar dressings of the Einstein frame metric in ordinary gravity. The tensorial calculus pertinent to examining the variational principles for $r = 2$ is well understood, whereas for $r > 2$ it remains to be investigated further. In particular, one may ask whether there exists any principle for singling out a preferred metric (possibly of rank two), using, for example, calibrations based on abelian p-forms of the type that will be discussed below. The application of these ideas to a geometrical characterization of the exact solutions of HS theory remains to be investigated.

8.5. Abelian p-Form Charges

Another type of intrinsically defined classical observables facilitated by the introduction of soldering one-forms are the charges of on shell closed abelian p-forms, viz.,

$$\mathcal{Q}[\Sigma, H] = \oint_\Sigma H(\mathcal{E}, B, S_\alpha, \bar{S}_{\dot\alpha}, r_{\alpha\beta}, \bar{r}_{\dot\alpha\dot\beta}) , \tag{8.21}$$

where Σ are closed p-cycles in $\mathcal{M}$ or open p-cycles with suitable boundary conditions, and H are globally defined differential forms that are cohomologically nontrivial on shell, namely $dH = 0$ and $\delta_\Lambda H = 0$ ($\Lambda \in \mathfrak{h}$), and H is not globally exact. We also recall that $r_{\alpha\beta}$ is defined in (6.11). The abelian p-forms were considered in [27]

$$H_{[p]} = \mathrm{Tr}\left[(\mathcal{E} \star \mathcal{E} + r^{(S)+})^{\star(p/2)} \star \kappa \right] , \qquad p = 2, 4, 6, \ldots \tag{8.22}$$

where

$$r^{(S)+} = \frac{1}{8i}(1 + \pi) r^{\alpha\beta} S_\alpha \star S_\beta - h.c. \tag{8.23}$$

The conservation of these charges can be checked directly by replacing the exterior derivative by the structure group covariant derivative $D(\Omega)$ inside the trace and using the fact that $D(\Omega)(\mathcal{E} \star \mathcal{E} + r^{(S)+}) = 0$ on shell.

8.6. On-Shell Actions

Actions have been proposed that imply the Vasiliev equations [16,17] upon applying the variational principle. Their on-shell evaluations involve subtleties stemming from their global formulation and the crucial role played by the boundary conditions. Thus, it remains unclear whether these actions vanish on-shell. Nonetheless, one can still employ the abelian p-form charges discussed above as off-shell topological deformations. To this end, one set of candidates that have been considered are [27]

$$S_{\rm top}[\Sigma_2] = {\rm Re}\left\{\tau_2 \oint_{\Sigma_2} {\rm Tr}\left[\kappa \star R\right]\right\}$$
$$S_{\rm top}[\Sigma_4] = {\rm Re}\left\{\oint_{\Sigma_4} {\rm Tr}\left[\kappa \star \left(\tau_4 R \star R + \tilde{\tau}_4 \left((\mathcal{E}\star\mathcal{E} + r^{(S)+}) \star R + \frac{1}{2}(\mathcal{E}\star\mathcal{E} + r^{(S)+})^{\star 2}\right)\right)\right]\right\}, \tag{8.24}$$

where τ_2, τ_4 and $\tilde{\tau}_4$ are complex constants, $\Sigma_{2,4}$ are submanifolds of $\mathcal{M}$ (where Lagrange multipliers vanish), and

$$R := \nabla W^+ + W^+ \star W^+ + \frac{1}{4i}(\omega^{\alpha\beta} M^{(0)}_{\alpha\beta} + \bar{\omega}^{\dot{\alpha}\dot{\beta}} \bar{M}^{(0)}_{\dot{\alpha}\dot{\beta}}), \tag{8.25}$$

which is the curvature of Ω. Using the field equation [27]

$$R + \mathcal{E}\star\mathcal{E} + r^{(S)+} = 0, \tag{8.26}$$

we can express $S_{\rm top}[\Sigma_4]$ on-shell as

$$S_{\rm top}[\Sigma_4] \approx {\rm Re}\left\{(\tau_4 - \frac{1}{2}\tilde{\tau}_4)\oint_{\Sigma_4} H_{[4]}\right\}, \tag{8.27}$$

where $H_{[4]}$ given in (8.22). Infinities arise from the integration over Σ_4 as well as twistor space. Assuming that the divergence from the AdS vacuum has a definite reality property, such that it can be removed by choosing τ_4 appropriately, one is left with an integral over perturbations that may in principle be finite modulo a prescription for integration contours; for related discussions, see [49,58]. Provided that one considers perturbations corresponding to boundary sources, it would be natural to interpret $S_{\rm top}[\Sigma_4]$ as the generating functional for the boundary correlation functions.

One may also construct topological two-forms given on-shell by

$$S_{\rm top}[\Sigma_2] \approx -{\rm Re}\left\{\tau_2 \oint_{\Sigma_2} H_{[2]}\right\}, \tag{8.28}$$

with $H_{[2]}$ given in (8.22). One application of this surface operator is to wrap Σ_2 around a point-like defect or singularity such as the center of the rotationally symmetric and static solution of [24]. In this case, the leading order contribution, which comes from the AdS vacuum, is a divergent integral over twistor space. If the divergence has a definite reality property, one can cancel it by choosing τ_2 appropriately. One would then be left with an integral over the perturbations. As the latter involve nontrivial functions in twistor space, the integral may be finite. It would be interesting to seek an interpretation of the resulting value of $S_{\rm top}[\Sigma_2]$ as some form of entropy of the black hole solutions reviewed above.

An alternative framework in which certain 2- and 4-forms in x-space, referred to as the Lagrangian forms, are introduced was proposed in [59], where their possible application for the computation of HS invariant charges and generating functional for the boundary correlations functions were discussed; see also [60] for the computation of such HS charges on black-hole solutions at the first order in the deformation parameters. Asymptotic charges in HS theories have also been described in [61–63].

Their relation to the HS charges[12] and their evaluation on certain exact solutions will be discussed elsewhere [26].

Acknowledgments: We thank David De Filippi, Gary Gibbons, Dmitry Ponomarev, Eugene Skvortsov and Xi Yin for helpful discussions in the course the writing of this review. The work of CI is supported in part by the Russian Science Foundation grant 14-42-00047 in association with the Lebedev Physical Institute in Moscow. The work of ES is supported in part by NSF grant PHY-1214344. The work of P.S. is supported by Fondecyt Regular grants N° 1140296 and N° 1151107 and Conicyt grant DPI 2014-0115.

Conflicts of Interest: The authors declare no conflict of interest.

References

1. Vasiliev, M.A. Consistent equations for interacting gauge fields of all spins in 3 + 1 dimensions. *Phys. Lett. B* **1990**, *243*, 378–382.
2. Vasiliev, M.A. Properties of equations of motion of interacting gauge fields of all spins in (3 + 1)-dimensions. *Class. Quantum Gravity* **1991**, *8*, 1387.
3. Vasiliev, M.A. More on equations of motion for interacting massless fields of all spins in 3 + 1 dimensions. *Phys. Lett. B* **1992**, *285*, 225–234.
4. Vasiliev, M.A. Higher spin gauge theories: Star product and AdS space. In *The Many Faces of the Superworld*; Shifman, M.A., Ed.; World Scientific Publishing Co. Inc.: Singapore, 1999; pp. 533–610.
5. Didenko, V.E.; Skvortsov, E.D. Elements of Vasiliev theory. *arXiv* **2014**, arXiv:1401.2975.
6. Iazeolla. C.; Sundell, P. Families of exact solutions to Vasiliev's 4D equations with spherical, cylindrical and biaxial symmetry. *J. High Energy Phys.* **2011**, *2011*, 84.
7. Iazeolla, C.; Sundell, P. 4D Higher Spin Black Holes with Nonlinear Scalar Fluctuations. *arXiv* **2017**, arXiv:1705.06713.
8. Prokushkin, S.F.; Vasiliev, M.A. Higher spin gauge interactions for massive matter fields in 3-D AdS space-time. *Nucl. Phys. B* **1999**, *545*, 385–433.
9. Vasiliev, M.A. Star-Product Functions in Higher-Spin Theory and Locality. *J. High Energy Phys.* **2015**, *2015*, 31
10. Bekaert, X.; Boulanger, N.; Sundell, P. How higher-spin gravity surpasses the spin two barrier: No-go theorems versus yes-go examples. *Rev. Mod. Phys.* **2012**, *84*, 987.
11. Giombi, S.; Yin, X. Higher Spin Gauge Theory and Holography: The Three-Point Functions. *J. High Energy Phys.* **2010**, *2010*, 115.
12. Boulanger, N.; Kessel, P.; Skvortsov, E.D.; Taronna, M. Higher spin interactions in four-dimensions: Vasiliev versus Fronsdal. *J. Phys. A* **2016**, *49*, 095402.
13. Bekaert, X.; Erdmenger, J.; Ponomarev, D.; Sleight, C. Towards holographic higher-spin interactions: Four-point functions and higher-spin exchange. *J. High Energy Phys.* **2015**, *2015*, 170.
14. Bekaert, X.; Erdmenger, J.; Ponomarev, D.; Sleight, C. Quartic AdS Interactions in Higher-Spin Gravity from Conformal Field Theory. *J. High Energy Phys.* **2015**, *2015*, 149.
15. Sleight, C.; Taronna, M. Higher Spin Interactions from Conformal Field Theory: The Complete Cubic Couplings. *Phys. Rev. Lett.* **2016**, *116*, 181602.
16. Boulanger, N.; Sundell, P. An action principle for Vasiliev's four-dimensional higher-spin gravity. *J. Phys. A* **2011**, *44*, 495402.

[12] It is worth noting that the HS charges described here, as well as those based on the above mentioned Lagrangian 2-form and evaluated in [60], present significant differences with respect to the asymptotic charges. For instance, the first ones are closed and gauge invariant everywhere in spacetime, which makes them proper classical observables also in the strong-coupling regions. Crucial for this to happen is that they contain contributions from all spins. On the other hand, each spin-s asymptotic charge is conserved per se, but is crucially defined only via an integration over a two-cycle at spatial infinity under the hypotesis of AdS asymptotics, i.e., only in the weak-coupling region. Indeed, it was shown in [60] that the HS charge based on the Lagrangian 2-form can be written as a combination of contributions from each spin-s field integrated over any two-cycle. Each of the latter contributions gives rise to a separately conserved spin-s charge only when the latter two-cycle is pushed to spatial infinity. Moreover, the asymptotic charges depend for their definition on the existence of asymptotic symmetries, whereas the HS charges, being purely master field constructs, are conserved even without assuming any symmetry. As stressed in [60], this can be seen as a consequence of the non-local expansion in derivatives of the physical fields that is naturally contained in the master fields on-shell.

17. Boulanger, N.; Sezgin, E.; Sundell, P. 4D Higher Spin Gravity with Dynamical Two-Form as a Frobenius-Chern-Simons Gauge Theory. *arXiv* **2015**, arXiv:1505.04957.
18. Vasiliev, M.A. Algebraic aspects of the higher spin problem. *Phys. Lett. B* **1991**, *257*, 111–118.
19. Bolotin, K.I.; Vasiliev, M.A. Star-product and massless free field dynamics in AdS(4). *Phys. Lett. B* **2000**, *479*, 421–428.
20. Sezgin, E.; Sundell, P. An Exact solution of 4-D higher-spin gauge theory. *Nucl. Phys. B* **2007**, *762*, 1–37.
21. Iazeolla, C.; Sezgin, E.; Sundell, P. Real forms of complex higher spin field equations and new exact solutions. *Nucl. Phys. B* **2008**, *791*, 231–264.
22. Iazeolla, C.; Sundell, P. Biaxially symmetric solutions to 4D higher-spin gravity. *J. Phys. A* **2013**, *46*, 214004.
23. Aros, R.; Iazeolla, C.; Noreña, J.; Sezgin, E.; Sundell, P.; Yin, Y. FRW and domain walls in higher spin gravity. *arXiv* **2017**, arXiv:1712.02401.
24. Didenko, V.E.; Vasiliev, M.A. Static BPS black hole in 4d higher-spin gauge theory. *Phys. Lett. B* **2009**, *682*, 305–315; Erratum in **2013**, *722*, 389.
25. Sezgin, E.; Sundell, P. Analysis of higher spin field equations in four-dimensions. *J. High Energy Phys.* **2002**, *2002*, 055.
26. Iazeolla, C. Boundary conditions and conserved charges of 4D higher-spin black holes, 2018. In preparation.
27. Sezgin, E.; Sundell, P. Geometry and Observables in Vasiliev's Higher Spin Gravity. *J. High Energy Phys.* **2012**, *2012*, 121.
28. Vasiliev, M.A. Nonlinear equations for symmetric massless higher spin fields in (A)dS(d). *Phys. Lett. B* **2003**, *567*, 139–151.
29. Bekaert, X.; Cnockaert, S.; Iazeolla, C.; Vasiliev, M.A. Nonlinear higher spin theories in various dimensions. *arXiv* **2005**, arXiv:hep-th/0503128.
30. Didenko, V.E.; Matveev, A.S.; Vasiliev, M.A. BTZ Black Hole as Solution of 3-D Higher Spin Gauge Theory. *Theor. Math. Phys.* **2007**, *153*, 1487–1510.
31. Iazeolla, C.; Raeymaekers, J. On big crunch solutions in Prokushkin-Vasiliev theory. *J. High Energy Phys.* **2016**, *2016*, 177.
32. Engquist, J.; Sundell, P. Brane partons and singleton strings. *Nucl. Phys. B* **2006**, *752*, 206–279.
33. Arias, C.; Sundell, P.; Torres-Gomez, A. Differential Poisson Sigma Models with Extended Supersymmetry. *arXiv* **2016**, arXiv:1607.00727.
34. Barrett, J.W.; Gibbons, G.W.; Perry, M.J.; Pope, C.N.; Ruback, P. Kleinian geometry and the N = 2 superstring. *Int. J. Mod. Phys. A* **1994**, *9*, 1457–1494.
35. Sezgin, E.; Sundell, P. On an exact cosmological solution of higher spin gauge theory. *arXiv* **2005**, arXiv:hep-th/0511296.
36. Plyushchay, M.S. R deformed Heisenberg algebra. *Mod. Phys. Lett. A* **1996**, *11*, 2953–2964.
37. Boulanger, N.; Sundell, P.; Valenzuela, M. Three-dimensional fractional-spin gravity. *J. High Energy Phys.* **2014**, *2014*, 52; Erratum in **2016**, *3*, 076.
38. Vasiliev, M.A. Higher Spin Algebras and Quantization on the Sphere and Hyperboloid. *Int. J. Mod. Phys. A* **1991**, *6*, 1115–1135.
39. Barabanshchikov, A.V.; Prokushkin, S.F.; Vasiliev, M.A. Free equations for massive matter fields in (2 + 1)-dimensional anti-de Sitter space from deformed oscillator algebra. *Theor. Math. Phys.* **1997**, *110*, 295–304.
40. Hertog, T.; Horowitz, G.T. Towards a big crunch dual. *J. High Energy Phys.* **2004**, *2014*, 073.
41. Hertog, T.; Horowitz, G.T. Holographic description of AdS cosmologies. *J. High Energy Phys.* **2005**, *2005*, 005.
42. Gubser, S.S.; Song, W. An axial gauge ansatz for higher spin theories. *J. High Energy Phys.* **2014**, *2014*, 36.
43. Iazeolla, C.; Sundell, P. A Fiber Approach to Harmonic Analysis of Unfolded Higher-Spin Field Equations. *J. High Energy Phys.* **2008**, *2008*, 022.
44. Sundell, P.; Yin, Y. New classes of bi-axially symmetric solutions to four-dimensional Vasiliev higher spin gravity. *J. High Energy Phys.* **2017**, *2017*, 43.
45. Bourdier, J.; Drukker, N. On Classical Solutions of 4d Supersymmetric Higher Spin Theory. *J. High Energy Phys.* **2015**, *2015*, 97.
46. Sezgin, E.; Sundell, P. Higher spin N = 8 supergravity. *J. High Energy Phys.* **1998**, *1998*, 016.
47. Didenko, V.E.; Misuna, N.G.; Vasiliev, M.A. Perturbative analysis in higher-spin theories. *J. High Energy Phys.* **2016**, *2016*, 146.

48. Vasiliev, M.A. On the Local Frame in Nonlinear Higher-Spin Equations. *arXiv* **2017**, arXiv:1707.03735.
49. Giombi, S.; Yin, X. Higher Spins in AdS and Twistorial Holography. *J. High Energy Phys.* **2011**, *2011*, 86.
50. Skvortsov, E.D.; Taronna, M. On Locality, Holography and Unfolding. *J. High Energy Phys.* **2015**, *2015*, 44.
51. Vasiliev, M.A. Current Interactions and Holography from the 0-Form Sector of Nonlinear Higher-Spin Equations. *J. High Energy Phys.* **2017**, *2017*, 117.
52. Sezgin, E.; Skvortsov, E.D.; Zhu, Y. Chern-Simons Matter Theories and Higher Spin Gravity. *arXiv* **2017**, arXiv:1705.03197.
53. Didenko, V.E.; Vasiliev, M.A. Test of the local form of higher-spin equations via AdS/CFT. *Phys. Lett. B* **2017**, *775*, 352–360.
54. Gelfond, O.A.; Vasiliev, M.A. Current Interactions from the One-Form Sector of Nonlinear Higher-Spin Equations. *arXiv* **2017**, arXiv:1706.03718.
55. Sleight, C.; Taronna, M. Higher spin gauge theories and bulk locality: A no-go result. *arXiv* **2017**, arXiv:1704.07859.
56. Ponomarev, D. A Note on (Non)-Locality in Holographic Higher Spin Theories. *arXiv* **2017**, arXiv:1710.00403.
57. Bonezzi, R.; Boulanger, N.; de Filippi, D.; Sundell, P. Noncommutative Wilson lines in higher-spin theory and correlation functions of conserved currents for free conformal fields. *J. Phys. A* **2017**, *50*, 475401.
58. Colombo, N.; Sundell, P. Twistor space observables and quasi-amplitudes in 4D higher spin gravity. *J. High Energy Phys.* **2011**, *2011*, 42.
59. Vasiliev, M.A. Invariant Functionals in Higher-Spin Theory. *Nucl. Phys. B* **2017**, *916*, 219–253.
60. Didenko, V.E.; Misuna, N.G.; Vasiliev, M.A. Charges in nonlinear higher-spin theory. *J. High Energy Phys.* **2017**, *2017*, 164.
61. Barnich, G.; Bouatta, N.; Grigoriev, M. Surface charges and dynamical Killing tensors for higher spin gauge fields in constant curvature spaces. *J. High Energy Phys.* **2005**, *2005*, 010.
62. Campoleoni, A.; Henneaux, M.; Hörtner, S.; Leonard, A. Higher-spin charges in Hamiltonian form. II. Fermi fields. *J. High Energy Phys.* **2017**, *2017*, 58.
63. Campoleoni, A.; Henneaux, M.; Hörtner, S.; Leonard, A. Higher-spin charges in Hamiltonian form. I. Bose fields. *J. High Energy Phys.* **2016**, *2016*, 146.

universe MDPI

Review

Higher Spins without (Anti-)de Sitter

Stefan Prohazka * and Max Riegler

International Solvay Institutes, Université Libre de Bruxelles, Physique Mathématique des Interactions Fondamentales, Campus Plaine—CP 231, B-1050 Bruxelles, Belgium; max.riegler@ulb.ac.be
* Correspondence: stefan.prohzaka@ulb.ac.be

Received: 31 October 2017 ; Accepted: 9 January 2018; Published: 19 January 2018

Abstract: Can the holographic principle be extended beyond the well-known AdS/CFT correspondence? During the last couple of years, there has been a substantial amount of research trying to find answers for this question. In this work, we provide a review of recent developments of three-dimensional theories of gravity with higher spin symmetries. We focus in particular on a proposed holographic duality involving asymptotically flat spacetimes and higher spin extended $\mathfrak{bms}_3$ symmetries. In addition, we also discuss developments concerning relativistic and nonrelativistic higher spin algebras. As a special case, Carroll gravity will be discussed in detail.

Keywords: higher spin; non-anti-de sitter; flat space; holography; nonrelativistic holography

1. Introduction

Higher spin theories on Anti-de Sitter (AdS) backgrounds provide many useful insights into various aspects of the holographic principle. Many of these works were inspired by the seminal work of Klebanov and Polyakov [1–3] who conjectured a holographic correspondence between the $O(N)$ vector model in three dimensions and Fradkin–Vasiliev higher spin gravity on AdS_4 [4–6] (see [7–9] for reviews and [10–16] for some key developments). There are many features of higher spin theories that make them interesting to study. In the context of holography, one of these features is that it is a weak/weak correspondence [17,18]. In contrast, the usual AdS/CFT correspondence [19–21] is a weak/strong correspondence that makes it useful for applications, but harder to check in detail since calculations are often feasible only on one side of the correspondence.

In particular, three-dimensional higher spin theories are useful in this context, since (in contrast to the higher-dimensional examples) one can truncate the otherwise infinite tower of higher spin excitations [22]. Furthermore, the equations that describe the propagation of a massless field of spin s in three dimensions imply that there are no local degrees of freedom when $s \geq 2$. Thus, one can also formulate three-dimensional higher spin theories as Chern–Simons theories [23] with specific boundary conditions [24–28]. This is a considerable simplification in comparison with the more complicated higher-dimensional case. Developments in three-dimensional higher spin theories in AdS include[1], e.g., the discovery of minimal model holography [44–46], higher spin black holes [31,47–49] and higher spin holographic entanglement entropy [50,51].

Since higher spin holography in AdS backgrounds has lead to many interesting insights, a natural question to ask is how to generalize this duality such that it involves other spacetimes or quantum field theories. Additionally, indeed, there are many applications where one has spacetimes that do not asymptote to AdS or do so in a weaker way compared to the Brown–Henneaux boundary conditions [52]. Some examples include Lobachevsky spacetimes [53–55], null warped AdS and their generalizations Schrödinger [56–60], Lifshitz spacetimes [61–63], flat space [64–67] and de Sitter

[1] Further examples can be found in [29–43].

holography [68–70]. Some of these spacetimes play an important role as gravity duals for nonrelativistic CFTs, which are a common occurrence in, e.g., condensed matter physics and thus may be able to provide new insight into these strongly interacting systems. Schrödinger spacetimes for example can be used as a holographic dual to describe cold atoms [56,57].

Even though non-AdS higher spin holography is a rather new field of research there has been quite a lot of research in this direction during the last couple of years. Our aim with this review is to give an overview of the results and ideas that have been accumulated over the years. A special focus of this review will be on higher spin theories in three-dimensional flat space as well as the construction of new higher spin theories using kinematical algebras as bulk isometries.

This review is organized as follows. In Section 2, we present an overview of non-AdS holography that makes use of non-AdS boundary conditions of certain higher spin gravity theories. In Section 3, we focus on a specific example of non-AdS higher spin holography, namely flat space higher spin gravity. Section 4 can be read independently of Sections 2 and 3 and explains a different approach of studying non-AdS higher spin theories not via boundary conditions, but rather by using different choices of gauge algebras realizing certain higher spin theories in the bulk.

2. Non-AdS through Boundary Conditions

A lot of the progress in non-AdS higher spin holography has been achieved by imposing suitable boundary conditions that in turn allow one to compare physical boundary observables with their bulk counterpart. In three dimensions, this can be done rather nicely using a first order formulation of gravity [71,72]. In order to set the stage for non-AdS higher spin holography, we give now a brief review[2] of this formulation for the case of Einstein gravity, as well as AdS higher spin gravity.

2.1. *The (Higher Spin) Chern–Simons Formulation of Gravity*

In many situations, it is advantageous to not describe gravity in terms of a metric formulation, but rather in terms of local orthonormal Lorentz frames. That is, one exchanges the metric $g_{\mu\nu}$ with a vielbein e and a spin connection ω. In three dimensions, the dreibein e and dualized spin connection ω can have the same index structure in their Lorentz indices. Thus, one can combine these two quantities into a single gauge field:

$$\mathcal{A} \equiv e^a \mathrm{P}_a + \omega^a \mathrm{J}_a, \tag{1}$$

where the generators P_a and J_a generate the following Lie algebra[3]:

$$[\mathrm{P}_a, \mathrm{P}_b] = \mp\frac{1}{\ell^2}\epsilon_{abc}\mathrm{J}^c, \quad [\mathrm{J}_a, \mathrm{J}_b] = \epsilon_{abc}\mathrm{J}^c, \quad [\mathrm{J}_a, \mathrm{P}_b] = \epsilon_{abc}P^c. \tag{2}$$

- For $-\frac{1}{\ell^2}$, i.e., de Sitter spacetimes, this gauge algebra is $\mathfrak{so}(3,1)$.
- For $\ell \to \infty$, i.e., flat spacetimes, this gauge algebra is $\mathfrak{isl}(2,\mathbb{R})$.
- For $+\frac{1}{\ell^2}$, i.e., anti-de Sitter spacetimes, this gauge algebra is $\mathfrak{so}(2,2) \sim \mathfrak{sl}(2,\mathbb{R}) \oplus \mathfrak{sl}(2,\mathbb{R})$.

It has been shown [71,72] that the Chern–Simons action:

$$S_{\text{CS}}[\mathcal{A}] = \frac{k}{4\pi}\int_{\mathcal{M}} \left\langle \mathcal{A} \wedge \mathrm{d}\mathcal{A} + \frac{2}{3}\mathcal{A} \wedge \mathcal{A} \wedge \mathcal{A} \right\rangle, \tag{3}$$

defined on a three-dimensional manifold $\mathcal{M} = \Sigma \times \mathbb{R}$, with the invariant nondegenerate symmetric bilinear form:

$$\langle \mathrm{J}_a, \mathrm{P}_b \rangle = \eta_{ab}, \quad \langle \mathrm{J}_a, \mathrm{J}_b \rangle = \langle \mathrm{P}_a, \mathrm{P}_b \rangle = 0, \tag{4}$$

---

2 Parts of this review are based on [73–75]. There is also a slight overlap with [76].
3 We raise and lower indices with $\eta = \mathrm{diag}(-,+,+)$ and $\epsilon_{012} = 1$.

is equivalent (up to boundary terms) to the Einstein–Hilbert–Palatini action[4] with vanishing cosmological constant, provided one identifies the Chern–Simons level k with Newton's constant G in three dimensions as:

$$k = \frac{1}{4G}. \tag{5}$$

The just mentioned bilinear form is also called an invariant metric. Its properties are important for each component of the Chern–Simons gauge field to have a kinematical term (non-degeneracy) and for the action to be invariant under gauge transformations (invariance). This is the general setup for Einstein gravity in three dimensions using the Chern–Simons formalism.

AdS spacetimes in particular have some very nice features in this formalism that allow for a very efficient treatment of many physical questions. Maybe the most convenient feature from a Chern–Simons perspective is that the isometry algebra of AdS $\mathfrak{so}(2,2)$ is a direct sum of two copies of $\mathfrak{sl}(2,\mathbb{R})$. This also means that one can split the gauge field $\mathcal{A}$ into two parts A and $\bar{A}$. On the level of the generators, this split can be made explicit by introducing the generators:

$$\mathtt{T}_a = \frac{1}{2}\left(\mathtt{J}_a + \ell \mathtt{P}_a\right), \qquad \bar{\mathtt{T}}_a = \frac{1}{2}\left(\mathtt{J}_a - \ell \mathtt{P}_a\right). \tag{6}$$

These new generators satisfy:

$$[\mathtt{T}_a, \bar{\mathtt{T}}_b] = 0, \qquad [\mathtt{T}_a, \mathtt{T}_b] = \epsilon_{abc}\mathtt{T}^c, \qquad [\bar{\mathtt{T}}_a, \bar{\mathtt{T}}_b] = \epsilon_{abc}\bar{\mathtt{T}}^c. \tag{7}$$

Both $\mathtt{T}_a$ and $\bar{\mathtt{T}}_a$ satisfy an $\mathfrak{sl}(2,\mathbb{R})$ algebra. From (4), one can immediately see that the invariant bilinear forms are given by:

$$\langle \mathtt{T}_a, \mathtt{T}_b \rangle = \frac{\ell}{2}\eta_{ab}, \qquad \langle \bar{\mathtt{T}}_a, \bar{\mathtt{T}}_b \rangle = -\frac{\ell}{2}\eta_{ab}. \tag{8}$$

The gauge field $\mathcal{A}$ in terms of this split can now be written as:

$$\mathcal{A} = A^a \mathtt{T}_a + \bar{A}^a \bar{\mathtt{T}}_a. \tag{9}$$

Thus, after implementing this explicit split of $\mathfrak{so}(2,2)$ into $\mathfrak{sl}(2,\mathbb{R}) \oplus \mathfrak{sl}(2,\mathbb{R})$, the Chern–Simons action (3) also splits into two contributions:

$$S_{\text{EH}}^{\text{AdS}}[A, \bar{A}] = S_{\text{CS}}[A] + S_{\text{CS}}[\bar{A}], \tag{10}$$

where the invariant bilinear forms appearing in the Chern–Simons action are given by (8). Since both $\mathtt{T}_a$ and $\bar{\mathtt{T}}_a$ satisfy an $\mathfrak{sl}(2,\mathbb{R})$ algebra, it is usually practical to not distinguish between the two generators, i.e., setting $\mathtt{T}_a = \bar{\mathtt{T}}_a$. This in turn also means that the invariant bilinear form in both sectors will be the same. From (8), however, one knows that the invariant bilinear form in both sectors should have the opposite sign. This is not a real problem since this relative minus sign can be easily introduced by hand by not taking the sum, but rather the difference of the two Chern–Simons actions:

$$S_{\text{EH}}^{\text{AdS}} = S_{\text{CS}}[A] - S_{\text{CS}}[\bar{A}] = \frac{1}{16\pi G}\left[\int_{\mathcal{M}} \mathrm{d}^3x\sqrt{|g|}\left(\mathcal{R} + \frac{2}{\ell^2}\right) - \int_{\partial\mathcal{M}} \omega^a \wedge e_a\right]. \tag{11}$$

4 For a nice and explicit calculation, see Appendix A in [51].

As the factor of ℓ in (8) only yields an overall factor of ℓ to the action (11), one can also absorb this factor simply in the Chern–Simons level as:

$$k = \frac{\ell}{4G}. \tag{12}$$

The form of the Chern–Simons connection (11) is usually the one discussed in the literature on AdS holography in three dimensions. The big advantage of this split into an unbarred and a barred part in the case of AdS holography is that usually one only has to explicitly treat one of the two sectors, as the other sector works in complete analogy, up to possible overall minus signs.

Aside from this technical simplification, there is another reason why the Chern–Simons formulation is very often used in AdS and non-AdS holography alike. While a generalization to higher-dimensional gravity is easier in the metric formulation, higher spin extensions are more straightforward in this setup. Since a Chern–Simons gauge theory with gauge algebra $\mathfrak{sl}(2,\mathbb{R}) \oplus \mathfrak{sl}(2,\mathbb{R})$ corresponds to spin-2 gravity with AdS isometries, it is natural to promote the gauge algebra to $\mathfrak{sl}(N,\mathbb{R}) \oplus \mathfrak{sl}(N,\mathbb{R})$[5] in order to describe gravity theories with additional higher spin symmetries. Indeed, in [23], it was shown that for $N \geq 3$, such a Chern–Simons theory describes the nonlinear interactions of gravity coupled to a finite tower of massless integer spin-$s \leq N$ fields.

From a holographic perspective, one point of interest is the asymptotic symmetries of these higher spin gravity theories for given sets of boundary conditions. The first set of consistent boundary conditions that lead to interesting higher spin extensions of the Virasoro algebra has been worked out in [24,25].

Aside from extending the gauge algebra, one also has to take care of the normalization of the Chern–Simons level k. This has to be done in such a way that the spin-2 part of the resulting higher spin theory coincides with Einstein gravity. In order to give the Chern–Simons description an interpretation in terms of a metric, one needs to re-extract the geometric information hidden in the gauge field $\mathcal{A}$. For AdS, as well as flat space higher spin theories (in the principal embedding), this can be done via:

$$g_{\mu\nu} = \#\langle e^z_\mu, e^z_\nu \rangle, \tag{13}$$

where # is some normalization constant and e^z_μ is the so-called zuvielbein that can be seen as a higher spin extension of the dreibein e_μ encountered previously. The expression zuvielbein is a German expression meaning "too many legs" to emphasize that the object e^z_μ now contains more geometric information than the usual dreibein found in spin-2 gravity. In the well-known AdS case, the previous equation can be equivalently written as [40,48]:

$$g_{\mu\nu} = \#\left\langle A_\mu - \bar{A}_\mu, A_\nu - \bar{A}_\nu \right\rangle. \tag{14}$$

2.2. Boundary Terms and Higher Spins

After this brief reminder about the Chern–Simons formulation of (AdS) higher spin theories in three dimensions, we now want to set the stage for the transition to non-AdS spacetimes. All of the interesting physics aside from global properties in three-dimensional gravity are governed by degrees of freedom at the boundary. Thus, it is of utmost importance to make sure that one can impose consistently fall off conditions on the gauge field[6] at the asymptotic boundary. Consistent in this context means that one still has a well-defined variational principle after imposing said boundary

---

[5] To be more precise: the spectrum of the higher spin gravity theory depends on the specific embedding of $\mathfrak{sl}(2,\mathbb{R}) \hookrightarrow \mathfrak{sl}(N,\mathbb{R})$. A very popular choice in the literature on AdS higher spin holography is the principal embedding of $\mathfrak{sl}(2,\mathbb{R}) \hookrightarrow \mathfrak{sl}(N,\mathbb{R})$. This is due to the fact that all generators in that particular embedding have a conformal weight greater or equal to two and thus can be interpreted as describing fields with spin $s \geq 2$.

[6] Or the metric in a second order formulation.

conditions. This is crucial since a consistent variational principle is the core principle underlying the definition of equations of motion of a physical system described by some action, as well as the one needed in the path integral. Thus, the necessity of having such a well-defined variational principle in turn also influences the possible set of boundary conditions that can be consistently imposed.

In order to see this, take a closer look at the variation of the Chern–Simons action (3):

$$\delta S_{\mathrm{CS}}[\mathcal{A}] = \frac{k}{2\pi}\int_{\mathcal{M}}\langle\delta\mathcal{A}\wedge F\rangle + \frac{k}{4\pi}\int_{\partial\mathcal{M}}\langle\delta\mathcal{A}\wedge\mathcal{A}\rangle\,. \tag{15}$$

This expression only vanishes on-shell, i.e., when $F = 0$, if the second term on the right-hand side vanishes as well. Assuming that the boundary $\partial\mathcal{M}$ is parametrized by a timelike coordinate t and an angular coordinate φ, this amounts to:

$$\frac{k}{4\pi}\int_{\partial\mathcal{M}}\langle\delta\mathcal{A}_t\mathcal{A}_\varphi - \delta\mathcal{A}_\varphi\mathcal{A}_t\rangle\,. \tag{16}$$

This term vanishes, for instance, if either $\mathcal{A}_\varphi$ or $\mathcal{A}_t$ are equal to zero at the boundary. This is quite a stringent condition on possible boundary conditions. Thus, it would be nice to have a way of enlarging the possible set of consistent boundary conditions. This can be most easily done by adding a boundary term $B[\mathcal{A}]$ to the Chern–Simons action (3). One could consider for example the following boundary term:

$$B[\mathcal{A}] = \frac{k}{4\pi}\int_{\partial\mathcal{M}}\langle\mathcal{A}_\varphi\mathcal{A}_t\rangle\,. \tag{17}$$

Including this boundary term, the total variation of the resulting action is on-shell:

$$\delta S_{\mathrm{CS}}[\mathcal{A}]^{\mathrm{Tot}} = \frac{k}{2\pi}\int_{\partial\mathcal{M}}\langle\delta\mathcal{A}_t\mathcal{A}_\varphi\rangle\,. \tag{18}$$

Vanishing of the total variation then can be achieved for example via:

$$\mathcal{A}_\varphi\Big|_{\partial\mathcal{M}} = 0 \quad\text{or}\quad \delta\mathcal{A}_t\Big|_{\partial\mathcal{M}} = 0. \tag{19}$$

Choosing $\delta\mathcal{A}_t\Big|_{\partial\mathcal{M}} = 0$, one is thus able to enlarge the possible set of boundary conditions by making sure that the variation of a part of the Chern–Simons connection vanishes.

2.3. Examples of Non-AdS Spacetimes Realized with Higher Spin Symmetries

Adding a suitable boundary term to the Chern–Simons connection in order to allow for a bigger set of possible boundary conditions is one of the necessary prerequisites for doing non-AdS holography. The second one is due to an observation first made explicit in [77]. That is, higher spin isometries, i.e., isometries based on $\mathfrak{sl}(N,\mathbb{R})$, can be used to realize certain non-AdS spacetimes asymptotically.

Take for example a direct product of maximally symmetric spacetimes such as $\mathrm{AdS}_2\times\mathbb{R}$ or $\mathbb{H}_2\times\mathbb{R}$, where $\mathbb{H}_2$ is the two-dimensional Lobachevsky plane. Then, assume that the gauge algebra of the Chern–Simons connection is given by a direct sum of an embedding of $\mathfrak{sl}(2,\mathbb{R})\hookrightarrow\mathfrak{sl}(N,\mathbb{R})$ that contains at least one singlet $\mathtt{S}$ with $\mathrm{tr}(\mathtt{S}^2)\neq 0$ and whose $\mathfrak{sl}(2,\mathbb{R})$ generators are labeled as $\mathtt{L}_n$. Furthermore, assume that the manifold $\mathcal{M}$ where the Chern–Simons theory is defined has the topology of a cylinder with radial coordinate ρ and boundary coordinates x^1 and x^2. Then, using (14) and the connection:

$$A = \mathtt{L}_0\,\mathrm{d}\rho + a_1 e^{\rho}\mathtt{L}_+\,\mathrm{d}x^1, \qquad \bar{A} = -\mathtt{L}_0\,\mathrm{d}\rho + e^{\rho}\mathtt{L}_-\,\mathrm{d}x^1 + \mathtt{S}\,\mathrm{d}x^2, \tag{20}$$

where a_1 is some non-zero constant, one obtains the following non-vanishing metric components:

$$g_{\rho\rho} = 2\mathrm{tr}(\mathtt{L}_0^2), \qquad g_{11} = -a_1\mathrm{tr}(\mathtt{L}_+\mathtt{L}_-)e^{2\rho}, \qquad g_{22} = \frac{1}{2}\mathrm{tr}(\mathtt{S}^2). \tag{21}$$

Depending on the sign of a_1, this metric is locally and asymptotically either $\mathrm{AdS}_2 \times \mathbb{R}$ or $\mathbb{H}_2 \times \mathbb{R}$.

This was a first indication that one can model Lobachevsky spacetimes using higher spin gauge-invariant Chern–Simons theories. Following up on this, a natural question to ask is whether or not one can introduce boundary conditions in this setup that lead to interesting boundary dynamics. In [53,55], it was shown that this is, indeed, possible using a very general algorithm[7]. This algorithm can roughly be summarized by the following steps:

Identify Bulk Theory and Variational Principle:

The first step in this algorithm consists of identifying the bulk theory[8] one wants to describe. After that, one has to propose a suitable generalized variational principle, i.e., add appropriate boundary terms that are consistent with the theory under consideration.

Impose Suitable Boundary Conditions:

After having chosen the bulk theory, the next step in this algorithm is choosing appropriate boundary conditions for the Chern–Simons connection $\mathcal{A}$. This is the most crucial step in the whole analysis as the boundary conditions essentially determine the physical content of the putative dual field theory at the boundary. Since one is dealing with a Chern–Simons gauge theory, one also has some gauge freedom left that can be used to simplify computations. Choosing a gauge:

$$\mathcal{A}_\mu = b^{-1}\left(\mathfrak{a}_\mu + a_\mu^{(0)} + a_\mu^{(1)}\right) b + b^{-1}\,\mathrm{d}b, \;\; b = b(\rho), \tag{22}$$

one can then identify the following three contributions to the Chern–Simons connections:

- $\mathfrak{a}_\mu$ denotes the (fixed) background that was chosen in the previous step.
- $a_\mu^{(0)}$ corresponds to state-dependent leading contributions in addition to the background that contains all the physical information about the field degrees of freedom at the boundary.
- $a_\mu^{(1)}$ are subleading contributions.

Choosing suitable boundary conditions in this context thus means choosing $a_\mu^{(0)}$ and $a_\mu^{(1)}$ in such a way that there exist gauge transformations that preserve these boundary conditions, i.e.,

$$\delta_\varepsilon \mathcal{A}_\mu = \mathcal{O}\left(b^{-1}a_\mu^{(0)}b\right) + \mathcal{O}\left(b^{-1}a_\mu^{(1)}b\right), \tag{23}$$

for some gauge parameter ε, which can also be written as:

$$\varepsilon = b^{-1}\left(\epsilon^{(0)} + \epsilon^{(1)}\right) b. \tag{24}$$

The transformations $\epsilon^{(0)}$ usually generate the asymptotic symmetry algebra, while $\epsilon^{(1)}$ are trivial gauge transformations.

Perform Canonical Analysis and Check the Consistency of Boundary Conditions:

Once the boundary conditions and the gauge transformations that preserve these boundary conditions have been fixed, one has to determine the canonical boundary charges. This is a standard

---

[7] See also, e.g., [52,78,79].

[8] This usually boils down to choosing an appropriate embedding of $\mathfrak{sl}(2,\mathbb{R}) \hookrightarrow \mathfrak{sl}(N,\mathbb{R})$ and then fixing the Chern–Simons connections A and $\bar{A}$ in such a way that they correctly reproduce the desired gravitational background.

procedure that is described in great detail for example in [78,79] and is based on the results of [80]. This procedure eventually leads to the variation of the canonical boundary charge:

$$\delta\mathcal{Q}[\varepsilon] = \frac{k}{2\pi}\int_{\partial\Sigma}\left\langle \epsilon^{(0)}\delta a_{\varphi}^{(0)}\right\rangle \mathrm{d}\varphi, \tag{25}$$

where φ parametrizes the cycle of the boundary cylinder. Of course, one also has to check whether or not the boundary conditions chosen at the beginning of the algorithm are actually physically admissible. For the three-dimensional higher spin gravity gravity examples that are treated in this review, that means that the variation of the canonical boundary charge is finite, conserved in time and integrable in field space. However, we want to stress that there are also other examples such as, e.g., [81], where one can also have physically interesting boundary conditions where the canonical boundary charges do not necessarily meet all of the previously stated conditions.

Determine Semiclassical Asymptotic Symmetry Algebra:

This step consists of working out the Dirac brackets between the canonical generators $\mathcal{G}$ that directly yield the semiclassical asymptotic symmetry algebra. There is a well-known trick that can be used to simplify calculations at this point. Assume that one has two charges with Dirac bracket $\{\mathcal{G}[\varepsilon_1],\mathcal{G}[\varepsilon_2]\}$. Then, one can exploit the fact that these brackets generate a gauge transformation as $\{\mathcal{G}[\varepsilon_1],\mathcal{G}[\varepsilon_2]\} = \delta_{\varepsilon_2}\mathcal{G}$ and read of the Dirac brackets by evaluating $\delta_{\varepsilon_2}\mathcal{G}$. This relation for the canonical gauge generators is on-shell equivalent to a corresponding relation only involving the canonical boundary charges:

$$\{\mathcal{Q}[\varepsilon_1],\mathcal{Q}[\varepsilon_2]\} = \delta_{\varepsilon_2}\mathcal{Q}, \tag{26}$$

which in most cases is straightforward to calculate. This directly leads to the semiclassical asymptotic symmetry algebra including all possible semiclassical central extensions.

Determine the Quantum Asymptotic Symmetry Algebra:

This part of the algorithm first appeared in [24]. One insight of this paper was that the asymptotic symmetry algebra derived in the previous steps is only valid for large values of the central charges. For non-linear algebras, such as $\mathcal{W}$-algebras that are frequently encountered in higher spin holography, that means in particular that one has to think about how normal ordering affects the algebra when passing from a semi-classical to a quantum description of the asymptotic symmetries. One particularly simple way of doing this is to take the semi-classical symmetry algebra, normal order non-linear terms and add all possible deformations to the commutation relations. Requiring that the resulting algebra satisfies the Jacobi identities (see, e.g., [82]) is usually enough to fix all the structure constants yielding the quantum asymptotic symmetry algebras.

Identify the Dual Field Theory:

With the results from all the previous steps, one can then proceed in trying to identify or put possible restrictions on a quantum field theory that explicitly realizes these quantum asymptotic symmetries. Once this dual field theory is identified, one can perform further nontrivial checks of the holographic conjecture.

2.3.1. Lobachevsky Spacetimes

As an explicit example of this algorithm, let us consider the Lobachevsky case worked out in [53,55]. In this work, the non-principal embedding of $\mathfrak{sl}(2,\mathbb{R}) \hookrightarrow \mathfrak{sl}(3,\mathbb{R})$ was used to describe fluctuations around the background:

$$\mathrm{d}s^2 = \mathrm{d}t^2 + \mathrm{d}\rho^2 + \sinh^2\rho\,\mathrm{d}\varphi^2. \tag{27}$$

In a Chern–Simons formulation, this means that one can consider a connection of the form:

$$A_t = 0, \qquad \bar{A}_t = \sqrt{3}\,\mathrm{S}, \qquad A_\rho = \mathrm{L}_0, \qquad \bar{A}_\rho = -\mathrm{L}_0, \tag{28a}$$

$$A_\varphi = -\frac{1}{4}\,\mathrm{L}_1\,e^\rho + \frac{2\pi}{k_{\text{cs}}}\left(\mathcal{J}(\varphi)\mathrm{S} + \mathcal{G}^\pm(\varphi)\psi^\pm_{-\frac{1}{2}}\,e^{-\rho/2} + \mathcal{L}(\varphi)\mathrm{L}_{-1}\,e^{-\rho}\right), \tag{28b}$$

$$\bar{A}_\varphi = -\mathrm{L}_{-1}\,e^\rho + \frac{2\pi}{k_{\text{cs}}}\,\bar{\mathcal{J}}(\varphi)\mathrm{S}, \tag{28c}$$

where the non-principal embedding of $\mathfrak{sl}(2,\mathbb{R}) \hookrightarrow \mathfrak{sl}(3,\mathbb{R})$ is characterized by three $\mathfrak{sl}(2)$ generators L_n $(n = -1, 0, 1)$, two sets of generators $\psi^\pm_n$ $(n = -\frac{1}{2}, \frac{1}{2})$ and one singlet S.

Performing the algorithm described previously, one finds that the (quantum) asymptotic symmetry algebra is given by a direct sum of the Polyakov–Bershadsky algebra [83,84] and a $\hat{u}(1)$ current algebra.

Defining $\hat{k} = -k - 3/2$ and denoting normal ordering with respect to a highest-weight representation by : :, the asymptotic symmetry algebra is given by:

$$[\mathrm{J}_n, \mathrm{J}_m] = \kappa\, n\, \delta_{n+m,0} = [\bar{\mathrm{J}}_n, \bar{\mathrm{J}}_m], \tag{29a}$$

$$[\mathrm{J}_n, \mathrm{L}_m] = n\,\mathrm{J}_{n+m}, \tag{29b}$$

$$[\mathrm{J}_n, \mathrm{G}^\pm_m] = \pm\mathrm{G}^\pm_{m+n}, \tag{29c}$$

$$[\mathrm{L}_n, \mathrm{L}_m] = (n-m)\mathrm{L}_{m+n} + \frac{c}{12}\,n(n^2-1)\,\delta_{n+m,0}, \tag{29d}$$

$$[\mathrm{L}_n, \mathrm{G}^\pm_m] = \left(\frac{n}{2} - m\right)\mathrm{G}^\pm_{n+m}, \tag{29e}$$

$$[\mathrm{G}^+_n, \mathrm{G}^-_m] = \frac{\lambda}{2}\left(n^2 - \frac{1}{4}\right)\delta_{n+m,0} - (\hat{k}+3)\mathrm{L}_{m+n} + \frac{3}{2}(\hat{k}+1)(n-m)\mathrm{J}_{m+n} + 3\sum_{p\in\mathbb{Z}} :\mathrm{J}_{m+n-p}\mathrm{J}_p:, \tag{29f}$$

with the $\hat{u}(1)$ level:

$$\kappa = \frac{2\hat{k}+3}{3}, \tag{30}$$

the Virasoro central charge:

$$c = 25 - \frac{24}{\hat{k}+3} - 6(\hat{k}+3), \tag{31}$$

and the central term in the $G^\pm$ commutator:

$$\lambda = (\hat{k}+1)(2\hat{k}+3)\,. \tag{32}$$

Looking at unitary representations of this algebra, one finds that there is only one value where there are negative norm states that are absent, and that is for $\hat{k} = -1$ and thus also $c = 1$. Hence, a natural guess for a dual quantum field theory is a free boson.

Applying the same logic to other non-principally embedded $\mathfrak{sl}(N,\mathbb{R})$ Chern–Simons theories, it became quickly clear that the requirement of having no negative norm states is a very simple tool in restricting possible values of the Chern–Simons level. Furthermore, one could also see that with increasing N, also the allowed values for the central charges started to grow. In [54,85], it was shown that a Chern–Simons theory with next-to-principally embedded $\mathfrak{sl}(2,\mathbb{R}) \hookrightarrow \mathfrak{sl}(N,\mathbb{R})$ allows for boundary conditions that yield a $\mathcal{W}_N^{(2)}$ Feigin–Semikhatov [86] algebra as an asymptotic symmetry algebra. Looking at negative norm states for these algebras, one finds again restrictions on the allowed values of the central charge c that depend on N in such a way that the central charge can take arbitrarily large (but finite[9]) values. This is quite an interesting result since this provides an example of a unitary theory of gravity whose boundary dynamics are covered by a dual quantum field theory that allows both for a semiclassical (large values of the central charge), as well as an ultra quantum (central charge of $\mathcal{O}(1)$) regime. Thus, this family of $\mathcal{W}_N^{(2)}$ models provides a novel class of models that may be good candidates for toy models of quantum gravity in three dimensions.

2.3.2. Lifshitz Spacetimes

Even though the Lobachevsky case was the first example where higher spin symmetries proved useful for describing asymptotics beyond AdS, it is by far not the only case considered in the literature so far. Another example that gained quite a bit of attention is the case of asymptotic Lifshitz spacetimes [61]:

$$\mathrm{d}s^2 = \ell^2 \left(\frac{\mathrm{d}r^2 + \mathrm{d}x^2}{r^2} - \frac{\mathrm{d}t^2}{r^{2z}} \right), \tag{33}$$

where $z \in \mathbb{R}$ is a scaling exponent.

The authors of [62] used the Chern–Simons higher spin formulation successfully to describe non-rotating black holes in three-dimensional Lifshitz spacetimes with $z = 2$. In addition, this allowed them also to study the thermodynamic properties of these black holes in detail.

Another very interesting aspect of describing Lifshitz spacetimes using higher spin symmetries has been explored in [63]. The starting point of the analysis was again an $\mathfrak{sl}(3,\mathbb{R}) \oplus \mathfrak{sl}(3,\mathbb{R})$ higher spin Chern–Simons theory with boundary conditions such that the corresponding metric asymptotes to the Lifshitz spacetime (33). Looking at the resulting form of the asymptotic symmetry algebra, the authors found two copies of a $\mathcal{W}_3$ algebra with a central charge $c = \frac{3\ell}{2G}$. This is quite an interesting result, since this is exactly what one would get starting with a spin-3 extension of AdS_3 [25,27]. It was then later argued in [60] that this may be due to the non-invertibility of the zuvielbein in the higher spin Lifshitz case, and thus, the metric interpretation of Lifshitz spacetimes in higher spin theories might be questioned.

These are not the only interesting features that have been explored in the context of Lifshitz holography using higher spin symmetries. Furthermore, very interesting relations to integrable systems have been discovered in [88–90].

2.3.3. Null Warped, Schrödinger Spacetimes

Null warped AdS:

$$\mathrm{d}s^2 = \ell^2 \left(\frac{\mathrm{d}r^2}{4r^2} + 2r\,\mathrm{d}t\,\mathrm{d}\varphi + f(r,z)\,\mathrm{d}\varphi^2 \right), \tag{34}$$

with $f(r,z) = r^z + \beta r + \alpha^2$ and where z is a real parameter and α and β constants of motion, is another case of spacetimes that have been linked to higher spin theories. In [59], the authors proposed boundary conditions that asymptote to null warped AdS and found a single copy of the $\mathcal{W}_3^{(2)}$ Polyakov–Bershadsky algebra (29) as asymptotic symmetries.

Last, but not least, we also want to mention that Schrödinger spacetimes:

$$\mathrm{d}s^2 = -r^{2z}\,\mathrm{d}t^2 - 2r^2\,\mathrm{d}t\,\mathrm{d}x^- + \frac{\mathrm{d}r^2}{r^2} + r^2\,\mathrm{d}x^i, \tag{35}$$

can be treated in a higher spin context, both in three dimensions, as well as in higher dimensions [91,92].

9 In [87], it has been shown that any embedding of $\mathfrak{sl}(2,\mathbb{R}) \hookrightarrow \mathfrak{sl}(N,\mathbb{R})$ that contains a singlet contains negative norm states for $c \to \infty$.

Even though this review is focused on higher spins without anti-de Sitter, we also want to point out some work on higher spins in de Sitter [68–70], as well an example of chiral higher spin theories in AdS[10] [94].

3. Flat Space Higher Spin Theories as Specific Examples

Besides the examples of non-AdS higher spin theories that have already been mentioned in the previous section, there is another quite prominent example of a holographic correspondence involving higher spins, that is flat space. Before we go into more details regarding higher spins in flat space, we want to give a brief overview of important developments regarding flat space holography in general.

The first indications that there might be a holographic correspondence in asymptotically flat spacetimes were worked out in [95–97]. In the last decade, there has been a lot of progress in that direction especially in three spacetime dimensions. In 2006, Barnich and Compère [98] presented a consistent set of boundary conditions for asymptotically flat spacetimes at null infinity[11] that extended previous considerations of [102]. Using these boundary conditions, Barnich and Compère were able to show that the corresponding asymptotic symmetry algebra is given by the three-dimensional Bondi–Metzner–Sachs algebra ($\mathfrak{bms}_3$) [103,104]. Since the discovery of the Barnich–Compère boundary conditions, many other boundary conditions in asymptotically flat spacetimes have been found leading to either extensions of the $\mathfrak{bms}_3$ algebra as asymptotic symmetry algebra such as [105–110] or to other algebras such as a warped conformal algebra [111], Heisenberg algebras [112] or an $\mathfrak{isl}(2)_k$ algebra [113].

In particular, the Barnich–Compère boundary conditions, and the associated $\mathfrak{bms}_3$ asymptotic symmetries were used quite extensively for various non-trivial checks of a putative holographic correspondence [114–132].

The previously mentioned developments were mainly focused on either pure Einstein gravity or supersymmetric extensions thereof. Now, what about (massless) higher spin theories in flat space? In four or higher dimensional flat space, there are in fact quite a number of no-go theorems that forbid non-trivial higher spin interactions such as the Coleman–Mandula theorem [133], its generalization by Pelc and Horwitz [134], the Aragone–Deser no-go result [135], the Weinberg–Witten theorem [136], and others. For a very nice overview of all these various no-go theorems, please refer to [137]. This seems like bad news for non-trivial interacting (massless) higher spin theories. However, every no-go theorem is only as good as its premises, and as such, there are various ways of circumventing these theorems such as, e.g., having a non-zero cosmological constant [6]. Interestingly enough, the no-go theorems mentioned previously do not apply in three dimensions, and thus, it seems possible to have non-trivial interacting (massless) higher spin theories in three dimensions[12].

Indeed, in [64,65], the first consistent boundary conditions for a higher spin extension of the Poincaré algebra were found.

3.1. Flat Space Spin-3 Gravity

Higher spin theories in three-dimensional flat space can be described in a very similar fashion as in the AdS_3 case, that is by a suitable Chern–Simons formulation. In the AdS_3 case, the basic gauge symmetries of the Chern–Simons gauge field are given by a direct sum of two copies of $\mathfrak{sl}(N,\mathbb{R})$ (or more general $\mathfrak{hs}[\lambda]$). In the flat space case, the corresponding connections take values in $\mathfrak{isl}(N,\mathbb{R})$.

---

10 The boundary conditions in this work can be seen as the spin-3 extension of the boundary conditions found in [93].

11 These are boundary conditions for either future or past null infinity. Thus, to be more precise, one obtains one copy of $\mathfrak{bms}_3$ on future and another copy on past null infinity. For successful efforts of connecting these two algebras, see [99–101].

12 Even though (A)dS backgrounds favor interactions of massless higher spin fields, higher spin interactions are not completely ruled out even in higher-dimensional flat space. For recent developments regarding higher spins in four or higher-dimensional flat space, see, e.g., [138–141].

The structure of $\mathfrak{isl}(N,\mathbb{R})$ is that of a semidirect sum of $\mathfrak{sl}(N,\mathbb{R})$ with an abelian ideal that is isomorphic to $\mathfrak{sl}(N,\mathbb{R})$ as a vector space. One nice thing about this structure is that it can be straightforwardly obtained by suitable İnönü–Wigner contractions [142], and thus, one has a direct way of obtaining these algebras from the well-known AdS_3 higher spin gauge symmetries[13]. These kinds of contractions have been used quite successfully to obtain new higher spin algebras in flat space (both isometries and asymptotic symmetries) [66,73,147], as well as flat space analogues of important formulas like the (spin-3) Cardy formula [125,148].

Thus, the starting point for a spin-3 theory in flat space is a Chern–Simons action with gauge algebra[14] $\mathfrak{isl}(3,\mathbb{R})$ equipped with an appropriate bilinear form. Then, one can choose boundary conditions as [64,65]:

$$\mathcal{A} = b^{-1}\,\mathrm{d}b + b^{-1}a(u,\varphi)b, \qquad b = e^{\frac{r}{2}\mathtt{M}_{-1}}, \tag{36}$$

with:

$$a(u,\varphi) = a_\varphi(u,\varphi)\,\mathrm{d}\varphi + a_u(u,\varphi)\,\mathrm{d}u, \tag{37}$$

where:

$$a_\varphi(u,\varphi) = \mathtt{L}_1 - \frac{\mathcal{M}}{4}\mathtt{L}_{-1} - \frac{\mathcal{N}}{2}\mathtt{M}_{-1} + \frac{\mathcal{V}}{2}\mathtt{U}_{-2} + \mathcal{Z}\mathtt{V}_{-2}, \tag{38a}$$

$$a_u(u,\varphi) = \mathtt{M}_1 - \frac{\mathcal{M}}{4}\mathtt{M}_{-1} + \frac{\mathcal{V}}{2}\mathtt{V}_{-2}. \tag{38b}$$

The operators $\mathtt{L}_n, \mathtt{M}_n$ with $n = \pm 1, 0$ and $\mathtt{U}_m, \mathtt{V}_m$ with $m = \pm 2, \pm 1, 0$ span the $\mathfrak{isl}(3,\mathbb{R})$ algebra[15] with invariant bilinear form:

$$\langle \mathtt{L}_n \mathtt{M}_m \rangle = -2\eta_{nm}, \qquad \langle \mathtt{U}_n \mathtt{V}_m \rangle = \frac{2}{3}\mathcal{K}_{nm}, \tag{39}$$

where $\eta_{nm} = \text{antidiag}(1, -\frac{1}{2}, 1)$ and $\mathcal{K}_{nm} = \text{antidiag}(12, -3, 2, -3, 12)$. The zuvielbein can be extracted from the gauge connection by using that $\mathcal{A} = e_n^{(2)}\mathtt{M}_n + e_n^{(3)}\mathtt{V}_n + \omega_n^{(2)}\mathtt{L}_n + \omega_n^{(3)}\mathtt{U}_n$. Using these ingredients, one can determine the metric[16]

$$g_{\mu\nu} = \eta_{ab} e_\mu^{(2)a} e_\nu^{(2)b} + \mathcal{K}_{ab} e_\mu^{(3)a} e_\nu^{(3)b}. \tag{40}$$

Thus, consequently, these boundary conditions describe the following metric and spin-3 field:

$$\mathrm{d}s^2 = \mathcal{M}\,\mathrm{d}u^2 - 2\,\mathrm{d}r\,\mathrm{d}u + 2\mathcal{N}\,\mathrm{d}u\,\mathrm{d}\varphi + r^2\,\mathrm{d}\varphi^2, \qquad \Phi_{\mu\nu\lambda}\,\mathrm{d}x^\mu\,\mathrm{d}x^\nu\,\mathrm{d}x^\lambda = 2\mathcal{V}\,\mathrm{d}u^3 + 4\mathcal{Z}\,\mathrm{d}u^2\,\mathrm{d}\varphi. \tag{41}$$

From a geometric point of view, the metric is nothing else than flat space in Eddington–Finkelstein coordinates. Working out the asymptotic symmetries, one finds that these boundary conditions lead to the following non-linear, centrally-extended asymptotic symmetry algebra:

---

13 Please refer to [117,119,126,132,143–146] for early, as well as recent work in flat space holography in three dimensions that rely on contractions.

14 To be more precise, it is the principal embedding of $\mathfrak{isl}(2,\mathbb{R}) \hookrightarrow \mathfrak{isl}(3,\mathbb{R})$.

15 The commutation relations are identical to the ones in (42) after restricting the mode numbers as already mentioned and in addition dropping all non-linear terms.

16 The spin-3 field can be determined in analogy by using the cubic Casimir of the $\mathfrak{sl}(3,\mathbb{R})$ subalgebra.

$$\begin{aligned}
[\mathtt{L}_n, \mathtt{L}_m] =&(n-m)\mathtt{L}_{n-m} + \frac{c_L}{12}n(n^2-1)\delta_{n+m,0}, &(42a)\\
[\mathtt{L}_n, \mathtt{M}_m] =&(n-m)\mathtt{M}_{n-m} + \frac{c_M}{12}n(n^2-1)\delta_{n+m,0}, &(42b)\\
[\mathtt{L}_n, \mathtt{U}_m] =&(2n-m)\mathtt{U}_{n+m}, &(42c)\\
[\mathtt{L}_n, \mathtt{V}_m] =&(2n-m)\mathtt{V}_{n+m}, &(42d)\\
[\mathtt{M}_n, \mathtt{U}_m] =&(2n-m)\mathtt{V}_{n+m}, &(42e)\\
[\mathtt{U}_n, \mathtt{U}_m] =&(n-m)(2n^2+2m^2-nm-8)\mathtt{L}_{n+m} + \frac{192}{c_M}(n-m)\Lambda_{n+m}\\
&- \frac{96c_L}{c_M^2}(n-m)\Theta_{n+m} + \frac{c_L}{12}n(n^2-1)(n^2-4)\delta_{n+m,0}, &(42f)\\
[\mathtt{U}_n, \mathtt{V}_m] =&(n-m)(2n^2+2m^2-nm-8)\mathtt{M}_{n+m} + \frac{96}{c_M}(n-m)\Theta_{n+m}\\
&+ \frac{c_M}{12}n(n^2-4)(n^2-1)\delta_{n+m,0}, &(42g)
\end{aligned}$$

with:

$$\Lambda_n = \sum_{p\in\mathbb{Z}} \mathtt{L}_p\mathtt{M}_{n-p}, \qquad \Theta_n = \sum_{p\in\mathbb{Z}} \mathtt{M}_p\mathtt{M}_{n-p}, \tag{43}$$

and:

$$c_L = 0, \qquad c_M = \frac{3}{G}. \tag{44}$$

This algebra is usually denoted by $\mathcal{FW}_3$ to denote its role similar to the $\mathcal{W}_3$ algebra in AdS_3 higher spin holography. Furthermore, this algebra can be obtained as a specific İnönü–Wigner contraction that can be interpreted as a limit of vanishing cosmological constant of the the AdS_3 spin-3 asymptotic $\mathcal{W}_3$ symmetries.

Assuming that one starts with two copies of a quantum[17] $\mathcal{W}_3$ algebra [149] whose generators are labeled as $\mathcal{L}_n$, $\bar{\mathcal{L}}_n$ and $\mathcal{W}_n$, $\bar{\mathcal{W}}_n$, then one can define the following linear combinations:

$$\mathtt{L}_n := \mathcal{L}_n - \bar{\mathcal{L}}_{-n}, \qquad \mathtt{M}_n := \frac{1}{\ell}\left(\mathcal{L}_n + \bar{\mathcal{L}}_{-n}\right), \tag{45a}$$

$$\mathtt{U}_n := \mathcal{W}_n - \bar{\mathcal{W}}_{-n}, \qquad \mathtt{V}_n := \frac{1}{\ell}\left(\mathcal{W}_n + \bar{\mathcal{W}}_{-n}\right), \tag{45b}$$

and in the limit $\ell \to \infty$ obtain exactly (42). It should also be noted that besides the contraction (45), one can also perform a so-called nonrelativistic contraction using the following alternative linear combination:

$$\mathtt{L}_n := \mathcal{L}_n + \bar{\mathcal{L}}_n, \qquad \mathtt{M}_n := -\epsilon\left(\mathcal{L}_n - \bar{\mathcal{L}}_n\right), \tag{46a}$$

$$\mathtt{U}_n := \mathcal{W}_n + \bar{\mathcal{W}}_n, \qquad \mathtt{V}_n := -\epsilon\left(\mathcal{W}_n - \bar{\mathcal{W}}_n\right), \tag{46b}$$

that in the limit $\epsilon \to 0$ yields another kind of non-linear, centrally-extended algebra [147] that can be seen as natural (quantum) higher spin extension of the Galilean conformal algebra $\mathfrak{gca}_2$. In the spin-2 case, these two limits yield two isomorphic algebras, namely the $\mathfrak{bms}_3$ and $\mathfrak{gca}_2$ algebra, respectively. However, as soon as one adds higher spins, these two limits do not yield isomorphic algebras anymore. The reason for this is basically that each limit favors different representations. The ultrarelativistic limit that leads to the $\mathfrak{bms}_3$ algebra favors so-called (unitary) induced representations, whereas the

[17] That means that all non-linear terms are normal ordered with respect to some highest-weight representation and the central terms have $\mathcal{O}(1)$ corrections that are necessary to satisfy the Jacobi identities when the non-linear terms are normal ordered.

nonrelativistic contraction favors (generically non-unitary [66]) highest-weight representations [147]. Since normal ordering requires a notion of vacuum, what is meant by normal ordering differs as soon as there are non-linear terms present in the algebra and as such also influences the structure constants.

3.2. Flat Space Cosmologies with Spin-3 Hair

Cosmological solutions in flat space [150,151] are well known and thoroughly studied objects. As such, another very interesting thing to study in the context of higher spin theories is comprised of cosmological solutions in flat space that also carry higher spin hair. This has been done successfully first in [67] and subsequently also in [152]. The basic idea of describing such cosmological solutions is by taking the φ-part of the connection (37) and extending the u-part by arbitrary, but fixed chemical potentials in such a way that the equations of motion $F = 0$ are satisfied. By imposing suitable holonomy conditions[18], one can then determine, the inverse temperature, angular potential and higher spin chemical potentials and subsequently also the thermal entropy of cosmological solutions with additional spin-3 hair. If one denotes the spin-2 charges by $\mathcal{N}$, $\mathcal{M}$, the spin-3 charges by $\mathcal{Z}$, $\mathcal{V}$ and introducing the dimensionless ratios:

$$\frac{\mathcal{R}-1}{4\mathcal{R}^{3/2}} = \frac{|\mathcal{V}|}{\mathcal{M}^{3/2}}, \qquad \mathcal{R} > 3, \qquad \text{and} \qquad \mathcal{P} = \frac{\mathcal{Z}}{\sqrt{\mathcal{M}\mathcal{N}}}, \tag{47}$$

one obtains the following formula for the thermal entropy of cosmological solutions with spin-3 hair:

$$S_{\text{Th}} = \frac{\pi}{2G}\frac{|\mathcal{N}|}{\sqrt{\mathcal{M}}}\frac{2\mathcal{R}-3-12\mathcal{P}\sqrt{\mathcal{R}}}{(\mathcal{R}-3)\sqrt{4-3/\mathcal{R}}}. \tag{48}$$

This result can also be understood in terms of a limiting procedure of the AdS spin-3 results for the thermal entropy of BTZsolutions with spin-3 hair [148]. In complete analogy to the limiting procedure of the BTZ black hole entropy, one has to consider the following expression that can be seen as a inner horizon entropy formula of spin-3 charged BTZ black holes:

$$S_{\text{inner}} = 2\pi\left|\sqrt{\frac{c\,\mathcal{L}}{6}}\sqrt{1-\frac{3}{4C}} - \sqrt{\frac{\bar{c}\,\bar{\mathcal{L}}}{6}}\sqrt{1-\frac{3}{4\bar{C}}}\right|, \tag{49}$$

where $c = \bar{c} = \frac{3\ell}{2G}$ and the dimensionless ratios C and $\bar{C}$ are given by:

$$\sqrt{\frac{c}{6\mathcal{L}^3}}\frac{\mathcal{W}}{4} = \frac{C-1}{C^{\frac{3}{2}}}, \qquad \sqrt{\frac{\bar{c}}{6\bar{\mathcal{L}}^3}}\frac{\bar{\mathcal{W}}}{4} = \frac{\bar{C}-1}{\bar{C}^{\frac{3}{2}}}. \tag{50}$$

In order to take the limit, one also has to introduce suitable relation of the AdS spin-2 and spin-3 charges $\mathcal{L}$, $\bar{\mathcal{L}}$, $\mathcal{W}$, $\bar{\mathcal{W}}$ with their flat space counterparts $\mathcal{N}$, $\mathcal{M}$ and $\mathcal{Z}$, $\mathcal{V}$. If one chooses the relations as:

$$\mathcal{M} = 12\left(\frac{\mathcal{L}}{c}+\frac{\bar{\mathcal{L}}}{\bar{c}}\right), \qquad \mathcal{N} = 6\ell\left(\frac{\mathcal{L}}{c}-\frac{\bar{\mathcal{L}}}{\bar{c}}\right), \tag{51a}$$

$$\mathcal{V} = 12\left(\frac{\mathcal{W}}{c}+\frac{\bar{\mathcal{W}}}{\bar{c}}\right), \qquad \mathcal{Z} = 6\ell\left(\frac{\mathcal{W}}{c}-\frac{\bar{\mathcal{W}}}{\bar{c}}\right), \tag{51b}$$

and in addition defines:

$$C = \mathcal{R} + \frac{2}{\ell}D(\mathcal{R},\mathcal{P},\mathcal{M},\mathcal{N}), \qquad \bar{C} = \mathcal{R} - \frac{2}{\ell}D(\mathcal{R},\mathcal{P},\mathcal{M},\mathcal{N}). \tag{52}$$

[18] Alternatively, one can also use a closed Wilson loop wrapped around the horizon [153] in order to determine the thermal entropy.

with:

$$D(\mathcal{R},\mathcal{P},\mathcal{M},\mathcal{N}) = \frac{\mathcal{N}}{\mathcal{M}} \frac{\mathcal{R}\left(\mathcal{R}^{\frac{3}{2}}\mathcal{P} + 3\mathcal{R} - 3\right)}{(\mathcal{R}-3)}, \tag{53}$$

then it is straightforward to show that, indeed, one reproduces the entropy formula (48) in the limit $\ell \to \infty$.

Having an entropy formula like (49) at hand also allows one to study possible phase transitions of these higher spin cosmological solutions in flat space by looking at the free energy. Indeed, one finds the usual phase transition to hot flat space first described in [118] plus additional phase transitions because of the additional spin-3 charges. Interestingly and in contrast to the possible phase transitions in AdS [154], there can also be first order phase transitions between various thermodynamical phases in the flat space case.

3.3. Higher Spin Soft Hair in Flat Space

Soft hair excitations of black holes as possible solutions[19] to the black hole information paradox have attracted quite some research interest lately; see, e.g., [157,158]. Especially, three-dimensional gravity proved to be quite an active playground to study soft hair on (higher spin) black holes in AdS [159–162], higher-derivative theories of gravity [163], as well as flat space [112,164]. What is most intriguing about all these near horizon boundary conditions is that they all lead to a (number of) $\hat{\mathfrak{u}}(1)$ current algebra(s), but the entropy is always given in a very simple form:

$$S_{\mathrm{Th}} = 2\pi\left(\mathtt{J}_0^+ + \mathtt{J}_0^-\right), \tag{54}$$

where $\mathtt{J}_0^\pm$ are the spin-2 zero modes of the near horizon symmetry algebras. In the following, we give a brief overview on how to obtain this result for the entropy for spin-3 gravity in flat space.

The starting point is again a Chern–Simons formulation of gravity with a gauge algebra $\mathfrak{isl}(3,\mathbb{R})$ as in Section 3.1[20]. However, one is now interested in describing near horizon boundary conditions of flat space cosmologies with additional spin-3 hair in contrast to the examples previously that focused on the asymptotic symmetries of such configurations. These near horizon boundary conditions can be described by:

$$\mathcal{A} = b^{-1}(a + \mathrm{d})\, b \tag{55}$$

where the radial dependence is encoded in the group element b as [112]:

$$b = \exp\left(\frac{1}{\mu_{\mathcal{P}}}\mathtt{M}_1\right)\exp\left(\frac{r}{2}\mathtt{M}_{-1}\right), \tag{56}$$

and the connection a reads:

$$a = a_v\,\mathrm{d}v + a_\varphi\,\mathrm{d}\varphi, \tag{57}$$

with:

$$a_\varphi = \mathcal{J}\,\mathtt{L}_0 + \mathcal{P}\,\mathtt{M}_0 + \mathcal{J}^{(3)}\,\mathtt{U}_0 + \mathcal{P}^{(3)}\,\mathtt{V}_0, \tag{58a}$$

$$a_v = \mu_{\mathcal{P}}\,\mathtt{L}_0 + \mu_{\mathcal{J}}\,\mathtt{M}_0 + \mu_{\mathcal{P}}^{(3)}\,\mathtt{U}_0 + \mu_{\mathcal{J}}^{(3)}\,\mathtt{V}_0. \tag{58b}$$

All the functions appearing in (58) are in principle arbitrary functions of the advanced time v and the angular coordinate φ. Based on these boundary conditions, it is straightforward to determine the near horizon symmetry algebra as:

---

19 For a contrasting view on the role of soft hair in solving the black hole paradox, see, e.g., [155,156] and the references therein.
20 Please note that instead of the retarded time coordinate u it is more natural to use the advanced time coordinate v.

$$[\mathrm{J}_n, \mathrm{P}_m] = k\, n\, \delta_{n+m,0} \qquad [\mathrm{J}^{(3)}_n, \mathrm{P}^{(3)}_m] = \frac{4k}{3}\, n\, \delta_{n+m,0} \tag{59}$$

where $k = \frac{1}{4G}$, which can also be brought into a different form by:

$$\mathrm{J}^{\pm}_{\pm n} = \frac{1}{2}(\mathrm{P}_n \pm \mathrm{J}_n) \qquad \mathrm{J}^{(3)\pm}_{\pm n} = \frac{1}{2}(\mathrm{P}^{(3)}_n \pm \mathrm{J}^{(3)}_n)\,. \tag{60}$$

The generators $\mathrm{J}^{\pm}_n$ and $\mathrm{J}^{(3)\pm}_n$ then satisfy:

$$[\mathrm{J}^{\pm}_n, \mathrm{J}^{\pm}_m] = \frac{k}{2} n\delta_{n+m,0} \qquad [\mathrm{J}^{+}_n, \mathrm{J}^{-}_m] = 0\,, \tag{61a}$$

$$[\mathrm{J}^{(3)\pm}_n, \mathrm{J}^{(3)\pm}_m] = \frac{2k}{3} n\delta_{n+m,0}\,, \qquad [\mathrm{J}^{(3)+}_n, \mathrm{J}^{(3)-}_m] = 0\,. \tag{61b}$$

Thus, the near horizon symmetries are given by two pairs of $\hat{\mathfrak{u}}(1)$ current algebras. Calculating both the Hamiltonian (in order to check that these excitations are, indeed, soft), as well as the thermal entropy, which is given by (54), is a straightforward exercise, and we refer the interested reader to [164] for more details.

With a simple result like (54) for the thermal entropy of a spin-3 charged flat space cosmology and a rather complicated one like (48), a natural question to ask is: How exactly are these two related? Is there a way to construct the asymptotic state-dependent functions $\mathcal{M}, \mathcal{N}, \mathcal{V}$ and $\mathcal{Z}$ in terms of the near-horizon state-dependent functions $\mathcal{J}, \mathcal{P}, \mathcal{J}^{(3)}$ and $\mathcal{P}^{(3)}$?

In order to answer these questions, one has to find a gauge transformation that maps these two connections into each other without changing the canonical boundary charges. Such a gauge transformation can, indeed, be found and gives the relations:

$$\mathcal{M} = \mathcal{J}^2 + \frac{4}{3}\left(\mathcal{J}^{(3)}\right)^2 + 2\mathcal{J}', \tag{62a}$$

$$\mathcal{N} = \mathcal{J}\mathcal{P} + \frac{4}{3}\mathcal{J}^{(3)}\mathcal{P}^{(3)} + \mathcal{P}', \tag{62b}$$

$$\mathcal{V} = \frac{1}{54}\left(18\mathcal{J}^2\mathcal{J}^{(3)} - 8\left(\mathcal{J}^{(3)}\right)^3 + 9\mathcal{J}'\mathcal{J}^{(3)} + 27\mathcal{J}\mathcal{J}^{(3)\prime} + 9\mathcal{J}^{(3)\prime\prime}\right), \tag{62c}$$

$$\mathcal{Z} = \frac{1}{36}\Big(6\mathcal{J}^2\mathcal{P}^{(3)} - 8\mathcal{P}^{(3)}\left(\mathcal{J}^{(3)}\right)^2 + 3\mathcal{P}^{(3)}\mathcal{J}' + 3\mathcal{J}^{(3)}\mathcal{P}' + 9\mathcal{J}\mathcal{P}^{(3)\prime} + 9\mathcal{P}\mathcal{J}^{(3)\prime} + 12\mathcal{P}\mathcal{J}\mathcal{J}^{(3)} + 3\mathcal{P}^{(3)\prime\prime}\Big) \tag{62d}$$

that are basically (twisted) Sugawara constructions for the spin-2 and spin-3 fields. One can use these relations and an appropriate Fourier expansion in order to solve for the zero modes $\mathrm{P}_0 = \mathrm{J}^{+}_0 + \mathrm{J}^{-}_0$, which gives:

$$P_0 = \mathrm{J}^{+}_0 + \mathrm{J}^{-}_0 = \frac{1}{4G}\frac{\mathcal{N}\left(4\mathcal{R} - 6 + 3\mathcal{P}\sqrt{\mathcal{R}}\right)}{4\sqrt{\mathcal{M}}(\mathcal{R}-3)\sqrt{1-\frac{3}{4\mathcal{R}}}} \tag{63}$$

and correctly reproduces (48). Thus, one sees that also for flat space cosmologies, there seems to be a much easier way to count the microscopical states contributing to the thermal entropy; that is, in terms of near horizon variables instead of asymptotic ones.

3.4. One Loop Higher Spin Partition Functions in Flat Space

One loop partition functions often provide very useful insights on the consistency of the spectrum for a possible interacting quantum field theory. On (A)dS backgrounds, this feature has been exploited quite successfully. In three bulk dimensions for example, the comparison between bulk and boundary partition functions [165–167] has been an important ingredient in defining the holographic correspondence between higher spin gauge theories and minimal model CFTs [45]. In spacetime dimensions higher than three, the analysis of one-loop partition functions of infinite sets of higher spin fields provided the first quantum checks [168–172] of analogous AdS/CFT dualities [1].

Since the study of one-loop higher spin partition functions has proven to be quite a fruitful line of research, a natural question to ask is whether or not one can extend such considerations also to higher spin theories in d-dimensional flat space. This venue has been successfully pursued in [173], where one-loop partition functions of (supersymmetric) higher spin fields in d-dimensional thermal flat space with angular potentials $\vec{\theta}$ and inverse temperature β have been computed for the first time using both a heat kernel, as well as a group theoretic approach. Also in this case, the three-dimensional case is of special interest since for $d = 3$, one can explicitly show that suitable products of massless one-loop partition functions:

$$Z[\beta,\vec{\theta}] = e^{\delta_{s,2}\frac{\beta c_M}{24}} \prod_{n=s}^{\infty} \frac{1}{|1-e^{in(\theta+i\epsilon)}|^2}, \quad c_M = 3/G, \tag{64}$$

coincide with vacuum characters of $\mathcal{FW}_N$ algebras:

$$\chi_{\mathcal{FW}_N} = e^{\beta c_M/24} \prod_{s=2}^{N} \left(\prod_{n=s}^{\infty} \frac{1}{|1-e^{in(\theta+i\epsilon)}|^2} \right). \tag{65}$$

3.5. Further Aspects of Higher Spins in 3D Flat Space

As some final remarks regarding higher spins in flat space, we want to describe a little bit more explicitly the content of the two works [174,175].

The first work [174] shows how higher spin symmetries could be used to get rid of the causal singularity in the Milne metric[21] in three dimensions [115,116]. The basic idea here is that one can reformulate the Milne metric equivalently in terms of a Chern–Simons connection and then enlarging the gauge algebra of the Chern–Simons connection from $\mathfrak{isl}(2,\mathbb{R})$ to $\mathfrak{isl}(3,\mathbb{R})$. Requiring that the holonomies of the higher spin connection match those of the original spin-2 connection does place some restrictions on the possible spin-3 extensions of the Chern–Simons gauge field; however, it still leaves enough freedom that can be used to get rid of the causal singularity that is present in the spin-2 case at the level of the Ricci scalar[22] and in addition have a non-singular spin-3 field supporting the geometry.

The second work in this context that we would like to mention explicitly is [175]. One very important ingredient in establishing a (higher-spin) holographic principle in asymptotically flat spacetimes is to find concrete theories that are invariant under the corresponding asymptotic symmetries. For Einstein gravity without a cosmological constant and Barnich–Compère boundary conditions, this would be the $\mathfrak{bms}_3$ algebra, and indeed, for this case, it has been suggested in [181] that a flat limit of Liouville theory would be a suitable candidate[23]. The work [175] extended the previous considerations accordingly to a two-dimensional action invariant under a spin-3 extension of the $\mathfrak{bms}_3$ algebra. The corresponding action can also be obtained as a suitable limit of $\mathfrak{sl}(3,\mathbb{R})$ Toda theory as expected.

4. Non-AdS through the Choice of Gauge Group

After the considerations of the preceding sections, it is natural to ask if there are higher spin theories based on Lie algebras beyond (A)dS and Poincaré. This is of interest because it became clear

21 See [176] for a work along similar lines, however, for a null-orbifold of flat space [177–180].

22 It should be noted that there is still the possibility that a possible spin-3 generalization of the Ricci scalar is singular. However, there is at the moment no full geometric interpretation of higher spin symmetries that would be necessary in order to check this.

23 See also [131] for a more group theoretic approach to the problem.

that for nonrelativistic holography, also nonrelativistic geometries play a fundamental role; for a review, see, e.g., [182]. In Section 2.3, the focus was on obtaining geometries beyond (A)dS using different backgrounds and boundary conditions than (A)dS, while still working in a theory given by the gauge group of AdS and its higher spin generalizations. In this section (like in Section 3), we are going to make non-(A)dS geometries manifest by changing the gauge algebra.

Since the tools that were used in the derivation of kinematical algebras and their Chern–Simons theories are the same for the spin-2 case and their spin-3 generalization, we will first focus on the former and comment afterwards on the latter.

4.1. Kinematical Algebras

A classification of interesting kinematical algebras, consisting of generators of time and spatial translations $\mathtt{H}$ and $\mathtt{P}_a$[24], rotations $\mathtt{J}$ and inertial transformations $\mathtt{G}_a$ has been given by Bacry and Levy-Leblond [183]. The classification was provided under the assumptions that:

1. Space is isotropic.
2. Parity and time-reversal are automorphisms of the kinematical groups.
3. Inertial transformations in any given direction form a non-compact subgroup.

This analysis led to other Lie algebras besides the already mentioned (A)dS and Poincaré algebras. Other prominent examples are the Galilei algebra and Carroll algebra and their cousins that appear in the context of spacetimes with non-vanishing cosmological constant. All of them can be conveniently summarized as a cube of İnönü–Wigner contractions[25] starting from the (A)dS algebras; see Figure 1. Since contractions are physically seen as approximations, they often automatically provide insights from the original to the contracted theory.

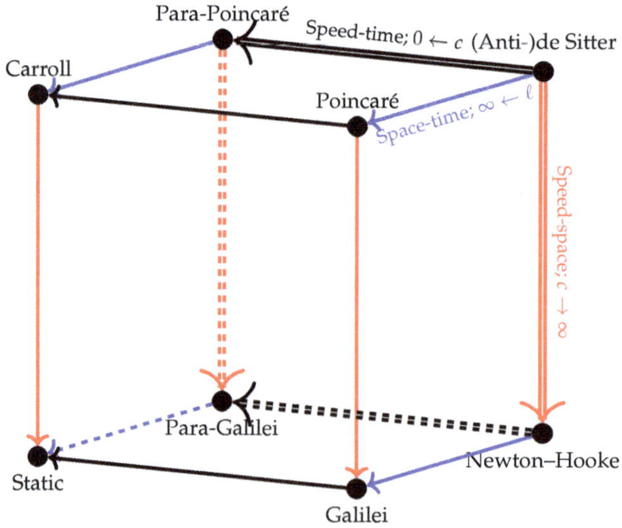

Figure 1. This cube summarizes the kinematical Lie algebras [183]. Each dot represents a kinematical Lie algebra, given explicitly in Appendix A, and each arrow represents an İnönü–Wigner contraction.

---

[24] The indices take now the values $a, b, m = (1, 2)$.

[25] We will use the term İnönü–Wigner contractions here to denote contractions of the form originally defined in [142], sometimes called simple İnönü–Wigner contractions. In contrast to some generalizations like generalized İnönü–Wigner contraction, they are linear in the contraction parameter.

The double arrows mean that there are two contractions since we start with two Lie algebras (AdS and dS). The algebras on the back surface and therefore of finite (A)dS radius ℓ can be considered as cosmological algebras. The top and bottom surfaces can be understood as relative and absolute time Lie algebras, connected by the nonrelativistic limit $c \to \infty$. Sending the speed of light c to zero, the ultrarelativistic limit, leads to absolute space Lie algebras. The parameters of the limits should not be taken too serious since, one of course cannot take the limit of $c \to 0$ and $c \to \infty$ simultaneously (there are actually three parameters involved). They should merely make intuitively clear that the light cone either closes in the ultrarelativistic limit or opens up as for the nonrelativistic one.

To define an İnönü–Wigner contraction, one starts with a Lie algebra $\mathfrak{g}$, which is a vector space direct sum of $\mathfrak{h}$ and $\mathfrak{i}$, i.e., $\mathfrak{g} = \mathfrak{h} \oplus \mathfrak{i}$. One then rescales $\mathfrak{i} \mapsto \frac{1}{\epsilon}\mathfrak{i}$. The commutation relations before and after the contraction are then explicitly given by:

$$[\mathfrak{h},\mathfrak{h}] = \mathfrak{h} + \cancel{\frac{1}{\epsilon}}\mathfrak{i}, \qquad \qquad [\mathfrak{h},\mathfrak{h}] = \mathfrak{h}, \tag{66}$$

$$[\mathfrak{h},\mathfrak{i}] = \epsilon\mathfrak{h} + \mathfrak{i}, \qquad \xrightarrow{\epsilon\to 0} \qquad [\mathfrak{h},\mathfrak{i}] = \mathfrak{i}, \tag{67}$$

$$[\mathfrak{i},\mathfrak{i}] = \epsilon\mathfrak{h} + \epsilon^2\mathfrak{i}, \qquad \qquad [\mathfrak{i},\mathfrak{i}] = 0. \tag{68}$$

The term on the right-hand side of the $[\mathfrak{h},\mathfrak{h}]$ commutator that has been crossed out basically shows that the contraction is convergent in the $\epsilon \to 0$ limes, and therefore well defined, if and only if $\mathfrak{h}$ is a Lie subalgebra of $\mathfrak{g}$ [142]. Specifying $\mathfrak{h}$ completely determines the İnönü–Wigner contraction (up to an isomorphism) [184], and one can therefore enumerate possible contractions by specifying a subalgebra. With this knowledge, we start with the three-dimensional (anti)-de Sitter algebra (the upper sign is the AdS algebra, the lower for dS),

$$[\mathtt{J},\mathtt{G}_a] = \epsilon_{am}\mathtt{G}_m, \qquad [\mathtt{J},\mathtt{P}_a] = \epsilon_{am}\mathtt{P}_m, \tag{69}$$

$$[\mathtt{G}_a,\mathtt{G}_b] = -\epsilon_{ab}\mathtt{J}, \qquad [\mathtt{G}_a,\mathtt{H}] = -\epsilon_{am}\mathtt{P}_m, \tag{70}$$

$$[\mathtt{G}_a,\mathtt{P}_b] = -\epsilon_{ab}\mathtt{H}, \qquad [\mathtt{H},\mathtt{P}_a] = \pm\epsilon_{am}\mathtt{G}_m, \tag{71}$$

$$[\mathtt{P}_a,\mathtt{P}_b] = \mp\epsilon_{ab}\mathtt{J}, \tag{72}$$

and specify the contractions according to Table 1. Consecutive İnönü–Wigner contractions then lead to the cube of Figure 1. For completeness, all the Lie algebras are explicitly given in Table A1 and A2 in Appendix A. For nontrivial contractions, i.e., $\mathfrak{g} \neq \mathfrak{h}$, the resulting algebra is always non-semisimple due to the abelian ideal spanned by $\mathfrak{i}$.

Table 1. The four different İnönü–Wigner contractions classified in [183].

Contraction	$\mathfrak{h}$	$\mathfrak{i}$
Space-time	$\{\mathtt{J},\mathtt{G}_a\}$	$\{\mathtt{H},\mathtt{P}_a\}$
Speed-space	$\{\mathtt{J},\mathtt{H}\}$	$\{\mathtt{G}_a,\mathtt{P}_a\}$
Speed-time	$\{\mathtt{J},\mathtt{P}_a\}$	$\{\mathtt{G}_a,\mathtt{H}\}$
General	$\{\mathtt{J}\}$	$\{\mathtt{H},\mathtt{P}_a,\mathtt{G}_a\}$

4.2. Carroll Gravity

As already discussed in Section 2.1, if one wants to write a Chern–Simons theory for the Lie algebras at hand, it is important for the Lie algebra to admit an invariant metric. While the three-dimensional Carroll algebra automatically admits an invariant metric, others like the Galilei algebra do not. We will discuss in the next section why this is no surprise, but first, we want to show how to construct a Chern–Simons theory with the Carroll algebra and impose boundary conditions (we will follow closely [185], to which we also refer to for more details). Carroll geometries were recently studied because of their relation to asymptotically flat spacetimes [186,187].

The Carroll algebra:

$$[\mathtt{J}, \mathtt{G}_a] = \epsilon_{am}\mathtt{G}_m, \qquad [\mathtt{J}, \mathtt{P}_a] = \epsilon_{am}\mathtt{P}_m, \qquad [\mathtt{G}_a, \mathtt{P}_b] = -\epsilon_{ab}\mathtt{H}, \tag{73}$$

has the invariant metric:

$$\langle \mathtt{H}, \mathtt{J}\rangle = -1, \qquad \langle \mathtt{P}_a, \mathtt{G}_b\rangle = \delta_{ab}\,. \tag{74}$$

The connection:

$$A = \tau\,\mathtt{H} + e^a\,\mathtt{P}_a + \omega\,\mathtt{J} + B^a\,\mathtt{G}_a, \tag{75}$$

is the one-form that takes values in the Carroll algebra, and the action is the usual Chern–Simons action (3). The next step is to construct Brown–Henneaux-like boundary conditions around the Carroll vacuum configuration:

$$e^1_\varphi = \rho, \qquad e^2_\rho = 1, \qquad e^1_\rho = e^2_\varphi = 0, \tag{76}$$

which one can also write as

$$ds^2_{(2)} = e^a e^b \delta_{ab} = \rho^2 d\varphi^2 + d\rho^2\,. \tag{77}$$

We assume that ρ is a radial coordinate and φ is an angular coordinate that is periodically identified by $\varphi \sim \varphi + 2\pi$. Moreover, on the background, the time-component should be fixed as:

$$\tau = dt\,. \tag{78}$$

This can be accomplished by the gauge transformation:

$$A = b^{-1}(\rho)\,(d + a(t,\varphi))\,b(\rho), \qquad b(\rho) = e^{\rho\mathtt{P}_2}, \tag{79}$$

and using [185]:

$$a_\varphi = -\mathtt{J} + h(t,\varphi)\,\mathtt{H} + p_a(t,\varphi)\,\mathtt{P}_a + g_a(t,\varphi)\,\mathtt{G}_a \tag{80}$$

$$a_t = \mu(t,\varphi)\,\mathtt{H}\,. \tag{81}$$

These off-shell boundary conditions lead to:

$$ds^2_{(2)} = \left[(\rho + p_1(t,\varphi))^2 + p_2(t,\varphi)^2\right] d\varphi^2 + 2p_2(t,\varphi)\,d\varphi d\rho + d\rho^2, \tag{82}$$

$$= \left(\rho^2 + \mathcal{O}(\rho)\right) d\varphi^2 + \mathcal{O}(1)\ d\rho d\varphi + d\rho^2, \tag{83}$$

and:

$$\tau = \mu(t,\varphi)\,dt + \left(h(t,\varphi) - \rho\, g_1(t,\varphi)\right) d\varphi\,. \tag{84}$$

$$= \mu(t,\varphi)\,dt + \mathcal{O}(\rho)\,d\varphi\,. \tag{85}$$

The analysis of the asymptotic symmetries leads to conserved, integrable and finite charges, and using a suitable Fourier decomposition, these lead to the asymptotic symmetry algebra:

$$[\mathtt{J}, \mathtt{P}^a_n] = \epsilon_{ab}\, \mathtt{P}^b_n \tag{86}$$
$$[\mathtt{J}, \mathtt{G}^a_n] = \epsilon_{ab}\, \mathtt{G}^b_n \tag{87}$$
$$[\mathtt{P}^a_n, \mathtt{G}^b_m] = -\left(\epsilon_{ab} + in\delta_{ab}\right) \mathtt{H}\, \delta_{n+m,0}\,. \tag{88}$$

It is interesting to note that the zero-mode generators J,H, $\mathtt{P}^a_0$ and $\mathtt{G}^a_0$ span a subalgebra that equals again the original Carroll algebra. Thus, this asymptotic symmetry algebra mirrors constructions from three-dimensional asymptotically flat and anti-de Sitter spacetimes. Two further sets of boundary conditions, which can be seen as a limit from AdS, were proposed in [113], one of which we will briefly describe in the following.

Assume again that Carroll gravity can be described by a Chern–Simons action with gauge algebra (73) and invariant metric (74). Choosing a connection like:

$$a_\varphi = \mathcal{K}(t,\varphi)\mathtt{J} + \mathcal{J}(t,\varphi)\mathtt{H} + \mathcal{G}^a(t,\varphi)(\mathtt{G}_a + \mathtt{P}_a) \qquad a_t = \mu(t,\, \varphi)\,\mathtt{H} \tag{89}$$

it is straightforward to determine the asymptotic symmetry algebra that reads in terms of the Fourier modes of the state dependent functions $\mathcal{K}$, $\mathcal{J}$ and $\mathcal{G}^a$:

$$[\mathtt{K}_n, \mathtt{J}_m] = kn\delta_{m+n,0}, \tag{90a}$$
$$[\mathtt{J}_n, \mathtt{G}^a_m] = \epsilon^a{}_b \mathtt{G}^b_{n+m}, \tag{90b}$$
$$[\mathtt{G}^a_n, \mathtt{G}^b_m] = -2\epsilon^{ab}\mathtt{K}_{n+m} - 2nk\,\delta^{ab}\delta_{n+m,0}\,. \tag{90c}$$

One puzzling aspect of these boundary conditions is that they appear to describe solutions that carry entropy despite (seemingly) having no horizon.

The spin-3 Carroll algebras (see Section 4.4), like their spin-2 subalgebras, admit an invariant metric and thus can be written, without obstructions, as a Chern–Simons theory. While they have been analyzed at a linear level, no boundary conditions were proposed so far.

4.3. Invariant Metrics and Double Extensions

The connection between Lie algebras and their invariant metrics has been greatly clarified by Mediny and Revoy [188] (here, we will follow [189]). They proved a structure theorem that explains how all Lie algebras permitting such an invariant nondegenerate symmetric bilinear form[26] can be constructed. This provides a useful guiding principle for the construction of Lie algebras with invariant metrics and, as will be shown later, also explains why the Carroll algebras inherit their invariant metric from the Poincaré algebra.

For that, one first restricts to indecomposable Lie algebras. We call a Lie algebra indecomposable if it cannot be decomposed as an orthogonal direct sum of two Lie algebras $\mathfrak{g}_1$ and $\mathfrak{g}_2$, i.e., it cannot be written in such a way that the two algebras commute $[\mathfrak{g}_1, \mathfrak{g}_2] = 0$ and that they are orthogonal $\langle \mathfrak{g}_1, \mathfrak{g}_2 \rangle = 0$. Additionally, one has to define double extensions [188]. The Lie algebra $\mathfrak{d} = D(\mathfrak{g}, \mathfrak{h})$ defined on the vector space direct sum $\mathfrak{g} \oplus \mathfrak{h} \oplus \mathfrak{h}^*$ (spanned by $\mathtt{G}_i$, $\mathtt{H}_\alpha$ and $\mathtt{H}^\alpha$, respectively) by:

---

[26] These Lie algebras are sometimes called symmetric self-dual or quadratic.

$$[\mathtt{G}_i, \mathtt{G}_j] = f_{ij}{}^k \mathtt{G}_k + f_{\alpha i}{}^k \Omega^{\mathfrak{g}}_{kj} \mathtt{H}^\alpha, \tag{91}$$
$$[\mathtt{H}_\alpha, \mathtt{G}_i] = f_{\alpha i}{}^j \mathtt{G}_j, \tag{92}$$
$$[\mathtt{H}_\alpha, \mathtt{H}_\beta] = f_{\alpha\beta}{}^\gamma \mathtt{H}_\gamma, \tag{93}$$
$$[\mathtt{H}_\alpha, \mathtt{H}^\beta] = -f_{\alpha\gamma}{}^\beta \mathtt{H}^\gamma, \tag{94}$$
$$[\mathtt{H}^\alpha, \mathtt{G}_j] = 0, \tag{95}$$
$$[\mathtt{H}^\alpha, \mathtt{H}^\beta] = 0, \tag{96}$$

is a double extension of $\mathfrak{g}$ by $\mathfrak{h}$. It has the invariant metric:

$$\Omega^{\mathfrak{d}}_{ab} = \begin{array}{c} \\ \mathtt{G}_i \\ \mathtt{H}_\alpha \\ \mathtt{H}^\alpha \end{array} \begin{array}{c} \begin{array}{ccc} \mathtt{G}_j & \mathtt{H}_\beta & \mathtt{H}^\beta \end{array} \\ \begin{pmatrix} \Omega^{\mathfrak{g}}_{ij} & 0 & 0 \\ 0 & h_{\alpha\beta} & \delta_\alpha{}^\beta \\ 0 & \delta^\alpha{}_\beta & 0 \end{pmatrix} \end{array}, \tag{97}$$

where $\Omega^{\mathfrak{g}}_{ij}$ is an invariant metric on $\mathfrak{g}$ and $h_{\alpha\beta}$ is some arbitrary (possibly degenerate) symmetric invariant bilinear form on $\mathfrak{h}$.

An example is the Poincaré algebra; see Equation (2) with $\ell \to \infty$. In this case, $\mathfrak{g}$ is trivial, and $\mathfrak{h}$ and $\mathfrak{h}^*$ is spanned by $\mathtt{J}_a$ and $\mathtt{P}_a$, respectively. The invariant metric is then given by (4). Similar considerations apply to the Carroll algebra of Section 4.2. These two algebra are actually related by a natural generalization of the İnönü–Wigner contractions to double extensions; see Section 5.3 in [75]. For that, one needs to apply the "dual contraction" on the dual part of a subspace of $\mathfrak{h}$. Explicitly, this means that one takes the Poincaré algebra (see Table A1) and rescales $\mathtt{G}_a \mapsto \frac{1}{c}\mathtt{G}_a$. Since $\mathtt{P}_a$ is in the dual part, this can be read off of Equation (4). One then applies the dual contraction $\mathtt{P}_a \mapsto c\mathtt{P}_a$. Using these rescalings leads to:

$$[\mathtt{J}, \mathtt{G}_a] = \epsilon_{am}\mathtt{G}_m, \qquad [\mathtt{J}, \mathtt{P}_a] = \epsilon_{am}\mathtt{P}_m, \qquad [\mathtt{G}_a, \mathtt{P}_b] = -\epsilon_{ab}\mathtt{H}, \tag{98}$$
$$[\mathtt{G}_a, \mathtt{G}_b] = -c^2\epsilon_{ab}\mathtt{J}, \qquad [\mathtt{G}_a, \mathtt{H}] = -c^2\epsilon_{am}\mathtt{P}_m, \tag{99}$$

and therefore to the Carroll algebra for $c \to 0$. The part that makes this new interpretation interesting is that it automatically leaves the invariant metric untouched since $\langle \mathtt{G}_a, \mathtt{P}_b\rangle \mapsto \langle \frac{1}{c}\mathtt{G}_a, c\mathtt{P}_b\rangle = \langle \mathtt{G}_a, \mathtt{P}_b\rangle$. That this is not just a coincidence, but that these non-simple Lie algebras actually have to be a double extension is explained by the following theorem.

Every indecomposable Lie algebra that permits an invariant metric, i.e., every indecomposable symmetric self-dual Lie algebra, is either [188,189]:

1. A simple Lie algebra.
2. A one-dimensional Lie algebra.
3. A double extended Lie algebra $D(\mathfrak{g}, \mathfrak{h})$ where:
   (a) $\mathfrak{g}$ has no factor $\mathfrak{p}$ for which the first and second cohomology group vanishes $H^1(\mathfrak{p}, \mathbb{R}) = H^2(\mathfrak{p}, \mathbb{R}) = 0$. This includes semisimple Lie algebra factors.
   (b) $\mathfrak{h}$ is either simple or one-dimensional.
   (c) $\mathfrak{h}$ acts on $\mathfrak{g}$ via outer derivations.

Since every decomposable Lie algebra can be obtained from the indecomposable ones, this theorem describes how all of them can be generated; see Figure 2 [75].

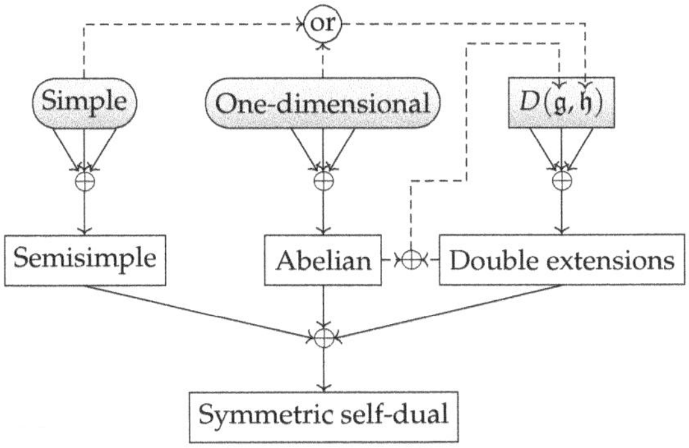

Figure 2. This diagram shows how all Lie algebras with an invariant metric, i.e., symmetric self-dual Lie algebras, are constructed. The fundamental indecomposable building blocks are the simple and the one-dimensional Lie algebras, and they need to be accompanied by the operations of direct sums ($\oplus$) and double extensions ($D(\mathfrak{g},\mathfrak{h})$). Direct sums of simple and one-dimensional Lie algebras lead to semisimple and abelian ones, respectively. To construct new indecomposable Lie algebras that admit an invariant metric, one needs to double extend an abelian or an already double extended Lie algebra (as explained, it should not have a simple factor).

4.4. Kinematical Higher Spin Algebras

We now want to investigate in which sense the kinematical Lie algebras can be generalized to higher spins; specifically, we will focus on spin-two and three fields. For that, it is again useful to start with the (semi)simple, (A)dS algebras[27] explicitly given in Table 2.

The spin-2 part is a subalgebra and is extended by the spin-3 generators $\mathtt{J}_a, \mathtt{H}_a, \mathtt{G}_{ab}, \mathtt{P}_{ab}$. For the generalization of the contractions to the higher spin algebra, the following restrictions are imposed [185]:

- The İnönü–Wigner contractions are restricted such that the contracted spin-2 Lie subalgebra of the contracted one coincides with the kinematical ones of Bacry and Levy-Leblond [183] (see Table 1 and Appendix A).
- The commutator of the spin-3 fields should be non-vanishing. This ensures that the spin-3 field also interacts with the spin-2 field.

Using these restrictions, one can systematically examine the possible contractions summarized in Table 3 [185][28]. These contractions can then be performed leading to the kinematical higher spin algebras. Consecutive contractions span the (higher spin) cube of Figure 3.

---

[27] Semisimple Lie algebras are a natural starting point for these kinds of considerations since no (nontrivial) contraction can lead to a semisimple Lie algebra.

[28] We ignore the traceless contractions in this review.

Table 2. Higher spin versions of the (A)dS algebras. The upper sign is for AdS and the lower sign for dS.

	$\mathfrak{hs}_3(\mathfrak{A})\mathfrak{dS}_{(\mp)}$
Equations (69) to (72)	
$[\mathtt{J}, \mathtt{J}_a]$	$\epsilon_{am}\mathtt{J}_m$
$[\mathtt{J}, \mathtt{G}_{ab}]$	$-\epsilon_{m(a}\mathtt{G}_{b)m}$
$[\mathtt{J}, \mathtt{H}_a]$	$\epsilon_{am}\mathtt{H}_m$
$[\mathtt{J}, \mathtt{P}_{ab}]$	$-\epsilon_{m(a}\mathtt{P}_{b)m}$
$[\mathtt{G}_a, \mathtt{J}_b]$	$-(\epsilon_{am}\mathtt{G}_{bm} + \epsilon_{ab}\mathtt{G}_{mm})$
$[\mathtt{G}_a, \mathtt{G}_{bc}]$	$-\epsilon_{a(b}\mathtt{J}_{c)}$
$[\mathtt{G}_a, \mathtt{H}_b]$	$-(\epsilon_{am}\mathtt{P}_{bm} + \epsilon_{ab}\mathtt{P}_{mm})$
$[\mathtt{G}_a, \mathtt{P}_{bc}]$	$-\epsilon_{a(b}\mathtt{H}_{c)}$
$[\mathtt{H}, \mathtt{J}_a]$	$\epsilon_{am}\mathtt{H}_m$
$[\mathtt{H}, \mathtt{G}_{ab}]$	$-\epsilon_{m(a}\mathtt{P}_{b)m}$
$[\mathtt{H}, \mathtt{H}_a]$	$\pm\epsilon_{am}\mathtt{J}_m$
$[\mathtt{H}, \mathtt{P}_{ab}]$	$\mp\epsilon_{m(a}\mathtt{G}_{b)m}$
$[\mathtt{P}_a, \mathtt{J}_b]$	$-(\epsilon_{am}\mathtt{P}_{bm} + \epsilon_{ab}\mathtt{P}_{mm})$
$[\mathtt{P}_a, \mathtt{G}_{bc}]$	$-\epsilon_{a(b}\mathtt{H}_{c)}$
$[\mathtt{P}_a, \mathtt{H}_b]$	$\mp(\epsilon_{am}\mathtt{G}_{bm} + \epsilon_{ab}\mathtt{G}_{mm})$
$[\mathtt{P}_a, \mathtt{P}_{bc}]$	$\mp\epsilon_{a(b}\mathtt{J}_{c)}$
$[\mathtt{J}_a, \mathtt{J}_b]$	$\epsilon_{ab}\mathtt{J}$
$[\mathtt{J}_a, \mathtt{G}_{bc}]$	$\delta_{a(b}\epsilon_{c)m}\mathtt{G}_m$
$[\mathtt{J}_a, \mathtt{H}_b]$	$\epsilon_{ab}\mathtt{H}$
$[\mathtt{J}_a, \mathtt{P}_{bc}]$	$\delta_{a(b}\epsilon_{c)m}\mathtt{P}_m$
$[\mathtt{G}_{ab}, \mathtt{G}_{cd}]$	$\delta_{(a(c}\epsilon_{d)b)}\mathtt{J}$
$[\mathtt{G}_{ab}, \mathtt{H}_c]$	$-\delta_{c(a}\epsilon_{b)m}\mathtt{P}_m$
$[\mathtt{G}_{ab}, \mathtt{P}_{cd}]$	$\delta_{(a(c}\epsilon_{d)b)}\mathtt{H}$
$[\mathtt{H}_a, \mathtt{H}_b]$	$\pm\epsilon_{ab}\mathtt{J}$
$[\mathtt{H}_a, \mathtt{P}_{bc}]$	$\pm\delta_{a(b}\epsilon_{c)m}\mathtt{G}_m$
$[\mathtt{P}_{ab}, \mathtt{P}_{cd}]$	$\pm\delta_{(a(c}\epsilon_{d)b)}\mathtt{J}$

Table 3. The contractions to the kinematical higher spin algebras. They can be summarized again as a (higher spin) cube; see Figure 3.

Contraction	**#**	$\mathfrak{h}$	$\mathfrak{i}$
Space-time	1	$\{\mathtt{J}, \mathtt{G}_a, \mathtt{J}_a, \mathtt{G}_{ab}\}$	$\{\mathtt{H}, \mathtt{P}_a, \mathtt{H}_a, \mathtt{P}_{ab}\}$
	2	$\{\mathtt{J}, \mathtt{G}_a, \mathtt{H}_a, \mathtt{P}_{ab}\}$	$\{\mathtt{H}, \mathtt{P}_a, \mathtt{J}_a, \mathtt{G}_{ab}\}$
Speed-space	3	$\{\mathtt{J}, \mathtt{H}, \mathtt{J}_a, \mathtt{H}_a\}$	$\{\mathtt{G}_a, \mathtt{P}_a, \mathtt{G}_{ab}, \mathtt{P}_{ab}\}$
	4	$\{\mathtt{J}, \mathtt{H}, \mathtt{G}_{ab}, \mathtt{P}_{ab}\}$	$\{\mathtt{G}_a, \mathtt{P}_a, \mathtt{J}_a, \mathtt{H}_a\}$
Speed-time	5	$\{\mathtt{J}, \mathtt{P}_a, \mathtt{J}_a, \mathtt{P}_{ab}\}$	$\{\mathtt{G}_a, \mathtt{H}, \mathtt{H}_a, \mathtt{G}_{ab}\}$
	6	$\{\mathtt{J}, \mathtt{P}_a, \mathtt{H}_a, \mathtt{G}_{ab}\}$	$\{\mathtt{G}_a, \mathtt{H}, \mathtt{J}_a, \mathtt{P}_{ab}\}$
General	7	$\{\mathtt{J}, \mathtt{J}_a\}$	$\{\mathtt{H}, \mathtt{P}_a, \mathtt{G}_a, \mathtt{H}_a, \mathtt{G}_{ab}, \mathtt{P}_{ab}\}$
	8	$\{\mathtt{J}, \mathtt{G}_{ab}\}$	$\{\mathtt{H}, \mathtt{P}_a, \mathtt{G}_a, \mathtt{J}_a, \mathtt{H}_a, \mathtt{P}_{ab}\}$
	9	$\{\mathtt{J}, \mathtt{H}_a\}$	$\{\mathtt{H}, \mathtt{P}_a, \mathtt{G}_a, \mathtt{J}_a, \mathtt{G}_{ab}, \mathtt{P}_{ab}\}$
	10	$\{\mathtt{J}, \mathtt{P}_{ab}\}$	$\{\mathtt{H}, \mathtt{P}_a, \mathtt{G}_a, \mathtt{J}_a, \mathtt{H}_a, \mathtt{G}_{ab}\}$

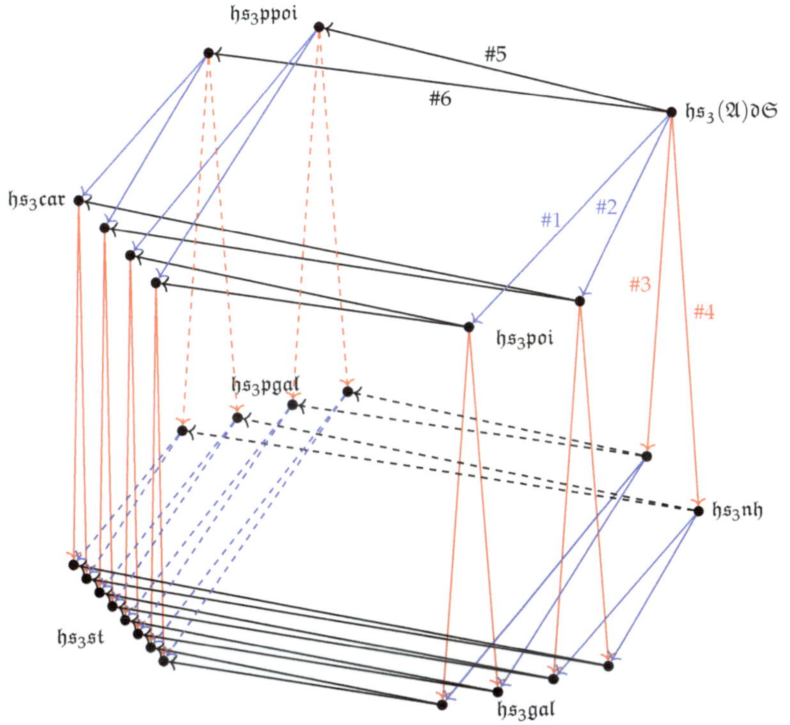

Figure 3. This figure [185] summarizes the contractions of Table 3. There are 2 space-time (blue; #1,#2), 2 speed-space (red; #3,#4) and 2 speed-time (black; #5,#6) contractions, and combining them leads to the full cube. The explicit commutators of all algebras can be found in the Appendix of [185]. In comparison to Figure 1, we have for clarity omitted the double lines.

Given the zoo of higher spin algebras, each corner of the higher spin cube representing one, the question arises if they permit an invariant metric. For the (semi)simple (A)dS algebras, this is obvious, but not so much for the other ones. For the case of the higher spin Poincaré algebras, the considerations of Section 2 generalize, and for the higher spin Carroll algebras, there are again invariant metric preserving contractions [75] analogous to the ones discussed in Section 4.4.

For the Galilei algebras, the situation is different, and the knowledge of double extensions proves to be useful. Already in the case of spin-2, the three-dimensional Galilei algebra has no invariant metric [190]. However, it is possible to centrally extend the Galilei algebra by two nontrivial central extensions (out of three possible ones [191]) to obtain a Lie algebra with an invariant metric. One of these central extensions is possible in any dimension and corresponds to the mass in the so-called Bargmann algebra. Due to the second extension that is a peculiarity of three spacetime dimensions, this algebra is called extended Bargmann algebra and makes a Chern–Simons formulation possible [190].

The higher spin Galilei generalizations also do not permit any invariant metric [185]. In contrast to the spin-2 case, central extensions are not sufficient to provide an invariant metric, but double extensions provide guidance. Interestingly, the double extension of the spin-3 Galilei algebras leads naturally to Lie algebras where the spin-2 part is exactly the just mentioned extended Bargmann algebra. Therefore, the higher spin generalization of the Galilei algebra can be considered the spin-3 extended Bargmann algebra. Furthermore, the properties of the higher spin Carroll and extended Bargmann theories have been studied in detail [185].

5. Conclusions and Outlook

Three-dimensional higher spin theories beyond (anti)-de Sitter can be roughly separated by the gauge algebra that is used in their Chern–Simons formulation. As reviewed in Section 2, using the higher spin (A)dS gauge algebras, one is able to construct backgrounds and boundary conditions for Lobachevsky, Lifshitz, null warped and Schrödinger spacetimes. Going beyond the (A)dS gauge algebra, the best understood cases so far are flat space higher spin theories that were discussed in Section 3. The interesting question if there are higher spin algebras beyond these cases has been answered by the construction of the kinematical higher spin algebras reviewed in Section 4.

While some progress has been made up until now, there are certainly interesting open problems that demand further investigation.

5.1. Boundary Conditions and Boundary Theories

While it was shown that AdS higher spin gauge algebras permit backgrounds and boundary conditions beyond the standard AdS choices, their asymptotic symmetry algebras often turned out to be related to already known ones. It would be interesting to further investigate if these boundary conditions and their asymptotic symmetry algebras can be further specialized in order to yield, e.g., Lifshitz-like asymptotic symmetries.

While for the Carroll case, boundary conditions have been proposed [113,185], for most of the other kinematical algebras, and especially the higher spin generalizations, no consistent boundary conditions have been established yet. Further examination is also needed for the mysterious result that it seems that one can assign entropy to Carroll geometries [113]. It would be interesting to see if this result can be generalized to the higher spin case. A generalization and interpretation of higher spin entanglement entropy [50,51] in these setups would be another intriguing option.

Another interesting generalization would be the calculation of one loop partition functions. Here, Newton–Hooke and Para-Poincaré seem to be intriguing options. This is due to the still non-vanishing cosmological constant, and one might therefore hope that they exhibit the "box-like" behavior of AdS.

For the higher spin cases, it would be interesting to see if the asymptotic symmetry algebras lead to nonlinear generalizations similar to the $\mathcal{W}_N$ algebras for AdS (some of them might be related to the ones discussed in [192]).

One interesting observation is related to possible dual theories of the Chern–Simons theories treated in this review to the Wess–Zumino–Novikov–Witten (WZNW) models [193,194]. Here, again, WZNW models based on a Lie algebra that admits an invariant metric play a distinguished role since they admit a (generalized) Sugawara construction [195]. Double extensions and the Medina–Revoy theorem are fundamental for the proof that the Sugawara construction factorizes in a semisimple and a non-semisimple one [196].

5.2. Kinematical (Higher Spin) Algebras

For the spin-2 extended Bargmann algebras, it was shown that they emerge as contractions of (anti)-de Sitter algebras that have been (trivially) centrally extended by two one-dimensional algebras [190][29]. It is also not clear as of yet which (semisimple) algebra can be naturally contracted to yield the double extended higher spin versions of the Bargmann algebra. This is interesting, because the deformed theories are often seen as more fundamental. We are not aware of a systematic discussion of contractions and double extensions; see however Section 5 in [75] for a start. Furthermore, it might be interesting to also look at the Chern–Simons theories based on the Lie algebras that have not been double extended.

[29] See also Section 9.2 in [75].

For many considerations, a generalization to the supersymmetric case seems possible; especially since the supersymmetric analog of double extensions exists [197] and an analog of the structure theorem of Section 4.3 has been proposed [198]. A Chern–Simons theory based on a supersymmetric version of the extended Bargmann algebra has already been investigated in [199].

Since the Chern–Simons theory based on the extended Bargmann algebra has been shown to be related to a specific version of Hořava-Lifshitz gravity [200], it would be interesting to see if the higher spin extended Bargmann theories lead to a spin-3 Hořava-Lifshitz theory.

Last, but not least, two obvious generalizations are to higher spins ($s > 3$), as well as to higher dimensions.

Acknowledgments: The authors want to thank Daniel Grumiller for collaboration during the start of this work. In addition, we want to thank all our collaborators, as well as all the other people that have been and are currently working on (non-)AdS higher spin holography for their valuable contributions to the field. Furthermore, the authors would like to thank Andrea Campoleoni and Nicolas Boulanger for inviting us to contribute to this Special Issue of Universe on higher spin gauge theories. The research of Stefan Prohazka is supported by the ERC Advanced Grant "High-Spin-Grav" and by FNRS (Fonds de la recherche scientifique) -Belgium (Convention FRFC PDR T.1025.14 and Convention IISN 4.4503.15). The research of Max Riegler is supported by the ERC Starting Grant 335146 "HoloBHC".

Author Contributions: Stefan Prohazka and Max Riegler contributed equally to this work.

Conflicts of Interest: The authors declare no conflict of interest.

Appendix A. Explicit Kinematical Algebra Relations

Table A1. (Anti-)de Sitter, Poincaré, Newton–Hooke and para-Poincaré algebras. The upper sign is for AdS (and contractions thereof) and the lower sign for dS (and contractions thereof).

	$(\mathfrak{A})\mathfrak{dS}_{(\mp)}$	$\mathfrak{poi}$	$\mathfrak{nh}$	$\mathfrak{ppoi}$
$[\mathrm{J},\mathrm{J}]$	0	0	0	0
$[\mathrm{J},\mathrm{G}_a]$	$\epsilon_{am}\mathrm{G}_m$	$\epsilon_{am}\mathrm{G}_m$	$\epsilon_{am}\mathrm{G}_m$	$\epsilon_{am}\mathrm{G}_m$
$[\mathrm{J},\mathrm{H}]$	0	0	0	0
$[\mathrm{J},\mathrm{P}_a]$	$\epsilon_{am}\mathrm{P}_m$	$\epsilon_{am}\mathrm{P}_m$	$\epsilon_{am}\mathrm{P}_m$	$\epsilon_{am}\mathrm{P}_m$
$[\mathrm{G}_a,\mathrm{G}_b]$	$-\epsilon_{ab}\mathrm{J}$	$-\epsilon_{ab}\mathrm{J}$	0	0
$[\mathrm{G}_a,\mathrm{H}]$	$-\epsilon_{am}\mathrm{P}_m$	$-\epsilon_{am}\mathrm{P}_m$	$-\epsilon_{am}\mathrm{P}_m$	0
$[\mathrm{G}_a,\mathrm{P}_b]$	$-\epsilon_{ab}\mathrm{H}$	$-\epsilon_{ab}\mathrm{H}$	0	$-\epsilon_{ab}\mathrm{H}$
$[\mathrm{H},\mathrm{P}_a]$	$\pm\epsilon_{am}\mathrm{G}_m$	0	$\pm\epsilon_{am}\mathrm{G}_m$	$\pm\epsilon_{am}\mathrm{G}_m$
$[\mathrm{P}_a,\mathrm{P}_b]$	$\mp\epsilon_{ab}\mathrm{J}$	0	0	$\mp\epsilon_{ab}\mathrm{J}$

Table A2. Carroll, Galilei, para-Galilei and static algebra. The upper sign is for AdS (and contractions thereof) and the lower sign for dS (and contractions thereof).

	$\mathfrak{car}$	$\mathfrak{gal}$	$\mathfrak{pgal}$	$\mathfrak{st}$
$[\mathrm{J},\mathrm{J}]$	0	0	0	0
$[\mathrm{J},\mathrm{G}_a]$	$\epsilon_{am}\mathrm{G}_m$	$\epsilon_{am}\mathrm{G}_m$	$\epsilon_{am}\mathrm{G}_m$	$\epsilon_{am}\mathrm{G}_m$
$[\mathrm{J},\mathrm{H}]$	0	0	0	0
$[\mathrm{J},\mathrm{P}_a]$	$\epsilon_{am}\mathrm{P}_m$	$\epsilon_{am}\mathrm{P}_m$	$\epsilon_{am}\mathrm{P}_m$	$\epsilon_{am}\mathrm{P}_m$
$[\mathrm{G}_a,\mathrm{G}_b]$	0	0	0	0
$[\mathrm{G}_a,\mathrm{H}]$	0	$-\epsilon_{am}\mathrm{P}_m$	0	0
$[\mathrm{G}_a,\mathrm{P}_b]$	$-\epsilon_{ab}\mathrm{H}$	0	0	0
$[\mathrm{H},\mathrm{P}_a]$	0	0	$\pm\epsilon_{am}\mathrm{G}_m$	0
$[\mathrm{P}_a,\mathrm{P}_b]$	0	0	0	0

References

1. Klebanov, I.; Polyakov, A. AdS dual of the critical O(N) vector model. *Phys. Lett. B* **2002**, *550*, 213–219.
2. Mikhailov, A. Notes on higher spin symmetries. *arXiv* **2002**, arXiv:hep-th/hep-th/0201019.
3. Sezgin, E.; Sundell, P. Massless higher spins and holography. *Nucl. Phys. B* **2002**, *644*, 303–370.
4. Fradkin, E.; Vasiliev, M.A. On the gravitational interaction of massless higher spin fields. *Phys. Lett. B* **1987**, *189*, 89–95.
5. Fradkin, E.S.; Vasiliev, M.A. Cubic interaction in extended theories of massless higher spin fields. *Nucl. Phys. B* **1987**, *291*, 141–171.
6. Vasiliev, M.A. Consistent equation for interacting gauge fields of all spins in (3 + 1)-dimensions. *Phys. Lett. B* **1990**, *243*, 378–382.
7. Sagnotti, A.; Taronna, M. String lessons for higher spin interactions. *Nucl. Phys. B* **2011**, *842*, 299–361.
8. Vasiliev, M.A. Holography, Unfolding and Higher Spin Theory. *J. Phys. A* **2013**, *46*, 214013.
9. Didenko, V.; Skvortsov, E. Elements of Vasiliev theory. *arXiv* **2014**, arXiv:hep-th/1401.2975.
10. Giombi, S.; Yin, X. Higher spin gauge theory and holography: The three-point functions. *J. High Energy Phys.* **2010**, *2010*, 115.
11. Giombi, S.; Yin, X. Higher Spins in AdS and Twistorial Holography. *J. High Energy Phys.* **2011**, *2011*, 86.
12. De Mello Koch, R.; Jevicki, A.; Jin, K.; Rodrigues, J.P. AdS_4/CFT_3 construction from collective fields. *Phys. Rev. D* **2011**, *83*, 025006.
13. Giombi, S.; Yin, X. On Higher Spin Gauge Theory and the Critical O(N) Model. *Phys. Rev. D* **2012**, *85*, 086005.
14. Douglas, M.R.; Mazzucato, L.; Razamat, S.S. Holographic dual of free field theory. *Phys. Rev. D* **2011**, *83*, 071701.
15. Giombi, S.; Yin, X. The Higher Spin/Vector Model Duality. *J. Phys. A* **2013**, *46*, 214003.
16. Sleight, C.; Taronna, M. Higher spin interactions from conformal field theory: The complete cubic couplings. *Phys. Rev. Lett.* **2016**, *116*, 181602.
17. Maldacena, J.; Zhiboedov, A. Constraining conformal field theories with a higher spin symmetry. *J. Phys. A* **2013**, *46*, 214011.
18. Maldacena, J.; Zhiboedov, A. Constraining conformal field theories with a slightly broken higher spin symmetry. *Class. Quantum Gravity* **2013**, *30*, 104003.
19. Maldacena, J.M. The Large N limit of superconformal field theories and supergravity. *Adv. Theor. Math. Phys.* **1998**, *2*, 231–252.
20. Gubser, S.; Klebanov, I.R.; Polyakov, A.M. Gauge theory correlators from noncritical string theory. *Phys. Lett. B* **1998**, *428*, 105–114.
21. Witten, E. Anti-de Sitter space and holography. *Adv. Theor. Math. Phys.* **1998**, *2*, 253–291.
22. Aragone, C.; Deser, S. Hypersymmetry in $D = 3$ of coupled gravity massless spin 5/2 system. *Class. Quantum Gravity* **1984**, *1*, L9.
23. Blencowe, M. A consistent interacting massless higher spin field theory in $D = (2 + 1)$. *Class. Quantum Gravity* **1989**, *6*, 443–452.
24. Henneaux, M.; Rey, S.J. Nonlinear W_∞ as asymptotic symmetry of three-dimensional higher spin anti-de sitter gravity. *J. High Energy Phys.* **2010**, *2010*, 7.
25. Campoleoni, A.; Fredenhagen, S.; Pfenninger, S.; Theisen, S. Asymptotic symmetries of three-dimensional gravity coupled to higher spin fields. *J. High Energy Phys.* **2010**, *2010*, 7.
26. Gaberdiel, M.R.; Hartman, T. Symmetries of Holographic Minimal Models. *J. High Energy Phys.* **2011**, *2011*, 31.
27. Campoleoni, A.; Fredenhagen, S.; Pfenninger, S. Asymptotic W-symmetries in three-dimensional higher spin gauge theories. *J. High Energy Phys.* **2011**, *2011*, 113.
28. Henneaux, M.; Perez, A.; Tempo, D.; Troncoso, R. Chemical potentials in three-dimensional higher spin anti-de Sitter gravity. *J. High Energy Phys.* **2013**, *2013*, 48.
29. Castro, A.; Lepage-Jutier, A.; Maloney, A. Higher spin theories in AdS_3 and a gravitational exclusion principle. *J. High Energy Phys.* **2011**, *2011*, 142.
30. Ammon, M.; Gutperle, M.; Kraus, P.; Perlmutter, E. Spacetime geometry in higher spin gravity. *J. High Energy Phys.* **2011**, *2011*, 53.

31. Castro, A.; Hijano, E.; Lepage-Jutier, A.; Maloney, A. Black holes and singularity resolution in higher spin gravity. *J. High Energy Phys.* **2012**, *2012*, 31.
32. Ammon, M.; Kraus, P.; Perlmutter, E. Scalar fields and three-point functions in D = 3 higher spin gravity. *J. High Energy Phys.* **2012**, *2012*, 113.
33. Henneaux, M.; Lucena Gómez, G.; Park, J.; Rey, S.J. Super- W(infinity) Asymptotic Symmetry of Higher Spin AdS_3 Supergravity. *J. High Energy Phys.* **2012**, *2012*, 37.
34. Campoleoni, A.; Fredenhagen, S.; Pfenninger, S.; Theisen, S. Towards metric-like higher spin gauge theories in three dimensions. *J. Phys. A* **2013**, *46*, 214017.
35. De Boer, J.; Jottar, J.I. Thermodynamics of higher spin black holes in AdS_3. *J. High Energy Phys.* **2014**, *2014*, 23.
36. Compère, G.; Jottar, J.I.; Song, W. Observables and microscopic entropy of higher spin black holes. *J. High Energy Phys.* **2013**, *2013*, 54.
37. De Boer, J.; Jottar, J.I. Boundary conditions and partition functions in higher spin AdS_3/CFT_2. *J. High Energy Phys.* **2016**, *2016*, 107.
38. Campoleoni, A.; Fredenhagen, S. On the higher spin charges of conical defects. *Phys. Lett. B* **2013**, *726*, 387–389.
39. Campoleoni, A.; Henneaux, M. Asymptotic symmetries of three-dimensional higher spin gravity: The metric approach. *J. High Energy Phys.* **2015**, *2015*, 143.
40. Castro, A.; Gopakumar, R.; Gutperle, M.; Raeymaekers, J. Conical defects in higher spin theories. *J. High Energy Phys.* **2012**, *2012*, 96.
41. Castro, A.; Llabrés, E. Unravelling Holographic Entanglement Entropy in Higher Spin Theories. *J. High Energy Phys.* **2015**, *2015*, 124.
42. De Boer, J.; Castro, A.; Hijano, E.; Jottar, J.I.; Kraus, P. Higher spin entanglement and $\mathcal{W}_N$ conformal blocks. *J. High Energy Phys.* **2015**, *2015*, 168.
43. Bañados, M.; Castro, A.; Faraggi, A.; Jottar, J.I. Extremal Higher Spin Black Holes. *J. High Energy Phys.* **2016**, *2016*, 77.
44. Gaberdiel, M.R.; Gopakumar, R. An AdS_3 Dual for Minimal Model CFTs. *Phys. Rev. D* **2011**, *83*, 066007.
45. Gaberdiel, M.R.; Gopakumar, R. Minimal Model Holography. *J. Phys. A* **2013**, *46*, 214002.
46. Candu, C.; Gaberdiel, M.R.; Kelm, M.; Vollenweider, C. Even spin minimal model holography. *J. High Energy Phys.* **2013**, *2013*, 185.
47. Gutperle, M.; Kraus, P. Higher Spin Black Holes. *J. High Energy Phys.* **2011**, *2011*, 22.
48. Ammon, M.; Gutperle, M.; Kraus, P.; Perlmutter, E. Black holes in three dimensional higher spin gravity: A review. *J. Phys. A* **2013**, *46*, 214001.
49. Bunster, C.; Henneaux, M.; Perez, A.; Tempo, D.; Troncoso, R. Generalized Black Holes in Three-dimensional Spacetime. *J. High Energy Phys.* **2014**, *2014*, 31.
50. Ammon, M.; Castro, A.; Iqbal, N. Wilson Lines and Entanglement Entropy in Higher Spin Gravity. *J. High Energy Phys.* **2013**, *2013*, 110.
51. De Boer, J.; Jottar, J.I. Entanglement Entropy and Higher Spin Holography in AdS_3. *J. High Energy Phys.* **2014**, *2014*, 89.
52. Brown, J.D.; Henneaux, M. Central charges in the canonical realization of asymptotic symmetries: An example from three-dimensional gravity. *Commun. Math. Phys.* **1986**, *104*, 207–226.
53. Riegler, M. Asymptotic symmetry algebras in non-anti-de-sitter higher spin gauge theories. *arXiv* **2012**, arXiv:hep-th/1210.6500.
54. Afshar, H.; Gary, M.; Grumiller, D.; Rashkov, R.; Riegler, M. Non-AdS holography in 3-dimensional higher spin gravity—General recipe and example. *J. High Energy Phys.* **2012**, *2012*, 99.
55. Afshar, H.; Gary, M.; Grumiller, D.; Rashkov, R.; Riegler, M. Semi-classical unitarity in 3-dimensional higher spin gravity for non-principal embeddings. *Class. Quantum Gravity* **2013**, *30*, 104004.
56. Son, D. Toward an AdS/cold atoms correspondence: A Geometric realization of the Schrodinger symmetry. *Phys. Rev. D* **2008**, *78*, 046003.
57. Balasubramanian, K.; McGreevy, J. Gravity duals for non-relativistic CFTs. *Phys. Rev. Lett.* **2008**, *101*, 061601.
58. Adams, A.; Balasubramanian, K.; McGreevy, J. Hot Spacetimes for Cold Atoms. *J. High Energy Phys.* **2008**, *2008*, 59.
59. Breunhölder, V.; Gary, M.; Grumiller, D.; Prohazka, S. Null warped AdS in higher spin gravity. *J. High Energy Phys.* **2015**, *2015*, 21.

60. Lei, Y.; Ross, S.F. Connection versus metric description for non-AdS solutions in higher spin theories. *Class. Quantum Gravity* **2015**, *32*, 185005.
61. Kachru, S.; Liu, X.; Mulligan, M. Gravity Duals of Lifshitz-like Fixed Points. *Phys. Rev. D* **2008**, *78*, 106005.
62. Gutperle, M.; Hijano, E.; Samani, J. Lifshitz black holes in higher spin gravity. *J. High Energy Phys.* **2014**, *2014*, 20.
63. Gary, M.; Grumiller, D.; Prohazka, S.; Rey, S.J. Lifshitz Holography with Isotropic Scale Invariance. *J. High Energy Phys.* **2014**, *2014*, 1.
64. Afshar, H.; Bagchi, A.; Fareghbal, R.; Grumiller, D.; Rosseel, J. Spin-3 Gravity in Three-Dimensional Flat Space. *Phys. Rev. Lett.* **2013**, *111*, 121603.
65. Gonzalez, H.A.; Matulich, J.; Pino, M.; Troncoso, R. Asymptotically flat spacetimes in three-dimensional higher spin gravity. *J. High Energy Phys.* **2013**, *2013*, 16.
66. Grumiller, D.; Riegler, M.; Rosseel, J. Unitarity in three-dimensional flat space higher spin theories. *J. High Energy Phys.* **2014**, *2014*, 15.
67. Gary, M.; Grumiller, D.; Riegler, M.; Rosseel, J. Flat space (higher spin) gravity with chemical potentials. *J. High Energy Phys.* **2015**, *2015*, 152.
68. Krishnan, C.; Raju, A.; Roy, S.; Thakur, S. Higher Spin Cosmology. *Phys. Rev. D* **2014**, *89*, 045007.
69. Krishnan, C.; Roy, S. Higher Spin Resolution of a Toy Big Bang. *Phys. Rev. D* **2013**, *88*, 044049.
70. Basu, R. Higher spin de sitter quantum gravity. *J. High Energy Phys.* **2015**, *2015*, 151.
71. Achucarro, A.; Townsend, P. A Chern–Simons action for three-dimensional anti-de sitter supergravity theories. *Phys. Lett. B* **1986**, *180*, 89–92.
72. Witten, E. (2 + 1)-Dimensional Gravity as an Exactly Soluble System. *Nucl. Phys. B* **1988**, *311*, 46–48.
73. Riegler, M. How General Is Holography? Ph.D. Thesis, Technische Universität Wien, Vienna, Austria, 2016.
74. Riegler, M.; Zwikel, C. Canonical Charges in Flatland. *arXiv* **2017**, arXiv:hep-th/1709.09871.
75. Prohazka, S. Chern–Simons Holography: Boundary Conditions, Contractions and Double Extensions for a Journey Beyond Anti-de Sitter. Ph.D. Thesis, Technische Universität Wien, Vienna, Austria, 2017.
76. Afshar, H.; Bagchi, A.; Detournay, S.; Grumiller, D.; Prohazka, S.; Riegler, M. Holographic Chern–Simons Theories. *Lect. Notes Phys.* **2015**, *892*, 311–329.
77. Gary, M.; Grumiller, D.; Rashkov, R. Towards non-AdS holography in 3-dimensional higher spin gravity. *J. High Energy Phys.* **2012**, *2012*, 22.
78. Henneaux, M.; Teitelboim, C. *Quantization of Gauge Systems*; Princeton University Press: Princeton, NJ, USA, 1992.
79. Blagojevic, M. *Gravitation and Gauge Symmetries*; CRC Press: Boca Raton, FL, USA, 2010.
80. Regge, T.; Teitelboim, C. Role of Surface Integrals in the Hamiltonian Formulation of General Relativity. *Ann. Phys.* **1974**, *88*, 286–318.
81. Barnich, G.; Troessaert, C. BMS charge algebra. *J. High Energy Phys.* **2011**, *2011*, 105.
82. Gaberdiel, M.R.; Gopakumar, R. Triality in Minimal Model Holography. *J. High Energy Phys.* **2012**, *2012*, 127.
83. Polyakov, A.M. Gauge Transformations and Diffeomorphisms. *Int. J. Mod. Phys. A* **1990**, *5*, 833.
84. Bershadsky, M. Conformal field theories via Hamiltonian reduction. *Commun. Math. Phys.* **1991**, *139*, 71–82.
85. Afshar, H.; Creutzig, T.; Grumiller, D.; Hikida, Y.; Ronne, P.B. Unitary W-algebras and three-dimensional higher spin gravities with spin one symmetry. *J. High Energy Phys.* **2014**, *2014*, 63.
86. Feigin, B.L.; Semikhatov, A.M. W(2)(n) algebras. *Nucl. Phys. B* **2004**, *698*, 409–449.
87. Castro, A.; Hijano, E.; Lepage-Jutier, A. Unitarity Bounds in AdS_3 Higher Spin Gravity. *J. High Energy Phys.* **2012**, *2012*, 1.
88. Gutperle, M.; Li, Y. Higher spin lifshitz theory and integrable systems. *Phys. Rev. D* **2015**, *91*, 046012.
89. Beccaria, M.; Gutperle, M.; Li, Y.; Macorini, G. Higher spin Lifshitz theories and the Korteweg-de Vries hierarchy. *Phys. Rev. D* **2015**, *92*, 085005.
90. Gutperle, M.; Li, Y. Higher Spin Chern–Simons Theory and the Super Boussinesq hierarchy. *arXiv* **2017**, arXiv:hep-th/1709.02345.
91. Lei, Y.; Peng, C. Higher spin holography with Galilean symmetry in general dimensions. *Class. Quantum Gravity* **2016**, *33*, 135008.
92. Lei, Y. Singularities in Holographic Non-Relativistic Spacetimes. Ph.D. Thesis, Durham University, Durham, UK, 2016.
93. Grumiller, D.; Riegler, M. Most general AdS_3 boundary conditions. *J. High Energy Phys.* **2016**, *2016*, 23.

94. Krishnan, C.; Raju, A. Chiral Higher Spin Gravity. *Phys. Rev. D* **2017**, *95*, 126004.
95. Polchinski, J. S matrices from AdS space-time. *arXiv* **1999**, arXiv:hep-th/hep-th/9901076.
96. Susskind, L. Holography in the flat space limit. *AIP Conf. Proc.* **1998**, 493, 98–112.
97. Giddings, S.B. Flat space scattering and bulk locality in the AdS/CFT correspondence. *Phys. Rev. D* **2000**, *61*, 106008.
98. Barnich, G.; Compere, G. Classical central extension for asymptotic symmetries at null infinity in three spacetime dimensions. *Class. Quantum Gravity* **2007**, *24*, F15–F23.
99. Strominger, A. On BMS Invariance of Gravitational Scattering. *J. High Energy Phys.* **2014**, *2014*, 152.
100. Kapec, D.; Lysov, V.; Pasterski, S.; Strominger, A. Higher-dimensional supertranslations and weinberg's soft graviton theorem. *Ann. Math. Sci. Appl.* **2017**, *2*, 69–94.
101. Prohazka, S.; Salzer, J.; Schöller, F. Linking Past and Future Null Infinity in Three Dimensions. *Phys. Rev. D* **2017**, *95*, 086011.
102. Ashtekar, A.; Bicak, J.; Schmidt, B.G. Asymptotic structure of symmetry reduced general relativity. *Phys. Rev. D* **1997**, *55*, 669–686.
103. Bondi, H.; van der Burg, M.G.J.; Metzner, A.W.K. Gravitational waves in general relativity. 7. Waves from axisymmetric isolated systems. *Proc. R. Soc. Lond. A* **1962**, *269*, 21–52.
104. Sachs, R. Asymptotic symmetries in gravitational theory. *Phys. Rev.* **1962**, *128*, 2851–2864.
105. Barnich, G.; Donnay, L.; Matulich, J.; Troncoso, R. Asymptotic symmetries and dynamics of three-dimensional flat supergravity. *J. High Energy Phys.* **2014**, *2014*, 71.
106. Barnich, G.; Lambert, P.H.; Mao, P.J. Three-dimensional asymptotically flat Einstein–Maxwell theory. *Class. Quantum Gravity* **2015**, *32*, 245001.
107. Detournay, S.; Riegler, M. Enhanced Asymptotic Symmetry Algebra of 2 + 1 Dimensional Flat Space. *Phys. Rev. D* **2017**, *95*, 046008.
108. Setare, M.R.; Adami, H. Enhanced asymptotic BMS_3 algebra of the flat spacetime solutions of generalized minimal massive gravity. *arXiv* **2017**, arXiv:hep-th/1703.00936.
109. Basu, R.; Detournay, S.; Riegler, M. Spectral Flow in 3D Flat Spacetimes. *arXiv* **2017**, arXiv:hep-th/1706.07438.
110. Fuentealba, O.; Matulich, J.; Troncoso, R. Asymptotic structure of $\mathcal{N} = 2$ supergravity in 3D: Extended super-BMS_3 and nonlinear energy bounds. *J. High Energy Phys.* **2017**, *2017*, 30.
111. Afshar, H.; Detournay, S.; Grumiller, D.; Oblak, B. Near-Horizon Geometry and Warped Conformal Symmetry. *J. High Energy Phys.* **2016**, *2016*, 187.
112. Afshar, H.; Grumiller, D.; Merbis, W.; Perez, A.; Tempo, D.; Troncoso, R. Soft hairy horizons in three spacetime dimensions. *arXiv* **2016**, arXiv:hep-th/1611.09783.
113. Grumiller, D.; Merbis, W.; Riegler, M. Most general flat space boundary conditions in three-dimensional Einstein gravity. *Class. Quantum Gravity* **2017**, *34*, 184001.
114. Bagchi, A.; Detournay, S.; Fareghbal, R.; Simón, J. Holography of 3D Flat Cosmological Horizons. *Phys. Rev. Lett.* **2013**, *110*, 141302.
115. Barnich, G. Entropy of three-dimensional asymptotically flat cosmological solutions. *J. High Energy Phys.* **2012**, *2012*, 95.
116. Barnich, G.; Gomberoff, A.; Gonzalez, H.A. The Flat limit of three dimensional asymptotically anti-de Sitter spacetimes. *Phys. Rev. D* **2012**, *86*, 024020.
117. Bagchi, A.; Fareghbal, R. BMS/GCA Redux: Towards Flatspace Holography from Non-Relativistic Symmetries. *J. High Energy Phys.* **2012**, *2012*, 92.
118. Bagchi, A.; Detournay, S.; Grumiller, D.; Simon, J. Cosmic evolution from phase transition of 3-dimensional flat space. *Phys. Rev. Lett.* **2013**, *111*, 181301.
119. Fareghbal, R.; Naseh, A. Flat-Space Energy-Momentum Tensor from BMS/GCA Correspondence. *J. High Energy Phys.* **2014**, *2014*, 5.
120. Krishnan, C.; Raju, A.; Roy, S. A Grassmann path from AdS_3 to flat space. *J. High Energy Phys.* **2014**, *2014*, 36.
121. Bagchi, A.; Basu, R. 3D Flat Holography: Entropy and Logarithmic Corrections. *J. High Energy Phys.* **2014**, *2014*, 20.
122. Detournay, S.; Grumiller, D.; Schöller, F.; Simón, J. Variational principle and one-point functions in three-dimensional flat space Einstein gravity. *Phys. Rev. D* **2014**, *89*, 084061.
123. Barnich, G.; Oblak, B. Notes on the BMS group in three dimensions: I. Induced representations. *J. High Energy Phys.* **2014**, *2014*, 129.

124. Bagchi, A.; Basu, R.; Grumiller, D.; Riegler, M. Entanglement entropy in Galilean conformal field theories and flat holography. *Phys. Rev. Lett.* **2015**, *114*, 111602.
125. Fareghbal, R.; Naseh, A. Aspects of Flat/CCFT Correspondence. *Class. Quantum Gravity* **2015**, *32*, 135013.
126. Fareghbal, R.; Hosseini, S.M. Holography of 3D Asymptotically Flat Black Holes. *Phys. Rev. D* **2015**, *91*, 084025.
127. Barnich, G.; Gonzalez, H.A.; Maloney, A.; Oblak, B. One-loop partition function of three-dimensional flat gravity. *J. High Energy Phys.* **2015**, *2015*, 178.
128. Barnich, G.; Oblak, B. Notes on the BMS group in three dimensions: II. Coadjoint representation. *J. High Energy Phys.* **2015**, *2015*, 033.
129. Bagchi, A.; Grumiller, D.; Merbis, W. Stress tensor correlators in three-dimensional gravity. *Phys. Rev. D* **2016**, *93*, 061502.
130. Asadi, M.; Baghchesaraei, O.; Fareghbal, R. Stress Tensor Correlators of $CCFT_2$ using Flat-Space Holography. *arXiv* **2016**, arXiv:hep-th/1701.00063.
131. Barnich, G.; Gonzalez, H.A.; Salgado-Rebolledo, P. Geometric actions for three-dimensional gravity. *arXiv* **2017**, arXiv:hep-th/1707.08887.
132. Fareghbal, R.; Karimi, P. Logarithmic Correction to BMSFT Entanglement Entropy. *arXiv* **2017**, arXiv:hep-th/1709.01804.
133. Coleman, S.R.; Mandula, J. All Possible Symmetries of the S Matrix. *Phys. Rev.* **1967**, *159*, 1251–1256.
134. Pelc, O.; Horwitz, L.P. Generalization of the Coleman-Mandula theorem to higher dimension. *J. Math. Phys.* **1997**, *38*, 139–172.
135. Aragone, C.; Deser, S. Consistency Problems of Hypergravity. *Phys. Lett. B* **1979**, *86*, 161–163.
136. Weinberg, S.; Witten, E. Limits on Massless Particles. *Phys. Lett. B* **1980**, *96*, 59–62.
137. Bekaert, X.; Boulanger, N.; Sundell, P. How higher spin gravity surpasses the spin two barrier: No-go theorems versus yes-go examples. *Rev. Mod. Phys.* **2012**, *84*, 987–1009.
138. Sleight, C.; Taronna, M. Higher Spin Algebras, Holography and Flat Space. *J. High Energy Phys.* **2017**, *2017*, 95.
139. Ponomarev, D.; Tseytlin, A.A. On quantum corrections in higher spin theory in flat space. *J. High Energy Phys.* **2016**, *2016*, 184.
140. Ponomarev, D.; Skvortsov, E.D. Light-Front Higher Spin Theories in Flat Space. *J. Phys. A* **2017**, *50*, 095401.
141. Campoleoni, A.; Francia, D.; Heissenberg, C. On higher spin supertranslations and superrotations. *J. High Energy Phys.* **2017**, *2017*, 120.
142. Inonu, E.; Wigner, E.P. On the Contraction of groups and their represenations. *Proc. Natl. Acad. Sci. USA* **1953**, *39*, 510–524.
143. Bagchi, A.; Gopakumar, R. Galilean Conformal Algebras and AdS/CFT. *J. High Energy Phys.* **2009**, *2009*, 37.
144. Bagchi, A.; Mandal, I. On representations and correlation functions of galilean conformal algebras. *Phys. Lett. B* **2009**, *675*, 393–397.
145. Bagchi, A.; Gopakumar, R.; Mandal, I.; Miwa, A. GCA in 2d. *J. High Energy Phys.* **2010**, *2010*, 4.
146. Bagchi, A. Correspondence between asymptotically flat spacetimes and nonrelativistic conformal field theories. *Phys. Rev. Lett.* **2010**, *105*, 171601.
147. Campoleoni, A.; Gonzalez, H.A.; Oblak, B.; Riegler, M. BMS modules in three dimensions. *Int. J. Mod. Phys. A* **2016**, *31*, 1650068.
148. Riegler, M. Flat space limit of higher spin Cardy formula. *Phys. Rev. D* **2015**, *91*, 024044.
149. Zamolodchikov, A. Infinite additional symmetries in two-dimensional conformal quantum field theory. *Theor. Math. Phys.* **1985**, *65*, 1205–1213.
150. Cornalba, L.; Costa, M.S. A New cosmological scenario in string theory. *Phys. Rev. D* **2002**, *66*, 066001.
151. Cornalba, L.; Costa, M.S. Time dependent orbifolds and string cosmology. *Fortschr. Phys.* **2004**, *52*, 145–199.
152. Matulich, J.; Perez, A.; Tempo, D.; Troncoso, R. Higher spin extension of cosmological spacetimes in 3D: Asymptotically flat behaviour with chemical potentials and thermodynamics. *J. High Energy Phys.* **2015**, *2015*, 25.
153. Basu, R.; Riegler, M. Wilson lines and holographic entanglement entropy in galilean conformal field theories. *Phys. Rev. D* **2016**, *93*, 045003.
154. David, J.R.; Ferlaino, M.; Kumar, S.P. Thermodynamics of higher spin black holes in 3D. *J. High Energy Phys.* **2012**, *2012*, 135.
155. Bousso, R.; Porrati, M. Soft Hair as a Soft Wig. *Class. Quantum Gravity* **2017**, *34*, 204001.

156. Bousso, R.; Porrati, M. Observable Supertranslations. *Phys. Rev. D* **2017**, *96*, 086016.
157. Hawking, S.W.; Perry, M.J.; Strominger, A. Soft Hair on Black Holes. *Phys. Rev. Lett.* **2016**, *116*, 231301.
158. Hawking, S.W.; Perry, M.J.; Strominger, A. Superrotation charge and supertranslation hair on black holes. *J. High Energy Phys.* **2017**, *2017*, 161.
159. Afshar, H.; Detournay, S.; Grumiller, D.; Merbis, W.; Perez, A.; Tempo, D.; Troncoso, R. Soft Heisenberg hair on black holes in three dimensions. *Phys. Rev. D* **2016**, *93*, 101503.
160. Afshar, H.; Grumiller, D.; Sheikh-Jabbari, M.M. Black hole horizon fluffs: Near horizon soft hairs as microstates of three dimensional black holes. *Phys. Rev. D* **2017**, *96*, 084032.
161. Grumiller, D.; Perez, A.; Prohazka, S.; Tempo, D.; Troncoso, R. Higher Spin Black Holes with Soft Hair. *J. High Energy Phys.* **2016**, *2016*, 119.
162. Grumiller, D.; Perez, A.; Tempo, D.; Troncoso, R. Log corrections to entropy of three dimensional black holes with soft hair. *J. High Energy Phys.* **2017**, *2017*, 107.
163. Setare, M.R.; Adami, H. The Heisenberg algebra as near horizon symmetry of the black flower solutions of Chern–Simons-like theories of gravity. *Nucl. Phys. B* **2017**, *914*, 220–233.
164. Ammon, M.; Grumiller, D.; Prohazka, S.; Riegler, M.; Wutte, R. Higher Spin Flat Space Cosmologies with Soft Hair. *J. High Energy Phys.* **2017**, *2017*, 031.
165. Gaberdiel, M.R.; Gopakumar, R.; Saha, A. Quantum W-symmetry in AdS_3. *J. High Energy Phys.* **2011**, *2011*, 4.
166. Gaberdiel, M.R.; Gopakumar, R.; Hartman, T.; Raju, S. Partition functions of holographic minimal models. *J. High Energy Phys.* **2011**, *2011*, 77.
167. Creutzig, T.; Hikida, Y.; Ronne, P.B. Higher spin AdS_3 supergravity and its dual CFT. *J. High Energy Phys.* **2012**, *2012*, 109.
168. Giombi, S.; Klebanov, I.R. One Loop Tests of Higher Spin AdS/CFT. *J. High Energy Phys.* **2013**, *2013*, 68.
169. Giombi, S.; Klebanov, I.R.; Safdi, B.R. Higher Spin $\mathrm{AdS}_{d+1}/\mathrm{CFT}_d$ at One Loop. *Phys. Rev. D* **2014**, *89*, 084004.
170. Giombi, S.; Klebanov, I.R.; Tseytlin, A.A. Partition functions and casimir energies in higher spin $\mathrm{AdS}_{d+1}/\mathrm{CFT}_d$. *Phys. Rev. D* **2014**, *90*, 024048.
171. Beccaria, M.; Tseytlin, A.A. Higher spins in AdS_5 at one loop: vacuum energy, boundary conformal anomalies and AdS/CFT. *J. High Energy Phys.* **2014**, *2014*, 114.
172. Beccaria, M.; Tseytlin, A.A. On higher spin partition functions. *J. Phys. A* **2015**, *48*, 275401.
173. Campoleoni, A.; Gonzalez, H.A.; Oblak, B.; Riegler, M. Rotating Higher Spin Partition Functions and Extended BMS Symmetries. *J. High Energy Phys.* **2016**, *2016*, 34.
174. Krishnan, C.; Roy, S. Desingularization of the Milne Universe. *Phys. Lett. B* **2014**, *734*, 92–95.
175. Gonzalez, H.A.; Pino, M. Boundary dynamics of asymptotically flat 3D gravity coupled to higher spin fields. *J. High Energy Phys.* **2014**, *2014*, 127.
176. Kiran, K.S.; Krishnan, C.; Saurabh, A.; Simón, J. Strings vs. Spins on the Null Orbifold. *J. High Energy Phys.* **2014**, *2014*, 2.
177. Horowitz, G.T.; Steif, A.R. Singular string solutions with nonsingular initial data. *Phys. Lett. B* **1991**, *258*, 91–96.
178. Figueroa-O'Farrill, J.M.; Simon, J. Generalized supersymmetric fluxbranes. *J. High Energy Phys.* **2001**, *2001*, 11.
179. Liu, H.; Moore, G.W.; Seiberg, N. Strings in time dependent orbifolds. *J. High Energy Phys.* **2002**, *2002*, 31.
180. Simon, J. The Geometry of null rotation identifications. *J. High Energy Phys.* **2002**, *2002*, 1.
181. Barnich, G.; Gomberoff, A.; González, H.A. Three-dimensional Bondi-Metzner-Sachs invariant two-dimensional field theories as the flat limit of Liouville theory. *Phys. Rev. D* **2013**, *87*, 124032.
182. Taylor, M. Lifshitz holography. *Class. Quantum Gravity* **2016**, *33*, 033001.
183. Bacry, H.; Levy-Leblond, J. Possible kinematics. *J. Math. Phys.* **1968**, *9*, 1605–1614.
184. Saletan, E.J. Contraction of Lie Groups. *J. Math. Phys.* **1961**, *2*, 1–21.
185. Bergshoeff, E.; Grumiller, D.; Prohazka, S.; Rosseel, J. Three-dimensional Spin-3 Theories Based on General Kinematical Algebras. *J. High Energy Phys.* **2017**, *2017*, 114.
186. Duval, C.; Gibbons, G.W.; Horvathy, P.A. Conformal Carroll groups and BMS symmetry. *Class. Quantum Gravity* **2014**, *31*, 092001.
187. Hartong, J. Gauging the Carroll Algebra and Ultra-Relativistic Gravity. *J. High Energy Phys.* **2015**, *2015*, 69.
188. Medina, A.; Revoy, P. Algèbres de Lie et produit scalaire invariant. *Ann. Sci. l'École Norm. Supér.* **1985**, *18*, 553–561.

189. Figueroa-O'Farrill, J.M.; Stanciu, S. On the structure of symmetric selfdual Lie algebras. *J. Math. Phys.* **1996**, *37*, 4121–4134.
190. Papageorgiou, G.; Schroers, B.J. A Chern–Simons approach to Galilean quantum gravity in 2 + 1 dimensions. *J. High Energy Phys.* **2009**, *2009*, 9.
191. Levy-Léblond, J.M. Galilei group and galilean invariance. In *Group Theory and Its Applications*; Loebl, E.M., Ed.; Academic Press: Cambridge, MA, USA, 1971; pp. 221–299.
192. Rasmussen, J.; Raymond, C. Galilean contractions of W-algebras. *Nucl. Phys. B* **2017**, *922*, 435–479.
193. Witten, E. Quantum Field Theory and the Jones Polynomial. *Commun. Math. Phys.* **1989**, *121*, 351–399.
194. Elitzur, S.; Moore, G.W.; Schwimmer, A.; Seiberg, N. Remarks on the Canonical Quantization of the Chern–Simons-Witten Theory. *Nucl. Phys. B* **1989**, *326*, 108–134.
195. Mohammedi, N. On bosonic and supersymmetric current algebras for non-semisimple groups. *Phys. Lett. B* **1994**, *325*, 371–376.
196. Figueroa-O'Farrill, J.M.; Stanciu, S. Nonsemisimple Sugawara constructions. *Phys. Lett. B* **1994**, *327*, 40–46.
197. Benamor, H.; Benayadi, S. Double extension of quadratic lie superalgebras. *Commun. Algebra* **1999**, *27*, 67–88.
198. Bajo, I.; Benayadi, S.; Bordemann, M. Generalized double extension and descriptions of qadratic Lie superalgebras. *arXiv* **2007**, arXiv:math-ph/0712.0228.
199. Bergshoeff, E.A.; Rosseel, J. Three-dimensional extended bargmann supergravity. *Phys. Rev. Lett.* **2016**, *116*, 251601.
200. Hartong, J.; Lei, Y.; Obers, N.A. Nonrelativistic Chern–Simons theories and three-dimensional Hořava-Lifshitz gravity. *Phys. Rev. D* **2016**, *94*, 065027.

universe

MDPI

Review

Higher Spin Fields in Hyperspace. A Review

Dmitri Sorokin [1] and Mirian Tsulaia [2,*]

1 Istituto Nazionale di Fisica Nucleare, Sezione di Padova, via F. Marzolo 8, 35131 Padova, Italy; dmitri.sorokin@pd.infn.it

2 School of Physics M013, The University of Western Australia, 35 Stirling Highway, Crawley, Perth, WA 6009, Australia

* Correspondence: mirian.tsulaia@gmail.com

Received: 29 October 2017; Accepted: 22 December 2017; Published: 3 January 2018

Abstract: We give an introduction to the so-called tensorial, matrix or hyperspace approach to the description of massless higher-spin fields.

Keywords: higher spin theories; supersymmetry; conformal field theory

1. Introduction

Every consistent theory of interacting higher spin fields necessarily includes an infinite number of such fields. For this reason, it is extremely important to develop a formalism which effectively includes an infinite number of fields into a simpler field-theoretical object. This formalism should yield correct field equations first of all at the free level and then be promoted to an interacting theory. An elegant geometrical approach to higher spin theories of this kind is known as the method of tensorial spaces. This approach was first suggested by Fronsdal [1]. Its explicit dynamical realization and further extensive developments have been carried out in [2–28].

In a certain sense, the method of tensorial spaces is reminiscent of the Kaluza–Klein theories. In such theories, one usually considers massless field equations in higher dimensions and then, assuming that the extra dimensions are periodic (compact), one obtains a theory in lower dimensions, which contains fields with growing masses. In the method of tensorial (super)spaces, one also considers theories in multi-dimensional space–times, but in this case the extra dimensions are introduced in such a way that they generate the fields with higher spins instead of the fields with increasing masses. A main advantage of the formulation of the higher spin theories on extended tensorial (super) spaces is that one can combine curvatures of an infinite number of bosonic and fermionic higher spin fields into a single "master" (or "hyper") scalar and spinor field which propagate through the tensorial supesrpaces (also called hyperspaces). The field equations in the tensorial spaces are invariant under the action of $Sp(2n)$ group, whereas the dimensions of the corresponding tensorial spaces are equal to $\frac{n(n+1)}{2}$. The case of four space–time dimensions $D = 4$ is of particular interest since the approach of tensorial (super)spaces comprises all massless higher spin fields from zero to infinity. The free field equations are invariant under the $Sp(8)$ group, which contains a four dimensional conformal group $SO(2,4)$ as a subgroup. In fact, the entire structure of the $Sp(8)$ invariant formulation of the higher spin fields is a straightforward generalization of the conformally invariant formulation of the four-dimensional scalar and spinor fields. This allows one to use the experience and intuition gained from the usual conformal field theories for studying the dynamics of higher spin fields on flat and AdS backgrounds, and to construct their correlation functions.

Being intrinsically related to the unfolded formulation [29–33] of higher-spin field theory, the hyperspace approach provides an extra and potentially powerful tool for studying higher spin AdS/CFT correspondence (for reviews on higher-spin holography, see, e.g., [34,35]). The origin of higher-spin holographic duality can be traced back to the work of Flato and Fronsdal [36] who showed that the tensor product of single-particle states of a 3D massless conformal scalar and spinor fields (singletons) produces the tower of all single-particle representations of 4D massless fields whose spectrum matches that of 4D higher spin gauge theories. The hyperspace formulation provides an explicit field theoretical realization of the Flato-Fronsdal theorem in which higher spin fields are embedded in a single scalar and spinor fields, though propagating in hyperspace. The relevance of the unfolded and hyperspace formulation to the origin of holography has been pointed out in [33]. In this interpretation, holographically dual theories share the same unfolded formulation in extended spaces which contains twistor-like variables and each of these theories corresponds to a different reduction, or "visualization", of the same "master" theory.

In what follows, we will review main features and latest developments of the tensorial space approach, and associated generalized conformal theories. It is mainly based on Papers [3,8,10,13,23,24,27]. We hope that this will be a useful complement to a number of available reviews on the higher-spin gauge theories which reflect other aspects and different approaches to the subject

- Frame-like approach in higher-spin field theory [37–42].
- Metric-like approach [43–55].
- Review that address the both approaches [56].
- Higher-spin Holography [34,35,57,58].
- Reviews which contain both the metric-like approach and the hyperspace approach [59,60].
- A short review on the hyperspace approach [61].
- A short review that contains frame-like approach, hyperspaces and higher-spin holography [62].

The review is organized as follows. In Section 2 we introduce a general concept of flat hyperspaces. To this end we use somewhat heuristic argument, which includes a direct generalization of the famous twistor-like representation of a light-light momentum of a particle to higher dimensional tensorial spaces i.e., to hyperspaces. The basic fields in this set up are one bosonic and one fermionic hyperfield, which contain infinite sets of bosonic and fermionic field strengths of massless fields with spins ranging from zero to infinity. Physically interesting examples are hyperspaces associated with ordinary space–times of dimensions $D = 3, 4, 6$ and 10. In what follows, we will always keep in mind these physical cases, though from the geometric perspective the tensorial spaces of any dimension have the same properties.

We demonstrate in detail that the solutions of wave equations in hyperspace are generating functionals for higher spin fields. These equations are nothing but a set of free conformal higher spin equations in $D = 3, 4, 6$ and 10. The case of $D = 3$ describes only scalar and spinor fields, the case of $D = 4$ comprises the all massless bosonic and fermionic higher spin fields with spins from 0 to ∞ and the cases of $D = 6$ and $D = 10$ describe infinite sets of fields whose field strengths are self-dual multiforms. These fields carry unitary irreducible representations of the higher-dimensional conformal group and are sometimes called "spinning singletons" [63].

We then describe a generalized conformal group $Sp(2n)$ which contains a convention conformal group $SO(2, D)$ as its subgroup (for $D = 3, 4, 6, 10$ and $n = 2, 4, 8, 16$, respectively) and show how the coordinates in hyperspace and the hyperfields transform under these generalized conformal transformations.

In Section 3, we consider an example of curved hyperspaces which are $Sp(n)$ group manifolds. An interesting property of these manifolds is that they are hyperspace generalizations of AdS_D spaces. Similarly to the AdS_D space which can be regarded as a coset space of the conformal group $SO(2, D)$, the $Sp(n)$ group manifold is a coset space of the generalized conformal group $Sp(2n)$. This results in the fact that the property of the conformal flatness of the AdS_D spaces (i.e., the existence of a basis in which the AdS metric is proportional to a flat metric) is also generalized to the case of hyperspaces. In particular, a metric on the $Sp(n)$ group manifold is flat up to a rotation of the $GL(n)$ group, the property that we call "GL–flatness".

In Section 4, we briefly discuss how the field equations given in the previous Sections can be obtained as a result of the quantization of (super)particle models on hyperspaces.

In Section 5, we derive the field equations on $Sp(n)$ group manifolds. We show that the field equations on flat hyperspaces and $Sp(n)$ group manifolds can be transformed into each other by performing a generalized conformal rescaling of the hyperfields. We discus plane wave solutions on generalized AdS spaces and present a generalized conformal (i.e., $Sp(2n)$) transformations of the hyperfields on the $Sp(n)$ group manifolds. In all these considerations, the property of $GL(n)$ flatness plays a crucial role.

Section 6 describes a supersymmetric generalization of the construction considered in Section 2 and Section 7 deals with the supersymmetric generalization of the field theory on $Sp(n)$ introduced in Section 3. The generalization is straightforward but nontrivial. Instead of hyperspace, we consider hyper-superspaces and instead of hyperfields we consider hyper-superfields. The generalized superconformal symmetry is the $OSp(1|2n)$ supergroup and the generalized super-AdS spaces are $OSp(1|n)$ supergroup manifolds. We show that all the characteristic features of the hyperspaces and hyperfield equations are generalized to the supersymmetric case as well.

The direct analogy with usual D-dimensional CFTs suggests a possibility of considering generalized conformal field theories in hyperspaces. Sections 8 and 9 deal with such a theory which is based on the invariance of correlation functions under the generalized conformal group $Sp(2n)$. The technique used in these Sections is borrowed from usual D-dimensional CFTs and the correlation functions are obtained via solving the generalized Ward identities in (super) hyperspaces.

In Section 8, we derive $OSp(1|2n)$ invariant two-, three- and four-point functions for scalar super-hyperfields. The correlation functions for component fields can be obtained by simply expanding the results in series of the powers of Garssmann coordinates. Therefore, we shall not consider the derivation of $Sp(2n)$ invariant correlation functions for the component fields separately.

Finally, in Section 9, we introduce generalized conserved currents and generalized stress-tensors. Their explicit forms and the transformation rules under $Sp(2n)$ can be readily obtained from the free field equations and the transformation rules of the free hyperfields.

Further, we show how one can compute $Sp(2n)$ invariant correlation functions which involve the basic hyperfields together with higher rank tensors such as conserved currents and the generalized stress tensor. We show that the $Sp(2n)$ invariance itself does not impose any restriction on the generalized conformal dimensions of the basic hyperfields even if the conformal dimensions of the current and stress tensor remains canonical.

However, the further requirements of the conservation of the generalized current and generalized stress tensor fixes also the conformal dimensions of the basic hyperfields, implying that the generalized conformal theory will not allow for nontrivial interactions.

We briefly discuss possibilities of avoiding these restrictions by considering spontaneously broken $Sp(2n)$ symmetry or local $Sp(2n)$ invariance, which may lead to an interacting hyperfield theory.

Appendices contain some technical details such as conventions used in the review, a derivation of the field equations on $Sp(n)$ group manifolds and some useful identities.

2. Flat Hyperspace

Let us formulate the basic idea behind the introduction of tensorial space. We shall mainly concentrate on a tensorial extension of four-dimensional Minkowski space–time. A generalization to higher dimensional $D = 6$ and $D = 10$ spaces will be given later in this Section.

Consider a four dimensional massless scalar field. Its light-like momentum $p_m p^m = 0$, $m = 0, 1, 2, 3$ can be expressed via the Cartan–Penrose (twistor) representation as a bilinear combination of a commuting Weyl spinor λ_A and its complex conjugate $\overline{\lambda}_{\dot{A}}$ $(A, \dot{A} = 1, 2)$

$$p^m = \lambda^A (\sigma^m)_{A\dot{A}} \bar{\lambda}^{\dot{A}}, \quad or \quad P_{A\dot{A}} = \lambda_A \overline{\lambda}_{\dot{A}}. \tag{1}$$

Obviously, since the spinors are commuting, one has $\lambda^A\lambda^B\varepsilon_{AB} \equiv \lambda^A\lambda_A = 0 = \bar{\lambda}^{\dot{A}}\bar{\lambda}_{\dot{A}}$ and therefore $P^{A\dot{A}}P_{A\dot{A}} = 0$, where the spinor indices are raised and lowered with the unit antisymmetric tensors ε^{AB} and ε_{AB}.

In order to generalize this construction to higher dimensions note that one can equivalently rewrite Equation (1) in terms of four-dimensional real Majorana spinors λ^α ($\alpha = 1, ..., 4$)

$$p^m = \lambda^\alpha \gamma^m_{\alpha\beta}\lambda^\beta. \tag{2}$$

Due to the Fierz identities

$$(\gamma^m)_{\alpha\beta}(\gamma_m)_{\gamma\delta} + (\gamma^m)_{\alpha\delta}(\gamma_m)_{\beta\gamma} + (\gamma^m)_{\alpha\gamma}(\gamma_m)_{\delta\beta} = 0 \tag{3}$$

satisfied by the Dirac matrices $(\gamma^m)_{\alpha\beta} = (\gamma^m)_{\beta\alpha}$ one has $p^m p_m = 0$. (The four-component spinor indices are raised and lowered by antisymmetric charge conjugation matrices $C^{\alpha\beta}$ and $C_{\alpha\beta}$ see the Appendix A.) Let us note that since identities similar to (3) hold also in $D = 3, 6$ and 10, the Cartan–Penrose relation (2) is valid in these dimensions as well.

Let us continue with the four-dimensional case. The momentum $P_{A\dot{A}}$ is canonically conjugate to coordinates $x^{A\dot{A}}$. One can easily solve the quantum analogue of Equation (1)

$$\left(\frac{\partial}{\partial x^{A\dot{A}}} - i\lambda_A\bar{\lambda}_{\dot{A}}\right)\Phi(x,\lambda) = 0 \tag{4}$$

to obtain a plane wave solution for the massless scalar particle

$$\Phi(x,\lambda,\bar{\lambda}) = \phi(\lambda,\bar{\lambda})e^{ix^{A\dot{A}}\lambda_A\bar{\lambda}_{\dot{A}}}, \tag{5}$$

or in terms of the Majorana spinors

$$\Phi(x,\lambda) = \phi(\lambda)e^{ix_m\lambda^\alpha\gamma^m_{\alpha\beta}\lambda^\beta}, \tag{6}$$

with $\phi(\lambda)$ being an arbitrary spinor function.

Let us now consider the equation

$$P_{\alpha\beta} = \lambda_\alpha\lambda_\beta, \tag{7}$$

which looks like a straightforward generalization of (1) and see its implications. A space–time described by the coordinates $X^{\alpha\beta}$ (conjugate to $P_{\alpha\beta}$) is now ten-dimensional, since $X^{\alpha\beta}$ is a 4×4 symmetric matrix. A basis of symmetric matrices is formed by the four Dirac matrices $\gamma^m_{\alpha\beta}$ and their six antisymmetric products $\gamma^{mn}_{\alpha\beta} = -\gamma^{mn}_{\alpha\beta}$. In this basis, $X^{\alpha\beta}$ has the following expansion

$$X^{\alpha\beta} = \frac{1}{2}x^m(\gamma_m)^{\alpha\beta} + \frac{1}{4}y^{mn}(\gamma_{mn})^{\alpha\beta}. \tag{8}$$

The analogue of the wave Equation (4) is now

$$\left(\frac{\partial}{\partial X^{\alpha\beta}} - i\lambda_\alpha\lambda_\beta\right)\Phi(X,\lambda) = 0, \tag{9}$$

whose solution is

$$\Phi(X,\lambda) = e^{iX^{\alpha\beta}\lambda_\alpha\lambda_\beta}\phi(\lambda). \tag{10}$$

At this point, one might ask the question what is the meaning of Equation (9) and of the extra coordinates y^{mn} and λ^α? As we shall see, the answer is that Equation (9) is nothing else but Vasiliev's unfolded equations for free massless higher-spin fields in four-dimensional Minkowski space–time [29]. The wave function $\Phi(X,\lambda)$ depends on the coordinates x^m, y^{mn} and λ^α. While x^m parameterize the conventional four-dimensional Minkowski space–time, the coordinates y^{mn} (and/or λ^α) are associated with integer and half-integer spin degrees of freedom of four-dimensional fields with spin values ranging from zero to infinity.

2.1. Higher Spin Content of the Tensorial Space Equations

To demonstrate the above statement let us first Fourier transform the wave function (10) into a conjugate representation with respect to the spinor variable λ_α considered in [4]

$$C(X,\mu) = \int d^4\lambda\, e^{-i\mu^\alpha\lambda_\alpha}\,\Phi(X,\lambda) = \int d^4\lambda\, e^{-i\mu^\alpha\lambda_\alpha + iX^{\alpha\beta}\lambda_\alpha\lambda_\beta}\phi(\lambda). \tag{11}$$

The function $C(X,\mu)$ obeys the equation

$$\left(\frac{\partial}{\partial X^{\alpha\beta}} - i\frac{\partial^2}{\partial\mu^\alpha\partial\mu^\beta}\right) C(X,\mu) = 0. \tag{12}$$

Let us expand the function $C(X,\mu)$ in series of the variables μ^α

$$C(X,\mu) = \sum_{n=0}^{\infty} C_{\alpha_1\cdots\alpha_n}(X)\,\mu^{\alpha_1}\cdots\mu^{\alpha_n} = b(X) + f_\alpha(X)\mu^\alpha + \cdots. \tag{13}$$

and insert this expansion into the Equation (12). Then one finds that all the components of $C(X,\mu)$ proportional to the higher powers of μ^α are expressed in terms of two fields the scalar $b(X)$ and the spinor $f_\alpha(X)$. As a result of (13), these fields satisfy the relations [4]

$$\partial_{\alpha\beta}\partial_{\gamma\delta}\, b(X) - \partial_{\alpha\gamma}\partial_{\beta\delta}\, b(X) = 0, \tag{14}$$
$$\partial_{\alpha\beta} f_\gamma(X) - \partial_{\alpha\gamma} f_\beta(X) = 0. \tag{15}$$

The basic fields $b(X)$ and $f_\alpha(X)$ depend on x^m and y^{mn}. Let us now expand these fields in series of the tensorial coordinates y^{mn}

$$\begin{aligned} b(x,y) &= \phi(x) + y^{m_1n_1}F_{m_1n_1}(x) + y^{m_1n_1}\,y^{m_2n_2}\,\hat{R}_{m_1n_1,m_2n_2}(x) \\ &\quad + \textstyle\sum_{s=3}^{\infty} y^{m_1n_1}\cdots y^{m_sn_s}\,\hat{R}_{m_1n_1,\cdots,m_sn_s}(x), \end{aligned} \tag{16}$$

$$\begin{aligned} f^\alpha(x,y) &= \psi^\alpha(x) + y^{m_1n_1}\,\hat{\mathcal{R}}^\alpha_{m_1n_1}(x) \\ &\quad + \textstyle\sum_{s=\frac{5}{2}}^{\infty} y^{m_1n_1}\cdots y^{m_{s-\frac{1}{2}}n_{s-\frac{1}{2}}}\,\hat{\mathcal{R}}^\alpha_{m_1n_1,\cdots,m_{s-\frac{1}{2}}n_{s-\frac{1}{2}}}(x). \end{aligned} \tag{17}$$

Each four-dimensional component field in this expansion is antisymmetric under the permutation of the indices m_i and n_i and is symmetric with respect to the permutation of the pairs (m_i,n_i) with (m_j,n_j). In order to answer the question about the physical meaning of these fields, let us first consider the scalar field Equation (14). Using the expression (8) for the tensorial coordinates and four-dimensional γ-matrix identities, one can decompose (14) as follows

$$\partial_p\partial^p\, b(x^l,y^{mn}) = 0, \quad \left(\partial_p\partial_q - 4\partial_{pr}\partial^r{}_q\right) b(x^l,y^{mn}) = 0, \quad \epsilon^{pqrt}\partial_{pq}\partial_{rs}\, b(x^l,y^{mn}) = 0,$$

$$\epsilon^{pqrt}\partial_q\partial_{rt}\, b(x^l,y^{mn}) = 0, \quad \partial_q^{\;p}\partial_p\, b(x^l,y^{mn}) = 0. \tag{18}$$

where $\partial_p = \frac{\partial}{\partial x^p}$ and $\partial_{pq} = \frac{\partial}{\partial y^{pq}}$. The meaning of Equations (18) is the following. The first equation is a Klein-Gordon equation. The second equation implies that the trace (with respect to the $4D$ Minkowski metric) of the tensor which comes with the s-th power of y^{mn} in the expansion (14) is expressed via the second derivative of the tensor which comes with the $(s-2)$-th power of y^{mn}. Therefore, traces are not independent degrees of freedom and the independent tensorial fields under consideration are effectively traceless. The third and fourth Equation in (18) imply that the tensor fields satisfy the four-dimensional Bianchi identities, and the last equation implies that they are co–closed. These are equations for massless higher-spin fields written in terms of their curvatures $\hat{\mathcal{R}}^{\alpha}_{m_1n_1,\cdots,m_{s-\frac{1}{2}}n_{s-\frac{1}{2}}}(x)$. In four dimensions these equations are conformally invariant. Therefore one can conclude that in the expansion (16) the field $\phi(x)$ is a conformal scalar, $F_{mn}(x)$ is the field strength of spin-1 Maxwell field, the field $\hat{R}_{m_1n_1,m_2n_2}(x)$ is a linearized Riemann tensor for spin-2 graviton, etc.

The treatment of Equation (15) which describes half-integer higher-spin fields in terms of corresponding curvatures is completely analogous to the bosonic one (14). The independent equations for the conformal half-integer spin fields are

$$\gamma^p \partial_p f(x^l, y^{mn}) = 0, \tag{19}$$

$$(\partial_p - 2\gamma^r \partial_{pr}) f(x^l, y^{mn}) = 0 \tag{20}$$

From (19)–(20) one can derive the equation

$$\partial_{mn} f(x,y) = \frac{1}{2}\gamma_{[m}\partial_{n]} f(x,y) + \frac{1}{2}(\partial_{mn} + \frac{1}{2}\varepsilon_{mnpq}\partial^{pq}\gamma_5) f(x,y). \tag{21}$$

This equation describes the decomposition of the spinor-tensor $\partial_{mn} f$ into the part which contains the $D = 4$ space–time derivative of f and the "physical" part which is self-dual and gamma-traceless, i.e.,

$$\begin{array}{l} \gamma^m(\partial_{mn} + \frac{1}{2}\varepsilon_{mnpq}\partial^{pq}\gamma_5) f(x,y) = 0 \\ (\partial_{mn} + \frac{1}{2}\varepsilon_{mnpq}\partial^{pq}\gamma_5) f(x^l.y^{mn}) = \frac{1}{2}\varepsilon_{mnrs}(\partial^{rs} + \frac{1}{2}\varepsilon^{rspq}\partial^{pq}\gamma_5) f(x,y) \end{array} \tag{22}$$

Therefore, one can conclude that due to Equations (19) and (20) the field $\psi^\alpha(x)$ in the expansion (17) is a spin-$\frac{1}{2}$ field, the field $\hat{\mathcal{R}}^{\alpha}_{m_1n_1}(x)$ corresponds to the field strength of the spin-$\frac{3}{2}$ Rarita–Schwinger field, while the other fields are the field strengths of the half-integer conformal higher-spin fields in $D = 4$.

Finally, let us define the hyperspaces associated with $D = 6$ and $D = 10$ space–time. The dynamics of the fields will be again determined by the equation (7) with the corresponding hyperspaces and the twistor-like variables λ_α defined as follows.

In $D = 10$ the twistor-like variable λ_α is a 16–component Majorana–Weyl spinor. The gamma–matrices $\gamma_m^{\alpha\beta}$ and $\gamma_{m_1\cdots m_5}^{\alpha\beta}$ form a basis of the symmetric 16×16 matrices, so the $n = 16$ tensorial manifold is parameterized by the coordinates

$$X^{\alpha\beta} = \frac{1}{16}\left(x^m \gamma_m^{\alpha\beta} + \frac{1}{2\cdot 5!} y^{m_1\ldots m_5}\gamma_{m_1\ldots m_5}^{\alpha\beta}\right) = X^{\beta\alpha}, \tag{23}$$

$$(m = 0,1,\ldots,9; \quad \alpha,\beta = 1,2,\ldots,16),$$

where $x^m = X^{\alpha\beta}\gamma^m_{\alpha\beta}$ are associated with the coordinates of the $D = 10$ space–time, while the anti-self-dual coordinates

$$y^{m_1\ldots m_5} = X^{\alpha\beta}\gamma_{\alpha\beta}^{m_1\ldots m_5} = -\frac{1}{5!}\epsilon^{m_1\ldots m_5 n_1\ldots n_5} y_{n_1\ldots n_5},$$

describe spin degrees of freedom.

The corresponding field Equations are again (14) and (15) and the entire discussion repeats as in the case of $D = 4$. The crucial difference is that now the expansions (16) and (17) is performed in terms of the coordinates $y^{m_1\cdots m_5}$. As a result one obtains a description of conformal fields whose curvatures are self-dual with respect to each set of indexes $(m_i n_i p_i q_i r_i)$. These traceless rank $5s$ tensors $R_{[5]_1\cdots[5]_s}$ are automatically irreducible under $GL(10,\mathbb{R})$ due to the self-duality property, and are thus associated with the rectangular Young diagrams (s,s,s,s,s) which are made of five rows of equal length s ("multi-five-forms"). The field equations, which are ten-dimensional analogues of the four-dimensional Equations (18), can be found in [13].

In $D = 6$ the commuting spinor λ_α is a symplectic Majorana–Weyl spinor. The spinor index can be decomposed as follows $\alpha = a \otimes i$ ($\alpha = 1, \ldots, 8; a = 1,2,3,4; i = 1,2$). The tensorial space coordinates $X^{\alpha\beta} = X^{ai\,bj}$ are decomposed into

$$\begin{aligned} X^{ai\,bj} &= \tfrac{1}{8}\, x^m\, \tilde{\gamma}^{ab}_m\, \epsilon^{ij} + \tfrac{1}{16\cdot 3!}\, y_I^{mnp}\, \tilde{\gamma}^{ab}_{mnp}\, \tau_I^{ij}\,, \\ & m,n,p = 0,\ldots,5; \quad a,b = 1,\ldots,4; \quad i,j = 1,2; \quad I = 1,2,3 \end{aligned} \tag{24}$$

where $\epsilon^{12} = -\epsilon_{12} = 1$, and τ_I^{ij} ($I = 1,2,3$) provide a basis of 2×2 symmetric matrices, They are related to the usual $SU(2)$-group Pauli matrices $\tau_{I\,ij} = \epsilon_{jj'}\,\sigma_{I\,i}{}^{j'}$. The matrices $\tilde{\gamma}^{ab}_m$ (where $\gamma^m_{ab} = 1/2\,\varepsilon_{abcd}\tilde{\gamma}^{m\,cd}$) form a complete basis of 4×4 antisymmetric matrices with upper (lower) indices transforming under an (anti)chiral fundamental representation of the non-compact group $SU^*(4) \sim Spin(1,5)$. For the space of 4×4 symmetric matrices with upper (lower) indices, a basis is provided by the set of self-dual and anti-self-dual matrices $(\tilde{\gamma}^{mnp})^{ab}$ and γ_{ab}^{mnp}, respectively,

$$(\tilde{\gamma}^{mnp})^{ab} = \frac{1}{3!}\epsilon^{mnpqrs}\tilde{\gamma}^{ab}_{qrs}\,, \qquad \gamma_{ab}^{mnp} = -\frac{1}{3!}\epsilon^{mnpqrs}(\gamma_{qrs})_{ab}\,. \tag{25}$$

The coordinates $x^m = x^{ai\,bj}\,\gamma^m_{ab}\,\epsilon_{ij}$ are associated with $D = 6$ space–time, while the self-dual coordinates

$$y_I^{mnp} = x^{ai\,bj}\,\gamma_{ab}^{mnp}\,\tau_{I\,ij} = -\frac{1}{3!}\,\epsilon^{mnpqrs}y^I_{qrs}\,, \tag{26}$$

describe spinning degrees of freedom.

The consideration proceeds as in the $D = 4$ and $D = 10$ case. Because of the form of the tensorial coordinates in (24) the six-dimensional analogue of the expansions (16) and (17) contains powers of y_i^{mnp}. Corresponding field strengths, which again describe conformal fields in six dimensions, are self-dual with respect to each set of the indexes $(m_i n_i p_i)$. In other words, one has an infinite number of conformally invariant (self-dual) "multi-3-form" higher-spin fields in the six-dimensional space–time which form the $(2[s]+1)$-dimensional representations of the group $SO(3)$.

In [9,16,21] Equation (12) has been generalized to include several commuting spinor variables $\mu^{p\alpha}$ ($p, q = 1, \ldots, r$)

$$\left(\frac{\partial}{\partial X^{\alpha\beta}} \pm i\eta^{pq}\frac{\partial^2}{\partial\mu^{p\alpha}\partial\mu^{q\beta}}\right) C^r_{\pm}(X,\mu) = 0. \tag{27}$$

where $\eta^{pq} = \eta^{qp}$ is a nondegenerate metric. The value of r is called the "rank". As we explained above, the free higher-spin fields in $D = 4$ are described by the rank-one equations in the ten-dimensional tensorial space. The higher-spin currents are fields of rank-two $r = 2$. These currents obey the equations with off-diagonal η^{pq} [19]. The currents $J(X, \mu^p)$ are bilinear in the higher-spin gauge fields C_+ and C_-, which obey the rank-one equation (27) $J = C_+C_-$.

On the other hand, when considering rank-two equations the corresponding tensorial space can be embedded in the higher-dimensional tensorial space. From the discussion above, it follows that a natural candidate for such higher-dimensional space is the tensorial extension of $D = 6$ space–time. In this way one effectively linearizes the problem since the conformal currents in four dimensions are identified with the fields in $D = 6$ [21].

2.2. Four Dimensional Unfolded Higher-Spin Field Equations from the Hyperspace Field Equations

Let us rewrite, in the case of the $D = 4$ theory, the hyperspace relations in terms of the Weyl spinors. The momenta (7) take the form

$$P_{AB} = \lambda_A\lambda_B\,, \quad \overline{P}_{\dot{A}\dot{B}} = \overline{\lambda}_{\dot{A}}\overline{\lambda}_{\dot{B}}\,, \quad P_{A\dot{A}} = \lambda_A\overline{\lambda}_{\dot{A}}\,, \tag{28}$$

while Equation (7) splits into

$$\left(\sigma^{mn}_{AB}\frac{\partial}{\partial y^{mn}}+i\frac{\partial^2}{\partial\mu^A\partial\mu^B}\right)C(x,y,\mu)=0,$$
$$\left(\bar{\sigma}^{mn}_{\dot{A}\dot{B}}\frac{\partial}{\partial y^{mn}}-i\frac{\partial^2}{\partial\bar{\mu}^{\dot{A}}\partial\bar{\mu}^{\dot{B}}}\right)C(x,y,\mu)=0 \tag{29}$$

and

$$\left(\sigma^{m}_{A\dot{A}}\frac{\partial}{\partial x^{m}}+i\frac{\partial^2}{\partial\mu^A\partial\bar{\mu}^{\dot{A}}}\right)C(x,y,\mu)=0\,. \tag{30}$$

Equations (29) relate the dependence of $C(x,y,\mu)$ on the coordinates y^{mn} to its dependence on μ^α. Thus, using this relation, one can regard the wave function $C(x^m,\mu^\alpha):=C(X^{\alpha\beta},\mu^\alpha)|_{y^{mn}=0}$ as the fundamental field.

The expansion of $C(x^m,\mu)$ in series of μ^A and $\bar{\mu}^{\dot{A}}$ is

$$C(x^p,\mu^A,\bar{\mu}^{\dot{A}})=\sum_{m,n=0}^{\infty}\frac{1}{m!n!}C_{A_1\dots A_m,\dot{B}_1\dots\dot{B}_n}(x^p)\,\mu^{A_1}\dots\mu^{A_m}\,\bar{\mu}^{\dot{B}_1}\dots\bar{\mu}^{\dot{B}_n}\,, \tag{31}$$

where the reality of the wave function implies $(C_{A_1\dots A_m,\dot{B}_1\dots\dot{B}_n})^*=C_{B_1\dots B_n,\dot{A}_1\dots\dot{A}_m}$, and by construction the spin-tensors are symmetric in the indices A_i and in $\dot{B}_i$.

The consistency of (30) implies the integrability conditions

$$\frac{\partial^2}{\partial\mu^{[A}\partial x^{B]\dot{B}}}C(x,\mu)=0,\quad \frac{\partial^2}{\partial\bar{\mu}^{[\dot{A}}\partial x^{\dot{B}]B}}C(x,\mu)=0\,. \tag{32}$$

We have thus obtained the equations of the Vasiliev's unfolded formulation of free higher spin fields in terms of zero–forms. In this formulation the $C_{0,0}$ component (a physical scalar), $C_{A_1\dots A_{2s},0}$ and $C_{0,\dot{A}_1,\dots\dot{A}_{2s}}$ components of the expansion (31) correspond to the physical fields, while the other fields are auxiliary. The latter two fields are the self-dual and anti-self-dual components of the spin–s field strength. The nontrivial equations on the dynamical fields are [38] the Klein–Gordon equation for the spin zero scalar field $\partial^m\partial_m C_{0,0}=0$ and the massless equations for spin $s>0$ field strengths

$$\partial^{B\dot{B}}C_{BA_1\dots A_{2s-1}}(x)=0\,,\quad \partial^{B\dot{B}}C_{\dot{B}\dot{A}_1\dots\dot{A}_{2s-1}}(x)=0\,, \tag{33}$$

which follow from (32). All the components of $C(x^m,\mu^A,\bar{\mu}^{\dot{A}})$ that depend on *both* μ^A and $\bar{\mu}^{\dot{A}}$ are auxiliary fields expressed by (30) in terms of space–time derivatives of the dynamical fields contained in the analytic fields $C(x^m,\mu^A,0)$ and $C(x^m,0,\mu^{\dot{A}})$ and thus one arrives at the unfolded formulation of [38].

Let us summarize what we have considered by now. To describe the dynamics of higher-spin fields in four dimensions we have introduced extended ten-dimensional tensorial space, hyperspace, parameterized by the coordinates $X^{\alpha\beta}$ (8). The main object is a generating functional for higher-spin fields described by $C(X,\mu)$ or by $\Phi(X,\lambda)$. The generating functional depends on the tensorial coordinates $X^{\alpha\beta}$ and on the commuting spinors μ^α or λ^α. The dynamics is described by the field Equation (9) or (12). To obtain from these the higher-spin field equations in the ordinary space–time parameterized by the coordinates x^m one can use two options. In the first approach one gets rid of the tensorial coordinates y^{mn} and arrives at Vasiliev's unfolded formulation in terms of the functional (31). Alternatively, one can first get rid of the commuting spinor variables and arrive at the equations for the bosonic (16) and fermionic (17) hyperfields. Both pictures provide the equations for the field strengths of the higher-spin potentials, the difference being that these field strengths are realized either as tensors or spin-tensors.

2.3. Generalized Conformal Group $Sp(2n)$

Let us consider in more detail the symmetries of Equation (7) in which now the Greek indices $\alpha, \beta, \ldots$ run from 1 to an arbitrary even integer $2n$. However, as we explained in the previous Section, the physically interesting cases are associated with $n = 2, 4, 8$, and 16, which correspond to the number of space–time dimensions equal to 3, 4, 6 and 10, respectively.

It turns out that Equation (7) is invariant under the transformations of the $Sp(2n)$ group [5,8]

$$\delta\lambda_\alpha = g_\alpha{}^\beta \lambda_\beta - k_{\alpha\beta} X^{\beta\gamma} \lambda_\gamma, \tag{34}$$

$$\delta X^{\mu\nu} = a^{\mu\nu} + (X^{\mu\rho} g_\rho{}^\nu + X^{\nu\rho} g_\rho{}^\mu) - X^{\mu\rho} k_{\rho\lambda} X^{\lambda\nu}. \tag{35}$$

The constant parameters $a^{\alpha\beta} = a^{\beta\alpha}$, g_γ^α and $k_{\alpha\beta} = k_{\beta\alpha}$ correspond to the generators of generalized translations $P_{\alpha\beta}$, generalized Lorentz transformations and dilatations G_β^α (generated by the $GL(n)$ algebra) and generalized conformal boosts $K_{\alpha\beta}$. The differential operator representation of these generators have the form

$$P_{\mu\nu} = -i\frac{\partial}{\partial X^{\mu\nu}} \equiv -i\partial_{\mu\nu}, \tag{36}$$

$$G_\nu{}^\mu = -2iX^{\mu\rho}\,\partial_{\rho\nu} \tag{37}$$

and

$$K^{\mu\nu} = iX^{\mu\rho}X^{\nu\lambda}\partial_{\rho\lambda} \tag{38}$$

These symmetries are the hyperspace counterparts of the conventional Poincaré translations, Lorentz rotations, dilatations and conformal boosts of Minkowski space–time. The generalized Lorentz rotations are generated by the traceless operators $L_\mu{}^\nu = G_\mu{}^\nu - \frac{1}{n}\delta_\mu^\nu G_\lambda{}^\lambda$, forming the $SL(n)$–algebra, whereas dilatations are generated by the trace of $G_\mu{}^\nu$. The generators (36), (37) and (38) form the $Sp(2n)$ algebra which plays the role of a generalized conformal symmetry in the hyperspace

$$\begin{gathered}
[P_{\mu\nu}, P_{\rho\lambda}] = 0, \qquad [K^{\mu\nu}, K^{\rho\lambda}] = 0, \qquad [G_\nu{}^\mu, G_\lambda{}^\rho] = i(\delta_\lambda^\mu\, G_\nu{}^\rho - \delta_\nu^\rho\, G_\lambda{}^\mu), \\
[P_{\mu\nu}, G_\lambda{}^\rho] = -i(\delta_\mu^\rho P_{\nu\lambda} + \delta_\nu^\rho P_{\mu\lambda}), \qquad [K^{\mu\nu}, G_\lambda{}^\rho] = i(\delta_\lambda^\mu K^{\nu\rho} + \delta_\lambda^\nu K^{\mu\rho}), \\
[P_{\mu\nu}, K^{\lambda\rho}] = \tfrac{i}{4}(\delta_\mu^\rho G_\nu{}^\lambda + \delta_\nu^\rho G_\mu{}^\lambda + \delta_\mu^\lambda G_\nu{}^\rho + \delta_\nu^\lambda G_\mu{}^\rho).
\end{gathered} \tag{39}$$

From the structure of this algebra, one can see that the flat hyperspace $\mathcal{M}_n$ can be realized as a coset manifold associated with the translations $P = \frac{Sp(2n)}{K\otimes GL(n)}$ where $K\otimes GL(n)$ is the semi–direct product of the Abelian group generated by the generalized conformal boosts $K_{\mu\nu}$ and the general linear group.

The generators of the translations, Lorentz rotations and conformal boosts of the conventional conformal group can be obtained from the $Sp(2n)$ generators as projections onto the x-space, for example $p_m = (\gamma_m)^{\mu\nu} P_{\mu\nu}$, etc.

Let us note that the $Sp(2n)$ algebra can be conveniently realized with the use of the twistor-like variables λ_α and their conjugate μ^α

$$[\mu^\alpha, \lambda_\beta] = \delta_\beta^\alpha. \tag{40}$$

In the twistor representation the generators of the $Sp(2n)$ group have the following form

$$P_{\alpha\beta} = \lambda_\alpha\lambda_\beta, \qquad G_\alpha^\beta = \lambda_\alpha\mu^\beta, \qquad K_{\alpha\beta} = \mu_\alpha\mu_\beta. \tag{41}$$

Equations (14) and (15) are invariant under the $Sp(2n)$ transformations (35), provided that the fields transform as follows

$$\delta b(X) = -(a^{\mu\nu}\partial_{\mu\nu} + \frac{1}{2}g_\mu{}^\mu + 2g_\nu{}^\mu X^{\nu\rho}\partial_{\mu\rho} - k_{\mu\nu}(\frac{1}{2}X^{\mu\nu} + X^{\mu\rho}X^{\nu\lambda}\partial_{\rho\lambda}))b(X), \quad (42)$$

$$\begin{aligned}\delta f_\rho(X) &= -(a^{\mu\nu}\partial_{\mu\nu} + \tfrac{1}{2}g_\mu{}^\mu + 2g_\nu{}^\mu X^{\nu\lambda}\partial_{\mu\lambda} - k_{\mu\nu}(\tfrac{1}{2}X^{\mu\nu} + X^{\mu\tau}X^{\nu\lambda}\partial_{\tau\lambda}))f_\rho(X) + \\ &\quad -(g_\rho{}^\nu - k_{\lambda\rho}X^{\lambda\nu})f_\nu(X).\end{aligned} \quad (43)$$

Note that these variations contain the term $\frac{1}{2}(g_\mu{}^\mu - k_{\mu\nu}X^{\mu\nu})$, implying that the fields have the canonical conformal weight $1/2$. A natural generalization of these transformations to fields of a generic conformal weight Δ is [4]

$$\delta b(X) = -(a^{\mu\nu}\partial_{\mu\nu} + \Delta\,(g_\mu{}^\mu - k_{\mu\nu}X^{\mu\nu}) + 2g_\nu{}^\mu X^{\nu\rho}\partial_{\mu\rho} - k_{\mu\nu}X^{\mu\rho}X^{\nu\lambda}\partial_{\rho\lambda})b(X), \quad (44)$$

$$\begin{aligned}\delta f_\rho(X) &= -(a^{\mu\nu}\partial_{\mu\nu} + \Delta\,(g_\mu{}^\mu - k_{\mu\nu}X^{\mu\nu}) + 2g_\nu{}^\mu X^{\nu\lambda}\partial_{\mu\lambda} - k_{\mu\nu}X^{\mu\tau}X^{\nu\lambda}\partial_{\tau\lambda})f_\rho(X) \\ &\quad -(g_\rho{}^\nu - k_{\lambda\rho}X^{\lambda\nu})f_\nu(X).\end{aligned} \quad (45)$$

3. Hyperspace Extension of AdS Spaces

A hyperspace extension of AdS_D spaces is another coset of the $Sp(2n)$ group. Recall that the usual AdS_D space can be realized as the coset space (Here, K and $\mathbb{D}$ denote the generalized conformal boosts and dilatation, respectively.) $\frac{SO(2,D)}{K\otimes(SO(1,D-1)\times\mathbb{D})}$ parameterized by the coset element $e^{\mathcal{P}_m x^m}$. The generators of the AdS_D boosts $\mathcal{P}_m$ can be singled out from the generators of the four dimensional conformal group $SO(2,D)$ by taking a linear combination of the generators of the Poincaré translations P_m and conformal boosts K_m as $\mathcal{P}_m = P_m - \xi^2 K_m$, where ξ is the inverse of the AdS_D radius.

Analogously, for the case of the hyperspace extension of the AdS_D space let us consider the generators

$$\mathcal{P}_{\alpha\beta} = P_{\alpha\beta} - \frac{\xi^2}{16}K_{\alpha\beta}, \quad [\mathcal{P},\mathcal{P}] \sim M, \quad [\mathcal{P},M] \sim \mathcal{P}, \quad (46)$$

where $K_{\alpha\beta} = C_{\alpha\gamma}C_{\beta\delta}K^{\gamma\delta}$, $M_{\alpha\beta}$ stands for the symmetric part of the $GL(n)$ transformations $M_{\alpha\beta} = G_{(\alpha}{}^\gamma C_{\gamma\beta)} \equiv \frac{1}{2}(G_\alpha{}^\gamma C_{\gamma\beta} + G_\beta{}^\gamma C_{\gamma\alpha})$ and $C_{\alpha\beta} = -C_{\beta\alpha}$ is the $Sp(n)$-invariant symplectic metric. One can see that the corresponding manifold is an $Sp(n)$ group manifold [8] which can be realized as a coset space $\frac{Sp(2n)}{K\otimes GL(n)}$ with the coset element $e^{(P-\frac{\xi^2}{16}K)_{\alpha\beta}X^{\alpha\beta}}$. Indeed, let us recall that $Sp(n)$ group is generated by $n\times n$ symmetric matrices $M_{\alpha\beta}$ which form the algebra

$$\left[M_{\alpha\beta}, M_{\gamma\delta}\right] = -\frac{i\xi}{2}\left[C_{\gamma(\alpha}M_{\beta)\delta} + C_{\delta(\alpha}M_{\beta)\gamma}\right], \quad \alpha,\beta = 1,...,n. \quad (47)$$

As a group manifold, $Sp(n)$ is the coset $[Sp(n)_L \times Sp(n)_R]/Sp(n)$ which has the isometry group $Sp(n)_L \times Sp(n)_R$, the latter being the subgroup of $Sp(2n)$ generated by

$$M^L_{\alpha\beta} = P_{\alpha\beta} - \frac{\xi^2}{16}K_{\alpha\beta} - \frac{\xi}{4}M_{\alpha\beta} \qquad M^R_{\alpha\beta} = P_{\alpha\beta} - \frac{\xi^2}{16}K_{\alpha\beta} + \frac{\xi}{4}M_{\alpha\beta}, \quad (48)$$

as one may see from the structure of the $Sp(2n)$ algebra (39). The generators $M_{\alpha\beta}$ form the diagonal $Sp(n)$ subalgebra of $Sp(n)_L \times Sp(n)_R$.

Let us note that, for the case of $n=4$, i.e., for the case of four space–time dimensions, AdS_4 space is a coset subspace of $Sp(4) \sim SO(2,3)$ of the maximal dimension. For $n>4$, an AdS_D space is also a subspace of $Sp(n)$ manifold but is no longer the maximal coset of this group.

3.1. GL-Flatness of $Sp(n)$ Group Manifolds

Let us describe a property of GL-flatness of the $Sp(n)$ group manifolds which is a generalization of the conformal flatness property of AdS_D spaces. By GL-flatness we mean that, in a local coordinate basis associated with $X^{\alpha\beta}$, the corresponding $Sp(n)$ Cartan form $\Omega^{\alpha\beta}$ has the form

$$\Omega^{\alpha\beta} = dX^{\mu\nu} G_\mu{}^\alpha(X) G_\nu{}^\beta(X) , \tag{49}$$

with the matrix $G_\mu{}^\alpha(X)$ being

$$G_\mu{}^\alpha(X) = \delta_\mu^\alpha + \sum_{k=1}^{\infty} \left(-\frac{\xi}{4}\right)^k (X^k)_\mu{}^\alpha . \tag{50}$$

This expression implies that the $Sp(n)$ Cartan form is obtained from the flat differential $dX^{\mu\nu}$ by a specific $GL(n)$ rotation of the latter.

This property can be demonstrated by showing that the Cartan forms (49) satisfy the $Sp(n)$-group Maurer–Cartan equations (see [8,23], for technical details)

$$d\Omega^{\alpha\beta} + \frac{\xi}{2} \Omega^{\alpha\gamma} \wedge \Omega_\gamma{}^\beta = 0 . \tag{51}$$

The matrix $G_\alpha^{-1\mu}(X)$ inverse to (50) depends linearly on $X_\alpha{}^\mu$ and has a very simple form

$$G_\alpha^{-1\mu}(X) = \delta_\alpha^\mu + \frac{\xi}{4} X_\alpha{}^\mu . \tag{52}$$

Note that the possibility of representing the Cartan forms in the form (49) is a particular feature of the $Sp(n)$ group manifold since, in general, it is not possible to decompose the components of the Cartan form into a "direct product" of components of some matrix $G_\mu{}^\alpha$.

3.2. An Explicit Form of the AdS_4 Metric

Let us now demonstrate that, for the case of $n = 4$ ($D = 4$), the pure x^m-dependent part of the matrix $G_\mu{}^\alpha(X)$ indeed generates the metric on AdS_4 in a specific parameterization. To this end, we should evaluate the expression

$$\Omega^{\alpha\beta}(x^m) = \frac{1}{2} dx^m (\gamma_m)^{\delta\sigma} G_\delta{}^\alpha G_\sigma{}^\beta = \frac{1}{2} dx^m e_m^a (\gamma_a)^{\alpha\beta} + \frac{1}{4} dx^m \omega_m^{ab} (\gamma_{ab})^{\alpha\beta} , \tag{53}$$

where the dependence of the matrices $X^{\alpha\beta}$ on the coordinates y^{mn} (see Equation (8)) was discarded, i.e., $X_\alpha{}^\beta = \frac{1}{2} x^n (\gamma_n)_\alpha{}^\beta$. Denoting

$$x^2 = x^m x^n \eta_{mn}, \quad x_m = \eta_{mn} x^n \tag{54}$$

and, using the explicit form (50) of $G_\mu{}^\alpha(X)$, one obtains

$$\Omega^{\alpha\beta}(x) = \frac{1}{2} \frac{dx^m}{[1-(\frac{\xi}{8})^2 x^2]^2} \left[(\gamma_\ell)^{\alpha\beta} \left([1+(\tfrac{\xi}{8})^2 x^2] \delta_m^\ell - 2(\tfrac{\xi}{8})^2 \eta_{mn} x^n x^\ell \right) - \tfrac{\xi}{4} x^n (\gamma_{mn})^{\alpha\beta} \right] . \tag{55}$$

In this way, we obtain a four-dimensional space vierbein and spin-connection

$$e_m^a = \frac{1}{[1-(\frac{\xi}{8})^2 x^2]^2} \left([1+(\tfrac{\xi}{8})^2 x^2] \delta_m^a - 2(\tfrac{\xi}{8})^2 x^a x_m \right) , \tag{56}$$

$$\omega_m^{ab} = \frac{-2\xi}{[1-(\frac{\xi}{8})^2 x^2]^2} \delta_m^{[a} x^{b]} = -\frac{8(\frac{\xi}{8})}{(1-(\frac{\xi}{8})^2 x^2)^2} (x^a \delta_m^b - x^b \delta_m^a) . \tag{57}$$

The corresponding metric is

$$g_{mn} = \frac{1}{[1-(\frac{\xi}{8})^2x^2]^4}\left([1+(\tfrac{\xi}{8})^2x^2]^2\eta_{mn} - 4(\tfrac{\xi}{8})^2x_mx_n\right), \tag{58}$$

It is well-known (see also Section 5.1) that the metric on AdS_D can be represented as an embedding in a flat $(D+1)$-dimensional space

$$ds^2 = \eta_{mn}dy^m dy^n - (dy^D)^2, \tag{59}$$

via the embedding constraint

$$\eta_{mn}y^m y^n - (y^D)^2 = -r^2. \tag{60}$$

Choosing the embedding coordinates for AdS_4 to be

$$y^m = \frac{1+(\frac{\xi}{8})^2x^2}{[1-(\frac{\xi}{8})^2x^2]^2}\,x^m, \qquad y^4 = \sqrt{r^2 + x^2\frac{1+(\frac{\xi}{8})^2x^2}{[1-(\frac{\xi}{8})^2x^2]^2}}, \tag{61}$$

one readily recovers the metric (58), with the parameter ξ being related to the AdS_4 radius r as follows

$$\xi = \frac{2}{r}. \tag{62}$$

Finally, computing the Riemann tensor

$$R^{ab}{}_{mn} = -32(\tfrac{\xi}{8})^2\frac{1+(\frac{\xi}{8})^2x^2}{[1-(\frac{\xi}{8})^2x^2]^4}\left([1+(\tfrac{\xi}{8})^2x^2]\delta^{[a}_m\delta^{b]}_n + 4(\tfrac{\xi}{8})^2x^{[a}\delta^{b]}_{[m}x_{n]}\right), \tag{63}$$

and the Ricci scalar

$$R = -192\left(\frac{\xi}{8}\right)^2 = -3\xi^2, \tag{64}$$

one verifies that the metric (58) indeed corresponds to a space with constant negative curvature, i.e., the AdS_4 space.

4. Particles in Hyperspaces

In this Section, we would like to explain the physical meaning of the tensorial space coordinates as spin degrees of freedom from the perspective of the dynamics of a particle in hyperspace.

Historically, the first dynamical system in which the Fronsdal hyperspace proposal for higher–spin fields was realized explicitly was the twistor-like superparticle model of Bandos and Lukierski [2] which, for $D=4$, possesses the generalized superconformal symmetry under $OSp(1|8)$. The original motivation behind this model was a geometric interpretation of commuting tensorial charges in an extended supersymmetry algebra. Its higher–spin content was found later in [3,64] where the quantum states of the superparticle were shown to form an infinite tower of massless higher–spin fields, and the relation of this model to the unfolded formulation was assumed. This relation was analyzed in detail in [4,5,8,10,13]. In addition to the relation to higher spins, the model of Bandos and Lukierski [2] has revealed other interesting features, such as the invariance under supersymmetry with tensorial charges (which are usually associated with brane solutions of Superstring and M–Theory). Moreover, it has provided the first example of a dynamical BPS system preserving more than half of the bulk supersymmetries. BPS states preserving $\frac{2n-1}{2n}$ supersymmetries (with $n = 16$ for $D = 10, 11$) were then shown to be building blocks of any BPS states, and this led to a natural conjecture that they can be elementary constituents or "preons" of M–theory [65].

Let us consider the generic case of a particle moving in an $Sp(2n)$-invariant hyperspace $\mathcal{M}$ described by the action

$$S[X,\lambda] = \int E^{\alpha\beta}\,(X(\tau))\,\lambda_\alpha(\tau)\,\lambda_\beta(\tau), \tag{65}$$

where $X^{\mu\nu}(\tau)$ are the hyperspace coordinates of the particle. The auxiliary commuting variables $\lambda_\alpha(\tau)$ $(\alpha = 1, \cdots, n)$ is a real spinor with respect to $Sp(n)$ and a vector with respect to $GL(n)$ (introduced in Section 2). Finally $E^{\alpha\beta}(X(\tau)) = E^{\beta\alpha}(X(\tau)) = dX^{\lambda\rho}(\tau)E^{\mu\nu}_{\alpha\beta}(X)$ is the pull–back on the particle worldline of the hyperspace vielbein. For flat hyperspace

$$E^{\alpha\beta}(X(\tau)) = d\tau\,\partial_\tau X^{\alpha\beta}\,(\tau) = dX^{\alpha\beta}\,(\tau), \tag{66}$$

and for the case of the $Sp(n)$ group manifold

$$E^{\alpha\beta}(X(\tau)) = \Omega^{\alpha\beta}(X), \tag{67}$$

where $\Omega^{\alpha\beta}$ is an $Sp(n)$ Cartan form. The latter can be taken in the GL-flat realization as in (49). The dynamics of particles on the $OSp(N|n)$ supergroup manifolds was considered for $N = 1$ in [8,10,66] and for generic values of N in [4,5], and, as we have already mentioned, the twistor-like superparticle in the $n = 32$ super-hyperspace was considered in [67] as a point-like model for BPS preons [65], the hypothetical $\frac{31}{32}$-supersymmetric constituents of M-theory.

The action (65) is manifestly invariant under global $GL(n)$ transformations and *implicitly* invariant under global $Sp(2n)$ transformations, acting linearly on λ_ρ and non-linearly on $X^{\rho\nu}$. Thus, the model possesses the symmetry that Fronsdal proposed as an underlying symmetry of higher–spin field theory in the case $n = 4$, $D = 4$ [1]. To make the $Sp(2n)$ invariance manifest, it is convenient to rewrite the action (65) in a twistor form (for simplicity we consider the flat case (66))

$$S[\lambda,\mu] = \int (d\mu^\alpha(\tau)\,\lambda_\alpha(\tau) - \mu^\alpha(\tau)\,d\,\lambda_\alpha(\tau)) = \int dZ^{\mathcal{A}} Z_{\mathcal{A}}, \tag{68}$$

where

$$\mu^\alpha = X^{\alpha\beta}\,\lambda_\beta, \tag{69}$$

and

$$Z_{\mathcal{A}} = (\lambda_\alpha, \mu^\beta) \qquad Z^{\mathcal{A}} = C^{\mathcal{A}\mathcal{B}}\,Z_{\mathcal{B}} = (\mu^\alpha, -\lambda_\beta), \quad \mathcal{A} = 1, \cdots, 2n, \tag{70}$$

form a linear representation of $Sp(2n)$

$$\delta Z_{\mathcal{A}} = S_{\mathcal{A}}{}^{\mathcal{B}}\,Z_{\mathcal{B}}, \qquad S_{\mathcal{A}}{}^{\mathcal{B}} = \begin{pmatrix} g_\alpha^\beta & k_{\alpha\gamma} \\ a^{\delta\beta} & -(g_\gamma^\delta)^T \end{pmatrix}. \tag{71}$$

Hence, the bilinear form $dZ^{\mathcal{A}}\,Z_{\mathcal{A}}$ is manifestly $Sp(2n)$ invariant. Note that, as it follows from the action (68), the variables μ^α and λ_β are canonically conjugate coordinates and momenta of the particle. Upon quantization, they become the operators introduced in Section 2.3, Equation (40).

Using the relation (69) one can easily recover the $Sp(2n)$ transformation (35) of $X^{\alpha\beta}$.

Applying the Hamiltonian analysis to the particle model described by (65) and (66), one finds that the momentum conjugate to $X^{\alpha\beta}$ is related to the twistor-like variable λ_α via the constraint

$$P_{\alpha\beta} = \lambda_\alpha\lambda_\beta\,. \tag{72}$$

As we have already mentioned, this expression, e.g., in the case $n = 4$ for which $X^{\alpha\beta}$ is given in (8), is the direct analog and the generalization of the Cartan–Penrose (twistor) relation for the particle momentum $P_m = \bar{\lambda}\,\gamma_m\,\lambda$. A difference is that in $D = 4$ the Penrose twistor relation is invariant under the phase transformation

$$\lambda_\alpha \;\rightarrow\; e^{i\varphi\gamma^5}\,\lambda_\alpha, \tag{73}$$

or in the two–component Weyl spinor notation $\lambda_A \;\rightarrow\; e^{i\varphi}\,\lambda_A$, while Equation (72) does not possess this symmetry. rather the symmetry of the model is $\mathbb{Z}_2$ ($\lambda_\alpha \;\rightarrow\; -\lambda_\alpha$) subgroup of $U(1)$ and as a result in the model under consideration the phase component φ of λ_α is a dynamical degree of freedom. It turns out that upon quantization it is associated with the infinite number of massless quantum states (particles) with increasing spin (helicity). This is in contrast to the conventional twistor-like (super) particle models with a finite number of quantum states, considered e.g., in [68–79].

To understand the physical meaning of the phase φ, let us notice that Equation (72) is a constraint on possible values of the canonical momenta of the particle in the hyperspace. In the case $n = 4$ the Majorana spinor λ_α has four independent components. One of these components can be associated with the phase φ. The momentum $P_m = \bar{\lambda}\,\gamma_m\,\lambda$ of the particle along the four conventional Minkowski directions $x^m \;=\; \frac{1}{2}\,X^{\mu\nu}\,\gamma^m_{\mu\nu}$ of the hyperspace (8) is light-like. Therefore, P_m depends on three components of λ_α. It does not depend on the phase φ of λ_α, since it is invariant under the phase transformation (73). The momentum $P_{mn} = \bar{\lambda}\,\gamma_{mn}\,\lambda$ of the particle along the six additional tensorial directions $y^{mn} = \frac{1}{4}\,X^{\alpha\beta}\,\gamma^{mn}_{\alpha\beta}$ is not invariant under the phase transformations and, hence, depends on the four components of λ_α. However, we have already associated three of them with the light-like momentum P_m in $D = 4$. Therefore, the only independent component of the momentum P_{mn} is associated with the $U(1)$ phase φ of λ_α, and as a result the motion of the particle along the six tensorial directions y^{mn} is highly constrained. This means that, effectively, the particle moves in the four-dimensional Minkowski space and along a single direction in the six additional dimensions whose coordinate is conjugate to the compact momentum–space direction parameterized by the periodic phase φ. As shown in [3,64], the coordinate conjugate to the compactified momentum φ takes, upon quantization, an infinite set of integer and half-integer values associated with the helicities of higher–spin fields. The half-integer and integer–spin states are distinguished by the discrete symmetry $\mathbb{Z}_2$ ($\lambda_\alpha \;\rightarrow\; -\lambda_\alpha$).

The resulting infinite tower of discrete higher–spin states can be regarded [3,64] as an alternative to the Kaluza–Klein compactification mechanism akin to Fronsdal's original proposal. In contrast to the conventional Kaluza–Klein theory, in the hyperspace particle model, the compactification occurs in momentum space and not in coordinate space. The phase φ in (73) can be regarded as a compactified component of the momentum (72), while the corresponding conjugate hyperspace coordinate is quantized and labels the discrete values of spin of fields in the effective conventional space–time.

As we have already seen by virtue of the Fierz identity (3) the twistor particle momentum is light-like ($P^m P_m = 0$) in $D = 3, 4, 6$ and 10. Therefore, in the hyperspaces corresponding to these space–time dimensions the first–quantized particles are massless [2,3,64]. Moreover, since the model is invariant under the generalized conformal group $Sp(2n)$, the quantum states of this particle in the hyperspaces containing the $D = 3, 4, 6$ and 10 Minkowski spaces as subspaces correspond to the conformal higher–spin fields introduced in Section 2.

Let us conclude this section with a brief comment on the model describing a particle propagating on the $Sp(n)$ group manifold. Its action has the form (65), with the corresponding Cartan form given by (67). The property of GL-flatness greatly simplifies the analysis of this case. Namely, since the Cartan forms of the $Sp(n)$ group manifold and the flat hyperspace are related as in Equation (49), one can simply reduce the classical $Sp(n)$ action to the flat one by redefining the spinor variables as follows $\lambda_\alpha \rightarrow G_\alpha^{-1\beta}(X)\lambda_\beta$. However, when quantizing this system we should work with variables that appropriately describe the geometry of the $Sp(n)$ background in which the particle propagates. Thus upon quantization one gets Equation (92) as explained in detail in [10].

5. Field Equations on $Sp(n)$ Group Manifold

5.1. Scalar Field on AdS_D. A Reminder

Before deriving the field equations of hyperfields on $Sp(n)$ group manifolds, let us recollect some well known facts about a scalar field propagating on AdS_D background. In the next subsection we will see that the form of the scalar field equation on $Sp(n)$ and its certain solutions are somewhat similar to those of the AdS scalar.

Conformally invariant scalar on AdS_4 is described by the field equation [80]

$$\left(D^m D_m + \frac{2}{r^2}\right)\phi(x) = 0, \tag{74}$$

here D_m is the usual covariant derivative on AdS_4.

Equation (74) can be written in a so-called ambient space formalism. The ambient space is obtained by introducing one more time-like dimension and considering AdS_D as a hyperboloid in this higher dimensional space (for applications of this formalism to the description of higher-spin fields on AdS_D see for example [81–87])

$$\eta_{AB} y^A y^B = -r^2, \quad \eta_{AB} = diag(-1, 1, .., 1, -1), \quad A = 0, 1, .., D\,. \tag{75}$$

The AdS_D ambient-space generalization of (74) has the form

$$\left(\nabla^A \nabla_A + \frac{2(D-3)}{r^2}\right)\phi(y) = 0, \tag{76}$$

where

$$\nabla^A = \theta^{AB}\frac{\partial}{\partial y^B} \tag{77}$$

and

$$\theta^{AB} = \eta^{AB} + \frac{y^A y^B}{r^2} \tag{78}$$

is a projector, since in view of the relation (75) one has

$$\theta^{AB}\theta^{BC} = \theta^{AC}, \quad y^A \theta_A^B = 0, \quad y^A \nabla_A = 0, \quad \nabla^A y_A = D, \tag{79}$$

where the indexes A, B are raised and lowered with the metric η^{AB} and η_{AB}.

One also has the following identities

$$[\nabla_A, \nabla_B] = -y_A \nabla_B + y_B \nabla_A, \quad [\nabla^C \nabla_C, y^A] = 2\nabla^A + D y^A, \tag{80}$$
$$[\nabla^C \nabla_C, \nabla^A] = (2-D)\nabla^A + 2y^A \nabla^D \nabla_D$$

where we have set $r^2 = 1$. The generators of the $SO(2, D-1)$ group can be expressed as

$$M^{AB} = y^A \nabla^B - y^B \nabla^A. \tag{81}$$

One can check that the generators (81) can also be represented as

$$M_{AB} = y_A \partial_B - y_B \partial_A, \quad \partial_A = \frac{\partial}{\partial y^A}. \tag{82}$$

To form the $SO(2, D)$ conformal algebra we need extra generators. These generators are

$$M_{(D+1)A} = \partial_A + y_A y^B \partial_B + l y_A \tag{83}$$

Here l is the conformal weight of a field. For the scalar $l = 1$.

One can derive (83) as follows. Obviously (75) is invariant under the $SO(2, D-1)$ rotations. In order to realize the conformal transformations in the ambient space one adds to it one more dimension i.e., considers $D+2$ dimensional space, parameterized by the coordinates z^M, where $M = 0, 1, .., D+1$. These coordinates are subject to the constraint

$$-\left(z^0\right)^2 + \left(z^1\right)^2 + \left(z^2\right)^2 + \cdots + \left(z^{D-1}\right)^2 - \left(z^D\right)^2 + \left(z^{D+1}\right)^2 = z^M z^N g_{MN} = 0 \tag{84}$$

which is invariant under the group of rotations $SO(2, D)$ with the generators

$$M_{MN} = z_M \partial_N - z_N \partial_M. \tag{85}$$

One can solve the constraint (84) by introducing

$$y^A = r\frac{z^A}{z^{D+1}}, \tag{86}$$

satisfying Equation (75).

The generators M_{MN} (85) contain the generators M_{AB} of the AdS_D isometry group $SO(2, D-1)$ and the generators $M_{(D+1),A}$ which extend the latter to the conformal group $SO(2, D)$ by taking the functions on the cone (84) to be homogeneous of degree $-l$

$$z^M \frac{\partial}{\partial z^M} f(z) = -l f(z). \tag{87}$$

In this way, one gets (83).

Then using the explicit realization of the generators (81), (83) as well as the commutation relations (80) between the operators it is straightforward to check invariance of the field Equation (76) under the conformal group $SO(2, D)$.

5.2. $Sp(n)$ Group-Manifold Equations

In the previous subsection we considered in detail a conformal scalar field on AdS_D. As we discussed in Section 3, the hyperspace generalization of AdS spaces are $Sp(n)$ group manifolds. We will now consider an $Sp(n)$ counterpart of the conformal scalar field Equation (74).

Let us start with an $Sp(n)$ analogue of Equation (9). To this end one should replace the flat derivative $\partial_{\alpha\beta}$ with the covariant derivative on $Sp(n)$ group manifold. The covariant derivatives $\nabla_{\alpha\beta}$ satisfy the $Sp(n)$ algebra

$$[\nabla_{\alpha\beta}, \nabla_{\gamma\delta}] = \frac{\xi}{2}\left(C_{\alpha(\gamma}\nabla_{\delta)\beta} + C_{\beta(\gamma}\nabla_{\delta)\alpha}\right). \tag{88}$$

Due to the GL-flatness these covariant derivatives have a simple form

$$\nabla_{\alpha\beta} = G_\alpha^{-1\mu}(X) G_\beta^{-1\nu}(X) \partial_{\mu\nu}, \tag{89}$$

where $G_\alpha^{-1\mu}(X)$ was defined in (52). Further, one should replace the spinor product $\lambda_\alpha\lambda_\beta$ in (8) with an expression which like the covariant derivatives $\nabla_{\alpha\beta}$ also satisfies the $Sp(n)$ algebra. This can be done by introducing new variables

$$\tilde{Y}_\alpha \equiv \lambda_\alpha + \frac{i\xi}{8}\frac{\partial}{\partial\lambda^\alpha} \tag{90}$$

Obviously, the spinorial variables Y_α do not commute among each other

$$[\tilde{Y}_\alpha, \tilde{Y}_\beta] = \frac{i\xi}{4} C_{\alpha\beta}. \tag{91}$$

Using the covariant derivatives $\nabla_{\alpha\beta}$ and the variables Y_α. one can write an $Sp(n)$ analogue of Equation (9) as

$$\left[\nabla_{\alpha\beta} - \frac{i}{2}(\tilde{Y}_\alpha \tilde{Y}_\beta + \tilde{Y}_\beta Y_\alpha)\right] \Phi(X,\lambda) = 0\,. \tag{92}$$

Similarly, one finds an $Sp(n)$ version of Equation (12)

$$\left[\nabla_{\alpha\beta} - \frac{i}{2}(Y_\alpha Y_\beta + Y_\beta Y_\alpha)\right] C(X,\mu) = 0, \quad Y_\alpha \equiv \frac{\xi}{8}\mu_\alpha + i\frac{\partial}{\partial\mu^\alpha}\,. \tag{93}$$

To obtain the equations for component fields one should expand, e.g. the functional $C(X,\mu)$ in power of μ^α

$$C(X,\mu) = \sum_{n=0}^{\infty} C_{\alpha_1\cdots\alpha_n}(X)\,\mu^{\alpha_1}\cdots\mu^{\alpha_n} = B(X) + F_\alpha(X)\mu^\alpha + \cdots\,. \tag{94}$$

Plugging this expansion into (92) one can show that similarly to the case of the flat hyperspace only zeroth and the first components in the expansion in terms of the variables μ^α are independent fields whereas the other fields are expressed in terms of derivatives of the independent ones. The independent hyperfields $B(X)$ and $F_\alpha(X)$ satisfy Equations [10]

$$\begin{aligned} \nabla_{\alpha[\beta}\nabla_{\gamma]\delta}B(X) &= \tfrac{\xi}{16}\left(C_{\alpha[\beta}\nabla_{\gamma]\delta} - C_{\delta[\gamma}\nabla_{\beta]\alpha} + 2C_{\beta\gamma}\nabla_{\alpha\delta}\right)B(X) \\ &+ \tfrac{\xi^2}{64}\left(2C_{\alpha\delta}C_{\beta\gamma} - C_{\alpha[\beta}C_{\gamma]\delta}\right)B(X), \end{aligned} \tag{95}$$

$$\nabla_{\alpha[\beta}F_{\gamma]}(X) = -\frac{\xi}{4}\left(C_{\alpha[\gamma}F_{\beta]}(X) + 2C_{\beta\gamma}F_\alpha(X)\right). \tag{96}$$

The derivation of these equations which are $Sp(n)$ versions of Equations (14) and (15) is straightforward and is given in the Appendix B.

Note that if one introduce the covariant derivatives $D_{\alpha\beta}$ acting on the spinors as follows (see [23] for more details)

$$D_{\alpha\beta}F_\gamma(X) = \nabla_{\alpha\beta}F_\gamma(X) + \frac{\xi}{4}C_{\gamma(\alpha}F_{\beta)}(X) \tag{97}$$

the form of Equations (95) and (96) simplifies to

$$D_{\alpha[\beta}D_{\gamma]\delta}B(X) = \frac{\xi^2}{8^2}\left(2C_{\alpha\delta}C_{\beta\gamma} - C_{\alpha[\beta}C_{\gamma]\delta}\right)B(X), \tag{98}$$

$$D_{\alpha[\beta}F_{\gamma]}(X) = 0\,. \tag{99}$$

We see that Equation (98) reminds that of the AdS scalar field (74), especially when we contract its indices.

5.2.1. Connection between the Fields in Flat Hyperspaces and $Sp(n)$ Group Manifolds

One can check [23] using the equations

$$\partial_{\mu\nu}G^{-1\alpha\beta}(X) = \frac{\xi}{8}(\delta^\alpha_\mu\delta^\beta_\nu + \delta^\beta_\mu\delta^\alpha_\nu)\,, \tag{100}$$

and

$$\partial_{\mu\nu}(\det G(X))^k = \frac{\xi k}{8}(\det G(X))^k(G_{\mu\nu}(X) + G_{\nu\mu}(X))\,, \tag{101}$$

that the fields $B(X)$ and $F_\alpha(X)$ satisfying Equations (95) and (96) are related to the fields $b(X)$ and $f_\mu(X)$ satisfying the flat hyperspace Equations (14) and (15) as follows

$$B(X) = (\det G(X))^{-\frac{1}{2}}\,b(X)\,, \tag{102}$$

$$F_\alpha(X) = (\det G(X))^{-\frac{1}{2}}\, G_\alpha^{-1\mu}(X) f_\mu(X). \tag{103}$$

These relations are similar to the relations between the conformally invariant scalar and spinor equations in the conventional flat and AdS spaces and reduce to them in the case of $n = 2, D = 3$.

5.2.2. Plane Wave Solutions

Equations (92) and (93) can be solved to obtain "plane-wave" solutions. Let us consider the case of the $Sp(4)$ group manifold. One can check that Equations (92) and (93) have the following solutions

$$\Phi(X,\lambda) = \int d^4\mu \sqrt{\det G^{-1}(X)}\, e^{iX^{\alpha\beta}(\lambda_\alpha+\frac{\xi}{8}\mu_\alpha)(\lambda_\beta+\frac{\xi}{8}\mu_\beta)+i\lambda_\alpha\mu^\alpha}\, \varphi(\mu)\,, \tag{104}$$

$$C(X,\mu) = \int d^4\lambda \sqrt{\det G^{-1}(X)}\, e^{iX^{\alpha\beta}(\lambda_\alpha+\frac{\xi}{8}\mu_\alpha)(\lambda_\beta+\frac{\xi}{8}\mu_\beta)-i\lambda_\alpha\mu^\alpha}\, \varphi(\lambda)\,. \tag{105}$$

These solutions describe plane-wave-like fields in the GL–flat parameterization of the metric [10]. They can be compared with the plane-wave solutions for the higher-spin curvatures on AdS_4 given in [8,88]. The latter can be found by solving the AdS_4 deformation of the field Equations (33)

$$\begin{aligned} &D_{M\dot{M}} C_{A_1,\ldots,A_{n+2s},\dot{A},\ldots,\dot{A}_n}(x) = \\ &e^{A\dot{A}}_{M\dot{M}} C_{A_1,\ldots,A_{n+2s},A,\dot{A},\ldots,\dot{A}_n\dot{A}}(x) - n(n+2s) e_{M\dot{M},\{A\dot{A}} C_{A_2,\ldots,A_{n+2s},\dot{A},\ldots,\dot{A}_n}(x) \end{aligned} \tag{106}$$

where $D_{M\dot{M}}$ is a covariant derivative on AdS_4 and $e^{A\dot{A}}_{M\dot{M}}$ are the corresponding vierbeins in the Weyl spinor representation. The physical higher-spin curvatures satisfy the equations

$$e^{M\dot{M}}_{A\dot{A}} D_{M\dot{M}} C^{A_1,\ldots,A_{2s}}(x) = 0 \tag{107}$$

whereas the auxiliary fields are expressed via derivatives of the physical fields with the help of Equation (106). Choosing the AdS_4 metric in the conformally flat form

$$e^{A\dot{A}}_{M\dot{M}} = e^{\frac{\rho(x)}{2}} \delta^A_M \delta^{\dot{A}}_{\dot{M}}, \quad \rho(x) = \ln \frac{4}{\left(1-(\frac{x}{r})^2\right)^2} \tag{108}$$

one can find the plane wave solutions of Equation (107)

$$C_{A_1,\ldots,A_{2s}}(x) = \frac{\partial}{\partial\mu^{A_1}} \cdots \frac{\partial}{\partial\mu^{A_{2s}}} C(x,\mu,\overline{\mu})|_{\mu=\overline{\mu}=0} \tag{109}$$

with

$$\begin{aligned} &C(x,\mu,\overline{\mu}) = \textstyle\int d^2\lambda d^2\overline{\lambda} \Phi(\lambda,\overline{\lambda}) \cdot \\ &\exp\left(i(\mu_A \overline{\mu}_{\dot{A}} + \lambda_A \overline{\lambda}_{\dot{A}}) x^{A\dot{A}} - \frac{\rho(x)}{2} + \left(1-(\frac{x}{r})^2\right)^{\frac{1}{2}} (\mu^A \lambda_A + \overline{\mu}^{\dot{A}} \overline{\lambda}_{\dot{A}}) \right). \end{aligned} \tag{110}$$

Comparing (110) with (105), one can see that the latter is a direct generalization of the AdS_4 plane-wave solution to the case of the $Sp(4)$ group manifold.

As a simplest example of this construction let us note that the conformal scalar on AdS_4 discussed in Section 5.1 admits a plane-wave solution [8] of the form

$$\phi(x) = \int d^2\lambda d^2\overline{\lambda} e^{\,ix^{A\dot{A}}\lambda_A\overline{\lambda}_{\dot{A}} - \frac{1}{2}\rho(x)} \phi_0(\lambda,\overline{\lambda}) \tag{111}$$

which can be checked substituting the expression (111) into the field Equation (74).

5.3. $Sp(2n)$ Transformations of the Fields

Using the relation between the fields of weight $\Delta = \frac{1}{2}$ on flat hyperspace and on $Sp(n)$ group manifold (102) we have the following relation between the $Sp(2n)$ transformations of the weight-$\frac{1}{2}$ fields on $Sp(n)$ and in flat hyperspace

$$\delta B(X) = (\det G(X))^{-\frac{1}{2}} \delta b(X)\,, \tag{112}$$

$$\delta F_\alpha(X) = (\det G(X))^{-\frac{1}{2}}\, G_\alpha^{-1\mu}(X)\, \delta f_\mu(X)\,. \tag{113}$$

Note that in the above expressions the matrix $G_\alpha{}^\mu(X)$ is not varied since it is form-invariant, i.e., $G(X')$ has the same form as $G(X)$.

Then, the $Sp(n)$-variations of $B(X)$ and $F_\alpha(X)$ have the following form [23]

$$\begin{aligned}\delta B(X) = &-(a^{\alpha\beta}\mathcal{D}_{\alpha\beta} + \tfrac{1}{2}(g_\alpha{}^\alpha - k_{\alpha\beta}X^{\alpha\beta}) + 2g_\beta{}^\alpha X^{\beta\gamma}\mathcal{D}_{\alpha\gamma} \\ &-k_{\alpha\beta}X^{\alpha\gamma}X^{\beta\delta}\mathcal{D}_{\gamma\delta})B(X)\,,\end{aligned} \tag{114}$$

$$\begin{aligned}\delta F_\sigma(X) = &-(a^{\alpha\beta}\mathcal{D}_{\alpha\beta} + \frac{1}{2}(g_\alpha{}^\alpha - k_{\alpha\beta}X^{\alpha\beta}) + 2g_\beta{}^\alpha X^{\beta\gamma}\mathcal{D}_{\alpha\gamma} \\ &-k_{\alpha\beta}X^{\alpha\gamma}X^{\beta\delta}\mathcal{D}_{\gamma\delta})F_\sigma(X) - (g_\sigma{}^\beta - k_{\sigma\alpha}X^{\alpha\beta})F_\beta(X),\end{aligned}$$

where the derivative $\mathcal{D}_{\alpha\beta}$ is defined as

$$\mathcal{D}_{\alpha\beta} = \partial_{\alpha\beta} + \frac{\xi}{16}(G_{\alpha\beta}(X) + G_{\beta\alpha}(X))\,. \tag{115}$$

Using

$$\partial_{\mu\nu}G_\rho{}^\sigma(X) = \frac{\xi}{8}(G_{\rho\mu}(X)G_\nu{}^\sigma(X) + G_{\rho\nu}(X)G_\mu{}^\sigma(X))\,, \tag{116}$$

one can check that these derivatives commute with each other $[\mathcal{D}_{\alpha\beta}, \mathcal{D}_{\gamma\delta}] = 0$ just as in the flat case.

Let us note that the relation between the flat and $Sp(n)$ hyperfields of an arbitrary weight Δ and the form of the corresponding $Sp(2n)$ transformations require additional study since for this one should know the form of $Sp(2n)$-invariant equations satisfied by these fields, which is still an open problem.

6. Supersymmetry

In this Section, we present a supersymmetric generalization of the $Sp(2n)$ invariant systems. We will mainly follow [24].

6.1. Flat Hyper-Superspace and Its Symmetries

The concept of hyperspaces, hyperfields and of the corresponding field equations can be generalized to construct supersymmetric $OSp(1|2n)$ invariant systems and the corresponding infinite-dimensional higher-spin supermultiplets. In this section we shall describe this generalization in detail.

The flat hyper–superspace (see e.g., [3,4,12]) is parameterized by $\frac{n(n+1)}{2}$ bosonic matrix coordinates $X^{\mu\nu} = X^{\nu\mu}$ and n real Grassmann–odd "spinor" coordinates θ^μ ($\mu = 1, \cdots, n$). The supersymmetry variation

$$\delta\theta^\mu = \epsilon^\mu, \qquad \delta X^{\mu\nu} = -i\epsilon^{(\mu}\theta^{\nu)}\,, \tag{117}$$

leaves invariant the Volkov-Akulov-type one-form

$$\Pi^{\mu\nu} = dX^{\mu\nu} + i\theta^{(\mu}d\theta^{\nu)}\,. \tag{118}$$

The supersymmetry transformations form a generalized super–translation algebra

$$\{Q_\mu, Q_\nu\} = 2P_{\mu\nu}, \qquad [Q_\mu, P_{\nu\rho}] = 0\,, \qquad [P_{\mu\nu}, P_{\rho\lambda}] = 0\,, \tag{119}$$

with $P_{\mu\nu}$ generating translations along $X^{\mu\nu}$.

The realization of $P_{\mu\nu}$ and Q_μ as differential operators is given by

$$P_{\mu\nu} = -i\frac{\partial}{\partial X^{\mu\nu}} \equiv -i\partial_{\mu\nu}\,, \qquad Q_\mu = \partial_\mu - i\theta^\nu \partial_{\nu\mu}\,, \qquad \partial_\mu \equiv \frac{\partial}{\partial \theta^\mu}\,, \tag{120}$$

The algebra (119) is invariant under rigid $GL(n)$ transformations

$$Q'_\mu = g_\mu{}^\nu\, Q_\nu\,, \qquad P'_{\mu\nu} = g_\mu{}^\rho\, g_\nu{}^\lambda\, P_{\rho\lambda}, \tag{121}$$

generated by

$$G_\mu{}^\nu = -2i(X^{\nu\rho} + \frac{i}{2}\theta^\nu\theta^\rho)\partial_{\rho\mu} - i\theta^\nu\, Q_\mu\,, \tag{122}$$

which act on $P_{\mu\nu}$ and Q_μ as follows

$$[P_{\mu\nu}, G_\lambda{}^\rho] = -i(\delta^\rho_\mu P_{\nu\lambda} + \delta^\rho_\nu P_{\mu\lambda})\,, \qquad [Q_\mu, G_\nu{}^\rho] = -i\delta^\rho_\mu\, Q_\nu\,, \tag{123}$$

and close into the $gl(n)$ algebra as in (39)

$$[G_\nu{}^\mu, G_\lambda{}^\rho] = i(\delta^\mu_\lambda\, G_\nu{}^\rho - \delta^\rho_\nu\, G_\lambda{}^\mu)\,. \tag{124}$$

The algebra (119), (123) and (124) is the hyperspace counterpart of the conventional super–Poincaré algebra enlarged by dilatations. That this is so can be most easily seen by taking $n = 2$ (i.e., $\mu = 1,2$), in which case this algebra is recognized as the $D = 3$ super–Poincaré algebra with $G_\mu{}^\nu - \frac{1}{2}\delta^\nu_\mu\, G_\rho{}^\rho = M_m(\gamma^m)_\mu{}^\nu$ ($m = 0,1,2$) generating the $SL(2,R) \sim SO(1,2)$ Lorentz rotations and $\mathbb{D} = \frac{1}{2}G_\rho{}^\rho$ being the dilatation generator. Note that the factor $\frac{1}{2}$ in the definition of the dilatation generator is required in order to have the canonical scaling of the momentum generator $P_{\mu\nu}$ with weight 1 and the supercharge Q_μ with weight $\frac{1}{2}$, as follows from Equation (123).

This algebra may be further extended to the $OSp(1|2n)$ algebra, generating generalized superconformal transformations of the flat hyper–superspace, by adding the additional set of supersymmetry generators

$$S^\mu = -(X^{\mu\nu} + \frac{i}{2}\theta^\mu\theta^\nu)Q_\nu\,, \tag{125}$$

and the generalized conformal boosts

$$K^{\mu\nu} = i(X^{\mu\rho} + \frac{i}{2}\theta^\mu\theta^\rho)(X^{\nu\lambda} + \frac{i}{2}\theta^\nu\theta^\lambda)\partial_{\rho\lambda} - i\theta^{(\mu}S^{\nu)}\,. \tag{126}$$

The generators S^μ and $K^{\mu\nu}$ form a superalgebra similar to (119)

$$\{S^\mu, S^\nu\} = -2K^{\mu\nu}, \qquad [S^\mu, K^{\nu\rho}] = 0\,, \qquad [K^{\mu\nu}, K^{\rho\lambda}] = 0\,, \tag{127}$$

while the non-zero (anti)commutators of S^μ and $K^{\mu\nu}$ with Q_μ, $P_{\mu\nu}$ and $G_\mu{}^\nu$ read

$$\begin{aligned} &\{Q_\mu, S^\nu\} = -G_\mu{}^\nu\,, \quad [S^\mu, P_{\nu\rho}] = i\delta^\mu_{(\nu}\, Q_{\rho)}, \\ &[Q_\mu, K^{\nu\rho}] = -i\delta^{(\nu}_\mu\, S^{\rho)}\,, \quad [S^\mu, G_\nu{}^\rho] = i\delta^\mu_\nu\, S^\rho\,. \end{aligned} \tag{128}$$

Let us note that in the case $n = 4$, in which the physical space–time is four-dimensional the generalized superconformal group $OSp(1|8)$ contains the $D = 4$ conformal symmetry group $SO(2,4) \sim SU(2,2)$ as a subgroup, but not the superconformal group $SU(2,2|1)$. The reason being that, although $OSp(1|8)$ and $SU(2,2|1)$ contain the same number of (eight) generators, the anticommutators of the former close on the generators of the whole $Sp(8)$, while those of the latter only close on an $U(2,2)$ subgroup of $Sp(8)$, and the same supersymmetry generators cannot satisfy the different anti-commutation relations simultaneously. In fact, the minimal OSp–supergroup containing $SU(2,2|1)$ as a subgroup is $OSp(2|8)$.

6.2. Scalar Superfields and Their $OSp(1|2n)$-Invariant Equations of Motion

Let us now consider a superfield $\Phi(X,\theta)$ transforming as a scalar under the super–translations (120)

$$\delta\Phi(X,\theta) = -(\epsilon^{\alpha} Q_{\alpha} + i a^{\mu\nu} P_{\mu\nu})\, \Phi(X,\theta)\,. \tag{129}$$

To construct equations of motion for $\Phi(X,\theta)$ which are invariant under (129) and comprise the equations of motion of an infinite tower of integer and half-integer higher-spin fields with respect to conventional space–time, we introduce the spinorial covariant derivatives

$$D_{\mu} = \partial_{\mu} + i\theta^{\nu}\partial_{\nu\mu}\,, \qquad \{D_{\mu}, D_{\nu}\} = 2i\partial_{\mu\nu}\,, \tag{130}$$

which (anti)commute with Q_{μ} and $P_{\mu\nu}$.

The Φ–superfield equations then take the form [12]

$$D_{[\mu}D_{\nu]}\Phi(X,\theta) = 0\,, \tag{131}$$

As was shown in [12], these superfield equations imply that all the components of $\Phi(X,\theta)$ except for the first and the second one in the θ^{μ}-expansion of $\Phi(X,\theta)$ should vanish

$$\Phi(X,\theta) = b(X) + i\theta^{\mu} f_{\mu}(X) + i\theta^{\mu}\theta^{\nu} A_{\mu\nu}(X) + \cdots\,, \tag{132}$$

(i.e., $A_{\mu_1\ldots\nu_k} = 0$ for $k > 1$) while the scalar and spinor fields $b(X)$ and $f_{\mu}(X)$ satisfy Equations (14) and (15).

The superfield Equations (131) are invariant under the generalized superconformal $OSp(1|2n)$ symmetry, provided that $\Phi(X,\theta)$ transforms as a scalar superfield with the "canonical" generalized scaling weight $\frac{1}{2}$, i.e.,

$$\begin{aligned} \delta\Phi(X,\theta) \; &= \; -(\epsilon^{\mu} Q_{\mu} + \xi_{\mu} S^{\mu} + i a^{\mu\nu} P_{\mu\nu} + i k_{\mu\nu} K^{\mu\nu} + i g_{\mu}{}^{\nu} G_{\nu}{}^{\mu})\, \Phi(X,\theta) \\ &\quad -\tfrac{1}{2}\left(g_{\mu}{}^{\mu} - k_{\mu\nu}(X^{\mu\nu} + \tfrac{i}{2}\theta^{\mu}\theta^{\nu}) + \xi_{\mu}\theta^{\mu}\right) \Phi(X,\theta)\,, \end{aligned} \tag{133}$$

where the factor $\frac{1}{2}$ in the second line is the generalized conformal weight and ϵ^{μ}, ξ_{μ}, $a^{\mu\nu}$, $k_{\mu\nu}$ and $g_{\mu}{}^{\nu}$ are the rigid parameters of the $OSp(1|2n)$ transformations.

Scalar superfields with anomalous generalized conformal dimension Δ transform under $OSp(1|2n)$ as

$$\begin{aligned} \delta\Phi(X,\theta) \; &= \; -(\epsilon^{\mu} Q_{\mu} + \xi_{\mu} S^{\mu} + i a^{\mu\nu} P_{\mu\nu} + i k_{\mu\nu} K^{\mu\nu} + i g_{\mu}{}^{\nu} G_{\nu}{}^{\mu})\, \Phi(X,\theta) \\ &\quad -\Delta\left(g_{\mu}{}^{\mu} - k_{\mu\nu}(X^{\mu\nu} + \tfrac{i}{2}\theta^{\mu}\theta^{\nu}) + \xi_{\mu}\theta^{\mu}\right) \Phi(X,\theta)\,. \end{aligned} \tag{134}$$

It is instructive to demonstrate how the generalized conformal dimension Δ, which is defined to be the same for all values of n in $OSp(1|2n)$, is related to the conventional conformal weight of scalar superfields in various space–time dimensions. As we have already mentioned in Section 6.1, the dilatation operator should be identified with $\mathbb{D} = \frac{1}{2}G_{\mu}{}^{\mu}$. Therefore, considering a $GL(n)$ transformation (134) with the parameter $g_{\mu}{}^{\nu}$

$$\delta\Phi(X,\theta) = -i g_{\mu}{}^{\nu} G_{\nu}{}^{\mu}\Phi(X,\theta),$$

the part of the transformation corresponding to the dilatation reads

$$\delta_{\mathbb{D}}\Phi(X,\theta) = -\frac{i}{n}g_\mu{}^\mu\, G_\nu{}^\nu\Phi(X,\theta) = -\frac{2i}{n}g_\mu{}^\mu\mathbb{D}\Phi(X,\theta) = -i\tilde{g}\mathbb{D}\Phi(X,\theta)\,, \tag{135}$$

where $\tilde{g} = \frac{2}{n}g_\mu{}^\mu$ is the genuine dilatation parameter. From (134), it then follows that the conventional conformal weight Δ_D of the scalar superfield is related to the generalized one Δ via

$$\Delta_{\mathbb{D}} = \frac{n}{2}\Delta, \quad D = \frac{n}{2}+2. \tag{136}$$

In the $n = 2$ case corresponding to the $\mathcal{N} = 1$, $D = 3$ scalar superfield theory the two definitions of the conformal dimension coincide, whereas in the case $n = 4$ describing conformal higher-spin fields in $D = 4$ one finds $\Delta_4 = 2\Delta$. Relation (136) indeed provides the correct conformal dimensions of scalar superfields (and consequently of their components) in the corresponding space–time dimensions. For instance, when $\Delta = \frac{1}{2}$, in $D = 3$ one finds $\frac{1}{2}$ as the canonical conformal dimension of the scalar superfield, while in the cases $D = 4$ and $D = 6$ ($n = 8$) it is found to be equal to one and two, respectively. For convenience, we shall henceforth associate the scaling properties of the fields to the universal D– and n–independent generalized conformal weight Δ.

6.3. Infinite-Dimensional Higher-Spin Representation of $\mathcal{N} = 1$, $D = 4$ Supersymmetry

Using the example of $n = 4$ ($D = 4$) we will now show that in four space–time dimensions, the fields of integer and half-integer spin $s = 0, \frac{1}{2}, 1, \cdots, \infty$ encoded in $b(X)$ and $f_\mu(X)$ (see Section 2.1) form an irreducible infinite-dimensional supermultiplet with respect to the supersymmetry transformations generated by the *generalized* super–Poincaré algebra (119). The hyperfields $b(X)$ and $f_\mu(X)$, satisfying (14) and (15), transform under the supertranslations (129) as follows

$$\delta b(X) = -i\epsilon^\mu f_\mu(X)\,, \qquad \delta f_\mu(X) = -\epsilon^\nu\,\partial_{\nu\mu}\,b(X)\,. \tag{137}$$

and their expansion in terms of the y^{mn} coordinates is given in (16) and (17).

The fact that the higher– spin fields should form an infinite-dimensional representation of the generalized $\mathcal{N} = 1$, $D = 4$ supersymmetry (119) is prompted by the observation that the spectrum of bosonic fields contains a single real scalar field $\phi(x)$, which alone cannot have a fermionic superpartner, while each field with $s > 0$ has two helicities $\pm s$. Indeed, from (137), we obtain an infinite entangled chain of supersymmetry transformations for the $D = 4$ fields

$$\begin{aligned}
&\delta\phi(x) = -i\epsilon^\mu\,\psi_\mu(x),\\
&\delta\psi_\mu(x) = \epsilon^\nu(\gamma^m_{\nu\mu}\,\partial_m\phi(x) + \gamma^{mn}_{\nu\mu}\,F_{mn}(x)),\\
&\delta F_{mn}(x) = -i\epsilon^\mu\left(R_{\mu\,mn}(x) - \frac{1}{2}\partial_{[m}(\gamma_{n]}\psi)_\mu(x)\right),\\
&\delta R_{\mu\,mn}(x) = \frac{1}{2}\partial_{[m}(\gamma_{n]}\delta\psi(x))_\mu - \frac{1}{2}\epsilon^\nu\,\gamma^p_{\nu\mu}\,\partial_p F_{mn}(x)\\
&-\epsilon^\nu\,\gamma^{pq}_{\nu\mu}\left(R_{pq,mn}(x) - \frac{1}{2}\partial_q\eta_{p[m}\partial_{n]}\phi(x)\right),
\end{aligned} \tag{138}$$

and so on.

The algebraic reason behind the appearance of the infinite-dimensional supermultiplet of the $D = 4$ higher–spin fields is related to the following fact. In the $n = 4$, $D = 4$ case the superalgebra (119) takes the following form

$$\{Q_\mu, Q_\nu\} = (\gamma^m)_{\mu\nu}P_m + (\gamma^{mn})_{\mu\nu}Z_{mn}\,, \tag{139}$$

where P_m is the momentum along the four-dimensional space–time and $Z_{mn} = -Z_{nm}$ are the tensorial charges associated with the momenta along the extra coordinates y^{mn}.

On the other hand, the conventional $N = 1$, $D = 4$ super–Poincaré algebra is

$$\{Q_\mu, Q_\nu\} = (\gamma^m)_{\mu\nu} P_m\,. \tag{140}$$

Though both algebras have the same number of the supercharges Q_μ, their anti-commutator closes on different sets of bosonic generators. Thus, the super–Poincaré algebra (140) is not a subalgebra of (139). Hence the representations of (139) do not split into (finite-dimensional) representations of the standard super–Poincaré algebra. In this sense the supersymmetric higher–spin systems under consideration differ from most of supersymmetric models of finite-dimensional super–Poincaré or AdS higher–spin supermultiplets considered in the literature (see e.g., [40,46,89–112]).

7. Hyperspace Extension of Supersymmetric AdS Spaces

In Section 3 we have seen that the hyperspace extension of AdS spaces are $Sp(n)$ group manifolds. In this section we consider their minimal supersymmetric extension, namely $OSp(1|n)$ supergroup manifolds.

The $OSp(1|n)$ superalgebra is formed by n anti commuting supercharges $\mathcal{Q}_\alpha$ and $\frac{n(n+1)}{2}$ generators $M_{\alpha\beta} = M_{\beta\alpha}$ of $Sp(n)$

$$\begin{aligned} \{\mathcal{Q}_\alpha, \mathcal{Q}_\beta\} &= 2M_{\alpha\beta}\,, \qquad [\mathcal{Q}_\alpha, M_{\beta\gamma}] = \tfrac{i\xi}{2} C_{\alpha(\beta}\,\mathcal{Q}_{\gamma)}\,, \\ [M_{\alpha\beta}, M_{\gamma\delta}] &= -\tfrac{i\xi}{2}(C_{\gamma(\alpha}M_{\beta)\delta} + C_{\delta(\alpha}M_{\beta)\gamma})\,, \end{aligned} \tag{141}$$

The $OSp(1|n)$ algebra (141) is recognized as a subalgebra of $OSp(1|2n)$ (see the Section 6.1) with the identifications

$$\mathcal{Q}_\alpha = (Q_\alpha + \frac{\xi}{4}S_\alpha), \qquad M_{\alpha\beta} = P_{\alpha\beta} - \frac{\xi^2}{16}K_{\alpha\beta} - \frac{\xi}{4}G_{(\alpha\beta)}\,. \tag{142}$$

The $OSp(1|n)$ manifold is parameterized by the coordinates $(X^{\mu\nu}, \theta^\mu)$ and its geometry is described by the Cartan forms

$$\Omega = \mathcal{O}^{-1}d\mathcal{O}(X,\theta) = -i\Omega^{\alpha\beta}M_{\alpha\beta} + iE^\alpha \mathcal{Q}_\alpha\,, \tag{143}$$

where $\mathcal{O}(X,\theta)$ is an $OSp(1|n)$ supergroup element. The Cartan forms satisfy the Maurer–Cartan equations associated with the $OSp(1|n)$ superalgebra (141)

$$d\Omega^{\alpha\beta} + \frac{\xi}{2}\Omega^{\alpha\gamma}\wedge\Omega_\gamma{}^\beta = -iE^\alpha\wedge E^\beta, \qquad dE^\alpha + \frac{\xi}{2}E^\gamma\wedge\Omega_\gamma{}^\alpha = 0\,, \tag{144}$$

with the external differential acting from the right.

7.1. GL Flatness of $OSp(1|n)$ Group Manifolds

There is a supersymmetric generalization of the $GL(n)$ flatness property of $Sp(n)$ group manifolds to the case of $OSp(1|n)$ supergroup manifolds [8]. In particular, the Maurer–Cartan Equations (144) are solved by the following forms

$$\Omega^{\alpha\beta} = dX^{\mu\nu}G_\mu{}^\alpha G_\nu{}^\beta(X) + \frac{i}{2}(\Theta^\alpha\mathcal{D}\Theta^\beta + \Theta^\beta\mathcal{D}\Theta^\alpha) = \Pi^{\mu\nu}\,\mathcal{G}_\mu{}^\alpha\,\mathcal{G}_\nu{}^\beta(X,\Theta), \tag{145}$$

$$E^\alpha = P(\Theta^2)\mathcal{D}\Theta^\alpha - \Theta^\alpha\mathcal{D}P(\Theta^2) \tag{146}$$

where Θ is related to θ as follows

$$\theta^\alpha = \Theta^\beta G_\beta^{-1\alpha}P^{-1}(\Theta^2), \qquad \Theta^2 = \Theta^\alpha\Theta_\alpha, \qquad P^2(\Theta^2) = 1 + \frac{i\xi}{8}\Theta^2, \tag{147}$$

while the covariant derivative

$$\mathcal{D}\Theta^{\alpha} = d\Theta^{\alpha} + \frac{\xi}{4}\Theta^{\beta}\,\omega_{\beta}{}^{\alpha}(X)\,, \tag{148}$$

contains the Cartan form of the $Sp(n)$ group manifold

$$\omega^{\alpha\beta}(X) = dX^{\mu\nu}G_{\mu}{}^{\alpha}(X)G_{\nu}{}^{\beta}(X), \tag{149}$$

and

$$\mathcal{G}_{\alpha}{}^{\beta}(X,\Theta) = G_{\alpha}{}^{\beta}(X) - \frac{i\xi}{8}(\Theta_{\alpha} - 2G_{\alpha}{}^{\gamma}(X)\Theta_{\gamma})\Theta^{\beta}, \tag{150}$$

where $G_{\alpha}{}^{\beta}(X)$ is given in (50). The inverse matrix of (150) is

$$\begin{array}{rcl}\mathcal{G}_{\alpha}^{-1\beta}(X,\Theta) & = & G_{\alpha}^{-1\beta}(X) - \frac{i\xi}{8}(\Theta^{\delta}G^{-1}_{\delta\alpha}(X))\,(\Theta^{\delta}\,G_{\delta}^{-1\beta}(X))P^{-2}(\Theta^{2}) \\ & = & G_{\alpha}^{-1\beta}(X) - \frac{i\xi}{8}\theta_{\alpha}\,\theta^{\beta} = \delta_{\alpha}^{\beta} + \frac{\xi}{4}(X_{\alpha}{}^{\beta} - \frac{i}{2}\theta_{\alpha}\,\theta^{\beta})\end{array} \tag{151}$$

with $G_{\alpha}^{-1\beta}(X)$ given in (52).

7.2. Field Equations on $OSp(1|n)$ Supergroup Manifold

The scalar superfield equation on $OSp(1|n)$ has the form [12]

$$\left(\nabla_{[\alpha}\nabla_{\beta]} - \frac{i\xi}{8}C_{\alpha\beta}\right)\Phi_{OSp}(X,\theta) = 0\,, \tag{152}$$

where the Grassmann–odd covariant derivatives ∇_{α} and their bosonic counterparts $\nabla_{\alpha\beta}$ satisfy the $OSp(1|n)$ superalgebra similar to (141), namely

$$\{\nabla_{\alpha},\nabla_{\beta}\} = 2i\nabla_{\alpha\beta} \tag{153}$$

$$[\nabla_{\gamma},\nabla_{\alpha\beta}] = \frac{\xi}{2}C_{\gamma(\alpha}\nabla_{\beta)}, \tag{154}$$

$$[\nabla_{\alpha\beta},\nabla_{\gamma\delta}] = \frac{\xi}{2}(C_{\alpha(\gamma}\nabla_{\delta)\beta} + C_{\beta(\gamma}\nabla_{\delta)\alpha})\,. \tag{155}$$

while the $OSp(1|n)$ covariant derivatives are obtained from the flat superspace ones by the following GL transformations

$$\begin{array}{rcl}\nabla_{\alpha} & = & \mathcal{G}_{\alpha}^{-1\mu}(X,\Theta)\,D_{\mu}\,, \\ \nabla_{\alpha\beta} & = & \mathcal{G}_{\alpha}^{-1\mu}(X,\Theta)\,\mathcal{G}_{\beta}^{-1\nu}(X,\Theta)\left(\partial_{\mu\nu} + 2iD_{(\mu}\ln\left((\det G(X))^{\frac{1}{2}}P^{-1}(\Theta^{2})\right)\,D_{\nu)}\right).\end{array} \tag{156}$$

Connection between Superfields on Flat Hyper-Superspace and on $OSp(1|n)$ Supergroup Manifolds

Using the relations given in Appendix C one can show that the superfield $\Phi_{OSp}(X,\theta)$ satisfying (152) is related to the superfield $\Phi(X,\theta)$ satisfying the flat superspace Equation (131) by the super–Weyl transformation

$$\begin{array}{l}\Phi_{OSp(1|n)}(X,\theta) = (\det\mathcal{G}(X,\Theta))^{-\frac{1}{2}}\,\Phi_{flat}(X,\theta) \\ = (\det G(X))^{-\frac{1}{2}}P(\Theta^{2})\,\Phi_{flat}(X,\theta),\end{array} \tag{157}$$

Substituting (132) into (157) and using the definition (147), together with the fact that on the mass shell all higher components in (132) vanish, we find

$$\begin{array}{l}\Phi_{OSp(n)}(X,\theta) = (\det G(X))^{-\frac{1}{2}}\,b(X) \\ +\Theta^{\alpha}(\det G(X))^{-\frac{1}{2}}\,G_{\alpha}^{-1\mu}(X)\,f_{\mu}(X) + O(\Theta^{2},b(X)),\end{array} \tag{158}$$

where the first two terms are the fields

$$B(X) = (\det G(X))^{-\frac{1}{2}}\, b(X), \qquad F_\alpha(X) = (\det G(X))^{-\frac{1}{2}}\, G_\alpha^{-1\mu}(X)\, f_\mu(X) \tag{159}$$

propagating on the $Sp(n)$ group manifold, and $O(\Theta^2, b(X))$ stands for higher order terms in Θ^2 which only depend on $b(X)$. The fields (159) satisfy the equations of motion on $Sp(n)$ group manifolds (95)–(96). Note that in these equations the covariant derivatives are restricted to the bosonic group manifold $Sp(n)$, i.e., $\nabla_{\alpha\beta} = G_\alpha^{-1\mu}(X)\, G_\beta^{-1\nu}(X)\, \partial_{\mu\nu}$.

7.3. $OSp(1|2n)$ Transformations of Superfields

Since the flat superspace field equation is invariant under the generalized superconformal $OSp(1|2n)$ transformations (133), the above relation leads us to conclude that also the $OSp(1|n)$ superspace Equations (152) are invariant under the $OSp(1|2n)$ transformations, under which the superfield $\Phi_{OSp}(X,\theta)$ varies as follows

$$\begin{aligned}\delta\Phi_{OSp}(X,\theta) &= -(\epsilon^\mu\, \mathbb{Q}_\mu + \xi_\mu\, \mathcal{S}^\mu + ia^{\mu\nu}\, \mathcal{P}_{\mu\nu} + ik_{\mu\nu}\, \mathcal{K}^{\mu\nu} + ig_\mu{}^\nu\, \mathcal{G}_\nu{}^\mu)\, \Phi_{OSp}(X,\theta) \\ &\quad -\tfrac{1}{2}\left(g_\mu{}^\mu - k_{\mu\nu}(X^{\mu\nu} + \tfrac{i}{2}\theta^\mu\theta^\nu) + \xi_\mu\, \theta^\mu\right)\, \Phi_{OSp}(X,\theta)\,.\end{aligned} \tag{160}$$

Here,

$$\mathcal{P}_{\mu\nu} = -i\mathcal{D}_{\mu\nu} = -i(\partial_{\mu\nu} + \frac{\xi}{8}\mathcal{G}_{(\mu\nu)}(X,\Theta))\,, \tag{161}$$

and

$$\mathbb{Q}_\mu = Q_\mu - \frac{i\xi}{8}\Theta_\mu P(\Theta)\,. \tag{162}$$

Using the relations given in the Appendix C one may check that the operators (161) and (162) obey the flat hyperspace supersymmetry algebra

$$[\mathcal{P}_{\mu\nu}, \mathcal{P}_{\rho\sigma}] = 0, \qquad \{\mathbb{Q}_\mu, \mathbb{Q}_\nu\} = -2\mathcal{P}_{\mu\nu}, \qquad [\mathcal{P}_{\mu\nu}, \mathbb{Q}_\rho] = 0\,. \tag{163}$$

The other generators of the $OSp(1|2n)$ are

$$\mathcal{S}^\mu = -(X^{\mu\nu} + \frac{i}{2}\theta^\mu\theta^\nu)\mathbb{Q}_\nu\,, \quad \mathcal{G}_\mu{}^\nu = -2i(X^{\nu\rho} + \frac{i}{2}\theta^\nu\theta^\rho)\mathcal{D}_{\rho\mu} - i\theta^\nu\, \mathbb{Q}_\mu\,, \tag{164}$$

and

$$\mathcal{K}^{\mu\nu} = i(X^{\mu\rho} + \frac{i}{2}\theta^\mu\theta^\rho)(X^{\nu\lambda} + \frac{i}{2}\theta^\nu\theta^\lambda)\mathcal{D}_{\rho\lambda} - i\theta^{(\mu}\mathcal{S}^{\nu)}\,. \tag{165}$$

Taking into account the commutation relations (163) we see that the operators $\mathbb{Q}_\mu, \mathcal{S}^\mu, \mathcal{P}_{\mu\nu}, \mathcal{G}_\mu{}^\nu$ and $\mathcal{K}^{\mu\nu}$ obey the same $OSp(1|2n)$ algebra as the operators Q_μ, S^μ, $P_{\mu\nu}$, $G_\mu{}^\nu$ and $K^{\mu\nu}$ considered in the Section 6.1.

8. Generalized CFT. Part I. Correlation Functions in $OSp(1|2n)$-Invariant Models

In the previous sections, we have described the generalized conformal group $Sp(2n)$ and generalized conformal supergroup $OSp(1|2n)$. We introduced the fundamental fields and superfields and showed how they transform under generalized conformal transformations.

In this Section we shall construct two-, three- and four-point correlation functions of these fields, by requiring the $Sp(2n)$ symmetry of the correlators, i.e., by solving the corresponding Ward identities. In other words we will follow the conventional approach adopted in multidimensional CFTs (see e.g., [113]). In particular, we will consider $OSp(1|2n)$ invariant correlation functions from which the $Sp(2n)$ invariant correlation functions can be recovered as components of the expansions of the former in series of the Grassman coordinates θ^μ. $Sp(2n)$-invariant correlation functions in the tensorial spaces have been studied in [11,23,24,27] and in the unfolded formulation in [114].

8.1. Two-Point Functions

Let us denote the two-point correlation function by

$$W(Z_1, Z_2) = \langle \Phi(X_1, \theta_1)\Phi(X_2, \theta_2)\rangle \,. \tag{166}$$

The invariance under supersymmetry transformation generated by the operators Q, Equation (120), requires that

$$\epsilon^\mu \left(\frac{\partial}{\partial\theta_1^\mu} - i\theta_1^\nu \frac{\partial}{\partial X_1^{\mu\nu}} + \frac{\partial}{\partial\theta_2^\mu} - i\theta_2^\nu \frac{\partial}{\partial X_2^{\mu\nu}} \right) W(Z_1, Z_2) = 0 \,, \tag{167}$$

which implies

$$\langle \Phi(X_1, \theta_1)\Phi(X_2, \theta_2)\rangle = W(\det|Z_{12}|), \tag{168}$$

where

$$Z_{12}^{\mu\nu} = X_1^{\mu\nu} - X_2^{\mu\nu} - \frac{i}{2}\theta_1^\mu\theta_2^\nu - \frac{i}{2}\theta_1^\nu\theta_2^\mu \tag{169}$$

is the interval between two points in hyper–superspace which is invariant under the rigid supersymmetry transformations (117).

We next require the invariance of the correlator under the S-supersymmetry (125)

$$\begin{aligned} &\xi_\mu \left[(X_1^{\mu\nu} + \frac{i}{2}\theta_1^\mu\theta_1^\nu) \left(\frac{\partial}{\partial\theta_1^\nu} - i\theta_1^\rho \frac{\partial}{\partial X_1^{\nu\rho}} \right) + (X_2^{\mu\nu} + \frac{i}{2}\theta_2^\mu\theta_2^\nu) \left(\frac{\partial}{\partial\theta_2^\nu} - i\theta_2^\rho \frac{\partial}{\partial X_2^{\nu\rho}} \right) \right] \cdot \\ &W(\det|Z_{12}|) \\ &+\xi_\mu \left(\frac{i}{2}\theta_1^\mu + \frac{i}{2}\theta_2^\mu \right) W(\det|Z_{12}|) = 0 \,, \end{aligned} \tag{170}$$

which is solved by

$$W(\det|Z_{12}|) = c_2(\det|Z_{12}|)^{-\frac{1}{2}} \quad \Rightarrow \quad \langle \Phi(X_1, \theta_1)\Phi(X_2, \theta_2)\rangle = c_2(\det|Z_{12}|)^{-\frac{1}{2}} \,. \tag{171}$$

The two-point function (171) reproduces the correlators of the component bosonic and fermionic hyperfields $b(X)$ and $f_\mu(X)$ after the expansion of the former in powers of the Grassmann coordinates $\theta_1^{(\mu}\theta_2^{\nu)}$. Since on the mass shell the superfield (132) has only two non-zero components, all terms in the θ-expansion of the two-point function (171), starting from the ones quadratic in $\theta_1^{(\mu}\theta_2^{\nu)}$, should vanish. This is indeed the case, as a consequence of the field equations.

To see this, let us recall that in the separated points the two-point function of the bosonic hyperfield of weight $\frac{1}{2}$ satisfies the free field equation. Therefore for $X^1_{\alpha\beta} \neq X^2_{\alpha\beta}$ one has (when the two points coincide, one can define an analog of the Dirac delta-function in the tensorial spaces, see [5] for the relevant discussion)

$$(\partial^1_{\mu\nu}\partial^1_{\rho\sigma} - \partial^1_{\mu\rho}\partial^1_{\nu\sigma})\langle b(X_1)b(X_2)\rangle = (\partial^1_{\mu\nu}\partial^1_{\rho\sigma} - \partial^1_{\mu\rho}\partial^1_{\nu\sigma})(\det|X_{12}|)^{-\frac{1}{2}} = 0 \,. \tag{172}$$

Similarly, for $X^1_{\alpha\beta} \neq X^2_{\alpha\beta}$ the fermionic two-point function satisfies the free field equation for the fermionic hyperfield. Written in terms of the superfields, these equations are encoded in the superfield equation (for $Z_{12} \neq 0$)

$$(D^1_\mu D^1_\nu - D^1_\nu D^1_\mu)\langle \Phi(X_1, \theta_1)\Phi(X_2, \theta_2)\rangle = (D^1_\mu D^1_\nu - D^1_\nu D^1_\mu)(\det|Z_{12}|)^{-\frac{1}{2}} = 0. \tag{173}$$

Expanding the two-point function $(\det|Z_{12}|)^{-\frac{1}{2}}$ in powers of the Grassmann variables

$$(\det|Z_{12}|)^{-\frac{1}{2}} = (\det|X_{12}|)^{-\frac{1}{2}} \\ -i\partial_{\alpha\beta}(\det|X_{12}|)^{-\frac{1}{2}}\theta_1^{(\alpha}\theta_2^{\beta)} - \frac{1}{2}\partial_{\gamma\delta}\partial_{\alpha\beta}(\det|X_{12}|)^{-\frac{1}{2}}\theta_1^{(\alpha}\theta_2^{\beta)}\theta_1^{(\gamma}\theta_2^{\delta)} + \dots , \tag{174}$$

one may see that the terms in the expansion starting from $(\theta_1^{(\mu}\theta_2^{\nu)})^2$ vanish due to the free field Equation (172). From Equations (171) and (174) and from the explicit form of the superfield (132), one may immediately reproduce the correlation functions for the component fields [11]

$$\langle b(X_1)b(X_2)\rangle = c_2(\det|X_{12}|)^{-\frac{1}{2}}, \tag{175}$$

$$\langle f_\mu(X_1)f_\nu(X_2)\rangle = \frac{ic_2}{2}(X_{12})^{-1}_{\mu\nu}(\det|X_{12}|)^{-\frac{1}{2}}. \tag{176}$$

The two-point functions on the $OSp(1|n)$ manifold may now be obtained from (171) via the rescaling (157), which relates the superfields in flat superspace and on the $OSp(1|n)$ group manifold

$$\langle \Phi_{OSp}(X_1,\theta_1)\Phi_{OSp}(X_2,\theta_2)\rangle = \\ (\det G(X_1))^{-\frac{1}{2}}P(\Theta_1^2)(\det G(X_2))^{-\frac{1}{2}}P(\Theta_2^2)\langle \Phi(X_1,\theta_1)\Phi(X_2,\theta_2)\rangle . \tag{177}$$

Finally, as in the $D = 3$ case, one may derive the superconformally invariant two-point function for superfields carrying an arbitrary generalized conformal weight Δ, which on flat hyper superspace has the form

$$\langle \Phi^{\Delta_1}(X_1,\theta_1)\Phi^{\Delta_2}(X_2,\theta_2)\rangle = c_2(\det|Z_{12}|)^{-\Delta}, \qquad \Delta_1 = \Delta_2 = \Delta . \tag{178}$$

8.2. Three-Point Functions

The three-point functions for the superfields with arbitrary generalized conformal dimensions Δ_i, $(i = 1,2,3)$

$$W(Z_1, Z_2, Z_3) = \langle \Phi(X_1,\theta_1)\Phi(X_2,\theta_2)\Phi(X_3,\theta_3)\rangle , \tag{179}$$

may be computed in a way similar to the two-point functions using the superconformal Ward identities. The invariance under Q–supersymmetry implies that they depend on the superinvariant intervals Z_{ij}, i.e.,

$$\langle \Phi(X_1,\theta_1)\Phi(X_2,\theta_2)\Phi(X_3,\theta_3)\rangle = W(Z_{12}, Z_{23}, Z_{31}) , \tag{180}$$

where

$$Z_{ij}^{\mu\nu} = X_i^{\mu\nu} - X_j^{\mu\nu} - \frac{i}{2}(\theta_i^\mu\theta_j^\nu + \theta_i^\nu\theta_j^\mu), \qquad i,j = 1,2,3. \tag{181}$$

Invariance under S–supersymmetry then fixes the form of the function W to be

$$\langle \Phi(X_1,\theta_1)\Phi(X_2,\theta_2)\Phi(X_3,\theta_3)\rangle \\ = c_3(\det Z_{12})^{-\frac{1}{2}(\Delta_1+\Delta_2-\Delta_3)}(\det Z_{23})^{-\frac{1}{2}(\Delta_2+\Delta_3-\Delta_1)}(\det Z_{31})^{-\frac{1}{2}(\Delta_3+\Delta_1-\Delta_2)} . \tag{182}$$

Let us note that the three-point function is not annihilated by the operator entering the free equations of motion (131) for generic values of the generalized conformal dimensions, including the case in which the values of all the generalized conformal dimensions are canonical

$$(D_\mu^1 D_\nu^1 - D_\nu^1 D_\mu^1)\langle \Phi(X_1,\theta_1), \Phi(X_2,\theta_2), \Phi(X_2,\theta_2)\rangle \\ = c_3(D_\mu^1 D_\nu^1 - D_\nu^1 D_\mu^1)\left((\det|Z_{12}|)^{-\frac{1}{4}}(\det|Z_{23}|)^{-\frac{1}{4}}(\det|Z_{31}|)^{-\frac{1}{4}}\right) \neq 0 . \tag{183}$$

Again, the three-point functions on the supergroup manifold $OSp(1|n)$ can be obtained via the Weyl rescaling (157), as in the case of the two-point functions (177)

$$
\begin{aligned}
&\langle \Phi_{OSp}(X_1,\theta_1)\Phi_{OSp}(X_2,\theta_2)\Phi_{OSp}(X_3,\theta_3)\rangle \\
&= (\det G(X_1))^{-\frac{1}{2}}P(\Theta_1^2)(\det G(X_2))^{-\frac{1}{2}}P(\Theta_2^2)(\det G(X_3))^{-\frac{1}{2}}P(\Theta_3^2)\cdot \\
&\langle \Phi(X_1,\theta_1)\Phi(X_2,\theta_2)\Phi(X_3,\theta_3)\rangle\,.
\end{aligned} \tag{184}
$$

8.3. Four-Point Functions

Finally, let us consider, first in flat hyper superspace, the correlation function of four real scalar superfields with arbitrary generalized conformal dimensions, Δ_i $(i = 1,2,3,4)$

$$
W(Z_1, Z_2, Z_3) = \langle \Phi(X_1,\theta_1)\Phi(X_2,\theta_2)\Phi(X_3,\theta_3)\Phi(X_4,\theta_4)\rangle\,. \tag{185}
$$

Invariance under Q–supersymmetry again implies that the correlation function depends only on the superinvariant intervals $Z_{ij}^{\mu\nu}$ (181). Following the analogy with conventional conformal field theory we find

$$
W(X_1,X_2,X_3,X_4) = c_4 \prod_{ij,i<j} \frac{1}{(\det|Z_{ij}|)^{k_{ij}}}\tilde{W}(z,z')\,, \tag{186}
$$

with W being an arbitrary function of the cross-ratios

$$
z = \det\left(\frac{|Z_{12}||Z_{34}|}{|Z_{13}||Z_{24}|}\right), \qquad z' = \det\left(\frac{|Z_{12}||Z_{34}|}{|Z_{23}||Z_{14}|}\right), \tag{187}
$$

subject to the crossing symmetry constraints

$$
\tilde{W}(z,z') = \tilde{W}\left(\frac{1}{z},\frac{z'}{z}\right) = \tilde{W}\left(\frac{z}{z'},\frac{1}{z'}\right). \tag{188}
$$

Furthermore, the k_{ij}'s are constrained by the invariance of the four-point function under the S–supersymmetry to satisfy

$$
\sum_{j\neq i} k_{ij} = \Delta_i\,. \tag{189}
$$

Similar to the case of two- and three-point functions, the four-point function of the scalar superfields on $OSp(1|n)$ can be obtained from (186) via the Weyl re-scaling (157).

8.4. An Example. $\mathcal{N} = 1$ $D = 3$ Superconformal Models

As we mentioned earlier, the case of $D = 3$ is the simplest example of "hyperspace" which in this case coincides with the three-dimensional space time itself, and the fundamental fields are just the scalar $b(x)$ and the two-component spinor $f_\alpha(x)$. All known results for three-dimensional (super)conformal theories are reproduced from the above generic formulas restricted to the case of $n = 2$ and $D = 3$, as we will show on the example of $\mathcal{N} = 1$ $D = 3$ superconformal two– and three-point functions.

The superconformally invariant two- and three-point correlation functions of the $\mathcal{N} = 1$, $D = 3$ scalar supermultiplet model have been constructed in [115].

Let us use the spinor–tensor representation for the description of the three-dimensional space–time coordinates

$$
x^{\alpha\beta} = x^{\beta\alpha} = x^m(\gamma_m)^{\alpha\beta}, \tag{190}
$$

where now $\alpha,\beta = 1,2$ are $D = 3$ spinorial indices and $m = 0,1,2$ is the vectorial one. Since (190) provides a representation of the symmetric 2×2 matrices $x^{\alpha\beta}$, no extra coordinates, like y^{mn}, are present and, hence, no higher-spin fields.

The inverse matrix of (190), $x^{-1}_{\alpha\beta}$

$$
x^{\alpha\beta}x^{-1}_{\beta\gamma} = \delta_\alpha^\gamma\,, \tag{191}
$$

takes the simple form

$$x^{-1}_{\alpha\beta} = -\frac{1}{x^m x_m} x^n (\gamma_n)_{\alpha\beta} = -\frac{1}{x^2} x_{\alpha\beta} \,. \tag{192}$$

We may now consider a real scalar superfield in $D = 3$

$$\Phi(x,\theta) = \phi(x) + i\theta^\alpha f_\alpha(x) + \theta^\alpha\theta_\alpha F(x) \,, \tag{193}$$

with $\phi(x)$ being a physical scalar, $f_\alpha(x)$ a physical fermion and $F(x)$ an auxiliary field.

If (193) satisfies the free equation of motion (131), which in the $D = 3$ case reduces to

$$D^\alpha D_\alpha \Phi(x,\theta) = 0 \,. \tag{194}$$

This equation implies that on the mass shell the auxiliary field $F(x)$ vanishes, the scalar field $\phi(x)$ satisfies the massless Klein–Gordon equation and $f_\alpha(x)$ satisfies the massless Dirac equation. The field Equation (194) is superconformally invariant if the superfield $\Phi(x,\theta)$ has the canonical conformal weight $\Delta = \frac{1}{2}$.

Let us consider a superconformal transformation of (193). The Poincaré supersymmetry transformations of Φ are

$$\delta\Phi(x,\theta) = \epsilon^\alpha \left(\frac{\partial}{\partial\theta^\alpha} - i\theta^\beta \frac{\partial}{\partial x^{\alpha\beta}} \right) \Phi(x,\theta) = \epsilon^\alpha Q_\alpha \Phi(x,\theta) \,. \tag{195}$$

They encode the supersymmetry transformations of the component fields

$$\delta\phi(x) = i\epsilon^\alpha f_\alpha(x) \,, \tag{196}$$

$$\delta f_\alpha(x) = -2i\epsilon_\alpha F(x) - \epsilon^\beta \partial_{\alpha\beta}\phi(x) \,, \tag{197}$$

$$\delta F(x) = \frac{1}{2}\epsilon^\alpha \partial_{\alpha\beta} f^\beta(x) \,, \tag{198}$$

where we have made use of the identity

$$\theta^\alpha\theta^\beta = \frac{1}{2} C^{\alpha\beta}(\theta^\gamma\theta_\gamma) \,. \tag{199}$$

Under conformal supersymmetry, $\Phi(x,\theta)$ transforms as follows

$$\delta\Phi(x,\theta) = \xi_\alpha(x^{\alpha\beta} + \frac{i}{2}\theta^\alpha\theta^\beta) Q_\beta \Phi(x,\theta) - i(\xi_\alpha\theta^\alpha)\Delta\Phi(x,\theta) \,, \tag{200}$$

where Δ is the conformal weight of the superfield. The superconformal transformations of the component fields are

$$\delta\phi(x) = i\xi_\alpha \, x^{\alpha\beta} f_\beta(x), \tag{201}$$

$$\delta f_\alpha(x) = -2i\xi_\beta \, x^\beta{}_\alpha F(x) + \xi_\beta \, x^{\beta\gamma} \, \partial_{\gamma\alpha}\phi(x) + \xi_\alpha \Delta\phi(x), \tag{202}$$

$$\delta F(x) = \frac{1}{2} \, \xi_\alpha \, x^{\alpha\beta}\partial_{\beta\gamma} f^\gamma(x) - \frac{1}{2}\xi_\alpha \left(\frac{1}{2} - \Delta\right) f^\alpha(x). \tag{203}$$

The conformal weights of ϕ, f_α and F are Δ, $\Delta + \frac{1}{2}$ and $\Delta + 1$, respectively.

As we have already seen, the two-point function for a superfield of an arbitrary noncannonical dimension has the form (178). Expanding the expression on the right hand side of (178) in powers of θ, we obtain

$$\begin{aligned} (\det|z_{12}|)^{-\Delta} &= (\det|x_{12}|)^{-\Delta} - i\partial_{\alpha\beta}(\det|x_{12}|)^{-\Delta}\, \theta_1^{(\alpha}\theta_2^{\beta)} \\ &\quad - \tfrac{1}{2}\partial_{\gamma\delta}\partial_{\alpha\beta}(\det|x_{12}|)^{-\Delta}\, \theta_1^{(\alpha}\theta_2^{\beta)}\theta_1^{(\gamma}\theta_2^{\delta)} \,. \end{aligned} \tag{204}$$

Using the identities

$$\partial_{\alpha\beta}(\det|x|)^{-\Delta} = -\Delta\, x^{-1}_{\alpha\beta}\ \det|x|^{-\Delta}\,, \tag{205}$$

and

$$\partial_{\alpha\beta}\partial_{\gamma\delta}(\det|x|)^{-\Delta} = \Delta\left(\Delta\, x^{-1}_{\alpha\beta}x^{-1}_{\gamma\delta} + \frac{1}{2}x^{-1}_{\alpha\gamma}x^{-1}_{\beta\delta} + \frac{1}{2}x^{-1}_{\beta\gamma}x^{-1}_{\alpha\delta}\right)(\det|x|)^{-\Delta}\,, \tag{206}$$

one may rewrite the expression (204) as

$$(\det|z_{12}|)^{-\Delta} \;=\; (\det|x_{12}|)^{-\Delta}\left(1 - i\Delta\frac{x^m_{12}(\gamma_m)_{\alpha\beta}}{x^2_{12}}\theta^\alpha_1\theta^\beta_2 - \frac{(2\Delta-1)\Delta}{4}\frac{1}{x^2_{12}}\theta^2_1\theta^2_2\right). \tag{207}$$

Thus, from Equation (204) or (207), one may immediately read off the expressions for the correlation functions of the component fields of the superfield (193)

$$\langle\phi(x_1)\phi(x_2)\rangle = c_2(\det|x_{12}|)^{-\frac{1}{2}}\,, \tag{208}$$

$$\langle f_\alpha(x_1)f_\beta(x_2)\rangle = -ic_2\partial_{\alpha\beta}(\det|x_{12}|)^{-\frac{1}{2}}\,, \tag{209}$$

$$\langle\phi(x_1)f_\alpha(x_2)\rangle = 0\,, \qquad \langle F(x_1)\phi(x_2)\rangle = 0\,, \qquad \langle F(x_1)f_\alpha(x_2)\rangle = 0\,, \tag{210}$$

$$\langle F(x_1)F(x_2)\rangle = -\frac{c_2}{8}\partial^{\alpha\beta}\partial_{\alpha\beta}(\det|x|)^{-\Delta}\,. \tag{211}$$

Let us note that when the superfield $\Phi(x,\theta)$ has the canonical conformal dimension $\Delta = \frac{1}{2}$, due to the identity

$$C^{\alpha\gamma}C^{\beta\delta}\partial^1_{\alpha\beta}\partial^1_{\gamma\delta}(\det|x_{12}|)^{-\frac{1}{2}} = -\frac{1}{2}\eta^{mn}\frac{\partial}{\partial x^m_1}\frac{\partial}{\partial x^n_1}(\det|x_{12}|)^{-\frac{1}{2}}, \tag{212}$$

the last term in (204) is proportional to the δ–function if one moves to the Euclidean signature. Then, one has for the two-point function for the auxiliary field

$$\langle F(x_1)F(x_2)\rangle = -\frac{\pi}{4}c_2\delta^{(3)}(x_1 - x_2). \tag{213}$$

Note that the correlation functions of the auxiliary field F with the physical fields and with itself (for $x^m_1 \neq x^m_2$) vanish.

On the other hand, if the conformal weight of the superfield (193) is anomalous, i.e., $\Delta \neq \frac{1}{2}$, the correlators of the auxiliary field with the physical ones still vanish (in agreement with the fact that their conformal weights are different), but the $\langle FF\rangle$ correlator is

$$\begin{aligned}\langle F(x_1)F(x_2)\rangle &= -c_2\frac{(2\Delta-1)\Delta}{4}\frac{1}{x^2_{12}}(\det|x_{12}|)^{-\Delta} \\ &= -c_2\frac{(2\Delta-1)\Delta}{4}(\det|x_{12}|)^{-\Delta-1}.\end{aligned} \tag{214}$$

This situation may correspond to an interacting quantum $\mathcal{N} = 1$ superconformal field theory [116], where the auxiliary field is non-zero, and fields acquire anomalous dimensions due to quantum corrections.

The consideration of three-point functions is analogous. Using the expression for the three-point function (182) and expanding it in series of the θ^μ_i variables, we get for the component fields whose labels of scaling dimension we skip for simplicity

$$\langle\phi(x_1)\phi(x_2)\phi(x_3)\rangle = c_3(\det|x_{12}|)^{-k_1}(\det|x_{23}|)^{-k_2}(\det|x_{31}|)^{-k_3}\,, \tag{215}$$

$$\begin{aligned}&\langle f_\alpha(x_1)f_\beta(x_2)\phi(x_3)\rangle \\ &= -ic_3\frac{k_1x^m_{12}(\gamma_m)_{\alpha\beta}}{x^2_{12}}(\det|x_{12}|)^{-k_1}(\det|x_{23}|)^{-k_2}(\det|x_{31}|)^{-k_3} \\ &= -ic_3k_1x^m_{12}(\gamma_m)_{\alpha\beta}(\det|x_{12}|)^{-k_1-1}(\det|x_{23}|)^{-k_2}(\det|x_{31}|)^{-k_3}\,,\end{aligned} \tag{216}$$

$$\begin{aligned}&\langle f_\alpha(x_1)F(x_2)f_\beta(x_3)\rangle\\&=c_3\frac{k_1k_2}{2x_{12}^2x_{23}^2}(\gamma_m)_\alpha{}^\delta(\gamma_n)_{\delta\beta}(x_{12}^m)(x_{23}^n)(\det|x_{12}|)^{-k_1}(\det|x_{23}|)^{-k_2}(\det|x_{31}|)^{-k_3}\\&=c_3\frac{k_1k_2}{2}(\gamma_m)_\alpha{}^\delta(\gamma_n)_{\delta\beta}(x_{12}^m)(x_{23}^n)(\det|x_{12}|)^{-k_1-1}(\det|x_{23}|)^{-k_2-1}(\det|x_{31}|)^{-k_3},\end{aligned} \tag{217}$$

$$\langle F(x_1)F(x_2)\phi(x_3)\rangle=-\frac{c_3}{8}\partial^m\partial_m((\det|x_{12}|)^{-k_1})(\det|x_{23}|)^{-k_2}(\det|x_{31}|)^{-k_3}. \tag{218}$$

The remaining three-point functions containing an odd number of fermions, as well as the correlator $\langle F\phi\phi\rangle$, vanish. Note that, dimensional arguments would allow for a non-zero $\langle F\phi\phi\rangle$ correlator, but supersymmetry forces it to vanish. The correlator $\langle F(x_1)F(x_2)F(x_3)\rangle$ is zero as well, since it is proportional to $(\gamma_m\gamma_n\gamma_p)x_{12}^m x_{23}^n x_{31}^p = 2i\epsilon_{mnp}x_{12}^m x_{23}^n x_{31}^p = 0$.

Moreover, from the above expressions we see that superconformal symmetry does not fix the values of the scaling dimensions Δ_i. This indicates that quantum operators may acquire anomalous dimensions and the quantum $\mathcal{N}=1$, $D=3$ superconformal theory of scalar superfields can be non-trivial, in agreement e.g., with the results of [116].

If the value of Δ were restricted by superconformal symmetry to its canonical value and no anomalous dimensions were allowed (for all the operators which are not protected by supersymmetry) one would conclude that the conformal fixed point is that of the free theory. This is the case, for instance, for the $\mathcal{N}=1$, $D=4$ Wess-Zumino model in which the chirality of $\mathcal{N}=1$ matter multiplets and their three-point functions restricts the scaling dimensions of the chiral scalar supermultiplets to be canonical. This implies that in the conformal fixed point the coupling constant is zero, i.e., the theory is free [117,118].

9. Generalized CFT. Part II

In this Section, we shall continue our consideration of the generalized CFT based on the symmetries of the generalized conformal group $Sp(2n)$. We shall mainly follow [27].

9.1. Conserved Currents

In Section 2, we introduced the bosonic and fermionic fields in hyperspace which play the role of the scalar and fermionic fields in ordinary conformal field theory. In order to continue the analogy with CFTs let us consider the fields $b^A_\Delta(X)$ and $f^A_{\mu\Delta}(X)$ where now $A=1,...N$ is an index of an internal $O(N)$ group (not to be confused with the Weyl spinor indices of the previous Sections) and Δ are corresponding generalized conformal weights.

The two point functions of these fields are similar to those obtained in the previous section, with an obvious generalization including the "color" indexes

$$\langle b^A_{\Delta_1}(X_1), b^B_{\Delta_2}(X_2)\rangle = c_{bb}(det|X_{12}|)^{-\Delta}\ \delta^{AB}, \tag{219}$$

$$\langle f^A_{\alpha(\Delta_1)}(X_1), f^B_{\beta(\Delta_2)}(X_2)\rangle = c_{ff}(det|X_{12}|)^{-\Delta}(X_{12})^{-1}_{\alpha\beta}\ \delta^{AB}, \tag{220}$$

where $\Delta_1=\Delta_2=\Delta$, and $(X_{12})_{\alpha\beta}=(X_1)_{\alpha\beta}-(X_2)_{\alpha\beta}$.

Having introduced global $O(N)$ symmetry, one can construct bosonic and fermionic bilinears

$$J^{AB}_{\mu\nu}(X)=b^A(X)\partial_{\mu\nu}b^B(X)-b^B(X)\partial_{\mu\nu}b^A(X), \tag{221}$$

$$J^{AB}_{\mu\nu}(X)=f^A_\mu(X)f^B_\nu(X)+f^A_\nu(X)f^B_\mu(X). \tag{222}$$

These bilinears correspond to conserved $O(N)$ currents. Indeed one can check that the currents (221) and (222) satisfy the generalized conservation conditions (first introduced in [6])

$$\partial_{\mu\nu}J^{AB}_{\alpha\beta}(X)-\partial_{\mu\alpha}J^{AB}_{\nu\beta}(X)-\partial_{\beta\nu}J^{AB}_{\alpha\mu}(X)+\partial_{\beta\alpha}J^{AB}_{\nu\mu}(X)=0 \tag{223}$$

provided that the fields $b^A(X)$ and $f^A_\mu(X)$ satisfy the free equations of motion (14) and (15).

Knowing the $Sp(2n)$ transformations (42) and (43) of the fields $b^A(X)$ and $f^A_\mu(X)$ and using Equations (221) and (222), one can derive the $Sp(2n)$ transformations of the conserved currents

$$\delta_a J^{AB}_{\mu\nu}(X) = -a^{\alpha\beta}\partial_{\alpha\beta} J^{AB}_{\mu\nu}(X) \tag{224}$$

$$\delta_g J^{AB}_{\mu\nu}(X) = -\left(g_\alpha{}^\alpha + 2g_\alpha{}^\beta X^{\alpha\gamma}\partial_{\beta\gamma}\right) J^{AB}_{\mu\nu}(X) - 2g_{(\mu}{}^\rho J^{AB}_{\rho\nu)}(X) \tag{225}$$

$$\delta_k J^{AB}_{\mu\nu}(X) = (k_{\alpha\beta}X^{\alpha\beta} + k_{\alpha\beta}X^{\alpha\gamma}X^{\beta\delta}\partial_{\gamma\delta}) J^{AB}_{\mu\nu}(X) + 2k_{(\mu\alpha}X^{\alpha\beta} J^{AB}_{\beta\nu)}(X) \tag{226}$$

From this transformation laws i.e, from the coefficients in front of the terms $g_\alpha{}^\alpha$ and $k_{\alpha\beta}X^{\alpha\beta}$ one can conclude that the generalized conformal dimension Δ_J of the currents (221) and (222) is equal to 1. The same conclusion can be reached from the fact that (221) and (222) correspond to free currents and the generalized conformal dimension of the fields $b(X)$ and $f_\mu(X)$ is equal to $\frac{1}{2}$. Using the general expression (136), one can see that the generalized conformal dimension is related to the usual scaling dimension as follows. Recall (see Section 2.3) that $SL(n)$ subalgebra of $GL(n)$ algebra is parameterized by $l_\mu{}^\nu = g_\mu{}^\nu - \frac{1}{n}\delta^\nu_\mu g_\rho{}^\rho$. Let us rewrite Equation (225) as

$$\delta_g J^{AB}_{\mu\nu}(X) = -\left(\frac{n+2}{n} g_\alpha{}^\alpha + 2g_\alpha{}^\beta X^{\alpha\gamma}\partial_{\beta\gamma}\right) J^{AB}_{\mu\nu}(X) - 2l_{(\mu}{}^\rho J^{AB}_{\rho\nu)}(X) \tag{227}$$

and define a weight Δ_1 as follows

$$\Delta_1 = 1 + \frac{2}{n}. \tag{228}$$

Then using the relations (136) one can see that

$$\Delta_{\mathbb{D},1} = D - 1 \tag{229}$$

which is the canonical conformal weight of a spin-1 field.

9.2. Stress Tensor

Since we are considering a generalized CFT it is natural to define a generalized stress tensor, which contains a usual CFT stress tensor when projected to the x-subspace. Taking

$$\tilde{T}_{\mu\nu,\rho\sigma}(X) = (\partial_{\mu\nu}b(X))(\partial_{\rho\sigma}b(X)) - \frac{1}{3}b(X)(\partial_{\mu\nu}\partial_{\rho\sigma}b(X)) \tag{230}$$

and

$$\tilde{T}_{\mu\nu,\rho\sigma}(X) = f_\rho(X)\partial_{\mu\nu}f_\sigma(X) \tag{231}$$

we define the generalized stress tensor as a symmetrized combination

$$T_{\mu\nu,\rho\sigma}(X) = \tilde{T}_{\mu\nu,\rho\sigma}(X) + \tilde{T}_{\mu\rho,\nu\sigma}(X) + \tilde{T}_{\mu\sigma,\nu\rho}(X) \tag{232}$$

The reason of taking the expression (232) as a definition for the generalized stress tensor instead of (230) and (231) is that (232) transforms properly under the $Sp(2n)$ transformations

$$\delta_a T_{\mu\nu\rho\sigma}(X) = -a^{\alpha\beta}\partial_{\alpha\beta} T_{\mu\nu,\rho\sigma}(X), \tag{233}$$

$$\begin{aligned} \delta_g T_{\mu\nu\rho\sigma}(X) &= -(g_\alpha{}^\alpha + 2g_{\alpha\beta}X^{\alpha\gamma}\partial_{\beta\gamma})T_{\mu\nu\rho\sigma}(X) \\ &\quad - g_\mu{}^\alpha T_{\alpha\nu\rho\sigma}(X) - \ldots - g_\sigma{}^\alpha T_{\mu\nu\rho\alpha}(X), \end{aligned} \tag{234}$$

$$\begin{aligned} \delta_k T_{\mu\nu\rho\sigma}(X) &= (k_{\alpha\beta}X^{\alpha\beta} + k_{\alpha\beta}X^{\alpha\gamma}X^{\beta\delta}\partial_{\gamma\delta})T_{\mu\nu\rho\sigma}(X) \\ &\quad + k_{\mu\alpha}X^{\alpha\beta}T_{\beta\nu\rho\sigma}(X) + \ldots + k_{\sigma\alpha}X^{\alpha\beta}T_{\mu\nu\rho\beta}(X). \end{aligned} \tag{235}$$

The transformations above are again derived using the transformations for the free fields (42) and (43) and the explicit form of the stress energy tensor (232). Again, using (136), one can see that the generalized conformal dimension of the stress tensor is $\Delta_T = 1$, whereas the conformal dimension Δ_2 (analogous to the expression (228) for $s = 1$ current) is

$$\Delta_2 = 1 + \frac{4}{n} \tag{236}$$

and the canonical spin-2 field weight is

$$\Delta_{D,2} = D$$

in compliance with the general formula $\Delta_{D,s} = D + s - 2$.

Like the conserved current $J^{AB}_{\mu\nu}$, the stress energy tensor satisfies the generalized conservation conditions

$$\partial_{\mu\nu} T_{\alpha\beta\gamma\delta}(X) - \partial_{\mu\alpha} T_{\nu\beta\gamma\delta}(X) - \partial_{\beta\nu} T_{\alpha\mu\gamma\delta}(X) + \partial_{\beta\alpha} T_{\nu\mu\gamma\delta}(X) = 0 \tag{237}$$

provided the fields satisfy the free equations of motion (14) and (15).

9.3. Higher Spin Conserved Currents

By analogy with $J_{\alpha\beta}(X)$ and $T_{\alpha\beta\gamma\delta}(X)$ one can introduce [6] higher-spin conserved currents $T_{\alpha_1\ldots\alpha_{2s}}(X)$ $(2s = 1, 2, 3, \ldots)$ which transform under $Sp(2n)$ as follows

$$\delta_a T_{\alpha_1\ldots\alpha_{2s}}(X) = -a^{\mu\nu}\partial_{\mu\nu} T_{\alpha_1\ldots\alpha_{2s}}(X), \tag{238}$$

$$\begin{aligned} \delta_g T_{\alpha_1\ldots\alpha_{2s}}(X) &= -(\Delta_s\, g_\mu{}^\mu + 2g_\nu{}^\mu X^{\nu\rho}\partial_{\mu\rho}) T_{\alpha_1\ldots\alpha_{2s}}(X) \\ &\quad -2s l_{(\alpha_1}{}^\mu T_{\alpha_2\ldots\alpha_{2s})\mu}(X), \end{aligned} \tag{239}$$

$$\begin{aligned} \delta_k T_{\alpha_1\ldots\alpha_{2s}}(X) &= (k_{\mu\nu} X^{\mu\nu} + k_{\mu\nu} X^{\mu\rho} X^{\nu\lambda}\partial_{\rho\lambda}) T_{\alpha_1\ldots\alpha_{2s}}(X) \\ &\quad +4k_{\mu(\alpha_1} X^{\mu\nu} T_{\alpha_2\ldots\alpha_{2s})\nu}(X), \end{aligned} \tag{240}$$

where

$$\Delta_s = 1 + \frac{2s}{n}. \tag{241}$$

Again, using the relations (136), one can see that

$$\Delta_{\mathbb{D},s} = D + s - 2 \tag{242}$$

which is a conventional expression for a canonical conformal weight for a field with spin s.

The higher spin currents obey $Sp(2n)$ conservation conditions [6]

$$\partial_{\mu\nu} T_{\alpha\beta\gamma(2s-2)}(X) - \partial_{\mu\alpha} T_{\nu\beta\gamma(2s-2)}(X) - \partial_{\beta\nu} T_{\alpha\mu\gamma(2s-2)}(X) + \partial_{\alpha\beta} T_{\mu\nu\gamma(2s-2)}(X) = 0. \tag{243}$$

9.4. Two-Point Correlation Functions of the Currents

We have already considered two-point functions for scalar and spinorial hyperfields (219) and (220). Using these expressions as well as the expressions for the generalized conserved currents (221) and (222), it is straightforward to compute the two-point functions of two currents

$$\langle J^{AB}_{\alpha\beta}(X_1), J^{CD}_{\mu\nu}(X_2)\rangle = C_{JJ}(\det|X_{12}|)^{-1}(P_{12})_{\alpha\beta,\mu\nu}(\delta^{AC}\delta^{BD} - \delta^{AD}\delta^{BC}). \tag{244}$$

Here, we introduced an $Sp(2n)$-invariant tensor structure (When checking the invariance under the generalized conformal boosts notice that the first pair of the indices of $(P_{12})_{\alpha\beta,\gamma\delta}$ gets rotated with the matrix $k_{\alpha\sigma} X_1^{\sigma\delta}$ and the second pair gets rotated with $k_{\mu\sigma} X_2^{\sigma\delta}$) (which we call P–structure)

$$(P_{ab})_{\alpha\beta,\mu\nu} = (X^{-1}_{ab})_{\mu\alpha}(X^{-1}_{ab})_{\nu\beta} + (X^{-1}_{ab})_{\nu\alpha}(X^{-1}_{ab})_{\mu\beta} \tag{245}$$

$a, b = 1, 2$ and $a \neq b$. which will be one of the building blocks for higher point correlation functions as well.

One more building block for the correlation functions is $(X_{12})^{-1}_{\alpha\beta}$ which is $Sp(2n)$ invariant when considered as a bilocal tensor

$$\begin{aligned}\delta_{tot}(X^{-1}_{12})_{\alpha\beta} &= -(X^{-1}_{12})_{\alpha\gamma}(\delta X_1 - \delta X_2)^{\gamma\delta}(X^{-1}_{12})_{\delta\beta} \\ &+ 2g_{(\alpha}{}^{\gamma}(X^{-1}_{12})_{\beta)\gamma}k_{\alpha\gamma}X_1^{\gamma\delta}(X^{-1}_{12})_{\delta\beta} - (X^{-1}_{12})_{\alpha\delta}X_2^{\delta\gamma}k_{\gamma\beta} = 0\,.\end{aligned}$$

Similarly, for the two stress tensors one finds

$$\langle T_{\alpha\beta\gamma\delta}(X_1), T_{\mu\nu\rho\sigma}(X_2)\rangle = C_{TT}\frac{1}{\det|X_{12}|}\left((P_{12})_{\alpha\beta,\mu\nu}(P_{12})_{\gamma\delta,\rho\sigma} + symm.\right), \tag{246}$$

where the total symmetrization of the both sets of indices $(\alpha\beta\gamma\delta)$ and $(\mu\nu\rho\sigma)$ is assumed.

It is instructive to recall the similar expressions for two-point functions in the usual CFT

$$\langle T^{(l)}_{\mu_1,\ldots,\mu_n}(x_1), T^{(l)}_{\nu_1,\ldots,\nu_n}(x_2)\rangle = c_{TT}\frac{g_{\mu_1\nu_1}(x_{12})\ldots g_{\mu_n\nu_n}(x_{12})}{(x_{12})^l} - traces \tag{247}$$

with

$$g_{\mu\nu} = \delta_{\mu\nu} - \frac{x_\mu x_\nu}{x^2}. \tag{248}$$

Obviously, the $Sp(2n)$-invariant structure $(P_{12})_{\alpha\beta,\gamma\delta}$ is a generalization of $g_{\mu\nu}$. Notice also that the expressions for two-point functions (244)–(246) can be obtained from solving generalized Ward identities, as it has been done for the case of scalar and spinor hyperfields. The generalized Ward identity for an n-point function

$$\langle \Phi^{\Delta^{(1)}}_{\alpha_1\ldots\alpha_{r_1}}(X_1)\ldots\Phi^{\Delta^{(k)}}_{\beta_1\ldots\beta_{r_k}}(X_k)\rangle \equiv G_{\alpha_1\ldots\alpha_{r_1},\ldots,\beta_1\ldots\beta_{r_k}}(X_1,\ldots,X_k)\,. \tag{249}$$

is as follows

$$\begin{aligned}&\Sigma_{i=1}^{k}\left[\Delta_i(g_\mu{}^\mu - k_{\mu\nu}X_i^{\mu\nu}) + \delta X_i^{\mu\nu}\frac{\partial}{\partial X_i^{\mu\nu}}\right]G_{\alpha_1\ldots\alpha_{r_1},\ldots,\beta_1\ldots\beta_{r_k}}(X_1,\ldots,X_k) \\ &+\Sigma_{j=1}^{1}(g_{\alpha_j}{}^{\mu_j} - k_{\alpha_j\nu}X_1^{\nu\mu_j})\,G_{\mu_1\ldots\mu_j\ldots\mu_{r_1},\ldots,\beta_1\ldots\beta_{r_k}}(X_1,\ldots,X_k) + \cdots \\ &+\Sigma_{j=1}^{r_k}(g_{\beta_j}{}^{\mu_j} - k_{\beta_j\nu}X_k^{\nu\mu_j})\,G_{\alpha_1\ldots\alpha_{r_k},\ldots,\mu_1\ldots\mu_j\ldots\mu_{r_k}}(X_1,\ldots,X_k) = 0,\end{aligned} \tag{250}$$

It is straightforward to check that the two-point functions solve Equations (250).

9.5. Three Point Functions: bbb and ffb

Three-point functions for three scalars and for two fermions and a scalar (computed firstly in [11]) have been given in Section 8.2 in the supersymmetric form and as a particular example for $D = 3$ were given in Section 8.4. The only difference with the case without supersymmetry is that the overall constants in front of the non-supersymmetric ones are independent of each other

$$\langle b_{\Delta_1}(X_1)b_{\Delta_2}(X_2)b_{\Delta_3}(X_3)\rangle = C_{bbb}\,(\det|X_{12}|)^{-k_3}\,(\det|X_{23}|)^{-k_1}\,(\det|X_{13}|)^{-k_2}, \tag{251}$$

$$\langle f_\alpha(X_1)f_\beta(X_2)b(X_3)\rangle = c_{ffb}\,(X^{-1}_{12})_{\alpha\beta}(\det|X_{12}|)^{-k_3}\,(\det|X_{23}|)^{-k_1}\,(\det|X_{13}|)^{-k_2}. \tag{252}$$

$$k_a = \frac{1}{2}(\Delta^{(a+1)} + \Delta^{(a+2)} - \Delta^{(a)}), \quad cycl. \quad (a = 1, 2, 3). \tag{253}$$

9.6. Three-Point Functions with J and T

Now, we would like to consider three-point functions which include the generalized conserved current $J^{AB}_{\alpha\beta}(X)$ and generalized stress tensor $T_{\alpha\beta\gamma\delta}(X)$. These can give us an answer whether an interacting generalized conformal field theory based on $Sp(2n)$ symmetry exists. As we shall see below, the answer to this question is negative.

Our strategy is as follows. As we have seen the generalized conformal weighs of $J^{AB}_{\alpha\beta}(X)$ and $T_{\alpha\beta\gamma\delta}(X)$ are equal to one, $\Delta_J = \Delta_T = 1$. If we assume that the corresponding symmetries are not broken by interactions, then the values of Δ_J and Δ_T will remain the same. Therefore, we would like to construct $Sp(2n)$-invariant three- and higher-order correlation functions which include $J^{AB}_{\alpha\beta}(X)$, $T_{\alpha\beta\gamma\delta}(X)$ and other operators $\mathcal{O}$ and see if the conservation conditions (223) and (237) along with $Sp(2n)$ invariance allow for the operators $\mathcal{O}$ to have anomalous dimensions. We will find that this is unfortunately not the case for $n > 2$.

First let us introduce one more $Sp(2n)$-invariant tensor structure (which we call Q–structure)

$$(Q^c_{ab})_{\alpha\beta} = (X^{-1}_{ac})_{\alpha\beta} - (X^{-1}_{bc})_{\alpha\beta}, \quad a,b,c = 1,2,3 \tag{254}$$

This structure, along with (245) and

$$(p_{ab})_{\alpha\beta} = (X^{\alpha\beta}_a - X^{\alpha\beta}_b)^{-1}, \quad a,b = 1,2, \quad a \neq b. \tag{255}$$

is a building block for all the $Sp(2n)$-invariant correlation functions. In other words, the most general multi-point function can be written as a sum over all possible polynomials of a required rank of the three structures $p_{ab} = X^{-1}_{ab}$, P_{ab} and Q^c_{ab} times a pre-factor

$$\langle \Phi ... \Phi \rangle = G(p_{ab}, P_{ab}, Q^c_{ab} | X_{ab}). \tag{256}$$

Following this prescription one can immediately write the simplest three-point function of two scalars (with generalized conformal dimensions $\Delta_1 = \Delta_2 = \Delta$) and a conserved current (with $\Delta_J = 1$)

$$\begin{aligned} &\langle b_{\Delta_1}(X_1) b_{\Delta_2}(X_2) J_{\alpha\beta}(X_3) \rangle = \\ &= C_{bbJ} (\det|X_{12}|)^{-k_3} (\det|X_{13}|)^{-k_2} (\det|X_{23}|)^{-k_1} (Q^3_{12})_{\alpha\beta}, \end{aligned} \tag{257}$$

and a three-point function of the two scalars (with $\Delta_1 = \Delta_2 = \Delta$) and the stress tensor (with $\Delta_T = 1$)

$$\begin{aligned} &\langle b(X_1) b(X_2) T_{\alpha\beta\gamma\delta}(X_3) \rangle = C_{bbT} (\det|X_{12}|)^{-k_3} (\det|X_{13}|)^{-k_2} \times \\ &(\det|X_{23}|)^{-k_1} ((Q^3_{12})_{\alpha\beta} (Q^3_{12})_{\gamma\delta} + (Q^3_{12})_{\alpha\gamma} (Q^3_{12})_{\beta\delta} + (Q^3_{12})_{\alpha\delta} (Q^3_{12})_{\beta\gamma}), \end{aligned} \tag{258}$$

where k_a are restricted according to (253). One can see that $Sp(2n)$ invariance alone does not impose any requirement on the generalized conformal dimension Δ of the scalar field.

The next step is to require the conservation of the current $J^{AB}_{\alpha\beta}(X)$ and the stress tensor $T_{\alpha\beta\gamma\delta}(X)$ according to Equations (223) and (237). This implies

$$k_1 = k_2 = \frac{1}{2}, \quad \text{and any} \quad k_3 . \tag{259}$$

Therefore, in this case, no restriction on generalized conformal dimension of the scalar field appears i.e., anomalous dimension and therefore interactions are allowed. At this, the current and the stress tensor remain conserved, and their dimensions remain canonical $\Delta_J = \Delta_T = 1$.

The next nontrivial example is a three point-function of two conserved currents and one scalar operator $\mathcal{O}(X)$ of dimension Δ. From the $Sp(2n)$-invariance condition we have

$$\begin{aligned}\langle J_{\mu\nu}(X_1)\mathcal{O}(X_2)J_{\alpha\beta}(X_3)\rangle = & (\det|X_{12}|)^{-\frac{\Delta}{2}}(\det|X_{13}|)^{-\frac{2-\Delta}{2}}(\det|X_{23}|)^{-\frac{\Delta}{2}} \\ & \times\left(\mathcal{A}[(Q^3_{12})_{\alpha\beta}(Q^1_{23})_{\mu\nu}]+\mathcal{B}(P_{13})_{\mu\nu,\alpha\beta}\right)\end{aligned} \tag{260}$$

where $\mathcal{A}$ and $\mathcal{B}$ are some constants. Again, one can see that $Sp(2n)$ symmetry alone does not impose any restriction on the generalized conformal dimension of $\mathcal{O}(X)$.

However, imposing the current conservation condition (223), one gets

$$\mathcal{A}=\mathcal{B}, \quad and \quad \Delta=1, \tag{261}$$

that is the dimension of the operator $\mathcal{O}(X)$ is fixed (Since the canonical dimension of the field $b(X)$ is equal to $\frac{1}{2}$ it is natural to assume that the operator $\mathcal{O}(X)$ is a composite one $O(X) = b^2(X)$.) by the current conservation condition. Let us note that from the point of view of the x-space the current $J^{AB}_{\alpha\beta}(X)$ contains higher spin currents as a result of its expansion in series of y coordinates. Therefore, this result is in accordance with the theorem of [119] stating that the conformal field theories which contain conserved higher-spin currents should be free.

Let us note, however, that in the simplest case of $n=2$, i.e., $D=3$ CFTs with the $Sp(4)$ conformal group the two conditions (261) are reduced to one (see [27] for technical details)

$$\mathcal{A}(D-1-\Delta)-\mathcal{B}\Delta=0\,. \tag{262}$$

This means that the conformal dimension Δ of the operator $\mathcal{O}(X)$ remains undetermined, and hence this analysis does not ban the existence of interacting $D=3$ CFTs, as is well known.

9.7. General Case

Let us now discuss the general structure of the three-point correlators of conserved currents which are symmetric tensors of rank $r=2s$ with s being an integer "spin". To this end, it is convenient to hide the tensor indices away by contracting them with auxiliary variables λ^α_a, where a refers to the point of the operator insertion:

$$(p_{ab})_{\alpha\beta}\Rightarrow p_{ab}=(X^{-1}_{ab})_{\alpha\beta}\,\lambda^\alpha_a\lambda^\beta_b \quad \text{no summation over } a,b\,. \tag{263}$$

$$(P_{bc})_{\alpha\beta,\gamma\delta}\Rightarrow P_{ab}=2p_{ab}p_{ba}=(P_{ab})_{\alpha\beta,\gamma\delta}\,\lambda^\alpha_a\lambda^\beta_a\lambda^\gamma_b\lambda^\delta_b \quad \text{no summation over } a,b\,, \tag{264}$$

$$(Q^a_{bc})_{\alpha\beta}\Rightarrow Q^a_{bc}=(Q^a_{bc})_{\alpha\beta}\,\lambda^\alpha_a\lambda^\beta_a \quad \text{no summation over } a\,. \tag{265}$$

For instance, the correlator of two scalar operators $\mathcal{O}$ of the same dimension Δ with a conserved current of an integer spin-s obeying (243) is

$$\langle O(X_1)O(X_2)J_s(X_3)\rangle=C(\det|X_{12}|)^{-\frac{2-\Delta}{2}}(\det|X_{13}|)^{-\frac{1}{2}}(\det|X_{23}|)^{-\frac{1}{2}}(Q^3_{12})^s\,. \tag{266}$$

The current conservation condition leads to the same result as for the case of $s=1,2$, i.e., $k_1=k_2=\frac{1}{2}$, which means that the dimensions of the scalar operators are arbitrary.

However, if we consider a three-point function of a scalar operator and two conserved currents

$$J_s(X)=J_{\alpha_1\ldots\alpha_{2s}}(X)\lambda^{\alpha_1}\cdots\lambda^{\alpha_{2s}} \tag{267}$$

of ranks $2s_1$ and $2s_2$ with $s \geq 1$, we will again find that, up to an overall factor, all the free parameters in the correlator are fixed. For example,

$$\langle J_3(X_1)J_1(X_2)O(X_3)\rangle = C\frac{(Q^1_{23})^3Q^2_{13} - 3(Q^1_{23})^2P_{12}}{\left(\det|X_{12}|\det|X_{13}|\det|X_{23}|\right)^{1/2}}. \quad (268)$$

From the discussion above, one can conclude that in order to describe the $Sp(2n)$-invariant three-point functions, we can borrow the generating functions of 3-point correlators of free symmetric higher-spin fields in conventional conformal theories [114,120–123] simply because the $Sp(2n)$ group contains the corresponding conformal group $SO(2, D)$ as a subgroup, or, in other words, the correlators in the free CFTs can be covariantly embedded into the $Sp(2n)$ invariant correlators. For example, a generating function of the three-point functions of currents built out of free scalars $b(X)$ is

$$\langle J(X_1)J(X_2)J(X_3)\rangle = \frac{\cos(p_{12})\cos(p_{13})\cos(p_{23})\,\exp\left(\frac{1}{2}[Q^1_{23} + Q^2_{13} + Q^3_{12}]\right)}{(\det|X_{12}|\det|X_{23}|\det|X_{13}|)^{1/2}}. \quad (269)$$

It contains the operators $J_s(X)$, $s = 0, 1, 2, ...$ and the correlator $\langle J_{s_1}J_{s_2}J_{s_3}\rangle$ is obtained as the coefficient in front of $(\lambda_1)^{2s_1}(\lambda_2)^{2s_2}(\lambda_3)^{2s_3}$.

The generating function obtained from the currents built out of the free fermions $f_\alpha(X)$ is

$$\langle J(X_1)J(X_2)J(X_3)\rangle = \frac{\sin(p_{12})\sin(p_{13})\sin(p_{23})\,\exp\left(\frac{1}{2}[Q^1_{23} + Q^2_{13} + Q^3_{12}]\right)}{(\det|X_{12}|\det|X_{23}|\det|X_{13}|)^{1/2}}. \quad (270)$$

The generating function of multi-point correlators can be found in [114,122–125].

The above expressions deal with the bosonic symmetric tensor currents of even rank. The generating function which produces three-point correlators involving two fermionic currents of odd ranks is similar, see e.g., [119].

As a further development of this subject, it would be of interest to carry out the study of other aspects of the $Sp(2n)$-invariant higher-spin systems, in particular, to explore their links to recent results on conformal higher-spin theories in AdS_D backgrounds (see e.g., [126–130]) and to $Sp(2n)$-invariant unfolded higher-spin structures discussed in [131].

9.8. Breaking $Sp(2n)$ Symmetry

As it follows from the discussion above, to have an interacting generalized conformal field theory based on $Sp(2n)$ symmetry, one has to break this symmetry down to a subgroup. Obviously, to still use $Sp(2n)$ symmetry as a symmetry of the theory, it should be broken spontaneously rather then explicitly. On the other hand, the question whether a symmetry is broken spontaneously or explicitly could be simpler to address if one had the corresponding Lagrangian, which would produce the field Equations (14) and (15) (and/or their possible nonlinear or massive deformations). Unfortunately, such a Lagrangian is still lacking.

In this respect, let us mention that the issue of breaking $Sp(8)$ symmetry via current interactions in the unfolded formulation has been addressed in [26]. In particular, analyzing the system of equations

$$DC(x,\mu,\bar{\mu}) = F(\omega, J(x,\mu,\bar{\mu})), \qquad D_2J(x,\mu,\bar{\mu}) = 0, \quad (271)$$

where $D = d + \omega$ is a spin connection, J is a current which is billinear in the higher-spin functional C and D_2 is the corresponding kinetic operator (see the discussion around the Equation (27)), the authors showed that the $Sp(8)$ symmetry is broken to the four-dimensional conformal group $SO(2,4)$.

In the hyperspace framework, one may try to approach this problem as follows. First, one should construct a nonlinear deformation of Equations (14) and (15)

$$\partial_{\alpha\beta}\partial_{\gamma\delta}\, b(X) - \partial_{\alpha\gamma}\partial_{\beta\delta}\, b(X) = F_b(b,f,A)\,, \tag{272}$$

$$\partial_{\alpha\beta} f_\gamma(X) - \partial_{\alpha\gamma} f_\beta(X) = F_f(b,f,A)\,. \tag{273}$$

with some unknown functions $F_b(b,f,A)$ and $F_f(b,f,A)$. It is natural to expect that these functions depend also on higher-spin potentials A, in addition to the higher-spin curvatures contained in the hyperfields $b(X)$ and $f_\mu(X)$. Note that, in the unfolded description of the $Sp(8)$-invariant system, higher-spin gauge potentials were introduced, at the linearized level, in [16]. As a necessary step forward, one should understand whether and how the Equations (272) may result from a (non-linear) generalization of the construction of [16].

The right hand sides of the Equation (272) should be chosen under the requirement that the analysis of the Equations (272) and (273), similar to the one carried out for the free equations in Section 2.1 leads to a physically meaningful nonlinear equations in the x–space. This is an interesting open problem for a future study.

10. Conclusions

The idea to formulate higher-spin theories in an extended (super) space, where extra coordinates generate higher spins (by analogy with the Kaluza–Klein theories where compact extra dimensions generate "higher masses") seems to be very attractive, especially taking into account a level of complexity of higher-spin theories formulated in an ordinary space–time.

The underlying symmetry of this formulation is the $Sp(2n)$ group, which contains the corresponding D-dimensional conformal group as a subgroup. This allows one to borrow, for the analysis of the $Sp(2n)$-invariant systems, an intuition and techniques from conventional Conformal Field Theories.

To summarize, the reviewed appraoch generalizes familiar concepts to higher-dimensional tensorial spaces and the correspondence looks schematically as follows

- Space–time coordinates x^m are extended to tensorial coordinates $X^{\alpha\beta}$.
- Cartan–Penrose relation $P_{A\dot{A}} = \lambda_A \bar{\lambda}_{\dot{A}}$ gets extended to the hyperspace twistor-like relation $P_{\alpha\beta} = \lambda_\alpha \lambda_\beta$ which determines free dynamics of fields in the tensorial space with the momentum $P_{\alpha\beta}$ conjugate to $X^{\alpha\beta}$.
- AdS_D space is extended to the $Sp(n)$ group manifold.
- Conformal scalar $\phi(x)$ and conformal spinor $\psi_\mu(x)$ become the "hyperscalar" $b(X)$ and the "hyperspinor" $f_\mu(X)$.
- D-dimensional conformal group $SO(2,D)$ is extended to the $Sp(2n)$ group which underlies the Generalized Conformal Field Theory of the fields $b(X)$ and $f_\mu(X)$.

We have shown that the hyperspace approach describes (in $D = 3,4,6$ and 10) free dynamics of an infinite set of massless conformal higher-spin fields in an elegant compact form. An important and non-trivial problem is to find a non-linear generalization of this formulation which would correspond to an interacting higher-spin theory. This problem has been addressed by several authors. As we have seen, it is related to the necessity to break the $Sp(2n)$ symmetry in an appropriate way. Attempts to construct such a generalization in the framework of hyperspace supergravity and a non-linear realization of the $OSp(1|8)$ supergroup were undertaken, respectively, in [12,14]. Obstacles encountered in these papers may be related to the fact that their constructions utilized only higher-spin field strengths but did not include couplings to higher-spin gauge potentials, while the consistent formulation of nonlinear equations of massless higher-spin fields contains both [37–39]. Therefore, to successfully address the problem of interactions it is important to incorporate higher-spin potentials in the hyperspace approach, e.g., by further elaborating on the construction of [16].

Another issue, which can be related to the previous one, is a question of consistent breaking $Sp(2n)$ symmetry. The manifestation of this breaking was observed e.g. in higher-spin current interactions [26]. As we have seen in Section 9, when considering generalized CFT based on global $Sp(2n)$ invariance (see [27]), the requirement of generalized current conservation turns out to be too strong to allow for the basic hyperfields to have anomalous conformal dimensions and again points at the necessity to (spontaneously) break $Sp(2n)$ invariance.

Theories with spontaneously broken $Sp(2n)$ symmetry might be also useful for studying massive higher-spin fields in hyperspaces. A consideration of theories with local $Sp(n)$ invariance i.e., some sort of generalized gravity is yet another interesting and widely unexplored area.

Finally, let us mention that field Equations (14) and (15) for the fields in hyperspaces remind (a part of) weak section conditions of exceptional field theories (see [132] for a review and references). This similarity can be relevant for higher-spin extensions of these theories, provided the section conditions can be properly relaxed (see e.g., [133,134] for a discussion of this point). It would be interesting to further elaborate on this issue, as a connection to the E_{11} framework [18].

Acknowledgments: We are grateful to I. Bandos, X. Bekaert, J.A. de Azcarraga, I. Florakis, J. Lukierski, P. Pasti, M. Plyushchay, E. Skvortsov and M. Tonin with whom we obtained the results reviewed in this article. We are thankful to P. Bouwknegt, O. Gelfond, A.R. Gover, S. Kuzenko, S. Sergeev, A.Tseytlin and especially to M. Vasiliev for many fruitful discussions. M.T. is grateful to the Department of Mathematics, the University of Auckland, New Zealand, where part of this work was done. Work of D.S. and M.T. was supported by the Australian Research Council grant DP160103633. Work of D.S. was also partially supported by the Russian Science Foundation grant 14-42-00047 in association with Lebedev Physical Institute.

Author Contributions: Both authors contributed to the article equally.

Conflicts of Interest: The authors declare no conflict of interest.

Appendix A. Conventions

The γ–matrices satisfy the following anti-commutation relations

$$(\gamma^m)^\alpha{}_\delta(\gamma^n)^\delta{}_\beta + (\gamma^n)^\alpha{}_\delta(\gamma^m)^\delta{}_\beta = 2\eta^{mn}\delta^\alpha_\beta, \tag{A1}$$

where m, n and other Latin letters are space–time vector indices, and α, β and other Greek letters label spinorial indices. Throughout the paper "$(,)$" denotes symmetrization and "$[,]$" denotes antisymmetrization with weight one. The symplectic matrix $C^{\alpha\beta} = -C^{\beta\alpha}$ is used to relate upper and lower spinorial indexes as follows

$$\mu^\alpha = C^{\alpha\beta}\mu_\beta, \quad \mu_\alpha = -C_{\alpha\beta}\mu^\beta, \quad C^{\alpha\gamma}C_{\gamma\beta} = -\delta^\alpha_\beta. \tag{A2}$$

The differentiation by hypercoordinates $X^{\alpha\beta}$ is as follows

$$\frac{dX^{\alpha\beta}}{dX^{\gamma\delta}} \equiv \partial_{\alpha\beta}X^{\gamma\delta} = \frac{1}{2}(\delta^\alpha_\gamma\delta^\beta_\delta + \delta^\beta_\gamma\delta^\alpha_\delta), \tag{A3}$$

$$\partial_{\mu\nu}X^{-1}_{\alpha\beta} = -\frac{1}{2}(X^{-1}_{\mu\alpha}X^{-1}_{\nu\beta} + X^{-1}_{\mu\beta}X^{-1}_{\nu\alpha}) \tag{A4}$$

and

$$\partial_{\mu\nu}(\det X) = X^{-1}_{\mu\nu}(\det X) \tag{A5}$$

where

$$X^{-1}_{\mu\nu}X^{\nu\alpha} = \delta^\alpha_\mu. \tag{A6}$$

Let us note that the product of an even number of $X^{\alpha\beta}$ matrices is antisymmetric in spinorial indexes, whereas the product of an odd number of $X^{\alpha\beta}$ is a symmetric matrix. For example,

$$X^{\alpha\gamma}X_\gamma{}^\beta = -X^{\beta\gamma}X_\gamma{}^\alpha, \quad X^\alpha{}_\gamma X^\gamma{}_\delta X^{\delta\beta} = +X^\beta{}_\delta X^\delta{}_\gamma X^{\gamma\alpha}, \quad \text{etc.} \tag{A7}$$

Appendix B. Derivation of the field Equations on $Sp(n)$

Let us evaluate the operator $Y_{(\alpha}Y_{\beta)}$ in (93):

$$\frac{1}{2}(Y_\alpha Y_\beta + Y_\beta Y_\alpha) \equiv Y_{(\alpha}Y_{\beta)} = (\tfrac{\xi}{8})^2 \mu_\alpha \mu_\beta + \tfrac{i\xi}{8}\left(\mu_\alpha \tfrac{\partial}{\partial\mu^\beta} + \mu_\beta \tfrac{\partial}{\partial\mu^\alpha}\right) - \tfrac{\partial}{\partial\mu^\alpha}\tfrac{\partial}{\partial\mu^\beta}\,. \tag{A8}$$

Appendix B.1. Fermionic Equation

Consider Equation (93). Substituting into it the expansion (94) one gets for the term linear in μ^α

$$\nabla_{\alpha\beta}F_\gamma(X)\,\mu^\gamma + \frac{\xi}{8}(C_{\gamma\alpha}F_\beta(X) + C_{\gamma\beta}F_\alpha(X))\,\mu^\gamma = 0 \tag{A9}$$

The second term comes from $-\frac{i}{2}(Y_\alpha Y_\beta + Y_\alpha Y_\beta)$ acting on $F_\gamma\mu^\gamma$. From this equation, one gets (96).

Appendix B.2. Bosonic Equation

Equation (93) to the zeroth order in μ^α becomes:

$$\nabla_{\alpha\beta}B(X) = iY_{(\alpha}Y_{\beta)}\cdot\tfrac{1}{2}B_{\gamma\delta}(X)\mu^\gamma\mu^\delta\,. \tag{A10}$$

Obviously, only the double μ-derivative in $Y_{(\alpha}Y_{\beta)}$ will contribute to this order. Thus, we have:

$$\nabla_{\alpha\beta}B(X) = -i\tfrac{\partial}{\partial\mu^\alpha}\tfrac{\partial}{\partial\mu^\beta}\cdot\tfrac{1}{2}B_{(\gamma\delta)}(X)\mu^\gamma\mu^\delta \tag{A11}$$

Therefore,

$$\nabla_{\alpha\beta}B(X) = -i\,B_{(\alpha\beta)}(X)\,, \tag{A12}$$

Which indicates that all the higher order components in the expansion (94) are expressed in terms of $B(X)$ and $F_\alpha(X)$.

To zeroth order in μ^α, we compute:

$$(\nabla_{\alpha\beta} - iY_{(\alpha}Y_{\beta)})(\nabla_{\gamma\delta} - iY_{(\gamma}Y_{\delta)})\left[B(X) + \tfrac{1}{2}B_{\rho\sigma}(X)\mu^\rho\mu^\sigma + \tfrac{1}{4!}B_{\rho\sigma\tau\lambda}(X)\mu^\rho\mu^\sigma\mu^\tau\mu^\lambda + \ldots\right] = 0\,. \tag{A13}$$

$$\begin{aligned}0 =& \nabla_{\alpha\beta}\nabla_{\gamma\delta}B(X) + (C_{\alpha\gamma}C_{\beta\delta} + C_{\beta\gamma}C_{\alpha\delta})B(X) + (\tfrac{\xi}{8})^2 B_{(\alpha\beta\gamma\delta)}(X)\\ &+ i(\tfrac{\xi}{8})\left[C_{\alpha\gamma}B_{(\beta\delta)}(X) + C_{\alpha\delta}B_{(\beta\gamma)}(X) + C_{\beta\gamma}B_{(\alpha\delta)}(X) + C_{\beta\delta}B_{(\alpha\gamma)}(X)\right]\\ &+ i\left[\nabla_{\gamma\delta}B_{(\alpha\beta)}(X) + \nabla_{\alpha\beta}B_{(\gamma\delta)}(X)\right].\end{aligned} \tag{A14}$$

Now, using (A12), this becomes:

$$\begin{aligned}0 =& \nabla_{\alpha\beta}\nabla_{\gamma\delta}B(X) + (\tfrac{\xi}{8})^2(C_{\alpha\gamma}C_{\beta\delta} + C_{\beta\gamma}C_{\alpha\delta})B(X) + B_{(\alpha\beta\gamma\delta)}(X)\\ &- \tfrac{\xi}{8}\left[C_{\alpha\gamma}\nabla_{\beta\delta} + C_{\alpha\delta}\nabla_{\beta\gamma} + C_{\beta\gamma}\nabla_{\alpha\delta} + C_{\beta\delta}\nabla_{\alpha\gamma}\right]B(X)\\ &-\left[\nabla_{\gamma\delta}\nabla_{\alpha\beta} + \nabla_{\alpha\beta}\nabla_{\gamma\delta}\right]B(X)\,.\end{aligned} \tag{A15}$$

Using the algebra (155) for the covariant derivatives $\nabla_{\alpha\beta}$, we can write:

$$\nabla_{\gamma\delta}\nabla_{\alpha\beta}B(X) = (\tfrac{\xi}{8})^2(C_{\alpha\gamma}C_{\beta\delta} + C_{\beta\gamma}C_{\alpha\delta})B(X) + B_{(\alpha\beta\gamma\delta)}(X) - \tfrac{1}{2}[\nabla_{\alpha\beta}, \nabla_{\gamma\delta}]B(X)\,. \tag{A16}$$

From this equation, we obtain the bosonic Equation (95). Let us note that exchange of indexes as $\alpha \leftrightarrow \gamma$ and $\beta \leftrightarrow \delta$:

$$\nabla_{\alpha\beta}\nabla_{\gamma\delta}B(X) =(\tfrac{\xi}{8})^2(C_{\alpha\gamma}C_{\beta\delta}+C_{\beta\gamma}C_{\alpha\delta})B(X)+B_{(\alpha\beta\gamma\delta)}(X)+\tfrac{1}{2}[\nabla_{\alpha\beta},\nabla_{\gamma\delta}]B(X)\,. \tag{A17}$$

and subtraction of (A16) and (A17) leads to an identity.

Appendix C. Some Identities for Supercoordinates on $OSp(1|n)$ Group Manifold

The supercoordinates on $OSp(1|n)$ group manifold obey some useful relations in particular

$$\theta^\alpha \mathcal{G}_\alpha{}^\beta = \Theta^\beta P(\Theta^2), \qquad \theta^\alpha = \Theta^\beta \mathcal{G}_\beta^{-1\alpha} P(\Theta^2)\,, \tag{A18}$$

$$\mathcal{Q}_\beta\Theta^\alpha = P^{-1}(\Theta^2)\left(G_\beta{}^\alpha+\frac{i\xi}{8}\Theta_\beta\Theta^\alpha+\frac{i\xi}{8}G_\beta{}^\sigma\Theta_\sigma\Theta^\alpha+\left(\frac{i\xi}{8}\right)^2\Theta^2\Theta_\beta\Theta^\alpha\right), \tag{A19}$$

$$(\mathcal{Q}_\beta\Theta^\alpha)\Theta_\alpha = P(\Theta^2)\left(G_\beta{}^\sigma+\frac{i\xi}{8}\Theta_\beta\Theta^\sigma\right)\Theta_\sigma, \tag{A20}$$

$$\partial_{\alpha\beta}\Theta^\gamma = \frac{\xi}{4}\Theta_{(\alpha}G_{\beta)}{}^\delta(\delta_\delta^\gamma+\frac{i\xi}{8}\Theta_\delta\Theta^\gamma)\,, \tag{A21}$$

$$D_\beta\mathcal{G}_\alpha{}^\gamma = \frac{i\xi}{4}P(\Theta^2)\,(\Theta_\alpha-2G_\alpha{}^\rho\Theta_\rho)\mathcal{G}_\beta{}^\gamma \tag{A22}$$

$$\partial_{\alpha\beta}\mathcal{G}_\gamma{}^\delta = \frac{\xi}{4}\mathcal{G}_{\gamma(\alpha}\,\mathcal{G}_{\beta)}{}^\delta\,, \tag{A23}$$

and

$$\mathcal{Q}_\alpha\mathcal{G}_{\mu\nu} = -\frac{i\xi}{4}P(\Theta^2)\Theta_\nu\mathcal{G}_{\mu\alpha}\,. \tag{A24}$$

References

1. Fronsdal, C. Massless particles, orthosymplectic symmetry and another type of Kaluza–Klein theory. In *Essays on Supersymmetry*; Frosndal, C., Ed.; Reidel: Dordrecht, The Netherlands, 1986; pp. 163–265, ISBN 978-9401085557.
2. Bandos, I.A.; Lukierski, J. Tensorial central charges and new superparticle models with fundamental spinor coordinates. *Mod. Phys. Lett. A* **1999** *14*, 1257–1272.
3. Bandos, I.A.; Lukierski, J.; Sorokin, D.P. Superparticle models with tensorial central charges. *Phys. Rev. D* **2000**, *61*, 045002.
4. Vasiliev, M.A. Conformal higher spin symmetries of 4-d massless supermultiplets and osp(L,2M) invariant equations in generalized (super)space. *Phys. Rev. D* **2002**, *66*, 066006.
5. Vasiliev, M.A. Relativity, causality, locality, quantization and duality in the S(p)(2M) invariant generalized space-time. In *Multiple Facets of Quantization and Supersymmetry*; World Scientific: Singapore, 2002; pp. 826–872.
6. Vasiliev, M.A. Higher spin conserved currents in Sp(2M) symmetric space-time. *Russ. Phys. J.* **2002**, *45*, 670–681.
7. Didenko, V.E.; Vasiliev, M.A. Free field dynamics in the generalized AdS (super)space. *J. Math. Phys.* **2004**, *45*, 197–215.
8. Plyushchay, M.; Sorokin, D.; Tsulaia, M. Higher spins from tensorial charges and OSp(N | 2n) symmetry. *J. High Energy Phys.* **2003**, *2003*, 013.
9. Gelfond, O.A.; Vasiliev, M.A. Higher rank conformal fields in the Sp(2M) symmetric generalized space-time. *Theor. Math. Phys.* **2005**, *145*, 1400–1424.
10. Plyushchay, M.; Sorokin, D.; Tsulaia, M. GL flatness of OSp(1 | 2n) and higher spin field theory from dynamics in tensorial spaces. *arXiv* **2003**, arXiv:hep-th/0310297.
11. Vasiliev, M.A.; Zaikin, V.N. On Sp(2M) invariant Green functions. *Phys. Lett. B* **2004**, *587*, 225–229.
12. Bandos, I.; Pasti, P.; Sorokin, D.; Tonin, M. Superfield theories in tensorial superspaces and the dynamics of higher spin fields. *J. High Energy Phys.* **2004**, *2004*, 023.

13. Bandos, I.; Bekaert, X.; de Azcarraga, J.A.; Sorokin, D.; Tsulaia, M. Dynamics of higher spin fields and tensorial space. *J. High Energy Phys.* **2005**, *2005*, 031.
14. Ivanov, E.; Lukierski, J. Higher spins from nonlinear realizations of OSp(1 | 8). *Phys. Lett. B* **2005**, *624*, 304–315.
15. Gelfond, O.A.; Skvortsov, E.D.; Vasiliev, M.A. Higher spin conformal currents in Minkowski space. *Theor. Math. Phys.* **2008**, *154*, 294–302.
16. Vasiliev, M.A. On Conformal, SL(4,R) and Sp(8,R) Symmetries of 4d Massless Fields. *Nucl. Phys. B* **2008**, *793*, 469–526.
17. Ivanov, E. Nonlinear Realizations in Tensorial Superspaces and Higher Spins. *arXiv* **2007**, arXiv:hep-th/0703056.
18. West, P.C. E(11) and higher spin theories. *Phys. Lett. B* **2007**, *650*, 197–202.
19. Gelfond, O.A.; Vasiliev, M.A. Higher Spin Fields in Siegel Space, Currents and Theta Functions. *J. High Energy Phys.* **2009**, *2009*, 125.
20. Gelfond, O.A.; Vasiliev, M.A. Sp(8) invariant higher spin theory, twistors and geometric BRST formulation of unfolded field equations. *J. High Energy Phys.* **2009**, *2009*, 021.
21. Gelfond, O.A.; Vasiliev, M.A. Unfolded Equations for Current Interactions of 4d Massless Fields as a Free System in Mixed Dimensions. *J. Exp. Theor. Phys.* **2015**, *120*, 484–508.
22. Bandos, I.A.; de Azcarraga, J.A.; Meliveo, C. Extended supersymmetry in massless conformal higher spin theory. *Nucl. Phys. B* **2011**, *853*, 760–776.
23. Florakis, I.; Sorokin, D.; Tsulaia, M. Higher Spins in Hyperspace. *J. High Energy Phys.* **2014**, *2014*, 105.
24. Florakis, I.; Sorokin, D.; Tsulaia, M. Higher Spins in Hyper-Superspace. *Nucl. Phys. B* **2014**, *890*, 279–301.
25. Fedoruk, S.; Lukierski, J. New spinorial particle model in tensorial space-time and interacting higher spin fields. *J. High Energy Phys.* **2013**, *2013*, 128.
26. Gelfond, O.A.; Vasiliev, M.A. Symmetries of higher-spin current interactions in four dimensions. *Theor. Math. Phys.* **2016**, *187*, 797–812.
27. Skvortsov, E.; Sorokin, D.; Tsulaia, M. Correlation Functions of Sp(2n) Invariant Higher-Spin Systems. *J. High Energy Phys.* **2016**, *2016*, 128.
28. Goncharov, Y.O.; Vasiliev, M.A. Higher-spin fields and charges in the periodic spinor space. *J. Phys. A* **2017**, *50*, 275401.
29. Vasiliev, M.A. Equations of Motion of Interacting Massless Fields of All Spins as a Free Differential Algebra. *Phys. Lett. B* **1988**, *209*, 491–497.
30. Vasiliev, M.A. Consistent Equations for Interacting Massless Fields of All Spins in the First Order in Curvatures. *Ann. Phys.* **1989**, *190*, 59–106.
31. Vasiliev, M.A. Progress in higher spin gauge theories. In Proceedings of the MGIX MM Meeting, Roma, Italy, 2–8 July 2000.
32. Vasiliev, M.A. Higher spin gauge theories in various dimensions. *Fortschr. Phys.* **2004**, *52*, 702–717.
33. Vasiliev, M.A. Holography, Unfolding and Higher-Spin Theory. *J. Phys. A* **2013**, *46*, 214013.
34. Giombi, S.; Yin, X. The Higher Spin/Vector Model Duality. *J. Phys. A* **2013**, *46*, 214003.
35. Gaberdiel, M.R.; Gopakumar, R. Minimal Model Holography. *J. Phys. A* **2013**, *46*, 214002.
36. Flato, M.; Fronsdal, C. One Massless Particle Equals Two Dirac Singletons: Elementary Particles in a Curved Space. 6. *Lett. Math. Phys.* **1978**, *2*, 421–426.
37. Vasiliev, M.A. Higher spin gauge theories in four-dimensions, three-dimensions, and two-dimensions. *Int. J. Mod. Phys. D* **1996**, *5*, 763–797.
38. Vasiliev, M.A. Higher spin gauge theories: Star product and AdS space. In *The Many Faces of the Superworld*; World Scientific: Singapore, 2000; pp. 533–610.
39. Bekaert, X.; Cnockaert, S.; Iazeolla, C.; Vasiliev, M.A. Nonlinear higher spin theories in various dimensions. *arXiv* **2005**, arXiv:hep-th/0503128.
40. Sezgin, E.; Sundell, P. Supersymmetric Higher Spin Theories. *J. Phys. A* **2013**, *46*, 214022.
41. Didenko, V.E.; Skvortsov, E.D. Elements of Vasiliev theory. *arXiv* **2014**, arXiv:hep-th1401.2975.
42. Arias, C.; Bonezzi, R.; Boulanger, N.; Sezgin, E.; Sundell, P.; Torres-Gomez, A.; Valenzuela, M. Action principles for higher and fractional spin gravities. In *Higher Spin Gauge Theories*; World Scientific: Singapore, 2017; pp. 213-253.
43. Bekaert, X.; Buchbinder, I.L.; Pashnev, A.; Tsulaia, M. On higher spin theory: Strings, BRST, dimensional reductions. *Class. Quantum Gravity* **2004**, *21*, S1457.

44. Bouatta, N.; Compere, G.; Sagnotti, A. An Introduction to free higher-spin fields. *arXiv* **2004**, arXiv:hep-th/0409068.
45. Francia, D.; Sagnotti, A. Higher-spin geometry and string theory. *J. Phys. Conf. Ser.* **2006**, *33*, 57.
46. Fotopoulos, A.; Tsulaia, M. Gauge Invariant Lagrangians for Free and Interacting Higher Spin Fields. A Review of the BRST formulation. *Int. J. Mod. Phys. A* **2009**, *24*, 1–60.
47. Campoleoni, A. Metric-like Lagrangian Formulations for Higher-Spin Fields of Mixed Symmetry. *Riv. Nuovo Cim.* **2010**, *33*, 123–253.
48. Francia, D. On the Relation between Local and Geometric Lagrangians for Higher spins. *J. Phys. Conf. Ser.* **2010**, *222*, 012002.
49. Bekaert, X.; Boulanger, N.; Sundell, P. How higher-spin gravity surpasses the spin two barrier: No-go theorems versus yes-go examples. *Rev. Mod. Phys.* **2012**, *84*, 987–1009.
50. Taronna, M. Higher Spins and String Interactions. *arXiv* **2010**, arXiv:hep-th1005.3061.
51. Sagnotti, A. Notes on Strings and Higher Spins. *J. Phys. A* **2013**, *46*, 214006.
52. Joung, E.; Lopez, L.; Taronna, M. Solving the Noether procedure for cubic interactions of higher spins in (A)dS. *J. Phys. A* **2013**, *46*, 214020.
53. Taronna, M. Higher-Spin Interactions: Three-point functions and beyond. *arXiv* **2012**, arXiv:hep-th1209.5755.
54. Lucena Gómez, G. Aspects of Higher-Spin Theory with Fermions. *arXiv* **2014**, arXiv:hep-th1406.5319.
55. Leonard, A. Aspects of higher spin Hamiltonian dynamics: Conformal geometry, duality and charges. *arXiv* **2017**, arXiv:math-ph1709.00719.
56. Rahman, R.; Taronna, M. From Higher Spins to Strings: A Primer. *arXiv* **2015**, arXiv:hep-th1512.07932.
57. Sleight, C. Interactions in Higher-Spin Gravity: A Holographic Perspective. *J. Phys. A* **2017**, *50*, 383001.
58. Sleight, C. Metric-like Methods in Higher Spin Holography. *arXiv* **2017**, arXiv:1701.08360.
59. Sorokin, D. Introduction to the classical theory of higher spins. *AIP Conf. Proc.* **2006**, *767*, 172–202.
60. Tsulaia, M. On Tensorial Spaces and BCFW Recursion Relations for Higher Spin Fields. *Int. J. Mod. Phys. A* **2012**, *27*, 1230011.
61. Bandos, I.A. BPS preons in supergravity and higher spin theories. An Overview from the hill of twistor appraoch. *AIP Conf. Proc.* **2005**, *767*, 141–171.
62. Vasiliev, M.A. Higher-Spin Theory and Space-Time Metamorphoses. *Lect. Notes Phys.* **2015**, *892*, 227–264.
63. Angelopoulos, E.; Laoues, M. Masslessness in n-dimensions. *Rev. Math. Phys.* **1998**, *10*, 271–300.
64. Bandos, I.A.; Lukierski, J.; Sorokin, D.P. The OSp(1 | 4) superparticle and exotic BPS states. *arXiv* **1999**, arXiv:hep-th/9912264.
65. Bandos, I.A.; de Azcarraga, J.A.; Izquierdo, J.M.; Lukierski, J. BPS states in M theory and twistorial constituents. *Phys. Rev. Lett.* **2001**, *86*, 4451–4454.
66. Bandos, I.A.; Lukierski, J.; Preitschopf, C.; Sorokin, D.P. OSp supergroup manifolds, superparticles and supertwistors. *Phys. Rev. D* **2000**, *61*, 065009.
67. Bandos, I.A.; de Azcarraga, J.A.; Izquierdo, J.M.; Picon, M.; Varela, O. On BPS preons, generalized holonomies and D = 11 supergravities. *Phys. Rev. D* **2004**, *69*, 105010.
68. Ferber, A. Supertwistors and Conformal Supersymmetry. *Nucl. Phys. B* **1978**, *132*, 55–64.
69. Shirafuji, T. Lagrangian Mechanics of Massless Particles With Spin. *Prog. Theor. Phys.* **1983**, *70*, 18–35.
70. Bengtsson, A.K.H.; Bengtsson, I.; Cederwall, M.; Linden, N. Particles, Superparticles and Twistors. *Phys. Rev. D* **1987**, *36*, 1766–1772.
71. Bengtsson, I.; Cederwall, M. Particles, Twistors and the Division Algebras. *Nucl. Phys. B* **1988**, *302*, 81–103.
72. Sorokin, D.P.; Tkach, V.I.; Volkov, D.V. Superparticles, Twistors and Siegel Symmetry. *Mod. Phys. Lett. A* **1989**, *4*, 901–908.
73. Volkov, D.V.; Zheltukhin, A.A. Extension of the Penrose Representation and Its Use to Describe Supersymmetric Models. *J. Exp. Theor. Phys. Lett.* **1988**, *48*, 63–66.
74. Plyushchay, M.S. Covariant Quantization of Massless Superparticle in Four-dimensional Space-time: Twistor Approach. *Mod. Phys. Lett. A* **1989**, *4*, 1827–1837.
75. Gumenchuk, A.I.; Sorokin, D.P. Relativistic superparticle dynamics and twistor correspondence. *Sov. J. Nucl. Phys.* **1990**, *51*, 350–355. (In Russian)
76. Sorokin, D.P. Double Supersymmetric Particle Theories. *Fortschr. Phys.* **1990**, *38*, 923–943.
77. Bandos, I.A. Superparticle in Lorentz harmonic superspace. *Sov. J. Nucl. Phys.* **1990**, *51*, 906–914. (In Russian)

78. Bandos, I.A. Multivalued action functionals, Lorentz harmonics, and spin. *J. Exp. Theor. Phys. Lett.* **1990**, *52*, 205–207.
79. Plyushchay, M.S. Lagrangian formulation for the massless (super)particles in (super)twistor approach. *Phys. Lett. B* **1990**, *240*, 133–136.
80. Fronsdal, C. Elementary Particles in a Curved Space. 4. Massless Particles. *Phys. Rev. D* **1975**, *12*, 3819–3830.
81. Metsaev, R.R. Lowest eigenvalues of the energy operator for totally (anti)symmetric massless fields of the n-dimensional anti-de Sitter group. *Class. Quantum Gravity* **1994**, *11*, L141–L145.
82. Metsaev, R.R. Massless mixed symmetry bosonic free fields in d-dimensional anti-de Sitter space-time. *Phys. Lett. B* **1995**, *354*, 78–84.
83. Metsaev, R.R. Arbitrary spin massless bosonic fields in d-dimensional anti-de Sitter space. *Lect. Notes Phys.* **1999**, *524*, 331–340.
84. Metsaev, R.R. Fermionic fields in the d-dimensional anti-de Sitter space-time. *Phys. Lett. B* **1995**, *419*, 49–56.
85. Fotopoulos, A.; Panigrahi, K.L.; Tsulaia, M. Lagrangian formulation of higher spin theories on AdS space. *Phys. Rev. D* **2006**, *74*, 085029.
86. Bekaert, X.; Grigoriev, M. Notes on the ambient approach to boundary values of AdS gauge fields. *J. Phys. A* **2013**, *46*, 214008.
87. Bekaert, X.; Grigoriev, M. Higher order singletons, partially massless fields and their boundary values in the ambient approach. *Nucl. Phys. B* **2013**, *876*, 667–714.
88. Bolotin, K.I.; Vasiliev, M.A. Star product and massless free field dynamics in AdS(4). *Phys. Lett. B* **2000**, *479*, 421–428.
89. Gates, S.J., Jr.; Koutrolikos, K. On 4D, N = 1 massless gauge superfields of arbitrary superhelicity. *J. High Energy Phys.* **2014**, *2014*, 098.
90. Gates, S.J., Jr.; Koutrolikos, K. On 4D, N = 1 Massless Gauge Superfields of Higher Superspin: Half-Odd-Integer Case. *arXiv* **2013**, arXiv:1310.7386.
91. Candu, C.; Peng, C.; Vollenweider, C. Extended supersymmetry in AdS_3 higher spin theories. *J. High Energy Phys.* **2014**, *2014*, 113.
92. Kuzenko, S.M.; Ogburn, D.X. Off-shell higher spin N = 2 supermultiplets in three dimensions. *Phys. Rev. D* **2016**, *94*, 106010.
93. Buchbinder, I.L.; Snegirev, T.V.; Zinoviev, Y.M. Unfolded equations for massive higher spin supermultiplets in AdS_3. *J. High Energy Phys.* **2016**, *2016*, 075.
94. Kuzenko, S.M.; Tsulaia, M. Off-shell massive N = 1 supermultiplets in three dimensions. *Nucl. Phys. B* **2017**, *914*, 160–200.
95. Buchbinder, I.L.; Snegirev, T.V.; Zinoviev, Y.M. Lagrangian description of massive higher spin supermultiplets in AdS_3 space. *J. High Energy Phys.* **2017**, *2017*, 021.
96. Buchbinder, I.L.; Gates, S.J.; Koutrolikos, K. Higher Spin Superfield interactions with the Chiral Supermultiplet: Conserved Supercurrents and Cubic Vertices. *arXiv* **2017**, arXiv:1708.06262.
97. Kuzenko, S.M.; Sibiryakov, A.G. Massless gauge superfields of higher integer superspins. *J. Exp. Theor. Phys. Lett.* **1993**, *57*, 539–542.
98. Kuzenko, S.M.; Sibiryakov, A.G. Free massless higher superspin superfields on the anti-de Sitter superspace. *Phys. Atom. Nucl.* **1994**, *57*, 1257–1267.
99. Buchbinder, I.L.; Kuzenko, S.M.; Sibiryakov, A.G. Quantization of higher spin superfields in the anti-De Sitter superspace. *Phys. Lett. B* **1995**, *352*, 29–36.
100. Gates, S.J., Jr.; Kuzenko, S.M.; Sibiryakov, A.G. N = 2 supersymmetry of higher superspin massless theories. *Phys. Lett. B* **1997**, *412*, 59–68.
101. Gates, S.J., Jr.; Kuzenko, S.M.; Sibiryakov, A.G. Towards a unified theory of massless superfields of all superspins. *Phys. Lett. B* **1997**, *394*, 343–353.
102. Sezgin, E.; Sundell, P. Higher spin N = 8 supergravity. *J. High Energy Phys.* **1998**, *1998*, 016.
103. Alkalaev, K.B.; Vasiliev, M.A. N = 1 supersymmetric theory of higher spin gauge fields in AdS(5) at the cubic level. *Nucl. Phys. B* **2003**, *655*, 57–92.
104. Engquist, J.; Sezgin, E.; Sundell, P. Superspace formulation of 4-D higher spin gauge theory. *Nucl. Phys. B* **2003**, *664*, 439–456.
105. Zinoviev, Y.M. Massive N = 1 supermultiplets with arbitrary superspins. *Nucl. Phys. B* **2007**, *785*, 98–114.
106. Curtright, T. Massless Field Supermultiplets With Arbitrary Spin. *Phys. Lett. B* **1979**, *85*, 219–224.

107. Vasiliev, M.A. 'Gauge' Form Of Description Of Massless Fields With Arbitrary Spin. *Yad. Fiz.* **1980**, *32*, 855–861. (In Russian)
108. Bellon, M.P.; Ouvry, S. $D = 4$ Supersymmetry for Gauge Fields of Any Spin. *Phys. Lett. B* **1987**, *187*, 93–96.
109. Fradkin, E.S.; Vasiliev, M.A. Superalgebra of Higher Spins and Auxiliary Fields. *Int. J. Mod. Phys. A* **1988**, *3*, 2983–3010.
110. Bergshoeff, E.; Salam, A.; Sezgin, E.; Tanii, Y. Singletons, Higher Spin Massless States and the Supermembrane. *Phys. Lett. B* **1988**, *205*, 237–244.
111. Konstein, S.E.; Vasiliev, M.A. Extended Higher Spin Superalgebras and Their Massless Representations. *Nucl. Phys. B* **1990**, *331*, 475499.
112. Kuzenko, S.M.; Sibiryakov, A.G.; Postnikov, V.V. Massless gauge superfields of higher half integer superspins. *J. Exp. Theor. Phys. Lett.* **1993**, *57*, 534–538.
113. Osborn, H.; Petkou, A.C. Implications of conformal invariance in field theories for general dimensions. *Ann. Phys.* **1994**, *231*, 311–362.
114. Didenko, V.E.; Skvortsov, E.D. Exact higher-spin symmetry in CFT: All correlators in unbroken Vasiliev theory. *J. High Energy Phys.* **2013**, *2013*, 158.
115. Park, J.H. Superconformal symmetry in three-dimensions. *J. Math. Phys.* **2000**, *41*, 7129–7161.
116. Synatschke, F.; Braun, J.; Wipf, A. N = 1 Wess Zumino Model in d = 3 at zero and finite temperature. *Phys. Rev. D* **2010**, *81*, 125001.
117. Ferrara, S.; Iliopoulos, J.; Zumino, B. Supergauge Invariance and the Gell-Mann—Low Eigenvalue. *Nucl. Phys. B* **1974**, *77*, 413–419.
118. Conlong, B.P.; West, P.C. Anomalous dimensions of fields in a supersymmetric quantum field theory at a renormalization group fixed point. *J. Phys. A* **1993**, *26*, 3325–3332.
119. Maldacena, J.; Zhiboedov, A. Constraining Conformal Field Theories with A Higher Spin Symmetry. *J. Phys. A* **2013**, *46*.
120. Giombi, S.; Yin, X. Higher Spins in AdS and Twistorial Holography. *J. High Energy Phys.* **2011**, *2011*, 86.
121. Colombo, N.; Sundell, P. Higher Spin Gravity Amplitudes From Zero-form Charges. *arXiv* **2012**, arXiv:1208.3880.
122. Gelfond, O.A.; Vasiliev, M.A. Operator algebra of free conformal currents via twistors. *Nucl. Phys. B* **2013**, *876*, 871–917.
123. Didenko, V.E.; Mei, J.; Skvortsov, E.D. Exact higher-spin symmetry in CFT: Free fermion correlators from Vasiliev Theory. *Phys. Rev. D* **2013**, *88*, 046011.
124. Sleight, C.; Taronna, M. Higher Spin Interactions from Conformal Field Theory: The Complete Cubic Couplings. *Phys. Rev. Lett.* **2016**, *116*, 181602.
125. Bonezzi, R.; Boulanger, N.; De Filippi, D.; Sundell, P. Noncommutative Wilson lines in higher-spin theory and correlation functions of conserved currents for free conformal fields. *arXiv* **2017**, arXiv:1705.03928.
126. Tseytlin, A.A. On partition function and Weyl anomaly of conformal higher spin fields. *Nucl. Phys. B* **2013**, *877*, 598–631.
127. Metsaev, R.R. Arbitrary spin conformal fields in (A)dS. *Nucl. Phys. B* **2014**, *885*, 734–771.
128. Metsaev, R.R. Mixed-symmetry fields in AdS(5), conformal fields, and AdS/CFT. *J. High Energy Phys.* **2015**, *2015*, 77.
129. Nutma, T.; Taronna, M. On conformal higher spin wave operators. *J. High Energy Phys.* **2014**, *2014*, 066.
130. Beccaria, M.; Tseytlin, A.A. On higher spin partition functions. *J. Phys. A* **2015**, *48*, 275401.
131. Sharapov, A.A.; Skvortsov, E.D. Formal higher-spin theories and Kontsevich–Shoikhet–Tsygan formality. *Nucl. Phys. B* **2017**, *921*, 538–584.
132. Baguet, A.; Hohm, O.; Samtleben, H. $E_{6(6)}$ Exceptional Field Theory: Review and Embedding of Type IIB. *arXiv* **2015**, arXiv:1506.01065.
133. Cederwall, M. Twistors and supertwistors for exceptional field theory. *J. High Energy Phys.* **2015**, *2015*, 123.
134. Bandos, I. Exceptional field theories, superparticles in an enlarged 11D superspace and higher spin theories. *Nucl. Phys. B* **2017**, *925*, 28–62.

universe

MDPI

Review

Frame- and Metric-Like Higher-Spin Fermions

Rakibur Rahman [1,2]

[1] Department of Physics, University of Dhaka, Dhaka 1000, Bangladesh; rahmanrakib@gmail.com
[2] Max-Planck-Institut für Gravitationsphysik (Albert-Einstein-Institut), Am Mühlenberg 1, D-14476 Potsdam-Golm, Germany

Received: 26 December 2017; Accepted: 5 February 2018; Published: 11 February 2018

Abstract: Conventional descriptions of higher-spin fermionic gauge fields appear in two varieties: the Aragone–Deser–Vasiliev frame-like formulation and the Fang–Fronsdal metric-like formulation. We review, clarify and elaborate on some essential features of these two. For frame-like free fermions in Anti-de Sitter space, one can present a gauge-invariant Lagrangian description such that the constraints on the field and the gauge parameters mimic their flat-space counterparts. This simplifies the explicit demonstration of the equivalence of the two formulations at the free level. We comment on the subtleties that may arise in an interacting theory.

Keywords: higher-spin gauge theory; fermions; frame-like formulation; metric-like formulation

1. Introduction

Arbitrary-spin massless particles are expected to play a crucial role in the understanding of quantum gravity. Lower-spin theories may be realized as low-energy limits of spontaneously-broken higher-spin gauge theories since lower-spin symmetries are subgroups of higher-spin ones. It is believed that the tensionless limit of string theory is a theory of higher-spin gauge fields. The study of fermionic fields is interesting in this regard because they are required by supersymmetry.

Higher-spin gauge fields can be described in the framework of two different formulations: frame-like and metric-like. The frame-like formulation generalizes the Cartan formulation of gravity where the gauge fields are described in terms of differential forms carrying irreducible representations of the fiber Lorentz group. This is available in Minkowski [1–3] as well as in Anti-de Sitter (AdS) [4–7] spaces. The metric-like formulation, on the other hand, is a generalization of the metric formulation of linearized gravity [8]. Originally developed by Fronsdal [9,10] and Fang–Fronsdal [11,12], it encodes the degrees of freedom of higher-spin particles in symmetric tensors and tensor-spinors. In this approach, the construction of a gauge-invariant action for a higher-spin field requires that the field and the gauge parameter obey some off-shell algebraic constraints (see [13,14] for a recent review). Note that the latter requirement can be avoided by recourse to other formulations [15–24] (see Appendix A).

Both of these approaches are geometric, albeit in different manners, in that the frame-like formulation extends Cartan geometry, whereas the metric-like formulation extends Riemannian geometry. The latter is however a particular gauge of the former just like in the case of gravity. The construction of interacting theories for higher-spin fields, fermions in particular, appears to be in dire need of the frame-like formulation. The metric-like formulation, in contrast, seems rather clumsy in managing the non-linearities required by gauge-theoretic consistency. Yet, it has the advantage of having a simplified field content that may make some features of the interactions more transparent. Understanding the connections between the two formulations may therefore provide valuable information [25–28].

In this article, we will focus exclusively on higher-spin gauge fermions. These fields appear naturally in the supersymmetric versions of Vasiliev theory [29–35] (see [36] for a recent review) and also in the tensionless limit of superstring theory compactified on $AdS_5 \times S^5$. The frame-like formulation of gauge fermions [1–3,6] has been discussed more recently by various authors [37–42]. The Fang–Fronsdal metric-like approach for higher-spin fermions, on the other hand, has been studied in arbitrary dimensions in [43–45]. We will consider the free theory of a spin $s = n + \frac{1}{2}$ massless fermionic field in flat and AdS spaces. Although we consider Majorana fermions for simplicity, our main results are valid almost verbatim for Dirac fermions in arbitrary spacetime dimensions. A crucial property of frame-like fermions in flat space is their shift symmetry w.r.t. a gauge parameter, which is an irreducible tensor-spinor in the fiber space with the symmetry property of the Young diagram $\mathbb{Y}(n-1,1)$. This symmetry makes it almost manifest that the free Lagrangian is equivalent to that of the metric-like formulation [1]. In AdS space, however, the constraints on this parameter may receive nontrivial corrections, which vanish in the flat limit [39,40]. This is tantamount to having no such corrections provided that some appropriate mass-like terms appear in the gauge transformation. In other words, one can have a gauge-invariant Lagrangian description for frame-like fermions in AdS space that does not deform the flat-space constraints on the field and the gauge parameters.

The organization of this article is as follows. In the remainder of this section, we spell out our notations and conventions. A review of frame-like higher-spin massless fermions in flat space appears in Section 2, where we write down the free Lagrangian [40,42] and discuss its gauge symmetries along with the constraints on the field and the gauge parameters. We also show how this theory simplifies in $D = 3, 4$. Section 3 formulates the free theory in AdS space with a trivial, but convenient modification of the well-known mass-like term [39,40]. By virtue of judiciously-chosen terms in the gauge transformation, we ensure that the constraints on the field and the gauge parameters mimic their flat-space counterparts. The value of the mass parameter, determined uniquely by gauge invariance, is in complete agreement with the known results [45,46]. In Section 4, we demonstrate explicitly the equivalence of the frame-like Lagrangian to the metric-like one at the free level. We conclude in Section 5 with some remarks, especially on the subtleties that may arise in an interacting theory. An appendix summarizes the essentials of the metric-like formulation of higher-spin gauge fermions.

Conventions and Notations

We adopt the conventions of [47], with mostly positive metric signature $(- + \cdots +)$. The expression $(i_1 \cdots i_n)$ denotes a totally symmetric one in all the indices $i_1, \cdots, i_n$ with no normalization factor, e.g., $(i_1 i_2) = i_1 i_2 + i_2 i_1$, etc. The totally antisymmetric expression $[i_1 \cdots i_n]$ has the same normalization. The number of terms appearing in the (anti-)symmetrization is assumed to be the possible minimum. A prime will denote a trace w.r.t. the background metric, e.g., $A' = \bar{g}^{\mu\nu} A_{\mu\nu} = A^{\mu}{}_{\mu}$. The Levi–Civita symbol is normalized as $\varepsilon_{01\ldots D-1} = +1$, where D is the spacetime dimension.

Fiber indices and world indices will respectively be denoted with lower case Roman letters and Greek letters. Repeated indices with the same name (appearing all as either covariant or contravariant ones) are (anti-)symmetrized with the minimum number of terms. This results in the following rules: $a(k)a = aa(k) = (k+1)a(k+1)$, $a(k)a(2) = a(2)a(k) = \binom{k+2}{2} a(k+2)$, $a(k)a(k') = a(k')a(k) = \binom{k+k'}{k} a(k+k')$, etc., where $a(k)$ has a unit weight by convention, and so, the proportionality coefficient gives the weight of the right-hand side.

The γ-matrices satisfy the Clifford algebra: $\{\gamma^a, \gamma^b\} = +2\eta^{ab}$ and $\gamma^{a\dagger} = \eta^{aa}\gamma^a$. Totally antisymmetric products of γ-matrices, $\gamma^{a_1 \ldots a_r} = \frac{1}{r!}\gamma^{[a_1}\gamma^{a_2} \cdots \gamma^{a_r]}$, have unit weight. A "slash" will denote a contraction with the γ-matrix, e.g., $\not{A} = \gamma^a A_a$.

A Majorana spinor χ obeys the reality condition: $\chi^C = \chi$. Two Majorana spinors $\chi_{1,2}$ follow the bilinear identity: $\bar{\chi}_1 \gamma^{a_1 \ldots a_r} \chi_2 = t_r \bar{\chi}_2 \gamma^{a_1 \ldots a_r} \chi_1$, where a "bar" denotes Majorana conjugation, and $t_r = \pm 1$, depending on the value of r and spacetime dimensionality [47].

2. Frame-Like Fermions in Flat Space

In the frame-like formulation, a fermion of spin $s = n + \frac{1}{2}$ is described by a vielbein-like one-form $\Psi^{a(n-1)}$, which is a symmetric rank-$(n-1)$ irreducible tensor-spinor in the fiber space:

$$\Psi^{a(n-1)} = \Psi_{\mu}{}^{a(n-1)} dx^{\mu}, \qquad \gamma_a \Psi^{ab(n-2)} = 0. \tag{1}$$

The Minkowski background is described by the vielbein $\bar{e}^a = \bar{e}^a_\mu dx^\mu$ that satisfies $\eta_{ab}\bar{e}^a_\mu \bar{e}^b_\nu = \eta_{\mu\nu}$, and the spin-connection $\bar{\omega}^{ab} = \bar{\omega}_\mu{}^{ab} dx^\mu = -\bar{\omega}_\mu{}^{ba} dx^\mu$, which fulfill the following equations:

$$T^a \equiv d\bar{e}^a + \bar{\omega}^a{}_b \bar{e}^b = 0, \qquad \rho^{ab} \equiv d\bar{\omega}^{ab} + \bar{\omega}^a{}_c \bar{\omega}^{cb} = 0. \tag{2}$$

In the Cartesian coordinates, in particular, the solution of Equation (2) is given by $\bar{e}^a_\mu = \delta^a_\mu$ and $\bar{\omega}_\mu{}^{ab} = 0$. We will however work with a generic coordinate system in order to facilitate the transition to AdS space. The following quantities will be useful in the subsequent discussion:

$$^*\bar{e}_{a_1} \dots \bar{e}_{a_p} \equiv \frac{1}{(D-p)!} \epsilon_{a_1 \dots a_p a_{p+1} \dots a_D} \bar{e}^{a_{p+1}} \dots \bar{e}^{a_D}, \tag{3}$$

$$\eta^{a_1 a_2 | b_1 b_2} \equiv \frac{1}{2} \left(\eta^{a_1 b_1} \eta^{a_2 b_2} - \eta^{a_1 b_2} \eta^{a_2 b_1} \right). \tag{4}$$

The frame-like free action for a Majorana gauge fermion, in arbitrary dimensions (Majorana fermions exist in $D = 3, 4, 8, 9, 10$ and 11. In dealing with such objects, it is important to assume the anti-commuting nature of fermions already at the classical level (before quantization)), reads [40,42]:

$$S = -\frac{1}{2} \int \left[\bar{\Psi}_{b_1 c(n-2)} \mathcal{A}^{a_1 a_2 a_3, b_1 b_2} \hat{D} \Psi_{b_2}{}^{c(n-2)} \right] {}^*\bar{e}_{a_1} \bar{e}_{a_2} \bar{e}_{a_3}, \tag{5}$$

where $\hat{D}$ denotes the Lorentz covariant derivative, and:

$$\mathcal{A}^{a_1 a_2 a_3, b_1 b_2} \equiv \frac{1}{6n} \left(\gamma^{a_1 a_2 a_3} \eta^{b_1 b_2} + 2(n-1) \eta^{b_1 b_2 | [a_1 a_2} \gamma^{a_3]} \right). \tag{6}$$

The action (5) enjoys the following gauge invariance:

$$\delta \Psi^{a(n-1)} = \hat{D} \zeta^{a(n-1)} + \bar{e}_b \lambda^{b, a(n-1)}, \tag{7}$$

where the zero-form gauge parameters $\zeta^{a(n-1)}$ and $\lambda^{b,a(n-1)}$ are irreducible tensor-spinors of rank $(n-1)$ and rank n respectively with the symmetry of the Young diagrams $\mathbb{Y}(n-1)$ and $\mathbb{Y}(n-1,1)$, i.e.,

$\zeta^{a(n-1)} \sim$ 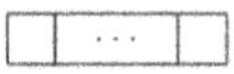 $n-1$, $\qquad \lambda^{b,a(n-1)} \sim$ 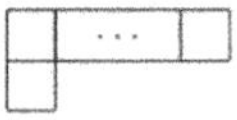 $n-1$. (8)

These irreducible tensor-spinors are subject to the following constraints:

$$\gamma_b \zeta^{ba(n-2)} = 0, \qquad \gamma_b \lambda^{b,a(n-1)} = 0, \qquad \gamma_c \lambda^{b,ca(n-2)} = 0, \qquad \lambda^{a,a(n-1)} = 0. \tag{9}$$

It is obvious that the action (5) is invariant, up to a total derivative term, under the gauge transformation of the parameter $\zeta^{a(n-1)}$, since $\hat{D}^2 = 0$ in flat space. To prove the shift symmetry w.r.t. the parameter $\lambda^{b,a(n-1)}$, let us make use of the identity: $\bar{e}^c\, {}^*\bar{e}_{a_1} \bar{e}_{a_2} \bar{e}_{a_3} = {}^*\bar{e}_{[a_1} \bar{e}_{a_2} \delta^c_{a_3]}$, so that the variation of the action can be written as:

$$\delta_\lambda S = -3 \int \left[\bar{\Psi}_{b_1}{}^{c(n-2)} \mathcal{A}^{a_1 a_2 a_3, b_1 b_2} \hat{D} \lambda_{a_3, b_2 c(n-2)} \right] {}^*\bar{e}_{a_1} \bar{e}_{a_2}. \tag{10}$$

Now, let us take a careful look at the identity:

$$6n\mathcal{A}^{a_1a_2a_3,\,b_1b_2} = \left(\gamma^{a_1a_2}\eta^{b_1b_2} + 2(n-1)\eta^{a_1a_2|b_1b_2}\right)\gamma^{a_3} + (n-1)\gamma^{[a_1}\eta^{a_2]b_1}\eta^{a_3b_2} \\ - \gamma^{[a_1}\eta^{a_2]a_3}\eta^{b_1b_2} - (n-1)\gamma^{[a_1}\eta^{a_2]b_2}\eta^{a_3b_1}. \quad (11)$$

When plugged into the gauge variation (10), the first line on the right-hand side of this identity gives a vanishing contribution on account of the γ-trace constraints (9) on the gauge parameter $\lambda^{b,a(n-1)}$. The two terms in the second line, on the other hand, cancel each other, thanks to the property $\lambda^{a,a(n-1)} = 0$. This proves the shift symmetry since $\delta_\lambda S = 0$.

Let us count the number of independent components of the parameters $\zeta^{a(n-1)}$ and $\lambda^{b,a(n-1)}$. Because the frame indices are γ-traceless, the number of possible values each index can take is essentially $(D-1)$. Then, it is easy to compute the number of components of the corresponding Young diagrams (8); they respectively turn out to be $\binom{D+n-3}{n-1}f_D$ and $(n-1)\binom{D+n-3}{n}f_D$, where:

$$f_D \equiv 2^{D/2+((-)^D-5)/4}, \quad (12)$$

for a Majorana fermion in D dimensions. On the other hand, one needs to take into account the vanishing of the trace when one contracts two indices from different rows of $\lambda^{b,a(n-1)}$, which removes $\binom{D+n-4}{n-2}f_D$ components. Therefore, the total numbers are given by:

$$\Delta_\zeta = \binom{D+n-3}{n-1}f_D, \qquad \Delta_\lambda = (n-1)\binom{D+n-3}{n}f_D - \binom{D+n-4}{n-2}f_D. \quad (13)$$

This counting will be useful later on.

Special Case: $D = 3$

The case of $D = 3$ is important in the context of hypergravity theories [3] (see also [48] for a recent discussion). In this case, note that the quantity ${}^*\bar{e}_{a_1}\bar{e}_{a_2}\bar{e}_{a_3}$ reduces to the Levi–Civita tensor $\epsilon_{a_1a_2a_3}$. Furthermore, one has at one's disposal the useful D-dimensional identity:

$$\mathcal{A}^{a_1a_2a_3,\,b_1b_2} = \tfrac{1}{6}\gamma^{a_1a_2a_3}\eta^{b_1b_2} + \left(\tfrac{n-1}{6n}\right)\gamma^{a_1a_2a_3b_1b_2} - \left(\tfrac{n-1}{12n}\right)\left(\gamma^{b_1}\gamma^{b_2}\gamma^{a_1a_2a_3} + \gamma^{a_1a_2a_3}\gamma^{b_1}\gamma^{b_2}\right). \quad (14)$$

The second term on the right-hand side in the above identity is zero in $D = 3$, whereas the last term gives a vanishing contribution because of the γ-trace condition on the field. On account of the relation: $\gamma^{a_1a_2a_3}\epsilon_{a_1a_2a_3} = (3!)\mathbb{I}$, therefore, the action (5) reduces to the well-known Aragone–Deser form [3]:

$$S_{D=3} = -\tfrac{1}{2}\int \bar{\Psi}_{a(n-1)}\hat{D}\Psi^{a(n-1)}. \quad (15)$$

On the other hand, the gauge symmetry (7)–(9) reduces to:

$$\delta\Psi^{a(n-1)} = \hat{D}\zeta^{a(n-1)}, \qquad \gamma_b\zeta^{ba(n-2)} = 0. \quad (16)$$

This is because in $D = 3$, the shift parameter $\lambda^{b,a(n-1)}$ is trivial, but $\zeta^{a(n-1)}$ is not,

$$\Delta_\lambda = 0, \qquad \Delta_\zeta = n, \quad (17)$$

as one can easily see from Equation (13).

Special Case: $D = 4$

In this case, the quantity ${}^*\bar{e}_{a_1}\bar{e}_{a_2}\bar{e}_{a_3}$ reduces to the one-form $\epsilon_{a_1a_2a_3b}\bar{e}^b$, while only the first piece on the right-hand side of the identity (14) contributes. Then, the dimension-dependent identity: $\gamma^{a_1a_2a_3} = -i\epsilon^{a_1a_2a_3b}\gamma_5\gamma_b$, reduces the action (5) to:

$$S_{D=4} = -\tfrac{i}{2}\int \bar{\Psi}_{a(n-1)}\gamma_5\gamma_b\bar{e}^b\hat{D}\Psi^{a(n-1)}. \tag{18}$$

Because $\Delta_\zeta = n(n+1) \neq 0$, $\Delta_\lambda = (n-1)(n+2) \neq 0$, both the parameters $\zeta^{a(n-1)}$ and $\lambda^{b,a(n-1)}$ are nontrivial, and so, the gauge symmetry has the full general form of (7). The Lagrangian (18) appeared in both [1,2], but only the former reference could correctly identify the gauge symmetries.

3. Frame-Like Fermions in AdS Space

The AdS background is described by the vielbein $\bar{e}^a = \bar{e}^a_\mu dx^\mu$ that satisfies $\eta_{ab}\bar{e}^a_\mu\bar{e}^b_\nu = \bar{g}_{\mu\nu}$, and the spin-connection $\bar{\omega}^{ab} = \bar{\omega}_\mu{}^{ab}dx^\mu = -\bar{\omega}_\mu{}^{ba}dx^\mu$, which fulfill the following equations:

$$T^a \equiv d\bar{e}^a + \bar{\omega}^a{}_b\bar{e}^b = 0, \qquad \rho^{ab} \equiv d\bar{\omega}^{ab} + \bar{\omega}^a{}_c\bar{\omega}^{cb} = -\frac{1}{l^2}\bar{e}^a\bar{e}^b, \tag{19}$$

where l is the AdS radius. Let us write the free action for a Majorana gauge fermion in AdS space by augmenting the kinetic term, already studied in the context of flat space, by a mass term:

$$S = -\tfrac{1}{2}\int \left[\bar{\Psi}_{b_1c(n-2)}\mathcal{A}^{a_1a_2a_3,b_1b_2}\hat{D}\Psi_{b_2}{}^{c(n-2)}\right]{}^*\bar{e}_{a_1}\bar{e}_{a_2}\bar{e}_{a_3} - \tfrac{1}{2}\mu\int \left[\bar{\Psi}_{b_1c(n-2)}\mathcal{B}^{a_1a_2,b_1b_2}\Psi_{b_2}{}^{c(n-2)}\right]{}^*\bar{e}_{a_1}\bar{e}_{a_2}, \tag{20}$$

where μ is some parameter with the dimensions of mass, to be specified later, and:

$$\mathcal{B}^{a_1a_2,b_1b_2} \equiv \tfrac{1}{2n}\left[\gamma^{a_1a_2}\eta^{b_1b_2} + 2(n-1)\eta^{a_1a_2|b_1b_2} - \tfrac{1}{2}\left(\tfrac{n-1}{D+2n-4}\right)\left(\gamma^{b_1}\gamma^{b_2}\gamma^{a_1a_2} + \gamma^{a_1a_2}\gamma^{b_1}\gamma^{b_2}\right)\right]. \tag{21}$$

Note that our choice of $\mathcal{B}^{a_1a_2,b_1b_2}$ differs from that of [39,40] by a trivial term, which vanishes upon implementing the constraint on the field. Yet, this term will be useful for our purpose.

It suffices to consider, invoking another mass parameter $\tilde{\mu}$, the gauge transformation:

$$\delta\Psi^{a(n-1)} = \hat{D}\zeta^{a(n-1)} + \tilde{\mu}\bar{e}_b\left[\gamma^b\zeta^{a(n-1)} - \left(\tfrac{2}{D+2n-4}\right)\gamma^a\zeta^{a(n-2)b}\right] + \bar{e}_b\lambda^{b,a(n-1)}, \tag{22}$$

which is compatible with the γ-trace constraint, $\gamma_a\Psi^{ab(n-2)} = 0$, on the field without requiring any modification of the properties (8) and (9) of the gauge parameters. In other words, the choice of this gauge transformation (22) is such that the field and the gauge parameters mimic their flat-space properties. This point is implicit in the choice made in [39,40].

To see that the shift transformation w.r.t. the parameter $\lambda^{b,a(n-1)}$ is a symmetry of the Lagrangian (20), let us first note that the invariance of the kinetic term follows exactly the flat-space logic. Then, from the variation of the mass term, we have:

$$\delta_\lambda S = -2\mu\int \left[\bar{\Psi}_{b_1c(n-2)}\mathcal{B}^{a_1a_2,b_1b_2}\lambda_{a_2,b_2}{}^{c(n-2)}\right]{}^*\bar{e}_{a_1}. \tag{23}$$

On account of the identity:

$$\begin{aligned} 2n\mathcal{B}^{a_1a_2,b_1b_2} &= \eta^{b_1b_2}\gamma^{a_1}\gamma^{a_2} + (n-1)\eta^{a_1b_1}\eta^{a_2b_2} - \tfrac{1}{2}\left(\tfrac{n-1}{D+2n-4}\right)\left(\gamma^{a_1a_2b_1}\gamma^{b_2} + \eta^{b_2[a_1}\gamma^{a_2]}\right) \\ &\quad - \eta^{a_1a_2}\eta^{b_1b_2} - (n-1)\eta^{a_1b_2}\eta^{a_2b_1}, \end{aligned} \tag{24}$$

we then see that $\delta_\lambda S = 0$. The cancellations happen in much the same way as the identity (11) eliminates contributions from the kinetic term.

The symmetry requirement of the Lagrangian (20) w.r.t. the ζ-transformation in (22) would relate the mass parameters μ and $\tilde{\mu}$ to each other and with the inverse AdS radius. There are a priori three kinds of contributions resulting from the ζ-transformation: 2-derivative, 1-derivative and 0-derivative ones. Not surprisingly, by virtue of the commutator formula:

$$\hat{D}^2\zeta_{a(n-1)} = -\frac{1}{l^2}\bar{e}^b\bar{e}^c\left[\eta_{ab}\zeta_{ca(n-2)} + \tfrac{1}{4}\gamma_{bc}\zeta_{a(n-1)}\right], \tag{25}$$

the two-derivative piece actually reduces to a zero-derivative piece. The explicit computation makes use of the identities: $\bar{e}^b\bar{e}^{c*}\bar{e}_{a_1}\bar{e}_{a_2}\bar{e}_{a_3} = {}^*\bar{e}_{[a_1}\delta^b_{a_2}\delta^c_{a_3]}$ and $\bar{e}^{b*}\bar{e}_{a_1}\bar{e}_{a_2} = {}^*\bar{e}_{[a_1}\delta^b_{a_2]}$, and leads straightforwardly to:

$$-\frac{(D+2n-3)(D+2n-4)}{4n}\frac{1}{l^2} - \frac{(D-2)(D+2n-3)}{n(D+2n-4)}\mu\tilde{\mu} = 0, \tag{26}$$

in order that the even-derivative terms cancel each other. Cancellation of the one-derivative terms, on the other hand, requires that the following condition be met:

$$-(D-2)\tilde{\mu} - \mu = 0. \tag{27}$$

Conditions (26) and (27) can be combined into the relation:

$$\mu^2 l^2 = \left(n + \tfrac{D-4}{2}\right)^2 > 0, \tag{28}$$

which gives, up to a sign, the real mass parameter μ in terms of the inverse AdS radius. The parameter $\tilde{\mu}$ is then also determined from Equation (27). This uniquely fixes the Lagrangian (20), as well as the gauge transformation (22), while the field and gauge parameters mimic their respective flat-space properties.

The physical significance of the mass parameter μ will be made clear in the next section as we work out the gauge fixed equations of motion. To proceed, let us forgo the language of differential forms and rewrite the action (20) as:

$$S = -\tfrac{1}{2}\int d^Dx\,\bar{e}\,\bar{\Psi}_{\mu,ac(n-2)}\left(6\mathcal{A}^{\mu\rho\nu,ab}\hat{D}_\rho + 2\mu\mathcal{B}^{\mu\nu,ab}\right)\Psi_{\nu,b}{}^{c(n-2)}, \tag{29}$$

where $\bar{e} \equiv \det \bar{e}^a_\mu$ is the determinant of the background AdS vielbein. The resulting Lagrangian equations of motion for the frame-like fermion field $\Psi_{\mu,a(n-1)}$ take the form:

$$\mathcal{R}^{\mu,a(n-1)} \equiv \left(\tfrac{6}{n-1}\right)\left(\mathcal{A}^{\mu\rho\nu,ab}\hat{D}_\rho + \tfrac{1}{3}\mu\,\mathcal{B}^{\mu\nu,ab}\right)\Psi_{\nu,b}{}^{a(n-2)} = 0. \tag{30}$$

Here, the normalization factor keeps the equations of motion well defined also for $n = 1$, as we will see. We emphasize that the equations of motion (30) are γ-traceless in the fiber indices, i.e.,

$$\gamma_b\mathcal{R}^{\mu,ba(n-2)} = 0, \tag{31}$$

as they should be. Actually, the very choices of $\mathcal{A}^{\mu\rho\nu,ab}$ and $\mathcal{B}^{\mu\nu,ab}$ made respectively in Equations (6) and (21) were such that the action (29) manifestly has the following form:

$$S = -\tfrac{1}{2}\int d^Dx\,\bar{e}\,\bar{\Psi}_{\mu,a(n-1)}\mathcal{R}^{\mu,a(n-1)}. \tag{32}$$

Clearly, the equations of motion (30) share the gauge symmetries (22) of the action:

$$\delta\Psi_{\mu,a(n-1)} = \hat{D}_\mu\zeta_{a(n-1)} + \tilde{\mu}\bar{e}^b_\mu\left[\gamma_b\zeta_{a(n-1)} - \left(\tfrac{2}{D+2n-4}\right)\gamma_a\zeta_{a(n-2)b}\right] + \bar{e}^b_\mu\lambda_{b,a(n-1)}. \tag{33}$$

In the next section, we will fix these gauge symmetries to find, among other things, the number of physical degrees of freedom, which should match with that of a Majorana fermion of spin $s = n + \frac{1}{2}$.

4. Equivalence of Frame- and Metric-Like Formulations

The first step to establish the equivalence of the frame- and metric-like descriptions of a gauge fermion is to find a match in the respective number of local degrees of freedom. To count this for a frame-like fermion [6], we rewrite the equations of motion (30) exclusively in terms of world indices:

$$\mathcal{R}^{\mu,\alpha(n-1)} \equiv \left(\gamma^{\mu\rho\nu}\nabla_\rho + \mu\gamma^{\mu\nu}\right)\Psi_{\nu,}{}^{\alpha(n-1)} + \tfrac{1}{2n}\mathcal{C}^{\mu\nu\beta,\alpha}\Psi_{\nu,\beta}{}^{\alpha(n-2)} = 0, \tag{34}$$

where $\mathcal{C}^{\mu\nu\beta,\alpha}$ is an operator antisymmetric in the μ,ν,β indices, given by:

$$\mathcal{C}^{\mu\nu\beta,\alpha} \equiv \left[\gamma^\alpha,\gamma^{\mu\rho\nu\beta}\right]\nabla_\rho - \mu\left\{\gamma^\alpha,\gamma^{\mu\nu\beta}\right\} - \left(\tfrac{2}{D+2n-4}\right)\mu\gamma^\alpha\gamma^{\mu\nu\beta}. \tag{35}$$

Some of the dynamical modes however are not physical because of gauge invariance. In order to exclude the correct number of pure gauge modes, let us rewrite the gauge transformations (33) as:

$$\delta\Psi_{\mu,\alpha(n-1)} = \nabla_\mu\zeta_{\alpha(n-1)} + \tilde{\mu}\left[\gamma_\mu\zeta_{\alpha(n-1)} - \left(\tfrac{2}{D+2n-4}\right)\gamma_\alpha\zeta_{\alpha(n-2)\mu}\right] + \lambda_{\mu,\alpha(n-1)}\,. \tag{36}$$

Now, one can use this freedom to choose the following covariant gauge:

$$\not{\Psi}_{\alpha(n-1)} \equiv \gamma^\mu\Psi_{\mu,\alpha(n-1)} = 0, \qquad \Longrightarrow \qquad \Psi'_{\alpha(n-2)} \equiv \bar{g}^{\mu\nu}\Psi_{\mu,\nu\alpha(n-2)} = 0. \tag{37}$$

As a consequence, the equations of motion (34) reduce to the following form:

$$\left(\not{\nabla} - \mu\right)\Psi^{\mu,}{}_{\alpha(n-1)} - \gamma^\mu\nabla^\nu\Psi_{\nu,\alpha(n-1)} + \tfrac{1}{2n}\mathcal{C}^{\mu\nu\rho,}{}_\alpha\,\chi_{\nu,\rho\alpha(n-2)} = 0, \tag{38}$$

where $\chi_{\mu,\alpha(n-1)}$ is the irreducible part of the field $\Psi_{\mu,\alpha(n-1)}$ with the symmetry of the Young diagram $\mathbb{Y}(n-1,1)$, i.e., it has exactly the same properties as the gauge parameter $\lambda_{\mu,\alpha(n-1)}$. Its appearance in the last term of Equation (38) is easy to understand. The antisymmetry property of $\mathcal{C}^{\mu\nu\rho,\alpha}$ removes the completely symmetric part of $\Psi_{\mu,\alpha(n-1)}$, while the γ-trace parts are trivial by the gauge choice (37).

The condition (37) is however not a complete gauge fixing. This can be seen by taking its gauge variation, which results in the Dirac equation for $\zeta_{\alpha(n-1)}$:

$$\delta\not{\Psi}_{\alpha(n-1)} \equiv \left[\not{\nabla} - \left(\tfrac{D+2n-2}{D+2n-4}\right)\mu\right]\zeta_{\alpha(n-1)} = 0. \tag{39}$$

Not only does this allow for nontrivial solutions for $\zeta_{\alpha(n-1)}$, but it also leaves $\lambda_{\mu,\alpha(n-1)}$ completely unaffected. Therefore, one can use to freedom of the shift parameter $\lambda_{\mu,\alpha(n-1)}$ to further gauge fix:

$$\chi_{\mu,\alpha(n-1)} = 0. \tag{40}$$

This finally reduces the equations of motion (38) to the Dirac form plus the divergence constraint:

$$\left(\not{\nabla} - \mu\right)\Psi_{\mu,\alpha(n-1)} = 0, \qquad \nabla^\mu\Psi_{\mu,\alpha(n-1)} = 0. \tag{41}$$

To exhaust the residual freedom of $\zeta_{\alpha(n-1)}$, let us choose the gauge:

$$\Psi_{0,\alpha(n-1)} = 0. \tag{42}$$

Its is easy to see that no residual freedom of $\zeta_{\alpha(n-1)}$ is left. A would-be residual parameter must obey some screened Poisson equation with no source term, which has no nontrivial solutions.

The count of local physical degrees of freedom is now immediate. The system (41) describes $(D-1)\Delta_\zeta$ many dynamical variables, where Δ_ζ is given in Equation (13). However, the gauge choices (37), (40) and (42) respectively remove Δ_ζ, Δ_λ and Δ_ζ degrees of freedom. Therefore, the total number of physical degrees of freedom is $(D-3)\Delta_\zeta - \Delta_\lambda$, which is the same as:

$$\Delta_{\text{Frame}} = \binom{D+n-4}{n} f_D. \tag{43}$$

This confirms, in view of Equation (A13), that the count matches in the two formulations: $\Delta_{\text{Frame}} = \Delta_{\text{Metric}}$.

The physical significance of the mass parameter μ is now clear from the Dirac equation in (41). While Equation (28) says that μ must be real, one may choose $\mu > 0$ without any loss of generality. Then,

$$\mu = \frac{1}{l}\left(n + \frac{D-4}{2}\right) > 0. \tag{44}$$

Our μ corresponds to the lowest value of the mass parameter m for a fermion carrying a unitary irreducible representation of the AdS isometry algebra:

$$(\slashed{\nabla} - m)\,\Psi_{\mu,\alpha(n-1)} = 0, \qquad m \geq \mu > 0. \tag{45}$$

The bound saturates for the massless representation [44–46], as we see.

Next, we will show that the two formulations are equivalent at the level of the free Lagrangian. With this end in view, let us decompose the fermion field $\Psi_{\mu,\alpha(n-1)}$ into totally symmetric, γ-traceless mixed-symmetric and γ-trace parts:

$$\Psi_{\mu,\alpha(n-1)} = \psi_{\mu\alpha(n-1)} + \chi_{\mu,\alpha(n-1)} + \gamma_{[\mu}\theta_{\alpha]\alpha(n-2)}, \tag{46}$$

where the fields appearing on the right-hand side have the symmetry of the following Young diagrams:

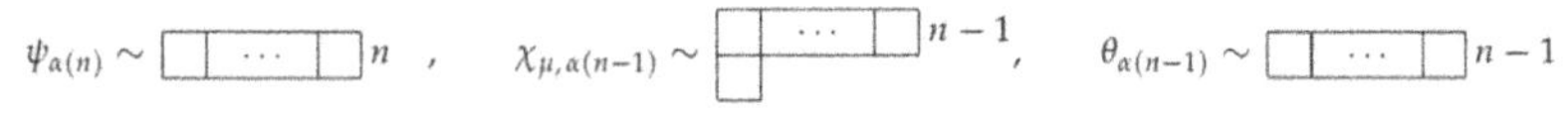

$$\psi_{\alpha(n)} \sim \boxed{\begin{array}{cccc}\ \ &\cdots&\ \ \end{array}}n \quad , \qquad \chi_{\mu,\alpha(n-1)} \sim \boxed{\begin{array}{cccc}&\cdots&\end{array}}n-1 \quad , \qquad \theta_{\alpha(n-1)} \sim \boxed{\begin{array}{cccc}\ &\cdots&\ \end{array}}n-1 \tag{47}$$

We have imposed irreducibility conditions on $\chi_{\mu,\alpha(n-1)}$, so that it is subject to the following constraints:

$$\gamma^\mu \chi_{\mu,\alpha(n-1)} = 0, \qquad \gamma^\beta \chi_{\mu,\alpha(n-2)\beta} = 0, \qquad \chi_{\alpha,\alpha(n-1)} = 0. \tag{48}$$

Of course, there will be additional constraints on the fields $\psi_{\alpha(n)}$ and $\theta_{\alpha(n-1)}$ coming from the γ-trace condition on the parent field $\Psi_{\mu,\alpha(n-1)}$ in the α-indices. To find them, let us first take a γ-trace of Equation (46) in an α-index. This results in:

$$\slashed{\psi}_{\mu\alpha(n-2)} - (D-2)\theta_{\mu\alpha(n-2)} - (n-1)\gamma_\mu \slashed{\theta}_{\alpha(n-2)} + \gamma_\alpha \slashed{\theta}_{\mu\alpha(n-3)} = 0. \tag{49}$$

Another γ-trace w.r.t. the μ-index gives:

$$\psi'_{\alpha(n-2)} - (Dn - 2n + 2)\,\slashed{\theta}_{\alpha(n-2)} - \gamma_\alpha \theta'_{\alpha(n-3)} = 0. \tag{50}$$

Now, a third γ-trace in an α-index yields:

$$\slashed{\psi}'_{\alpha(n-3)} - (Dn + D - 4)\theta'_{\alpha(n-3)} + \gamma_\alpha \slashed{\theta}'_{\alpha(n-4)} = 0. \tag{51}$$

On the other hand, one could also have obtained a triple γ-trace by first contracting the μ index with an α index in Equation (46) and then taking a γ trace. This however produces a different result:

$$\slashed{\psi}'_{\alpha(n-3)} - (D+n-4)\theta'_{\alpha(n-3)} + \gamma_\alpha \slashed{\theta}'_{\alpha(n-4)} = 0. \tag{52}$$

Equations (51) and (52) impose the following constraints:

$$\not\psi'_{\alpha(n-3)} = 0, \qquad \theta'_{\alpha(n-3)} = 0, \tag{53}$$

i.e., the symmetric rank-n field $\psi_{\alpha(n)}$ must be triply γ-traceless, whereas the symmetric rank-$(n-1)$ field $\theta_{\alpha(n-1)}$ must be traceless. This in turn results, from Equations (49) and (50), in the following relation:

$$\theta_{\alpha(n-1)} = \left(\tfrac{1}{D-2}\right)\left[\not\psi_{\alpha(n-1)} - \left(\tfrac{1}{nD-2n+2}\right)\gamma_\alpha \psi'_{\alpha(n-2)}\right]. \tag{54}$$

Finally, plugging the above expression into the decomposition (46), we obtain:

$$\begin{aligned}\Psi_{\mu,\alpha(n-1)} &= \psi_{\mu\alpha(n-1)} + \chi_{\mu,\alpha(n-1)} + \left(\tfrac{1}{D-2}\right)\left[\gamma_{[\mu}\not\psi_{\alpha]\alpha(n-2)} - \left(\tfrac{2}{Dn-2n+2}\right)\gamma_{\mu\alpha}\psi'_{\alpha(n-2)}\right] \\ &\quad + \tfrac{1}{(D-2)(Dn-2n+2)}\left[(n-2)\gamma_\alpha\gamma_\mu\psi'_{\alpha(n-2)} - 2\bar{g}_{\alpha(2)}\psi'_{\mu\alpha(n-3)}\right].\end{aligned} \tag{55}$$

This decomposition generalizes that of [1] to arbitrary dimensions.

It will be convenient to write the covariant equations of motion (34) in the following form:

$$\mathcal{R}^{\mu,\alpha(n-1)} \equiv \mathcal{O}^{\mu\nu,\alpha(n-1)\beta(n-1)}\Psi_{\nu,\beta(n-1)} = 0, \tag{56}$$

where we have defined the operator $\mathcal{O}$ as:

$$\mathcal{O}^{\mu\nu,\alpha(n-1)\beta(n-1)} \equiv \left(\gamma^{\mu\rho\nu}\nabla_\rho + \mu\gamma^{\mu\nu}\right)\bar{g}^{\alpha(n-1),\beta(n-1)} + \tfrac{1}{2n(n-1)}C^{\mu\nu\beta,\alpha}\bar{g}^{\alpha(n-2),\beta(n-2)}, \tag{57}$$

with $\bar{g}^{\alpha(k),\beta(k)} \equiv \frac{1}{k^2}\bar{g}^{\alpha\beta}\bar{g}^{\alpha\beta}\ldots\bar{g}^{\alpha\beta}$ (multiplicity k) denoting the unit-strength symmetric tensor product of k background metric tensors. This enables us to present the corresponding Lagrangian as:

$$\tfrac{1}{\sqrt{-\bar{g}}}\mathcal{L} = -\tfrac{1}{2}\bar{\Psi}_{\mu,\alpha(n-1)}\mathcal{O}^{\mu\nu,\alpha(n-1)\beta(n-1)}\Psi_{\nu,\beta(n-1)}. \tag{58}$$

When the decomposition (55) is plugged into the above Lagrangian, the irreducible mixed-symmetric part $\chi_{\mu,\alpha(n-1)}$ completely drops out, thanks to the shift symmetry. The fact that the parameter $\lambda_{\mu,\alpha(n-1)}$ enjoys exactly the same properties as $\chi_{\mu,\alpha(n-1)}$ plays a crucial role in this regard. The resulting Lagrangian contains only the completely symmetric part $\psi_{\alpha(n)}$ and can be viewed as a gauge-fixed version of the original Lagrangian (58) with the gauge fixing: $\chi_{\mu,\alpha(n-1)} = 0$. The explicit derivation of this Lagrangian is tedious, but straightforward. The calculations can however be simplified by noting that, on account of the γ-tracelessness of the equations of motion (56) in the α-indices, the Lagrangian splits into the sum of two pieces:

$$\tfrac{1}{\sqrt{-\bar{g}}}\mathcal{L} = -\tfrac{1}{2}\bar{\Xi}_{\mu,\alpha(n-1)}\mathcal{O}^{\mu\nu,\alpha(n-1)\beta(n-1)}\Xi_{\nu,\beta(n-1)} + \tfrac{1}{2}\bar{\zeta}_{\mu,\alpha(n-2)}\gamma_\alpha\mathcal{O}^{\mu\nu,\alpha(n-1)\beta(n-1)}\gamma_\beta\zeta_{\nu,\beta(n-2)}, \tag{59}$$

where the tensor-spinors $\Xi_{\mu,\alpha(n-1)}$ and $\zeta_{\mu,\alpha(n-2)}$ are given by:

$$\begin{aligned}\Xi_{\mu,\alpha(n-1)} &= \psi_{\mu\alpha(n-1)} + \left(\tfrac{1}{D-2}\right)\left[(n-1)\gamma_\mu\not\psi_{\alpha(n-1)} - \left(\tfrac{2}{Dn-2n+2}\right)\bar{g}_{\mu\alpha}\psi'_{\alpha(n-2)}\right], \\ \zeta_{\mu,\alpha(n-2)} &= \left(\tfrac{1}{D-2}\right)\left[-\not\psi_{\mu\alpha(n-2)} + \left(\tfrac{1}{Dn-2n+2}\right)\left(n\gamma_\mu\psi'_{\alpha(n-2)} - \gamma_\alpha\psi'_{\mu\alpha(n-3)}\right)\right].\end{aligned} \tag{60}$$

One can explicitly carry out the calculations to get to the following result:

$$\begin{aligned}-\tfrac{2}{\sqrt{-\bar{g}}}\mathcal{L} &= \bar{\psi}_{\alpha(n)}\left(\not\nabla - \mu\right)\psi^{\alpha(n)} + n\bar{\not\psi}_{\alpha(n-1)}\left(\not\nabla + \mu\right)\not\psi^{\alpha(n-1)} - \tfrac{1}{4}n(n-1)\bar{\psi}'_{\alpha(n-2)}\left(\not\nabla - \mu\right)\psi'^{\alpha(n-2)} \\ &\quad -2n\bar{\not\psi}_{\alpha(n-1)}\nabla_\mu\psi^{\mu\alpha(n-1)} - n(n-1)\bar{\psi}'_{\alpha(n-2)}\nabla_\mu\not\psi^{\mu\alpha(n-2)}.\end{aligned} \tag{61}$$

This indeed coincides with the Lagrangian (A7) for a metric-like gauge fermion in AdS space. Because only the symmetric part of the parent field $\Psi_{\mu,\alpha(n-1)}$ appears in this Lagrangian, the corresponding gauge symmetry is obtained simply by a total symmetrization of the indices in Equation (36). The result is:

$$\delta\psi_{\alpha(n)} = \tfrac{1}{n}\left(\nabla_\alpha\zeta_{\alpha(n-1)} - \tfrac{1}{2l}\gamma_\alpha\zeta_{\alpha(n-1)}\right), \tag{62}$$

which also matches perfectly with the metric-like gauge symmetry (A11).

This hardly comes as a surprise. The symmetric part of $\Psi_{\mu,\alpha(n-1)}$ has all the characteristics of a metric-like gauge fermion; in particular, it is triple γ-traceless, as we have shown in Equation (53). Moreover, it transforms w.r.t. a symmetric γ-traceless gauge parameter $\zeta_\alpha(n-1)$. The gauge-invariant Lagrangian description for such a system is unique [43–45]. Therefore, $\psi_{\alpha(n)}$ is a metric-like gauge fermion in every sense.

5. Remarks

In this article, we have elaborated on some key features of higher-spin gauge fermions and the connections between their frame- and metric-like formulations at the free level. A gauge-invariant frame-like Lagrangian description in AdS space, with the constraints on the fields and the gauge parameters resembling their flat-space cousins, facilitates the explicit derivation of the corresponding metric-like Lagrangian as a gauge fixing. This derivation generalizes that of [1] to AdS space and arbitrary dimensions. Although the equivalence of the frame- and metric-like formulations at the free level may not come as a surprise, our work fills a gap in the literature.

As is well known, the frame-like formulation packages the non-linearities in an interacting theory in a very efficient way. For higher-spin fermions, this can be seen in a very simple setup: the Aragone–Deser hypergravity [3]—a consistent gauge theory of a spin $s = n + \frac{1}{2}$ massless Majorana fermion coupled to Einstein gravity in 3D flat space. While only fermion bilinears appear in the frame-like formulation [3], the metric-like formulation will also include four-fermion couplings that originate from integrating out the spin-connection, just like in supergravity [47]. Moreover, the fermion-bilinear terms will look more complicated in the metric-like variables. To see this, note that with frame-like fermions, the cubic cross-coupling in the covariant language has the simple form [49]:

$$\mathcal{L}_3 \sim \bar{\Psi}_{\mu,\alpha(n-1)}\gamma^{\mu\nu\rho}\gamma^{\sigma\lambda}\Psi_{\nu,}{}^{\alpha(n-1)}\partial_\sigma h_{\rho\lambda}, \tag{63}$$

where $h_{\mu\nu}$ is the metric perturbation. Because the irreducible hook part $\chi_{\mu,\alpha(n-1)}$ of the frame-like fermion is trivial in $D = 3$, the decomposition (55) amounts to a complicated field redefinition:

$$\Psi_{\mu,\alpha(n-1)} = \psi_{\mu\alpha(n-1)} + \gamma_{[\mu}\not{\psi}_{\alpha]\alpha(n-2)} + \left(\tfrac{1}{n+2}\right)\left[n\gamma_\alpha\gamma_\mu\psi'_{\alpha(n-2)} - 2\eta_{\mu\alpha}\psi'_{\alpha(n-2)} + 2\eta_{\alpha(2)}\psi'_{\mu\alpha(n-3)}\right], \tag{64}$$

where $\psi_{\alpha(n)}$ is the metric-like fermion. After this redefinition is performed, the cubic coupling (63) will look cumbersome in terms of the metric-like fermion. Within the metric-like formulation, it would be more difficult to construct or to prove the consistency of this cubic coupling, say using the techniques of [50,51]. The fermion-bilinear cross-couplings do not stop at any finite order in the graviton fluctuations, and the situation gets only worse at higher orders, while the frame-like formulation captures all the non-linearities in a very neat way [3].

In higher dimensions, the difference between the two formulations becomes more drastic. The hook part of the frame-like fermion never shows up in the interacting Lagrangian because of the deformed shift symmetry. However, there appear the so-called "extra" fields: a set of additional fields that arises when one tries to construct a complete set of gauge-invariant objects (curvatures) [52] (The extra fields are generalizations of the spin-connection. The number of extra fields depends on the spin; the higher the spin, the more are the extra fields needed for constructing curvatures. The extra fields however do not enter the free action, and so, they are not expressed in terms of physical fields via equations of motion) . To understand the role of these extra fields that are absent in the free Lagrangian, one may express them in terms of the physical fields by means of appropriate constraints implemented via Lagrange multipliers [4,6,52–54]. Then, up to pure gauge parts, the extra fields are given by derivatives of the physical fields. The extra fields therefore induce higher-derivative terms in the interactions, while their absence in the free Lagrangian merely reflects the absence of higher-derivative kinetic terms. Explicit solutions of the aforementioned constraints are difficult and actually not needed. The main idea of the so-called Fradkin–Vasiliev formalism [52–54] is that one can treat the extra fields as independent variables since most of the gauge-invariant curvatures vanish on shell.

Acknowledgments: The author is grateful to N. Boulanger, A. Campoleoni, G. Lucena Gómez, M. Henneaux and especially to E. D. Skvortsov for valuable inputs and useful comments. He would like to thank the organizers of the 4th Mons Workshop on Higher Spin Gauge Theories (2017), during which this study was initiated.

Conflicts of Interest: The author declares no conflict of interest.

Appendix A. Metric-Like Formulation

The metric-like formulation of gauge fermions originated in the work of Fang and Fronsdal [11,12], who studied the massless limit of the Lagrangian for massive higher-spin fermions. The Fang–Fronsdal Lagrangian can be derived uniquely by considering gauge invariance and supersymmetry transformations for a massless system involving the pair of spins $\left(s, s+\frac{1}{2}\right)$ [55]. The construction was later generalized for maximally-symmetric spaces with arbitrary dimension in [43–45]. In the metric-like formulation, a spin $s = n + \frac{1}{2}$ gauge fermion is described by a completely symmetric rank-n tensor-spinor $\psi_{\mu(n)}$ in the world indices. It satisfies the triple γ-trace condition:

$$\not{\psi}'_{\mu(n-3)} = 0. \tag{A1}$$

It is convenient to describe metric-like theories in the operator formalism, where contraction and symmetrization of indices are realized through auxiliary variables and tensor operations are simplified in terms of operator calculus. Symmetric tensor-spinor fields are represented by:

$$\psi(x,u) = \tfrac{1}{n!}\,\psi_{\mu_1\ldots\mu_n}(x)\,\bar{e}^{\mu_1}_{a_1}(x)u^{a_1}\,\ldots\,\bar{e}^{\mu_n}_{a_n}(x)u^{a_n}, \tag{A2}$$

where $\bar{e}^{\mu}_{a}(x)$ is the background vielbein and u^a is an auxiliary tangent variable. The action of the covariant derivative is defined as a differential operation involving both x and u:

$$\nabla_\mu = \bar{\nabla}_\mu + \bar{\omega}_\mu{}^{ab} u_a \tfrac{\partial}{\partial u^b}, \tag{A3}$$

where $\bar{\nabla}_\mu$ is the standard covariant derivative acting on naked tensorial indices and $\bar{\omega}_\mu{}^{ab}$ the background spin connection. In what follows, we work only with the contracted auxiliary variable and the associated derivative:

$$u^\mu \equiv \bar{e}^{\mu}_{a}(x)u^a, \quad \partial^\mu_u \equiv \bar{e}^{\mu a}(x)\tfrac{\partial}{\partial u^a}. \tag{A4}$$

The vielbein postulate then implies that $[\nabla_\mu, u^\nu] = 0$, as well as $[\nabla_\mu, \partial^\nu_u] = 0$. The commutator of covariant derivatives on a spinor function of u and ∂_u will be given by:

$$[\nabla_\mu, \nabla_\nu] = R_{\mu\nu\rho\sigma}(x)u^\rho\partial^\sigma_u + \tfrac{1}{4}R_{\mu\nu\rho\sigma}(x)\gamma^{\rho\sigma}. \tag{A5}$$

One would have to use the following set of operators [43–45]:

$$\mathbb{G} = \left\{ \not{\nabla},\, \partial_u\cdot\nabla,\, u\cdot\nabla,\, \not{\partial}_u,\, \not{u},\, \partial^2_u,\, u^2,\, u\cdot\partial_u \right\}. \tag{A6}$$

The set comprises eight operators: the Dirac operator $\not{\nabla}$, divergence $\partial_u\cdot\nabla$, symmetrized-gradient $u\cdot\nabla$, γ-trace $\not{\partial}_u$, symmetrized-γ $\not{u}$, trace ∂^2_u, symmetrized-metric u^2 and rank $u\cdot\partial_u$. These operators have nontrivial commutation relations because of $[\partial_u^\mu, u^\nu] = \bar{g}^{\mu\nu}$ and the non-commutativity (A5) of the covariant derivatives if the background is non-flat.

Then, the Lagrangian for a massless fermionic field in AdS space can be written as (for a Majorana fermion, certain terms in the Lagrangian are equivalent up to total derivatives) [45]:

$$\begin{aligned} \tfrac{1}{\sqrt{-\bar{g}}}\mathcal{L} &= -\tfrac{1}{2}\bar{\psi}(*_n)\left(\not\nabla - u\cdot\nabla\not\partial_u - \not u\,\partial_u\cdot\nabla + \not u\not\nabla\not\partial_u + \tfrac{1}{2}\not u\,u\cdot\nabla\,\partial_u^2 + \tfrac{1}{2}u^2\,\partial_u\cdot\nabla\,\not\partial_u - \tfrac{1}{4}u^2\not\nabla\,\partial_u^2\right)\psi \\ &\quad + \tfrac{1}{2}\mu\,\bar{\psi}(*_n)\left(1 - \not u\,\not\partial_u - \tfrac{1}{4}u^2\,\partial_u^2\right)\psi, \end{aligned} \tag{A7}$$

where the operation: $(*_k) \equiv \left(\overleftarrow{\partial_u}\cdot\overrightarrow{\partial_u}\right)^k$ enables contraction between two rank-k tensor-spinors and has the properties: $(*_k)u^\mu = k\overleftarrow{\partial_u}^\mu(*_{k-1})$ and $(*_k)\overrightarrow{\partial_u}^\mu = (k+1)^{-1}u^\mu(*_{k+1})$. The mass parameter:

$$\mu = \frac{1}{l}\left(n + \tfrac{D-4}{2}\right), \tag{A8}$$

is uniquely fixed by gauge invariance [44,45], where l is the AdS radius. The gauge symmetry of the Lagrangian (A7) is w.r.t. a symmetric γ-traceless rank-$(n-1)$ tensor-spinor parameter:

$$\varepsilon = \tfrac{1}{(n-1)!}\,\varepsilon_{\mu_1\ldots\mu_{n-1}}u^{\mu_1}\ldots u^{\mu_{n-1}}, \qquad \not\partial_u\varepsilon = 0, \tag{A9}$$

while the triple γ-tracelessness condition (A1) on the field translates in the operator formalism to:

$$\not\partial_u\partial_u^2\psi = \partial_u^2\not\partial_u\psi = 0. \tag{A10}$$

Explicitly, the gauge transformations are given by:

$$\delta\psi = u\cdot\nabla\varepsilon - \frac{1}{2l}\not u\,\varepsilon. \tag{A11}$$

This can be verified by using the commutator (A5), which reduces in AdS space to:

$$[\nabla_\mu, \nabla_\nu] = -\frac{1}{l^2}\left(u_{[\mu}d_{\nu]} + \tfrac{1}{2}\gamma_{\mu\nu}\right), \tag{A12}$$

and the various commutators of the operators in $\mathbb{G}$ given the properties (A9) and (A10).

The metric-like description of higher-spin gauge fermions in flat-space is easily obtained by taking the limit $l \to \infty$ of the gauge invariant system (A7)–(A12). The degrees of freedom count in flat [13] and AdS [14] spaces are of course the same and given by:

$$\Delta_{\text{Metric}} = \binom{D+n-4}{n} f_D, \tag{A13}$$

where f_D for a Majorana fermion is given in Equation (12), while for a Dirac fermion, the value is twice as much. Note that Equation (A13) counts the number of physical dynamical fields plus their conjugate momenta. In AdS space, one of course gets the same number since the counting of dynamical equations, constraints and gauge freedom works in the same way.

As already mentioned in the Introduction, the γ-trace constraints (A9)–(A10) on the gauge parameter and the higher-spin fermionic field can be avoided by recourse to other formulations. These include the non-local formulation [15], the Becchi–Rouet–Stora–Tyutin (BRST) formulation [17,22], the higher-derivative compensator formulation [20], the quartet formulation [21] and the non-minimal formulation with no higher derivatives [24].

References

1. Vasiliev, M.A. 'gauge' Form Of Description Of Massless Fields With Arbitrary Spin. *Sov. J. Nucl. Phys.* **1980**, *32*, 439. (In Russian)
2. Aragone, C.; Deser, S. Higher Spin Vierbein Gauge Fermions and Hypergravities. *Nucl. Phys. B* **1980**, *170*, 329–352.
3. Aragone, C.; Deser, S. Hypersymmetry in $D = 3$ of Coupled Gravity Massless Spin 5/2 System. *Class. Quantum Gravity* **1984**, *1*, L9.
4. Vasiliev, M.A. Free Massless Fields of Arbitrary Spin in the De Sitter Space and Initial Data for a Higher Spin Superalgebra. *Fortschr. Phys.* **1987**, *35*, 741–770.
5. Lopatin, V.E.; Vasiliev, M.A. Free Massless Bosonic Fields of Arbitrary Spin in d-dimensional De Sitter Space. *Mod. Phys. Lett. A* **1988**, *3*, 257–270.
6. Vasiliev, M.A. Free Massless Fermionic Fields of Arbitrary Spin in d-dimensional De Sitter Space. *Nucl. Phys. B* **1988**, *301*, 26–68.
7. Vasiliev, M.A. Cubic interactions of bosonic higher spin gauge fields in AdS_5. *Nucl. Phys. B* **2001**, *616*, 106–162; Erratum in **2003**, *652*, 407.
8. De Wit, B.; Freedman, D.Z. Systematics of Higher Spin Gauge Fields. *Phys. Rev. D* **1980**, *21*, 358.
9. Fronsdal, C. Massless Fields with Integer Spin. *Phys. Rev. D* **1978**, *18*, 3624.
10. Fronsdal, C. Singletons and Massless, Integral Spin Fields on de Sitter Space. *Phys. Rev. D* **1979**, *20*, 848–856.
11. Fang, J.; Fronsdal, C. Massless Fields with Half Integral Spin. *Phys. Rev. D* **1978**, *18*, 3630.
12. Fang, J.; Fronsdal, C. Massless, Half Integer Spin Fields in De Sitter Space. *Phys. Rev. D* **1980**, *22*, 1361.
13. Rahman, R.; Taronna, M. From Higher Spins to Strings: A Primer. *arXiv* **2015**, arXiv:1512.07932.
14. Campoleoni, A.; Henneaux, M.; Hörtner, S.; Leonard, A. Higher-spin charges in Hamiltonian form. II. Fermi fields. *J. High Energy Phys.* **2017**, *2017*, 58.
15. Francia, D.; Sagnotti, A. Free geometric equations for higher spins. *Phys. Lett. B* **2002**, *543*, 303–310.
16. Bekaert, X.; Boulanger, N. On geometric equations and duality for free higher spins. *Phys. Lett. B* **2003**, *561*, 183–190.
17. Buchbinder, I.L.; Krykhtin, V.A.; Pashnev, A. BRST approach to Lagrangian construction for fermionic massless higher spin fields. *Nucl. Phys. B* **2005**, *711*, 367–391.
18. Francia, D.; Sagnotti, A. Minimal local Lagrangians for higher-spin geometry. *Phys. Lett. B* **2005**, *624*, 93–104.
19. Bekaert, X.; Boulanger, N. Tensor gauge fields in arbitrary representations of GL(D,R). II. Quadratic actions. *Commun. Math. Phys.* **2007**, *271*, 723–773.
20. Francia, D.; Mourad, J.; Sagnotti, A. Current Exchanges and Unconstrained Higher Spins. *Nucl. Phys. B* **2007**, *773*, 203–237.
21. Buchbinder, I.L.; Galajinsky, A.V.; Krykhtin, V.A. Quartet unconstrained formulation for massless higher spin fields. *Nucl. Phys. B* **2007**, *779*, 155–177.
22. Buchbinder, I.L.; Krykhtin, V.A.; Reshetnyak, A.A. BRST approach to Lagrangian construction for fermionic higher spin fields in (A)dS space. *Nucl. Phys. B* **2007**, *787*, 211–240.
23. Francia, D. Geometric Lagrangians for massive higher-spin fields. *Nucl. Phys. B* **2008**, *796*, 77–122.
24. Campoleoni, A.; Francia, D.; Mourad, J.; Sagnotti, A. Unconstrained Higher Spins of Mixed Symmetry. II. Fermi Fields. *Nucl. Phys. B* **2010**, *828*, 405–514.
25. Campoleoni, A.; Fredenhagen, S.; Pfenninger, S.; Theisen, S. Towards metric-like higher-spin gauge theories in three dimensions. *J. Phys. A* **2013**, *46*, 214017.
26. Fredenhagen, S.; Kessel, P. Metric- and frame-like higher-spin gauge theories in three dimensions. *J. Phys. A* **2015**, *48*, 035402.
27. Campoleoni, A.; Henneaux, M. Asymptotic symmetries of three-dimensional higher-spin gravity: The metric approach. *J. High Energy Phys.* **2015**, *2015*, 143.
28. Boulanger, N.; Kessel, P.; Skvortsov, E.D.; Taronna, M. Higher spin interactions in four-dimensions: Vasiliev versus Fronsdal. *J. Phys. A* **2016**, *49*, 095402.
29. Konstein, S.E.; Vasiliev, M.A. Extended Higher Spin Superalgebras and Their Massless Representations. *Nucl. Phys. B* **1990**, *331*, 475–499.
30. Vasiliev, M.A. Properties of equations of motion of interacting gauge fields of all spins in (3+1)-dimensions. *Class. Quantum Gravity* **1991**, *8*, 1387–1417.

31. Vasiliev, M.A. More on equations of motion for interacting massless fields of all spins in (3+1)-dimensions. *Phys. Lett. B* **1992**, *285*, 225–234.
32. Sezgin, E.; Sundell, P. Higher spin N = 8 supergravity. *J. High Energy Phys.* **1998**, *1998*, 16.
33. Sezgin, E.; Sundell, P. Higher spin N = 8 supergravity in AdS(4). In Proceedings of the Johns Hopkins Workshop on Current Problems in Particle Theory 22, Göteborg, Sweden, 20–22 August 1998; p. 241
34. Sezgin, E.; Sundell, P. Analysis of higher spin field equations in four-dimensions. *J. High Energy Phys.* **2002**, *2002*, 55.
35. Engquist, J.; Sezgin, E.; Sundell, P. On N = 1, N = 2, N = 4 higher spin gauge theories in four-dimensions. *Class. Quantum Gravity* **2002**, *19*, 6175–6196.
36. Sezgin, E.; Sundell, P. Supersymmetric Higher Spin Theories. *J. Phys. A* **2013**, *46*, 214022.
37. Alkalaev, K.B. Free fermionic higher spin fields in AdS(5). *Phys. Lett. B* **2001**, *519*, 121–128.
38. Alkalaev, K.B. Mixed-symmetry massless gauge fields in AdS(5). *Theor. Math. Phys.* **2006**, *149*, 1338–1348.
39. Sorokin, D.P.; Vasiliev, M.A. Reducible higher-spin multiplets in flat and AdS spaces and their geometric frame-like formulation. *Nucl. Phys. B* **2009**, *809*, 110–157.
40. Zinoviev, Y.M. Frame-like gauge invariant formulation for massive high spin particles. *Nucl. Phys. B* **2009**, *808*, 185–204.
41. Zinoviev, Y.M. Frame-like gauge invariant formulation for mixed symmetry fermionic fields. *Nucl. Phys. B* **2009**, *821*, 21–47.
42. Skvortsov, E.D.; Zinoviev, Y.M. Frame-like Actions for Massless Mixed-Symmetry Fields in Minkowski space. Fermions. *Nucl. Phys. B* **2011**, *843*, 559–569.
43. Hallowell, K.; Waldron, A. Constant curvature algebras and higher spin action generating functions. *Nucl. Phys. B* **2005**, *724*, 453–486.
44. Metsaev, R.R. Gauge invariant formulation of massive totally symmetric fermionic fields in (A)dS space. *Phys. Lett. B* **2006**, *643*, 205–212.
45. Metsaev, R.R. CFT adapted approach to massless fermionic fields, AdS/CFT, and fermionic conformal fields. *arXiv* **2013**, arXiv:1311.7350.
46. Metsaev, R.R. Massive totally symmetric fields in AdS(d). *Phys. Lett. B* **2004**, *590*, 95–104.
47. Freedman, D.Z.; Van Proeyen, A. *Supergravity*; Cambridge University Press: Cambridge, UK, 2012; ISBN-10: 0521194016.
48. Fuentealba, O.; Matulich, J.; Troncoso, R. Extension of the Poincaré group with half-integer spin generators: Hypergravity and beyond. *J. High Energy Phys.* **2015**, *2015*, 3.
49. Henneaux, M.; Lucena Gómez, G.; Rahman, R. The uniqueness of hypergravity. Unpublished work, 2018.
50. Henneaux, M.; Gómez, G.L.; Rahman, R. Higher-Spin Fermionic Gauge Fields and Their Electromagnetic Coupling. *J. High Energy Phys.* **2012**, *2012*, 93.
51. Henneaux, M.; Gómez, G.L.; Rahman, R. Gravitational Interactions of Higher-Spin Fermions. *J. High Energy Phys.* **2014**, *2014*, 87.
52. Fradkin, E.S.; Vasiliev, M.A. Candidate to the Role of Higher Spin Symmetry. *Ann. Phys.* **1987**, *177*, 63–112.
53. Fradkin, E.S.; Vasiliev, M.A. On the Gravitational Interaction of Massless Higher Spin Fields. *Phys. Lett. B* **1987**, *189*, 89–95.
54. Fradkin, E.S.; Vasiliev, M.A. Cubic Interaction in Extended Theories of Massless Higher Spin Fields. *Nucl. Phys. B* **1987**, *291*, 141–171.
55. Curtright, T. Massless Field Supermultiplets With Arbitrary Spin. *Phys. Lett. B* **1979**, *85*, 219–224.

universe

MDPI

Article

Exploring Free Matrix CFT Holographies at One-Loop

Jin-Beom Bae [1], Euihun Joung [2,*] and Shailesh Lal [3,*]

[1] Korea Institute for Advanced Study, 85 Hoegiro, Dongdaemun-Gu, Seoul 02455, Korea; jinbeom@kias.re.kr
[2] Department of Physics and Research Institute of Basic Science, Kyung Hee University, Seoul 02447, Korea
[3] LPTHE-UMR 7589, UPMC Paris 06, Sorbonne Universites, 75005 Paris, France
* Correspondence: euihun.joung@khu.ac.kr (E.J.); shailesh@lpthe.jussieu.fr (S.L.)

Received: 18 August 2017; Accepted: 11 October 2017; Published: 9 November 2017

Abstract: We extend our recent study on the duality between stringy higher spin theories and free conformal field theories (CFTs) in the $SU(N)$ adjoint representation to other matrix models, namely the free $SO(N)$ and $Sp(N)$ adjoint models as well as the free $U(N) \times U(M)$ bi-fundamental and $O(N) \times O(M)$ bi-vector models. After determining the spectrum of the theories in the planar limit by Polya counting, we compute the one loop vacuum energy and Casimir energy for their respective bulk duals by means of the Character Integral Representation of the Zeta Function (CIRZ) method, which we recently introduced. We also elaborate on possible ambiguities in the application of this method.

Keywords: string theory, higher-spin gauge theory, partition functions

1. Introduction

The AdS/CFT duality [1] is a remarkable conjecture proposing the equivalence between a quantum gravity in Anti de Sitter (AdS) space and a conformal field theory (CFT) defined on the boundary of the same AdS space (see [2] for a review).

These dualities were observed in string theory, building on the observation [3] that the D-branes of string theory and the black branes of supergravity are essentially complementary descriptions of the same system, being valid respectively at weak and strong string coupling. The AdS theory is the closed string theory—a theory of quantum gravity—that the black branes are embedded in, and the CFT is the field theory which describes the low energy dynamics of the world volume of the D-branes. We therefore expect that the AdS/CFT dualities would share two very common features. First, fields in CFT should be matrix valued because of the CFT is the low energy effective theory of a stack of D-branes. Second, since closed string theories contain supergravity as a low-energy limit, there should be a regime in the parameter space of the duality where the AdS theory is described well by supergravity. It turns out that, in field theory, this corresponds to taking the strongly coupled limit. We also observe that supersymmetry is almost ubiquitous in these dualities, and is an important ingredient for ensuring that the weak and strong string coupling descriptions can be extrapolated to each other to lead to the duality.

It is interesting to contrast this situation with the case of AdS/CFT dualities involving higher-spin theories and vector model CFTs [4,5]. These dualities are counterexamples to the above expectations in almost every way. Firstly, they are non-supersymmetric, or at least there is no apparent benefit in working with their supersymmetric extensions. Secondly, the CFT is a typically a vector model rather than a matrix model, this leads to important simplifications in the spectrum of the bulk and boundary theories and may also have important implications on black hole physics in these theories [6]. Thirdly, there is no obvious point in the parameter space of the duality where we obtain bulk General Relativity.

Nonetheless, the study of these dualities might have important insights into the physics of 'stringy' AdS/CFT dualities. This turns out to be the case from two *a priori* distinct motivations. Firstly, while it is clearly desirable to use AdS/CFT dualities to probe bulk quantum gravity using the dual CFT, in practice it is somewhat more difficult as the dynamics of a CFT at a generic coupling is extremely complicated in itself. From this point of view, it is natural to consider AdS/CFT dualities involving *free* CFTs to examine how the CFT repackages itself into a theory of quantum gravity[1]. Secondly, taking the free 't-Hooft coupling limit on the CFT side corresponds to a particularly interesting limit on the AdS side as well [7,8].

For definiteness, let us consider the case of the duality between Type IIB superstrings on $AdS_5 \times S^5$ and $\mathcal{N} = 4$ super Yang-Mills, where the dictionary between the bulk and boundary parameters reads as

$$N^2 \sim \left(\frac{\ell_{AdS}}{\ell_P}\right)^8, \qquad \lambda \sim \frac{\ell_{AdS}^4}{(\alpha')^2}. \tag{1}$$

This dictionary indicates, as is familiar, that taking the planar limit of $\mathcal{N} = 4$ super Yang-Mills corresponds to taking the classical limit in the bulk, where the radii of AdS_5 and S^5 are much larger than the Planck length. Further, now setting λ to zero corresponds to setting the string tension α'^{-1} to zero or, equivalently, taking the string length to be much larger than the AdS_5 radius. In either way of thinking about this limit in the bulk, it should be clear that this is a very stringy limit as it corresponds to working at an energy scale much larger than the string tension, at which point the string no longer looks like a point object as it would to a low energy observer, which is essentially the supergravity approximation, corresponding to taking λ to infinity.

Moreover, the tensionless limit is a window of string theory about which much remains to be understood, however there are important hints that new symmetries should manifest themselves in this phase [9–14] and indeed that higher-spin symmetry may be one such symmetry [12–14]. It is therefore natural to explore this window of AdS/CFT duality both for gaining a foothold into tensionless string theory and also for a more general program of extracting bulk physics from CFT data.

The approach we adopt in this paper is to assume that a CFT with an't-Hooft expansion admits an AdS dual in the planar limit, and then compute $1/N$ corrections in the duality. This approach also provides an interesting point of view regarding a different but related question. In particular, how does one couple massive representations of the higher-spin algebra to the Vasiliev system? Although the coupling of massive and massless higher-spin fields in AdS has been studied at the cubic level in [15–17], directly constructing the bulk theory is still quite difficult. However, since the single-trace operator spectrum of a free matrix model CFT contains the conserved currents found in the vector model along with conformal primaries lying above the unitarity bound, we expect its AdS dual to be a theory of massless higher spins coupled to massive higher spins. Further, by varying the content of the CFT, the operator spectrum can be quite easily varied. Hence this setting is expectedly useful for generating a zoo of theories with massless and massive higher-spins coupled to each other in AdS.

As a preliminary exploration of the duality between tensionless strings in AdS and free matrix model CFTs, it is particularly appealing to focus on one-loop quantum effects in the bulk, especially the vacuum energy in AdS with sphere boundary and thermal AdS with torus boundary. We shall refer these two quantities as one-loop vacuum energy and Casimir energy. The spectral problem for arbitrary spin tensor (and spinor) fields has been almost completely solved for Laplacians on hyperboloids [18–21], and this provides the vacuum energy of the corresponding particle. The full result is expectedly determined by summing over contributions from every particles in the spectrum of the bulk theory. This was very explicitly carried out for the higher spin theories in [22–33] and the resulting computation matched. We refer the reader to [34] for a review higher spin holography in general, including the one-loop computations mentioned here.

[1] We consider 'free' CFTs as being obtained from a zero 't-Hooft coupling limit of the large-N expansion of a given CFT. Hence the bulk theory still admits a semi-classical expansion, identified to the 't-Hooft expansion of the dual CFT, and single trace conformal primaries in the CFT correspond to fields in the bulk.

It is clearly of interest to explore how these computations can be extended to the case of the tensionless string, but there is an obvious complication. The higher-spin computations of one-loop free energies rely on an explicit knowledge of the bulk spectrum. Further, summing over free energy contributions of each particle leads to naively divergent sums that need to be regulated. Meanwhile, an independent formulation of even the classical bulk theory for the tensionless string is lacking. Even if we were to attempt to use the CFT data to reconstruct the bulk by identifying the CFT single trace operator spectrum to the spectrum of bulk particles, there is so far no simple closed form expression for the operator spectrum of a matrix valued CFT [35,36].

For these reasons, an alternate approach was adopted in [37–40] which bypasses both these problems by expressing the one-loop vacuum energy of a given field in terms of a linear operator acting on the conformal algebra character corresponding to the field. For technical reasons, this was referred to as the *Character Integral Representation of the Zeta Function (CIRZ) method*. This method completely reproduces the previous results for Vasiliev's higher spin theory as well as readily extracts the answers for the tensionless string as well as its bosonic cousins, the bulk duals of the free $SU(N)$-adjoint scalar CFT and free $SU(N)$ Yang-Mills. In particular, it was found that the one-loop free energies of these bulk dual theories are non-zero, and equal to minus of the one-loop free energy of the corresponding boundary conformal field (scalar, spin-1, etc.). Further, the computations involved undergo simplifications for the *maximally* supersymmetric case which are seemingly quite miraculous [40].

In this paper we shall discuss the extension of these results to the free CFTs in the adjoint representation of $SO(N)$ and $Sp(N)$, as well as the bi-fundamental and the bi-vector representation of $U(N) \times U(M)$ and $O(N) \times O(M)$, respectively. We concentrate our consideration on the AdS_5/CFT_4 dualities, but all our analysis can be generalized to any even d in a straightforward manner and to odd d with a bit more effort (see [37] for AdS_4/CFT_3 case, and [33] for a generalization to arbitrary dimensions.). Another aim of the current paper is to provide a concise summary of the series of our recent works [37–39] and to append more details on the relevant technicalities such as the spectral analysis of AdS space and the operator counting problem.

1.1. Organization of Paper

A brief overview of this paper is as follows. In Sections 2 and 3 we shall review the formalism for one-loop computations in AdS_5, recollecting the essential results for computing the Casimir Energy and vacuum energy at one-loop. Section 4 provides a few more details about the duality between the tensionless string and free matrix models, focusing strongly on a pedagogical treatment of Polya counting, an essential tool for many of the computations presented here. Section 5 then presents the applications of this formalism to adjoint $Sp(N)$ and $SO(N)$ CFTs, namely free scalar, Yang Mills and $\mathcal{N} = 4$ SYM, and bi-fundamental and bi-vector scalar and fermion models. Finally, some more technical details are reviewed in the Appendices. Appendix A contains a review of some facts of harmonic analysis on AdS spaces which are useful to these computations, while Appendix B reviews key features of unitary representations of $so(2,4)$. Finally, Appendix C contains an overview of the applications of the methods of Sections 2 and 3 to the higher-spin/CFT dualities.

2. Casimir Energy in Thermal AdS_5

We begin with how the one-loop AdS/CFT Casimir energy may be computed in thermal AdS_5. In particular, we will review the observation of [25] that 'naive' computation of the AdS/CFT Casimir energy for higher-spin theories yields a divergent answer which may be suitably regularized to obtain a result consistent with CFT expectations. Importantly, the latter regularization also does not require us to know the precise spectrum of the theory, except in some implicit way through the thermal partition function of the theory, computed in the *canonical ensemble*. This is discussed below. Equally importantly, the computations here contain the same key idea which is very useful for the analysis presented later, but in a simpler setting.

We begin with the Vasiliev Type A theory in AdS_5. Its duality is discussed at somewhat greater length in Appendix C but for now it is sufficient to note that the *non-minimal* Vasiliev theory contains massless spins from spin equal to 1 to infinity appearing once each in the spectrum, along with a scalar with $\Delta = 2$, and is dual to the $U(N)$ vector model. Further, there is a *minimal* Vasiliev system arrived at by truncating the non-minimal one to even spins only, and this is dual to the $O(N)$ vector model. Next, we note that the Casimir energy of a massless spin-s field in AdS_5 is given by [25]

$$E_c^{(s)} = -\frac{1}{1440} s(s+1)\left[18s^2(s+1)^2 - 14s(s+1) - 11\right]. \tag{2}$$

While the scalar of the theory is not massless, its Casimir energy can be determined from the formula (2) by setting $s = 0$ in it. Therefore the Casimir energy for the *non-minimal* AdS theory is given by $E_c = \sum_{s=0}^{\infty} E_c^{(s)}$, which is clearly divergent. This divergence can be regularized by means of an appropriate zeta function, or by inserting an exponential damping $e^{-\epsilon\left(s+\frac{1}{2}\right)}$ when evaluating the sum and discarding all terms divergent in ϵ in the limit $\epsilon \to 0$. We thus obtain [25]

$$E_c = \sum_{s=0}^{\infty} E_c^{(s)} e^{-\epsilon\left(s+\frac{1}{2}\right)}|_{\text{finite}} = 0. \tag{3}$$

As is apparent from the above analysis, carrying out this computation requires knowledge of the precise spectrum of the theory, along with a prescription for regulating the divergence for summing over the infinite number of fields in the spectrum of the theory. This data is unavailable for the bulk duals of matrix CFTs at present. We will now show in below how this requirement may be evaded[2]. Our starting point is the relation between the (blind) character

$$\chi_{\mathcal{V}}(\beta) = \text{Tr}\left(e^{-\beta H}\right) = \sum_n d_n e^{-\beta E_n}, \tag{4}$$

computed over the UIR $\mathcal{V}$ of $so(2,4)$ and the Casimir energy in AdS of the corresponding field. Here (E_n, d_n) are the eigenvalues and degeneracies of the hamiltonian H. Given $\chi_{\mathcal{V}}(\beta)$ we may take its Mellin transform to obtain $\tilde{\chi}_{\mathcal{V}}(s)$ as

$$\tilde{\chi}_{\mathcal{V}}(z) = \mathcal{L}_{\text{Mellin}}[\chi_{\mathcal{V}}; z] \equiv \int_0^{\infty} d\beta \frac{\beta^{z-1}}{\Gamma(z)} \chi_{\mathcal{V}}(\beta) = \sum_n d_n E_n^{-z}, \tag{5}$$

which implies

$$\tilde{\chi}_{\mathcal{V}}(-1) = \sum_n d_n E_n = E_c, \tag{6}$$

where E_c is the Casimir energy [41]. Anticipating future developments, in the above we have defined a linear functional $\mathcal{L}_{\text{Mellin}}$ which acts on the character $\chi_{\mathcal{V}}$ to return the Mellin transform. Note that it is straightforward to apply (6) to the case where $\mathcal{V}$ is the short representation $\left(s+2, \frac{s}{2}, \frac{s}{2}\right)$ to obtain the expression (2) for the Casimir energy of a massless spin-s field. For fermions, the Casimir energy is defined with an overall minus sign, so we insert the fermion number operator into the character and define a partition function $Z_{\mathcal{V}}(\beta) = \text{Tr}_{\mathcal{V}}\left((-1)^F e^{-\beta H}\right)$ in terms of which we obtain E_c as (5) and (6).

2 Somewhat related arguments are also implicit in some computations of [25]. In particular, note their computations from Equations (5.16)–(5.21) which are essentially a 'one-shot' computation of the full bulk Casimir energy from the thermal partition function, much as we present here in (8).

Now we use the fact that the Hilbert space $\mathcal{H}$ of one-particle excitations of the AdS_5 theory decomposes by definition into UIRs of the conformal algebra $so(2,4)$. Further, again by definition

$$\mathcal{Z}_{\mathcal{H}}(\beta) = \text{Tr}_{\mathcal{H}}\left((-1)^F e^{-\beta H}\right) = \sum_{\{b\}} n_b \chi_b(\beta) - \sum_{\{f\}} n_f \chi_f(\beta), \tag{7}$$

where $\{b\}$ and $\{f\}$ denote respectively the sets of bosonic and fermionic fields in the theory. Then we may use the linearity property of $\mathcal{L}_{\text{Mellin}}$ and act with it on the total partition function $\mathcal{Z}_{\mathcal{H}}(\beta)$ to find the total Casimir energy

$$E_c = \tilde{\mathcal{Z}}_{\mathcal{H}}(-1); \quad \text{where} \quad \tilde{\mathcal{Z}}_{\mathcal{H}}(z) = \mathcal{L}_{\text{Mellin}}[\mathcal{Z}_{\mathcal{H}}; z]. \tag{8}$$

It turns out that in all cases of which we are aware, this definition of the Casimir energy perfectly reproduces the expressions found by the regularots such as (3) that are used in the literature. Further, often it is possible to evaluate the full partition function $\chi_{\mathcal{H}}$ without knowing the explicit spectrum of the theory. Indeed matrix CFTs are an example of this possibility, as we review below. Therefore, the definition (8) is particularly useful to apply to the cases of matrix model CFTs which we encounter in 'stringy' AdS/CFT dualities.

Finally, we also note that (5) may be efficiently evaluated by deforming the contour of β integration, which originally stretches along the positive real β axis from 0 to ∞, to Figure 1 to get

$$\tilde{\mathcal{Z}}_{\mathcal{H}}(z) = \frac{i}{2\sin(\pi z)} \oint_C d\beta \frac{\beta^{z-1}}{\Gamma(z)} \mathcal{Z}_{\mathcal{H}}(\beta). \tag{9}$$

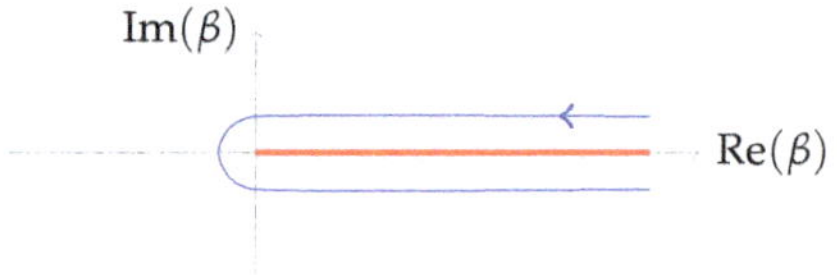

Figure 1. Integration contour for the zeta function.

Further, if the partition function $\chi_{\mathcal{H}}(\beta)$ has no singularities on the positive axis of β except for poles at $\beta = 0$, then the contour C can be shrunk to a small circle around $\beta = 0$ to give

$$E_c = -\frac{1}{2} \oint_C \frac{d\beta}{2\pi i \beta^2} \mathcal{Z}_{\mathcal{H}}(\beta), \tag{10}$$

which may be evaluated by the residue theorem. We therefore find that the AdS Casimir energy is simply $-\frac{1}{2}$ of the $\mathcal{O}(\beta)$ term in the Laurent series expansion of $\mathcal{Z}_{\mathcal{H}}(\beta)$ about $\beta = 0$.

3. Formalism for One-Loop Computations in AdS_5

In this section we review the formalism and techniques for carrying out one-loop computations in AdS_5. The techniques are more generally applicable and extend to arbitrary odd-dimensional AdS spaces straightforwardly and even-dimensional AdS spaces with a bit more effort.

3.1. *Vacuum Energy in AdS_5*

Evaluating the one-loop partition function of a quantum theory reduces to the problem of evaluating functional determinants:

$$Z^{(1)} = \int \mathcal{D}\Psi\, e^{-\frac{1}{2}\langle\Psi,\mathcal{K}\Psi\rangle} = \frac{1}{\sqrt{\det \mathcal{K}}}, \qquad \mathcal{K}(x_1, x_2) = \frac{\delta^2 S[\Phi]}{\delta\Phi(x_1)\delta\Phi(x_2)}\Big|_{\Phi=\bar{\Phi}}, \tag{11}$$

where $S[\Phi]$ is the classical action and $\Phi = \bar{\Phi} + \Psi$. The $\bar{\Phi}$ and Ψ are a background field and the fluctuation over it, respectively. The bracket $\langle \cdot, \cdot \rangle$ is the scalar product defined by

$$\langle \Psi_1, \Psi_2 \rangle = \int d^D x \sqrt{g}\, \Psi_1(x)^* \cdot \Psi_2(x) . \tag{12}$$

Here, we suppressed all the indices for simplicity but one should understand that the fields Φ or Ψ are tensors in general. '$\cdot$' is a Lorentz invariant scalar product that contracts the (suppressed) spin indices of the fields Ψ.

The one-loop free energy $\Gamma^{(1)}$ or the vacuum energy is simply

$$\Gamma^{(1)} = -\ln Z^{(1)} = \frac{1}{2} \operatorname{Tr} \ln \mathcal{K} , \tag{13}$$

hence, we need to evaluate the Tr ln (or functional determinant) of the operator $\mathcal{K}$. If we treat the operator $\mathcal{K}$ as if it is a finite dimensional diagonalizable matrix with eigenvalues κ_n, then we would get

$$\operatorname{Tr} \ln \mathcal{K} = \sum_n d_n \ln \kappa_n , \tag{14}$$

where n parametrizes the eigenvalue and d_n is the degeneracy. Defining the zeta function $\zeta(z)$ to be

$$\zeta(z) = \sum_n \frac{d_n}{\kappa_n{}^z} , \tag{15}$$

it is easy to see that

$$\zeta'(0) = -\operatorname{Tr} \ln \mathcal{K} = -2\,\Gamma^{(1)} . \tag{16}$$

However, expressions such as (15) are not ideally suited for a direct evaluation in the case of a differential operator $\mathcal{K}$. Typically, the naive degeneracy corresponding to a given eigenvalue is infinite. We shall therefore use the fact that given an orthonormal set of eigenvectors[3] $\left\{ \Psi_m^{(n)} \right\}$ belonging to the eigenvalue κ_n, the degeneracy may be defined as

$$d_n = \sum_m \langle \Psi_m^{(n)}, \Psi_m^{(n)} \rangle . \tag{17}$$

We emphasize that though (17) is a tautology for compact spaces, for non-compact spaces it is essentially a non-trivial definition. When evaluated explicitly, the answer is still divergent but may be regulated in accordance with general principles of AdS/CFT. We shall be applying these methods to the typical kinetic operators $\mathcal{K} = -\Box + c$ in AdS$_5$, so it is useful to specialize a little to that case. It turns out that the spectrum of eigenvalues is continuous, labeled by a positive real number u, and is given by

$$\kappa_u = u^2 + c' . \tag{18}$$

3 The operators we are interested in will be of the form $-\Box + c$ where c is a constant and $-\Box = g^{\mu\nu}\nabla_\mu \nabla_\nu$. The spectral problem for operators of this form has been explicitly solved for a wide class of spin fields in AdS space. In contrast, if we wish to compute the same determinants over quotients of AdS, in principle we have to impose quantization conditions over $\Psi_m^{(n)}$. This may prove easy or difficult depending on the orbifold at hand. Nonetheless, for the quotients we are interested in, it is possible to compute the determinants on the quotient space by the method of images. We review these facts in Appendix.

Here, c' is a constant number, essentially encoded in the parameter c appearing in the bulk kinetic operator. For physical fields it will be related to Δ, the conformal dimension of the dual operator on the boundary. The zeta function for the operator $\mathcal{K}$ is then given by

$$\zeta(z) = \int du \, \frac{\sum_m \langle \Psi_m^{(u)}, \Psi_m^{(u)} \rangle}{(u^2 + c')^z}, \tag{19}$$

where the wave functions $\Psi_m^{(u)}$ now obey the orthonormality conditions

$$\langle \Psi_m^{(u)}, \Psi_{m'}^{(u')} \rangle = \delta(u - u')\, \delta_{m,m'} \,. \tag{20}$$

We now specialize to the case of global AdS_5, to develop general expressions useful for the forthcoming analysis. In this case, for a wide class of fields, the eigenfunctions of the Laplace operator $\Box = g^{\mu\nu}\nabla_\mu \nabla_\nu$ have been explicitly computed [18–21,42]. Further, using the homogeneity of AdS, it follows that $\sum_m \Psi_m^{(u)}(x)^* \cdot \Psi_m^{(u)}(x)$ is independent of $x \in AdS_5$,[4] and we define

$$\sum_m \Psi_m^{(u)}(x)^* \cdot \Psi_m^{(u)}(x) = \sum_m \Psi_m^{(u)}(0)^* \cdot \Psi_m^{(u)}(0) \equiv \mu(u) \,. \tag{21}$$

Here $x = 0$ is a point on AdS_5 which may be arbitrarily chosen. In practice, it is chosen so that all but a finite number of eigenfunctions $\Psi_m^{(u)}(x)$ vanish at that point, and the sum over m may be easily evaluated. Finally, we see that (19) reduces to

$$\zeta(z) = \mathrm{Vol}_{\mathrm{AdS}_5} \int du \, \frac{\mu(u)}{(u^2 + c')^z}, \tag{22}$$

where $\mu(u)$ plays the role of measure over the parameter u which indexes the eigenvalues of the Laplacian. It is known as the Plancherel measure.

For the operator $\mathcal{K}$ corresponding the irreducible representation $\mathcal{D}(\Delta, (\ell_1, \ell_2))$, the constant c' is given by

$$c' = (\Delta - 2)^2, \tag{23}$$

and the measure $\mu(u)$ by

$$\mu(u) = \frac{1}{3\pi^2} \frac{\ell_1 + \ell_2 + 1}{2} \frac{\ell_1 - \ell_2 + 1}{2} \left(u^2 + (\ell_1 + 1)^2\right) \left(u^2 + \ell_2^2\right). \tag{24}$$

The derivation of the c' and $\mu(u)$ is provided in Appendix A. The factor of the volume $\mathrm{Vol}_{\mathrm{AdS}_5}$ is infinity due to the non-compactness nature of AdS space. This IR divergence can be also regularized as

$$\mathrm{Vol}_{AdS_5} = \pi^2 \log(\mu R), \tag{25}$$

where R is the raduis of AdS space and μ the renormalization scale. See [43] for the details and discussions. With the above result, suppressing the μ dependence, the AdS_5 vacuum energy is given always proportional to $\log R$.

---

4 This statement is a generalization of the addition theorem for spherical harmonics on S^2 to general spin fields on symmetric spaces, and is particularly transparent when the group theory underlying harmonic analysis on symmetric spaces is used. These facts are reviewed in Appendix A.

3.2. Character Integral Representation of Zeta Function

We have seen in Section 2 that the Casimir energy for a theory in thermal AdS_5 is naturally encoded in the thermal partition function, or the blind character in other words, via a linear operator acting on it. This provides a natural resummation of the Casimir energies of the individual fields in the spectrum. In [37], we have shown how the character may be similarly used to resum the one-loop free energies in *global* AdS_5. That is, there exists a linear operator $\mathcal{L}$ which, like $\mathcal{L}_{\text{Mellin}}$ of Section 2, acts on the character over a UIR $\mathcal{V}$ of $so(2,4)$ and returns the one-loop vacuum energy of the corresponding field, now in global AdS. $\mathcal{L}$ again takes the form of a β integral over the character, now with additional operations included, and returns the zeta function corresponding to the one-loop determinant, as defined in Section 3.1. For this reason, we refer to this method as Character Integral Representation of Zeta function (CIRZ).

Let us provide a brief summary of the result of [37]. The zeta function for a Hilbert space $\mathcal{H}$—which might be a single UIR space or any collection of them—can be written as the sum of three pieces:

$$\zeta_{\mathcal{H}}(z) = \zeta_{\mathcal{H}|2}(z) + \zeta_{\mathcal{H}|1}(z) + \zeta_{\mathcal{H}|0}(z)\,, \tag{26}$$

where $\zeta_{\mathcal{H}|n}$ are the Mellin transforms,

$$\frac{\Gamma(z)\,\zeta_{\mathcal{H}|n}(z)}{\log R} = \int_0^\infty d\beta\, \frac{\left(\frac{\beta}{2}\right)^{2(z-1-n)}}{\Gamma(z-n)}\, f_{\mathcal{H}|n}(\beta)\,, \tag{27}$$

of the functions $f_{\mathcal{H}|n}(\beta)$ given by

$$\begin{aligned}
f_{\mathcal{H}|2}(\beta) &= \frac{\sinh^4 \frac{\beta}{2}}{2}\, \chi_{\mathcal{H}}(\beta,0,0)\,,\\
f_{\mathcal{H}|1}(\beta) &= \sinh^2 \tfrac{\beta}{2} \left[\frac{\sinh^2 \frac{\beta}{2}}{3} - 1 - \sinh^2 \tfrac{\beta}{2} \left(\partial_{\alpha_1}^2 + \partial_{\alpha_2}^2\right)\right] \chi_{\mathcal{H}}(\beta,\alpha_1,\alpha_2)\Big|_{\alpha_i=0}\,,\\
f_{\mathcal{H}|0}(\beta) &= \left[1 + \frac{\sinh^2 \frac{\beta}{2}\left(3 - \sinh^2 \frac{\beta}{2}\right)}{3}\left(\partial_{\alpha_1}^2 + \partial_{\alpha_2}^2\right)\right.\\
&\qquad \left. - \frac{\sinh^4 \frac{\beta}{2}}{3}\left(\partial_{\alpha_1}^4 - 12\,\partial_{\alpha_1}^2\partial_{\alpha_2}^2 + \partial_{\alpha_2}^4\right)\right] \chi_{\mathcal{H}}(\beta,\alpha_1,\alpha_2)\Big|_{\alpha_i=0}\,.
\end{aligned} \tag{28}$$

Here $\chi_{\mathcal{H}}$ is the character defined by

$$\chi_{\mathcal{H}}(\beta,\alpha_1,\alpha_2) = \text{Tr}_{\mathcal{H}}\left(e^{-\beta H + i\alpha_1 M_{12} + i\alpha_2 M_{34}}\right). \tag{29}$$

Since both of the relations (27) and (28) are linear, they define a linear map $\mathcal{L}$ between the zeta function and the character: $\zeta_{\mathcal{H}}(z) = \mathcal{L}[\chi_{\mathcal{H}}; z]$.

One can recast the β integral (27) with sufficently large $\text{Re}(z)$ into an integral over the contour which runs from the positive real infinity and encircles the branch cut generated by $\beta^{2(z-1-n)}$ in the counter-clockwise direction (see Figure 1) as

$$\frac{\Gamma(z)\,\zeta_{\mathcal{H}|n}(z)}{\log R} = \frac{i\,(-2)^{2(n+1-z)}}{2\,\sin(2\pi z)\,\Gamma(z-n)} \oint_C d\beta\, \frac{f_{\mathcal{H}|n}(\beta)}{\beta^{2(n+1-z)}}\,. \tag{30}$$

Now the right hand side of the above equation is well defined in the $z \to 0$ limit. Defining

$$\gamma_{\mathcal{H}|n} = -(-4)^n\, n! \oint \frac{d\beta}{2\pi i}\, \frac{f_{\mathcal{H}|n}(\beta)}{\beta^{2(n+1)}}\,, \tag{31}$$

the total one-loop vacuum energy of the AdS_5 theory is given by the sum

$$\Gamma^{(1)}_{\mathcal{H}} = \log R \left(\gamma_{\mathcal{H}|2} + \gamma_{\mathcal{H}|1} + \gamma_{\mathcal{H}|0}\right). \tag{32}$$

When the function $f_{\mathcal{H}|n}$ does not have any singularities in positive real axis of β the contour can be eventually shrunken to a small circle around $\beta = 0$. The functions $f_{\mathcal{H}|n}$ for any one particle state in AdS_5 as well as for the spectrum of Vasiliev's theory indeed satisfy this property. However, quite generically, the functions $f_{\mathcal{H}|n}$ for the AdS dual to a matrix model CFT do have additional poles or branch cuts. Physically, this corresponds to the fact that higher-spin theories do not have a Hagedorn transition [6] while string theory does [7].

When there are bulk fermionic degrees of freedom to be summed over, as in the case of supersymmetric theories, it is sufficient to use the CIRZ method as presented for bosons, but instead of using the thermal partition function, use the weighted partition function

$$Z_{\mathcal{H}}(\beta, \alpha_1, \alpha_2) = \mathrm{Tr}_{\mathcal{H}}\left((-1)^F\, e^{-\beta\, H + i\,\alpha_1\, M_{12} + i\,\alpha_2\, M_{34}}\right), \tag{33}$$

introduced in [11,44].

4. Computing Partition Functions by Polya Counting

The large N expansion qualitatively works in the same way in the vector model as the matrix one: all the correlation functions can be organized in terms of the color loop number. At the leading order of N, it is sufficient to consider the single trace operators, i.e., those made with a single color loop. On these single trace operators, there is important difference between vector models and matrix models. In vector models, the single trace operators are the scalar product of two fields in vector representation,

$$\partial\partial\cdots\partial\vec{\phi}_1\cdot\partial\partial\cdots\vec{\phi}_2\,, \tag{34}$$

whereas in matrix models they are the traces of arbitrary number of fields in a matrix valued representation,

$$\mathrm{Tr}\Big[(\partial\partial\cdots\partial\boldsymbol{\phi}_1)(\partial\partial\cdots\partial\boldsymbol{\phi}_2)\cdots(\partial\partial\cdots\partial\boldsymbol{\phi}_n)\Big]. \tag{35}$$

In (34) and (35), we suppressed all the indices and the field operators ϕ_n can be either scalar, spinor or vector in four dimensions. Therefore, even though the number of single trace operators is infinite in both cases, the number is infinitely larger in matrix models than vector models.

One of difficulties in matrix models is the control or organization of infinitely many single trace operators. For this reason, one often focuses on a certain class of single trace operators, such as BPS operators, whose study does not invoke the knowledge of the rest of operators. However, in studying the total one-loop Casimir or vacuum energy, we need the full operator spectrum of the theory. This can in principle be identified by decomposing the operators (35) into irreducible $so(2,4)$ representations. The decomposition requires a particular symmetrization of indices which projects the operator (35) to the irreducible representation. Finding out the exact forms of these projections is not easy, but this process can be cast as a standard group theoretical problem. For the scalar product or inside of a trace, we put derivatives of field operator up to its equation. They form a basis for the Hilbert space V of the conformal field ϕ carrying a short-representation of $so(2,4)$:

$$V = \mathrm{Span}\{\,\phi,\,\partial\phi,\,\partial\partial\phi,\ldots\} = \mathrm{Span}\{v_i\}\,, \tag{36}$$

where the i is the index indicating one of descedant (or primary) states of ϕ, hence it is infinite dimensional.

In vector models, we construct single trace operators by using two elements of V as (34), hence the vector space of single trace operators is the tensor product $V \otimes V$. When the field $\vec{\phi}_1$ and $\vec{\phi}_2$ are the same, which is the case in the $O(N)$ vector model, single trace operators are symmetric in the exchange of 1 and 2 as the latter label is dummy. Then the corresponding vector space of single trace operators are the symmetrized tensor product of two V's, denoted by $V \vee V$. In order to find out the operator spectrum, we need to decompose $V \vee V$ into $so(2,4)$ UIRs and it can be conveniently done in terms of the $so(2,4)$ characters. In addition, the symmetrizations of the tensor product, or the plethysm, can be also handily treated at the level of characters. Suppose that for an element $g \in SO(2,4)$, V has eigenvalues $\{\lambda_i\}$ (since only the conjugacy class of g matters due to the trace, we can focus on the Cartan subgroup as in (A43)), then the character reads $\chi_V(g) = \mathrm{Tr}_V(g) = \sum_i \lambda_i$. Then, the character for $V \vee V$ is

$$\chi_{V\vee V}(g) = \sum_{i\leq j} \lambda_i\,\lambda_j = \frac{(\sum_i \lambda_i)^2 + \sum_i \lambda_i^2}{2}\,. \tag{37}$$

Since $\sum_i \lambda_i^2 = \mathrm{Tr}_V(g^2)$, we get the relation

$$\chi_{V\vee V}(g) = \frac{\chi_V(g)^2 + \chi_V(g^2)}{2}\,. \tag{38}$$

In the case of matrix models, we need to consider n tensor product of V and impose appropriate symmetrization compatible with trace and also the gauge group.

4.1. $SU(N)$ Adjoint Models

When the gauge group is $SU(N)$, the fields $\partial\partial\cdots\partial\phi_p$ have the cyclic symmetry in $p \to p+1$ ($p = 1,\ldots n, p+1 \equiv 1$) due to the trace operation. Then, the projection to the cyclic invariant requires only some combinatorial consideration. For intuitive understanding let us consider a few lower n's where n is the number of operators in the trace. First, the $n = 2$ cyclic symmetry is nothing but the permutation symmetry. For $n = 3$, the character of cyclic 3 tensor product of V, denoted by $\mathrm{Cyc}_3(V)$, is

$$\begin{aligned}\chi_{\mathrm{Cyc}_3(V)}(g) &= \sum_{i=j=k\,\vee\, i=j\neq k\,\vee\, i<j<k\,\vee\, i<k<j} \lambda_i\,\lambda_j\,\lambda_k \\ &= \sum_i \lambda_i^3 + \sum_{i\neq j} \lambda_i^2\,\lambda_j + 2\sum_{i<j<k} \lambda_i\,\lambda_j\,\lambda_k\,.\end{aligned} \tag{39}$$

where the summation over (i,j,k) is chosen for the proper counting of elements with the cyclicity $(i,j,k) \equiv (j,k,i)$. Since

$$\left(\sum_i \lambda_i\right)^3 = \sum_i \lambda_i^3 + 3\sum_{i\neq j} \lambda_i^2\,\lambda_j + 6\sum_{i<j<k} \lambda_i\,\lambda_j\,\lambda_k\,, \tag{40}$$

We find that

$$\chi_{\mathrm{Cyc}_3(V)}(g) = \frac{(\sum_i \lambda_i)^3 + 2\,\sum_i (\lambda_i)^3}{3} = \frac{\chi_V(g)^3 + 2\,\chi_V(g^3)}{3}\,. \tag{41}$$

As one can see from this example, the point is the counting of $(i_1,\ldots,i_n)$ taking into account the cyclic equivalence. This is well-know problem of counting inequivalent necklaces with n beads (see Figure 2). The index i_p indicates the type or color of beads (which, in our context, corresponds to the descendant state of the conformal field ϕ).

The solution to this problem is provided by Polya's enumeration theorem as

$$\chi_{\mathrm{Cyc}_n(V)}(g) = \sum_{\text{cyclic }(i_1,\ldots,i_n)} \lambda_{i_1}\cdots\lambda_{i_n} = \frac{1}{n}\sum_{k=1}^{n}\left(\sum_i \lambda_i^{\frac{n}{\gcd(k,n)}}\right)^{\gcd(k,n)}\,. \tag{42}$$

Here, $\gcd(k,n)$ is the greatest common divisor of k and n. Alternatively, the above can be written as

$$\chi_{\mathrm{Cyc}_n(V)}(g) = \frac{1}{n}\sum_{k|n}\varphi(k)\left(\sum_i \lambda_i^k\right)^{\frac{n}{k}} = \frac{1}{n}\sum_{k|n}\varphi(k)\,\chi_V(g^k)^{\frac{n}{k}}, \tag{43}$$

where $k|p$ denotes the divisor k of p and the Euler totient function $\varphi(p)$ is the number of relative primes of p in $\{1,\ldots,p\}$.

The total partition function is the sum of the above from $n=2$ to infinity. Hence, we are temped to sum $\chi_{\mathrm{Cyc}_n(V)}$ over n. It turns out that it is possible to at least partially sum over n in (43) via [7,45,46]

$$\begin{aligned}\chi_{\mathrm{Cyc}(V)}(g) &= \sum_{n=2}^{\infty}\chi_{\mathrm{Cyc}_n(V)}(g) = \sum_{n=2}^{\infty}\frac{1}{n}\sum_{k|n}\varphi(k)\,[\chi_V(g^k)]^{\frac{n}{k}}\\ &= -\chi_V(g) + \sum_{k=1}^{\infty}\sum_{m=1}^{\infty}\frac{1}{m\,k}\,\varphi(k)\,[\chi_V(g^k)]^m\\ &= -\chi_V(g) + \sum_{k=1}^{\infty}\frac{\varphi(k)}{k}\,\chi_{\log,k(V)}(g),\end{aligned} \tag{44}$$

where $\chi_{\log,k(V)}$ are given by

$$\chi_{\log,k(V)}(\beta,\alpha_1,\alpha_2) = -\log\left[1-\chi_V(k\beta,k\alpha_1,k\alpha_2)\right]. \tag{45}$$

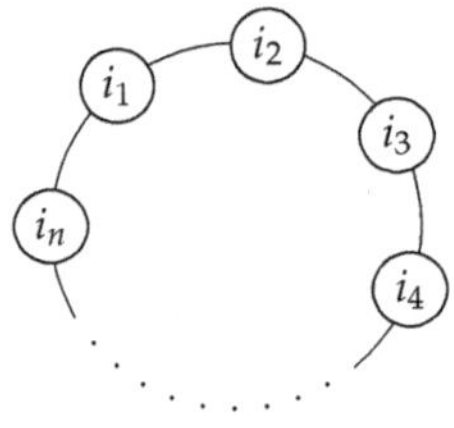

Figure 2. Necklace with n beads.

4.2. $Sp(N)$ and $SO(N)$ Adjoint Models

Now let us turn to the cases where the field $\boldsymbol{\phi}$ takes value in the adjoint representation $Sp(N)$. Then the field is symmetric: $\boldsymbol{\phi}^t = \boldsymbol{\phi}$. Consequently, the single-trace operators (35) admit the (anti-)symmetry,

$$\mathrm{Tr}\left(\partial^{k_1}\boldsymbol{\phi}\,\partial^{k_2}\boldsymbol{\phi}\cdots\partial^{k_n}\boldsymbol{\phi}\right) = \mathrm{Tr}\left(\partial^{k_n}\boldsymbol{\phi}\cdots\partial^{k_2}\boldsymbol{\phi}\,\partial^{k_1}\boldsymbol{\phi}\right), \tag{46}$$

under the flip of the $\partial^k\boldsymbol{\phi}$ ordering inside the trace. In terms of indices,

$$(i_1,i_2,\ldots,i_n) \equiv (i_n,i_{n-1},\ldots,i_1). \tag{47}$$

Hence, the space of independent single-trace operators corresponds now to the subspace invariant under the actions of the dihedral group Dih_n (which includes also reflections on top of the cyclic rotations).

Let us again start the discussion with $n=3$ case:

$$\chi_{\mathrm{Dih}_3^+(V)}(g) = \sum_{i=j=k\,\vee\, i=j,k\,\vee\, i<j<k}\lambda_i\,\lambda_j\,\lambda_k = \sum_i \lambda_i^3 + \sum_{i\neq j}\lambda_i^2\,\lambda_j + \sum_{i<j<k}\lambda_i\,\lambda_j\,\lambda_k. \tag{48}$$

where the summation over (i,j,k) is chosen for the proper counting of elements with the cyclicity $(i,j,k) \equiv (j,k,i) \equiv (i,k,j)$. Note that for $n = 3$, the dihedral group coincides with the symmetric group. The above can be written as

$$\chi_{\mathrm{Dih}_3^+(V)}(g) = \frac{(\sum_i \lambda_i)^3 + 3\left(\sum_i \lambda_i^2\right)\left(\sum_j \lambda_j\right) + 2\sum_i \lambda_i^3}{6}, \tag{49}$$

using (40) and

$$\left(\sum_i \lambda_i^2\right)\left(\sum_j \lambda_j\right) = \sum_i \lambda^3 + \sum_{i\neq j} \lambda_i^2 \lambda_j. \tag{50}$$

In terms of the basic character χ_V, it reads

$$\chi_{\mathrm{Dih}_3^+(V)}(g) = \frac{\chi_V(g)^3 + 2\chi_V(g^3) + 3\chi_V(g^2)\chi_V(g)}{6} = \frac{1}{2}\chi_{\mathrm{cyc}^3(V)}(g) + \frac{1}{2}\chi_V(g)\chi_V(g^2). \tag{51}$$

As one can see from this $n = 3$ example, the character of dihedrial tensor-product space $\mathrm{Dih}_n^+(V)$ is roughly half of the cyclic one, but not exactly. The precise formula is

$$\chi_{\mathrm{Dih}_n^+(V)}(g) = \frac{1}{2}\chi_{\mathrm{Cyc}_n(V)}(g) + \begin{cases} \frac{1}{2}\chi_V(g)\chi_V(g^2)^{\frac{n-1}{2}} & [n \text{ odd}] \\ \frac{1}{4}\left(\chi_V(g)^2\chi_V(g^2)^{\frac{n-2}{2}} + \chi_V(g^2)^{\frac{n}{2}}\right) & [n \text{ even}] \end{cases}. \tag{52}$$

If the field $\boldsymbol{\phi}$ takes value in the adjoint representation of $SO(N)$, it is antisymmetric: $\boldsymbol{\phi}^t = -\boldsymbol{\phi}$. Then, the single trace operators have the following reflection property,

$$(i_1, i_2, \ldots, i_n) \equiv (-1)^n\,(i_n, i_{n-1}, \ldots, i_1), \tag{53}$$

in addition to the cyclicity. Due to the factor $(-1)^n$ we have less number of operators in the $SO(N)$ case compared to the $Sp(N)$ case.

Again, let us consider the $n = 3$ example. Since $(i,j,k) \equiv -(k,j,i)$, any repeated index vanish: $(i,i,j) \equiv (i,j,i) \equiv -(i,j,i)$. In the end, only strictly ordered set (i,j,k) with $i < j < k$ survive. Hence, the character is

$$\begin{aligned} \chi_{\mathrm{Dih}_3^-(V)}(g) &= \sum_{i<j<k} \lambda_i \lambda_j \lambda_k = \frac{(\sum_i \lambda_i)^3 - 3\left(\sum_i \lambda_i^2\right)\left(\sum_j \lambda_j\right) + 2\sum_i \lambda_i^3}{6} \\ &= \frac{\chi_V(g)^3 + 2\chi_V(g^3) - 3\chi_V(g^2)\chi_V(g)}{6} = \frac{1}{2}\chi_{\mathrm{cyc}^3(V)}(g) - \frac{1}{2}\chi_V(g)\chi_V(g^2). \end{aligned} \tag{54}$$

One can notice that compared to the $Sp(N)$ case of (51), we have minus sign after the last equality. This pattern extends to an arbitrary odd n:

$$\chi_{\mathrm{Dih}_n^-(V)}(g) = \frac{1}{2}\chi_{\mathrm{Cyc}_n(V)}(g) + \begin{cases} -\frac{1}{2}\chi_V(g)\chi_V(g^2)^{\frac{n-1}{2}} & [n \text{ odd}] \\ \frac{1}{4}\left(\chi_V(g)^2\chi_V(g^2)^{\frac{n-2}{2}} + \chi_V(g^2)^{\frac{n}{2}}\right) & [n \text{ even}] \end{cases}, \tag{55}$$

because, for odd n, cyclic operators can be split into either symmetric or anti-symmetric ones under reflection.

Finally, one may attempt to sum $\chi_{\mathrm{Dih}_n^\pm(V)}$ over n. The term half of cyclic character can be treated as the cyclic case, whereas the additional contributions can be summed as they are geometric series.

It will be useful to consider the two partition functions obtained by summing separately over the even and odd values of n. In particular

$$\begin{aligned} \chi_{\text{even}(V)}(g) &= \frac{1}{4}\sum_{n=2,4,6,\cdots}^{\infty}\left(\chi_V(g)^2\chi_V(g^2)^{\frac{n-2}{2}}+\chi_V(g^2)^{n/2}\right)=\frac{1}{4}\frac{\chi_V(g)^2+\chi_V(g^2)}{1-\chi_V(g^2)}, \\ \chi_{\text{odd}(V)}(g) &= \frac{1}{2}\sum_{n=3,5,7,\cdots}^{\infty}\chi_V(g)\,\chi_V(g^2)^{\frac{n-1}{2}}=\frac{1}{2}\frac{\chi_V(g)\,\chi_V(g^2)}{1-\chi_V(g^2)}. \end{aligned} \tag{56}$$

In the end we get

$$\chi_{\text{Dih}^{\pm}(V)}(g)=\frac{1}{2}\chi_{\text{Cyc}(V)}(g)+\chi_{\text{even}(V)}(g)\pm\chi_{\text{odd}(V)}(g)\,. \tag{57}$$

4.3. $U(N)\times U(M)$ Bi-Fundamental and $O(N)\times O(M)$ Bi-Vector Models

If the conformal fields $\boldsymbol{\phi}$ carry bi-fundamental representations with respect to $U(N)$ and $U(M)$, hence taking value in $M\times N$ complex matrix, then the single trace operators will take the form of

$$\text{Tr}\left(\partial^{k_1}\boldsymbol{\phi}\,\partial^{k_2}\boldsymbol{\phi}^{\dagger}\,\cdots\,\partial^{k_{2n}}\boldsymbol{\phi}^{\dagger}\right). \tag{58}$$

Note here that the operators involve always even number of fields in $\boldsymbol{\phi}\,\boldsymbol{\phi}^{\dagger}$ form, and the operators are invariant under the cyclic rotation by 2. This means that the basic vector space in this case is not V but $V\otimes V$ and the single trace operators with $2n$ operators is governed by the cyclic group C_n. The character can be constructed in an analogous manner and reads

$$\chi_{\text{Bf}_{2n}(V)}(g)=\frac{1}{n}\sum_{k|n}\varphi(k)\,\chi_V(g^k)^{\frac{2n}{k}}\,. \tag{59}$$

We can collect the above for $n=1,\ldots,\infty$ to get

$$\chi_{\text{Bf}(V)}(g)=\sum_{n=1}^{\infty}\chi_{\text{Bf}_{2n}(V)}(g)=-\sum_{k=1}^{\infty}\frac{\varphi(k)}{k}\log\left[1-\chi_V(g^k)^2\right]. \tag{60}$$

The final case is the $O(N)\times O(M)$ bi-vector models where the scalar fields $\boldsymbol{\phi}$ are real as opposed to the $U(N)\times U(M)$ bi-fundamental models. Hence, the space of its single-trace operators are spanned by

$$\text{Tr}\left(\partial^{k_1}\boldsymbol{\phi}\,\partial^{k_2}\boldsymbol{\phi}^{t}\cdots\partial^{k_{2n}}\boldsymbol{\phi}^{t}\right), \tag{61}$$

which has the reflection symmetry:

$$\text{Tr}\left(\partial^{k_1}\boldsymbol{\phi}\,\partial^{k_2}\boldsymbol{\phi}^{t}\cdots\partial^{k_{2n}}\boldsymbol{\phi}^{t}\right)=\text{Tr}\left(\partial^{k_{2n}}\boldsymbol{\phi}\cdots\partial^{k_2}\boldsymbol{\phi}\,\partial^{k_1}\boldsymbol{\phi}^{t}\right), \tag{62}$$

On top of the cyclic rotation by two. Again, the character of the bi-vector models is the half of the bi-fundamental ones up to the contribution from the reflection symmetries. This time, the latter is simpler and we end up with

$$\chi_{\text{Bv}_{2n}(V)}(g)=\frac{1}{2}\left(\chi_{\text{Bf}_{2n}(V)}(g)+\chi_V(g^2)^n\right). \tag{63}$$

The character for all single-trace operator is again the sum of the latter over all positive integer n and reads

$$\chi_{\text{Bv}(V)}(g)=\sum_{n=1}^{\infty}\chi_{\text{Bv}_{2n}(V)}(g)=\frac{1}{2}\left(\chi_{\text{Bf}(V)}(g)+\frac{\chi_V(g^2)}{1-\chi_V(g^2)}\right). \tag{64}$$

4.4. Symmetric Group

Finally, we can consider the operators made by conformal fields which are fully symmetric in any permutations. This is the symmetric group and the corresponding character is given by

$$\chi_{\mathrm{Sym}_n(V)}(g) = \sum_{j_1+2j_2+\cdots+nj_n=n} \prod_{k=1}^{n} \frac{\chi_V(g^k)^{j_k}}{k^{j_k}\, j_k!}, \tag{65}$$

or in terms of Bell polynomial as

$$\chi_{\mathrm{Sym}_n(V)}(g) = \frac{1}{n!} B_n(0!\,\chi_V(g), 1!\,\chi_V(g^2), \ldots, (n-1)!\,\chi_V(g^n)). \tag{66}$$

The generating function or the full partition function has rather simple form,

$$\chi_{\mathrm{Sym}(V)}(g) = \exp\left(\sum_{k=1}^{\infty} \frac{1}{k} \chi_V(g^k)\right), \tag{67}$$

sometimes referred to as plethystic exponential (PE). Notice that here we do not have the notion of large N expansion (and single trace, multi trace etc) hence we have also included the $n = 1$ operator, that is ϕ itself.

In order to see the implication of the above formula, let us consider a few toy examples. We first take the one-particle partition function of free scalar in two dimensions:

$$\chi_V(q, \bar{q}) = \frac{q}{1-q} + \frac{\bar{q}}{1-\bar{q}}. \tag{68}$$

One can evaluate the sum over k by expanding first $1/(1-q)$ and $1/(1-\bar{q})$ as

$$\sum_{k=1}^{\infty} \frac{1}{k} \frac{q^k}{1-q^k} = \sum_{k=1}^{\infty} \frac{1}{k} q^k \sum_{n=0}^{\infty} q^{kn} = \sum_{n=1}^{\infty} \log\left(\frac{1}{1-q^n}\right), \tag{69}$$

hence we get

$$\chi_{S(V)}(q, \bar{q}) = \left(\prod_{n=1}^{\infty} \frac{1}{1-q^n}\right)\left(\prod_{n=1}^{\infty} \frac{1}{1-\bar{q}^n}\right). \tag{70}$$

This differs from the partition function of free boson by $(q\bar{q})^{-1/24}/\log(q\bar{q})$. The $(q\bar{q})^{-1/24}$ factor is missing because our character did not include the $q^{-c/24}$. The $\log(q\bar{q})$ factor is due to zero mode contribution.

Let us consider the following toy partition functions inspired by the two-dimensional free boson,

$$\chi_V(q) = \frac{q}{(1-q)^d}, \tag{71}$$

which captures certain aspects of the scalar character in higher dimensions. By using

$$\frac{q}{(1-q)^d} = \frac{1}{(d-1)!} \left(\frac{\partial}{\partial s}\right)^{d-1} \frac{s^{d-1} q}{1-sq}\Big|_{s=1}, \tag{72}$$

we get

$$\begin{aligned} \log \chi_{S(V)}(q) &= \frac{1}{(d-1)!} \left(\frac{\partial}{\partial s}\right)^{d-1} \sum_{n=1}^{\infty} s^{d+n-2} \log\left(\frac{1}{1-q^n}\right)\Big|_{s=1} \\ &= \sum_{n=1}^{\infty} \binom{d+n-2}{d-1} \log\left(\frac{1}{1-q^n}\right). \end{aligned} \tag{73}$$

In the end, the full partition function,

$$\chi_{S(V)}(q) = \prod_{n=1}^{\infty} \frac{1}{(1-q^n)^{\binom{d+n-2}{d-1}}}, \tag{74}$$

gives the MacMahon's (unsuccessful) guess formula for the generating function of d-dimensional partitions.

4.5. Fermions

So far, our consideration was only on the Hilbert space V of bosonic conformal fields ϕ. Let us now include the Hilbert space W of fermionic fields ψ:

$$W = \mathrm{Span}\{\psi, \partial\psi, \partial\partial\psi, \ldots\} = \mathrm{Span}\{w_p\}. \tag{75}$$

The total Hilbert space of conformal fields is then $H = V \oplus W$. We generalize the character to cover the fermionic case as

$$Z(g) = \mathrm{Tr}\left((-1)^F g\right), \tag{76}$$

where F is the fermionic number operator.

Let us reconsider the $SU(N)$ adjoint partition function for single trace operators of lower n's. For $n=2$, we have two additional class of operators. First, we have fermionic operators,

$$Z_{VW}(g) = -\mathrm{Tr}_{V\otimes W}(g) = -\sum_{i,p} \lambda_i \lambda_p = Z_V(g)\, Z_W(g). \tag{77}$$

Second, there is the bosonic one made by two fermions,

$$Z_{WW}(g) = \mathrm{Tr}_{W\wedge W}(g) = \sum_{p<q} \lambda_p \lambda_q = \frac{\left(\sum_p \lambda_p\right)^2 - \sum_p \lambda_p^2}{2} = \frac{1}{2}\left(Z_W(g)^2 + Z_W(g^2)\right). \tag{78}$$

In the end, we get the same form as (38):

$$Z_{HH}(g) = \frac{1}{2}\left(Z_H(g)^2 + Z_H(g^2)\right). \tag{79}$$

Moving to $n=3$, we have three more classes of operators, VVW, VWW and WWW. The first and second are simply

$$Z_{VVW}(g) = Z_V(g)^2\, Z_W(g), \qquad Z_{VWW}(g) = Z_V(g)\, Z_W(g)^2, \tag{80}$$

and the last is

$$Z_{WWW}(g) = \frac{Z_W(g)^3 + 2\, Z_W(g^3)}{3}. \tag{81}$$

Note that the fermionic nature does not play any role in WWW as the cyclic permutation can be viewed as the commutation of bosonic WW and fermionic W space. In the end, we get

$$Z_{HHH}(g) = \frac{Z_H(g)^3 + 2\, Z_H(g^3)}{3}. \tag{82}$$

In this way, one can convince her/himself that the partition fuction of single trace operators made by both of bosonic and fermionic conformal fields has the same form as the character in the pure bosonic case:

$$Z_{\mathrm{cyc}^n(H)}(g) = \frac{1}{n} \sum_{k|n} \varphi(k)\, Z_H(g^k)^{\frac{n}{k}}. \tag{83}$$

One can include fermions in the dihedral, bi-fundamental and bi-vector models in the same way.

5. One Loop Tests of Free Matrix CFT Holographies

In the previous section, we have computed the partition function of all single trace operators in various free CFTs with fields in various different representations of the internal symmetry group. Using the AdS/CFT dictionary, these operator spectrum can be identified to the spectrum of AdS fields in the dual theory. Hence, the partition function for single trace operators computed above can be simply interpreted as the partition function of the dual AdS theory. Then, the CIRZ formalism presented in Section 3 can be readily applied to computing the one-loop vacuum energy. Analogously, the methods of Section 2 can be used to compute the Casimir energy of such AdS theories. We now turn to these computations.

A natural starting point is to evaluate the one-loop vacuum and Casimir energies of the AdS fields dual to the single trace operators made by n boundary fields. The corresponding partition functions are given by χ_{Cyc_n} (43), $\chi_{\mathrm{Dih}_n^+}$ (52), $\chi_{\mathrm{Dih}_n^-}$ (55), χ_{Bf_n} (59), χ_{Bv_n} (63) and χ_{Sym_n} (65). The set of AdS fields dual to the single trace operators appearing in each tensor product above will be referred to as comprising the $(n-1)^{\mathrm{th}}$ order Regge trajectory, following [37,38,40].

In order to obtain the full vacuum energy we need analytic expressions for the vacuum energy of the fields in a given Regge trajectory. As we shall show shortly, this is not available except for the $\mathcal{N}=4$ theory. However, it is still possible to calculate these quantities for n's large enough to observe a certain pattern. Either from the pattern or from the analytic expression in the $\mathcal{N}=4$ case, we can see that the one-loop vacuum and Casimir energies seem to diverge with increasing n.

Given this, we need an alternative prescription to sum over fields and one-loop energies to cure this divergence. For intution, let us consider the corresponding computations for the higher-spin theories dual to vector models [23–25,28]. In that case, divergence in the total energy can be traced back to the fact that contributions from individual fields are computed *first* and summed over *second*. Indeed, our computations of [37,38] reviewed in Appendix C show that this divergence is cured by summing over states first, by computing the thermal partition function, and evaluating the vacuum energy second. Hence it is natural to attempt to cure the divergence arising in matrix model CFTs by applying the CIRZ technique directly to the full partition functions—χ_{Cyc} (44), $\chi_{\mathrm{Dih}^\pm}$ (57), χ_{Bf} (60), χ_{Bv} (64) and χ_{Sym} (67). We shall also carry out these computations here.

The rest of the section is organized as follows. Section 5.1 contains the computation of one-loop vacuum and casimir energies for the bulk duals of the free $Sp(N)$ and $SO(N)$ scalar matrix models as well as the free Yang Mills theories. Next, in Section 5.2 we turn to the corresponding computations for the bulk dual of $\mathcal{N}=4$ super Yang-Mills. Finally in Section 5.3 we study the bulk duals of CFTs with bifundamental matter, in particular, scalars and Majorana fermions.

Finally, some reminders of notation. In what follows, the partition function of the boundary scalar, spin-$\frac{1}{2}$, and spin-1 fields is respectively denoted by χ_0, $\chi_{\frac{1}{2}}$ and χ_1, and their explicit forms are given in (A49), (A50) and (A51). The Casimir energy is denoted as E and the one-loop vacuum energy is denoted as $\Gamma^{(1)}$. Often these quantities will have subscripts which indicate which fields or set of fields they correspond to. For example, E_0 is the Casimir energy for a boundary scalar, while $\Gamma^{(1)}_{\mathrm{Dih}_n^+}$ is the vacuum energy summed over all fields contained in the cyclic character for $Sp(N)$ at some fixed n.

5.1. Non-Supersymmetric $Sp(N)$ and $SO(N)$ Adjoint Models

5.1.1. Bulk Dual of Free Scalar

At fixed values of n, the Casimir energies $E_{\mathrm{Dih}_n^\pm}$ and vacuum energy $\Gamma^{(1)}_{\mathrm{Dih}_n^\pm}$ for the free $Sp(N)$ and $SO(N)$ adjoint scalar CFTs can be obtained by applying CIRZ methods to (52) and (55), where χ_V is taken to be χ_0. The results obtained up to $n=32$ are exhibited graphically in Figure 3. We make the following observations at this stage. Firstly, we note that the $Sp(N)$ and $SO(N)$ plots almost overlap each other. Secondly, upon normalizing by $\varphi(n)\,E_0$ or $\varphi(n)\,\Gamma_0^{(1)}$, the chaotic pattern of the results in Figure 3 maps to the constant $\frac{1}{2}$ with very tiny fluctuations: see Figure 4. This is because the correction terms in (52) and (55) very quickly decay as n increases. Further, it was also observed for the $SU(N)$ adjoint model in [37] that when $n = 2^m$ for an integer m, then the fluctuations exactly vanish. This turns out to be no longer true for the dihedral models due to the presence of the correction terms (52) and (55).

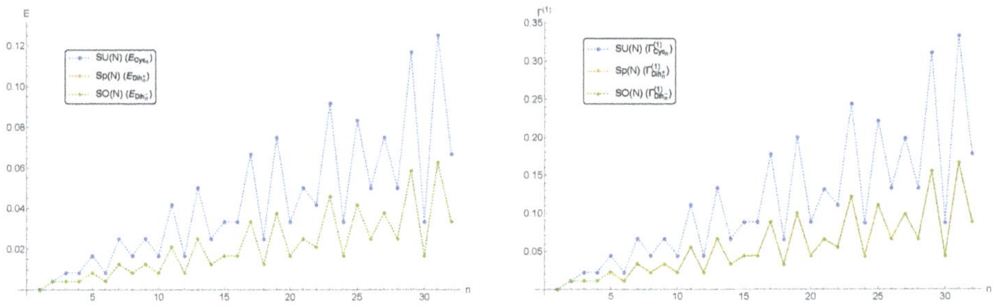

(a) (b)

Figure 3. Plot of the first 32 results for $E_{\mathrm{Dih}_n^\pm}$ and $\Gamma^{(1)}_{\mathrm{Dih}_n^\pm}$ (in the unit of $\log R$). (**a**) Casimir Energy; (**b**) Vacuum Energy.

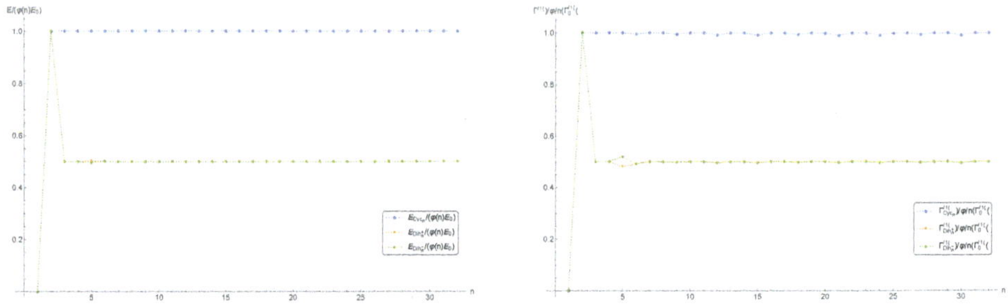

(a) (b)

Figure 4. Plot of $E_{\mathrm{Dih}_n^\pm}/(\varphi(n)\,E_0)$ and $\Gamma^{(1)}_{\mathrm{Dih}_n^\pm}/(\varphi(n)\,\Gamma_0^{(1)})$. (**a**) Casimir Energy; (**b**) Vacuum Energy.

In order to obtain the full Casimir or vacuum energy, we need to sum these results over n . In the $N \to \infty$ limit, the summation is from $n = 2$ to ∞. Since we do not have an analytic expression for n at our disposal, we cannot evaluate this sum. However, if the pattern of Figure 4 persists, one can expect from the pattern that the total results are divergent.

We will now examine if the CIRZ method can again be used to regulate this divergence by summing over the spectrum first and evaluating the free energy afterwards. It is quickly apparent that the CIRZ method, when applied to (57), returns a finite value for the Casimir energy for both $Sp(N)$ and $SO(N)$ models. In particular,

$$E_{\mathrm{Dih}^+} = \frac{27}{240}, \qquad E_{\mathrm{Dih}^-} = \frac{28}{240} \qquad \left(E_0 = \frac{1}{240}\right), \tag{84}$$

where we have also presented the Casimir energy of the boundary scalar for comparison. We now finally turn to the one-loop vacuum energy computation in the $N \to \infty$ case, where we provide a few more details. Firstly, again from examining (57) we see that it is useful to focus on the correction terms

χ_{even} and χ_{odd} computed in (56). Applying the CIRZ formalism, we see that the one-loop vacuum energies receive the following contributions

$$\Gamma^{(1)}_{\text{odd}} = -\frac{1}{180}\log R, \qquad \Gamma^{(1)}_{\text{even}} = \frac{101}{180}\log R. \tag{85}$$

Using these results, as well as the result [37]

$$\Gamma^{(1)}_{\text{Cyc}} = -\frac{1}{90}\log R. \tag{86}$$

We see that for the bulk dual of the scalar matrix model

$$\Gamma^{(1)}_{\text{Dih}^+} = \frac{11}{20}\log R\,, \qquad \Gamma^{(1)}_{\text{Dih}^-} = \frac{101}{180}\log R \qquad \left(\Gamma^{(1)}_0 = \frac{1}{90}\log R\right). \tag{87}$$

The vacuum energy for the boundary scalar is also presented above for comparison. We postpone discussions of how to interpret these results to the conclusions.

5.1.2. Bulk Dual of Free Yang Mills

In contrast to the free scalar case, for free Yang Mills—we take $\chi_V = \chi_1$—we see that the one-loop vacuum and Casimir energies plotted in Figure 5a,b shows a runaway behavior. Further, though the scale of the graph hides it, the contribution to the vacuum energy flips sign as n is increased. This may be readily inferred from the fact that for $n = 2$ the $Sp(N)$ and $SO(N)$ partition functions (52) and (55) are equal to the $SU(N)$ partition function, for which the vacuum energy contribution was computed in [38] and found to be $+\frac{62}{45}\log R$, and the fact that for larger values of n the vacuum energy contribution takes negative values.

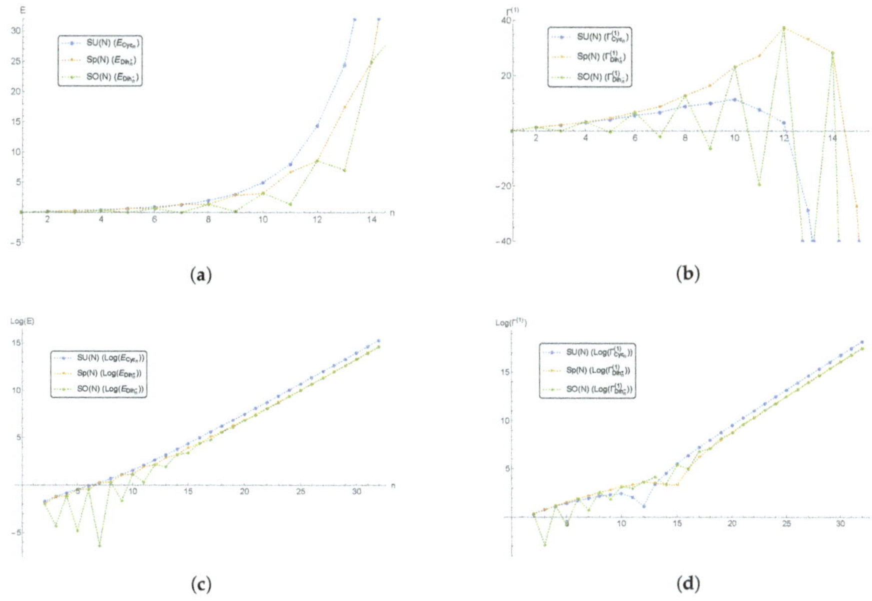

Figure 5. Plots of the Casimir energy and the one-loop vacuum energy of the bulk dual of free Yang-Mills theory in $SO(N)$ and $Sp(N)$ adjoint representation up to $n = 32$. (**a**) Casimir Energy for free Yang-Mills; (**b**) Vacuum Energy for free Yang-Mills; (**c**) Log plot of the Casimir Energy; (**d**) Log plot of the Vacuum Energy.

Hence, even in the Yang Mills case the one-loop free energies do not appear to converge as we increase the value of n. It is therefore again natural to regulate the result by directly using the character of the full partition function (57). In this way we obtain the total Casimir energy as

$$E_{\text{Dih}^+} = \frac{1377}{120}, \qquad E_{\text{Dih}^-} = \frac{1388}{120} \qquad \left(E_1 = \frac{11}{120}\right). \tag{88}$$

While for the total vacuum energy we find

$$\Gamma^{(1)}_{\text{Dih}^+} = \frac{791}{10} \log R, \qquad \Gamma^{(1)}_{\text{Dih}^-} = \frac{7181}{90} \log R \qquad \left(\Gamma^{(1)}_1 = \frac{31}{45} \log R\right). \tag{89}$$

5.2. Bulk Dual of Free $\mathcal{N} = 4$ SYM

We now turn to the maximally supersymmetric case of the AdS_5 dual of free planar $\mathcal{N} = 4$ super Yang-Mills with gauge group $SO(N)$ or $Sp(N)$. The partition function to use is $Z_V = \chi_1 - 4\chi_{\frac{1}{2}} + 6\chi_0$, as in the $SU(N)$ case studied in [40] we again find that the computation can be analytically carried out for arbitrary values of n, in contrast to the non-supersymmetric cases studied above.

We remind the reader here that the result for the one-loop vacuum and Casimir energies from the n-th order Regge Trajectory in the $SU(N)$ (i.e. cyclic) case was given by

$$\Gamma^{(1)}_{\text{Cyc}_n} = n \log R, \qquad E_{\text{Cyc}_n} = \frac{3}{16} n. \tag{90}$$

The partition function for the n-th order Regge Trajectories in the $Sp(N)$ and $SO(N)$ adjoint models is given in (52) and (55) respectively. We focus on the correction term in both expressions. The contributions to the one-loop free energies from these terms readily be evaluated and summarized as

$$\Gamma^{(1)}_{\text{Dih}^\pm_n} - \frac{1}{2}\Gamma^{(1)}_{\text{Cyc}_n} = \log R \begin{cases} \pm\frac{n}{2} & [n \text{ odd}] \\ \frac{n}{2} & [n \text{ even}] \end{cases}, \qquad E_{\text{Dih}^\pm_n} - \frac{1}{2} E_{\text{Cyc}_n} = \begin{cases} \pm\frac{3n}{32} & [n \text{ odd}] \\ \frac{3n}{32} & [n \text{ even}] \end{cases}. \tag{91}$$

Combining with the cyclic result, we obtain the one-loop free energies for the n-th order Regge Trajectory in $Sp(N)$ as

$$\Gamma^{(1)}_{\text{Dih}^+_n} = n \log R, \qquad E_{\text{Dih}^+_n} = \frac{3n}{16}, \tag{92}$$

while for $SO(N)$ they are given by

$$\Gamma^{(1)}_{\text{Dih}^-_n} = \begin{cases} 0 & [n \text{ odd}] \\ n \log R & [n \text{ even}] \end{cases}, \qquad E_{\text{Dih}^-_n} = \begin{cases} 0 & [n \text{ odd}] \\ \frac{3n}{16} & [n \text{ even}] \end{cases}. \tag{93}$$

The total one-loop free energies are given formally as

$$\begin{aligned} \Gamma^{(1)}_{\text{Dih}^+} &= \log R \sum_{n=2}^{\infty} n, & E_{\text{Dih}^+} &= \frac{3}{16} \sum_{n=2}^{\infty} n, \\ \Gamma^{(1)}_{\text{Dih}^-} &= 2 \log R \sum_{p=1}^{\infty} p, & E_{\text{Dih}^-} &= \frac{3}{8} \sum_{p=1}^{\infty} p, \end{aligned} \tag{94}$$

where for the $SO(N)$ case we used $n = 2p$. The above involve clearly divergent sum $\sum_{p=1}^{\infty} p$, which has been regularized to zero in [27]. Hence, if we use the same regularization scheme, we would obtain

$$\Gamma^{(1)}_{\text{Dih}^+} = -\log R, \qquad \Gamma^{(1)}_{\text{Dih}^-} = 0, \qquad E_{\text{Dih}^+} = -\frac{3}{16}, \qquad E_{\text{Dih}^-} = 0. \tag{95}$$

We can rederive the same result applying the CIRZ directly to the full partition functions $\chi_{\mathrm{Dih}^{\pm}}$ (57). However, we see that at $\beta = 0$, the $\mathcal{N} = 4$ singleton partition function Z_V equals 1. As a result, the geometric series in (56) are divergent at that point. To avoid this, we introduce a factor r^{n-2} in summing $\chi_{\mathrm{Dih}_n^{\pm}}$ from $n = 2$ to ∞. With the regulator r, we find that the correction terms (56) give

$$\Gamma^{(1)}_{\mathrm{even}} = \tfrac{1}{2}\, q(r)\, \log R\,, \qquad \Gamma^{(1)}_{\mathrm{odd}} = \frac{1}{2}\,(q(r)-1)\,\log R\,, \tag{96}$$

$$E_{\mathrm{even}} = \tfrac{3}{32}\, q(r)\,, \qquad E_{\mathrm{odd}} = \frac{3}{32}\,(q(r)-1)\,, \tag{97}$$

with

$$q(r) = \frac{r+1}{(r-1)^2}\,. \tag{98}$$

This immediately yields (95) in the $SO(N)$ case as the function $q(r)$ simply cancels out. In the $Sp(N)$ case, this function survives and becomes singular in the $r \to 1$ limit. By adopting the scheme '$q(1) = 0$' we can again recover (95). We would however like to emphasize that the result obtained by naively using (56) as the partition functions for $\mathcal{N} = 4$ is finite and different from the above. In particular, for the $Sp(N)$ matrix model the Casimir and vacuum energies are $-\frac{371}{1152}$ and $-\frac{1049}{648}\log R$ respectively, while for the $SO(N)$ matrix model they are $-\frac{3}{128}$ and $-\frac{1}{8}\log R$.

5.3. $U(N) \times U(M)$ Bi-Fundamental and $O(N) \times O(M)$ Bi-Vector Models

We now turn to a computation of the one-loop vacuum and Casimir energies for the bulk duals of the free $U(N) \times U(M)$ bi-fundamental model, and next, the $O(N) \times O(M)$ bi-vector model. We will consider the cases where the fundamental field is either a scalar ($\chi_V = \chi_0$) or a Majorana fermion ($Z_V = -\chi_{\frac{1}{2}}$).

For the case of the $U(N) \times U(M)$ bi-fundamental model, it turns out that the one-loop vacuum and Casimir energies are almost trivially zero. This is because both $\chi_0(g)$ and $\chi_{\frac{1}{2}}(g)$ are odd in β, and hence $\chi_0(g^k)^2$ and $\chi_{\frac{1}{2}}(g^k)^2$ are even in β. Hence the functions $f_{\mathcal{H}|0}, f_{\mathcal{H}|1}$ and $f_{\mathcal{H}|2}$ are also even in β and therefore possess no odd powers of β in the small β expansion. Hence the one-loop vacuum energy for the corresponding AdS theories are trivially zero. By a similar reasoning, the one-loop Casimir energy also vanishes.

Let us turn to the computation of the vacuum and Casimir energies for $O(N) \times O(M)$ bivector model. Working first at fixed values of n (here $2n$ is the number of fields in single trace operators), we observe that as n grows, the absolute value of the one-loop free energies E_{Bv_n} and $\Gamma^{(1)}_{\mathrm{Bv}_n}$ rapidly decay for both scalar and fermion cases. To understand better this decaying behavior, we depict the log-plots of the Casimir and vacuum energies for scalar and fermion bi-vector models in n at Figure 6: the linear behavior implies that the one-loop free energies exponentially decay to zero in n, for both scalar and fermion cases.

When $N, M \to \infty$, the full one-loop free energies are the sum of all these results from $n = 1$ to ∞. But, due to the absence of analytic expressions, we cannot evaluate this sum. Instead, we can again apply the CIRZ method directly to the full partition functions (64). In the end, we find that the one-loop free energies of the bulk duals of $O(N) \times O(M)$ bivector model are simply zero:

$$E_{\mathrm{Bv}} = 0\,, \qquad \Gamma^{(1)}_{\mathrm{Bv}} = 0\,, \tag{99}$$

both for the scalar and fermion cases. As a final comment, we note that if only $N \to \infty$ while M is kept finite, the single trace operators can involve only finite number of fields in a trace, hence possible value of n is bounded above. An extreme case is the vector model with $M = 1$ where the only allowed value of n is 2 (remind that $2n$ is the number of fields in a trace). Therefore, for finite M, the bulk dual theory will involve only finite number of Regge trajectories and the full one-loop free energies will be a finite sum of E_{Bv_n} and $\Gamma^{(1)}_{\mathrm{Bv}_n}$.

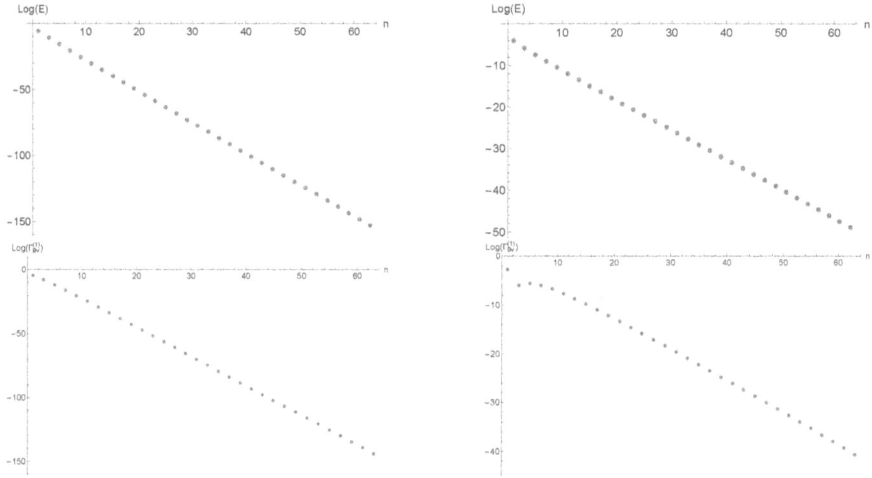

Figure 6. The vacuum energy of the scalar bivector model (**lower left**) and fermion bivector model (**lower right**); and the Casimir energy of the scalar bivector model (**upper left**) and fermion bivector model (**upper right**).

5.4. *Symmetric Group*

Let us consider a toy AdS/CFT model based on free scalar with symmetric group, even though it does not fit well in the standard picture on holography in many respects. Putting the interpretation issues aside, let us simply provide the result of Casimir energy computations. Using the partition functions (65), we calculate first 32 Casimir energies E_{Sym_n} and plot the result in Figure 7. We find the Casimir energy has an oscillating behaviour with exponentially growing oscillation amplitude.

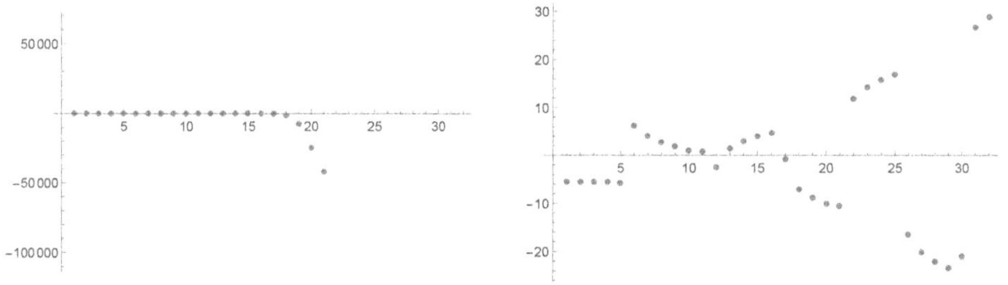

(**a**) (**b**)

Figure 7. Casimir energies for the first 32 results. (**a**) E_{Sym_n}; (**b**) $f(E_{\mathrm{Sym}_n})$ with $f(x) = \mathrm{sign}(x)\,\log|x|$.

6. Summary and Concluding Remarks

In this paper we have computed the one-loop free energies for the holographic duals of a number of free CFTs. These results are summarized in Table 1.

Table 1. Summary of one-loop free energies computed for bulk duals of free matrix conformal field theories (CFTs) in the $N \to \infty$ limit. $\Gamma^{(1)}$ is in unit of $\log R$. The fermion is Majorana.

Boundary CFT	Symmetry Group	Casimir Energy E	Vacuum Energy $\Gamma^{(1)}$
Adjoint Scalar	$Sp(N)$	$\frac{27}{240}$	$\frac{11}{20}$
	$SO(N)$	$\frac{28}{240}$	$\frac{101}{180}$
Yang Mills	$Sp(N)$	$\frac{1377}{120}$	$\frac{791}{10}$
	$SO(N)$	$\frac{1388}{120}$	$\frac{7181}{90}$
$\mathcal{N}=4$ SYM	$Sp(N)$	$-\frac{3}{16}$	-1
	$SO(N)$	0	0
Bi-fundamental Scalar	$U(N)\times U(M)$	0	0
Bi-vector Fermion	$U(N)\times U(M)$	0	0
Bi-vector Scalar	$O(N)\times O(M)$	0	0
Bi-vector Fermion	$O(N)\times O(M)$	0	0

We now briefly discuss the physical interpretation of these results in terms of matching the free energies across the bulk and the boundary theories. For definiteness, we will focus on vacuum energy in the scalar $SO(N)$ adjoint model, though the discussion readily generalizes to the other dualities discussed above. The CFT free energy on S^4 has a logarithmic divergence corresponding to the a anomaly, and is given by

$$F_{\text{CFT}} = \frac{N(N-1)}{2}\frac{1}{90}\log\Lambda\,, \tag{100}$$

while the AdS free energy takes the form

$$F_{\text{AdS}} = \left(g^{-1}\mathcal{L}_0 + \frac{101}{180} + \mathcal{O}(g)\right)\log R\,, \tag{101}$$

where g is the bulk coupling constant. Then by matching the free energies and using the correspondence between $\log\Lambda$ and $\log R$ we see that

$$g^{-1} = \frac{N(N-1)}{2} - \frac{101}{2}, \quad \mathcal{L}_0 = \frac{1}{90}\,. \tag{102}$$

Hence, in contrast to the SU(N) case, the one-loop shift in the definition of the coupling constant is by a half-integer, and not an integer. Further, if we apply a similar process to interpret the Casimir energy, we obtain

$$g^{-1} = \frac{N(N-1)}{2} + 28. \tag{103}$$

Note that this shift is different from that obtained for global AdS$_5$ with S^4 boundary in (102). This is an interesting counterpoint to the situation for higher-spin CFT dualities as well as SU(N) matrix CFT dualities the shift was the same in both backgrounds and was always by an integer amount. It would be interesting to have a better understanding of (102) and (103). While it is true that quantum effects are sensitive to topologies, it is puzzling that the AdS/CFT dictionary itself can get altered in a background dependent way. The results (102) and (103) might be indicating that putative AdS/CFT dualities involving the free $SO(N)$ and $Sp(N)$ adjoint scalar model and free Yang Mills hold only in the planar limit. It might also be possible that higher loop corrections can alter this discussion.

In this regard the situation for $\mathcal{N}=4$ super Yang Mills is perhaps more satisfactory, once an appropriate regularization is adopted. In the $Sp(N)$ case we find the shift

$$g^{-1} = \frac{N(N+1)}{2} - 1 = \frac{(N-1)(N+2)}{2}\,. \tag{104}$$

While for the $SO(N)$ case it does not shift at all from the tree-level identification

$$g^{-1} = \frac{N(N-1)}{2}. \tag{105}$$

An additional curiosity regarding the dihedral matrix models is that for all the matter content that we have examined, be it the scalar, spin-1 or $\mathcal{N}=4$ supersymmetric, the following relations hold:

$$E_{\text{Dih}^-} = E_{\text{Dih}^+} + E_{\text{singleton}}, \qquad \Gamma^{(1)}_{\text{Dih}^-} = \Gamma^{(1)}_{\text{Dih}^+} + \Gamma^{(1)}_{\text{singleton}}. \tag{106}$$

Or in other words, the contribution of $\chi_{\text{odd}(V)}$ to the one-loop free energies is equal to $-1/2$ times the contribution of χ_V.

We finally turn to a summary of possible ambiguities in the application of the CIRZ method that we have so far only briefly discussed. These have to do with the presence of additional poles in the CIRZ integrands $f_{\mathcal{H}|n}$ and the partition function Z in the complex-β plane when we work with the full partition function. Firstly, we note that due to the presence of terms of the form

$$\log\left[1-\chi_V\left(g^k\right)\right], \quad \text{and} \quad \log\left[1-\chi_V\left(g^k\right)^2\right] \tag{107}$$

the partition functions (57), (60) and (64) contain branch points in β, one of which lies on the positive real axis[5].

Since the reduction of the contours in (5) and (31) to small loops around $\beta = 0$ relies on the integrands having no additional singularities in the complex β plane, the contribution of these branch points to the β integrals may need to be separately accounted for. Next, as for the $SU(N)$ case we also choose to apply the CIRZ partition function on the partition functions $\chi_{\log,k}$ defined in (45) *at fixed k*. This is again because of the singular points in β contained in (45) at fixed k. In particular, if β_c is a singular point for (45) at $k=1$, then $\frac{\beta_c}{k}$ is a singular point of χ_{Cyc} in (57) at arbitrary integer values of k. Similar remarks apply to the singular points of (60) and (64). These singularities would tend to cluster around $\beta = 0$, making the partition function highly non-analytic in that neighbourhood.

In addition, the presence of this pole in the complex β plane also introduces an additional ambiguity which we have so far not discussed in much detail. In particular, the summation of the geometric series carried out to evaluate (56) and (64) assumes that the absolute value of the character $\chi_V(g)$ is less than 1. If the absolute value is greater than 1, the summed expression may be used as an analytic continuation of the divergent geometric series. However, this breaks down when $|\chi_V(g)|$ is equal to 1, and the resulting expressions for the partition functions (56) such as diverge. This is what happens for the character of the $\mathcal{N}=4$ Maxwell multiplet at $\beta = 0$, and the role of the regulation with r that we carried out above is essentially to discard the contribution of this singularity to the contour integrals (5) and (31). Since the characters $\chi_0(g)$ and $\chi_1(g)$ become 1 at some positive real β, and not at $\beta = 0$ we have implicitly carried out this prescription for the scalar matrix models and for free Yang Mills. The inclusion of this additional pole leads to contributions to the free energy which are not rational numbers.

Acknowledgments: We would like to thank Rajesh Gopakumar, Karapet Mkrtchyan, Zhenya Skvortsov and Mikhail Vasiliev for helpful discussions regarding various technical and conceptual points during the course of this and previous related work. We also thank the participants of the workshop "New Ideas in Higher Spin Gravity and Holography" at Kyung-Hee University for helpful discussions and comments. SL thanks Kyung-Hee University for hospitality during the course of this work. The work of EJ was supported by the National Research Foundation of Korea through the grant NRF2014R1A6A3A04056670. SL's research is supported by a Marie Sklodowska Curie Individual Fellowship 2014.

---

5 This branch point gives rise to the Hagedorn transition in CFTs with adjoint and bifundamental matter, see e.g., [7,47]. This transition is absent in pure higher-spin/CFT dualities and is potentially an important distinguishing feature between higher-spin and tensionless string(y) theories [6].

Author Contributions: All authors contrbuted equally to this work.

Conflicts of Interest: The authors declare no conflict of interest.

Appendix A. Harmonic Analysis on Spheres, Hyperboloids, and Their Quotients

In this section we will review some essential facts about the spectrum of the Laplacian $\Box = -g^{\mu\nu}\nabla_\mu\nabla_\nu$ on Anti-de Sitter space, focusing mostly on Euclidean AdS_5 and more generally AdS_{2n+1}. The extension to even dimensional subspaces has some subtleties which we mention below. Our starting point is the observation that Euclidean AdS_5 is the symmetric space $SO(5,1)/SO(5)$. From this it follows, see [18,48] for a review of these facts, that the spectrum of the Laplacian over arbitrary spin fields is determined in terms of the representation theory of $SO(5,1)$ and $SO(5)$. Further, since the results are more generally valid for all homogeneous spaces, we shall present the results for the general case, giving concrete examples along the way.

Firstly, given Lie groups G and $H \subset G$, we define the coset space G/H to be the set of equivalence classes of elements in G obtained by the right action of the subgroup H, i.e.,

$$g_1 \equiv g_2 \quad \text{if} \quad \exists \quad h \in H \quad \text{s.t.} \quad g_1 = g_2 \cdot h. \tag{A1}$$

The set of all g equivalent to an element g_o under this relation is denoted by g_oH. Now G is the principle bundle over G/H with fibre isomorphic to H, and we can define the projection map from the bundle to the base space

$$\pi : G \to G/H, \quad \pi(g) = gH, \quad \forall \quad g \in G. \tag{A2}$$

Further we can also define a *section* $\sigma(x)$, $x \in G/H$, through

$$\sigma : G/H \to G, \quad \sigma(x) \in xH, \tag{A3}$$

i.e., $\sigma(x)$ is an element of the coset which contains x. Clearly, there is no canonical choice of section, and all sections are equivalent to each other upto right multiplication by H. Therefore, given two sections σ_1 and σ_2, at every $x_o \in G/H$, there exists an $h \in H$ such that $\sigma_1(x_o) = \sigma_2(x_o)\, h$. Then in this case the spin of a given field is fixed by specifying a UIR S of H. Secondly, given a UIR S of H, let us choose the set of all UIRs R of G that contain S. With these inputs, the eigenvalues of the Laplacian for a spin-S field are given by

$$E_R^{(S)} = -\frac{1}{a^2}\left(C_2(R) - C_2(S)\right), \tag{A4}$$

where a is the AdS radius. The corresponding eigenfunctions are given by

$$\psi_a^I(x) = \frac{1}{N_R^{(s)}}\left[\mathcal{U}^R\left(\sigma(x)^{-1}\right)\right]_a^I. \tag{A5}$$

Here I is an index for the vector space carrying the representation R and a is an index for the vector space carrying the representation S. $N_R^{(s)}$ is a normalization constant, which we shall fix subsequently. Notice that eigenfuctions carrying different values of I for the same R are necessarily degenerate. Hence the degeneracy of the eigenvalue $E_R^{(S)}$ is at least [6] d_R. With these inputs, we may write the zeta function for the operator $-\Box + \nu^2$ over a spin-S field on the space G/H as

[6] The actual degeneracy can in principle be more as many representations R can carry the same quadratic Casimir $C_2(R)$. We neglect this possibility below as it does not affect the subsequent analysis.

$$\zeta_{\nu,S}(z) = \sum_R \sum_I \sum_a \frac{1}{\left(E_R^{(s)} + \nu^2\right)^z} \int_{G/H} \sqrt{g}\, d^d x \left[\psi_a^I(x)\right]^* \cdot \psi_a^I(x)\,. \tag{A6}$$

Here the sum over a is the dot product over local spin indices and the sum over I is the sum over degenerate eigenvalues while the sum over R is the sum over non-degenerate eigenvalues. Now, using (A5), we have

$$\sum_I \left[\psi_b^I(x)\right]^* \cdot \psi_a^I(x) = \frac{1}{|N_R^{(s)}|^2}\delta_a^b, \tag{A7}$$

which is independent of the point x on the coset space[7]. Further, using the definition of degeneracy

$$\int d^d x \sqrt{g} \sum_{I,a} \left[\psi_a^I(x)\right]^* \cdot \psi_a^I(x) = d_R, \tag{A8}$$

we must have

$$\frac{1}{|N_R^{(s)}|^2} = \frac{d_R}{d_s}\frac{1}{V_{G/H}}. \tag{A9}$$

The zeta function (A6) is therefore perfectly consistent with the original definition

$$\zeta_{\nu,S}(z) = \sum_R \frac{d_R}{\left(E_R^{(s)} + \nu^2\right)^z}. \tag{A10}$$

It should be immediately apparent, however, that the above procedure is at first sight ill-defined for the case of hyperboloids like AdS_5. In this specific example G is $SO(5,1)$, and its unitary representations are necessarily infinite dimensional. Further, the volume $V_{G/H}$ is infinite. For this reason, a slight modification of the above computation is adopted. We note from (A7) that the quantity $\sum_I \left[\psi_b^I(x)\right]^* \cdot \psi_a^I(x)$ is independent of the point x. As a result, it is possible to define the *coincident* zeta function

$$\zeta_{\nu,S}^{coin}(z) = \sum_R \sum_I \sum_a \frac{1}{\left(E_R^{(s)} + \nu^2\right)^z} \left[\psi_a^I(x)\right]^* \cdot \psi_a^I(x)\,, \tag{A11}$$

such that

$$\zeta_{\nu,S}(z) = V_{G/H}\, \zeta_{\nu,S}^{coin}(z)\,. \tag{A12}$$

We shall now recapitulate the evaluation of the coincident zeta function of fields on AdS_5 by means of analytic continuation from S^5. For definiteness we shall focus on the Laplace operator $-\Box$ but the discussion easily generalized to arbitrary ν. Firstly, we specify the spin S of the field by the UIR of $SO(5)$ that it carries. This representation is in turn specified by the quantum numbers $S = (s_1, s_2)$, where

$$s_1 \geq s_2 \geq 0. \tag{A13}$$

To solve for the zeta function on S^5, we consider all UIRs R of $SO(6)$ that contain S when restricted to $SO(5)$. Firstly, UIRs of $SO(6)$ are labelled by the triplet (ℓ, m_1, m_2) where

$$\ell \geq m_1 \geq |m_2|, \tag{A14}$$

[7] While we are working here with compact groups, this statement should hold equally well when we consider AdS_5 for which $G = SO(5,1)$. This is the group theoretic origin of the Equation (21) which exploited the homogeneity of AdS_5 to define the Plancherel measure.

and ℓ, m_1, m_2 are all simultaneously integers or half-integers. The dimension of such a representation R is given by

$$d_R = \frac{(\ell+2)^2-(m_1+1)^2}{3}\frac{(\ell+2)^2-m_2^2}{4}\frac{(m_1+1)^2-m_2^2}{1}. \tag{A15}$$

This representation R contains S provided

$$\ell \geq s_1 \geq m_1 \geq s_2 \geq |m_2|. \tag{A16}$$

We shall, however, consider the case that the given field satisfies irreducibility conditions, such as transversality and tracelessness, where some of the inequalities in (A16) saturate to yield

$$\ell \geq s_1 = m_1 \geq s_2 = |m_2|. \tag{A17}$$

Further, the eigenvalue of the Laplacian is determined from (A4) to be

$$E_\ell^{(s_1,s_2)} = (\ell+2)^2 - (s_1+s_2) - 4, \tag{A18}$$

and as a result, the coincident zeta function on a five-sphere of unit radius is given by

$$\zeta^{coin}_{(s_1,s_2)} = \frac{1}{12V_{S^5}}\sum_{\ell \geq s_1} \frac{\left[(\ell+2)^2-(s_1+1)^2\right]\left[(\ell+2)^2-s_2^2\right]\left[(s_1+1)^2-s_2^2\right]}{\left((\ell+2)^2-(s_1+s_2)-4\right)^z}. \tag{A19}$$

Next, we move to the case of AdS_5 where to carry out the above procedure we need to enumerate all UIRs of $SO(5,1)$ which contain S when restricted to $SO(5)$. UIRs of $SO(5,1)$ are labelled by the triplet $R = (i\lambda, m_1, m_2)$, where $\lambda \in \mathbb{R}_+$ and $m_1 \geq |m_2| \geq 0$ and contain S provided that

$$s_1 \geq m_1 \geq s_2 \geq |m_2|. \tag{A20}$$

Further, we shall apply the same irreducibility conditions on the field as we did on S^5 to saturate some inequalities in (A20). In particular, we take $m_1 = s_1$ and $|m_2| = s_2$. It was explicitly demonstrated in [18–21,42] that for a wide class of fields, the coincident zeta function in AdS_5 may be computed from the corresponding S^5 answer by means of the following analytic continutation

$$\ell + 2 \mapsto i\lambda, \quad \lambda \in \mathbb{R}_+. \tag{A21}$$

As a result the coincident zeta function on AdS_5 is given by

$$\zeta^{coin}_{(s_1,s_2)} = \frac{1}{12V_{S^5}}\int_0^\infty d\lambda\, \frac{\left[\lambda^2+(s_1+1)^2\right]\left[\lambda^2+s_2^2\right]\left[(s_1+1)^2-s_2^2\right]}{\left(\lambda^2+(s_1+s_2)+4\right)^z}. \tag{A22}$$

We now provide an explicit example of computing the coincident zeta function on AdS_5 for a scalar field without using the analytic continuation proposed above. Also, as the analytic continuation as presented above seems somewhat abstract, we shall use this example as an explicit setting to demonstrate how this continuation works in practice.

Appendix A.1. The Scalar on AdS_5 and S^5

We begin with noting that the metric on S^N of unit radius in spherical polar coordinates

$$ds^2_{S^N} = d\chi^2 + \sin^2\chi\, ds^2_{S^{N-1}} \tag{A23}$$

is related to the metric on the corresponding hyperbolic space AdS_N or H^N

$$ds^2_{H^N} = dy^2 + \sinh^2 y\, ds^2_{S^{N-1}} \tag{A24}$$

via $\chi = iy$. The Laplace eigenvalue equation for scalar fields on S^N is given by

$$\Box \varphi = \ell\,(\ell + N - 1)\,\varphi, \tag{A25}$$

and its solutions are given in terms of hypergeometric functions as

$$\varphi_{\ell m\sigma} = (\sin\chi)^m\, {}_2F_1\left(\ell + m + N - 1, m - \ell, m + \tfrac{N}{2}; \sin^2\tfrac{\chi}{2}\right) Y_{m\sigma}. \tag{A26}$$

The hypergeometric function of the second kind is ruled out by requiring smoothness at $\chi = 0$ while requiring the eigenfunctions to be smooth at $\chi = \pi$ restricts $m = 0, 1, \ldots, \ell$. The $Y_{m\sigma}$ solve the Laplace equation on S^{N-1} with eigenvalue $m\,(m + N - 2)$. Then the eigenfunctions of the scalar Laplace equation on H^N

$$\Box \varphi = \left(\lambda^2 + \rho^2\right)\varphi, \quad \rho = \tfrac{N-1}{2}, \tag{A27}$$

are obtained by making the replacements $\chi = iy$ and $\ell + \rho = i\lambda$ in (A26). We therefore have

$$\varphi_{\lambda m\sigma} = N_{\lambda m}\,(i\sinh y)^m\, {}_2F_1\left(i\lambda + \rho\ell + m, -i\lambda + m + \rho, m + \tfrac{N}{2}; -\sinh^2\tfrac{y}{2}\right) Y_{m\sigma}, \tag{A28}$$

where $N_{\lambda m}$ is an overall constant. Notice that these are purely local solutions on which the only boundary condition that has been imposed is regularity at $y = 0$. Next, demanding that the eigenfunctions be square integrable and a complete set fixes λ to be real and positive, while demanding that they be Dirac delta normalized fixes

$$N_{\lambda m} = \left(\frac{2^{N-2}}{\pi}\right)^{1/2} \left|\frac{\sqrt{\pi}\,\Gamma\left(i\lambda + (N-1)/2 + m\right)}{2^{N+m-2}\,\Gamma(i\lambda)\,\Gamma(m + N/2)}\right|. \tag{A29}$$

Note that $\ell \mapsto i\lambda - \rho$ for $N = 5$ is precisely the analytic continuation used above from AdS_5 to S^5. The coincident zeta function is therefore given by

$$\zeta^{coin}(z) = \int_0^\infty d\lambda \sum_{m\sigma} \frac{\varphi^*_{\lambda m\sigma}\varphi_{\lambda m\sigma}}{(\lambda^2 + \rho^2)^z}. \tag{A30}$$

We choose to evaluate this quantity at $y = 0$ where the eigenfunction $\varphi_{\lambda m\sigma}$ vanishes unless $m = 0$. Further, $Y_{m\sigma}$ for $m = 0$ is just the constant mode on S^{N-1}, given by $|V_{S^{N-1}}|^{-1/2}$ for reasons of normalization, and the sum over σ is also then trivial. We finally obtain

$$\sum_{m\sigma} \varphi^*_{\lambda m\sigma}\varphi_{\lambda m\sigma} = \frac{1}{2^{N-2}} \left|\frac{\Gamma\left(i\lambda + (N-1)/2\right)}{\Gamma(i\lambda)\,\Gamma(N/2)}\right|^2 \left|\frac{\Gamma(N/2)}{2\pi^{N/2}}\right|. \tag{A31}$$

For the case of $N = 5$ we obtain

$$\sum_{m\sigma} \varphi^*_{\lambda m\sigma}\varphi_{\lambda m\sigma} = \frac{1}{12\pi^3}\lambda^2\left(\lambda^2 + 1\right). \tag{A32}$$

Hence the coincident zeta function for scalar fields on AdS_5 is given by

$$\zeta^{coin}(z) = \frac{1}{12\pi^3}\int_0^\infty d\lambda\, \frac{\lambda^2\left(\lambda^2 + 1\right)}{(\lambda^2 + 4)^z}. \tag{A33}$$

This matches perfectly with (A22) which was obtained by analytic continuation if we set $s_1 = 0$ and $s_2 = 0$ there.

Appendix A.2. Zeta Functions of AdS_5 Fields

To evaluate the partition function of the bulk theory to one-loop order it is sufficient to consider the quadratic action for the fields about AdS_5. For the case of massless symmetric rank-s fields, this is given by the Fronsdal action

$$S\left[\phi_{(s)}\right] = \int d^D x \sqrt{g} \phi^{\mu_1 \dots \mu_s} \left(\hat{\mathcal{F}}_{\mu_1,\dots,\mu_s} - \frac{1}{2} g_{(\mu_1 \mu_2} \hat{\mathcal{F}}_{\mu_3 \dots \mu_s)\lambda}{}^{\lambda} \right), \tag{A34}$$

where

$$\hat{\mathcal{F}}_{\mu_1 \dots \mu_s} = \mathcal{F}_{\mu_1 \dots \mu_s} - \frac{s^2 + (D-6)s - 2(D-3)}{\ell^2} \phi_{\mu_1 \dots \mu_s} - \frac{2}{\ell^2} g_{(\mu_1 \mu_2} \phi_{\mu_3 \dots \mu_s)\lambda}{}^{\lambda}, \tag{A35}$$

and

$$\mathcal{F}_{\mu_1 \dots \mu_s} = \Delta \phi_{\mu_1 \dots \mu_s} - \nabla_{(\mu_1} \nabla^{\lambda} \phi_{\mu_2 \dots \mu_s)\lambda} + \frac{1}{2} \nabla_{(\mu_1} \nabla_{\mu_2} \phi_{\mu_3 \dots \mu_s)\lambda}{}^{\lambda}. \tag{A36}$$

The expressions are true for AdS_D though we shall explicitly consider the case of $D = 5$ only. It may be shown that this action is invariant under the gauge transformation

$$\phi_{\mu_1 \dots \mu_s} \mapsto \phi_{\mu_1 \dots \mu_s} + \nabla_{(\mu_1} \xi_{\mu_2 \dots \mu_s)}. \tag{A37}$$

For a consistent description, it turns out that the fields ϕ necessarily satisfy a double-tracelessness constraint $\phi_{\mu_1 \dots \mu_{s-4} \nu \rho}{}^{\nu\rho} = 0$ for $s \geq 4$ and ξ satisfies a tracelessness constraint $\xi_{\mu_1 \dots \mu_{s-3} \nu}{}^{\nu} = 0$. It is then straightforward to evaluate the functional integral

$$Z^{(s)} = \frac{1}{\text{Vol(gauge group)}} \int \left[D\phi_{(s)} \right] e^{-S[\phi_{(s)}]}, \tag{A38}$$

to obtain the partition function as a ratio of one loop determinants evaluated over symmetric transverse traceless (STT) fields

$$\mathcal{Z}^{(s)} = \frac{\left[\det\left(-\Box - \frac{(s-1)(3-D-s)}{\ell^2} \right)_{(s-1)} \right]^{\frac{1}{2}}}{\left[\det\left(-\Box + \frac{s^2+(D-6)s-2(D-3)}{\ell^2} \right)_{(s)} \right]^{\frac{1}{2}}}. \tag{A39}$$

The subscripts indicate that the numerator is evaluated over rank $s-1$ STT fields and the denominator is evaluated over rank s STT fields. The numerator is the ghost determinant that arises from gauge fixing the freedom (A37). These determinants may be evaluated over quotients of AdS space using the techniques of [49]. In the specific case of AdS_5 where we turn on a temperature β as well as chemical potentials α_1 and α_2 for the $SO(4)$ Cartans, we find that the partition function is given by

$$\log \mathcal{Z}^{(s)} = \sum_{m=1}^{\infty} \frac{1}{m} \frac{e^{-m\beta(s+2)}}{|1 - e^{-m(\beta - i\phi_1)}|^2 |1 - e^{-m(\beta - i\phi_2)}|^2} \left[\chi^{SO(4)}_{(s,0)} - \chi^{SO(4)}_{(s-1,0)} e^{-m\beta} \right], \tag{A40}$$

and $\phi_1 = \alpha_1 + \alpha_2$, $\phi_2 = \alpha_1 - \alpha_2$. The reader will recognize this as the partition function $\mathrm{Tr}\left(e^{-\beta H - \alpha^i J_i} \right)$ computed in the *grand canonical* ensemble for the conformal primary with highest weights $\left(s+2, \frac{s}{2}, \frac{s}{2} \right)$. It was observed in [50] that it is possible to formally invert this procedure, i.e., given a character of the conformal algebra, one may infer the spectrum of the corresponding kinetic operator which gives rise to the corresponding grand canonical partition function.

Appendix B. Unitary Irreducible Representations of the $so(2,4)$ Algebra

Fields in AdS_5 carry quantum numbers under $so(2,4)$, the isometry algebra of AdS_5, and fall into its Unitary Irreducible Representations (UIRs). A necessary condition for any AdS/CFT duality is that for every field in AdS_5, there is a 'dual' operator in the CFT_4 carrying the same $so(2,4)$ quantum numbers as the AdS_5 field. We therefore review very briefly the set of UIRs of $so(2,4)$. These have been extensively explored in [51,52] and particularly accessible accounts are available in [35,53].

The starting point is the Verma module $\mathcal{V}(\Delta,(\ell_1,\ell_2))$ of $so(2,4)$. The numbers Δ and (ℓ_1,ℓ_2) label the irreps of $so(2)\oplus so(4)$ subalgebra carried by the lowest weight state of the module. Since $so(4)\simeq su(2)\oplus su(2)$, we shall also use the $su(2)\oplus su(2)$ label $[j_+,j_-]$. The two labels are simply related by $j_\pm=(\ell_1\pm\ell_2)/2$. The Verma module $\mathcal{V}(\Delta,(\ell_1,\ell_2))$ is unitary and irreducible when Δ is greater than the critical dimension Δ_{ℓ_1,ℓ_2}, which will be introduced shortly in below.

- The UIR belonging to the interior the unitary region of Δ is referred as to *long representations*. They can be realized as higher-spin operators in CFT_4 or as massive higher-spin fields in AdS_5 with the mass-squared given by $M^2=[\Delta(\Delta-4)-\ell_1-\ell_2]/L^2$ (L is the radius of AdS_5) [54–56].

In the critical cases lying on the boundary of the unitary region, the Verma module develops an invariant subspace and an UIR can be obtained by quotienting the Verma module with the invariant subspace. These 'critical' representations are again divided into two groups, *semi-short* and *short*, depending on (ℓ_1,ℓ_2).

- The semi-short UIR appears when $\ell_1\neq\pm\ell_2$ and Δ reaches its critical value $\Delta_{\ell_1,\ell_2}=\ell_1+2$, and the UIR is given by the quotient,

$$\mathcal{D}(\ell_1+2,(\ell_1,\pm\ell_2))=\mathcal{V}(\ell_1+2,(\ell_1,\pm\ell_2))/\mathcal{V}(\ell_1+3,(\ell_1-1,\pm\ell_2))\,. \tag{A41}$$

The semi-short representations can be realized as conserved current operator in CFT_4 or massless (mixed-symmetry) higher-spin fields in AdS_5.

- The short representation arises when $|\ell_2|=\ell_1$ and $\Delta_{\ell_1,\pm\ell_1}=\ell_1+1$. The invariant subspace of the Verma module appearing in this case is a semi-short representation, hence again contains an invariant subspace. Therefore, the UIR is given by a 'double' quotient,

$$\begin{aligned}\mathcal{D}(\ell+1,(\ell,\pm\ell)) &= \mathcal{V}(\ell+1,(\ell,\pm\ell))/\mathcal{D}(\ell+2,(\ell,\pm(\ell-1)))\,,\\ \mathcal{D}(\ell+2,(\ell,\pm(\ell-1))) &= \mathcal{V}(\ell+2,(\ell,\pm(\ell-1)))/\mathcal{V}(\ell+3,(\ell-1,\pm(\ell-1)))\,.\end{aligned} \tag{A42}$$

Differently from the long and semi-short representations, the short representations cannot be realized as a propagating AdS_5 field but only as a conformal field operator in CFT_4.

A convenient way to treat the representation spaces, in particular their tensor products and decompositions, is to use the Lie algebra character. In case of $so(2,4)$, it is given by the trace,

$$\chi_{\mathcal{R}}(q,x_+,x_-)=\mathrm{Tr}_{\mathcal{R}}\left(q^D\,{x_+}^{J_+}\,{x_-}^{J_-}\right), \tag{A43}$$

over a representation space $\mathcal{R}$. Here D, $J_\pm$ are the Cartan subalgebra of $so(2,4)$ and corresponds to the subalgebra $so(2)$ and $su(2)\oplus su(2)$. Since all the UIRs are given in term of the Verma module. Their characters can be also expressed by the Verma module character. It is given by the following simple function,

$$\chi_{\Delta,[j_+,j_-]}(q,x_+,x_-)=q^\Delta\,P(q,x_+,x_-)\,\chi_{j_+}(x_+)\,\chi_{j_-}(x_-)\,, \tag{A44}$$

where χ_j is the character of the spin-j representation of $su(2)$ taking the form,

$$\chi_j(x)=\frac{x^{j+\frac{1}{2}}-x^{-j-\frac{1}{2}}}{x^{\frac{1}{2}}-x^{-\frac{1}{2}}}\,, \tag{A45}$$

and $P(q, x_+, x_-)$ is given by

$$P(q, x_+, x_-) = \frac{1}{\left(1 - q\, x_+^{\frac{1}{2}} x_-^{\frac{1}{2}}\right)\left(1 - q\, x_+^{-\frac{1}{2}} x_-^{\frac{1}{2}}\right)\left(1 - q\, x_+^{\frac{1}{2}} x_-^{-\frac{1}{2}}\right)\left(1 - q\, x_+^{-\frac{1}{2}} x_-^{-\frac{1}{2}}\right)}. \tag{A46}$$

In the case of long representations, its character is simply that of Verma module. About the semi-short and short representations, the character is given by the difference,

$$\chi_{\mathcal{D}(\ell_1+2,(\ell_1,\ell_2))} = \chi_{\ell_1+2,(\ell_1,\ell_2)} - \chi_{\ell_1+3,(\ell_1-1,\ell_2)}\,, \tag{A47}$$

$$\chi_{\mathcal{D}(\ell+1,(\ell,\pm\ell))} = \chi_{\ell+1,(\ell,\pm\ell)} - \chi_{\ell+2,(\ell+1,\pm(\ell-1))} + \chi_{\ell+3,(\ell-1,\pm(\ell-1))}\,. \tag{A48}$$

It is often useful to work with $(\beta, \alpha_+, \alpha_-)$ variables which are related to (q, x_+, x_-) as $q = e^{-\beta}$ and $x_\pm = e^{i\,\alpha_\pm}$. Interpreting the character as a partition function, the variables β and $\alpha_\pm$ would correspond to the inverse temperature and two angular chemical potentials.

As a final remark we explicitly evaluate (A48) for the case of the the boundary singletons that we need. We obtain

$$\chi_0(g) = \chi_{\mathcal{D}(1,(0,0))} = \frac{e^{\beta} - e^{-\beta}}{4\,(\cos\alpha_1 - \cosh\beta)\,(\cos\alpha_2 - \cosh\beta)} \tag{A49}$$

for the scalar,

$$\chi_{\frac{1}{2}}(g) = \chi_{\mathcal{D}(\frac{3}{2},(\frac{1}{2},\frac{1}{2}))} + \chi_{\mathcal{D}(\frac{3}{2},(\frac{1}{2},-\frac{1}{2}))} = \frac{\left(e^{\frac{\beta}{2}} - e^{-\frac{\beta}{2}}\right)\cos\frac{\alpha_1}{2}\,\cos\frac{\alpha_2}{2}}{(\cos\alpha_1 - \cosh\beta)\,(\cos\alpha_2 - \cosh\beta)} \tag{A50}$$

for the fermion, and

$$\chi_1(g) = \frac{e^{-2\beta}\left(-2e^{\beta}\,(\cos\alpha_1 + \cos\alpha_2) + e^{2\beta}(2\cos\alpha_1\cos\alpha_2 + 1) + 1\right)}{2\,(\cos\alpha_1 - \cosh\beta)\,(\cos\alpha_2 - \cosh\beta)} \tag{A51}$$

for the spin-1 field. These expressions are useful for the computations in the main text.

Appendix C. Vector Models

Let us first consider the one-loop vacuum energy of the higher spin gravities in AdS_5, which are dual to free vector model CFTs in four dimensions. In even boundary dimensions, we have more possibilities of free CFT as there are infinitely many singleton representations. In four boundary dimensions, they correspond simply to the massless spin j representations. Here, we consider the $j = 0, 1/2$ and 1 cases whose AdS duals are referred to as the type A, B and C higher spin theories, respectively. The other CFTs with $j > 1$ are rather exotic as they do not have local stress tensor (hence the dual higher spin theory contains gravity inside). See [39] for the general j cases.

In each of these type A, B, C cases, we have again two different higher spin theories depending on their spectrum contains only even spin fields or all integer spin fields. And these models are referred to as minimal or non-minimal and correspond in boundary to $O(N)$ or $U(N)$ free CFT, respectively. Practically they are distinguished whether the two fields in a bilinear operator are symmetric. The non-minimal model does not enjoy such symmetry whereas the minimal model does so. This symmetry of the bilinear CFT operators can be simply reflected at the level of character. The space of bilinear operators without any symmetry, hence giving the $U(N)$-model operator spectrum, can be obtained from

$$\chi_{U(N)}(\beta, \alpha_1, \alpha_2) = \left[\chi_{sing}(\beta, \alpha_1, \alpha_2)\right]^2. \tag{A52}$$

About the $O(N)$ model, the space of bilinear operators symmetric in two fields can be obtained from

$$\chi_{O(N)}(\beta, \alpha_1, \alpha_2) = \tfrac{1}{2}\left[\chi_{sing}(\beta, \alpha_1, \alpha_2)^2 + (-1)^{F_{sing}}\,\chi_{sing}\,(2\beta, 2\alpha_1, 2\alpha_2)\right]. \tag{A53}$$

Here F_{sing} is the fermion number associated with the singleton representation. Following standard conventions, it is 0 for bosonic fields, and 1 for fermionic fields. By AdS/CFT dualities, the contents in (A52) and (A53) correspond to AdS_5 fields in non-minimal and minimal theories, respectively.

With the above inputs, we will now use the techniques reviewed in Section 3 to compute the one-loop vacuum energy for higher-spin theories in global AdS_5 and the one-loop Casimir energy in thermal AdS_5. The singletons for type A, type B and type C dualities have the lowest weights $[\Delta, j_+, j_-]$,

$$A : [1,0,0], \qquad B : [\tfrac{3}{2}, \tfrac{1}{2}, 0] \oplus [\tfrac{3}{2}, 0, \tfrac{1}{2}], \qquad C : [2,1,0] \oplus [2,0,1], \tag{A54}$$

respectively, and the corresponding characters can be determined from (A48). We will first focus on the simpler case of Casimir energies.

Appendix C.1. Casimir Energies

Firstly, we note that though singletons do not represent propagating degrees of freedom, formally their partition function may be evaluated through a one-loop determinant in AdS_5 and the answer matched with the CFT result. In particular, using the prescription of (10) and Laurent expanding the characters $(-1)^{F_{sing}} \chi_{sing}(\beta, 0, 0)$ about $\beta = 0$ and picking $-\frac{1}{2}$ times the $\mathcal{O}(\beta)$ coefficient, we find

$$E_{c;(0)} = \frac{1}{240}, \quad E_{c;(\frac{1}{2})} = \frac{17}{960}, \quad E_{c;(1)} = \frac{11}{120}, \tag{A55}$$

where the subscript (s) reminds us that this is the Casimir energy of a spin-s singleton. Next, the Casimir energies of the corresponding bulk non-minimal and minimal higher-spin theories may be found using the partition functions (A52) and (A53) respectively. We find that for the non-minimal version of all three dualities the Casimir energy vanishes [38]

$$E^{\text{non-min}}_{c;(A/B/C)} = 0. \tag{A56}$$

In contrast, for the minimal cases [38]

$$E^{\text{min}}_{c;(A/B/C)} = E_{c;(0/\frac{1}{2}/1)} \quad \text{respectively}. \tag{A57}$$

We now review the argument of [25] for interpreting the non-vanishing result as a shift in the dictionary between the bulk coupling constant g and N. In particular, it turns out that the total Casimir energy in the boundary theory scales as [25]

$$F_{O(N)\,\text{sing}}(\beta) = N\,\beta\,E_{sing} + \hat{F}_{O(N)\,\text{sing}}. \tag{A58}$$

On the bulk side we find a non-vanishing result at one-loop. The bulk answer therefore has the structure

$$\Gamma_{A/B/C,\text{min}} = \frac{1}{g} S_{A/B/C,\text{min}} + \beta E^{\text{min}}_{c;(A/B/C)} + \hat{\mathcal{F}}_{A/B/C,\text{min}}(\beta) + \ldots, \tag{A59}$$

where $S_{A/B/C,\text{min}}$ is the corresponding classical on-shell action. With $\hat{F}_{O(N)\,\text{sing}} = \hat{\mathcal{F}}_{A/B/C,\text{min}}$, (A58) and (A59) are consistent provided

$$g^{-1} = N - 1, \quad S_{A/B/C,\text{min}} = \beta\, E_{c;(0/\frac{1}{2}/1)}. \tag{A60}$$

Appendix C.2. Vacuum Energies

We now turn to the computation of one-loop vacuum energies in global AdS_5. The computations are a bit more invloved in practice but conceptually they are very similar to the Casimir energy computations presented above. The only difference is that instead of expanding the characters and corresponding partition functions, we shall first be computing the functions $f_{\mathcal{H}|n}$ defined in (28) and Laurent series expanding those to extract the corresponding $\gamma_{\mathcal{H}|n}$s defined in (32). As in the Casimir energy case, we have picked the computationally most convenient prescription to work in. We remind the reader that all three prescriptions are equivalent for higher-spin partition functions, and emphasize that *a priori* this will not be true for bulk duals of matrix CFTs. We also mention that while we are focussed here on the cases of Higher-Spin/CFT dualities involving singletons carrying spins 0, $\frac{1}{2}$ and 1, the results presented here are valid for more general spin s [39]. For the singleton cases, it is straightforward to compute the $f_{sing|n}$ and expand in β to extract the coefficients

$$\gamma_{(s)|2} = \frac{15\,s^4 - 1}{30}\,, \qquad \gamma_{(s)|1} = \frac{6\,s^4 - 3\,s^2 + 1}{18}\,, \qquad \gamma_{(s)|0} = \frac{s^4 - s^2}{2}\,. \tag{A61}$$

Finally, summing these three numbers, we obtain the vacuum energy as

$$\Gamma^{(1)\,\text{ren}}_{(s)} = (-1)^{2s}\left(1 - \tfrac{1}{2}\delta_{s,0}\right)\frac{60\,s^4 - 30\,s^2 + 1}{45}\,\log R\,, \tag{A62}$$

where $(-1)^{2s}$ arises from the possibly fermionic statistics of the singleton. The reader would recognize, for the $s = 0,\ \frac{1}{2},\ 1$ instances, the coefficient of the $\log R$ term as the conformal anomaly of the spin-s singleton on S^4.

We next turn to the vacuum energy of the non-minimal theory, whose partition function is given by (A52). Again we may evaluate the $f_{\mathcal{H}|n}$ and expand in β to obtain

$$\begin{aligned}\gamma^{\text{non-min}}_{(s)|2} &= \frac{2}{105}\,n_s\left(72\,s^4 - 24\,s^2 + 1\right), \quad \gamma^{\text{non-min}}_{(s)|1} = \frac{4}{315}\,n_s\left(60\,s^4 - 27\,s^2 + 2\right),\\ \gamma^{\text{non-min}}_{(s)|0} &= \frac{8}{15}\,n_s\,(s^4 - s^2)\,,\end{aligned} \tag{A63}$$

where $n_s = \frac{(2s-1)2s(2s+1)}{6}$ is an integer. The terms in (A63) add to give the full one-loop vacuum energy. Using (A62), we find that

$$\Gamma^{(1)\,\text{ren}}_{A/B/C,\text{non-min}} = 2\,n_{0/\frac{1}{2}/1}\,\Gamma^{(1)\,\text{ren}}_{(0/\frac{1}{2}/1)}\,. \tag{A64}$$

We thus reproduce the results that the one-loop vacuum energy vanishes for type A and B theories [24] and equals twice that of the singleton for the type C theory [28].

Having thus reproduced the results for the non-minimal dualities, we now turn to the minimal ones, where the thermal partition function is given by (A53). The contribution to the one-loop vacuum energy of the first term there has already been evaluated in the minimal case, but for the overall factor of $\frac{1}{2}$. It turns out that the second term contributes $\Gamma^{(1)\,\text{ren}}_{(s)}$ to the vacuum energy [37–39]. We finally obtain

$$\Gamma^{(1)\,\text{ren}}_{A/B/C,\text{min}} = \frac{1}{2}\,\Gamma^{(1)\,\text{ren}}_{s,\text{non-min}} + \Gamma^{(1)\,\text{ren}}_{(s)} = \left[(-1)^{2s}\,n_s + 1\right]\Gamma^{(1)\,\text{ren}}_{(s)}\,. \tag{A65}$$

This reproduces using CIRZ, the results of [24] for the Type A, B dualities and of [28] for the Type C duality.

We conclude this section with a discussion of the proposal of [23,24] for interpreting the non-zero one-loop answer derived above for the bulk higher-spin theory. For definiteness, we will work with a spin-0 field, but the discussion readily generalizes, and we will state the final result for all cases discussed above. Firstly, the logarithmically divergent part of the free energy for a theory with N scalar fields defined on S^4 is [57,58]

$$F_{CFT_4} = \frac{N}{90}\log\Lambda_{CFT} + \mathcal{O}\left(\Lambda^0_{CFT}\right). \tag{A66}$$

Next, the computation of the one-loop vacuum energy above indicates that the bulk free energy has the expansion

$$F_{AdS_5} = \left(\frac{\mathcal{L}_0}{g} + \frac{1}{90}\right) \log R + \mathcal{O}(g), \tag{A67}$$

where $\mathcal{L}_0 \log R$ is the on-shell action of the AdS_5 theory. Then equating (A66) with (A67) while using $\log \Lambda_{CFT} \sim \log R$ and identifying $\mathcal{L}_0 = \frac{1}{90}$ yields [24]

$$g^{-1} = N - 1. \tag{A68}$$

This discussion generalizes for the Type B and Type C cases also. In summary, for the non-minimal case, we find [24,28]

$$\text{Type A, B:} \quad g^{-1} = N, \qquad \text{Type C:} \quad g^{-1} = N - 1, \tag{A69}$$

while for the minimal case we find [24,28]

$$\text{Type A, B:} \quad g^{-1} = N - 1, \qquad \text{Type C:} \quad g^{-1} = N - 2. \tag{A70}$$

The corresponding expressions for spin-s singletons is available in [39].

References

1. Maldacena, J.M. The Large N limit of superconformal field theories and supergravity. *Int. J. Theor. Phys.* **1999**, *484*, 51–63.
2. Aharony, O.; Gubser, S.S.; Maldacena, J.; Ooguri, H.; Oz, Y. Large N field theories, string theory and gravity. *Phys. Rep.* **2000**, *323*, 183–386.
3. Polchinski, J. Dirichlet Branes and Ramond-Ramond charges. *Phys. Rev. Lett.* **1995**, *75*, 4724–4727.
4. Klebanov, I.R.; Polyakov, A.M. AdS dual of the critical O(N) vector model. *Phys. Lett. B* **2002**, *550*, 213–219.
5. Gaberdiel, M.R.; Gopakumar, R. An AdS_3 Dual for Minimal Model CFTs. *Phys. Rev. D* **2011**, *83*, 066007.
6. Shenker, S.H.; Yin, X. Vector Models in the Singlet Sector at Finite Temperature. *arXiv* **2011**, arXiv:1109.3519.
7. Sundborg, B. The Hagedorn transition, deconfinement and N=4 SYM theory. *Nucl. Phys. B* **2000**, *573*, 349–363.
8. Witten, E. Spacetime reconstruction. In Proceedings of the Conference in Honor of John Schwarz's 60th Birthday, Pasadena, CA, USA, 3–4 November 2001.
9. Gross, D.J. High-Energy Symmetries of String Theory. *Phys. Rev. Lett.* **1988**, *60*, 1229.
10. Gross, D.J.; Periwal, V. String Perturbation Theory Diverges. *Phys. Rev. Lett.* **1988**, *60*, 2105.
11. Beisert, N.; Bianchi, M.; Morales, J.F.; Samtleben, H. On the spectrum of AdS/CFT beyond supergravity. *J. High Energy Phys.* **2004**, *2004*, 001.
12. Bianchi, M.; Morales, J.F.; Samtleben, H. On stringy $AdS_5 \times S^5$ and higher spin holography. *J. High Energy Phys.* **2003**, *2003*, 062.
13. Beisert, N.; Bianchi, M.; Morales, J.F.; Samtleben, H. Higher spin symmetry and N=4 SYM. *J. High Energy Phys.* **2004**, *2004*, 058.
14. Bianchi, M.; Heslop, P.J.; Riccioni, F. More on "La Grande Bouffe": Towards higher spin symmetry breaking in AdS. *J. High Energy Phys.* **2005**, *2005*, 088.
15. Joung, E.; Lopez, L.; Taronna, M. Solving the Noether procedure for cubic interactions of higher spins in (A)dS. *J. Phys. A Math. Theor.* **2013**, *46*, 214020.
16. Joung, E.; Lopez, L.; Taronna, M. On the cubic interactions of massive and partially-massless higher spins in (A)dS. *J. High Energy Phys.* **2012**, *2012*, 041.
17. Joung, E.; Lopez, L.; Taronna, M. Generating functions of (partially-)massless higher-spin cubic interactions. *J. High Energy Phys.* **2013**, *2013*, 168.
18. Camporesi, R. Harmonic analysis and propagators on homogeneous spaces. *Phys. Rep.* **1990**, *196*, 1–134.
19. Camporesi, R.; Higuchi, A. Arbitrary spin effective potentials in anti-de Sitter space-time. *Phys. Rev. D* **1993**, *47*, 3339–3344.

20. Camporesi, R.; Higuchi, A. The plancherel measure for p-forms in real hyperbolic spaces. *J. Geom. Phys.* **1994**, *15*, 57–94.
21. Camporesi, R.; Higuchi, A. Spectral functions and zeta functions in hyperbolic spaces. *J. Math. Phys.* **1994**, *35*, 4217–4246.
22. Giombi, S.; Yin, X. The Higher Spin/Vector Model Duality. *J. Phys. A Math. Theor.* **2013**, *46*, 214003.
23. Giombi, S.; Klebanov, I.R. One Loop Tests of Higher Spin AdS/CFT. *J. High Energy Phys.* **2013**, *12*, 068.
24. Giombi, S.; Klebanov, I.R.; Safdi, B.R. Higher Spin AdS_{d+1}/CFT_d at One Loop. *Phys. Rev. D* **2014**, *89*, 084004.
25. Giombi, S.; Klebanov, I.R.; Tseytlin, A.A. Partition Functions and Casimir Energies in Higher Spin AdS_{d+1}/CFT_d. *Phys. Rev. D* **2014**, *90*, 024048.
26. Beccaria, M.; Bekaert, X.; Tseytlin, A.A. Partition function of free conformal higher spin theory. *J. High Energy Phys.* **2014**, *2014*, 113.
27. Beccaria, M.; Tseytlin, A.A. Higher spins in AdS_5 at one loop: Vacuum energy, boundary conformal anomalies and AdS/CFT. *J. High Energy Phys.* **2014**, *2014*, 114.
28. Beccaria, M.; Tseytlin, A.A. Vectorial AdS_5/CFT_4 duality for spin-one boundary theory. *J. Phys. A Math. Theor.* **2014**, *47*, 492001.
29. Beccaria, M.; Macorini, G.; Tseytlin, A.A. Supergravity one-loop corrections on AdS_7 and AdS_3, higher spins and AdS/CFT. *Nucl. Phys. B* **2015**, *892*, 211–238.
30. Giombi, S.; Klebanov, I.R.; Tan, Z.M. The ABC of Higher-Spin AdS/CFT. *arXiv* **2016**, arXiv:1608.07611.
31. Gunaydin, M.; Skvortsov, E.D.; Tran, T. Exceptional F(4) Higher-Spin Theory in AdS(6) at One-Loop and other Tests of Duality. *J. High Energy Phys.* **2016**, *2016*, 168.
32. Pang, Y.; Sezgin, E.; Zhu, Y. One Loop Tests of Supersymmetric Higher Spin AdS_4/CFT_3. *arXiv* **2016**, arXiv:1608.07298.
33. Skvortsov, E.D.; Tran, T. AdS/CFT in Fractional Dimension and Higher Spin Gravity at One Loop. *arXiv* **2017**, arXiv:1707.00758.
34. Giombi, S. TASI Lectures on the Higher Spin - CFT duality. *arXiv* **2016**, arXiv:1607.02967.
35. Barabanschikov, A.; Grant, L.; Huang, L.L.; Raju, S. The Spectrum of Yang Mills on a sphere. *J. High Energy Phys.* **2006**, *2006*, 160.
36. Newton, T.H.; Spradlin, M. Quite a Character: The Spectrum of Yang-Mills on S3. *Phys. Lett. B* **2009**, *672*, 382–385.
37. Bae, J.-B.; Joung, E.; Lal, S. One-loop test of free SU(N) adjoint model holography. *J. High Energy Phys.* **2016**, *2016*, 061.
38. Bae, J.-B.; Joung, E.; Lal, S. On the Holography of Free Yang-Mills. *J. High Energy Phys.* **2016**, *2016*, 074.
39. Bae, J.-B.; Joung, E.; Lal, S. A Note on Vectorial AdS_5/CFT_4 Duality for Spin-j Boundary Theory. *J. High Energy Phys.* **2016**, *2016*, 077.
40. Bae, J.-B.; Joung, E.; Lal, S. One-Loop Free Energy of Tensionless Type IIB String in $AdS_5 \times S^5$. *arXiv* **2016**, arXiv:1701.01507.
41. Gibbons, G.W.; Perry, M.J.; Pope, C.N. Partition functions, the Bekenstein bound and temperature inversion in anti-de Sitter space and its conformal boundary. *Phys. Rev. D* **2006**, *74*, 084009.
42. Camporesi, R.; Higuchi, A. On the Eigen functions of the Dirac operator on spheres and real hyperbolic spaces. *J. Geom. Phys.* **1996**, *20*, 1–18.
43. Diaz, D.E.; Dorn, H. Partition functions and double-trace deformations in AdS/CFT. *J. High Energy Phys.* **2007**, *2007*, 046.
44. Bianchi, M.; Dolan, F.A.; Heslop, P.J.; Osborn, H. N=4 superconformal characters and partition functions. *Nucl. Phys. B* **2007**, *767*, 163–226.
45. Polyakov, A.M. Gauge fields and space-time. *Int. J. Mod. Phys. A* **2002**, *17*, 119–136.
46. Aharony, O.; Marsano, J.; Minwalla, S.; Papadodimas, K.; Van Raamsdonk, M. The Hagedorn-deconfinement phase transition in weakly coupled large N gauge theories. *Adv. Theor. Math. Phys.* **2004**, *8*, 603–696.
47. Nishioka, T.; Takayanagi, T. On Type IIA Penrose Limit and N=6 Chern-Simons Theories. *J. High Energy Phys.* **2008**, *2008*, 001.
48. Jaroszewicz, T.; Kurzepa, P.S. Polyakov spin factors and Laplacians on homogeneous spaces. *Ann. Phys.* **1992**, *213*, 135–165.
49. Gopakumar, R.; Gupta, R.K.; Lal, S. The Heat Kernel on AdS. *J. High Energy Phys.* **2011**, *2011*, 010.
50. Lal, S. CFT(4) Partition Functions and the Heat Kernel on AdS(5). *Phys. Lett. B* **2013**, *727*, 325–329.

51. Mack, G. All Unitary Ray Representations of the Conformal Group SU(2,2) with Positive Energy. *Commun. Math. Phys.* **1977**, *55*, 1–28.
52. Dolan, F.A. Character formulae and partition functions in higher dimensional conformal field theory. *J. Math. Phys.* **2006**, *47*, 062303.
53. Minwalla, S. Restrictions imposed by superconformal invariance on quantum field theories. *Adv. Theor. Math. Phys.* **1998**, *2*, 783–851.
54. Metsaev, R.R. Massless mixed symmetry bosonic free fields in d-dimensional anti-de Sitter space-time. *Phys. Lett. B* **1995**, *354*, 78–84.
55. Metsaev, R.R. Mixed symmetry massive fields in AdS(5). *Class. Quant. Grav.* **2005**, *22*, 2777–2796.
56. Metsaev, R.R. Mixed-symmetry fields in AdS(5), conformal fields, and AdS/CFT. *J. High Energy Phys.* **2015**, *2015*, 077.
57. Duff, M.J. Observations on Conformal Anomalies. *Nucl. Phys. B* **1977**, *125*, 334–348.
58. Christensen, S.M.; Duff, M.J. New Gravitational Index Theorems and Supertheorems. *Nucl. Phys. B* **1979**, *154*, 301–342.

universe

MDPI

Article

A Note on Rectangular Partially Massless Fields

Thomas Basile

Department of Physics and Research Institute of Basic Science, Kyung Hee University, Seoul 02447, Korea; thomas.basile@khu.ac.kr

Received: 30 October 2017; Accepted: 26 December 2017; Published: 1 January 2018

Abstract: We study a class of non-unitary $\mathfrak{so}(2,d)$ representations (for even values of d), describing mixed-symmetry partially massless fields which constitute natural candidates for defining higher-spin singletons of higher order. It is shown that this class of $\mathfrak{so}(2,d)$ modules obeys of natural generalisation of a couple of defining properties of unitary higher-spin singletons. In particular, we find out that upon restriction to the subalgebra $\mathfrak{so}(2,d-1)$, these representations branch onto a sum of modules describing partially massless fields of various depths. Finally, their tensor product is worked out in the particular case of $d=4$, where the appearance of a variety of mixed-symmetry partially massless fields in this decomposition is observed.

Keywords: singletons; higher spin gauge theory

Contents

1. Introduction

The completion of the Bargmann-Wigner program in anti-de Sitter (AdS) spacetime [1] lead to some surprising lessons concerning the definition of masslessness in other backgrounds than Minkowski space. If nowadays, the most common way to discriminate between massless and massive fields in AdS is whether or not they enjoy some gauge symmetry, other proposals which involve a particular kind of $\mathfrak{so}(2,d)$ representations known as "singletons", were put forward [4,5]. Indeed, the proposed notions of "conformal masslessness" and "composite masslessness" both rely on two crucial properties of singletons, namely [2]:

- They are unitary and irreducible representations (UIRs) of $\mathfrak{so}(2,d)$ that remain irreducible when restricted to UIRs of $\mathfrak{so}(2,d-1)$, or in other words they correspond to the class of elementary particles in d-dimensional anti-de Sitter space which are conformal. This property is the very definition of a conformally massless UIR, and it turns out that the singletons are precisely the $\mathfrak{so}(2,d)$ UIRs to which conformally massless $\mathfrak{so}(2,d-1)$ UIRs can be lifted.
- The tensor product of two $\mathfrak{so}(2,3)$ singletons contains all conformally massless fields in AdS_4 [7]. In any dimensions however, the representations appearing in the decomposition of the tensor product of two $\mathfrak{so}(2,d)$ singletons (of spin 0 or $\frac{1}{2}$) are no longer conformally massless but make up, by definition, all of the composite massless UIRs of $\mathfrak{so}(2,d)$ [6,8–10]. In other words, composite massless UIRs are those modules which appear in the decomposition of the tensor product of two singletons.

This tensor product decomposition, called the Flato-Frønsdal theorem, is crucial in the context of Higher-Spin Gauge Theories and can be summed up as follows (in the case of two scalar singletons):

$$\text{Rac} \otimes \text{Rac} = \text{Massive scalar} \oplus \bigoplus_{s=1}^{\infty} \text{Gauge field of spin } s\,, \tag{1}$$

where Rac denotes the $\mathfrak{so}(2,d)$ scalar singleton. Notice that the spin-$s \geqslant 1$ gauge fields are both "composite massless" by definition, as well as massless in the modern sense (as they enjoy some gauge symmetry), whereas the scalar field is considered massive in the sense of being devoid of said gauge symmetries (despite the fact that it is also "composite massless"). The $\mathfrak{so}(2,d)$ UIRs on the right hand side make up the spectrum of fields of Vasiliev's higher-spin gravity [11–14] (see e.g., [15–17] for, respectively, non-technical and technical reviews). This decomposition can also be interpreted in terms of operators of a free d-dimensional Conformal Field Theory (CFT) as on the left hand side, the tensor product of two scalar singletons can be thought of as a bilinear operator in the fundamental scalar field and the right hand side as the various conserved currents that this CFT possesses. This dual interpretation of Equation (1) is by now regarded as a first evidence in favor of the AdS/CFT correspondence [18–20] in the context of Higher-Spin theory [21,22]. This duality relates (the type A) Vasiliev's bosonic (minimal) higher-spin gravity to the free $U(N)$ ($O(N)$) vector model and has passed several non-trivial checks since it has been proposed, from the computation and matching of the one-loop partition functions [23,24] to three point functions [25–27] on both sides of the duality (see e.g., [28–31] and references therein for reviews of this duality). The possible existence of such an equivalence between the type-A Higher-Spin (HS) theory and the free vector model opened the possibility of probing interactions in the bulk using the knowledge gathered on the CFT side, a program which was tackled in [32] (improving the earlier works [33–35]). This lead to the derivation

[1] In the sense that for every $\mathfrak{so}(2,d)$ UIRs of the lowest energy type, i.e., in the discrete series of representations, a realisation on a space of solutions of a wave equation is known. It should be noted however that the recent works [1,2] uncovered the existence of "continuous spin" fields on AdS which do not fall into the discrete series, but might correspond to $\mathfrak{so}(2,d)$ module induced from the non-compact subalgebra $\mathfrak{so}(1,1) \oplus \mathfrak{so}(1,d-1)$ (instead of the maximal compact subalgebra $\mathfrak{so}(2) \oplus \mathfrak{so}(d)$) as noticed in [3].

[2] See e.g., [6] where those properties were studied using the harmonic expansion of fields in (A)dS spacetimes.

of all cubic vertices and the quartic vertex for four scalar fields in the bulk [36,37], as dictated by the holographic duality, while [38] also raising questions on the locality properties of the bulk HS theory (see e.g., [39–43] and references therein for more details).

The fact that the prospective CFT dual to a HS theory in AdS_{d+1} is free can be understood retrospectively thanks to the Maldacena-Zhiboedov theorem [44] and its generalisation [45,46]. Indeed, it was shown in [44] that if a 3-dimensional CFT which is unitary, obeys the cluster decomposition axiom and has a (unique) Lorentz covariant stress-tensor plus at least one higher-spin current, then this theory is either a CFT of free scalars or free spinors. This was generalised to arbitrary dimensions in [45,46][3], where the authors showed that this result holds in dimensions $d \geqslant 3$, up to the additional possibility of a free CFT of $(\frac{d-2}{2})$-forms in even dimensions. These free conformal fields precisely correspond to the singleton representations of spin 0, $\frac{1}{2}$ and 1 in arbitrary dimensions [5,48][4]. Due to the fact that, according to the standard AdS/CFT dictionary, the higher-spin gauge field making up the spectrum of the Higher-Spin theory in the bulk are dual to higher-spin conserved current on the CFT side, this CFT should be free[5] as it falls under the assumption of the previously recalled results of [44–46]. Hence, the algebra generated by the set of charges associated with the conserved currents of the CFT whose fundamental field is a spin-s singletons corresponds to the HS algebra of the HS theory in the bulk with a spectrum of field given by the decomposition of the tensor product of two spin-s singletons. These HS algebras can be defined as follows:

$$\mathfrak{hs}_s^{(d)} \cong \frac{\mathcal{U}\big(\mathfrak{so}(2,d)\big)}{\mathrm{Ann}\big(\mathcal{D}_s^{\text{sing.}}\big)}\,, \tag{2}$$

where $\mathfrak{hs}_s^{(d)}$ stands for the HS algebra associated with the spin-s singleton in AdS_{d+1}, and $\mathcal{U}\big(\mathfrak{so}(2,d)\big)$ denotes the universal enveloping algebra of $\mathfrak{so}(2,d)$ whereas $\mathcal{D}_s^{\text{sing.}}$ denotes the spin-s singleton module of $\mathfrak{so}(2,d)$ and $\mathrm{Ann}\big(\mathcal{D}_s^{\text{sing.}}\big)$ the annihilator of this module. For more details, see e.g., [6,52,53] where the construction of HS algebras (and their relation with minimal representations of simple Lie algebras [54]) is reviewed and [55] where HS algebras associated with HS singletons were studied. Although Vasiliev's Higher-Spin theory is based on the HS algebra $\mathfrak{hs}_0^{(d)}$ and the HS theory based on $\mathfrak{hs}_s^{(d)}$ with $s = \frac{1}{2}, 1, \ldots,$ is unknown[6], the latter algebras are quite interesting as they all describe a spectrum containing mixed-symmetry fields. Even though this last class of massless field is well understood at the free level (in flat space as well as in AdS) [56–74], little is known about their interaction (see e.g., [52,75–78] on cubic vertices and [79–81] where mixed-symmetry fields have been studied in the context of the AdS/CFT correspondence).

---

3 See also [47] where the authors derived the bulk counterpart of this theorem.

4 Singletons of spin-$s > 1$ do not appear here due to the fact that the free CFT based on these fields do not possess a conserved current of spin-2, i.e., a stress-energy tensor and therefore are not covered by the theorem derived in [46]. This can be seen for instance from the decomposition of the tensor product of two spin-s singletons spelled out in [10], where one can check that there are no currents of spin lower than $2s$.

5 Notice that by changing the boundary condition of fields in the bulk opens the possibility of having an HS theory dual to an interacting CFT, see for instance [22,25,28,49,50]. From the CFT point of view, this corresponds to the modification of the previously mentioned Maldacena-Zhiboedov theorem studied by the same authors in [51]. Instead of the existence of at least one conserved higher spin current, it is assumed that the CFT possesses a parameter N together with a tower of single trace, approximately conserved currents of all even spin $s \geqslant 4$, such that the conservation law gets corrected by terms of order $1/N$. As a consequence, the anomalous dimensions of these higher spin currents are of order $1/N$, which translates into the fact that the dual higher spin fields in the bulk acquire masses through radiative corrections, thereby leading to changes in their boundary conditions.

6 Notice that this is true only for $d > 3$. For the AdS_4 case however, due to the fact that the annihilator of the scalar and the spin-$\frac{1}{2}$ singleton are isomorphic (i.e., $\mathrm{Ann}(\mathcal{D}_0^{\text{sing.}}) \cong \mathrm{Ann}(\mathcal{D}_{1/2}^{\text{sing.}})$, see e.g., [6] for more details), the HS algebra $\mathfrak{hs}_{1/2}^{(3)}$ is isomorphic to $\mathfrak{hs}_0^{(3)}$ and is therefore known. This translates into the fact that the two corresponding HS theory have almost the same spectrum of fields, the only difference being the mass of the bulk scalar field.

A possible extension of the HS algebras associated with singletons can be obtained by applying the above construction (2) with a generalisation of the singleton representations $\mathcal{D}_s^{\text{sing.}}$, referred to as "higher-order" singletons. The latter are also irreducible representations of $\mathfrak{so}(2,d)$, which share the property of describing fields "confined" to the boundary of AdS with the usual singletons but which are non-unitary (as detailed in [82]). This class of higher-order singletons, which are of spin 0 or $\frac{1}{2}$, is labelled by a (strictly) positive integer ℓ. In the case of the scalar singleton of order ℓ, such a representation describes a conformal scalar ϕ obeying the polywave equation:

$$\Box^{\ell}\phi = 0\,. \tag{3}$$

When $\ell = 1$ one recovers the usual singleton (free, unitary conformal scalar field), whereas $\ell > 1$ leads to non-unitary CFT. Such CFTs were studied in [83] for instance, and were proposed to be dual to HS theories [82] whose spectrum consists, on top of the infinite tower of (totally symmetric) higher-spin massless fields, also partially massless (totally symmetric) fields of arbitrary spin (theories which have been studied recently in [84–86]), thereby extending the HS holography proposal of Klebanov-Polyakov-Sezgin-Sundell to the non-unitary case. The corresponding HS algebras were studied in [87] for the simplest case $\ell = 2$ (as the symmetry algebra of the Laplacian square, thereby generalising the previous characterisation of $\mathfrak{hs}_0^{(d)}$ as the symmetry algebra of the Laplacian [88]) and for general values of ℓ in [89–91]. As we already mentionned, the interesting feature of such HS algebras is that their spectrum, i.e., the set of fields of the bulk theory, contains partially massless (totally symmetric) higher-spin fields [82] (introduced originally in [92–95], and whose free propagation was described in the unfolded formalism in [96]). Although non-unitary in AdS background, partially massless fields of arbitrary spin are unitary in de Sitter background [97], and hence constitute a particularly interesting generalisation of HS gauge fields to consider[7]. Partially massless fields, both totally symmetric and of mixed-symmetry, also appear in the spectrum of the HS algebra based on the order-ℓ spinor singleton [100]. It seems reasonable to expect that the known spectrum of the HS algebras based on a spin-s singleton is enhanced, when considering the HS algebras based on their higher-order extension, with partially massless fields of the same symmetry type as already present in the case of the original singleton. Therefore, a natural question is whether or not there exist higher-order higher-spin singletons. This question is adressed in the present note, in which we study a class of $\mathfrak{so}(2,d)$ modules which is a natural candidate for defining a higher-order higher-spin singleton.

This paper is organised as follows: in Section 2 we start by introducing the various notations that will be used throughout this note, then in Section 3 we first review the defining properties of the well-known (unitary) higher-spin singletons before introducing their would-be higher-order extension and spelling out the counterpart of the previously recalled characteristic properties. Finally, the tensor product of two such representations is decomposed in the low-dimensional case $d = 4$ in Section 4. Technical details on the branching and tensor product rule of $\mathfrak{so}(d)$ can be found in Appendix A while details of the proofs of Propositions 2 and 3 and are relegated to Appendix B.

2. Notation and Conventions

In the rest of this paper, we will use the following symbols:

- A $\mathfrak{so}(2,d)$ (generalised Verma) module is characterised by the $\mathfrak{so}(2) \oplus \mathfrak{so}(d)$ lowest weight $[\Delta; \boldsymbol{\ell}]$, where Δ is the $\mathfrak{so}(2)$ weight (in general a real number, and more often in this paper, a positive integer) corresponding to the minimal energy of the AdS field described by this representation and $\boldsymbol{\ell}$ is a dominant integral $\mathfrak{so}(d)$ weight corresponding to the spin of the representation. If irreducible,

[7] Notice that partially massless HS fields have been shown to also appear in the spectrum of fields resulting from various compactifications of anti-de Sitter spacetime [98,99].

those modules will be denoted $\mathcal{D}(\Delta;\boldsymbol{\ell})$, whereas if reducible (or indecomposable), they will be denoted $\mathcal{V}(\Delta;\boldsymbol{\ell})$.

- The spin $\boldsymbol{\ell} \equiv (\ell_1,\ldots,\ell_r)$, with $r := \text{rank}(\mathfrak{so}(d)) \equiv [\frac{d}{2}]$ (and where $[x]$ is the integer part of x), is a $\mathfrak{so}(d)$ integral dominant weight. The property that the weight $\boldsymbol{\ell}$ is integral means that its components ℓ_i, $i = 1,\ldots,r$ are either *all* integers or *all* half-integers. The fact that $\boldsymbol{\ell}$ is dominant means that the components are ordered in decreasing order, and all positive except for the component ℓ_r when $d = 2r$. More precisely,

$$\ell_1 \geqslant \ell_2 \geqslant \cdots \geqslant \ell_{r-1} \geqslant \ell_r \geqslant 0\,, \quad \text{for} \quad \mathfrak{so}(2r+1)\,, \tag{4}$$

and

$$\ell_1 \geqslant \ell_2 \geqslant \cdots \geqslant \ell_{r-1} \geqslant |\ell_r|\,, \quad \text{for} \quad \mathfrak{so}(2r)\,. \tag{5}$$

- In order to deal more efficiently with weight having several identical components, we will use the notation:

$$(\underbrace{\ell_1,\ldots,\ell_1}_{h_1\text{ times}},\underbrace{\ell_2,\ldots,\ell_2}_{h_2\text{ times}},\ldots,\underbrace{\ell_k,\ldots,\ell_k}_{h_k\text{ times}}) \equiv (\ell_1^{h_1},\ell_2^{h_2},\ldots,\ell_k^{h_k})\,, \quad \text{with} \quad k \leqslant r\,, \tag{6}$$

in other words the number h of components with the same value ℓ appears as the exponent of the latter. For the special cases where all components of the highest weight are equal either to 0 or to $\frac{1}{2}$, we will use bold symbols (and forget about the brackets), i.e.,

$$\mathbf{0} := (0,\ldots,0)\,, \quad \text{and} \quad \boldsymbol{\tfrac{1}{2}} := (\tfrac{1}{2},\ldots,\tfrac{1}{2})\,. \tag{7}$$

We will also write only the non-vanishing components of the various $\mathfrak{so}(d)$ weight encountered in this paper, i.e.,

$$(s_1,\ldots,s_k) := (s_1,\ldots,s_k,\underbrace{0,\ldots,0}_{r-k})\,, \quad \text{for} \quad 1 \leqslant k \leqslant r\,. \tag{8}$$

- If the spin is given by an irrep of an even dimensional orthogonal algebra, i.e., $\mathfrak{so}(d)$ for $d = 2r$, the last component ℓ_r of this highest weight (if non-vanishing) can either be positive or negative. Whenever the statements involving such a weight does not depend on this sign, we will write $\boldsymbol{\ell} = (\ell_1,\ldots,\ell_r)$ with the understanding that ℓ_r can be replaced by $-\ell_r$. However, if the sign of the component ℓ_r matters, we will distinguish the two cases by writting:

$$\boldsymbol{\ell}_\epsilon \equiv (\ell_1,\ldots,\ell_{r-1},\ell_r)_\epsilon := (\ell_1,\ldots,\ell_{r-1},\epsilon\ell_r)\,, \quad \text{with} \quad \epsilon = \pm 1\,. \tag{9}$$

It will also be convenient to consider the direct sum of two modules labelled by $\boldsymbol{\ell}_+$ and $\boldsymbol{\ell}_-$, which we will denote by:

$$\mathcal{D}(\Delta;\boldsymbol{\ell}_0) := \mathcal{D}(\Delta;\boldsymbol{\ell}_+) \oplus \mathcal{D}(\Delta;\boldsymbol{\ell}_-)\,. \tag{10}$$

- Finally, a useful tool that we will use throughout this paper is the character $\chi^{\mathfrak{so}(2,d)}_{\mathcal{V}(\Delta;\boldsymbol{\ell})}(q,\vec{x})$ of a (possibly reducible) generalised Verma module $\mathcal{V}(\Delta;\,\boldsymbol{\ell})$:

$$\chi^{\mathfrak{so}(2,d)}_{\mathcal{V}(\Delta;\boldsymbol{\ell})}(q,\vec{x}) = q^\Delta \chi^{\mathfrak{so}(d)}_{\boldsymbol{\ell}}(\vec{x})\mathcal{P}^{(d)}(q,\vec{x})\,, \tag{11}$$

where $q := e^{-\mu_0}$ and $x_i := e^{\mu_i}$ for $i = 1,\ldots,r$ with $\{\mu_0,\ldots,\mu_r\}$ a basis of the weight space of $\mathfrak{so}(2,d)$, and

$$\mathcal{P}^{(d)}(q,\vec{x}) := \frac{1}{(1-q)^{d-2r}} \prod_{i=1}^{r} \frac{1}{(1-qx_i)(1-qx_i^{-1})}\,, \tag{12}$$

i.e., the prefactor $\frac{1}{1-q}$ is absent for $d = 2r$, and $\chi_{\ell}^{\mathfrak{so}(d)}(\vec{x})$ is the character of the irreducible $\mathfrak{so}(d)$ representation ℓ. Any irreducible generalised Verma module $\mathcal{D}(\Delta\,;\,\ell)$ can be defined as the quotient of the (freely generated) generalised Verma module $\mathcal{V}(\Delta\,;\,\ell)$ by its maximal submodule $\mathcal{D}(\Delta'\,;\,\ell')$ (for some $\mathfrak{so}(2)\oplus\mathfrak{so}(d)$ weight $[\Delta'\,;\,\ell']$ related to $[\Delta\,;\,\ell]$). An important property that will be used extensively in this work is the fact that given two representation spaces V and W of the same algebra with respective characters χ_V and χ_W, the characters of the tensor product, direct sum or quotient of these two spaces obey:

$$\chi_{V\otimes W} = \chi_V \cdot \chi_W\,, \quad \chi_{V\oplus W} = \chi_V + \chi_W\,, \quad \chi_{V/W} = \chi_V - \chi_W\,. \tag{13}$$

As a consequence, the character of an irreducible generalised Verma module $\mathcal{D}(\Delta;\ell)$ takes the form:

$$\chi_{[\Delta;\ell]}^{\mathfrak{so}(2,d)}(q,\vec{x}) = \chi_{\mathcal{V}(\Delta;\ell)}^{\mathfrak{so}(2,d)}(q,\vec{x}) - \chi_{\mathcal{D}(\Delta';\ell')}^{\mathfrak{so}(2,d)}(q,\vec{x})\,, \tag{14}$$

whenever $\mathcal{V}(\Delta;\ell)$ possesses a submodule $\mathcal{D}(\Delta';\ell')$. For more details on characters of $\mathfrak{so}(2,d)$ generalised Verma modules, see e.g., [10,101].

3. Higher-Order Higher-Spin Singletons

In this section, we start by reviewing the definition of the usual (unitary) higher-spin singletons (about which more details can be found in the pedagogical review [102], and in [103] where they were studied from the point of view of minimal representations), before moving on to the proposed higher-order extension which is the main focus of this paper.

3.1. Unitary Higher-Spin Singletons

Higher-spin singletons have been first considered by Siegel [48], as making up the list of unitary and irreducible representations of the conformal algebra $\mathfrak{so}(2,d)$ which can lead to a free conformal field theory. They were later identified by Angelopoulos and Laoues [4,5,8,104] as being part of the same class of particular representations first singled out by Dirac [105], which is what is understood by singletons nowadays. Initially, what lead Dirac to single out the $\mathfrak{so}(2,3)$ representations $\mathcal{D}(\frac{1}{2};(0))$ and $\mathcal{D}(1;(\frac{1}{2}))$ studied in [105] as "remarkable" is the fact that, contrarily to the usual UIRs of *compact* orthogonal algebras, the former are labelled by an highest weight whose components are not both integers or both half-integers but rather one is an integer and the other is an half-integer. In other words, the highest weight defining this representation is not integral dominant. On top of that, the other intriguing feature of these representations, which was later elaborated on significantly by Flato and Frønsdal, is the fact they correspond respectively to a scalar and a spinor field in AdS which do not propagate any local degree of freedom in the bulk. This last property is the most striking from a field theoretical point of view. Indeed, the fact that representations of the $\mathfrak{so}(2,d)$ algebra can be interpreted both as fields in AdS_{d+1}, i.e., the bulk, and as conformal fields on d-dimensional Minkowski, i.e., the (conformal) boundary of AdS_{d+1} is at the core of the AdS/CFT correspondence. This last characteristic translates into a defining property of the $\mathfrak{so}(2,d)$ singleton modules, namely that they remain irreducible when restricted to either one of the subalgebras $\mathfrak{so}(2,d-1)$, $\mathfrak{so}(1,d)$ or $\mathfrak{iso}(1,d-1)$. This is reviewed below after we define the singletons as unitary and irreducible $\mathfrak{so}(2,d)$ modules.

First, let us recall that the unitarity conditions for generalised Verma modules of $\mathfrak{so}(2,d)$ (i.e., in its discrete series of representations) induced from the compact subalgebra $\mathfrak{so}(2)\oplus\mathfrak{so}(d)$ were derived independently in [57–59] and in [106] (where the more general result of [107] giving unitarity conditions for highest weight modules of Hermitian algebras was applied to $\mathfrak{so}(2,d)$). The outcome of these analyses is that the irreducible modules $\mathcal{D}(\Delta\,;\,\ell)$ which are unitary are:

- $\ell = 0$: modules with $\Delta \geqslant \frac{d-2}{2}$;
- $\ell = \frac{1}{2}$: modules with $\Delta \geqslant \frac{d-1}{2}$;
- $\ell = (s^p, s_{p+1}, \ldots, s_r)$ with $1 \geqslant s > s_{p+1} \geqslant \cdots \geqslant s_r$: modules with $\Delta \geqslant s + d - p - 1$.

With these unitarity bounds in mind for $\mathfrak{so}(2,d)$ generalised Verma modules, let us move onto the definition of *unitary* singletons:

Definition 1 (Singleton). *A spin-s singleton is defined as the $\mathfrak{so}(2,d)$ module:*

$$\mathcal{D}(s+\tfrac{d}{2}-1;(s^r)), \tag{15}$$

for $s=0,\frac{1}{2}$ in arbitrary dimensions, and $s\in\mathbb{N}$ when $d=2r$. Introducing the minimal energies of the scalar and spinor singleton

$$\epsilon_0 := \frac{d-2}{2}, \quad and \quad \epsilon_{1/2} := \frac{d-1}{2} \equiv \epsilon_0 + \tfrac{1}{2}, \tag{16}$$

all of the above modules can be denoted as $\mathcal{D}(\epsilon_0+s\,;\,(s^r))$. Depending on the value of s the structure of the the above module changes drastically:

- *If $s=0$ or $\frac{1}{2}$, then*

$$\text{Rac} := \mathcal{D}(\epsilon_0;\mathbf{0}) \cong \frac{\mathcal{V}(\epsilon_0;\mathbf{0})}{\mathcal{D}(d-\epsilon_0;\mathbf{0})}, \quad \text{Di} := \mathcal{D}(\epsilon_{1/2};\tfrac{\mathbf{1}}{\mathbf{2}}) \cong \frac{\mathcal{V}(\epsilon_{1/2};\frac{\mathbf{1}}{\mathbf{2}})}{\mathcal{D}(d-\epsilon_{1/2};\frac{\mathbf{1}}{\mathbf{2}})}, \tag{17}$$

where $\mathcal{D}(d-\epsilon_0;\mathbf{0}) = \mathcal{V}(d-\epsilon_0;\mathbf{0})$ and $\mathcal{D}(d-\epsilon_{1/2};\frac{\mathbf{1}}{\mathbf{2}}) = \mathcal{V}(d-\epsilon_{1/2};\frac{\mathbf{1}}{\mathbf{2}})$. Their character read [7,10]:

$$\chi^{\mathfrak{so}(2,d)}_{\text{Rac}}(q,\vec{x}) = q^{\epsilon_0}\,(1-q^2)\,\mathcal{P}^{(d)}(q,\vec{x}), \quad and \quad \chi^{\mathfrak{so}(2,d)}_{\text{Di}}(q,\vec{x}) = q^{\epsilon_{1/2}}\,(1-q)\,\chi^{\mathfrak{so}(d)}_{\frac{1}{2}}(\vec{x})\,\mathcal{P}^{(d)}(q,\vec{x}). \tag{18}$$

- *If $s\geqslant 1$ (and $d=2r$), then:*

$$\mathcal{D}(s+\tfrac{d}{2}-1\,;\,(s^r)) \cong \frac{\mathcal{V}(s+\frac{d}{2}-1\,;\,(s^r))}{\mathcal{D}(s+\frac{d}{2}\,;\,(s^{r-1},s-1))}. \tag{19}$$

In this case, the structure of the maximal submodule $\mathcal{D}(s+\frac{d}{2}\,;\,(s^{r-1},s-1))$ is more involved, the maximal submodule $\mathcal{D}(s+\frac{d}{2}\,;\,(s^{r-1},s-1))$ can be defined through the sequence of quotients of generalized Verma modules:

$$\mathcal{D}(s+\tfrac{d}{2}-1+k\,;\,(s^{r-k},(s-1)^k)) := \frac{\mathcal{V}(s+\frac{d}{2}-1+k\,;\,(s^{r-k},(s-1)^k))}{\mathcal{D}(s+\frac{d}{2}+k\,;\,(s^{r-k-1},(s-1)^{k+1}))}, \tag{20}$$

with $k=1,\dots,r-1$ and $\mathcal{D}(s+\frac{d}{2}+r-1\,;\,((s-1)^r)) \equiv \mathcal{V}(s+\frac{d}{2}+r-1\,;\,((s-1)^r))$ is an irreducible module. For more details on the structure of irreducible generalised $\mathfrak{so}(2,d)$ Verma module, see the classification displayed in [108]. Their character read [10]:

$$\chi^{\mathfrak{so}(2,d)}_{[s+\frac{d}{2}-1;(s^r)]}(q,\vec{x}) = q^{s+\frac{d}{2}-1}\Big(\chi^{\mathfrak{so}(d)}_{(s^r)}(\vec{x}) + \sum_{k=1}^{r}(-)^k q^k \chi^{\mathfrak{so}(d)}_{(s^{r-k},(s-1)^k)}(\vec{x})\Big)\mathcal{P}^{(d)}(q,\vec{x}). \tag{21}$$

Remark 1. *As advertised, all of the above modules corresponding to singletons are unitary. One can notice that they actually saturate the unitarity bound and are all irreps of twist ϵ_0 (the twist τ being the absolute value of the difference between the minimal energy Δ and the spin of s of a $\mathfrak{so}(2,d)$ irrep, $\tau := |\Delta - s|$).*

All of the above $\mathfrak{so}(2,d)$ modules share a couple of defining properties recalled below:

Theorem 1 (Properties of the singletons [4,5]). *A singleton on AdS_{d+1} is a module $\mathcal{D}(s+\frac{d}{2}-1;(s^r))$ of $\mathfrak{so}(2,d)$ with $s=0,\frac{1}{2}$ for any values d and $s\in\mathbb{N}$ for $d=2r$, enjoying the following properties:*

(i) *It decomposes into a (infinite) single direct sum of $\mathfrak{so}(2) \oplus \mathfrak{so}(d)$ (finite-dimensional) modules in which each irrep of $\mathfrak{so}(d)$ appears only once (is multiplicity free) and with a different $\mathfrak{so}(2)$ weight, i.e.,*

$$\mathcal{D}\big(s+\tfrac{d}{2}-1;(s^r)\big) \cong \bigoplus_{\sigma=0}^{\infty} \mathcal{D}_{\mathfrak{so}(2)\oplus\mathfrak{so}(d)}\big(s+\tfrac{d}{2}-1+\sigma;(s+\sigma,s^{r-1})\big)\,. \tag{22}$$

This was proven originally in [109] for the $d=3$ case where only the spin-0 and spin-$\frac{1}{2}$ singletons exist, and extended to arbitrary dimensions and for singletons of arbitrary spin in [5].

(ii) *It branches into a single irreducible module*[8] *of the subalgebras $\mathfrak{iso}(1,d-1)$, $\mathfrak{so}(1,d)$ or $\mathfrak{so}(2,d-1)$, in which case this branching rule reads:*

$$\mathcal{D}\big(s+r-1;\,(s^r)\big) \quad \underset{\mathfrak{so}(2,d-1)}{\overset{\mathfrak{so}(2,d)}{\downarrow}} \quad \mathcal{D}\big(s+r-1;\,(s^{r-1})\big)\,, \quad \text{for} \quad d=2r \quad \text{and} \quad s\geqslant 1\,, \tag{23}$$

and where the $\mathfrak{so}(2,d-1)$ module $\mathcal{D}(s+r-1;\,(s^{r-1}))$ correspond to a massless field of spin (s^{r-1}) in AdS_d. Conversely, singletons can be seen as the only $\mathfrak{iso}(1,d-1)$, $\mathfrak{so}(1,d)$ or $\mathfrak{so}(2,d-1)$ modules that can be lifted to a module of $\mathfrak{so}(2,d)$ (which is $\mathcal{D}(s+\frac{d}{2}-1;(s^r))$). From this point of view, this property can be restated as "Singletons are the only (massless) particles, or gauge fields, in d-dimensional Minkowski, de Sitter or anti-de Sitter spacetime which also admit conformal symmetries", as they are the only representations of the isometry algebra of the d-dimensional maximally symmetric spaces that can be lifted to a representation of the conformal algebra in d-dimensions. Again, this property was first proven in $d=3$ in [4] and later extended to arbitrary dimensions in [5,110]. This was revisited recently in [111].

Proof. Considering that the couple of defining properties of the singletons are already known, we will only sketch the idea of their proofs—that can be found in the original papers—by focusing on the simpler, low-dimensional, case of $\mathfrak{so}(2,4)$ spin-s singletons, leaving the general case in arbitrary dimensions to Appendix B.1.

(i) This decomposition can be proven by showing that the character of the module $\mathcal{D}(s+r-1;(s^r))$ can be rewritten in the form:

$$\chi^{\mathfrak{so}(2,d)}_{[s+r-1;(s^r)]}(q,\vec{x}) = \sum_{\sigma=0}^{\infty} q^{s+r-1+\sigma}\,\chi^{\mathfrak{so}(d)}_{(s+\sigma,s^{r-1})}(\vec{x})\,, \tag{24}$$

which is indeed the character of the direct sum of $\mathfrak{so}(2)\oplus\mathfrak{so}(d)$ modules displayed in (22). This was proven in [10], and in practice the idea is simply to use the property of the "universal" function $\mathcal{P}^{(d)}(q,\vec{x})$ that it can be rewritten as:

$$\mathcal{P}^{(d)}(q,\vec{x}) = \sum_{\sigma,n=0}^{\infty} q^{\sigma+2n}\,\chi^{\mathfrak{so}(d)}_{(\sigma)}(\vec{x})\,, \tag{25}$$

and then perform the tensor product between the $\mathfrak{so}(d)$ characters appearing in the character $\chi^{\mathfrak{so}(2,d)}_{[s+r-1;(s^r)]}(q,\vec{x})$ with $\chi^{\mathfrak{so}(d)}_{(s)}(\vec{x})$. Let us to do that explicitely for $d=4$, where we can take

8 Or at most two, which is the case of the scalar singleton as will be illustrated later (in Proposition 1).

advantage of the exceptional isomorphism $\mathfrak{so}(4) \cong \mathfrak{so}(3) \oplus \mathfrak{so}(3)$ to deal with $\mathfrak{so}(4)$ tensor products:

$$\chi^{\mathfrak{so}(2,4)}_{[s+1;(s,s)]}(q,\vec{x}) = q^{s+1}\Big(\chi^{\mathfrak{so}(4)}_{(s,s)}(\vec{x}) - q\chi^{\mathfrak{so}(4)}_{(s,s-1)}(\vec{x}) + q^2\chi^{\mathfrak{so}(4)}_{(s-1,s-1)}(\vec{x})\Big)\mathcal{P}^{(4)}(q,\vec{x}) \tag{26}$$

$$= \sum_{n=0}^{\infty} q^{s+1+2n}\Big(\sum_{\sigma=0}^{2s} q^{\sigma}\sum_{k=0}^{\sigma}\chi^{\mathfrak{so}(4)}_{(s+k,s+k-\sigma)}(\vec{x}) + \sum_{\sigma=2s+1}^{\infty} q^{\sigma}\sum_{k=0}^{2s}\chi^{\mathfrak{so}(4)}_{(\sigma+k-s,k-s)}(\vec{x}) \tag{27}$$

$$-\sum_{\sigma=0}^{2s-1} q^{\sigma+1}\sum_{k=0}^{\sigma}\chi^{\mathfrak{so}(4)}_{(s+k,s+k-\sigma-1)}(\vec{x}) - \sum_{\sigma=1}^{2s-1} q^{\sigma+1}\sum_{k=0}^{\sigma}\chi^{\mathfrak{so}(4)}_{(s+k-1,s+k-\sigma)}(\vec{x}) \tag{28}$$

$$-\sum_{\sigma=2s}^{\infty} q^{\sigma+1}\sum_{k=0}^{2s-1}\Big[\chi^{\mathfrak{so}(4)}_{(\sigma+k+1-s,k-s)}(\vec{x}) + \chi^{\mathfrak{so}(4)}_{(\sigma+k-s,k+1-s)}(\vec{x})\Big] \tag{29}$$

$$+\sum_{\sigma=0}^{2s-2} q^{\sigma+2}\sum_{k=0}^{\sigma}\chi^{\mathfrak{so}(4)}_{(s+k-1,s+k-\sigma-1)}(\vec{x}) + \sum_{\sigma=2s-1}^{\infty} q^{\sigma+2}\sum_{k=0}^{2s-2}\chi^{\mathfrak{so}(4)}_{(\sigma+k+1-s,k+1-s)}(\vec{x})\Big)$$

$$= \sum_{n=0}^{\infty} q^{s+1+2n}\Big(\sum_{\sigma=0}^{\infty} q^{\sigma}\chi^{\mathfrak{so}(4)}_{(s+\sigma,s)}(\vec{x}) - \sum_{\sigma=1}^{\infty} q^{\sigma+1}\chi^{\mathfrak{so}(4)}_{(s+\sigma-1,s)}(\vec{x})\Big) \tag{30}$$

$$= \sum_{\sigma=0}^{\infty} q^{s+1+\sigma}\chi^{\mathfrak{so}(4)}_{(s+\sigma,s)}(\vec{x})\,. \tag{31}$$

This decomposition can be illustrated by drawing a "weight diagram", representing the $\mathfrak{so}(2)$ weight of $\mathfrak{so}(2) \oplus \mathfrak{so}(d)$ modules as a function of the first component of their $\mathfrak{so}(d)$ weights, see Figure 1 below.

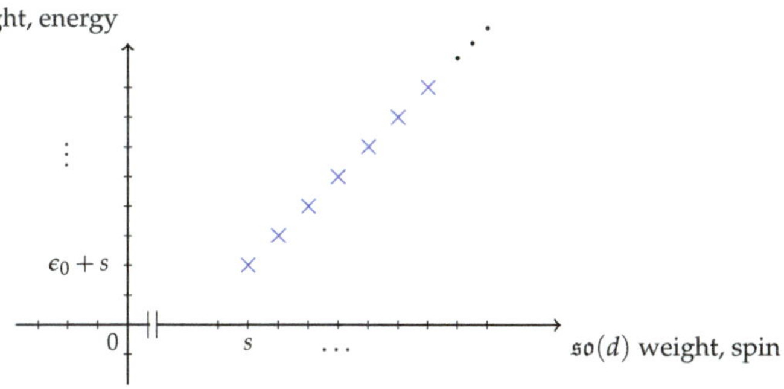

Figure 1. Weight diagram of the spin-s singleton.

The fact the weight diagram of singletons is made out of a single line, noticed in the case of the Dirac singletons of $\mathfrak{so}(2,3)$ in [109] and later extended to singletons in arbitrary dimensions [9] in [5], is the reason for the name "singletons" [102].

(ii) In order to prove the branching rule from $\mathfrak{so}(2,d)$ to $\mathfrak{so}(2,d-1)$, we will compare the $\mathfrak{so}(2) \oplus \mathfrak{so}(d-1)$ decomposition of the $\mathfrak{so}(2,d)$ spin-s singleton on the one hand, obtained by branching [10] the $\mathfrak{so}(d)$ components of the $\mathfrak{so}(2) \oplus \mathfrak{so}(d)$ of these modules displayed in the previous item onto $\mathfrak{so}(d-1)$, to the $\mathfrak{so}(2) \oplus \mathfrak{so}(d-1)$ decomposition of the $\mathfrak{so}(2,d-1)$ module $\mathcal{D}(s+r-1;(s^{r-1}))$ describing a massless field with spin (s^{r-1}). For the sake of brevity, we will only detail the low

[9] Actually, it was even adopted as a definition of singletons in [8].
[10] The branching rules for $\mathfrak{so}(d)$ irreps are recalled in Appendix A.1.

dimensional case of $\mathfrak{so}(2,4)$ spin-s singletons which captures the idea of the proof, and leave the treatment of the arbitrary dimension case to Appendix B.1.

Let us start by deriving the $\mathfrak{so}(2)\oplus\mathfrak{so}(3)$ decomposition of the $\mathfrak{so}(2,4)$ spin-s singleton module $\mathcal{D}(s+1;(s,s))$:

$$\mathcal{D}(s+1;(s,s)) \cong \bigoplus_{\sigma=0}^{\infty} \mathcal{D}_{\mathfrak{so}(2)\oplus\mathfrak{so}(4)}(s+1+\sigma;(s+\sigma,s)) \tag{32}$$

$$\underset{\mathfrak{so}(2,3)}{\overset{\mathfrak{so}(2,4)}{\downarrow}} \bigoplus_{\sigma=0}^{\infty}\bigoplus_{k=0}^{\sigma} \mathcal{D}_{\mathfrak{so}(2)\oplus\mathfrak{so}(3)}(s+1+\sigma;(s+k)) \tag{33}$$

$$\cong \bigoplus_{\sigma,n=0}^{\infty} \mathcal{D}_{\mathfrak{so}(2)\oplus\mathfrak{so}(3)}(s+1+\sigma+n;(s+\sigma)) \tag{34}$$

Next, we need to derive the $\mathfrak{so}(2)\oplus\mathfrak{so}(3)$ of a massless spin-s field corresponding to the $\mathfrak{so}(2,3)$ module $\mathcal{D}(s+1;(s))$. To do so, we will rewrite its character in a way that makes this decomposition explicit:

$$\chi^{\mathfrak{so}(2,3)}_{[s+1;(s)]}(q,x) = q^{s+1}\left(\chi^{\mathfrak{so}(3)}_{(s)}(x) - q\chi^{\mathfrak{so}(3)}_{(s-1)}(x)\right)\mathcal{P}^{(d)}(q,x) \tag{35}$$

$$= \sum_{\sigma,n=0}^{\infty} q^{s+1+\sigma+2n}\Big(\sum_{\tau=|s-\sigma|}^{s+\sigma}\chi^{\mathfrak{so}(3)}_{(\tau)}(x) - q\sum_{\tau=|s-1-\sigma|}^{s-1+\sigma}\chi^{\mathfrak{so}(3)}_{(\tau)}(x)\Big) \tag{36}$$

$$= \sum_{n=0}^{\infty} q^{s+1+2n}\Big(\chi^{\mathfrak{so}(3)}_{(s)} + \sum_{\sigma=1}^{\infty} q^{\sigma}\Big[\sum_{\tau=|s-\sigma|}^{s+\sigma}\chi^{\mathfrak{so}(3)}_{(\tau)}(x) - \sum_{\tau=|s-\sigma|}^{s+\sigma-2}\chi^{\mathfrak{so}(3)}_{(\tau)}(x)\Big]\Big) \tag{37}$$

$$= \sum_{n=0}^{\infty} q^{s+1+2n}(1+q)\sum_{\sigma=0}^{\infty} q^{\sigma}\chi^{\mathfrak{so}(3)}_{(s+\sigma)}(x) = \sum_{\sigma,n=0}^{\infty} q^{s+1+\sigma+n}\chi^{\mathfrak{so}(3)}_{(s+\sigma)}(x)\,, \tag{38}$$

where we used the property (25) of the function $\mathcal{P}^{(4)}(q,\vec{x})$, namely

$$\mathcal{P}^{(4)}(q,\vec{x}) = \sum_{s,n=0}^{\infty} q^{s+2n}\chi^{\mathfrak{so}(4)}_{(s)}(\vec{x})\,. \tag{39}$$

This proves that the decomposition of the $\mathfrak{so}(2,3)$ module of a massless spin-s field in AdS_4 reads:

$$\mathcal{D}(s+1;(s)) \cong \bigoplus_{\sigma,n=0}^{\infty} \mathcal{D}_{\mathfrak{so}(2)\oplus\mathfrak{so}(3)}(s+1+\sigma+n;(s+\sigma))\,, \tag{40}$$

which coincide with the $\mathfrak{so}(2)\oplus\mathfrak{so}(3)$ decomposition obtained after branching the $\mathfrak{so}(2,4)$ spin-s singleton module onto $\mathfrak{so}(2,3)$, i.e., we indeed have

$$\mathcal{D}(s+1;(s,s)) \underset{\mathfrak{so}(2,3)}{\overset{\mathfrak{so}(2,4)}{\downarrow}} \mathcal{D}(s+1;(s))\,. \tag{41}$$

This can be graphically seen by implementing the branching rule of the weight diagram in Figure 2. Indeed, the branching rule for the $\mathfrak{so}(2r)$ irrep $(s+\sigma,s^{r-1})$ is:

$$(s+\sigma,s^{r-1}) \underset{\mathfrak{so}(2,d-1)}{\overset{\mathfrak{so}(2,d)}{\downarrow}} \bigoplus_{k=0}^{\sigma}(s+k,s^{r-2})\,, \tag{42}$$

which means that one should add on each line of the weight diagram (representing the $\mathfrak{so}(d)$ modules appearing at fixed energy, or $\mathfrak{so}(2)$ weight) in Figure 2 a dot at each value of ℓ_1 to the left of the orignal one until $\ell_1 = s$ is reached. By doing so, an infinite wedge whose tip has coordinates $(E = s + \epsilon_0, \ell_1 = s)$ precisely corresponding to the weight diagram of a massless field of spin given by a rectangular Young diagram on maximal height and length s as can be seen from (40) for $d = 3$ and in Appendix B.1 for arbitrary odd values of d.

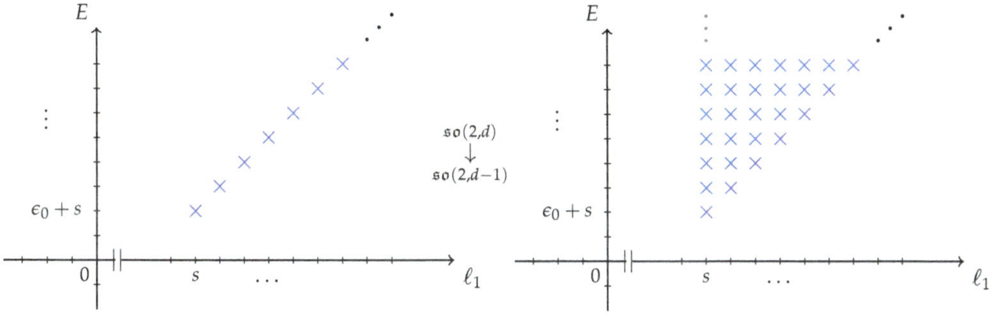

Figure 2. (**Left**) $\mathfrak{so}(2) \oplus \mathfrak{so}(d)$ weight diagram of the spin-s singletons (with in abscisse the first component of the $\mathfrak{so}(d)$ weights, denoted ℓ_1); (**Right**) $\mathfrak{so}(2) \oplus \mathfrak{so}(d-1)$ weight diagram of the $\mathfrak{so}(2, d-1)$ module $\mathcal{D}(s + \frac{d}{2} - 1;\, (s^{r-1}))$ (with in abscisse the first component of the $\mathfrak{so}(d-1)$ weights, denoted ℓ_1 as well). Lighter blue crosses × for a given $\mathfrak{so}(2)$ weight E represent the $\mathfrak{so}(d-1)$ representations coming from the branching rule of the $\mathfrak{so}(d)$ representations in the $\mathfrak{so}(2) \oplus \mathfrak{so}(d)$ module (with the same $\mathfrak{so}(2)$ weight E) of the singleton decomposition represented by a darker blue cross ×.

□

We did not, in the previous review of the proofs of the listed properties in Theorem 1, cover the branching rule of the $\mathfrak{so}(2, d)$ singletons onto $\mathfrak{iso}(1, d-1)$ or $\mathfrak{so}(1, d)$ for the following reasons:

- **From $\mathfrak{so}(2, d)$ to $\mathfrak{iso}(1, d-1)$.** As far as the branching rule from $\mathfrak{so}(2, d)$ to $\mathfrak{so}(1, d-1)$ are concerned, it can be recovered, assuming that the following diagram is commutative:

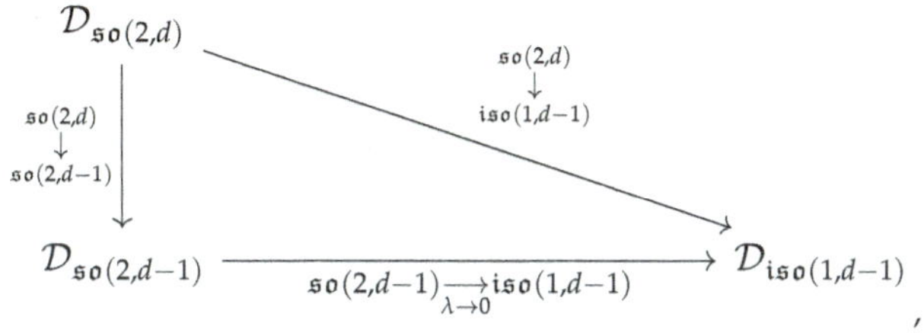

, (43)

i.e., by combining the branching rule from $\mathfrak{so}(2, d)$ to $\mathfrak{so}(2, d-1)$ and an Inönü-Wigner contraction. That is to say, it is equivalent (i) to branch a representation $\mathcal{D}_{\mathfrak{so}(2,d)}$ from $\mathfrak{so}(2, d)$ onto $\mathfrak{so}(2, d-1)$ and then perform a Inönü-Wigner contraction by sending the cosmological constant λ to zero to obtain a representation $\mathcal{D}_{\mathfrak{iso}(1,d-1)}$ of $\mathfrak{iso}(1, d-1)$, and (ii) to branch the $\mathfrak{so}(2, d)$ module $\mathcal{D}_{\mathfrak{iso}(1,d-1)}$ onto $\mathfrak{iso}(1, d-1)$ to obtain the same module $\mathcal{D}_{\mathfrak{iso}(1,d-1)}$ than previously. Under this assumption, we can use the branching rule (23) of the $\mathfrak{so}(2, d)$ singleton module onto $\mathfrak{so}(2, d-1)$ and then contracting it to a $\mathfrak{iso}(1, d-1)$ instead of deriving the branching rule from $\mathfrak{so}(2, d)$ onto $\mathfrak{iso}(1, d-1)$. The Inönü-Wigner contraction for massless fields in AdS_{d+1} (i.e., $\mathfrak{so}(2, d)$ modules) is known as the Brink-Metsaev-Vasiliev mechanism [112], which was proven in [65,66,72]. This mechanism states that massless $\mathfrak{so}(2, d)$ UIRs of spin given by a $\mathfrak{so}(d)$ Young diagram $\mathbb{Y}$ contracts to the direct sum of massless UIRs of the Poincaré algebra with spin given by all of the Young diagrams obtained from the branching rule of $\mathbb{Y}$ *except* those where boxes in the first block of $\mathbb{Y}$ have been removed. Higher-spin singleton, as well as the massless $\mathfrak{so}(2,d-1)$ module onto which they branch being labelled by a rectangular Young diagram, the BMV mechanism implies that they contract to a single $\mathfrak{iso}(1,d-1)$ i.e.,

$$\mathcal{D}_{\mathfrak{so}(2,d)}\big(s+r-1\,;\,(s^r)\big) \underset{\mathfrak{so}(2,d-1)}{\overset{\mathfrak{so}(2,d)}{\downarrow}} \mathcal{D}_{\mathfrak{so}(2,d-1)}\big(s+r-1\,;\,(s^{r-1})\big) \underset{\lambda\to 0}{\longrightarrow} \mathcal{D}_{\mathfrak{iso}(1,d-1)}\big(m=0\,;\,(s^{r-1})\big)\,, \tag{44}$$

as shown in [5].

- **From $\mathfrak{so}(2,d)$ to $\mathfrak{so}(1,d)$.** The $\mathfrak{so}(1,d)$ generalised Verma modules are induced by, and decompose into, $\mathfrak{so}(1,1)\oplus\mathfrak{so}(d-1)$ modules instead of $\mathfrak{so}(2)\oplus\mathfrak{so}(d-1)$ in the case of $\mathfrak{so}(2,d-1)$. As a consequence, the method used previously consisting in relying on the common $\mathfrak{so}(2)\oplus\mathfrak{so}(d-1)$ cannot be applied here and we will therefore refer to the original paper [5] for the proof of that branching rule.

3.2. Non-Unitary, Higher-Order Extension

Higher-order extension of the Dirac singletons (i.e., the scalar and spinor ones) are non-unitary $\mathfrak{so}(2,d)$ modules that share the crucial field theoretical property of singletons mentioned above, namely they correspond to AdS (scalar and spinor) field that do not propagate local degree of freedom in the bulk. They have been considered in [6] as well as in [82] where the confinement to the conformal boundary of these remarkable fields was highlighted, but were excluded from the exhaustive work [11] [5] because they fall below the unitary bound for representations of $\mathfrak{so}(2,d)$ (recalled in Subsection 3.1).

Definition 2 (Higher-order Dirac singletons). *The scalar and spinor, order-ℓ Dirac singletons are the $\mathfrak{so}(2,d)$ modules $\mathcal{D}(\epsilon_0^{(\ell)};\mathbf{0})$ and $\mathcal{D}(\epsilon_{1/2}^{(\ell)};\tfrac{1}{2})$ respectively, where*

$$\epsilon_0^{(\ell)} := \frac{d-2\ell}{2}\,, \quad \textit{and} \quad \epsilon_{1/2}^{(\ell)} := \frac{d+1-2\ell}{2} \equiv \epsilon_0^{(\ell)}+\tfrac{1}{2}\,, \tag{45}$$

and which are defined as the quotient:

$$\mathrm{Rac}_\ell := \mathcal{D}(\epsilon_0^{(\ell)}\,;\mathbf{0}) \cong \frac{\mathcal{V}(\epsilon_0^{(\ell)}\,;\mathbf{0})}{\mathcal{D}(d-\epsilon_0^{(\ell)}\,;\mathbf{0})}\,, \quad \textit{and} \quad \mathrm{Di}_\ell := \mathcal{D}(\epsilon_{1/2}^{(\ell)};\tfrac{1}{2}) \cong \frac{\mathcal{V}(\epsilon_{1/2}^{(\ell)};\tfrac{1}{2})}{\mathcal{D}(d-\epsilon_{1/2}^{(\ell)};\tfrac{1}{2})}\,. \tag{46}$$

Their character read:

$$\chi^{\mathfrak{so}(2,d)}_{\mathrm{Rac}_\ell}(q,\vec{x}) = q^{\epsilon_0^{(\ell)}}\,(1-q^{2\ell})\,\mathcal{P}^{(d)}(q,\vec{x})\,, \quad \textit{and} \quad \chi^{\mathfrak{so}(2,d)}_{\mathrm{Di}_\ell}(q,\vec{x}) = q^{\epsilon_{1/2}^{(\ell)}}\,(1-q^{2\ell-1})\,\chi^{\mathfrak{so}(d)}_{\frac{1}{2}}(\vec{x})\,\mathcal{P}^{(d)}(q,\vec{x})\,. \tag{47}$$

These modules are non-unitary for $\ell>1$, whereas they correspond to the original (unitary) Dirac singletons of Definition 1 for $\ell=1$.

On top of the confinement property, the Rac_ℓ and Di_ℓ singletons also possess properties analogous to those of their unitary counterparts reviewed in Theorem 1. Specifically, they can be decomposed as several direct sum of $\mathfrak{so}(2)\oplus\mathfrak{so}(d)$ modules, making up not only one but now several lines in the weight diagram, and they obey a branching rule (from $\mathfrak{so}(2,d)$ to $\mathfrak{so}(2,d-1)$) similar to that of Rac and Di. The properties of the higher-order Dirac singletons are summed up below.

---

[11] Although mentionned briefly in [8] as "multipleton".

Proposition 1 (Properties of Rac_ℓ and Di_ℓ). *The $\mathfrak{so}(2)\oplus\mathfrak{so}(d)$ decomposition of the order-ℓ scalar and spinor singletons respectively read* [12]*:*

$$\mathcal{D}(\epsilon_0^{(\ell)}\,;\mathbf{0}) \cong \bigoplus_{k=0}^{\ell-1}\bigoplus_{\sigma=0}^{\infty}\mathcal{D}_{\mathfrak{so}(2)\oplus\mathfrak{so}(d)}(\epsilon_0^{(\ell)}+\sigma+2k\,;\,(\sigma))\,, \tag{48}$$

and

$$\mathcal{D}(\epsilon_{1/2}^{(\ell)};\tfrac{\mathbf{1}}{\mathbf{2}}) \cong \bigoplus_{k=0}^{2(\ell-1)}\bigoplus_{\sigma=0}^{\infty}\mathcal{D}_{\mathfrak{so}(2)\oplus\mathfrak{so}(d)}(\epsilon_{1/2}^{(\ell)}+\sigma+k\,;\,(\sigma+\tfrac{1}{2},(\tfrac{1}{2})^{r-1}))\,. \tag{49}$$

These two modules obey the following branching rules [13]*:*

$$\mathcal{D}(\epsilon_0^{(\ell)}\,;\mathbf{0}) \quad \underset{\mathfrak{so}(2,d-1)}{\overset{\mathfrak{so}(2,d)}{\downarrow}} \quad \bigoplus_{k=0}^{2\ell-1}\mathcal{D}(\epsilon_0^{(\ell)}+k\,;\mathbf{0})\,, \tag{50}$$

and

$$\mathcal{D}(\epsilon_{1/2}^{(\ell)};\tfrac{\mathbf{1}}{\mathbf{2}}) \quad \underset{\mathfrak{so}(2,d-1)}{\overset{\mathfrak{so}(2,d)}{\downarrow}} \quad \bigoplus_{k=0}^{2(\ell-1)}\mathcal{D}(\epsilon_{1/2}^{(\ell)}+k\,;\tfrac{\mathbf{1}}{\mathbf{2}})\,. \tag{51}$$

Proof. As previsouly, we will use the property (25) of the function $\mathcal{P}^{(d)}(q,\vec{x})$ to rewrite the characters of the order-ℓ scalar and spinor singletons (47) as a sum of $\mathfrak{so}(2)\oplus\mathfrak{so}(d)$ characters, starting with the scalar Rac_ℓ:

$$\begin{aligned}
\chi^{\mathfrak{so}(2,d)}_{\mathrm{Rac}_\ell}(q,\vec{x}) &= q^{\epsilon_0^{(\ell)}}(1-q^{2\ell})\,\mathcal{P}^{(d)}(q,\vec{x}) = \sum_{\sigma,n=0}^{\infty} q^{\epsilon_0^{(\ell)}+\sigma+2n}(1-q^{2\ell})\,\chi^{\mathfrak{so}(d)}_{(\sigma)}(\vec{x}) && (52)\\
&= \sum_{k=0}^{\ell-1}\sum_{\sigma=0}^{\infty} q^{\epsilon_0^{(\ell)}+2k+\sigma}\,\chi^{\mathfrak{so}(d)}_{(\sigma)}(\vec{x}) && (53)\\
\Leftrightarrow \quad \mathrm{Rac}_\ell &= \bigoplus_{k=0}^{\ell-1}\bigoplus_{\sigma=0}^{\infty}\mathcal{D}_{\mathfrak{so}(2)\oplus\mathfrak{so}(d)}(\epsilon_0^{(\ell)}+2k+\sigma\,;\,(\sigma))\,. && (54)
\end{aligned}$$

For the Di_ℓ singleton we will also need the $\mathfrak{so}(d)$ tensor product rule:

$$(\sigma)\otimes\tfrac{\mathbf{1}}{\mathbf{2}} = (\sigma+\tfrac{1}{2},(\tfrac{1}{2})^{r-1})\oplus(\sigma-\tfrac{1}{2},(\tfrac{1}{2})^{r-1})\,, \qquad \text{for} \quad \sigma \geqslant 1\,. \tag{55}$$

Using the above identity and proceeding similarly to the scalar case, we end up with:

[12] Notice that the $\mathfrak{so}(2)\oplus\mathfrak{so}(d)$ decomposition (48) of the Rac_ℓ module was originally derived in [6] (where these singletons are refered to as "scalar p-linetons")

[13] Notice that the module on the right hand side of the branching rules (50) and (51) for $k=0$ are not the order ℓ singleton, due to the fact that $\epsilon_0^{(\ell)} = \frac{d-2\ell}{2}$ is not the critical energy of the Rac_ℓ singleton of $\mathfrak{so}(2,d-1)$.

$$\chi^{\mathfrak{so}(2,d)}_{\mathrm{Di}_\ell}(q,\vec{x}) = q^{\epsilon^{(\ell)}_{1/2}}(1-q^{2\ell-1})\,\chi^{\mathfrak{so}(d)}_{\frac{1}{2}}(\vec{x})\,\mathcal{P}^{(d)}(q,\vec{x}) \tag{56}$$

$$= \sum_{n=0}^{\infty} q^{\epsilon^{(\ell)}_{1/2}+2n}(1-q^{2\ell-1})\Big(\sum_{\sigma=0}^{\infty} q^\sigma \chi^{\mathfrak{so}(d)}_{(\sigma+\frac{1}{2},(\frac{1}{2})^{r-1})}(\vec{x}) + \sum_{\sigma=1}^{\infty} q^\sigma \chi^{\mathfrak{so}(d)}_{(\sigma-\frac{1}{2},(\frac{1}{2})^{r-1})}(\vec{x})\Big) \tag{57}$$

$$= \sum_{n,\sigma=0}^{\infty} q^{\epsilon^{(\ell)}_{1/2}+2n+\sigma}(1-q^{2\ell-1})\,(1+q)\,\chi^{\mathfrak{so}(d)}_{(\sigma+\frac{1}{2},(\frac{1}{2})^{r-1})}(\vec{x}) \tag{58}$$

$$= \sum_{k=0}^{2(\ell-1)}\sum_{\sigma=0}^{\infty} q^{\epsilon^{(\ell)}_{1/2}+k+\sigma}\chi^{\mathfrak{so}(d)}_{(\sigma+\frac{1}{2},(\frac{1}{2})^{r-1})}(\vec{x}) \tag{59}$$

$$\Leftrightarrow \quad \mathrm{Di}_\ell = \bigoplus_{k=0}^{2(\ell-1)}\bigoplus_{\sigma=0}^{\infty}\mathcal{D}_{\mathfrak{so}(2)\oplus\mathfrak{so}(d)}\big(\epsilon^{(\ell)}_{1/2}+\sigma+k\,;\,(\sigma+\tfrac{1}{2},(\tfrac{1}{2})^{r-1})\big)\,. \tag{60}$$

To prove the branching rule (50) and (51), we will follow the same strategy as previously, namely we will compare the $\mathfrak{so}(2)\oplus\mathfrak{so}(d-1)$ decomposition of the two sides of these identities. This decomposition reads, for the Rac_ℓ singleton:

$$\mathcal{D}(\epsilon^{(\ell)}_0\,;\mathbf{0}) \cong \bigoplus_{k=0}^{\ell-1}\bigoplus_{\sigma=0}^{\infty}\mathcal{D}_{\mathfrak{so}(2)\oplus\mathfrak{so}(d)}\big(\epsilon^{(\ell)}_0+\sigma+2k\,;\,(\sigma)\big) \tag{61}$$

$$\underset{\mathfrak{so}(2,d-1)}{\overset{\mathfrak{so}(2,d)}{\downarrow}} \bigoplus_{k=0}^{\ell-1}\bigoplus_{\sigma=0}^{\infty}\bigoplus_{n=0}^{\sigma}\mathcal{D}_{\mathfrak{so}(2)\oplus\mathfrak{so}(d-1)}\big(\epsilon^{(\ell)}_0+\sigma+2k\,;\,(n)\big) \tag{62}$$

$$\cong \bigoplus_{k=0}^{\ell-1}\bigoplus_{\sigma=0}^{\infty}\bigoplus_{n=0}^{\infty}\mathcal{D}_{\mathfrak{so}(2)\oplus\mathfrak{so}(d-1)}\big(\epsilon^{(\ell)}_0+\sigma+2k+n\,;\,(\sigma)\big)\,, \tag{63}$$

whereas for the Di_ℓ singleton:

$$\mathcal{D}(\epsilon^{(\ell)}_{1/2}\,;\tfrac{\mathbf{1}}{\mathbf{2}}) \cong \bigoplus_{k=0}^{2(\ell-1)}\bigoplus_{\sigma=0}^{\infty}\mathcal{D}_{\mathfrak{so}(2)\oplus\mathfrak{so}(d)}\big(\epsilon^{(\ell)}_{1/2}+\sigma+k\,;\,(\sigma+\tfrac{1}{2},(\tfrac{1}{2})^{r-1})\big) \tag{64}$$

$$\underset{\mathfrak{so}(2,d-1)}{\overset{\mathfrak{so}(2,d)}{\downarrow}} \bigoplus_{k=0}^{2(\ell-1)}\bigoplus_{\sigma=0}^{\infty}\bigoplus_{n=0}^{\sigma}\mathcal{D}_{\mathfrak{so}(2)\oplus\mathfrak{so}(d-1)}\big(\epsilon^{(\ell)}_{1/2}+\sigma+k\,;\,(n+\tfrac{1}{2},(\tfrac{1}{2})^{r-1})\big) \tag{65}$$

$$\cong \bigoplus_{k=0}^{2(\ell-1)}\bigoplus_{\sigma=0}^{\infty}\bigoplus_{n=0}^{\infty}\mathcal{D}_{\mathfrak{so}(2)\oplus\mathfrak{so}(d-1)}\big(\epsilon^{(\ell)}_{1/2}+\sigma+k+n\,;\,(\sigma+\tfrac{1}{2},(\tfrac{1}{2})^{r-1})\big)\,. \tag{66}$$

On the other hand, the character of an irreducible $\mathfrak{so}(2,d-1)$ module $\mathcal{D}(\Delta\,;\mathbf{0})\equiv\mathcal{V}(\Delta\,;\mathbf{0})$, i.e., a generalised Verma module which does not contain a submodule [14] can be rewritten as:

$$\chi^{\mathfrak{so}(2,d-1)}_{[\Delta\,;\mathbf{0}]}(q,\vec{x}) = q^\Delta\,\mathcal{P}^{(d-1)}(q,\vec{x}) = \sum_{\sigma,n=0}^{\infty} q^{\Delta+2n+\sigma}\chi^{\mathfrak{so}(d-1)}_{(\sigma)}(\vec{x}) \tag{67}$$

$$\Leftrightarrow \quad \mathcal{D}(\Delta\,;\mathbf{0}) \cong \bigoplus_{\sigma=0}^{\infty}\bigoplus_{n=0}^{\infty}\mathcal{D}_{\mathfrak{so}(2)\oplus\mathfrak{so}(d-1)}\big(\Delta+\sigma+2n\,;\,(\sigma)\big) \tag{68}$$

---

14 A scalar $\mathfrak{so}(2,d-1)$ module $\mathcal{V}(\Delta\,;\mathbf{0})$ possesses a submodule only if $\Delta=\frac{d-1-2\ell}{2}$, whereas a spin one-half module $\mathcal{V}(\Delta\,;\frac{\mathbf{1}}{\mathbf{2}})$ possesses a submodule only if $\Delta=\frac{d-2\ell}{2}$. In other words, only the Rac_ℓ and Di_ℓ modules are defined as quotients, see the classification in [108].

As a consequence,

$$\mathcal{D}(\epsilon_0^{(\ell)}+2k\,;\,\mathbf{0})\oplus\mathcal{D}(\epsilon_0^{(\ell)}+2k+1\,;\,\mathbf{0})\cong\bigoplus_{\sigma=0}^{\infty}\bigoplus_{n=0}^{\infty}\mathcal{D}_{\mathfrak{so}(2)\oplus\mathfrak{so}(d-1)}\big(\epsilon_0^{(\ell)}+\sigma+n+2k\,;\,(\sigma)\big)\,, \tag{69}$$

which proves (50). Finally, an irreducible $\mathfrak{so}(2,d-1)$ module $\mathcal{D}(\Delta\,;\,\frac{1}{2})\equiv\mathcal{V}(\Delta\,;\,\frac{1}{2})$ admits the following $\mathfrak{so}(2)\oplus\mathfrak{so}(d-1)$ decomposition:

$$\chi^{\mathfrak{so}(2,d-1)}_{[\Delta\,;\,\frac{1}{2}]}(q,\vec{x}) = q^{\Delta}\chi^{\mathfrak{so}(d-1)}_{\frac{1}{2}}(\vec{x})\mathcal{P}^{(d-1)}(q,\vec{x}) \tag{70}$$

$$= \sum_{n=0}^{\infty}q^{\Delta+2n}\Big(\sum_{\sigma=0}^{\infty}q^{\sigma}\chi^{\mathfrak{so}(d-1)}_{(\sigma+\frac{1}{2},(\frac{1}{2})^{r-1})}(\vec{x})+\sum_{\sigma=1}^{\infty}q^{\sigma}\chi^{\mathfrak{so}(d-1)}_{(\sigma-\frac{1}{2},(\frac{1}{2})^{r-1})}(\vec{x})\Big) \tag{71}$$

$$= \sum_{\sigma=0}^{\infty}\sum_{n=0}^{\infty}q^{\Delta+\sigma+n}\chi^{\mathfrak{so}(d-1)}_{(\sigma+\frac{1}{2},(\frac{1}{2})^{r-1})}(\vec{x}) \tag{72}$$

$$\Leftrightarrow\quad \mathcal{D}(\Delta\,;\,\tfrac{1}{2}) \cong \bigoplus_{\sigma=0}^{\infty}\bigoplus_{n=0}^{\infty}\mathcal{D}_{\mathfrak{so}(2)\oplus\mathfrak{so}(d-1)}\big(\Delta+\sigma+n\,;\,(\sigma+\tfrac{1}{2},(\tfrac{1}{2})^{r-1})\big)\,, \tag{73}$$

thereby proving (51). □

The branching rule (50) and (51) reproduce that given in [5,8] (and rederived in [113]) for the Rac and Di singletons upon setting $\ell=1$, and extend them to the higher-order Dirac singletons Rac_ℓ and Di_ℓ.

From a CFT point of view, the order-ℓ scalar and spinor singletons correspond to respectively a non-unitary fundamental scalar or spinor fields of respective conformal weight $\epsilon_0^{(\ell)}$ and $\epsilon_{1/2}^{(\ell)}$, and respectively subject to an order 2ℓ and $2\ell-1$ wave equation (see e.g., [82] for more details). The spectrum of current of these CFT contains an infinite tower of partially conserved totally symmetric currents of arbitrary spin, which should be dual to partially massless gauge fields in the bulk [114].

3.3. Candidates for Higher-Spin Higher-Order Singletons

The extension we will be concerned with corresponds to the $\mathfrak{so}(2,d)$ module, for $d=2r$:

$$\mathcal{D}(s+\tfrac{d}{2}-t\,;\,(s^r))\cong\frac{\mathcal{V}(s+\frac{d}{2}-t\,;\,(s^r))}{\mathcal{D}(s+\frac{d}{2}\,;\,(s^{r-1},s-t))}\,,\quad\text{for}\quad 1\leqslant t\leqslant s\,, \tag{74}$$

whose structure is similar to the unitary spin-s singletons for $s\geqslant 1$ in the sense that the various submodule to be modded out of $\mathcal{V}(s+\frac{d}{2}-t\,;\,(s^r))$ are defined throught the sequence:

$$\mathcal{D}(s+\tfrac{d}{2}+k\,;\,(s^{r-k-1},(s-1)^k,s-t)):=\frac{\mathcal{V}(s+\frac{d}{2}+k\,;\,(s^{r-k-1},(s-1)^k,s-t))}{\mathcal{D}(s+\frac{d}{2}+k+1\,;\,(s^{r-k-2},(s-1)^{k+1},s-t))}\,, \tag{75}$$

for $0\leqslant k\leqslant r-2$ and $\mathcal{D}(s+d-1\,;\,((s-1)^{r-1},s-t))\equiv\mathcal{V}(s+d-1\,;\,((s-1)^{r-1},s-t))$. In other words, except for the first submodule which is obtained by increasing the $\mathfrak{so}(2)$ weight of t units and removing t boxes from the last row of the rectangular Young diagram (s^r) labelling the irreducible module, the sequence of nested submodules are related to one another by adding one unit to the $\mathfrak{so}(2)$ weight of the previous submodule and removing one box in the row above the previously amputated row. Correspondingly, the character of this module reads:

$$\chi^{\mathfrak{so}(2,d)}_{[s+\frac{d}{2}-t;(s^r)]}(q,\vec{x})=q^{s+\frac{d}{2}-t}\Big(\chi^{\mathfrak{so}(d)}_{(s^r)}(\vec{x})+\sum_{k=0}^{r-1}(-)^{k+1}q^{t+k}\chi^{\mathfrak{so}(d)}_{(s^{r-1-k},(s-1)^k,s-t)}(\vec{x})\Big)\mathcal{P}^{(d)}(q,\vec{x})\,. \tag{76}$$

This definition encompasses the unitary spin-s singletons, which correspond to the case $t = 1$ saturating the unitarity bound. For $t > 1$ (but always $t \leqslant s$), the module (74) is non-unitary and describes a depth-t partially-massless field of spin (s^r). The spin being given by a rectangular Young diagram, we will refer to this class of module as "rectangular" partially massless (RPM) fields of spin s and depth t. From the boundary point of view, the modules (74) correspond to the curvature a conformal field of spin (s^{r-1}) (hence the curvature is given by a tensor of symmetry described by a rectangular Young diagram of length s and height r) obeying a partial conservation law of order t, i.e., taking t symmetrised divergences of this curvature identically vanishes on-shell (see e.g., [115] where the $d = 4$ and $t = 1$ case was discussed, and [116] for a more details on mixed symmetry conformal field in arbitrary dimensions).

Remark 2. *Notice that formally, the modules of the* Rac_ℓ *and* Di_ℓ *singletons, as well as the module* (74) *that we propose here as a higher-spin generalisation of the higher-order scalar and spinor singletons, can be denoted as:*

$$\mathcal{D}(s+\epsilon_0^{(t)}\,;\,(s^r))\,, \quad with \quad s=0,\tfrac{1}{2}\,, \quad and \quad s\in\mathbb{N} \quad if \quad d=2r \tag{77}$$

with $\epsilon_0^{(t)} = \frac{d-2t}{2}$ *as defined in Definition 2. On top of being notationally convenient, this coincidence is actually the reason why the modules* (74) *are "natural" generalisations of the unitary higher-spin singletons: by introducing the parameter* t *in this way, one considers a family of modules whose first representative is the unitary singletons whereas for* $t > 1$ *the modules are non-unitary but their structure is almost the same than in the unitary case.*

Let us now study what are the counterpart of the properties displayed in Theorem 1 for unitary singletons and Proposition 1 for the Rac_ℓ and Di_ℓ singletons, starting with the $\mathfrak{so}(2)\oplus\mathfrak{so}(d)$ decomposition of (74).

Proposition 2 ($\mathfrak{so}(2)\oplus\mathfrak{so}(d)$ decomposition). *The* $\mathfrak{so}(2,d)$ *module* $\mathcal{D}(s+r-t;(s^r))$ *for* $d = 2r$*, describing a depth-*t *and spin-*s *RPM field, admits the following* $\mathfrak{so}(2)\oplus\mathfrak{so}(d)$ *decomposition:*

$$\mathcal{D}(s+r-t;(s^r)) \quad \cong \quad \bigoplus_{\ell=0}^{t-1}\bigoplus_{n=0}^{t-1-\ell}\bigoplus_{\sigma=\ell}^{\infty}\mathcal{D}_{\mathfrak{so}(2)\oplus\mathfrak{so}(d)}(s+r-t+\sigma+2n;(s-\ell+\sigma,s^{r-2},s-\ell))\,. \tag{78}$$

Equivalently, this property means that the character (76) *can bewritten as:*

$$\chi^{\mathfrak{so}(2,d)}_{[s+r-t;(s^r)]}(q,\vec{x}) = \sum_{\ell=0}^{t-1}\sum_{n=0}^{t-1-\ell}\sum_{\sigma=\ell}^{\infty} q^{s+r-t+\sigma+2n}\chi^{\mathfrak{so}(d)}_{(s+\sigma-\ell,s^{r-2},s-\ell)}(\vec{x})\,. \tag{79}$$

Proof. As previously, we will only focus on the simpler $d = 4$ case and leave the proof of this property in arbitrary dimension to Appendix B.2. We will proceed in the exact same way as we did for unitary higher-spin singleton, that is we will use (25) in the character formula (76), so as to rewrite it in the following way:

$$\chi^{\mathfrak{so}(2,4)}_{[s+2-t;(s,s)]}(q,\vec{x}) = q^{s+2-t}\Big(\chi^{\mathfrak{so}(4)}_{(s,s)}(\vec{x}) - q^t\chi^{\mathfrak{so}(4)}_{(s,s-t)}(\vec{x}) + q^{t+1}\chi^{\mathfrak{so}(4)}_{(s-1,s-t)}(\vec{x})\Big)\mathcal{P}^{(4)}(q,\vec{x}) \tag{80}$$

$$= \sum_{n=0}^{\infty} q^{s+2-t+2n}\Big(\sum_{\sigma=0}^{2s} q^{\sigma}\sum_{k=0}^{\sigma}\chi^{\mathfrak{so}(4)}_{(s+k,s+k-\sigma)}(\vec{x}) + \sum_{\sigma=2s+1}^{\infty} q^{\sigma}\sum_{k=0}^{2s}\chi^{\mathfrak{so}(4)}_{(\sigma+k-s,k-s)}(\vec{x}) \tag{81}$$

$$-\sum_{m=0}^{t}\Big[\sum_{\sigma=m}^{2s-t} q^{\sigma+t}\sum_{k=0}^{\sigma}\chi^{\mathfrak{so}(4)}_{(s+k-m,s+k-\sigma-t+m)}(\vec{x}) + \sum_{\sigma=2s-t+1}^{\infty} q^{\sigma+t}\sum_{k=0}^{2s-t}\chi^{\mathfrak{so}(4)}_{(\sigma+k+t-s-m,k+m-s)}(\vec{x})\Big]$$

$$+\sum_{m=0}^{t-1}\Big[\sum_{\sigma=m}^{2s-t-1} q^{\sigma+t+1}\sum_{k=0}^{\sigma}\chi^{\mathfrak{so}(4)}_{(s+k-m-1,s+k-\sigma-t+m)}(\vec{x}) \tag{82}$$

$$+\sum_{\sigma=2s-t}^{\infty} q^{\sigma+t+1}\sum_{k=0}^{2s-t-1}\chi^{\mathfrak{so}(4)}_{(\sigma+k+t-s-m,k+m+1-s)}(\vec{x})\Big]\Big)$$

$$= \sum_{n=0}^{\infty} q^{s+2-t+2n}\Big(\sum_{\sigma=0}^{t-1} q^{\sigma}\sum_{k=0}^{\sigma}\chi^{\mathfrak{so}(4)}_{(s+k,s+k-\sigma)}(\vec{x}) + \sum_{m=0}^{t-1}\sum_{\sigma=t}^{\infty} q^{\sigma}\chi^{\mathfrak{so}(4)}_{(s+\sigma-m,s-m)}(\vec{x}) \tag{83}$$

$$-\sum_{m=1}^{t}\sum_{\sigma=m}^{\infty} q^{\sigma+t}\chi^{\mathfrak{so}(4)}_{(s+\sigma-m,s-t+m)}(\vec{x})\Big) \tag{84}$$

$$= \sum_{n=0}^{\infty} q^{s+2-t+2n}\sum_{m=0}^{t-1}\Big(\sum_{\sigma=m}^{\infty} q^{\sigma}\chi^{\mathfrak{so}(4)}_{(s+\sigma-m,s-m)}(\vec{x}) - \sum_{\sigma=t-m}^{\infty} q^{\sigma+t}\chi^{\mathfrak{so}(4)}_{(s+\sigma-t+m,s-m)}(\vec{x})\Big) \tag{85}$$

$$= \sum_{n=0}^{\infty} q^{s+2-t+2n}\sum_{m=0}^{t-1}\sum_{\sigma=m}^{\infty} q^{\sigma}(1-q^{2(t-m)})\chi^{\mathfrak{so}(4)}_{(s+\sigma-m,s-m)}(\vec{x}) \tag{86}$$

$$= \sum_{\ell=0}^{t-1}\sum_{n=0}^{t-1-\ell}\sum_{\sigma=\ell}^{\infty} q^{\sigma+s+2-t+2n}\chi^{\mathfrak{so}(4)}_{(s-\ell+\sigma,s-\ell)}(\vec{x}), \tag{87}$$

where we used

$$\sum_{\sigma=0}^{t-1} q^{\sigma}\sum_{k=0}^{\sigma}\chi^{\mathfrak{so}(4)}_{(s+k,s+k-\sigma)}(\vec{x}) = \sum_{m=0}^{t-1}\sum_{\sigma=m}^{t-1} q^{\sigma}\chi^{\mathfrak{so}(4)}_{(s+\sigma-m,s-m)}(\vec{x}), \tag{88}$$

between (83) and (85). Expression (87) shows that the depth-t PM module $\mathcal{D}(s+2-t;(s,s))$ decomposes as the direct sum of $\mathfrak{so}(2)\oplus\mathfrak{so}(4)$ modules:

$$\mathcal{D}(s+2-t;\,(s,s)) \cong \bigoplus_{\ell=0}^{t-1}\bigoplus_{n=0}^{t-1-\ell}\bigoplus_{\sigma=\ell}^{\infty}\mathcal{D}_{\mathfrak{so}(2)\oplus\mathfrak{so}(4)}\big(s+2-t+\sigma+2n;\,(s+\sigma-\ell,s-\ell)\big). \tag{89}$$

□

With the previous $\mathfrak{so}(2)\oplus\mathfrak{so}(d)$ decomposition at hand, we can now derive the branching rule of the spin-s depth-t RPM module.

Proposition 3 (Branching rule). *The $\mathfrak{so}(2,d)$ module $\mathcal{D}(s+r-t;\,(s^r))$ for $d=2r$, describing a depth-t and spin-s RPM field, branches onto the direct sum of $\mathfrak{so}(2,d-1)$ modules $\mathcal{D}(s+r-\tau;\,(s^{r-1}))$ with $\tau=1,\dots,t$ describing partially massless fields in AdS_d of spin (s^{r-1}) and with depth-τ:*

$$\mathcal{D}(s+r-t;\,(s^r)) \;\underset{\mathfrak{so}(2,d-1)}{\overset{\mathfrak{so}(2,d)}{\downarrow}}\; \bigoplus_{\tau=1}^{t}\mathcal{D}(s+r-\tau;\,(s^{r-1})). \tag{90}$$

Proof. Here again we will only display the proof for the low dimensional case $d=4$ in order to illustrate the general mechanism while being not too technically involved, and we leave the treatment in arbitrary dimensions to the Appendix B.2.

In order to prove the branching rule (90) for $\mathfrak{so}(2,4)$, we will compare the $\mathfrak{so}(2)\oplus\mathfrak{so}(3)$ decomposition of the $\mathfrak{so}(2,4)$ spin-s and depth-t singleton (obtained by first branching it onto $\mathfrak{so}(2,3)$) to the $\mathfrak{so}(2)\oplus\mathfrak{so}(3)$ decomposition of the $\mathfrak{so}(2,3)$ spin-s and depth-τ partially massless fields. Let us start with the latter, i.e., derive the $\mathfrak{so}(2)\oplus\mathfrak{so}(3)$ decomposition of the $\mathfrak{so}(2,3)$ module $\mathcal{D}(s+2-\tau\,;\,(s))$ using its character:

$$\chi^{\mathfrak{so}(2,3)}_{[s+2-\tau;(s)]}(q,x) = q^{s+2-\tau}\big(\chi^{\mathfrak{so}(3)}_{(s)}(x) - q^{\tau}\chi^{\mathfrak{so}(3)}_{(s-\tau)}(x)\big)\mathcal{P}^{(3)}(q,x) \quad (91)$$

$$= \sum_{n,\sigma=0}^{\infty} q^{s+2-\tau+\sigma+2n}\Big(\sum_{k=|s-\sigma|}^{s+\sigma}\chi^{\mathfrak{so}(3)}_{(k)}(x) - q^{\tau}\sum_{k=|s-\sigma-\tau|}^{s+\sigma-\tau}\chi^{\mathfrak{so}(3)}_{(k)}(x)\Big) \quad (92)$$

$$= \sum_{n=0}^{\infty} q^{s+2-\tau+2n}\Big(\sum_{\sigma=0}^{\infty} q^{\sigma}\sum_{k=|s-\sigma|}^{s+\sigma}\chi^{\mathfrak{so}(3)}_{(k)}(x) - \sum_{\sigma=\tau}^{\infty} q^{\sigma}\sum_{k=|s-\sigma|}^{s+\sigma-2\tau}\chi^{\mathfrak{so}(3)}_{(k)}(x)\Big) \quad (93)$$

$$= \sum_{n=0}^{\infty} q^{s+2-\tau+2n}\Big(\sum_{\sigma=0}^{\tau-1} q^{\sigma}\sum_{k=|s-\sigma|}^{s+\sigma}\chi^{\mathfrak{so}(3)}_{(k)}(x) + \sum_{\sigma=\tau}^{\infty} q^{\sigma}\sum_{k=s+\sigma-2\tau+1}^{s+\sigma}\chi^{\mathfrak{so}(3)}_{(k)}(x)\Big) \quad (94)$$

$$= \sum_{\sigma,n=0}^{\infty}\sum_{k=0}^{\tau-1} q^{s+2-\tau+n+\sigma+k}\chi^{\mathfrak{so}(3)}_{(s+\sigma-k)}(x)\,. \quad (95)$$

Hence, the $\mathfrak{so}(2)\oplus\mathfrak{so}(3)$ decomposition of a $\mathfrak{so}(2,3)$ spin-s and depth-τ partially massless field reads:

$$\mathcal{D}(s+2-\tau\,;\,(s)) \cong \bigoplus_{\sigma=0}^{\infty}\bigoplus_{n=0}^{\infty}\bigoplus_{k=0}^{\tau-1}\mathcal{D}_{\mathfrak{so}(2)\oplus\mathfrak{so}(3)}(s+2-\tau+n+\sigma+k\,;\,(s+\sigma-k))\,. \quad (96)$$

This can be represented graphically by the weight diagram displayed in Figure 3 for $\tau=3$.

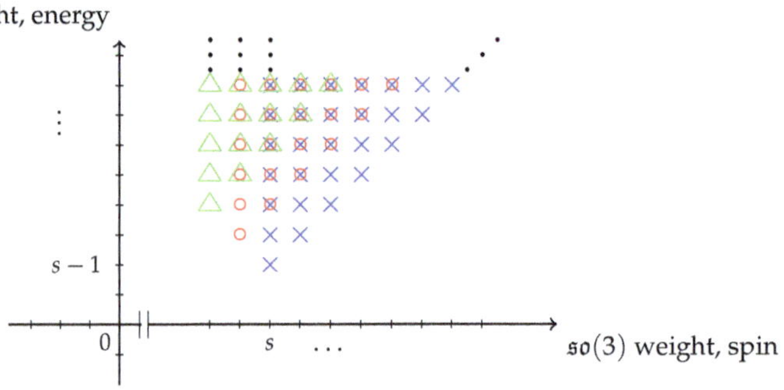

Figure 3. Weight diagram of a spin-s partially massless field of depth $\tau=3$. The contribution of the sum over k in (96) are represented by: blue crosses × for $k=0$, red circles ○ for $k=1$ and green triangles △ for $k=2$.

Now starting with the $\mathfrak{so}(2)\oplus\mathfrak{so}(4)$ decomposition (78) of the spin-s depth-t PM $\mathfrak{so}(2,4)$ module, we can derive its $\mathfrak{so}(2)\oplus\mathfrak{so}(3)$ decomposition:

$$\mathcal{D}(s+2-t\,;\,(s,s)) \cong \bigoplus_{\ell=0}^{t-1}\bigoplus_{n=0}^{t-1-\ell}\bigoplus_{\sigma=\ell}^{\infty}\mathcal{D}_{\mathfrak{so}(2)\oplus\mathfrak{so}(4)}(s+2-t+\sigma+2n\,;\,(s+\sigma-\ell,s-\ell)) \quad (97)$$

$$\overset{\mathfrak{so}(2,4)}{\underset{\mathfrak{so}(2,3)}{\downarrow}} \bigoplus_{\ell=0}^{t-1}\bigoplus_{n=0}^{t-1-\ell}\bigoplus_{\sigma=\ell}^{\infty}\bigoplus_{k=0}^{\sigma}\mathcal{D}_{\mathfrak{so}(2)\oplus\mathfrak{so}(3)}(s+2-t+\sigma+2n\,;\,(s+k-\ell)) \quad (98)$$

$$\cong \bigoplus_{\tau=1}^{t}\bigoplus_{k=0}^{\tau-1}\bigoplus_{\sigma=0}^{\infty}\bigoplus_{n=0}^{\infty}\mathcal{D}_{\mathfrak{so}(2)\oplus\mathfrak{so}(3)}(s+2-\tau+n+\sigma+k\,;\,(s+\sigma-k)) \quad (99)$$

which matches the direct sum of the $\mathfrak{so}(2) \oplus \mathfrak{so}(3)$ decomposition of the spin-s partially massless modules of depth $\tau = 1, \ldots, t$, i.e.,

$$\mathcal{D}(s+2-t;\,(s,s)) \underset{\mathfrak{so}(2,3)}{\overset{\mathfrak{so}(2,4)}{\downarrow}} \bigoplus_{\tau=1}^{t} \mathcal{D}(s+2-\tau;\,(s))\,. \tag{100}$$

This branching rule can also be represented graphically, by drawing on the one hand the $\mathfrak{so}(2) \oplus \mathfrak{so}(3)$ weight diagram of the spin-s and depth-t RPM field as read from (98) and on the other hand by drawing the $\mathfrak{so}(2) \oplus \mathfrak{so}(3)$ weight diagrams of the partially massless spin-s modules of depth $\tau = 1, \ldots, t$, and comparing the two diagrams. This is done for the $t = 2$ case in Figure 4 below. □

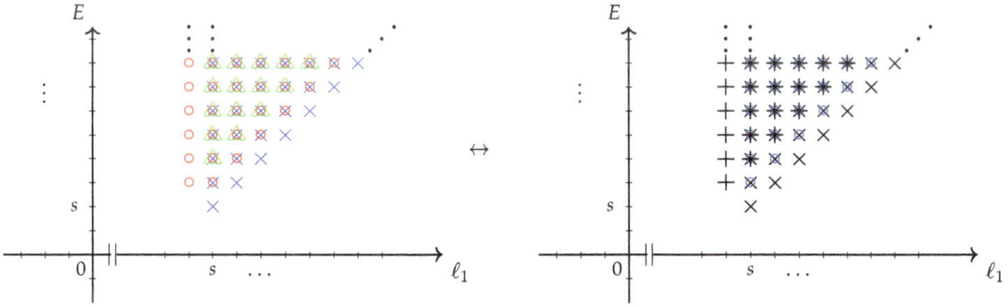

Figure 4. **(Left)** $\mathfrak{so}(2) \oplus \mathfrak{so}(3)$ weight diagram of the spin-s and depth $t = 2$ PM field. The contribution to the sum over ℓ and n in (98) are represented by: blue crosses × for $\ell = n = 0$, green triangles △ for $\ell = 0, n = 1$ and red circles ∘ for $\ell = 1, n = 0$. The lowest energy/$\mathfrak{so}(2)$ weight in this diagram is $s + 2 - t = s$ for $t = 2$; **(Right)** Superimposed $\mathfrak{so}(2) \oplus \mathfrak{so}(3)$ weight diagrams of the spin-s and depth $\tau = 1$ (blue circles ∘) and $\tau = 2$ (black crosses × and +) partially massless modules. The lowest energy/$\mathfrak{so}(2)$ weight in this diagram is $s + 2 - \tau = s$ for $\tau = 2$.

Remark 3. *Notice that the previous Proposition 3 encompasses the case of unitary higher-spin singleton, corresponding to $t = 1$. The above decomposition reduce, in this special case $t = 1$ to those previously derived and summed up in Theorem 1.*

From $\mathfrak{so}(2,d)$ to $\mathfrak{iso}(1,d-1)$.

Again assuming that the diagram (43) is commutative, the branching of the spin-s and depth-t RPM can be obtained by performing an Inönü-Wigner contraction of the $\mathfrak{so}(2,d-1)$ modules. Applying the BMV mechanism to a $\mathfrak{so}(2,d-1)$ partially massless fields of depth-t and spin given by a maximal height rectangular Young diagram yields [65,66,112]:

$$\mathcal{D}_{\mathfrak{so}(2,d-1)}(s+d-r-t;\,(s^{r-1})) \underset{\lambda\to 0}{\longrightarrow} \bigoplus_{\tau=0}^{t-1} \mathcal{D}_{\mathfrak{iso}(1,d-1)}(m=0;\,(s^{r-2},s-\tau))\,. \tag{101}$$

As a consequence, the branching rule of the $\mathfrak{so}(2,d)$ spin-s and depth-t RPM module onto $\mathfrak{iso}(1,d-1)$ reads:

$$\mathcal{D}_{\mathfrak{so}(2,d)}(s+r-t;\,(s^r)) \underset{\mathfrak{iso}(1,d-1)}{\overset{\mathfrak{so}(2,d)}{\downarrow}} \bigoplus_{\tau=0}^{t-1} (t-\tau)\,\mathcal{D}_{\mathfrak{iso}(1,d-1)}(m=0;\,(s^{r-2},s-\tau))\,. \tag{102}$$

At this point, a few comments are in order. As emphasised in the first part of this section, the crucial properties of unitary singletons is that (i) they constitute the class of representations that can be lifted from $\mathfrak{so}(2,d-1)$ to $\mathfrak{so}(2,d)$, i.e., they are AdS fields that are also conformal, and (ii) they describe AdS fields which are "confined" to its (conformal) boundary. The first property translates, for unitary singletons, into the fact that these $\mathfrak{so}(2,d)$ modules remain irreducible when restricted to $\mathfrak{so}(2,d-1)$—except in the case of the scalar singleton whose branching rule actually contains two modules. The second property is related to the fact that the singleton modules also remain irreducible when further contracting to the Poincaré algebra $\mathfrak{iso}(1,d-1)$ (thereby indicating that the AdS_{d+1} field does not propagate degrees of freedom in the bulk).

In the case of the RPM fields of spin-s and depth-t studied in the present note, it seems difficult to consider them as a suitable higher-order (i.e., non-unitary) extension of higher-spin singletons due to the fact that their branching rule (90) shows the appearance of t modules. Indeed, the presence of multiple modules in (90) for $t > 1$ prevent us from reading this decomposition "backward" (from right to left) as the property for a *single* field in AdS_d corresponding to a $\mathfrak{so}(2,d-1)$ module that can be lifted to a $\mathfrak{so}(2,d)$ module thereby illustrating that this AdS_d field is also conformal. Notice that this is in accordance with [110] where conformal AdS fields were classified, and confirmed in the more recent analysis [111] where, without insisting on unitarity, the authors were lead to rule out partially massless fields from the class of AdS fields which can be lifted to conformal representations. On top of that, the contraction of (90) to $\mathfrak{iso}(1,d-1)$ given in (102) produces several modules, some of them even appearing with a multiplicity greater than one, which seems to indicate that the "confinement" property of unitary singletons is also lost when relaxing the unitarity condition in the way proposed here (i.e., considering the modules $\mathcal{D}(s+\frac{d}{2}-t;(s^r))$ with $t>1$). It would nevertheless be interesting to study a field theoretical realisation of these modules to explicitely see how this property is lost when passing from $t=1$ to $t>1$.

4. Flato-Frønsdal Theorem

Let us now particularise the discussion to the $d=4$ case, where we can take advantage of the low dimensional isomorphism $(4) \cong \mathfrak{so}(3) \oplus \mathfrak{so}(3)$ to decompose the tensor product of two spin-s and depth-t RPM fields.

The tensor product of two higher-spin unitary singletons was considered (in arbitrary dimensions) in [10], and reads in the special case $d=4$:

$$\mathcal{D}(s+1;(s,s)_0)^{\otimes 2} \cong \bigoplus_{\sigma=0}^{2s} \mathcal{D}(2s+2;(\sigma,\sigma)_0) \oplus \bigoplus_{\sigma=2s+1}^{\infty} \mathcal{D}(\sigma+2;(\sigma,2s)_0) \oplus \bigoplus_{\sigma=2s}^{\infty} 2\,\mathcal{D}(\sigma+2;(\sigma))\,. \quad (103)$$

Considering singletons of fixed chirality, i.e., $\mathcal{D}(s+1;(s,s)_\epsilon)$, the decomposition of their tensor product then reads:

$$\mathcal{D}(s+1;(s,s)_\epsilon)^{\otimes 2} \cong \bigoplus_{\sigma=0}^{2s} \mathcal{D}(2s+2;(\sigma,\sigma)_\epsilon) \oplus \bigoplus_{\sigma=2s+1}^{\infty} \mathcal{D}(\sigma+2;(\sigma,2s)_\epsilon)\,, \quad (104)$$

i.e., it contributes to the above tensor product by producing the infinite tower of mixed symmetry massless fields $\mathcal{D}(\sigma+2;(\sigma,2s)_\epsilon)$ and the finite tower of massive fields $\mathcal{D}(2s+2;(\sigma,\sigma)_\epsilon)$. The tensor product of two spin-s singletons of opposite chirality, on the other hand, contribute to (103) by producing the infinite tower of totally symmetric massless fields $\mathcal{D}(\sigma+2;(\sigma))$:

$$\mathcal{D}(s+1;(s,s)_+) \otimes \mathcal{D}(s+1;(s,s)_-) \cong \bigoplus_{\sigma=2s}^{\infty} \mathcal{D}(\sigma+2;(\sigma))\,. \quad (105)$$

Remark 4. *The Higher-Spin algebra on which such a theory is based [115] can be decomposed as:*

$$\mathfrak{hs}_s^{(4)} \cong \bigoplus_{\sigma=2s}^{\infty} \begin{array}{|c|}\hline \sigma-1 \\ \hline \sigma-1 \\ \hline \end{array} \oplus \bigoplus_{\sigma=2s+1}^{\infty} \begin{array}{|c|}\hline \sigma-1 \\ \hline \sigma-1 \\ \hline 2s \\ \hline \end{array}\,, \tag{106}$$

In other words, it is composed of all the Killing tensor of the massless fields appearing in the decomposition of two spin-s singletons.

The tensor product of two higher-order Dirac singletons was worked out in arbitrary dimensions in [82,100], and hereafter we give the decomposition for the tensor product of two spin-s and depth-t RPM fields, considered as a possible generalisation of those higher-order singletons, in the special case $d = 4$.

Theorem 2 (Flato-Frønsdal theorem for rectangular partially massless fields). *The tensor product of two* $\mathfrak{so}(2,4)$ *rectangular partially massless fields of spin-s and depth-t decomposes as:*

- *If they are of the same chirality* ϵ*:*

$$\mathcal{D}(s+2-t;(s,s)_\epsilon)^{\otimes 2} \cong \bigoplus_{\tau=1}^{t} \bigoplus_{m=0}^{2(\tau-1)} \bigoplus_{n=0}^{\nu_m^\tau} \bigoplus_{\sigma=2s+2\tau-1-m-n}^{\infty} \mathcal{D}(\sigma+4-2\tau+m;(\sigma,2s-m)_\epsilon)$$
$$\oplus \bigoplus_{\tau=1}^{t} \bigoplus_{m=0}^{2(\tau-1)} \bigoplus_{n=0}^{\nu_m^\tau} \bigoplus_{k=-\mu_{m,n}^\tau}^{2s-n} \mathcal{D}(2s+4-2\tau+m;(k+m,k)_\epsilon)\,, \tag{107}$$

where $\nu_m^\tau := \min(m, 2(\tau-1)-m)$ *and*

$$\mu_{m,n}^\tau := \begin{cases} \min(n, \nu_m^\tau - n) & , \text{ if } m < \tau\,, \\ m-\tau+1+\min(n,\nu_m^\tau-n)\,, & \text{if } m \geqslant \tau\,. \end{cases} \tag{108}$$

- *If they are of opposite chirality:*

$$\mathcal{D}(s+2-t;(s,s)_+) \otimes \mathcal{D}(s+2-t;(s,s)_-) \cong \bigoplus_{\tau=1}^{t} \bigoplus_{m=0}^{t-\tau} \bigoplus_{n=0}^{t-\tau-m} \bigoplus_{\sigma=2s-m}^{\infty} \mathcal{D}(\sigma+4-2\tau-n;(\sigma,n)_0)\,. \tag{109}$$

Notice that in the above decomposition (109) *of two singletons of opposite chirality, the irreps describing* totally symmetric *partially massless fields, i.e., of spin given by a single row Young diagram, only appear once despite what the notation* $(\sigma)_0$ *would normally suggests.*

Proof. In order to prove the above decomposition, we will use the two expressions of the character of a spin-s and depth-t RPM:

$$\chi^{\mathfrak{so}(2,4)}_{[s+2-t;(s,s)]}(q,\vec{x}) = q^{s+2-t}\Big(\chi^{\mathfrak{so}(4)}_{(s,s)}(\vec{x}) - q^t \chi^{\mathfrak{so}(4)}_{(s,s-t)}(\vec{x}) + q^{t+1}\chi^{\mathfrak{so}(4)}_{(s-1,s-t)}(\vec{x})\Big)\mathcal{P}^{(4)}(q,\vec{x}) \tag{110}$$
$$= \sum_{\ell=0}^{t-1}\sum_{n=0}^{t-1-\ell}\sum_{\sigma=\ell}^{\infty} q^{s+2-t+\sigma+2n}\chi^{\mathfrak{so}(4)}_{(s-\ell+\sigma,s-\ell)}(\vec{x})\,, \tag{111}$$

and will decompose their product as the sum of the characters of the different modules appearing in (107) and (109). To do so, the idea is simply to look at the product of (110) and (111), decompose the tensor product of the $\mathfrak{so}(4)$ characters, and finally recognize the resulting expression as a sum of characters of:

- Partially massless fields of depth-τ and spin given by a two-row Young diagram (σ, n) which read:

$$\chi^{\mathfrak{so}(2,4)}_{[\sigma+3-\tau;(\sigma,n)]}(q,\vec{x}) = q^{\sigma+3-\tau}\mathcal{P}^{(4)}(q,\vec{x})\left(\chi^{\mathfrak{so}(4)}_{(\sigma,n)}(\vec{x}) - q^{\tau}\chi^{\mathfrak{so}(4)}_{(\sigma-\tau,n)}(\vec{x})\right), \tag{112}$$

- Massive fields of minimal energy Δ and spin given a two-row Young diagram (k,l) which read:

$$\chi^{\mathfrak{so}(2,4)}_{[\Delta;(k,l)]}(q,\vec{x}) = q^{\Delta}\mathcal{P}^{(4)}(q,\vec{x})\chi^{\mathfrak{so}(4)}_{(k,l)}(\vec{x}). \tag{113}$$

We will not display here the full computations for the sake of conciseness. □

5. Conclusions

In this note, we considered a class of non-unitary $\mathfrak{so}(2,d)$ modules (for $d = 2r$) parametrised by an integer t, as possible extensions of the higher-spin singletons. These $\mathfrak{so}(2,d)$ modules describe partially massless fields of spin (s^r) and depth-t, and restrict (for $d = 2r$) to a sum of partially massless $\mathfrak{so}(2,d-1)$ modules of spin (s^{r-1}) and depth $\tau = 1,\ldots,t$, thereby naturally generalising the case of unitary singletons corresponding to $t = 1$. Due to the fact that the branching rule (90) shows that these modules cannot be considered as AdS$_d$ field preserved by *conformal* symmetries, and that the branching rule (102) onto $\mathfrak{iso}(1,d-1)$ (deduced from (90) after a Inönü-Wigner contraction) seems to indicate that those fields are not "confined" to the boundary of AdS, the family of $\mathfrak{so}(2,d)$ module $\mathcal{D}(s+\frac{d}{2}-t;(s^r))$ does not appear to share the defining properties of singletons for $t > 1$.

The decomposition of their tensor product in the low-dimensional $\mathfrak{so}(2,4)$ case contains partially massless fields of the same type than in the unitary ($t = 1$) case, i.e., fields of spin (σ) with $\sigma \geqslant 2s$ and spin $(\sigma, 2s)$ with $\sigma \geqslant 2s+1$, as could be expected from comparison with what happens for the Rac$_\ell$ and Di$_\ell$ singletons. However, for $t > 1$, partially massless fields with a different spin also appear, namely of the type (σ, n) with n either taking the values $1,2,\ldots,t-1$ or $2(s-t+1),\ldots,2s$. It is also worth noticing that only for $s = 1$ the decomposition in Theorem 2 contains a conserved spin-2 current (i.e., the module $\mathcal{D}(4;\,(2))$). It would be interesting to extend this tensor product decomposition to arbitrary dimensions.

Acknowledgments: I am grateful to Xavier Bekaert and Nicolas Boulanger for suggesting this work in the first place, as well as for various discussions on the properties of higher-spin singletons and their comments on a previous version of this paper. I am also grateful for the insightful comments of an anonymous referee. This work was supported by a joint grant "50/50" Université François Rabelais Tours—Région Centre/UMONS.

Conflicts of Interest: The authors declare no conflict of interest.

Appendix A. Branching Rules and Tensor Products of $\mathfrak{so}(d)$

In this appendix we recall the branching and tensor product rules for $\mathfrak{so}(d)$ irreps, as well as detail the proofs of the branching rules (23) and (90) and the decomposition (78).

Appendix A.1. Branching Rules for $\mathfrak{so}(d)$

For $d = 2r+1$, the $\mathfrak{so}(d)$ irrep $(s_1,\ldots,s_r)$ branches onto $\mathfrak{so}(d-1)$ as:

$$(s_1,\ldots,s_r) \quad \underset{\mathfrak{so}(d-1)}{\overset{\mathfrak{so}(d)}{\downarrow}} \quad \bigoplus_{t_1=s_2}^{s_1}\cdots\bigoplus_{t_{r-1}=s_r}^{s_{r-1}}\bigoplus_{t_r=-s_r}^{s_r}(t_1,\ldots,t_r), \tag{A1}$$

whereas for $d = 2r$, the branching rule reads:

$$(s_1,\ldots,s_r) \quad \underset{\mathfrak{so}(d-1)}{\overset{\mathfrak{so}(d)}{\downarrow}} \quad \bigoplus_{t_1=s_2}^{s_1}\cdots\bigoplus_{t_{r-1}=s_r}^{s_{r-1}}(t_1,\ldots,t_{r-1}). \tag{A2}$$

Appendix A.2. Computing $\mathfrak{so}(d)$ Tensor Products

In order to prove the decomposition (22) and (78), as well as the similar decomposition for (partially) massless fields with spin given by a rectangular Young diagram of arbitrary length s, we first need to know know how to decompose the tensor product of two $\mathfrak{so}(d)$ Young diagrams, one of which being a single row of arbitrary length and the other one being an "almost" rectangular diagram, i.e., of the form:

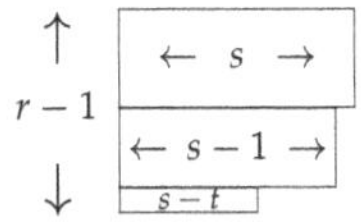

To do so, it is convenient to express the tensor product rule for $\mathfrak{so}(d)$ in terms of that of $\mathfrak{gl}(d)$ which is considerably simpler. The rule for decomposing a tensor product of two $\mathfrak{gl}(d)$ irreps labelled by two Young diagrams $\boldsymbol{\lambda} := (\lambda_1, \ldots, \lambda_d)$ and $\boldsymbol{\mu} = (\mu_1, \ldots, \mu_d)$, known as the Littlewood-Richardson rule, goes as follows (see e.g., [117]):

- First, assign to the boxes of each rows of one of the Young diagrams (say $\boldsymbol{\mu}$) a label which keeps track of the order of the rows (for instance, if the labels are letters of the alphabet, then each of the μ_1 boxes of the first row of $\boldsymbol{\mu}$ are assigned the label "a", each of the μ_2 boxes of the second row of $\boldsymbol{\mu}$ are assigned the label "b", etc.);
- Then, glue the boxes of $\boldsymbol{\mu}$ to $\boldsymbol{\lambda}$ in all possible ways such that the resulting diagram obey the following constraints:
  - Boxes in the same column should *not* have the same label;
  - When reading the row of the obtained Young diagram from right to left, and its columns from top to bottom, the number of boxes encountered should be decreasing with their label (i.e., less boxes of the second label are encountered than with the first label, less with the third than the second, etc.);
  - The resulting diagram should always be a legitimate Young diagram, i.e., the length of the rows is decreasing from top to bottom, and it is composed of at most d rows.

For orthogonal algebras $\mathfrak{so}(2r+1)$, the tensor product of two irreps, $\boldsymbol{\ell} = (\ell_1, \ldots, \ell_r)$ and (σ) can be computed as follows:

(i) Branch each of the two Young diagrams into $\mathfrak{so}(2r)$ and pair them by number of boxes removed from the original ones until the products of one of these diagrams are exhausted;
(ii) Compute the tensor product between these pairs of diagrams using the Littlewood-Richardson rule recalled above;
(iii) Discard the Young diagrams which are not acceptable for $\mathfrak{so}(d)$, i.e., those for which the sum of the height of their first two columns is *stricly* greater than d.

The tensor product $\boldsymbol{\ell} \otimes (\sigma)$ can therefore be represented as follows:

$$\boldsymbol{\ell} \otimes (\sigma) = \bigoplus_{m=0}^{\min(\sigma, \ell_1 - \ell_r)} \bigoplus_{n_1=0}^{\ell_1-\ell_2} \bigoplus_{n_2=0}^{\ell_2-\ell_3} \cdots \bigoplus_{n_r=0}^{\ell_r} (\ell_1 - n_1, \ldots, \ell_r - n_r) \underset{\mathfrak{gl}(d-1)}{\otimes} (\sigma - m) \Big|_{\mathfrak{so}(d)\text{ OK}} . \tag{A3}$$

Example A1. *Consider the tensor product between the $\mathfrak{so}(5)$ irreps*  *and*  *branching rule for these representations is:*

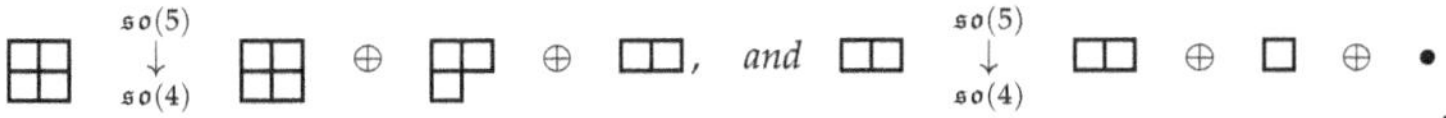

(A4)

Now computing the tensor products between those product paired by number of boxes removed, and using the Littlewood-Richardson rule yields:

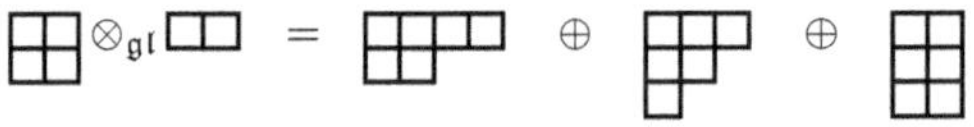

(A5)

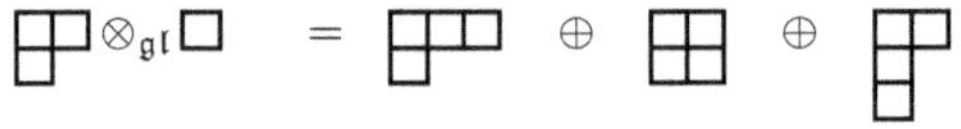

(A6)

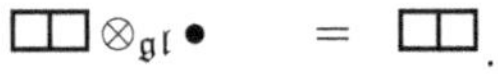

(A7)

Among the diagrams obtained above, one is not a legitimate $\mathfrak{so}(5)$ *Young diagram (as the sum of the height of its two first columns is*

(A8)

the fact that two $\mathfrak{so}(d)$ *Young diagrams whose first column is of height c and* $d-c$ *are equivalent, the initial tensor product finally reads:*

Notice that the well-known tensor product rule for $\mathfrak{so}(3)$ can be recovered from the above algorithm. Given two $\mathfrak{so}(3)$ irreps (s) and (s'), i.e., two one-row Young diagram of respective length s and s', their tensor product decomposes as:

$$(s)\otimes(s') = \bigoplus_{m=0}^{\min(s,s')} (s-m)\otimes_{\mathfrak{gl}(3)}(s'-m)\Big|_{\mathfrak{so}(3)\ \mathrm{OK}} = \bigoplus_{m=0}^{\min(s,s')}\bigoplus_{k=0}^{\min(s'-m,s-m)}(s+s'-2m-k,k)\Big|_{\mathfrak{so}(3)\ \mathrm{OK}}. \tag{A9}$$

The only Young diagrams that are $\mathfrak{so}(3)$ acceptable in the previous equation are those for which $k=0$ or 1, i.e., the second row contains no more than one box. In the latter case, such a Young diagram is equivalent to the one where the second row is absent, i.e. $(s,1)\cong(s)$. As a consequence, the decomposition reads:

$$(s)\otimes(s') = \bigoplus_{m=0}^{2\min(s,s')}(s+s'-m) = \bigoplus_{k=|s-s'|}^{s+s'}(k)\,, \tag{A10}$$

which is indeed the tensor product rule for $\mathfrak{so}(3)$.

Appendix B. Technical Proofs

Appendix B.1. Proof of the Branching Rule for Unitary HS Singletons

From now on, we will set $d=2r$. In order to prove the branching rule:

$$\mathcal{D}(s+r-1\,;\,(s^r)) \quad \underset{\mathfrak{so}(2,d-1)}{\overset{\mathfrak{so}(2,d)}{\downarrow}} \quad \mathcal{D}(s+r-1\,;\,(s^{r-1}))\,, \tag{A11}$$

we will need to derive the $\mathfrak{so}(2)\oplus\mathfrak{so}(d-1)$ decomposition of the $\mathfrak{so}(2,d)$ singleton module $\mathcal{D}(s+r-1\,;\,(s^r))$ and of the $\mathfrak{so}(2,d-1)$ spin (s^{r-1}) massless field module $\mathcal{D}(s+r-1\,;\,(s^{r-1}))$.

Decomposition of the Massless Modules.

To obtain the $\mathfrak{so}(2)\oplus\mathfrak{so}(d-1)$ decomposition of the $\mathfrak{so}(2,d-1)$ spin (s^{r-1}) massless field module $\mathcal{D}(s+r-1;\,(s^{r-1}))$, we will use its character and rewrite it as a sum of $\mathfrak{so}(2)\oplus\mathfrak{so}(d-1)$ characters. Using the property (25) of the function $\mathcal{P}^{(d-1)}(q,\vec{x})$, the character of this module becomes:

$$\begin{aligned}\chi^{\mathfrak{so}(2,d-1)}_{[s+r-1;(s^{r-1})]}(q,\vec{x}) &= q^{s+r-1}\,\mathcal{P}^{(d-1)}(q,\vec{x})\Big(\chi^{\mathfrak{so}(d-1)}_{(s^{r-1})}(\vec{x})+\sum_{k=1}^{r-1}(-1)^k q^k\chi^{\mathfrak{so}(d-1)}_{(s^{r-1-k},(s-1)^k)}(\vec{x})\Big) \quad \text{(A12)}\\ &= \sum_{\sigma,n=0}^{\infty} q^{s+r-1+\sigma+2n}\,\chi^{\mathfrak{so}(d-1)}_{(\sigma)}(\vec{x})\sum_{k=0}^{r-1}(-q)^k\chi^{\mathfrak{so}(d-1)}_{(s^{r-1-k},(s-1)^k)}(\vec{x})\,.\end{aligned}$$

Now we can use the tensor product rule recalled previously for $\mathfrak{so}(d-1)$ with $d-1=2r-1$ odd to reduce the above expression. It turns out that most of the terms in its alternating sum cancel one another. To see that, let us have a look at three consecutive terms in the above sum, that we will denote by "RHS". In order to make the expression more readable, we will also write $\boldsymbol{\ell}=(\ell_1,\ldots,\ell_r)$ for the character of the $\mathfrak{so}(d-1)$ representation $\boldsymbol{\ell}$. A typical triplet of terms in the alternating sum composing the character (A12) reads:

$$\begin{aligned}\text{RHS} &= \sum_{\sigma=0}^{\infty}\Big(q^{\sigma+k}(s^{r-k-1},(s-1)^k)-q^{\sigma+k+1}(s^{r-k-2},(s-1)^{k+1})+q^{\sigma+k+2}(s^{r-k-3},(s-1)^{k+2})\Big)\otimes(\sigma)\\ &= \sum_{\sigma=0}^{\infty} q^{\sigma+k}\sum_{m=0}^{\min(\sigma,s-1)}(s^{r-k-1},(s-1)^{k-1},s-1-m)\otimes_{\mathfrak{gl}(d)}(\sigma-m) \quad \text{(A13)}\\ &\quad +\sum_{\sigma=1}^{\infty} q^{\sigma+k}\sum_{m=0}^{\min(\sigma-1,s-1)}(s^{r-k-2},(s-1)^k,s-1-m)\otimes_{\mathfrak{gl}(d)}(\sigma-1-m) \quad \text{(A14)}\\ &\quad -\sum_{\sigma=0}^{\infty} q^{\sigma+k+1}\sum_{m=0}^{\min(\sigma,s-1)}(s^{r-k-2},(s-1)^k,s-1-m)\otimes_{\mathfrak{gl}(d)}(\sigma-m) \quad \text{(A15)}\\ &\quad -\sum_{\sigma=1}^{\infty} q^{\sigma+k+1}\sum_{m=0}^{\min(\sigma-1,s-1)}(s^{r-k-3},(s-1)^{k+1},s-1-m)\otimes_{\mathfrak{gl}(d)}(\sigma-1-m) \quad \text{(A16)}\\ &\quad +\sum_{\sigma=0}^{\infty} q^{\sigma+k+2}\sum_{m=0}^{\min(\sigma,s-1)}(s^{r-k-3},(s-1)^{k+1},s-1-m)\otimes_{\mathfrak{gl}(d)}(\sigma-m) \quad \text{(A17)}\\ &\quad +\sum_{\sigma=1}^{\infty} q^{\sigma+k+2}\sum_{m=0}^{\min(\sigma-1,s-1)}(s^{r-k-4},(s-1)^{k+2},s-1-m)\otimes_{\mathfrak{gl}(d)}(\sigma-1-m)\,. \quad \text{(A18)}\end{aligned}$$

The second term (corresponding to the fourth and fifth line, i.e., (A15) and (A16) above) can be rewritten as:

$$\begin{aligned}\text{Second term} &= -\sum_{\sigma=1}^{\infty} q^{\sigma+k}\sum_{m=1}^{\min(\sigma,s)}(s^{r-k-2},(s-1)^k,s-m)\otimes_{\mathfrak{gl}(d)}(\sigma-m) \quad \text{(A19)}\\ &\quad -\sum_{\sigma=1}^{\infty} q^{\sigma+k+1}\sum_{m=1}^{\min(\sigma,s)}(s^{r-k-3},(s-1)^{k+1},s-m)\otimes_{\mathfrak{gl}(d)}(\sigma-m)\,. \quad \text{(A20)}\end{aligned}$$

Both of these terms are compensated, the first line above by rewritting (A14) as:

$$+\sum_{\sigma=1}^{\infty} q^{\sigma+k}\sum_{m=1}^{\min(\sigma,s)}(s^{r-k-2},(s-1)^k,s-m)\otimes_{\mathfrak{gl}(d)}(\sigma-m)\,, \quad \text{(A21)}$$

and the second line by rewritting (A17) as:

$$+\sum_{\sigma=1}^{\infty} q^{\sigma+k+1}\sum_{m=1}^{\min(\sigma,s)}(s^{r-k-3},(s-1)^{k+1},s-m)\otimes_{\mathfrak{gl}(d)}(\sigma-m)\,. \quad \text{(A22)}$$

Hence it appears that the second term in this triplet is completely compensated, in part by the first terms and in part by the third one. Because this succession of three terms repeats itself in the alternating sum of the character (A12), and that the last term is completely cancelled by the previous one, only the part of the first term that is not compensated by the second one remains:

$$\begin{aligned}
\text{First two terms} &= \sum_{\sigma=0}^{\infty} q^{\sigma}(s^{r-1})\otimes(\sigma) - \sum_{\sigma=0}^{\infty} q^{\sigma+1}(s^{r-2},s-1)\otimes(\sigma) && \text{(A23)}\\
&= \sum_{\sigma=0}^{\infty} q^{\sigma} \sum_{m=0}^{\min(\sigma,s)} (s^{r-2},s-m)\otimes_{\mathfrak{gl}(d)}(\sigma-m) && \text{(A24)}\\
&\quad - \sum_{\sigma=0}^{\infty} q^{\sigma+1} \sum_{m=0}^{\min(\sigma,s-1)} (s^{r-2},s-1-m)\otimes_{\mathfrak{gl}(d)}(\sigma-m)\\
&\quad - \text{term that will be compensated} && \text{(A25)}\\
&= \sum_{\sigma=0}^{\infty} q^{\sigma} \sum_{m=0}^{\min(\sigma,s)} (s^{r-2},s-m) \underset{\mathfrak{gl}(d-1)}{\otimes} (\sigma-m) && \text{(A26)}\\
&\quad - \sum_{\sigma=1}^{\infty} q^{\sigma} \sum_{m=1}^{\min(\sigma,s)} (s^{r-2},s-m) \underset{\mathfrak{gl}(d-1)}{\otimes} (\sigma-m) && \text{(A27)}\\
&= \sum_{\sigma=0}^{\infty} q^{\sigma}(s^{r-1}) \underset{\mathfrak{gl}(d-1)}{\otimes} (\sigma) = \sum_{\sigma=0}^{\infty} q^{\sigma}(s+\sigma,s^{r-2}) \oplus \sum_{\sigma=1}^{\infty} q^{\sigma}(s+\sigma-1,s^{r-2}) && \text{(A28)}\\
&= \sum_{\sigma=0}^{\infty} (1+q)\, q^{\sigma}(s+\sigma,s^{r-2})\,. && \text{(A29)}
\end{aligned}$$

Notice that only two terms in the $\mathfrak{gl}(d-1)$ tensor product survived in (A28). Indeed, in full generality, it would produce:

$$(s^{r-1}) \underset{\mathfrak{gl}(d-1)}{\otimes} (\sigma) = \bigoplus_{k=0}^{\min(s,\sigma)} (s+\sigma-k, s^{r-2}, k)\,, \tag{A30}$$

however, only the first two terms are $\mathfrak{so}(d-1)$ acceptable Young diagrams, as they are the only ones for which the sum of the heights of their first two columns is lower than $d-1$. The character of the $\mathfrak{so}(2,d-1)$ module $\mathcal{D}(s+r-1;\,(s^{r-1}))$ therefore reads:

$$\chi^{\mathfrak{so}(2,d-1)}_{[s+r-1;(s^{r-1})]}(q,\vec{x}) = \sum_{n=0}^{\infty}\sum_{\sigma=0}^{\infty} q^{s+r-1+2n+\sigma}(1+q)\,\chi^{\mathfrak{so}(d-1)}_{(s+\sigma,s^{r-2})}(\vec{x}) = \sum_{n=0}^{\infty}\sum_{\sigma=0}^{\infty} q^{s+r-1+n+\sigma}\,\chi^{\mathfrak{so}(d-1)}_{(s+\sigma,s^{r-2})}(\vec{x})\,, \tag{A31}$$

which proves that the $\mathfrak{so}(2,d-1)$ module of a spin (s^{r-1}) massless field reads:

$$\mathcal{D}(s+r-1;(s^{r-1})) \cong \bigoplus_{\sigma,n=0}^{\infty} \mathcal{D}_{\mathfrak{so}(2)\oplus\mathfrak{so}(d-1)}(s+r-1+\sigma+n;(s+\sigma,s^{r-2}))\,. \tag{A32}$$

Decomposition of the Singleton Module.

The spin-s singleton module can be decomposed as an infinite direct sum of (finite-dimensional) $\mathfrak{so}(2)\oplus\mathfrak{so}(d)$ modules as [10]:

$$\mathcal{D}(s+r-1;(s^{r})) \cong \bigoplus_{\sigma=0}^{\infty} \mathcal{D}_{\mathfrak{so}(2)\oplus\mathfrak{so}(d)}(s+r-1+\sigma;(s+\sigma,s^{r-1}))\,. \tag{A33}$$

In order to branch this module from $\mathfrak{so}(2,d)$ to $\mathfrak{so}(2,d-1)$, one can simply branch the $\mathfrak{so}(d)$ part of the above decomposition onto $\mathfrak{so}(d-1)$, thereby yielding:

$$\mathcal{D}(s+r-1;(s^r)) \underset{\mathfrak{so}(2,d-1)}{\overset{\mathfrak{so}(2,d)}{\downarrow}} \bigoplus_{\sigma=0}^{\infty}\bigoplus_{\tau=0}^{\sigma} \mathcal{D}_{\mathfrak{so}(2)\oplus\mathfrak{so}(d-1)}(s+r-1+\sigma;(s+\tau,s^{r-2})) \tag{A34}$$

$$\cong \bigoplus_{\sigma=0}^{\infty}\bigoplus_{n=0}^{\infty} \mathcal{D}_{\mathfrak{so}(2)\oplus\mathfrak{so}(d-1)}(s+r-1+\sigma+n;(s+\sigma,s^{r-2}))\,. \tag{A35}$$

This $\mathfrak{so}(2)\oplus\mathfrak{so}(d-1)$ decomposition matches that of the $\mathfrak{so}(2,d-1)$ module of a spin (s^{r-1}) massless field, hence we proved:

$$\mathcal{D}(s+r-1;\,(s^r)) \underset{\mathfrak{so}(2,d-1)}{\overset{\mathfrak{so}(2,d)}{\downarrow}} \mathcal{D}(s+r-1;\,(s^{r-1}))\,. \tag{A36}$$

Appendix B.2. Proof of the Branching Rule for Rectangular Partially Massless Fields

Following the same steps as in the previous section, we will now proceed to proving the branching rule:

$$\mathcal{D}(s+r-t;\,(s^r)) \underset{\mathfrak{so}(2,d-1)}{\overset{\mathfrak{so}(2,d)}{\downarrow}} \bigoplus_{\tau=1}^{t} \mathcal{D}(s+r-\tau;\,(s^{r-1}))\,. \tag{A37}$$

Decomposition of the Partially Massless Modules for $d=2r+1$.

Let us start with the $\mathfrak{so}(2)\oplus\mathfrak{so}(d-1)$ decomposition of the $\mathfrak{so}(2,d-1)$ module $\mathcal{D}(s+r-\tau;(s^{r-1}))$ corresponding to a partially massless field of spin (s^{r-1}) and depth-τ. Using the same notational shortcuts than in the previous section, its character can be rewritten as:

$$\chi^{\mathfrak{so}(2,d-1)}_{[s+r-\tau;(s^{r-1})]}(q,\vec{x}) = \sum_{\sigma,n=0}^{\infty} q^{s+r-\tau+2n+\sigma}(\sigma)\otimes\Big((s^{r-1})+\sum_{k=0}^{r-2}(-1)^{k+1}q^{\tau+k}(s^{r-2-k},(s-1)^k,s-\tau)\Big) \tag{A38}$$

$$= \sum_{n=0}^{\infty} q^{s+r-\tau+2n}\Big(\sum_{\sigma=0}^{\infty} q^{\sigma}\sum_{m=0}^{\min(s,\sigma)}(\sigma-m)\underset{\mathfrak{gl}(d-1)}{\otimes}(s^{r-2},s-m) \tag{A39}$$

$$-\sum_{p=0}^{\tau}\sum_{\sigma=p}^{\infty} q^{\sigma+\tau}\sum_{m=0}^{\min(s-\tau,\sigma-p)}(\sigma-p-m)\underset{\mathfrak{gl}(d-1)}{\otimes}(s^{r-3},s-p,s-\tau-m)$$

$$+\sum_{k=1}^{r-2}(-1)^{k+1}\sum_{p=0}^{\tau-1}\sum_{\sigma=p}^{\infty} q^{\sigma+\tau+k}\times \tag{A40}$$

$$\sum_{m=0}^{\min(s-\tau,\sigma-p)}(\sigma-p-m)\underset{\mathfrak{gl}(d-1)}{\otimes}(s^{r-2-k},(s-1)^{k-1},s-1-p,s-\tau-m)$$

$$+\sum_{k=1}^{r-3}(-1)^{k+1}\sum_{p=0}^{\tau-1}\sum_{\sigma=p+1}^{\infty} q^{\sigma+\tau+k}\times \tag{A41}$$

$$\sum_{m=0}^{\min(s-\tau,\sigma-p-1)}(\sigma-p-1-m)\underset{\mathfrak{gl}(d-1)}{\otimes}(s^{r-3-k},(s-1)^k,s-1-p,s-\tau-m)\Big)$$

$$= \sum_{n=0}^{\infty} q^{s+r-\tau+2n}\Big(\sum_{\sigma=0}^{\infty} q^{\sigma}\sum_{m=0}^{\min(s,\sigma)}(\sigma-m)\underset{\mathfrak{gl}(d-1)}{\otimes}(s^{r-2},s-m) \tag{A42}$$

$$-\sum_{p=0}^{\tau}\sum_{\sigma=p}^{\infty} q^{\sigma+\tau}\sum_{m=0}^{\min(s-\tau,\sigma-p)}(\sigma-p-m)\underset{\mathfrak{gl}(d-1)}{\otimes}(s^{r-3},s-p,s-\tau-m) \tag{A43}$$

$$+\sum_{p=0}^{\tau-1}\sum_{\sigma=p}^{\infty} q^{\sigma+\tau+1}\sum_{m=0}^{\min(s-\tau,\sigma-p)}(\sigma-p-m)\underset{\mathfrak{gl}(d-1)}{\otimes}(s^{r-3},s-1-p,s-\tau-m)\Big) \tag{A44}$$

$$= \sum_{n=0}^{\infty} q^{s+r-\tau+2n}\Big(\sum_{\sigma=0}^{\infty} q^{\sigma}\sum_{m=0}^{\min(s,\sigma)}(\sigma-m)\underset{\mathfrak{gl}(d-1)}{\otimes}(s^{r-2},s-m) \tag{A45}$$

$$-\sum_{\sigma=0}^{\infty} q^{\sigma+\tau}\sum_{m=0}^{\min(s-\tau,\sigma)}(\sigma-m)\underset{\mathfrak{gl}(d-1)}{\otimes}(s^{r-2},s-\tau-m)\Big) \tag{A46}$$

$$= \sum_{\sigma,n=0}^{\infty} q^{s+r-\tau+\sigma+2n}\sum_{m=0}^{\min(\sigma,\tau-1)}(\sigma-m)\underset{\mathfrak{gl}(d-1)}{\otimes}(s^{r-2},s-m)\,. \tag{A47}$$

Leaving aside the sum of n, the above equation can be re-expressed as:

$$\sum_{\sigma=0}^{\infty} q^{\sigma}\sum_{m=0}^{\min(\sigma,\tau-1)}(\sigma-m)\underset{\mathfrak{gl}(d-1)}{\otimes}(s^{r-2},s-m) = \sum_{k=0}^{\tau-1}\sum_{\sigma=k}^{\infty} q^{\sigma}(\sigma-k)\underset{\mathfrak{gl}(d-1)}{\otimes}(s^{r-2},s-k)\,. \tag{A48}$$

Using the Littlewood-Richardson rule yields:

$$\sum_{\sigma=k}^{\infty} q^{\sigma}(\sigma-k)\underset{\mathfrak{gl}(d-1)}{\otimes}(s^{r-2},s-k) = \sum_{p=0}^{k}\sum_{\sigma=k+p}^{\infty} q^{\sigma}(1+q)(s+\sigma-k-p,s^{r-3}) \tag{A49}$$

$$= \sum_{p=0}^{k}\sum_{\sigma=2k-p}^{\infty} q^{\sigma}(1+q)(s+\sigma-2k+p,s^{r-3},s-p) \tag{A50}$$

$$= \sum_{p=0}^{k}\sum_{\sigma=p}^{\infty} q^{\sigma+2(k-p)}(1+q)(s+\sigma-p,s^{r-3},s-p)\,, \tag{A51}$$

hence (A48) becomes:

$$\sum_{k=0}^{\tau-1}\sum_{p=0}^{k}\sum_{\sigma=p}^{\infty} q^{\sigma+2(k-p)}(1+q)(s+\sigma-p,s^{r-3},s-p) = \sum_{k=0}^{\tau-1}\sum_{p=0}^{\tau-1-k}\sum_{\sigma=p}^{\infty} q^{\sigma+2p}(1+q)(s+\sigma-k,s^{r-3},s-k)\,. \tag{A52}$$

Finally, the character of the $\mathfrak{so}(2,d-1)$ module $\mathcal{D}(s+r-\tau;(s^{r-1}))$ can be expressed as:

$$\chi^{\mathfrak{so}(2,d-1)}_{[s+r-\tau;(s^{r-1})]}(q,\vec{x}) = \sum_{n=0}^{\infty}\sum_{k=0}^{\tau-1}\sum_{p=0}^{\tau-1-k}\sum_{\sigma=k}^{\infty} q^{s+r-\tau+2p+\sigma+n}\chi^{\mathfrak{so}(d-1)}_{(s+\sigma-k,s^{r-3},s-k)}(\vec{x}), \tag{A53}$$

which proves that this module decomposes as:

$$\mathcal{D}(s+r-\tau;(s^{r-1})) \cong \bigoplus_{k=0}^{\tau-1}\bigoplus_{p=0}^{\tau-1-k}\bigoplus_{n=0}^{\infty}\bigoplus_{\sigma=k}^{\infty}\mathcal{D}_{\mathfrak{so}(2)\oplus\mathfrak{so}(d-1)}(s+r-\tau+\sigma+n+2p;(s+\sigma-k,s^{r-3},s-k)). \tag{A54}$$

Decomposition of the Rectangular Partially Massless Field Module for $d=2r$.

Before deriving the $\mathfrak{so}(2)\oplus\mathfrak{so}(d-1)$ decomposition of the spin-s and depth-t RPM module, we will need to prove its $\mathfrak{so}(2)\oplus\mathfrak{so}(d)$ decomposition:

$$\mathcal{D}(s+r-t;(s^r)) \cong \bigoplus_{\ell=0}^{t-1}\bigoplus_{n=0}^{t-1-\ell}\bigoplus_{\sigma=\ell}^{\infty}\mathcal{D}_{\mathfrak{so}(2)\oplus\mathfrak{so}(d)}(s+r-t+\sigma+2n;(s-\ell+\sigma,s^{r-2},s-\ell)). \tag{A55}$$

To do so, we will need to use the Weyl character formula:

$$\chi^{\mathfrak{so}(d)}_{\boldsymbol{\ell}}(\vec{x}) = \frac{\sum_{w\in\mathcal{W}_{\mathfrak{so}(d)}}\varepsilon(w)e^{(w(\boldsymbol{\ell}+\boldsymbol{\rho})-\boldsymbol{\rho},\boldsymbol{\mu})}}{\prod_{\boldsymbol{\alpha}\in\Phi_+}(1-e^{(\boldsymbol{\alpha},\boldsymbol{\mu})})}, \tag{A56}$$

where Φ_+ is the set of positive roots of $\mathfrak{so}(d)$, $\mathcal{W}_{\mathfrak{so}(d)}$ its Weyl group, $\boldsymbol{\rho} := \frac{1}{2}\sum_{\boldsymbol{\alpha}\in\Phi_+}\boldsymbol{\alpha}$ the Weyl vector, $\varepsilon(w)$ the signature of the Weyl group element w, $(\,,)$ is inner product on the root space inherited from the Killing form and $\boldsymbol{\mu}=(\mu_1,\dots,\mu_r)$ an arbitrary root used to define the variables x_i on which depends the character via $x_i := e^{\mu_i}$. Defining

$$C^{\mathfrak{so}(d)}_{\boldsymbol{\ell}}(\vec{x}) := e^{(\boldsymbol{\ell},\boldsymbol{\mu})}\prod_{\boldsymbol{\alpha}\in\Phi_+}\frac{1}{1-e^{(\boldsymbol{\alpha},\boldsymbol{\mu})}} = \prod_{i=1}^{r}x_i^{\ell_i}\prod_{\boldsymbol{\alpha}\in\Phi_+}\frac{1}{1-e^{(\boldsymbol{\alpha},\boldsymbol{\mu})}}, \tag{A57}$$

one can show

$$w(C^{\mathfrak{so}(d)}_{\boldsymbol{\ell}}(\vec{x})) := e^{(w(\boldsymbol{\ell}),\boldsymbol{\mu})}\prod_{\boldsymbol{\alpha}\in\Phi_+}\frac{1}{1-e^{(w(\boldsymbol{\alpha}),\boldsymbol{\mu})}} = \varepsilon(w)C^{\mathfrak{so}(d)}_{w\cdot\boldsymbol{\ell}}(\vec{x}), \tag{A58}$$

with $w\cdot\boldsymbol{\ell} = w(\boldsymbol{\ell}+\boldsymbol{\rho})-\boldsymbol{\rho}$. Hence, the Weyl character formula (A57) can be rewritten as:

$$\chi^{\mathfrak{so}(d)}_{\boldsymbol{\ell}}(\vec{x}) = \sum_{w\in\mathcal{W}_{\mathfrak{so}(d)}} w(C^{\mathfrak{so}(d)}_{\boldsymbol{\ell}}(\vec{x})) := \mathfrak{W}_{\mathfrak{so}(d)}\Big(C^{\mathfrak{so}(d)}_{\boldsymbol{\ell}}(\vec{x})\Big). \tag{A59}$$

The Weyl group for $\mathcal{W}_{\mathfrak{so}(d)}$ for $d=2r$ is the semi-direct product $\mathcal{S}_r\ltimes(\mathbb{Z}_2)^{r-1}$ where $\mathcal{S}_r$ is the permutation group of r elements. Any element of $\mathcal{W}_{\mathfrak{so}(d)}$ acts on the $\mathfrak{so}(d)$ characters as a combination of permutation of the variables x_i and (pair of) sign flip of their exponents (for more details, see e.g., [10] or the classical textbooks [118,119]). As a consequence, any function invariant under such operation can go in and out of the Weyl symmetrizer $\mathfrak{W}_{\mathfrak{so}(d)}$. In particular, $\mathcal{P}^{(d)}(q,\vec{x})$ being invariant under any permutation and any inversion ($x_i\to x_i^{-1}$) of the variables x_i, it has the property:

$$\mathfrak{W}_{\mathfrak{so}(d)}(C^{\mathfrak{so}(d)}_{\boldsymbol{\ell}}(\vec{x})\mathcal{P}^{(d)}(q,\vec{x})) = \mathfrak{W}_{\mathfrak{so}(d)}(C^{\mathfrak{so}(d)}_{\boldsymbol{\ell}}(\vec{x}))\mathcal{P}^{(d)}(q,\vec{x}). \tag{A60}$$

Using this property, as well as the fact that $C^{\mathfrak{so}(d)}_{\boldsymbol{\ell}}(\vec{x})$ verifies:

$$C^{\mathfrak{so}(d)}_{(\ell_1,\dots,\ell_{j-1},\ell_j+a,\ell_{j+1},\dots,\ell_r)}(\vec{x}) = x_j^a\,C^{\mathfrak{so}(d)}_{(\ell_1,\dots,\ell_{j-1},\ell_j,\ell_{j+1},\dots,\ell_r)}(\vec{x}), \tag{A61}$$

we can now prove that the character of the $\mathfrak{so}(2) \oplus \mathfrak{so}(d)$ decomposition (A55) can be rewritten [15] as the character of the spin-s and depth-t RPM module (76):

$$\begin{aligned}
\text{Decomposition} &= \sum_{\ell=0}^{t-1}\sum_{n=0}^{t-1-\ell}\sum_{\sigma=\ell}^{\infty} q^{s+r-t+2n+\sigma}\chi^{\mathfrak{so}(d)}_{(s+\sigma-\ell,s^{r-2},s-\ell)}(\vec{x}) && \text{(A62)}\\
&= \sum_{\ell=0}^{t-1} q^{s+r-t+\ell}\frac{1-q^{2(t-\ell)}}{1-q^2}\mathfrak{W}_{\mathfrak{so}(d)}\Big(\sum_{\sigma=0}^{\infty} q^{\sigma}x_1^{\sigma}C^{\mathfrak{so}(d)}_{(s^{r-1},s-\ell)}(\vec{x})\Big) && \text{(A63)}\\
&= \frac{q^{s+r}}{1-q^2}\mathfrak{W}_{\mathfrak{so}(d)}\Big(\sum_{\ell=0}^{t-1}\left[q^{-t+\ell}-q^{t-\ell}\right]\frac{x_r^{-\ell}}{1-qx_1}C^{\mathfrak{so}(d)}_{(s^r)}(\vec{x})\Big) && \text{(A64)}\\
&= \frac{q^{s+r-t}}{1-q^2}\mathfrak{W}_{\mathfrak{so}(d)}\Big(\frac{1-(1-q^2)q^t x_r^{-t}-(1-q^{2t})qx_r-q^{2t+2}}{(1-qx_1)(1-qx_r)(1-qx_r^{-1})}C^{\mathfrak{so}(d)}_{(s^r)}(\vec{x})\Big) && \text{(A65)}\\
&= \frac{q^{s+r-t}}{1-q^2}\mathcal{P}^{(d)}(q,\vec{x})\times && \text{(A66)}\\
&\quad \mathfrak{W}_{\mathfrak{so}(d)}\Big(\left[1-q^{2t+2}-(1-q^{2t})qx_r\right. && \text{(A67)}\\
&\quad \left.-(1-q^2)q^t x_r^{-t}\right](1-qx^{-1})\prod_{i=2}^{r-1}(1-qx_i)(1-qx_i^{-1})C^{\mathfrak{so}(d)}_{(s^r)}(\vec{x})\Big)\,. && \text{(A68)}
\end{aligned}$$

Each terms of the above sum as some power of q times a $\mathfrak{so}(d)$ character: using (A57), (A60) and (A61) they are all of the form

$$\mathfrak{W}_{\mathfrak{so}(d)}\Big(q^{\beta}\prod_{i=1}^{r}x_i^{\alpha_i}C^{\mathfrak{so}(d)}_{(s^r)}(\vec{x})\Big) = q^{\beta}\,\chi^{\mathfrak{so}(d)}_{(s+\alpha_1,s+\alpha_2,\ldots,s+\alpha_r)}(\vec{x})\,. \qquad \text{(A69)}$$

- The first piece, proportional to $(1-q^{2t+2})$ reads:

$$\begin{aligned}
\mathfrak{W}_{\mathfrak{so}(d)}\Big((1-qx^{-1})\ \textstyle\prod_{i=2}^{r-1}\ (1-qx_i)(1-qx_i^{-1})\,C^{\mathfrak{so}(d)}_{(s^r)}(\vec{x})\Big) && \text{(A70)}\\
= \sum_{\substack{0\leqslant n\leqslant 2r-3\\ n:=n_{1,-}+\sum_{i=2}^{r-1}n_{i,+}+n_{i,-}}} (-q)^n\chi^{\mathfrak{so}(d)}_{(s-n_{1,-},s+n_{2,+}-n_{2,-},\ldots,s+n_{r-1,+}-n_{r-1,-},s)}(\vec{x})\,.
\end{aligned}$$

Using the symmetry property

$$\chi^{\mathfrak{so}(d)}_{(\ell_1,\ldots,\ell_j,\ell-1,\ell+1,\ell_{j+3},\ldots,\ell_r)}(\vec{x}) = -\chi^{\mathfrak{so}(d)}_{(\ell_1,\ldots,\ell_j,\ell,\ell,\ell_{j+3},\ldots,\ell_r)}(\vec{x})\,, \qquad \text{(A71)}$$

of the $\mathfrak{so}(d)$ characters, one can check that the previous sum reduces to a single contribution:

$$\mathfrak{W}_{\mathfrak{so}(d)}\Big((1-qx^{-1})\prod_{i=2}^{r-1}(1-qx_i)(1-qx_i^{-1})\,C^{\mathfrak{so}(d)}_{(s^r)}(\vec{x})\Big) = \chi^{\mathfrak{so}(d)}_{(s^r)}(\vec{x})\,. \qquad \text{(A72)}$$

For instance, for $r=3$:

$$\begin{aligned}
\mathfrak{W}_{\mathfrak{so}(d)}\Big((1-qx^{-1})(1-qx_2)(1-qx_2^{-1})\,C^{\mathfrak{so}(d)}_{(s^r)}(\vec{x})\Big) &= (1+q^2)\chi^{\mathfrak{so}(d)}_{(s,s,s)}(\vec{x}) - q(\chi^{\mathfrak{so}(d)}_{(s,s+1,s)}(\vec{x})\\
&\quad +\chi^{\mathfrak{so}(d)}_{(s,s-1,s)}(\vec{x}))\\
&\quad -q(1+q^2)\chi^{\mathfrak{so}(d)}_{(s-1,s,s)}(\vec{x})\\
&\quad +q^2(\chi^{\mathfrak{so}(d)}_{(s-1,s+1,s)}(\vec{x})+\chi^{\mathfrak{so}(d)}_{(s-1,s-1,s)}(\vec{x}))\,.
\end{aligned}$$

---

15 This technique was originally used in [10] (see appendix D) in order to derive the $\mathfrak{so}(2) \oplus \mathfrak{so}(d)$ decomposition of the unitary singleton modules. We merely adapt it here to the non-unitary case.

The symmetry property (A71) implies that:

$$\chi^{\mathfrak{so}(d)}_{(s,s+1,s)}(\vec{x}) = \chi^{\mathfrak{so}(d)}_{(s,s-1,s)}(\vec{x}) = \chi^{\mathfrak{so}(d)}_{(s-1,s-1,s)}(\vec{x}) = \chi^{\mathfrak{so}(d)}_{(s-1,s,s)}(\vec{x}) = 0\,, \quad \text{and} \quad \chi^{\mathfrak{so}(d)}_{(s-1,s+1,s)}(\vec{x}) = -\chi^{\mathfrak{so}(d)}_{(s,s,s)}(\vec{x})\,,$$

hence we indeed obtain:

$$\mathfrak{W}_{\mathfrak{so}(d)}\Big((1-qx^{-1})(1-qx_2)(1-qx_2^{-1})\,C^{\mathfrak{so}(d)}_{(s^r)}(\vec{x})\Big) = \chi^{\mathfrak{so}(d)}_{(s,s,s)}(\vec{x})\,. \tag{A73}$$

- The second piece, proportional to $(1-q^{2t})$, also reduces to a single contribution upon using the same symmetry property (A71):

$$\mathfrak{W}_{\mathfrak{so}(d)}\Big(qx_r(1-qx^{-1})\ \prod_{i=2}^{r-1}\ (1-qx_i)(1-qx_i^{-1})\,C^{\mathfrak{so}(d)}_{(s^r)}(\vec{x})\Big) \tag{A74}$$

$$= \sum_{\substack{0\leqslant n\leqslant 2r-3\\ n:=n_{1,-}+\sum_{i=2}^{r-1} n_{i,+}+n_{i,-}}} (-)^n q^{n+1}\chi^{\mathfrak{so}(d)}_{(s-n_{1,-},s+n_{2,+}-n_{2,-},\dots,s+n_{r-1,+}-n_{r-1,-},s+1)}(\vec{x})$$

$$= q^2\,\chi^{\mathfrak{so}(d)}_{(s^r)}(\vec{x}) \tag{A75}$$

- Finally, the third piece (proportional to $(1-q^2)$) contains more contributions:

$$\mathfrak{W}_{\mathfrak{so}(d)}\Big(q^t x_r^{-t}(1-qx^{-1})\ \prod_{i=2}^{r-1}\ (1-qx_i)(1-qx_i^{-1})\,C^{\mathfrak{so}(d)}_{(s^r)}(\vec{x})\Big) \tag{A76}$$

$$= \sum_{\substack{0\leqslant n\leqslant 2r-3\\ n:=n_{1,-}+\sum_{i=2}^{r-1} n_{i,+}+n_{i,-}}} (-1)^n q^{t+n}\chi^{\mathfrak{so}(d)}_{(s-n_{1,-},s+n_{2,+}-n_{2,-},\dots,s+n_{r-1,+}-n_{r-1,-},s-t)}(\vec{x})$$

$$= \sum_{k=0}^{r-1}(-1)^k q^{t+k}\chi^{\mathfrak{so}(d)}_{(s^{r-1-k},(s-1)^k,s-t)}(\vec{x})\,. \tag{A77}$$

Putting those three piece together yields:

$$\sum_{\ell=0}^{t-1}\sum_{n=0}^{t-1-\ell}\sum_{\sigma=\ell}^{\infty} q^{s+r-t+2n+\sigma}\chi^{\mathfrak{so}(d)}_{(s+\sigma-\ell,s^{r-2},s-\ell)}(\vec{x}) = \frac{q^{s+r-t}}{1-q^2}\Big((1-q^{2t+2})\,\chi^{\mathfrak{so}(d)}_{(s^r)}(\vec{x}) - (1-q^{2t})\,q^2\,\chi^{\mathfrak{so}(d)}_{(s^r)}(\vec{x}) \tag{A78}$$

$$-(1-q^2)\sum_{k=0}^{r-1}(-)^k q^{t+k}\,\chi^{\mathfrak{so}(d)}_{(s^{r-1-k},(s-1)^k,s-t)}(\vec{x})\Big)\mathcal{P}^{(d)}(q,\vec{x})$$

$$= q^{s+r-t}\,\mathcal{P}^{(d)}(q,\vec{x})\,\Big(\chi^{\mathfrak{so}(d)}_{(s^r)}(\vec{x}) \tag{A79}$$

$$-\sum_{k=0}^{r-1}(-)^k q^{t+k}\,\chi^{\mathfrak{so}(d)}_{(s^{r-1-k},(s-1)^k,s-t)}(\vec{x})\Big) \tag{A80}$$

$$= \chi^{\mathfrak{so}(2,d)}_{[s+r-t;(s^r)]}(q,\vec{x})\,, \tag{A81}$$

thereby proving the $\mathfrak{so}(2)\oplus\mathfrak{so}(d)$ decomposition (A55) of the spin-s and depth-t RPM module. Finally, the $\mathfrak{so}(2)\oplus\mathfrak{so}(d-1)$ of this module reads:

$$\mathcal{D}(s+r-t;(s^r)) \cong \bigoplus_{\ell=0}^{t-1}\bigoplus_{n=0}^{t-1-\ell}\bigoplus_{\sigma=\ell}^{\infty}\mathcal{D}_{\mathfrak{so}(2)\oplus\mathfrak{so}(d)}(s+r-t+\sigma+2n;(s-\ell+\sigma,s^{r-2},s-\ell)) \tag{A82}$$

$$\underset{\mathfrak{so}(2,d-1)}{\overset{\mathfrak{so}(2,d)}{\downarrow}} \bigoplus_{\ell=0}^{t-1}\bigoplus_{n=0}^{t-1-\ell}\bigoplus_{\sigma=\ell}^{\infty}\bigoplus_{\tau=0}^{\sigma-\ell}\bigoplus_{k=0}^{\ell}\mathcal{D}_{\mathfrak{so}(2)\oplus\mathfrak{so}(d-1)}(s+r-t+\sigma+2n;(s+\tau,s^{r-3},s-k)) \tag{A83}$$

$$\cong \bigoplus_{\tau=1}^{t}\bigoplus_{k=0}^{\tau-1}\bigoplus_{p=0}^{\tau-1-k}\bigoplus_{n=0}^{\infty}\bigoplus_{\sigma=k}^{\infty}\mathcal{D}_{\mathfrak{so}(2)\oplus\mathfrak{so}(d-1)}(s+r-\tau+\sigma+n+2p;(s+\sigma-k,s^{r-3},s-k))\,,$$

thereby proving the branching rule:

$$\mathcal{D}(s+r-t;(s^r)) \underset{\mathfrak{so}(2,d-1)}{\overset{\mathfrak{so}(2,d)}{\downarrow}} \bigoplus_{\tau=1}^{t}\mathcal{D}(s+r-\tau;(s^{r-1}))\,. \tag{A84}$$

References

1. Metsaev, R.R. Continuous spin gauge field in (A)dS space. *Phys. Lett. B* **2017**, *767*, 458–464.
2. Metsaev, R.R. Fermionic continuous spin gauge field in (A)dS space. *Phys. Lett. B* **2017**, *773*, 135–141.
3. Bekaert, X.; Skvortsov, E.D. Elementary particles with continuous spin. *Int. J. Mod. Phys. A* **2017**, *32*, 1730019.
4. Angelopoulos, E.; Flato, M.; Fronsdal, C.; Sternheimer, D. Massless Particles, Conformal Group and De Sitter Universe. *Phys. Rev. D* **1981**, *23*, 1278.
5. Angelopoulos, E.; Laoues, M. Masslessness in n-dimensions. *Rev. Math. Phys.* **1998**, *10*, 271–299.
6. Iazeolla, C.; Sundell, P. A Fiber Approach to Harmonic Analysis of Unfolded Higher-Spin Field Equations. *J. High Energy Phys.* **2008**, *2008*, 022.
7. Flato, M.; Fronsdal, C. One Massless Particle Equals Two Dirac Singletons. *Lett. Math. Phys.* **1978**, *2*, 421–426.
8. Angelopoulos, E.; Laoues, M. Singletons on AdS(n). In Proceedings of the Conference Moshe Flato, Dijon, France, 5–8 September 1999; pp. 3–23.
9. Vasiliev, M.A. Higher spin superalgebras in any dimension and their representations. *J. High Energy Phys.* **2004**, *2004*, 46.
10. Dolan, F.A. Character formulae and partition functions in higher dimensional conformal field theory. *J. Math. Phys.* **2006**, *47*, 062303.
11. Vasiliev, M.A. Equations of Motion of Interacting Massless Fields of All Spins as a Free Differential Algebra. *Phys. Lett. B* **1988**, *209*, 491–497.
12. Vasiliev, M.A. Consistent equation for interacting gauge fields of all spins in (3+1)-dimensions. *Phys. Lett. B* **1990**, *243*, 378–382.
13. Vasiliev, M.A. More on equations of motion for interacting massless fields of all spins in (3+1)-dimensions. *Phys. Lett. B* **1992**, *285*, 225–234.
14. Vasiliev, M.A. Nonlinear equations for symmetric massless higher spin fields in (A)dS(d). *Phys. Lett. B* **2003**, *567*, 139–151.
15. Bekaert, X.; Boulanger, N.; Sundell, P. How higher-spin gravity surpasses the spin two barrier. *Rev. Mod. Phys.* **2012**, *84*, 987–1009.
16. Bekaert, X.; Cnockaert, S.; Iazeolla, C.; Vasiliev, M.A. Nonlinear higher spin theories in various dimensions. In Proceedings of the 1st Solvay Workshop Higher Spin Gauge Theories, Brussels, Belgium, 12–14 May 2004; pp. 132–197.
17. Didenko, V.E.; Skvortsov, E.D. Elements of Vasiliev theory. *arXiv* **2014**, arXiv:1401.2975.
18. Maldacena, J.M. The Large N limit of superconformal field theories and supergravity. *Int. J. Theor. Phys.* **1999**, *38*, 1113–1133.
19. Witten, E. Anti-de Sitter space and holography. *Adv. Theor. Math. Phys.* **1998**, *2*, 253–291.
20. Gubser, S.S.; Klebanov, I.R.; Polyakov, A.M. Gauge theory correlators from noncritical string theory. *Phys. Lett. B* **1998**, *428*, 105–114.
21. Sezgin, E.; Sundell, P. Massless higher spins and holography. *Nucl. Phys. B* **2002**, *644*, 303–370.
22. Klebanov, I.R.; Polyakov, A.M. AdS dual of the critical O(N) vector model. *Phys. Lett. B* **2002**, *550*, 213–219.
23. Giombi, S.; Klebanov, I.R. One Loop Tests of Higher Spin AdS/CFT. *J. High Energy Phys.* **2013**, *2013*, 68.
24. Giombi, S.; Klebanov, I.R.; Tseytlin, A.A. Partition Functions and Casimir Energies in Higher Spin AdS_{d+1}/CFT_d. *Phys. Rev. D* **2014**, *90*, 024048.
25. Sezgin, E.; Sundell, P. Holography in 4D (super) higher spin theories and a test via cubic scalar couplings. *J. High Energy Phys.* **2005**, *2015*, 44.
26. Giombi, S.; Yin, X. Higher Spin Gauge Theory and Holography: The Three-Point Functions. *J. High Energy Phys.* **2010**, *2010*, 115.
27. Giombi, S.; Yin, X. The Higher Spin/Vector Model Duality. *J. Phys. A* **2013**, *46*, 214003.
28. Giombi, S. Higher spin—CFT duality. In Proceedings of the Theoretical Advanced Study Institute in Elementary Particle Physics: New Frontiers in Fields and Strings (TASI 2015), Boulder, CO, USA, 1–26 June 2015; pp. 137–214.
29. Giombi, S.; Klebanov, I.R.; Tan, Z.M. The ABC of Higher-Spin AdS/CFT. *arXiv* **2016**, arXiv:1608.07611.
30. Sleight, C. Interactions in Higher-Spin Gravity: A Holographic Perspective. Ph.D. Thesis, Ludwig Maximilian University of Munich, München, Germany, 2016.
31. Sleight, C. Metric-like Methods in Higher Spin Holography. *arXiv* **2017**, arXiv:1701.08360.

32. Bekaert, X.; Erdmenger, J.; Ponomarev, D.; Sleight, C. Towards holographic higher-spin interactions: Four-point functions and higher-spin exchange. *J. High Energy Phys.* **2015**, *2015*, 170.
33. Ruhl, W. The Masses of gauge fields in higher spin field theory on AdS(4). *Phys. Lett. B* **2005**, *605*, 413–418.
34. Manvelyan, R.; Ruhl, W. The Masses of gauge fields in higher spin field theory on the bulk of AdS(4). *Phys. Lett. B* **2005**, *613*, 197–207.
35. Manvelyan, R.; Ruhl, W. The Off-shell behaviour of propagators and the Goldstone field in higher spin gauge theory on AdS(d+1) space. *Nucl. Phys. B* **2005**, *717*, 3–18.
36. Sleight, C.; Taronna, M. Higher Spin Interactions from Conformal Field Theory: The Complete Cubic Couplings. *Phys. Rev. Lett.* **2016**, *116*, 181602.
37. Bekaert, X.; Erdmenger, J.; Ponomarev, D.; Sleight, C. Quartic AdS Interactions in Higher-Spin Gravity from Conformal Field Theory. *J. High Energy Phys.* **2015**, *2015*, 149.
38. Boulanger, N.; Kessel, P.; Skvortsov, E.D.; Taronna, M. Higher spin interactions in four-dimensions: Vasiliev versus Fronsdal. *J. Phys. A* **2016**, *49*, 095402.
39. Skvortsov, E.D.; Taronna, M. On Locality, Holography and Unfolding. *J. High Energy Phys.* **2015**, *2015*, 44.
40. Vasiliev, M.A. Current Interactions and Holography from the 0-Form Sector of Nonlinear Higher-Spin Equations. *J. High Energy Phys.* **2017**, *2017*, 111.
41. Taronna, M. A note on field redefinitions and higher-spin equations. *J. Phys. A* **2017**, *50*, 075401.
42. Sleight, C.; Taronna, M. Higher spin gauge theories and bulk locality: A no-go result. *arXiv* **2017**, arXiv:1704.07859.
43. Bonezzi, R.; Boulanger, N.; de Filippi, D.; Sundell, P. Noncommutative Wilson lines in higher-spin theory and correlation functions of conserved currents for free conformal fields. *J. Phys. A* **2017**, *50*, 475401.
44. Maldacena, J.; Zhiboedov, A. Constraining Conformal Field Theories with A Higher Spin Symmetry. *J. Phys. A* **2013**, *46*, 214011.
45. Alba, V.; Diab, K. Constraining conformal field theories with a higher spin symmetry in d = 4. *arXiv* **2013**, arXiv:1307.8092.
46. Alba, V.; Diab, K. Constraining conformal field theories with a higher spin symmetry in $d > 3$ dimensions. *J. High Energy Phys.* **2016**, *2016*, 44.
47. Boulanger, N.; Ponomarev, D.; Skvortsov, E.D.; Taronna, M. On the uniqueness of higher-spin symmetries in AdS and CFT. *Int. J. Mod. Phys. A* **2013**, *28*, 1350162.
48. Siegel, W. All Free Conformal Representations in All Dimensions. *Int. J. Mod. Phys. A* **1989**, *4*, 2015–2020.
49. Leigh, R.G.; Petkou, A.C. Holography of the N = 1 higher spin theory on AdS(4). *J. High Energy Phys.* **2003**, *2003*, 11.
50. Skvortsov, E.D. On (Un)Broken Higher-Spin Symmetry in Vector Models. In *Higher Spin Gauge Theories*; World Scientific: Singapore, 2015; pp. 103–137.
51. Maldacena, J.; Zhiboedov, A. Constraining conformal field theories with a slightly broken higher spin symmetry. *Class. Quant. Gravity* **2013**, *30*, 104003.
52. Boulanger, N.; Skvortsov, E.D. Higher-spin algebras and cubic interactions for simple mixed-symmetry fields in AdS spacetime. *J. High Energy Phys.* **2011**, *2011*, 63.
53. Joung, E.; Mkrtchyan, K. Notes on higher-spin algebras: Minimal representations and structure constants. *J. High Energy Phys.* **2014**, *2014*, 103.
54. Joseph, A. The minimal orbit in a simple Lie algebra and its associated maximal ideal. *Ann. Sci. École Norm. Super.* **1976**, *9*, 1–29.
55. Bekaert, X.; Grigoriev, M. Manifestly conformal descriptions and higher symmetries of bosonic singletons. *Symmetry Integr. Geom. Methods Appl.* **2010**, *6*, 38.
56. Labastida, J.M.F. Massless Particles in Arbitrary Representations of the Lorentz Group. *Nucl. Phys. B* **1989**, *322*, 185–209.
57. Metsaev, R.R. Massless mixed symmetry bosonic free fields in d-dimensional anti-de Sitter space-time. *Phys. Lett. B* **1995**, *354*, 78–84.
58. Metsaev, R.R. Arbitrary spin massless bosonic fields in d-dimensional anti-de Sitter space. In *Supersymmetries and Quantum Symmetries*; Springer: Berlin/Heidelberg, Germany, 1999; pp. 331–340.
59. Metsaev, R.R. Fermionic fields in the d-dimensional anti-de Sitter space-time. *Phys. Lett. B* **1998**, *419*, 49–56.
60. Burdik, C.; Pashnev, A.; Tsulaia, M. The Lagrangian description of representations of the Poincare group. *Nucl. Phys. Proc. Suppl.* **2001**, *102*, 285–292.

61. Burdik, C.; Pashnev, A.; Tsulaia, M. On the Mixed symmetry irreducible representations of the Poincare group in the BRST approach. *Mod. Phys. Lett. A* **2001**, *16*, 731–746.
62. Alkalaev, K.B.; Shaynkman, O.V.; Vasiliev, M.A. On the frame—Like formulation of mixed symmetry massless fields in (A)dS(d). *Nucl. Phys. B* **2004**, *692*, 363–393.
63. Bekaert, X.; Boulanger, N. On geometric equations and duality for free higher spins. *Phys. Lett. B* **2003**, *561*, 183–190.
64. Bekaert, X.; Boulanger, N. Mixed symmetry gauge fields in a flat background. In Proceedings of the 5th International Workshop on Supersymmetries and Quantum Symmetries (SQS'03), Dubna, Russia, 24–29 July 2003; pp. 37–42.
65. Boulanger, N.; Iazeolla, C.; Sundell, P. Unfolding Mixed-Symmetry Fields in AdS and the BMV Conjecture: I. General Formalism. *J. High Energy Phys.* **2009**, *2009*, 13.
66. Boulanger, N.; Iazeolla, C.; Sundell, P. Unfolding Mixed-Symmetry Fields in AdS and the BMV Conjecture: II. Oscillator Realization. *J. High Energy Phys.* **2009**, *2009*, 14.
67. Skvortsov, E.D. Mixed-Symmetry Massless Fields in Minkowski space Unfolded. *J. High Energy Phys.* **2008**, *2008*, 4.
68. Alkalaev, K.B.; Grigoriev, M.; Tipunin, I.Y. Massless Poincare modules and gauge invariant equations. *Nucl. Phys. B* **2009**, *823*, 509–545.
69. Skvortsov, E.D. Gauge fields in (A)dS(d) within the unfolded approach: Algebraic aspects. *J. High Energy Phys.* **2010**, *2010*, 106.
70. Skvortsov, E.D. Gauge fields in (A)dS(d) and Connections of its symmetry algebra. *J. Phys. A* **2009**, *42*, 385401.
71. Campoleoni, A. Metric-like Lagrangian Formulations for Higher-Spin Fields of Mixed Symmetry. *arXiv* **2010**, arXiv:0910.3155v2.
72. Alkalaev, K.B.; Grigoriev, M. Unified BRST description of AdS gauge fields. *Nucl. Phys. B* **2010**, *835*, 197–220.
73. Alkalaev, K.B.; Grigoriev, M. Unified BRST approach to (partially) massless and massive AdS fields of arbitrary symmetry type. *Nucl. Phys. B* **2011**, *853*, 663–687.
74. Campoleoni, A.; Francia, D. Maxwell-like Lagrangians for higher spins. *J. High Energy Phys.* **2013**, *2013*, 168.
75. Metsaev, R.R. Cubic interaction vertices of massive and massless higher spin fields. *Nucl. Phys. B* **2006**, *759*, 147–201.
76. Metsaev, R.R. Cubic interaction vertices for fermionic and bosonic arbitrary spin fields. *Nucl. Phys. B* **2012**, *859*, 13–69.
77. Alkalaev, K. FV-type action for AdS_5 mixed-symmetry fields. *J. High Energy Phys.* **2011**, *2011*, 31.
78. Boulanger, N.; Skvortsov, E.D.; Zinoviev, Y.M. Gravitational cubic interactions for a simple mixed-symmetry gauge field in AdS and flat backgrounds. *J. Phys. A* **2011**, *44*, 415403.
79. Alkalaev, K. Mixed-symmetry tensor conserved currents and AdS/CFT correspondence. *J. Phys. A* **2013**, *46*, 214007.
80. Alkalaev, K. Massless hook field in AdS(d+1) from the holographic perspective. *J. High Energy Phys.* **2013**, *2013*, 18.
81. Chekmenev, A.; Grigoriev, M. Boundary values of mixed-symmetry massless fields in AdS space. *Nucl. Phys. B* **2016**, *913*, 769–791.
82. Bekaert, X.; Grigoriev, M. Higher order singletons, partially massless fields and their boundary values in the ambient approach. *Nucl. Phys. B* **2013**, *876*, 667–714.
83. Brust, C.; Hinterbichler, K. Partially Massless Higher-Spin Theory. *J. High Energy Phys.* **2017**, *2017*, 086.
84. Alkalaev, K.B.; Grigoriev, M.; Skvortsov, E.D. Uniformizing higher-spin equations. *J. Phys. A* **2015**, *48*, 015401.
85. Brust, C.; Hinterbichler, K. Partially massless higher-spin theory II: One-loop effective actions. *J. High Energy Phys.* **2017**, *2017*, 126.
86. Brust, C.; Hinterbichler, K. Free $\Box^k$ scalar conformal field theory. *J. High Energy Phys.* **2017**, *2017*, 66.
87. Eastwood, M.; Leistner, T. Higher symmetries of the square of the Laplacian. In *Symmetries and Overdetermined Systems of Partial Differential Equations*; Springer: Berlin, Germany, 2008; pp. 319–338.
88. Eastwood, M.G. Higher symmetries of the Laplacian. *Ann. Math.* **2005**, *161*, 1645–1665.
89. Gover, A.R.; Silhan, J. Higher symmetries of the conformal powers of the Laplacian on conformally flat manifolds. *J. Math. Phys.* **2012**, *53*, 032301.
90. Michel, J.-P. Higher symmetries of the Laplacian via quantization. *Annales de L'institut Fourier* **2014**, *64*, 1581–1609.

91. Joung, E.; Mkrtchyan, K. Partially-massless higher-spin algebras and their finite-dimensional truncations. *J. High Energy Phys.* **2016**, *2016*, 3.
92. Deser, S.; Nepomechie, R.I. Gauge Invariance Versus Masslessness in De Sitter Space. *Ann. Phys.* **1984**, *154*, 396–420.
93. Deser, S.; Nepomechie, R.I. Anomalous Propagation of Gauge Fields in Conformally Flat Spaces. *Phys. Lett. B* **1983**, *132*, 321–324.
94. Higuchi, A. Symmetric Tensor Spherical Harmonics on the N Sphere and Their Application to the De Sitter Group SO(N,1). *J. Math. Phys.* **1987**, *28*, 1553–1566.
95. Deser, S.; Waldron, A. Partial masslessness of higher spins in (A)dS. *Nucl. Phys. B* **2001**, *607*, 577–604.
96. Skvortsov, E.D.; Vasiliev, M.A. Geometric formulation for partially massless fields. *Nucl. Phys. B* **2006**, *756*, 117–147.
97. Basile, T.; Bekaert, X.; Boulanger, N. Mixed-symmetry fields in de Sitter space: A group theoretical glance. *J. High Energy Phys.* **2017**, *2017*, 81.
98. Gwak, S.; Kim, J.; Rey, S.-J. Massless and Massive Higher Spins from Anti-de Sitter Space Waveguide. *J. High Energy Phys.* **2016**, *2016*, 24.
99. Gwak, S.; Kim, J.; Rey, S.-J. Higgs Mechanism and Holography of Partially Massless Higher Spin Fields. In Proceedings of the International Workshop on Higher Spin Gauge Theories, Singapore, 4–6 November 2015; pp. 317–352.
100. Basile, T.; Bekaert, X.; Boulanger, N. Flato-Fronsdal theorem for higher-order singletons. *J. High Energy Phys.* **2014**, *2014*, 131.
101. Beccaria, M.; Bekaert, X.; Tseytlin, A.A. Partition function of free conformal higher spin theory. *J. High Energy Phys.* **2014**, *2014*, 113.
102. Bekaert, X. Singletons and their maximal symmetry algebras. In Proceedings of the 6th Modern Mathematical Physics Meeting: Summer School and Conference on Modern Mathematical Physics, Belgrade, Serbia, 14–23 September 2010; pp. 71–89.
103. Fernando, S.; Günaydin, M. Massless conformal fields, AdS_{d+1}/CFT_d higher spin algebras and their deformations. *Nucl. Phys. B* **2016**, *904*, 494–526.
104. Laoues, M. Some Properties of Massless Particles in Arbitrary Dimensions. *Rev. Math. Phys.* **1998**, *10*, 1079–1109.
105. Dirac, P.A.M. A Remarkable representation of the 3 + 2 de Sitter group. *J. Math. Phys.* **1963**, *4*, 901–909.
106. Ferrara, S.; Fronsdal, C. Conformal fields in higher dimensions. In *Recent Developments in Theoretical and Experimental General Relativity, Gravitation and Relativistic Field Theories, Proceedings of the 9th Marcel Grossmann Meeting, MG'9, Rome, Italy, 2–8 July 2000*; World Scientific: Singapore, 2000; pp. 508–527.
107. Enright, T.; Howe, R.; Wallach, N. A classification of unitary highest weight modules. In *Representation Theory of Reductive Groups*; Springer: Berlin, Germany, 1983; pp. 97–143.
108. Shaynkman, O.V.; Tipunin, I.Y.; Vasiliev, M.A. Unfolded form of conformal equations in M dimensions and o (M + 2) modules. *Rev. Math. Phys.* **2006**, *18*, 823–886.
109. Ehrman, J.B. On the unitary irreducible representations of the universal covering group of the 3 + 2 deSitter group. In *Mathematical Proceedings of the Cambridge Philosophical Society*; Cambridge University Press: Cambridge, UK, 1957; Volume 53, pp. 290–303.
110. Metsaev, R.R. All conformal invariant representations of d-dimensional anti-de Sitter group. *Mod. Phys. Lett. A* **1995**, *10*, 1719–1731.
111. Barnich, G.; Bekaert, X.; Grigoriev, M. Notes on conformal invariance of gauge fields. *J. Phys. A* **2015**, *48*, 505402.
112. Brink, L.; Metsaev, R.R.; Vasiliev, M.A. How massless are massless fields in AdS(d). *Nucl. Phys. B* **2000**, *586*, 183–205.
113. Artsukevich, A.Y.; Vasiliev, M.A. Dimensional Degression in AdS(d). *Phys. Rev. D* **2009**, *79*, 045007.
114. Dolan, L.; Nappi, C.R.; Witten, E. Conformal operators for partially massless states. *J. High Energy Phys.* **2001**, *2001*, 16.
115. Bae, J.-B.; Joung, E.; Lal, S. A note on vectorial AdS_5/CFT_4 duality for spin-j boundary theory. *J. High Energy Phys.* **2016**, *2016*, 77.
116. Vasiliev, M.A. Bosonic conformal higher-spin fields of any symmetry. *Nucl. Phys. B* **2010**, *829*, 176–224.

117. Bekaert, X.; Boulanger, N. The Unitary representations of the Poincare group in any spacetime dimension. In Proceedings of the 2nd Modave Summer School in Theoretical Physics, Modave, Belgium, 6–12 August 2006.
118. Fulton, W.; Harris, J. *Representation Theory: A First Course;* Graduate Texts in Mathematics; Springer: New York, NY, USA, 1991.
119. Fuchs, J.; Schweigert, C. *Symmetries, Lie Algebras and Representations: A Graduate Course for Physicists;* Cambridge Monographs on Mathem; Cambridge University Press: Cambridge, UK, 2003.

universe

MDPI

Article

Higher Spin Extension of Fefferman-Graham Construction

Xavier Bekaert [1,*], Maxim Grigoriev [2] and Evgeny Skvortsov [2,3]

1 Laboratoire de Mathématiques et Physique Théorique Unité Mixte de Recherche 7350 du CNRS Fédération de Recherche 2964 Denis Poisson Université François Rabelais, Parc de Grandmont, Tours 37200, France

2 Lebedev Institute of Physics, Leninsky ave. 53, Moscow 119991, Russia; grig@lpi.ru (M.G.); evgeny.skvortsov@aei.mpg.de (E.S.)

3 Albert Einstein Institute, Am Mühlenberg 1, D-14476 Potsdam-Golm, Germany

* Correspondence: Xavier.Bekaert@lmpt.univ-tours.fr

Received: 31 October 2017; Accepted: 12 January 2018 ; Published: 29 January 2018

Abstract: Fefferman-Graham ambient construction can be formulated as $\mathfrak{sp}(2)$-algebra relations on three Hamiltonian constraint functions on ambient space. This formulation admits a simple extension that leads to higher-spin fields, both conformal gauge fields and usual massless fields on anti-de Sitter spacetime. For the bulk version of the system, we study its possible on-shell version which is formally consistent and reproduces conformal higher-spin fields on the boundary. Interpretation of the proposed on-shell version crucially depends on the choice of the functional class. Although the choice leading to fully interacting higher-spin theory in the bulk is not known, we demonstrate that the system has a vacuum solution describing general higher-spin flat backgrounds. Moreover, we propose a functional class such that the system describes propagation of higher-spin fields over any higher-spin flat background, reproducing all the structures that determine the known nonlinear higher-spin equations.

Keywords: higher-spin theory; AdS/CFT; Fefferman-Graham

1. Introduction

The theory of higher-spin gravity is intimately tied to Anti de Sitter/Conformal Field theory (AdS/CFT) correspondence [1–3] in the exotic regime of strong curvature/weak coupling [4–6]. Historically, the discovery of the deep relationship between AdS massless fields and elementary fields living on the conformal boundary (aka "singletons") by Flato and Fronsdal [7] anticipated some ideas of AdS/CFT correspondence. Most presumably, both subjects (higher spins and holography) still hold important insights worth exploring for their mutual benefit. Particular examples where the connection between both subjects might deserve to be explored further are the ambient construction of Fefferman and Graham and its relation to effective actions and higher-spin gauge fields that we investigate in this work.

First of all, the Fefferman-Graham (FG) ambient construction (The seminal paper is [8] but see e.g., [9] (and refs therein) for a comprehensive overview on the subject.) is one of the most important mathematical pillar sustaining the AdS/CFT correspondence since its very birth. In fact, it was instrumental in the holographic prescription, see e.g., [2,10,11]. In its simplest version it amounts to the flat ambient space approach [12,13] whose underlying idea is to make conformal and/or AdS symmetries manifest: conformal algebra acts on the projective hypercone, while AdS algebra acts on the hyperboloid.

The FG construction can be understood as the curved generalization of the naive ambient space approach. It can be used in several different ways: to study curved conformal geometry with the tools of the Riemannian geometry by extending the conformal structure off the hypercone; to study

Einstein equations with cosmological constant on the hyperboloid in terms of the Ricci flat ambient space geometry. Also, for odd bulk dimensions the conformal gravity equations on the boundary arise as an obstruction to extending the conformal structure to the FG ambient metric and can also be seen as arising from holographic Weyl anomaly [11]. In a similar way, equations of motion/Lagrangians of conformal gauge fields arise as an obstruction/holographic anomaly of the respective AdS gauge fields [11,14–17].

Typically, it is assumed that in the FG construction the ambient metric is subject to Ricci flatness condition. With this condition omitted the FG construction is equivalent to Hamiltonian constraints: three functions on ambient phase space (with Poisson bracket $\{X^A, P_B\} = \delta^A_B$) that obey the $\mathfrak{sp}(2)$ algebra (The fact that there is a triplet of operators that forms $\mathfrak{sp}(2)$ and it is closely related to the FG construction was pointed out in [18] (see also [19–22]). In the context of quantizing particle models this system also appeared in [23]),

$$\{F_+, F_-\} = F_0 \,, \qquad \{F_0, F_\pm\} = \pm 2\, F_\pm \,. \tag{1}$$

and is referred to as "off-shell FG construction" in what follows. These three constraints have a specific form: they contain no more than two powers of momenta. More precisely, one constraint is independent of the momenta, F_-, one is linear, F_0, and one is quadratic, F_+. The constraints can be shown to imply the existence of an ambient metric $G_{AB}(X)$ and a homothety vector field (closely related to what is known as "compensator field" in physics literature) $V^A(X)$ that satisfy the FG conditions, which can be summarized as the relation $G_{AB} = \nabla_A V_B$.

When the off-shell FG construction is realized as $\mathfrak{sp}(2)$ constraints, a higher-spin (HS) extension is naturally obtained by allowing the constraints to be arbitrary functions on the phase space (The $\mathfrak{sp}(2)$ algebra (or its extensions) plays a prominent role in the unfolded approach [24] to HS theory at the nonlinear level [25–27], but within a rather different framework). Since the HS extension is done in the ambient space, one can consider it either in the vicinity of the (curved) hypercone $V^2 = 0$ or in the vicinity of the (curved) hyperboloid $V^2 = -1$, which leads to two interpretations in terms of HS gauge fields on the conformal space (the projectivization of the hypercone) and HS fields on the hyperboloid.

The formulations of HS fields based on the ambient space $\mathfrak{sp}(2)$-system have originally appeared in the literature independently of FG ambient construction. The idea to describe a tower of HS fields on projective hypercone as an $\mathfrak{sp}(2)$-system in ambient space was proposed in [28] and developed further in [20,21]. In the context of HS fields on AdS an $\mathfrak{sp}(2)$-system was proposed in [29] where it was shown to describe off-shell HS fields upon linearizing over the AdS vacuum solution. The same system describes [17,30] the off-shell theory of conformal higher-spin (CHS) fields [31–34] provided one subjects the system to certain extra algebraic gauge symmetries.

CHS fields arise as natural sources for conserved higher-rank tensors, likewise the conformal graviton is a source for the stress-tensor in a CFT. Infinite multiplets of CHS fields are sources for HS currents that are present in free CFT's [35–39]. When CHS fields are viewed as infinitesimal sources, the effective action is simply a generating functional of the free CFT correlators. While it is trivial to couple the free scalar field to an arbitrary gravitational background, a remarkable fact [32] is that one can couple the free scalar field to an arbitrary CHS background as well, i.e., to extend CHS fields beyond infinitesimal sources. This requires an intricate structure of non-abelian symmetries that make the effective action of the free scalar field gauge-invariant on an arbitrary CHS background. It is these gauge symmetries for the CHS sources that the $\mathfrak{sp}(2)$-system describes, thereby encoding information about the effective action on any background. Moreover, for even boundary dimension the effective action has a local log-divergent part: it is the conformal gravity action if the background is gravitational and the action of CHS gravity if a HS background is turned on [31–34].

On the hyperboloid, $V^2 = -1$, the same $\mathfrak{sp}(2)$-system describes off-shell nonlinear bulk HS gauge fields. More precisely, the $\mathfrak{sp}(2)$-system linearized over the AdS background can be put on-shell, giving the ambient space description of the Fronsdal fields [40]. However, it is not clear how to extend this beyond the free approximation: a natural suggestion to put the system on-shell at higher orders is to introduce extra gauge symmetry factoring out the ideal generated by the $\mathfrak{sp}(2)$-fields themselves.

This extra gauge symmetry is precisely the one needed to describe CHS fields on the boundary and can also be seen as a natural gauge symmetry of the constrained system with constraints F_i, which is related to a redefinition of the constraints (A somewhat similar factorization procedure is employed in the Vasiliev system [27] in general space-time dimension. Precisely this gauge symmetry and its constrained system interpretation was proposed in [41]).

On general grounds, putting the system on-shell by gauging away off-shell modes crucially depends on the choice of the functional class. For instance, with naive but natural choice the procedure yields an empty system. When formulated in such terms the problem seems to be closely related to the issue of locality in field theory, more specifically in higher-spin theories (The degree of non-locality was quantified in [42–44] by reconstructing the quartic vertex, which revealed that the vertex is highly non-local. It should always be possible to manufacture interactions in AdS that would give the expected CFT correlation functions [42,45–48], but it is unclear how to fix such interactions in the bulk without having to invoke the AdS/CFT argument, which is due to a high degree of non-locality [43,44], the problem being similar to that in flat spacetime). Indeed, the choice of a functional class controls the derivative expansion of interactions, which is always strongly-coupled in higher-spin theories due to the dimensionless coupling constant and unbounded number of derivatives starting from the quartic order, see e.g., discussion in [49].

Instrumental in investigating various properties of gauge theories (in particular the $\mathfrak{sp}(2)$-system we are interested in) is the parent approach [50–53]. One of the advantages of the parent approach in the context of the AdS/CFT correspondence is that one can jump directly between bulk and boundary simply by changing the compensator field from timelike ($V^2 = -1$) to null ($V^2 = 0$) [16,17]. The parent equations of motion then rearrange themselves in accordance with the representation structure of the AdS/conformal algebra.

In this work, with the help of the parent approach we demonstrate that the $\mathfrak{sp}(2)$-system can be pushed one step further: the system has a class of exact solutions—higher-spin flat backgrounds. We also show that one can put the HS fields on-shell over a HS-flat background, which is not necessarily AdS and thereby probes interactions. More specifically, we propose a suitable functional class in the auxiliary space of the parent formulation, which allows one to put the system on-shell. The resulting equations have the correct form of a flatness condition deformed by a two-cocycle of the HS algebra [54]. These are the data that completely determine the Vasiliev equations [24,27].

It has been known for decades that HS fields are hard to make propagate consistently on anything but constant curvature backgrounds [55]. Nevertheless, we observe that special backgrounds, those given by flat connections of HS algebras, allow for propagation of HS fields. The flatness condition is hard to interpret from the vantage point of Fronsdal fields [56]. It is worth mentioning that HS-flat backgrounds were shown to describe rich physics of HS black holes in three-dimensions, see e.g., [57] and references therein/thereon.

The outline of the paper is as follows. In Section 2, we discuss the off-shell FG construction and show that it is equivalent to the $\mathfrak{sp}(2)$ constraints (with some details delegated to Appendix A.4). In Section 3, we review and discuss the on-shell FG construction. In Section 4, a HS extension is proposed and it is discussed how it is related to the known HS systems. In Section 5, we show that the HS extension can describe fluctuations of massless HS fields over any HS-flat background. Conclusions and discussion are in Section 6.

2. Off-Shell Fefferman-Graham Theory

By "off-shell gravity" in $d+1$ dimensions, we understand a gauge theory whose fields are the components of the metric tensor $g_{\mu\nu}$ ($\mu, \nu = 0, 1, \ldots, d$), which is assumed invertible and only (i.e., it is "off-shell" in the sense that the fields are *not* subject to field equations.) subject to the usual gauge transformations (infinitesimal diffeomorphisms):

$$\delta_\xi g_{\mu\nu} = \mathcal{L}_\xi g_{\mu\nu} := \xi^\rho \partial_\rho g_{\mu\nu} + \partial_\mu \xi^\rho g_{\rho\nu} + \partial_\nu \xi^\rho g_{\mu\rho}\,, \tag{2}$$

where ξ^μ are the components of a gauge parameter which is assumed unconstrained.

Similarly one defines "off-shell conformal gravity" in d dimensions via an invertible metric tensor g_{ab} $(a, b = 0, 1, \ldots, d-1)$ by introducing the extra gauge transformations (infinitesimal Weyl transformations):

$$\delta_{\xi,\omega} g_{ab} = \mathcal{L}_\xi g_{ab} + 2\,\omega\, g_{ab}\,, \tag{3}$$

where ξ^a and ω are, respectively, parameters of the infinitesimal diffeomorphisms and Weyl transformations.

The "off-shell FG theory" defined in $(d+2)$-dimensional ambient space (see Appendix A for historical overview and technical details) parameterized by the coordinates X^A requires two ingredients: a nondegenerate ambient metric G_{AB} $(A, B = 0, 0', 1, \ldots, d)$ and a nowhere-vanishing homothety vector field V^A that are assumed to obey

$$\mathcal{L}_V G_{AB} = 2\, G_{AB}\,, \qquad\qquad \partial_A V_B - \partial_B V_A = 0\,, \tag{4}$$

from which it follows that $V_A = \partial_A(V^2/2)$ and $G_{AB} = \nabla_A V_B$. The fields G_{AB} and V_A are subject to usual diffeomorphisms as gauge transformations.

Equivalently, as is shown in Appendix A.4, the same relations can be defined in terms of three totally symmetric polyvector fields of rank 2, 1 and 0: $G^{AB}(X)$, $V^A(X)$, and $F(X)$ defined on the $(d+2)$-dimensional ambient space. The equations of motion can be reformulated as the $\mathfrak{sp}(2)$ algebra relations:

$$\{F_+, F_-\} = F_0\,, \qquad \{F_0, F_\pm\} = \pm 2\, F_\pm\,, \tag{5}$$

where

$$F_+(X,P) = \frac{1}{2}\, G^{AB}(X)\, P_A P_B\,, \qquad F_0(X,P) = V^A(X)\, P_A\,, \qquad F_-(X,P) = F(X)\,, \tag{6}$$

i.e., we encoded the tensor fields in the three generating functions F_i $(i = +, -, 0)$ using extra variables P_A which are ambient momenta, conjugate to the coordinates X^A. The Poisson bracket $\{\,,\}$ is defined by

$$\left\{X^B, P_A\right\} = \delta^B_A\,. \tag{7}$$

The gauge symmetries in the FG ambient theory are given by

$$\delta_\xi F_i = \{\xi, F_i\}\,, \tag{8}$$

where $\xi = \xi^A(X)\, P_A$ is the generating function of the gauge parameters. It is clear that these gauge transformations are nothing but infinitesimal diffeomorphisms of the ambient space. At the same time, these are particular canonical transformations of the phase space X, P.

As explained in Appendix A.4, when G^{AB} is nondegenerate (and can thus be seen as the inverse of a metric G_{AB}), it follows from the three Equation (5) on the three functions (6) that

$$F(X) = -\frac{1}{2}\, G_{AB}(X) V^A(X) V^B(X) \tag{9}$$

and

$$\nabla_A V^B = \delta^B_A\,, \tag{10}$$

where ∇ is the Levi-Civita connection of the ambient metric $G_{AB}(X)$.

The off-shell FG theory in $d+2$ dimensions is equivalent to off-shell gravity in $d+1$ dimensions provided one disregards the direction along $V^A(x)$ as a genuine spacetime dimension. Other way around, any metric $g_{\mu\nu}$ in $d+1$ dimensions can be lifted to an ambient metric G_{AB} and homothety vector field V^A defined on the $(d+2)$-dimensional ambient space such that the original space is a "curved hyperboloid" determined by $G_{AB} V^A V^B = -1$, while the original metric is a pullback of G_{AB} to this level surface.

3. On-Shell Fefferman-Graham Theory

We begin with the $\mathfrak{sp}(2)$ system, i.e., (5) and (8):

$$\{F_i\,, F_j\} = C^k_{ij} F_k\,, \qquad\qquad \delta_\epsilon F_i = \{\epsilon, F_i\}\ , \tag{11}$$

which we write in a compact way using $\mathfrak{sp}(2)$ structure constants C^k_{ij}. Following [8], let us impose an extra condition on the ambient metric entering $F_\pm$:

$$R_{AB} = 0\,, \tag{12}$$

i.e., one requires the ambient metric G_{AB} to be Ricci flat. The system (11)–(12) defines "on-shell FG theory". More precisely, the ambient system should be understood within a certain expansion scheme, called the FG expansion.

There are two interpretations of the on-shell FG theory in $d+2$ dimensions:

- This system is equivalent to on-shell gravity in $d+1$ dimensions with a nonvanishing cosmological constant (in other words the metric $g_{\mu\nu}$ is Einstein). The spacetime manifold can be identified with the curved hyperboloid $V^2 = -1$;
- This system describes conformal gravity in d dimensions. For d odd it is off-shell, while for d even it is on-shell, the field equations resulting from the conformal anomaly (For d even, in the original FG approach the Ricci flatness was imposed only up to a certain power of the defining function so that the conformal gravity was always off-shell. Another point of view is to require Ricci flatness at all orders which results in conformal gravity equations. Note that in this case the system also describes subleading solutions). The spacetime manifold can be identified with the projectivization of the curved hypercone $V^2 = 0$;

There is another way to describe off-shell conformal gravity by introducing in place of (12) the following gauge equivalence (A version of this description was proposed in [17,30]).

$$G_{AB} \sim G_{AB} + \lambda\, G_{AB} + \lambda_{(A} V_{B)} + \lambda_{AB}\, V^2\,, \tag{13}$$

where λ, λ_A and λ_{AB} are gauge parameters. The parameter λ is related to the usual Weyl symmetry while λ_A and λ_{AB} implement the equivalence relation up to components along the homothety vector field and up to terms vanishing on the null-cone $V^2 = 0$ (Note that this formulation (more precisely, its parent version) of the off-shell conformal gravity gives a manifestly $\mathfrak{so}(d,2)$-covariant description of the respective jet-space and BRST complex employed [58,59] in classifying conformal invariants). Both interpretations of the system (11)–(12) as well the system (11) and (13) have simple toy model counterparts in the context of the scalar field in ambient space, described in Appendix C.

The condition (12) of Ricci flatness on the ambient metric in the FG construction can be understood as a gauge-fixing condition for the arbitrariness (13) in the extension of the metric from the projective null cone to the whole ambient space. This gives a field-theoretic explanation of the relation between the two equivalent ambient descriptions of off-shell conformal gravity.

One can even try to exploit the relation further and to interpret the ambient system (11) and (13) as defining a field theory on the curved hyperboloid $V^2 = -1$. We postpone detailed discussion of this approach till Section 5.2 and only mention that there is a simple example illustrating this idea. Consider the following ambient system:

$$(V^A \partial_A + \Delta)\Phi = 0\,, \qquad \Phi \sim \Phi + V^2 \lambda\,, \tag{14}$$

where Φ is a scalar field on ambient space. Interpret this system as defining a scalar field on the curved hyperboloid $V^2 = -1$ rather than on the projective hypercone $V^2 = 0$. In so doing one can try to assume Φ harmonic by adding terms proportional to V^2, i.e., try to pick a representative which obeys

$\nabla^2\Phi = 0$. Of course, this is a subtle procedure and for it to work properly one needs to be careful with functional issues.

To complete the discussion of the gauge equivalence (13) let us note that (13) can be compactly written in terms of the generating functions F_i :

$$F_+ \sim F_+ + \lambda F_+ + \lambda^A P_A \, F_0 + \frac{1}{2} \lambda^{AB} P_A P_B \, F_- \,. \tag{15}$$

In this form it is clear that this equivalence corresponds to the usual equivalence of constrained systems related to an infinitesimal redefinition of the constraints. This system (more precisely, its BRST extension) was considered in [30,41] in the context of AdS HS gauge theory.

4. Higher-Spin Extension of Fefferman-Graham Theory

The idea to describe HS fields by allowing all powers of momenta in the $\mathfrak{sp}(2)$ constraints was at the core of Bars' proposal in [28]. Accordingly, a bold guess for a HS extension of the FG construction is to remove the restriction on F_i to contain no more than two powers of ambient momentum P_A:

$$\begin{aligned} F_+(X,P) &= \Phi(X) + \Phi^A(X)P_A + \tfrac{1}{2} G^{AB}(X)P_A P_B + \Phi^{ABC}(X)P_A P_B P_C + \dots\,, \\ F_0(X,P) &= V(X) + V^A(X)P_A + \tfrac{1}{2} V^{AB}(X)P_A P_B + \dots\,, \\ F_-(X,P) &= \widetilde{G}(X) + \widetilde{G}^A(X)P_A + \dots\,, \end{aligned} \tag{16}$$

where the dots denote possible terms of higher order in the momenta. Moreover, one should require $G^{AB}(X)$ and $V^A(X)$ to remain, respectively, nondegenerate and nowhere vanishing. This requirement, or the choice of a particular vacuum, will break the apparent democracy between the three constraints $F_i(X,P)$ even if one allows for arbitrary dependence on the momentum for all the three constraints.

We assume that the equations of motion remain the same, which makes the system consistent with an arbitrary gravitational background (i.e., G_{AB} and V^C):

$$\{F_i, F_j\} = C_{ij}^k F_k \,. \tag{17}$$

Here C_{ij}^k are $\mathfrak{sp}(2)$ structure constants. The gauge symmetries are

$$\delta_\epsilon F_i = \{\epsilon, F_i\} \,. \tag{18}$$

In what follows we analyze this system and relate it to other formulations of HS gauge fields. As we are going to show, the same system with the Poisson bracket replaced with the Weyl star-commutator is more appropriate in the context of HS fields. Although the above system has something to do with both conformal HS fields in d dimensions and HS fields in $d+1$ dimensions, here we are mostly interested in the $(d+1)$-dimensional interpretation.

The full system (17) and (18) with F_i as in (16) is known in the literature in various versions (see e.g., [28,41]). In the context of HS fields in $d+1$ dimensions, precisely this system was proposed in [29] where it was proved to describe off-shell HS fields upon linearizing over the AdS background solution. Note also that a version of this system with two generators fixed and F_+ involving higher powers in P was also suggested in [20] to describe HS fields in d dimensions and, moreover, the $\mathfrak{sp}(2)$ system was cast into an action principle in [21].

The above construction can also be phrased in terms of tractor fields which can be identified with certain tensor fields in ambient space restricted to the hypercone [60–62]. In particular, this language was employed to described HS fields in [63–67].

4.1. Off-Shell Higher-Spin Fields on Gravitational Backgrounds

Suppose we are given a triplet of ambient functions $F^0_+(X,P)$, $F^0_0(X,P)$, $F^0_-(X,P)$ that are of degree $2,1,0$ in P_A, respectively, and whose Poisson brackets satisfy the algebra $\mathfrak{sp}(2)$. We assume that $F^0_+(X,P)$ defines a nondegenerate metric and $F^0_0(X,P)$ defines a nowhere-vanishing vector field. Then, as was mentioned before, it follows that they are of the form (6) with the scalar field $F(X)$ given by (9) and the vector field $V^A(X)$ satisfying $\nabla_A V^B = \delta^A_B$.

As was shown by Fefferman and Graham, at least locally any metric $g_{\mu\nu}$ in $d+1$ dimensions can be lifted to an ambient metric G_{AB} defined on the $(d+2)$-dimensional ambient space such that the original spacetime is a "curved hyperboloid" determined by $G_{AB}V^AV^B = -1$, while the original metric $g_{\mu\nu}$ is a pullback of G_{AB} to this surface.

We are going to interpret $F_i = F^0_i$ as a background solution of the full HS system, where only gravitational fields are nonvanishing. In order to see that the full system indeed describes HS fields, let us consider its linearization around $F_i = F^0_i$. The linearized equations and gauge symmetries read:

$$\left\{f_i, F^0_j\right\} + \left\{F^0_i, f_j\right\} = C^k_{ij} f_k\,, \qquad \delta_\epsilon f_i = \left\{\epsilon, F^0_i\right\}, \tag{19}$$

where f_i is a perturbation. In more detail, for the gauge transformations of f_0 and f_- one gets

$$\delta_\epsilon f_0 = \left\{\epsilon, F^0_0\right\} = (V\cdot D - P\cdot\partial_P)\epsilon\,, \qquad \delta_\epsilon f_- = \left\{\epsilon, F^0_-\right\} = (V\cdot\partial_P)\epsilon\,, \tag{20}$$

where D_M denotes the covariant derivative acting on generating functions: $D_M = \frac{\partial}{\partial X^M} + \Gamma^K_{MN} P_K \frac{\partial}{\partial P_N}$ (see also Appendix B). Since V^A is nowhere vanishing, this implies that the gauge $f_- = f_0 = 0$ is reachable. Indeed, by picking a suitable coordinate system one can always assume that the homothety vector field reads $V^A\frac{\partial}{\partial X^A} = \frac{\partial}{\partial\rho}$, where ρ is one of the coordinates. In such a coordinate system it is clear that using suitable ϵ one can set both f_- and f_0 to zero.

Although the above argument applies to the linearized system, it extends in a usual way to the nonlinear level provided one restricts oneself to solutions which are "sufficiently close" to F^0_i. For such solutions, the two remaining $\mathfrak{sp}(2)$ relations involving the perturbation f_+ (not necessarily small) imply:

$$(P\cdot\partial_P - V\cdot D - 2)f_+ = 0\,, \qquad (V\cdot\partial_P)f_+ = 0\,, \tag{21}$$

where $F_+ = F^0_+ + f_+$. While the first equation uniquely determines f_+ in terms of its value on the level surface $V^2 =$ constant, the second one implies that f_+ does not depend on the components of P along V. Finally, such f_+ are in one-to-one correspondence with totally symmetric tensor fields on the level surface $V^2 =$ constant.

Similarly, in the gauge $F_0 = F^0_0$ and $F_- = F^0_-$ the residual gauge symmetries read as

$$\delta_\epsilon f_+ = \{\epsilon, F_+\} = (P\cdot D)\epsilon + \{\epsilon, f_+\}\,, \quad \text{with} \quad (V\cdot\partial_P)\epsilon = 0\,, \quad (P\cdot\partial_P - V\cdot D)\epsilon = 0\,.$$

The global reducibility parameters ϵ_0 for the linearized symmetries in (19) are described by the $\mathfrak{sp}(2)$ centralizer equations $\{\epsilon_0, F^0_i\} = 0$ which read more explicitly:

$$(P\cdot D)\epsilon_0 = (P\cdot\partial_P - V\cdot D)\epsilon_0 = (V\cdot\partial_P)\epsilon_0 = 0\,. \tag{22}$$

These global symmetries (= global reducibility parameters) define an "off-shell HS algebra", by which we simply mean the $\mathfrak{sp}(2)$ centralizer (22) above, canonically equipped with a Lie algebra structure via the Poisson bracket. More invariantly, the off-shell HS algebra arises here in a usual way (see e.g., [68] for more details) as an algebra of gauge transformations preserving a given vacuum solution. Its Lie bracket comes from the commutator of the respective gauge transformations. In particular, for the flat vacuum $V^A = X^A$ we get a version of the standard off-shell HS algebra.

We comment more on various off-shell HS algebras below when the Poisson bracket is replaced by the star-product. It is worth stressing that the off-shell HS algebra is defined on any gravitational background, but it is generically trivial. In fact, the system of Equation (22) describes totally symmetric Killing tensors in $d+1$ dimensions, which may not admit nontrivial solutions on a generic background.

To summarize, the natural HS extension of the FG off-shell system provides an elegant description of totally symmetric tensor gauge fields in $d+1$-dimension subject to nonlinear gauge symmetries in an arbitrary gravitational background (This does not contradict the usual no-go theorems (such as [55]) on the propagation of HS gauge fields in generic gravitational backgrounds since here the tensor gauge fields are *off-shell*. For instance, at linearized level the covariantization of Fronsdal gauge transformations is perfectly well defined. It is only its compatibility with the field equations which is problematic).

4.2. Poisson Bracket vs. Star-Product

If we are interested in the *on-shell* fields, then the construction of the previous section is not entirely satisfactory. Moreover, even at the off-shell level the corresponding algebra of global symmetries coincides with the familiar off-shell HS algebra as a linear space, but is a Lie algebra with respect to the Poisson bracket, rather than the Weyl commutator as it should. This suggests that the construction has to be modified by replacing the Poisson bracket with the Weyl $\star$-commutator determined by

$$[X^A, P_B]_\star = \delta^A_B \,. \tag{23}$$

Another reason for considering the "quantum" version has to do with the interpretation of the off-shell nonlinear system as describing background conformal HS fields in d dimensions to which a scalar field can consistently couple (for more details, see e.g., [17,32–34,69,70]). Note that from this perspective the Poisson bracket version naturally corresponds to background fields for the point particle. This in turn is a mechanical model whose wave function is the above scalar field. The $\star$-product is crucial for the scalar field to couple to arbitrary HS background fields. It arises directly as a quantization of the phase space $\{x^a, p_b\} = \delta^a_b$ where HS background fields correspond to arbitrary functions $f(x,p)$. The $\mathfrak{sp}(2)$-system provides an ambient space extension of this construction [17,41,71].

Therefore, we pass to the $\star$-commutator instead of the Poisson bracket and the corresponding version of the basic system (17) reads as

$$[F_i, F_j]_\star = C^k_{ij} F_k \,, \qquad \delta_\epsilon F_i = [\epsilon, F_i]_\star \,. \tag{24}$$

Note that if one restricts to the spin-two version of this $\mathfrak{sp}(2)$-system, then the same solution (6), (9) and (10) that satisfies the Poisson bracket version (11) also solves (24). This system of operators plays an important role in the ambient description of scalar fields [18] (see also [72,73]), in particular for the singleton (see Appendix C).

Let us now restrict ourselves to the flat vacuum solution: $G^{AB} = \eta^{AB}$, $V^A = X^A$. It is easy to see that in this case F^0_i solves the above equations and hence gives a vacuum solution. Linearizing the equations and the gauge symmetries around F^0_i one gets

$$[F^0_i, f_j]_\star + [f_i, F^0_j]_\star = C^k_{ij} f_k \qquad\qquad \delta_\epsilon f_i = [\epsilon, F^0_i]_\star \,. \tag{25}$$

It follows again that f_0 and f_- can be gauged away, resulting in the linearized system (Note that had we linearized the system around general gravitational background the linearized gauge transformations for a spin s-field would in general not only involve contributions from parameters of spin $s-1$ but also those with lower spin. This mixing can be traced back to nonlinear in X terms in F^0_+ involved in the star-commutator with the parameter. It is similar to the analogous mixing observed [69] in the case of conformal HS fields on the boundary and is consistent with the fact that CHS fields are boundary values of the bulk ones. We are grateful to A. Tseytlin for discussion of this point.)

$$(X\cdot\partial_P)f_+=0\,,\qquad (P\cdot\partial_P-X\cdot\partial_X-2)f_+=0\,,$$
$$\delta_\epsilon f_+=[\epsilon,F_+^0]_\star=(P\cdot\partial_X)\epsilon\,,\qquad (X\cdot\partial_P)\epsilon=0\,,\quad (P\cdot\partial_P-X\cdot\partial_X)\epsilon=0\,. \tag{26}$$

To see the relation with Fronsdal fields, let us recall their ambient space formulation. In terms of the generating function $\Phi(X,P)=\sum_s\frac{1}{s!}\,\Phi^{A_1\ldots A_s}(X)\,P_{A_1}\ldots P_{A_s}$, the equations of motion and gauge symmetries read as

$$(\partial_X\cdot\partial_X)\Phi=(\partial_P\cdot\partial_X)\Phi=(\partial_P\cdot\partial_P)\Phi=0\,,\qquad \delta_\epsilon\Phi=(P\cdot\partial_X)\epsilon\,, \tag{27}$$

$$(X\cdot\partial_X-P\cdot\partial_P+2)\Phi=(X\cdot\partial_P)\Phi=0\,, \tag{28}$$

which is equivalent to

$$(\nabla^2-m_s^2)\phi_{\mu_1\ldots\mu_s}=\nabla^\mu\phi_{\mu\mu_2\ldots\mu_s}=\phi^\mu{}_{\mu\mu_3\ldots\mu_s}=0\,,\qquad \delta_\epsilon\phi_{\mu_1\ldots\mu_s}=\nabla_{(\mu_1}\epsilon_{\mu_2\ldots\mu_s)}\,, \tag{29}$$

in terms of the fields defined on the hyperboloid, where m_s is the mass of spin-s Fronsdal field on $(A)dS_{d+1}$.

It is clear from (27) that, in order to put the linearized off-shell system (26) on-shell, one has to impose the equation of motion $(\partial_X\cdot\partial_X)f_+=0$ (the remaining divergence and trace constraints arise automatically as consistency conditions). The above analysis of the system (24) and its relation to Fronsdal fields has been originally performed in [29] using the parent formulation technique.

To draw an analogy with the FG description of gravity note that $(\partial_X\cdot\partial_X)f_+=0$ is a linearized HS analogue of the Ricci flatness $R_{AB}=0$ condition. To find a nonlinear HS version of the Ricci flatness remains a tantalizing open problem in the metric-like formulation. We pursue a different approach in the next section.

5. Towards on-Shell Higher-Spin Theory

The $\mathfrak{sp}(2)$-system captures off-shell backgrounds, both gravitational and HS ones. In the case of gravity, the $(d+1)$-dimensional Einstein equations with cosmological constant result from Ricci flatness in the $(d+2)$-dimensional ambient space. A natural question is whether it is possible to directly put the off-shell HS system (24) on-shell.

One possible way would be to find nonlinear corrections to the constraints in the first line of (27). However, it is not clear which structures may control such deformation and in any case in this way there is no obvious way to reconstruct the system nonperturbatively.

An alternative is to further exploit the analogy with constrained systems. The $\mathfrak{sp}(2)$ relations imposed on F_i can be interpreted as a condition that F_i are first-class constraints while gauge transformations $\delta_\epsilon F_i=[\epsilon,F_i]_\star$ correspond to canonical transformations. The general first-class condition (= closure of the algebra)

$$[F_i,F_j]_\star=U_{ij}^k\star F_k \tag{30}$$

is preserved by the following gauge transformations

$$\delta F_i=\lambda_i^j\star F_j\,, \tag{31}$$

in addition to $\delta_\epsilon F_i=[\epsilon,F_i]_\star$. The transformations (31) correspond to infinitesimal redefinitions of the constraints (at classical level such symmetries preserve the constraint surface). The $\mathfrak{sp}(2)$ system (24) can be seen as a partial gauge-fixing of the system (30)–(31) (Strictly speaking one needs to prove that such gauge is reachable. Although this is easy to see at the linearized level and hence this is also true for the field configurations that are sufficiently close to the vacuum, a general statement is not known).

To explain what symmetries (31) can be useful for, let us consider the linearization of the system about the flat vacuum solution F_i^0 corresponding to $G^{AB} = \eta^{AB}$ and $V^A = X^A$, i.e., $F_+^0 = \frac{1}{2}P \cdot P$, $F_0^0 = X \cdot P$ and $F_-^0 = -\frac{1}{2}X \cdot X$. The linearization of (31) reads as

$$\delta f_i = \lambda_i^j \star F_j^0 \,. \tag{32}$$

At the formal level, these symmetries can be employed to make f_i satisfy the constraints in the first line of (27), i.e., $(\partial_X \cdot \partial_X) f_i = (\partial_P \cdot \partial_X) f_i = (\partial_P \cdot \partial_P) f_i = 0$. For instance, in the space of polynomials in X, P variables this is exactly the case.

As we have seen f_0 and f_- can be set to zero by the gauge symmetries $\delta f_i = [\epsilon, F_i^0]_\star$ (for $i = 0, -$). As a result, there is only one field left, f_+. It satisfies (27)–(28) and hence describes the Fronsdal system (29). The problem with this formal argument is that the space of polynomials in X, P is not the one where one can set to zero f_0, f_- (since one somehow has to "invert" the operators $X \cdot \partial_P$ and $X \cdot \partial_X - P \cdot \partial_P$). Moreover, it is not the functional space relevant for describing field theory configurations. Other way around, the space (natural from the field theory perspective) of polynomials in P and smooth functions in X defined in the vicinity of $X^2 = -1$ allows one to eliminate f_0, f_- but, in this space, the gauge transformations (31) can be used to set to zero all the fields f_i.

It turns out that one can nevertheless use this system to describe on-shell fields by reformulating the system in parent form (see Section 5.1) and requiring the fields to belong to a special functional class.

A heuristic explanation for why the extended system (30)–(31) is capable of describing on-shell HS gauge fields in the bulk employs boundary analysis. More precisely, the extended system in the vicinity of the hypercone $V^2 = 0$ is known [17,30] to describe off-shell conformal HS gauge fields in d dimensions. In their turn, these off-shell conformal fields are boundary values of the on-shell bulk fields in $d + 1$ dimensions. However, in the ambient approach bulk fields and their boundary values are described by exactly the same ambient system, considered either near $V^2 = -1$ or near $V^2 = 0$ (for more details see [16,17]). This justifies that the above extended system has something to do with on-shell bulk fields provided one considers it near $V^2 = -1$.

To give the above considerations a precise meaning it is useful to reformulate the system using the parent approach which has proved to be instrumental in analyzing boundary values [16,17,74].

5.1. Parent Reformulation

The system (24) can be equivalently reformulated in parent form. The underlying idea is to put the constrained system under consideration in an auxiliary space where genuine space-time coordinates are replaced by formal variables, typically denoted by Y^A, so that the equations of motion and gauge symmetries become purely algebraic. The parent formulation is then constructed as a gauge field theory whose target space is the space of fields on the auxiliary space while the source space is the original space-time. Moreover, in so doing the gauge parameters are promoted to 1-form fields of the parent system while reducibility parameters to p-form fields. In other words, parent formulation is an AKSZ sigma model [75] whose target space is the jet-space BRST complex of the intial theory. The equivalence with the original formulation is maintained by imposing free differential algebra relations and their associated gauge symmetries, on top of the auxiliary space version of the original equations of motion and gauge symmetries. More details and references can be found in [50–53].

Being a sigma model of AKSZ type (or equivalently a free differential algebra with constraints) the parent formulation can be consistently pulled back to any space-time submanifold or even put (at least formally) on a totally different space-time manifold. This property is extremely useful in analyzing the relation between the Hamiltonian and Lagrangian description [76], manifest realization of symmetries [51,71] boundary values of gauge fields [16,77], and, more generally, holographic dualities [17,77].

In the case at hand the parent reformulation is constructed by first extending X^A and P_B with additional variables Y^A (The geometric meaning is that Y^A are coordinates on the fibres of the tangent bundle over the ambient space). Now X^A are the usual space-time coordinate, while Y^A and P_B are auxiliary variables needed to conveniently pack fields into generating functions. Note however, that Weyl $\star$-product is now in the space of Y, P variables so that spacetime coordinates are not explicitly involved. The field content consists of the original constraints $F_i(X, Y, P)$ and a new field, the connection one-form $A = dX^B A_B(X, Y, P)$ associated to the original gauge parameter. The fields F_i, A are interpreted as generating functions for the component fields identified as the respective coefficients in the expansion over Y, P.

Due to the fact that the parent formulation contains an infinite number of fields the specification of the functional class of F_i and A in the Y space is part of the definition of the system. The minimal choice to begin with is that of formal power series in Y. With this choice the parent formulation is equivalent to (24). Let us for definiteness and simplicity also restrict ourselves to polynomials in P, so that we are dealing with polynomials in P with coefficients in formal power series in Y.

The parent form of the $\mathfrak{sp}(2)$ system, the equations of motion we are going to study, read [29]

$$dA - \frac{1}{2}[A, A]_\star = 0, \qquad dF - [A, F]_\star = 0, \qquad [F_i, F_j]_\star - C_{ij}^k F_k = 0, \\ \delta_\epsilon A = d\epsilon - [A, \epsilon]_\star, \qquad \delta_\epsilon F_i = [\epsilon, F_i]_\star, \tag{33}$$

where from now on $[\cdot, \cdot]_\star$ denotes the Weyl $\star$-commutator (in the Y, P-space). The classical limit where the star-product commutator is replaced by the Poisson bracket in the Y, P-space also makes sense but, as we argued above, its interpretation from the effective action point of view is unclear.

The above parent system (33) is background independent and can be considered on $(d+1)$-dimensional manifolds. This can be done by pulling-back the system (33) to the curved hyperboloid described by $V^2 = -1$. The advantage of the parent formulation is that V^A is non-dynamical and can be conveniently manipulated. In particular, in the resulting theory defined on $V^2 = -1$ one can gauge fix V^A such that V^A is constant. Furthermore, in contrast to the X, P-space of the previous sections, the auxiliary Y, P-space is flat and it is easier to impose algebraic conditions in order to put the system on-shell.

There is a parent realization for the anti-de Sitter solution of the HS extension of the FG construction, which reads: [29,41]

$$F_+^0 = \tfrac{1}{2} P \cdot P, \qquad F_0^0 = \tfrac{1}{2}\{(V+Y)^A, P_A\}_\star, \qquad F_-^0 = \tfrac{1}{2}(V+Y)\cdot(V+Y), \tag{34}$$

$$A^0 = \omega_B^A\,(V^B + Y^B) P_A, \tag{35}$$

where V^A is constant, $\{\ ,\ \}_\star$ denotes the $\star$-anticommutator and ω is a standard $\mathfrak{so}(d,2)$ flat connection such that $\nabla_\mu V^B$ has rank $d+1$. Note that there are more general solutions where, rather than (35), A_0 is taken to be any flat connection taking values in the off-shell HS algebra, i.e., $dA_0 - \frac{1}{2}[A_0, A_0]_\star = 0$ and $[F_i^0, A^0]_\star = 0$, so that A^0 is polynomial in Y. Note that although it is easy to check that (34)–(35) is a solution by redefining variables $Y^A + V^A \to Y^A$, this redefinition is not well-defined for formal power series in Y. In particular, the shift by V crucially affects the content of the theory.

The gauge symmetries leaving the vacuum solution (34)–(35) intact,

$$d\epsilon^0 - [A^0, \epsilon^0]_\star = 0, \qquad [F_i^0, \epsilon^0]_\star = 0, \tag{36}$$

are 1:1 with the off-shell HS algebra defined as the algebra of elements $\star$-commuting with all F_i^0's. In contrast to Section 4.1, this off-shell HS algebra is not just a Lie algebra, but is an associative one because the $\star$-product of two $\mathfrak{sp}(2)$-singlets is a singlet again. This off-shell algebra is directly related to the symmetries of the conformal Laplacian described by Eastwood [78], i.e., it has a two-sided ideal that can be quotiented out as to get the on-shell HS algebra.

Let us consider the linearization of the parent system around (34). The linearized fluctuations may be required to be totally traceless in which case one finds an on-shell version. This is analogous to imposing $(\partial_X \cdot \partial_X) f_+ = 0$ in Section 4.2. More precisely, it can be shown [29] that requiring the linearized fluctuations of (33) to be in the kernel of $\partial_Y \cdot \partial_Y$, $\partial_Y \cdot \partial_P$, and $\partial_P \cdot \partial_P$ results in the free Fronsdal equations. The problem is to extend this beyond the linearized approximation. For spin-two, the Ricci flatness provides a nonlinear completion of these equations but its HS analogue is still missing.

5.2. Factorization

We can push the on-shell HS extension of the FG construction a bit further and get equations that describe propagation of HS fields on any HS-flat background. Indeed, there is a formally consistent factorization of the system by the ideal generated by F_i. The factorization is obtained by imposing the extra gauge symmetry:

$$\delta A = \lambda^i \star F_i\,, \qquad \delta F_i = \lambda_i^j \star F_j\,, \tag{37}$$

where λ^i, λ_j^i are gauge parameters, and requiring equations of motion to hold modulo terms proportional to F_i. The full system of equations that is suitable for describing on-shell fields together with gauge transformations is [30] (see also [41] for earlier versions)

$$dA - \frac{1}{2}[A, A]_\star = u^i \star F_i\,, \qquad \delta A = d\epsilon - [A, \epsilon]_\star + \lambda^j \star F_j\,, \tag{38a}$$

$$dF_i - [A, F_i]_\star = u_i^j \star F_j\,, \qquad \delta F_i = [\epsilon, F_i]_\star + \lambda_i^j \star F_j\,, \tag{38b}$$

$$[F_i, F_j]_\star - C_{ij}^k F_k = u_{ij}^k \star F_k\,. \tag{38c}$$

Here u's are non-dynamical fields that transform under ϵ, λ_i^j, λ^j in an obvious way. Note that u_{jk}^i are not unconstrained and have to obey the relations following from the Jacobi identities. This system is a candidate for the on-shell HS-extended FG theory.

Now we are going to study the above system perturbatively over a HS-flat vacuum solution, where $F_i = F_i^0$ as in (34) while A^0 is more general. To this end we introduce an appropriate functional class that allows one to have gauge symmetry (37) without trivializing the solution space and such that the off-shell HS algebra gets reduced to the correct on-shell HS algebra. The functional class $\mathfrak{C}$ is that of polynomials in P with coefficients that are formal power series in Y (It is important to distinguish $\mathfrak{C}$ from the space of formal power series in Y with coefficients that are polynomials in P). Having formal power series in Y is important for being able to gauge away fluctuations of F_-, F_0. We also require that $\mathfrak{C}$ is of finite trace order, i.e., for any $f(Y, P) \in \mathfrak{C}$ there exists an $\ell \in \mathbb{N}$ such that

$$(\partial_Y \cdot \partial_Y)^\ell f = 0\,. \tag{39}$$

Note that the space of functions of finite trace order is a module over polynomials in Y, P, i.e., we can $\star$-multiply $f(Y, P) \in \mathfrak{C}$ by a polynomial $p(Y, P)$ and the result, $f(Y, P) \star p(Y, P)$, is still in $\mathfrak{C}$. It then follows that any function in $\mathfrak{C}$ can be decomposed as

$$f = f_0 + f_1^i \star F_i^0 + f_2^{ij} \star F_i^0 \star F_j^0 + \dots\,, \qquad f_n - \text{totally traceless}\,, \tag{40}$$

such that the number of terms is finite. Having an element f decomposed as above, we define a projector onto the traceless part: (Note that the trace decomposition is defined with respect to $F_-^0 = -\frac{1}{2}(Y+V)\cdot(Y+V)$, $F_0^0 = P \cdot (Y+V)$, and $F_+^0 = \frac{1}{2}P \cdot P$ i.e., this is not the usual decomposition into traceless tensors $f = f_0 + Y \cdot Y f_1 + \ldots$ due to the shift by V and due to the $\star$-product).

$$\Pi f = f_0\,. \tag{41}$$

Now we can linearize the system and put it on-shell. As anticipated above, we take a slightly more general vacuum, where A^0 does not have to be just the flat connection (35) linear in P, describing AdS_{d+1}. The nontrivial part of the vacuum equations reads

$$dA^0 - \frac{1}{2}[A^0, A^0]_\star = u^i \star F_i^0\,, \tag{42a}$$

$$[A^0, F_i^0]_\star = u_i^j \star F_j^0\,. \tag{42b}$$

It follows that A^0 is equivalent to a flat connection of the on-shell HS algebra. Indeed, the on-shell HS algebra of Eastwood [78] can be defined within the present framework as follows: (An oscillator realization of this algebra was given in [27]).

$$\chi \in \text{on-shell HS-algebra}: \qquad [\chi, F_i^0]_\star = 0\,, \qquad \chi \sim \chi + \lambda^i \star F_i^0\,, \tag{43}$$

where χ, λ^i are in $\mathfrak{C}$. The $\mathfrak{sp}(2)$-singlet constraints solved for χ in $\mathfrak{C}$ imply that χ is a polynomial in Y.

Next, for A^0 entering (42a)–(42b), we can use gauge symmetry (37) on the vacuum solution

$$\delta A^0 = \lambda^i \star F_i^0\,, \tag{44}$$

as to gauge away all traces and arrive at $A^0(Y,P)$ satisfying $\partial_Y \cdot \partial_Y A^0 = \partial_Y \cdot \partial_P A^0 = \partial_P \cdot \partial_P A^0 = 0$ (i.e., A^0 is a collection of traceless tensors in Y and P, or equivalently $\Pi A^0 = A^0$). Then the traceless part of (42a)–(42b) implies that A^0 is a flat connection of the on-shell HS algebra:

$$dA^0 = \frac{1}{2}\Pi\big([A^0, A^0]_\star\big)\,. \tag{45}$$

The gauge symmetries preserving the vacuum solution are determined by

$$d\epsilon^0 = [A^0, \epsilon^0]_\star + \lambda^i \star F_i^0\,, \qquad [F_i^0, \epsilon^0]_\star + \lambda_i^j \star F_j^0 = 0\,. \tag{46}$$

We can again decompose ϵ^0 into the trace part that is proportional to F_i^0 and the traceless part. The trace part unambiguously fixes the λ's, while the traceless part is covariantly constant with respect to A^0. Therefore, the global symmetry parameters are parameterized by the on-shell HS algebra, as it should be.

Now we can study fluctuations over A^0 and F_i^0 and determine the general structure of the equations. Let us expand

$$A = A^0 + a\,, \qquad F_i = F_i^0 + f_i\,, \tag{47}$$

where a and f_i are assumed to belong to the functional class $\mathfrak{C}$. The linearized equations read

$$da - [A^0, a]_\star = u^i \star F_i^0\,, \qquad \delta a = d\epsilon - [A^0, \epsilon]_\star + \lambda^i \star F_i^0\,, \tag{48a}$$

$$df_i - [A^0, f_i]_\star - [a, F_i^0]_\star = u_i^j \star F_j^0\,, \qquad \delta f_i = [\epsilon, F_i^0]_\star + \lambda_i^j \star F_j^0\,, \tag{48b}$$

$$[F_i^0, f_j]_\star - (i \leftrightarrow j) - C_{ij}^k f_k = u_{ij}^k \star F_k^0\,. \tag{48c}$$

We choose $\Pi a = a$, $\Pi f_i = f_i$ as a legitimate gauge condition. Just like in X, P-space the residual gauge symmetry $\delta f_i = [\epsilon, F_i^0]_\star$ can be employed to gauge away f_0, f_- in Y, P-space. The only subtlety is that now both f_i and ϵ are traceless and to see this one needs extra technique (see [51] for details).

With f_0, f_- set to zero the equations for a, f_+ imply

$$da = \Pi([A^0, a]_\star)\,, \tag{49a}$$
$$df_+ = \Pi([A^0, f_+]_\star) + (P\cdot\partial_Y)a\,, \tag{49b}$$
$$(Y+V)\cdot\frac{\partial}{\partial P}a = (P\cdot\frac{\partial}{\partial P} - (Y+V)\cdot\frac{\partial}{\partial Y})a = 0\,, \tag{49c}$$
$$(Y+V)\cdot\frac{\partial}{\partial P}f_+ = (P\cdot\frac{\partial}{\partial P} - (Y+V)\cdot\frac{\partial}{\partial Y} - 2)f_+ = 0\,. \tag{49d}$$

Note that for A^0 a flat $\mathfrak{so}(d,2)$-connection, the Π projector is not needed and Equation (49) are known [51] to describe Fronsdal fields on AdS space.

The Equation (49) can be reduced even further down to the unfolded form using a suitable generalization of the reduction put forward in [51]. Indeed, the procedure is purely algebraic and allows to eliminate all the components in f_+ which are in the image of the operator $P\cdot\frac{\partial}{\partial Y}$ and all the components in a which are not in the kernel of $P\cdot\frac{\partial}{\partial Y}$. The remaining fields $\bar{f}$ and $\bar{a}$ satisfy (49c) and (49d) and are such that $(P\cdot\partial_Y)\bar{a} = 0$ and $\bar{f}$ parametrizes the quotient $f_+ \sim f_+ + (P\cdot\partial_Y)\epsilon$. The reduced equations have the following structure

$$\begin{aligned} d\bar{a} &= \Pi(\,[A^0, \bar{a}]_\star) + \mu(A^0, A^0, \bar{f})\,,\\ d\bar{f} &= \Pi(\,[A^0, \bar{f}]_\star)\,, \end{aligned} \tag{50}$$

where $\mu(A^0, A^0, \bar{f})$ is a trilinear form.

The structure of Equations (49) and (50) becomes more clear if one reformulates them in terms of certain modules of the on-shell HS algebra. To this end let us consider the following two modules:

$$M^0 = \{f\in\mathfrak{C}\,:\,[F^0_-, f]_\star = [F^0_0, f]_\star = 0\}\,, \tag{51a}$$
$$M^1 = \{f\in\mathfrak{C}\,:\,[F^0_-, f]_\star = [F^0_0, f]_\star - 2f = 0\}\,. \tag{51b}$$

Note that these are precisely the spaces (49c) and (49d) where a and f_+ belong to. It is easy to see that for any $f\in M^{0,1}$ and B from the on-shell HS algebra (43), $B\star f$ and $f\star B$ belong to $f\in M^{0,1}$ so that $f\in M^{0,1}$ are bimodules over the HS algebra seen as an associative algebra.

Furthermore, it is clear that the operator $S^\dagger = [\cdot, F^0_+]_\star = P\cdot\frac{\partial}{\partial Y}$ defines a map $M^0\to M^1$ and moreover $S^\dagger(B\star f) = B\star(S^\dagger f)$ for any B in the HS algebra. It follows, that both $m^0\equiv \ker S^\dagger\subset M^0$ and $m^1\equiv \operatorname{coker} S^\dagger\subset M^1$ (i.e., the quotient of M^1 modulo $\operatorname{Im} S^\dagger$) are also modules of the HS algebra as an associative algebra.

In what follows we consider $M^{0,1}$ and $m^{0,1}$ as modules of the HS algebra seen as a Lie algebra, with the action being:

$$\rho(B)f = [B, f]_\star \tag{52}$$

In particular, $M^{0,1}$ and $m^{0,1}$ are modules over $\mathfrak{so}(d,2)$ which is a Lie subalgebra of the HS algebra. It is now clear that Equations (49) and (50) are, up to an extra term, nothing but covariant-constancy conditions with respect to the HS algebra connection A_0.

Strictly speaking the above construction applies to the off-shell version of the algebra and modules. The on-shell version is obtained by requiring all the elements to be totally traceless and applying the projector Π when necessary. The modules $m^{0,1}$ of the on-shell HS algebra are known in the literature as, respectively, the adjoint and the twisted-adjoint modules [24] (also as, respectively, gauge and Weyl modules). The above realization of the modules originates from [51] (see also [74]), where they were considered as $\mathfrak{so}(d,2)$-modules only.

System (50) describes propagation of free HS fields, encoded in $\bar{a}$ and $\bar{f}$, over a background described by the flat connection A^0 taking values in the on-shell HS algebra. The most nontrivial part of the system comes from the term $\mu(A^0, A^0, \bar{f})$ which is trilinear in the fields and cannot be reduced to a product in the HS algebra. In the next section, we explain that this is a correct structure which is completely fixed by the HS algebra. Therefore, the parent form of the HS extension of the FG construction solves an important problem of how to make HS fields propagate on backgrounds that differ from pure AdS space. The derivation above was quite abstract and we do not aim at deriving the explicit form of $\mu(A^0, A^0, \bar{f})$ (its free approximation when A^0 is an $\mathfrak{so}(d,2)$ connection was discussed many times, see e.g., [27,51,79]). This is due to the fact that the explicit form of $\mu(A^0, A^0, \bar{f})$ can be changed by field redefinitions and it is difficult to compare with vertices in the usual weak-field expansion.

Let us discuss if the on-shell conditions entering the full system (38) considered over the flat vacuum A^0, F_i^0 can be extended beyond the linear order. The general feature of the formulations that are based on jet bundles is that the field space contains infinitely many auxiliary fields that, as a consequence of equations of motion, encode derivatives of the fields of arbitrarily high order. The advantage is that interaction terms can then be written in an algebraic form. However, this does not come for free and non-linear expressions can easily contain infinitely many derivatives and make locality properties obscure. This problem becomes visible when nonlinearities are at least bilinear in fluctuations $\bar{f}$, i.e., are of order $\mathcal{O}(\bar{f}^2)$. This is due to the fact that the HS algebra has a well-defined grading that is mapped to polynomial degree in $Y+V$, P. Therefore, expressions of order $\mathcal{O}(\bar{a}\bar{f})$ or $\mathcal{O}(\bar{a}\bar{a}\bar{f})$ are always local once we fix the spins (i.e., homogeneity in P) in $\bar{a}$ and $\bar{f}$. However, expressions of order $\mathcal{O}(\bar{f}^2)$ can have unbounded number of derivatives (Indeed, $\bar{a}$ contains a finite number of derivatives per spin and $\bar{f}$ contains derivatives of unbounded order. Looking at the possible contribution of the interaction vertices to a given equation we see that infinite sums over derivatives require at least two $\bar{f}$'s on the right-hand-side). Such non-localities arise at the quartic order in weak-field expansion [42,43], i.e., $\mathcal{O}(\bar{a}\bar{f}^3)$ $\mathcal{O}(\bar{a}\bar{a}\bar{f}^2)$ in the parent formulation (38). In the framework offered by the nonlinear system (38) understood perturbatively over the flat vacuum A^0, F_i^0 the locality problem manifest itself in that the functional class $\mathfrak{C}$ is not closed under star-multiplication. Although this does not affect the linearized system (49), at higher orders either the functional class or even the system itself has to be amended.

5.3. Relation to Unfolded Equations

The system (50) can also be understood as a specific free differential algebra [80–82], unfolded Equation [24] or AKSZ sigma model [75] associated to a certain target-space Q-manifold (see also [83]). The underlying Q-manifold can be directly related to the deformation procedure that is relevant for higher-spin theories [24]. Our aim is to show how (50) arises. The field content, or the coordinates of the Q-manifold, consists of a connection of the on-shell HS algebra ω and a zero-form C that takes values in the on-shell HS algebra as well. The deformation procedure starts with the flatness condition for ω and a covariant constancy condition for C:

$$d\omega = \omega * \omega\,, \tag{53a}$$

$$dC = \omega * C - C * \pi(\omega)\,, \tag{53b}$$

where the automorphism π is induced by an automorphism of $\mathfrak{so}(d,2)$ that flips the sign of the transvection generators and leaves Lorentz generators intact. Note that in contrast to the previous section ω and C take values in the on-shell HS algebra from the very beginning and the associative product is denoted $*$. As we discussed in the previous section the twisted-adjoint covariant constancy equation for C can be systematically derived from the parent system (49), which in turn arises from the HS extended FG theory upon factorization.

System (53) involves a HS-flat background ω and a linear fluctuation C. When considered over AdS background, given by ω belonging to the $\mathfrak{so}(d,2)$ subalgebra, the equation for C describes a scalar field and $s = 1, 2, 3, \ldots$ massless fields in terms of their gauge-invariant field strengths contained in C: Faraday tensor, Weyl tensor and higher-spin generalizations thereof [24].

The deformation procedure has C as an expansion parameter. Therefore, the first order deformation of (53) involves a vertex that violates the flatness of ω:

$$d\omega = \omega * \omega + \mathcal{V}(\omega,\omega,C) + \mathcal{O}(C^2), \tag{54a}$$
$$dC = \omega * C - C * \pi(\omega) + \mathcal{O}(C^2). \tag{54b}$$

The consistency conditions on $\mathcal{V}$ follow from $d\,d \equiv 0$. The solution for $\mathcal{V}$ can be cast into the form [54] (A crucial assumption needed to reduce a complicated problem of Chevalley-Eilenberg cohomology of higher-spin algebra to a much simpler problem of Hochschild cohomology is to assume that higher-spin theory should allow for Yang-Mills gaugings $u(M)$ for any M. This assumption is justified by AdS/CFT correspondence where higher-spin theories are dual to free CFT's and it is always possible to take a number of free fields as to have $u(M)$ (times higher-spin algebra) as a global symmetry on the CFT side and hence HS algebra tensored with matrices is a gauge symmetry on the AdS side. Having this matrix factor allows one to reduce the problem [54] to the Hochschild cohomology).

$$\mathcal{V}(\omega,\omega,C) = \Phi(\omega,\omega) * \pi(C), \tag{55}$$

where $\Phi(\bullet,\bullet)$ is a Hochschild two-cocycle of the higher-spin algebra in the twisted-adjoint representation:

$$a * \Phi(b,c) + \Phi(a * b,c) - \Phi(a,b * c) + \Phi(a,b) * \pi(c) = 0. \tag{56}$$

The deformation does not stop at $\mathcal{O}(C^2)$ order and higher orders are needed. For a large class of algebras it can be shown [54,84] that there are no obstructions at higher orders. Therefore, any consistent system that has (54) as the order-$\mathcal{O}(C)$ approximation can be completed to a solution to the full non-linear deformation problem. The conclusion is that the full nonlinear system is determined by HS algebra, its twisted-adjoint module and the vertex $\mathcal{V}(\omega,\omega,C)$.

The system that describes fluctuations of HS fields over any HS-flat background is obtained from (54) by taking any flat connection A^0 and expanding the system $\omega \to A^0 + \omega$ to the first order in C and ω:

$$dA^0 = A^0 * A^0, \tag{57a}$$
$$d\omega = A^0 * \omega + \omega * A^0 + \mathcal{V}(A^0, A^0, C), \tag{57b}$$
$$dC = A^0 * C - C * \pi(A^0). \tag{57c}$$

These equations are fully consistent, gauge invariant and do not require any higher order corrections. They can be identified with Equations (45) and (50) obtained as a linearization of the HS-extension of the FG-construction (38).

To sum up, we observe that all the structures governing HS theories within the unfolded approach are already present in the linearization over a HS-flat background. In its turn this system and the respective structures can be extracted from the nonlinear system (38), which in turn can be related to the HS-extension of the FG construction. Equations (45) and (50) from the previous section are exactly of this form. Fluctuations over a HS-flat background seems to be the farthest one can get without facing the locality problem in this approach.

6. Conclusions and Discussion

In this work, we have shown that the $\mathfrak{sp}(2)$-constraints on the ambient phase space are equivalent to the off-shell Fefferman-Graham theory, if the three constraints are of degree 0, 1, 2 in the momentum. The HS extension then follows by letting the three constraints to have arbitrary powers of momenta. The HS extended $\mathfrak{sp}(2)$-system has already been studied in the past, both in the context of bulk HS theories [29] (see also [28,41]) and conformal HS fields on the boundary [17,20,30].

It is more convenient to analyze the equations after reformulating the system in parent form, i.e., by moving the ambient space-time to the fiber and introducing an extra gauge field associated with ambient diffeomorphisms. In so doing the original space time coordinates X^A are promoted to the components of the compensator field V^A. In particular, the bulk theory can now be formulated in terms of intrinsic geometry of AdS space by pulling back the ambient parent system to $d+1$ dimensions and setting $V^2 = -1$.

Although it was known that the linearized parent formulation of $\mathfrak{sp}(2)$-system can be put on-shell by supplementing the system with extra conditions, thereby giving rise to the infinite multiplet of on-shell Fronsdal fields, it was not clear how to implement this beyond the linearized approximation over the AdS vacuum. In this work, we propose a procedure that allows to go one step further such approximation. More precisely, we consider a factorized version of the parent $\mathfrak{sp}(2)$-system. Although with the simplest functional class the system is empty when considered over AdS background, we have succeeded to find another functional class (in auxiliary space) such that the modified system admits HS-flat vacuum solution and the respective linearized system is nonempty and properly describes propagation of HS fields over a HS-flat background. Even though the extension to higher orders still remains an open problem, this linearized system already contains all the structures determining the Vasiliev equations.

It has been known for many years that free massless HS fields cannot be put on nontrivial gravitational backgrounds as it leads to breakdown of gauge invariance. However, HS gauge fields *can* propagate on nontrivial backgrounds that have other higher-spin gauge fields turned on—backgrounds described by a flat connection of a HS algebra (In three dimensions HS fields do not have propagating degrees of freedom but HS-flat backgrounds were found to contain many interesting solutions, e.g., black holes and conical defects). The resulting equations have a clear algebraic meaning of flatness condition deformed by a Hochschild two-cocycle of the relevant HS algebra. This probes the structure of interactions, even though it is hard to directly make a link to the vertices in the weak field expansion over anti-de Sitter space.

It can be argued [54] that, at least within the formal deformation procedure, a HS algebra and its Hochschild cocycle determine the full non-linear completion so that the knowledge of free fluctuations over sufficiently general backgrounds is enough to recover the full structure of interactions. We showed that all this information is already present in the factorized version of the parent $\mathfrak{sp}(2)$-system. The problem of putting the system on-shell at higher orders is clearly related to the subtlety of higher-spin interactions in anti-de Sitter space: the interactions are known to be non-local starting from the quartic order [42–44] and the precise characterization of the appropriate functional class is not yet known.

The factorized version of the parent $\mathfrak{sp}(2)$-system employed in the paper also has a natural interpretation in the context of CHS theories. Indeed, the same system considered on the boundary, by setting $V^2 = 0$, describes nonlinear CHS fields at the off-shell level. More specifically, it is equivalent (at least perturbatively) to the system from [32] underlying the nonlinear theory of CHS fields.

It is not surprising that that the system describing off-shell conformal HS fields on the boundary has something to do with on-shell HS fields in the bulk. Indeed, the former are the boundary values of the later while in the ambient space formulation bulk fields and their boundary values are typically described by one and the same ambient system considered, respectively, around the hyperboloid and around the hypercone (see e.g., [17,85] for a parent formulation). Therefore, the parent $\mathfrak{sp}(2)$-system provides a direct link between the symmetries of the effective action of the scalar field on CHS background and HS gravity in the bulk.

From this perspective, the approach advocated in this paper can be viewed as a purely classical version of the holographic reconstruction (see e.g., [10,42,43] for somewhat related approaches). Indeed, the candidate system describing nonlinear on-shell HS fields in the bulk is obtained by pulling to the bulk the boundary system describing off-shell conformal fields. It is a remarkable feature of the ambient space formulation that it does not only allows one to go from bulk to boundary but also to reverse the procedure in order to reconstruct bulk theory from the boundary values. Strictly speaking, to make it work one also employs the parent formulation as to fine-tune the system by picking specific functional classes in the auxiliary space.

Possible generalizations and extensions of the above construction involve a number of cases: theories of partially-massless higher-spin fields [17,30,86], which result from the same $\mathfrak{sp}(2)$-system but making use of GJMS operators [18] for the higher-order singleton and a slightly different factorization procedure [30,87]. It would be also interesting to consider other algebras of constraints. For example, $\mathfrak{osp}(1|2)$-constraints should describe the yet-unknown Type-B theory that is dual to free massless fermions and Gross-Neveu model. This problem was recently discussed in [21,88].

Finally, let us note that there exists an alternative approach to described (HS) fields in AdS_{d+1} in terms of $\mathfrak{sp}(2)$-system in $(d+3)$-dimensional ambient space. In so doing one needs an additional ingredient, the scale tractor, that breaks the conformal invariance of the tractor formulation [89,90].

Acknowledgments: We thank the organizers of the very stimulating workshop on "Higher-spin gauge theories" (26-28 of April, 2017, Mons, Belgium) for providing us the opportunity to present this work in a warm atmosphere. This research was supported by the Russian Science Foundation grant 14-42-00047 in association with the Lebedev Physical Institute.

Author Contributions: All the authors contributed equally to this work.

Conflicts of Interest: The authors declare no conflict of interest.

Appendix A. FG Ambient Construction as $\mathfrak{sp}(2)$ Algebra of Constraints

Appendix A.1. Klein Flat Ambient Model

In modern conformal field theory and its holographic dual interpretation, a celebrated technique for performing computations is the ambient formalism. Its principle goes back to Dirac's cone reformulation of conformal fields and their wave equations [13] corresponding to the case of flat Lorentzian conformal geometry.

In turn, Dirac's approach goes back to Klein's model of flat Euclidean d-dimensional conformal geometry, that is the round sphere $\mathbb{S}^d$ with conformal isometry algebra $\mathfrak{so}(d+1,1)$. The main idea of the ambient approach is to make conformal symmetry manifest via an embedding of this geometry inside an "ambient" $(d+2)$-dimensional Minkowski space $\mathbb{R}^{d+1,1}$. The $(d+1)$-dimensional upper null cone $N \subset \mathbb{R}^{d+1,1}$, generated by light-like rays through the origin, plays a crucial role. The conformal sphere $\mathbb{S}^d$ is realized as the projective future light-cone $\mathbb{P}N := N/\mathbb{R}^+$ (with $\mathbb{R}^+$ the multiplicative group of positive real numbers) inside the (d+1)-dimensional projective ambient space $\mathbb{PR}^{d+1,1} := \mathbb{R}^{d+1,1}/\mathbb{R}^+$. The interior of the projective future light-cone $\mathbb{P}N \subset \mathbb{PR}^{d+1,1}$ is a (d+1)-dimensional hyperbolic ball $\mathbb{H}_{d+1} \cong \mathbb{B}^{d+1}$, of which the d-dimensional sphere $\mathbb{S}^d$ is the conformal boundary.

Appendix A.2. Fefferman-Graham Ambient Construction

In 1985, Fefferman and Graham generalized the ambient construction of Klein to *curved* conformal geometry (of any signature) [8].

Conformal space: The basic data of conformal geometry is a manifold M and a conformal metric $[g_{ab}]$ on M, i.e., an equivalence class of Riemanian metrics for the equivalence relation

$$\widetilde{g}_{ab} = \Omega^2 g_{ab} \sim g_{ab} \tag{A1}$$

with Ω a nowhere vanishing function on M (which we will assume strictly positive $\Omega > 0$).

Example. *In the case of flat Euclidean conformal geometry, $M = \mathbb{S}^d$ and g_{ab} is the standard metric of the unit sphere (which is conformally flat since the flat metric δ_{ab} belongs to the same equivalence class $[g_{ab}]$).*

Metric bundle: The first step in the construction of Fefferman and Graham is the introduction of a principal $\mathbb{R}^+$-bundle N over M, whose fiber at a point (of coordinates x^a with $a = 1, \cdots, d$) is the collection of values at this point of all representatives $\widetilde{g}_{ab}(x) = \Omega^2(x) g_{ab}(x)$ inside the conformal class. The base manifold is recovered as the quotient $M = N/\mathbb{R}^+$. One may take (x^a, t) as local coordinates on N with $t := \Omega(x)$. This bundle is called the *metric bundle* because its sections are the representatives $\widetilde{g}_{ab}$ of the given conformal class $[g_{ab}]$. The fundamental vector field of the principal $\mathbb{R}^+$-bundle N will be denoted v ($= t\partial_t$ in local coordinates). The conformal class $[g_{ab}]$ of metrics defines a degenerate metric on N by pullback along the projection map $\pi : N \to M$,

$$ds^2_N = t^2 g_{ab}(x)\, dx^a dx^b \,. \tag{A2}$$

This metric is homogeneous of degree 2 under the action of $\mathbb{R}^+$ on N. However, this metric is degenerate, it annihilates for instance the fundamental vector field v.

Example. *In the case of flat Euclidean conformal geometry the metric bundle N, as it was introduced, has the topology of a cylinder $\mathbb{S}^d \times \mathbb{R}^+$, which can be interpreted as an upper cone (but it can be extended to a complete cone if we extend the range of values of t to all real numbers). Together with its degenerate metric, it is indeed a null cone, as mentioned in Section A.1.*

Ambient space: The second step in the construction of Fefferman and Graham is to embed the $(d+1)$-dimensional manifold N inside a "slightly thicker" $(d+2)$-dimensional *ambient space*, e.g., $\widetilde{N} = N \times I$, where $I \subset \mathbb{R}$ is an open interval around zero. The natural extension to $\widetilde{N}$ of the fundamental vector field v on N is a vector field V on $\widetilde{N}$ which one might call the *homothety vector field*. A *defining function* ρ is a function on $\widetilde{N}$ with homogeneity degree zero under the homothety vector field V, and such that $\rho = 0$ but $d\rho \neq 0$ on N. The ambient space $\widetilde{N}$ is endowed with local coordinates $Y^M = (x^a, t, \rho)$ where ρ is a defining function. An *ambient metric* G_{MN} is a metric on ambient space $\widetilde{N}$ such that:

- its signature has one more timelike and one more spacelike direction with respect to g_{ab},
- it is homogeneous of degree two with respect to the homotheties: $\mathcal{L}_V G_{MN} = 2\, G_{MN}$,
- it is an extension to $\widetilde{N}$ of the degenerate metric (A2) on N,
- the one-form $V_M = G_{MN} V^N$ is closed.

There exists a local choice of coordinates such that the ambient metric reads [9]

$$ds^2_{\widetilde{N}} = t^2 g_{ab}(x)\, dx^a dx^b + 2\rho\, dt^2 + 2\, t\, dt\, d\rho \tag{A3}$$

and the vector field $\partial/\partial\rho$ is geodesic. The square of the homothety vector field is proportional to the defining function ρ since, from (A3), one has $G_{MN} V^M V^N = t^2 \rho$. Taking (A3) into account gives $V_M dY^M = 2\rho\, t\, dt + t^2\, d\rho = d(t^2\rho)$ in local coordinates.

Up to the well-known subtleties in the holographic reconstruction related to conformal anomalies, the ambient metric is essentially uniquely specified (in an infinitesimally thick neighborhood around N or, more precisely, as a formal power series in the variable ρ) if one further requires G_{MN} to be Ricci flat: $R_{MN} = 0$.

Example. *In the case of flat Euclidean conformal geometry, the ambient space is the Minkowski spacetime $\widetilde{N} = \mathbb{R}^{d+1,1}$ with ambient metric $\eta_{MN} = diag(-1,+1,...,+1)$ in the Cartesian coordinates X^M with $M = 0, 1, \cdots, d+1$. The homothety vector field reads $V = X^M \partial_M$ and the metric bundle N is embedded as the light-cone through the origin $V^2 = 0 \Leftrightarrow \eta_{MN} X^M X^N = 0$. The relation between the Cartesian coordinates X^M and the FG coordinates $Y^M = (x^a, t, \rho)$ is as follows: $X^a = t\, x^a$, $X^0 - X^{d+1} = t$ and $X^0 + X^{d+1} = t\,(\,\delta_{ab} x^a x^b - 2\rho\,)$. In such case, $ds^2_{\widetilde{N}} = \eta_{MN} dX^M dX^N$ reproduces* (A3) *with $g_{ab} = \delta_{ab}$.*

Bulk: The third step in the construction of Fefferman and Graham provided its holographic interpretation, cherished by theoretical physicists. The *bulk space* is the (d+1)-dimensional manifold $\widetilde{M} := (\widetilde{N} - N)/\mathbb{R}^+$. One may take (x^m, ρ) as local coordinates on the bulk space $\widetilde{M}$ which can be realized as the level hypersurface $t^2\rho$ = constant. It is endowed with a metric $g_{\mu\nu}$ which has $[g_{ab}]$ as conformal class at conformal infinity (i.e., at $\rho = 0$). If the ambient metric G_{MN} is Ricci flat ($R_{MN} = 0$), then the bulk metric is Einstein ($R_{\mu\nu} = \frac{1}{d+1} R\, g_{\mu\nu}$).

Example. *In the case of flat Euclidean conformal geometry, to be more precise the $(d+2)$-dimensional ambient space is the interior future light-cone $\widetilde{N}$ inside the Minkowski spacetime $\mathbb{R}^{d+1,1}$. Then the bulk space is the $(d+1)$-dimensional hyperbolic space $\mathbb{H}_{d+1}$ realized as one sheet of a two-sheeted hyperboloid $\eta_{MN}X^M X^N = -1$ of time-like vectors of constant square in ambient space. This corresponds to the level hypersurface $t^2\rho = -\frac{1}{2}$ in the FG coordinates. Up to the change of coordinate $z = 1/t$, this leads to the standard form $ds^2_{\widetilde{M}} = z^{-2}\,(\,\delta_{ab}\,dx^a dx^b + dz^2\,)$ of the hyperbolic space metric in Poincaré coordinates.*

Appendix A.3. Properties of the Ambient Metric and of the Homothety Vector Field

The two main ingredients of Fefferman-Graham construction are the ambient metric G_{AB} and the homothety vector field V^A. They are closely related to each other due to the two following properties:

(I) The ambient metric is of homogeneity degree two with respect to the homothety vector field:

$$\mathcal{L}_V G_{AB} = 2\,G_{AB}\,. \tag{A4}$$

(II) The homothety one-form is closed:

$$\partial_{[A}V_{B]} = 0\,. \tag{A5}$$

In particular, the property (II) implies that, locally, $V_A = \partial_A f$. In particular, the homothety vector field V^A is hypersurface orthogonal to the level surfaces f = constant.

The properties (I) and (II) are equivalent to the following useful property:

(III) The ambient metric is equal to the covariant derivative of the homothety one-form:

$$G_{AB} = \nabla_A V_B = \nabla_B V_A\,. \tag{A6}$$

Equivalently,

$$\nabla_A V^B = \delta_A^B\,. \tag{A7}$$

Proof. While the relation (A4) is equivalent to

$$\nabla_A V_B + \nabla_B V_A = 2\,G_{AB}\,, \tag{A8}$$

the relation (A5) is equivalent to $\nabla_{[A}V_{B]} = 0$, i.e.,

$$\nabla_A V_B - \nabla_B V_A = 0\,. \tag{A9}$$

Summing Equations (A8) and (A9) gives (A6). □

In turn, the property (III) implies the following corollary:

(IV) The homothety one-form is equal to half the gradient of the homothety vector field squared:

$$V_A = \partial_A\Big(\frac{V^2}{2}\Big)\,. \tag{A10}$$

Proof. Contracting $G_{AB} = \nabla_A V_B$ with V^B gives the relation $V_A = V^B \nabla_A V_B$ which implies (A10). □

In particular, the homothety vector field is hypersurface orthogonal to the level surfaces V^2 = constant.

Appendix A.4. Hypersurface Orthogonality and Homogeneity as $\mathfrak{sp}(2)$ *Algebra*

Consider the cotangent bundle $T^*\widetilde{N}$ of the ambient space. Local coordinates on $T^*\widetilde{N}$ read (X^M, P_N) and the canonical Poisson bracket on the algebra of functions on $T^*\widetilde{N}$ is such that $\{X^M, P_N\} = \delta^M_N$. In the main text, we need the relation between the Fefferman-Graham ambient construction and the $\mathfrak{sp}(2)$ algebra of constraints.

Consider as data, three Hamiltonian constraints which are respectively quadratic, linear, independent, of the momenta:

$$F_+ = \frac{1}{2}P_M P_N G^{MN}(X) \tag{A11}$$

$$F_0 = P_M V^M(X) \tag{A12}$$

$$F_- = F(X) \tag{A13}$$

The first Hamiltonian constraint is quadratic in the momenta, thus its coefficients define a covariant symmetric tensor $G^{MN}(X)$, which can be interpreted as an inverse metric if it is nondegenerate. This fact will be called the *nondegeneracy condition*. The second Hamiltonian constraint is linear in the momenta, thence its coefficients define a vector field $V^M(X)$. The third Hamiltonian constraint is a scalar field $F(X)$.

Proposition. Under the assumption of nondegeneracy, the property that the three functions (A11)–(A12) on the ambient phase space form the $\mathfrak{sp}(2)$ algebra

$$\{F_+, F_-\} = F_0\,, \quad \{F_0, F_+\} = +2\,F_+\,, \quad \{F_0, F_-\} = -2\,F_-\,, \tag{A14}$$

under the Poisson bracket is equivalent to the fact that

- the scalar field is equal to $F(X) = -\frac{1}{2}\,V^M(X)\,G_{MN}(X)\,V^N(X)$,
- the symmetric tensor field $G_{MN}(X)$ and the vector field $V^M(X)$ obey the properties (I)-(IV).

Proof. Since $h = P \cdot V(X)$ is linear in the momenta, the adjoint action of the constraint h via the Poisson bracket is related to the Lie derivative along the homothety vector field, $\{h, \cdot\} = -\mathcal{L}_V$. Therefore, the second relation in (6) is equivalent to the condition $\mathcal{L}_V G^{MN} = -2\,G^{MN} \Leftrightarrow \mathcal{L}_V G_{MN} = 2\,G_{MN}$, i.e., it is equivalent to (A4). The first relation in (A14) gives the relation $\partial_M F = -V_M$, i.e., it implies the property (II) in Section 2. Together with property (I), it implies the properties (III)–(IV) of Section 2. In particular, the relation (A10) implies $\partial_M F = -\partial_M(\frac{V^2}{2})$. This leads to $F(X) = C - \frac{1}{2}\,V^M(X) G_{MN}(X) V^N(X)$, where $C \in \mathbb{R}$ an arbitrary constant which is enforced to vanish by the third relation in (A14). Then, the third relation in (A14) is equivalent to the homogeneity property $\mathcal{L}_V G_{MN} = 2\,G_{MN}$. □

Appendix B. Covariant Derivatives

Suppose we are given with the ambient metric $g_{MN}(X)$ and the compensator $V^M(X)$ such that $\nabla_M V_N = G_{MN}$, where ∇ denotes the covariant derivative determined by the Levi-Civita connection.

Introduce the covariant derivative acting on functions in X, P as follows:

$$D_M f(X,P) := \partial_M + \Gamma^R_{MN} P_R \frac{\partial}{\partial P_N}$$

so that e.g.,

$$D_M(V^N P_N) = (\nabla_M V^N) P_N\,.$$

It is easy to check that the Poisson bracket of functions in X, P can be written in terms of the covariant derivative:

$$\{F, G\} := \frac{\partial F}{\partial X^M}\frac{\partial G}{\partial P_M} - \frac{\partial F}{\partial P_M}\frac{\partial G}{\partial X^M} = D_M F \frac{\partial G}{\partial P_M} - \frac{\partial F}{\partial P_M} D_M G\,.$$

For instance, for the adjoint action of $\mathfrak{sp}(2)$ generators we have:

$$\begin{aligned} \{V^M P_M, f\} &= (\partial_N V^M) P_M \left(\tfrac{\partial}{\partial P_N} f\right) - V^M \partial_M f \\ &= (\nabla_N V^M) P_M \left(\tfrac{\partial}{\partial P_N} f\right) - \Gamma^M_{NR} V^R P_M \left(\tfrac{\partial}{\partial P_N} f\right) - V^M \partial_M f \\ &= (\nabla_N V^M) P_M \left(\tfrac{\partial}{\partial P_N} f\right) - V^A D_A f \\ &= \left(P_M \tfrac{\partial}{\partial P_M} - V^A D_A\right) f . \end{aligned} \tag{A15}$$

Similarly for the other generators.

Appendix C. Off-shell vs On-shell, Boundary vs. Bulk

The different physical interpretations of the FG ambient construction, mentioned in Sections 2 and 3, can be illustrated in the simple case of a scalar field, which is an inspiring toy model (For more details on the ambient formulation of the scalar field and its holographic interpretation (in the case of the flat ambient space) see e.g., [16,17]). This simple example is actually of interest on its own, since it corresponds to the first-quantization of the $\mathfrak{sp}(2)$ Hamiltonian constraints. To be more precise, we consider an ambient scalar field in a background of off-shell FG theory.

Appendix C.1. Off-Shell Boundary Scalar Field

The null cone $V^2(X) = 0$ quotiented by the integral lines of the vector field $V = V^A \partial_A$ is a d-dimensional conformal space. A scalar primary conformal field of conformal weight Δ on this d-dimensional conformal space can be lifted uniquely to a scalar field on the null cone $V^2 = 0$ with homogeneity degree $-\Delta$. However, the latter does not determine a unique scalar field $\Phi(X)$ in the vicinity of the null cone, but only up to the following equivalence relation $\Phi \sim \Phi + V^2 \lambda$. This can be summarized by saying that an off-shell conformal scalar field in d dimensions is equivalently described by an ambient scalar field obeying to the following set of one equation and one equivalence relation:

$$(V^A \partial_A + \Delta)\Phi(X) = 0 , \qquad \Phi(X) \sim \Phi(X) + V^2(X)\, \lambda(X) . \tag{A16}$$

The consistency of the above equations amounts to the fact that the operators $V^A \partial_A$ and V^2 form a Lie algebra (the lower-triangular subalgebra of $\mathfrak{sp}(2)$).

Spin-two analogue: Off-shell conformal gravity determines the value of the ambient metric on the null cone. The ambient metric in off-shell FG theory has homogeneity degree two and its extension beyond the null cone is completely undetermined.

Appendix C.2. On-Shell Bulk Scalar Field

An on-shell scalar field on the $d + 1$ dimensional level manifold $V^2(X) = -1$, i.e., a scalar field obeying to Klein-Gordon equation in the bulk, can be lifted uniquely to a scalar field $\Phi(X)$ on the region $V^2 < 0$ of the ambient space. The lift is unique if the scalar field is of fixed homogeneity degree, say $-\Delta \in \mathbb{R}$, along the homothety vector field V, i.e., $\mathcal{L}_V \Phi = -\Delta\, \Phi$. More precisely, the ambient scalar field obeys the following two equations:

$$(V^A \partial_A + \Delta)\Phi(X) = 0 , \qquad \nabla^2 \Phi(X) = 0 , \tag{A17}$$

where $\nabla^2 = G^{AB} \nabla_A \nabla_B$ is the Laplacian for the ambient metric. The consistency of the two conditions in (A17) can be checked by using $\nabla_A V^B = \delta^B_A$ and its consequence $R_{AB|CD} V^D = 0$. Moreover, these two operators form the upper-triangular subalgebra of $\mathfrak{sp}(2)$.

When the scalar field $\Phi(X)$ admits a regular extension to the whole region $V^2 \leqslant 0$ of the ambient space, then its restriction to the null cone $V^2 = 0$ corresponds to the asymptotic boundary data of the on-shell bulk scalar field, with scaling behavior prescribed by Δ (see e.g., [16] for more details). To be more precise, two scaling behaviours are actually possible ($\Delta_+ = \Delta$ and $\Delta_- = d - \Delta$), corresponding to the two branches of solutions of the bulk Klein-Gordon equation.

Spin-two analogue: The Ricci flatness of the ambient metric is the analogue of the harmonicity of the ambient scalar field, while Einstein equation with a cosmological constant of bulk gravity is the analogue of Klein-Gordon equation of the bulk scalar field. The holographic reconstruction of the $(d+1)$-dimensional spacetime metric from the d-dimensional conformal boundary data via the $(d+2)$-dimensional ambient space is the essence of FG construction.

Appendix C.3. On-Shell Boundary Scalar Field (Aka Singleton)

For generic values of the conformal dimension Δ, the harmonicity $\nabla^2\Phi = 0$ is incompatible with the gauge equivalence $\Phi \sim \Phi + V^2\,\lambda$. In such cases, the harmonicity can be interpreted as a gauge-fixing condition for the gauge equivalence. This can be summarized by saying that the off-shell boundary scalar field can be equivalently described as the ambient scalar satisfying (A16) or (A17). However, when the conformal dimension takes the value $\Delta = (d-2)/2$ one can impose consistently

$$\nabla^2\Phi = 0\,, \qquad \left(V^A\partial_A + \frac{d-2}{2}\right)\Phi = 0\,, \qquad \Phi \sim \Phi + V^2\,\lambda\,. \tag{A18}$$

The consistency of the system follows from the fact that (as was originally observed in [18]) the operators ∇^2, $V^A\partial_A + \frac{d+2}{2}$, V^2 form the algebra $\mathfrak{sp}(2)$. Note that these operators can be thought as first-quantized versions of the $\mathfrak{sp}(2)$-constraints F_+, F_0, F_-.

Identifying an equivalence class determined by the second and the third relations in (A18) with an off-shell boundary scalar field, the first equation imposes the Yamabe equation for this conformal scalar field in d dimensions. Another interpretation of the same fact is that the first two constraints describe the bulk Klein-Gordon equation with critical mass. Considered in the vicinity of the boundary, the equivalence relation described by the third relation in (A18) then eliminates the subleading solutions. From this perspective, the Yamabe equation appears as an obstruction in extending the unconstrained boundary value to an on-shell bulk field (or, equivalently, as an obstruction in the near-boundary expansion of the on-shell bulk scalar field).

Spin-two analogue: Consider $d = 4$ for simplicity (but similar discussion holds for any even d). The Bach tensor appears as an obstruction (related to the holographic anomaly) in the FG expansion of bulk gravity in five dimensions. This obstruction is absent if and only if four-dimensional conformal gravity is on-shell.

References

1. Maldacena, J.M. The large N limit of superconformal field theories and supergravity. *Adv. Theor. Math. Phys.* **1998**, *2*, 231–252.
2. Witten, E. Anti-de Sitter space and holography. *Adv. Theor. Math. Phys.* **1998**, *2*, 253–291.
3. Gubser, S.S.; Klebanov, I.R.; Polyakov, A.M. Gauge theory correlators from non-critical string theory. *Phys. Lett. B* **1998**, *428*, 105–114.
4. Sundborg, B. Stringy gravity, interacting tensionless strings and massless higher spins. *Nucl. Phys. Proc. Suppl.* **2001**, *102*, 113–119.
5. Sezgin, E.; Sundell, P. Massless higher spins and holography. *Nucl. Phys. B* **2002**, *644*, 303–370.
6. Klebanov, I.; Polyakov, A. AdS dual of the critical O(N) vector model. *Phys. Lett. B* **2002**, *550*, 213–219.
7. Flato, M.; Fronsdal, C. One Massless Particle Equals Two Dirac Singletons: Elementary Particles in a Curved Space. *Lett. Math. Phys.* **1978**, *2*, 421–426.
8. Fefferman, C.; Graham, C. Conformal Invariants. *Astérisque Numero Hors. Ser.* **1985**, 95–116.
9. Fefferman, C.; Graham, C.R. The ambient metric. *arXiv* **2007**, arXiv:0710.0919.
10. De Haro, S.; Solodukhin, S.N.; Skenderis, K. Holographic reconstruction of space-time and renormalization in the AdS/CFT correspondence. *Commun. Math. Phys.* **2001**, *217*, 595–622.
11. Henningson, M.; Skenderis, K. The Holographic Weyl anomaly. *J. High Energy Phys.* **1998**, *1998*, 23.

12. Dirac, P. The Electron Wave Equation in de Sitter Space. *Ann. Math.* **1935**, *36*, 657–669.
13. Dirac, P.A.M. Wave equations in conformal space. *Ann. Math.* **1936**, *37*, 429–442.
14. Liu, H.; Tseytlin, A.A. D = 4 superYang-Mills, D = 5 gauged supergravity, and D = 4 conformal supergravity. *Nucl. Phys. B* **1998**, *533*, 88–108.
15. Metsaev, R.R. Gauge invariant two-point vertices of shadow fields, AdS/CFT, and conformal fields. *Phys. Rev. D* **2010**, *81*, 106002.
16. Bekaert, X.; Grigoriev, M. Notes on the ambient approach to boundary values of AdS gauge fields. *J. Phys. A* **2013**, *46*, 214008.
17. Bekaert, X.; Grigoriev, M. Higher order singletons, partially massless fields and their boundary values in the ambient approach. *Nucl. Phys. B* **2013**, *876*, 667–714.
18. Graham, C.R.; Jenne, R.; Mason, L.J.; Sparling, G.A.J. Conformally invariant powers of the laplacian, I: Existence. *J. Lond. Math. Soc.* **1992**, *2*, 557–565.
19. Bars, I. Survey of two-time physics. *Class. Quantum Gravity* **2001**, *18*, 3113–3130.
20. Bonezzi, R.; Latini, E.; Waldron, A. Gravity, Two Times, Tractors, Weyl Invariance and Six Dimensional Quantum Mechanics. *Phys. Rev. D* **2010**, *82*, 064037.
21. Bonezzi, R.; Corradini, O.; Waldron, A. Is Quantum Gravity a Chern-Simons Theory? *Phys. Rev. D* **2014**, *90*, 084018.
22. Bonezzi, R.; Corradini, O.; Latini, E.; Waldron, A. Quantum Gravity and Causal Structures: Second Quantization of Conformal Dirac Algebras. *Phys. Rev. D* **2015**, *91*, 121501.
23. Marnelius, R. Manifestly Conformal Covariant Description of Spinning and Charged Particles. *Phys. Rev. D* **1979**, *20*, 2091–2095.
24. Vasiliev, M.A. Consistent Equations for Interacting Massless Fields of All Spins in the First Order in Curvatures. *Ann. Phys.* **1989**, *190*, 59–106.
25. Vasiliev, M.A. *Higher Spin Gauge Theories: Star-Product and AdS Space*; World Scientific: Singapore, 2000.
26. Sezgin, E.; Sundell, P. 7-D bosonic higher spin theory: Symmetry algebra and linearized constraints. *Nucl. Phys. B* **2002**, *634*, 120–140.
27. Vasiliev, M.A. Nonlinear equations for symmetric massless higher spin fields in (A)dS(d). *Phys. Lett. B* **2003**, *567*, 139–151.
28. Bars, I.; Deliduman, C. High spin gauge fields and two-time physics. *Phys. Rev. D* **2001**, *64*, 247–256.
29. Grigoriev, M. Parent formulations, frame-like Lagrangians, and generalized auxiliary fields. *J. High Energy Phys.* **2012**, *2012*, 48.
30. Alkalaev, K.; Grigoriev, M.; Skvortsov, E. Uniformizing higher-spin equations. *J. Phys. A* **2015**, *48*, 015401.
31. Fradkin, E.S.; Tseytlin, A.A. Conformal Supergravity. *Phys. Rep.* **1985**, *119*, 233–362.
32. Segal, A.Y. Conformal higher spin theory. *Nucl. Phys. B* **2003**, *664*, 59–130.
33. Tseytlin, A.A. On limits of superstring in AdS(5) x S**5. *Theor. Math. Phys.* **2002**, *133*, 1376–1389.
34. Bekaert, X.; Joung, E.; Mourad, J. Effective action in a higher-spin background. *J. High Energy Phys.* **2011**, *2011*, 48.
35. Maldacena, J.; Zhiboedov, A. Constraining Conformal Field Theories with A Higher Spin Symmetry. *arXiv* **2011**, arXiv:1112.1016.
36. Alba, V.; Diab, K. Constraining conformal field theories with a higher spin symmetry in d = 4. *arXiv* **2013**, arXiv:1307.8092.
37. Boulanger, N.; Ponomarev, D.; Skvortsov, E.; Taronna, M. On the uniqueness of higher-spin symmetries in AdS and CFT. *Int. J. Mod. Phys. A* **2013**, *28*, 1350162.
38. Stanev, Y.S. Constraining conformal field theory with higher spin symmetry in four dimensions. *Nucl. Phys. B* **2013**, *876*, 651–666.
39. Alba, V.; Diab, K. Constraining conformal field theories with a higher spin symmetry in $d > 3$ dimensions. *arXiv* **2015**, arXiv:1510.02535.
40. Fronsdal, C. Massless Fields with Integer Spin. *Phys. Rev. D* **1978**, *18*, 3624–3629.
41. Grigoriev, M. Off-shell gauge fields from BRST quantization. *arXiv* **2006**, arXiv:hep-th/0605089.
42. Bekaert, X.; Erdmenger, J.; Ponomarev, D.; Sleight, C. Quartic AdS Interactions in Higher-Spin Gravity from Conformal Field Theory. *J. High Energy Phys.* **2015**, *2015*, 149.
43. Sleight, C.; Taronna, M. Higher spin gauge theories and bulk locality: A no-go result. *arXiv* **2017**, arXiv:1704.07859.

44. Ponomarev, D. A Note on (Non)-Locality in Holographic Higher Spin Theories. *arXiv* **2017**, arXiv:1710.00403.
45. Bekaert, X.; Erdmenger, J.; Ponomarev, D.; Sleight, C. Towards holographic higher-spin interactions: Four-point functions and higher-spin exchange. *J. High Energy Phys.* **2015**, *2015*, 170.
46. Kessel, P.; Gomez, G.L.; Skvortsov, E.; Taronna, M. Higher Spins and Matter Interacting in Dimension Three. *arXiv* **2015**, arXiv:1505.05887.
47. Skvortsov, E.D. On (Un)Broken Higher-Spin Symmetry in Vector Models. In Proceedings of the International Workshop on Higher Spin Gauge Theories, Singapore, 4–6 November 2015; pp. 103–137.
48. Sleight, C.; Taronna, M. Higher Spin Interactions from Conformal Field Theory: The Complete Cubic Couplings. *Phys. Rev. Lett.* **2016**, *16*, 181602.
49. Bekaert, X.; Boulanger, N.; Sundell, P. How higher-spin gravity surpasses the spin two barrier: No-go theorems versus yes-go examples. *arXiv* **2010**, arXiv:1007.0435.
50. Barnich, G.; Grigoriev, M.; Semikhatov, A.; Tipunin, I. Parent field theory and unfolding in BRST first-quantized terms. *Commun. Math. Phys.* **2005**, *260*, 147–181.
51. Barnich, G.; Grigoriev, M. Parent form for higher spin fields on anti-de Sitter space. *J. High Energy Phys.* **2006**, *2006*, 747–748.
52. Barnich, G.; Grigoriev, M. First order parent formulation for generic gauge field theories. *J. High Energy Phys.* **2011**, *2011*, 122.
53. Grigoriev, M. Parent formulation at the Lagrangian level. *J. High Energy Phys.* **2011**, *2011*, 1–24.
54. Sharapov, A.A.; Skvortsov, E.D. Formal higher-spin theories and Kontsevich–Shoikhet–Tsygan formality. *Nucl. Phys. B* **2017**, *921*, 538–584.
55. Aragone, C.; Deser, S. Consistency Problems of Hypergravity. *Phys. Lett. B* **1979**, *86*, 161–163.
56. Campoleoni, A.; Fredenhagen, S.; Pfenninger, S.; Theisen, S. Towards metric-like higher-spin gauge theories in three dimensions. *J. Phys. A* **2013**, *46*, 214017.
57. Ammon, M.; Gutperle, M.; Kraus, P.; Perlmutter, E. Black holes in three dimensional higher spin gravity: A review. *J. Phys. A* **2013**, *46*, 214001.
58. Boulanger, N. A Weyl-covariant tensor calculus. *J. Math. Phys.* **2005**, *46*, 053508.
59. Boulanger, N.; Erdmenger, J. A Classification of local Weyl invariants in D = 8. *Class. Quantum Gravity* **2004**, *21*, 4305–4316.
60. Čap, A.; Gover, A.R. Standard Tractors and the Conformal Ambient Metric Construction. *Ann. Glob. Anal. Geom.* **2003**, *24*, 231–259.
61. Gover, A.R.; Peterson, L.J. The ambient obstruction tensor and the conformal deformation complex. *Pac. J. Math.* **2006**, *226*, 309–351.
62. Gover, A. Spin-polarized free electron beam interaction with radiation and superradiant spin-flip radiative emission. *Phys. Rev. Spec. Top. Accel Beams* **2006**, *9*, 060703.
63. Gover, A.R.; Shaukat, A.; Waldron, A. Tractors, Mass and Weyl Invariance. *Nucl. Phys. B* **2009**, *812*, 424–455.
64. Gover, A.R.; Shaukat, A.; Waldron, A. Weyl Invariance and the Origins of Mass. *Phys. Lett. B* **2009**, *675*, 93–97.
65. Gover, A.; Waldron, A. The so(d+2,2) Minimal Representation and Ambient Tractors: The Conformal Geometry of Momentum Space. *Adv. Theor. Math. Phys.* **2009**, *13*, 1875–1894.
66. Gover, A.R.; Silhan, J. Higher symmetries of the conformal powers of the Laplacian on conformally flat manifolds. *arXiv* **2012**, arXiv:0911.5265.
67. Grigoriev, M.; Waldron, A. Massive Higher Spins from BRST and Tractors. *Nucl. Phys. B* **2011**, *853*, 291–326.
68. Barnich, G.; Brandt, F. Covariant theory of asymptotic symmetries, conservation laws and central charges. *Nucl. Phys. B* **2002**, *633*, 3–82.
69. Grigoriev, M.; Tseytlin, A.A. On conformal higher spins in curved background. *J. Phys. A* **2017**, *50*, 125401.
70. Bonezzi, R. Induced Action for Conformal Higher Spins from Worldline Path Integrals. *arXiv* **2017**, arXiv:1709.00850.
71. Bekaert, X.; Grigoriev, M. Manifestly conformal descriptions and higher symmetries of bosonic singletons. *Symmetry Integrability Geom. Method Appl.* **2010**, *6*, 038.
72. Gover, A.R.; Peterson, L.J. Conformally invariant powers of the Laplacian, Q-curvature, and tractor calculus. *Commun. Math. Phys.* **2002**, *235*, 339–378.
73. Manvelyan, R.; Mkrtchyan, K.; Mkrtchyan, R. Conformal invariant powers of the Laplacian, Fefferman-Graham ambient metric and Ricci gauging. *Phys. Lett. B* **2007**, *657*, 112–119.

74. Chekmenev, A.; Grigoriev, M. Boundary values of mixed-symmetry massless fields in AdS space. *Nucl. Phys. B* **2016**, *913*, 769–791.
75. Alexandrov, M.; Kontsevich, M.; Schwartz, A.; Zaboronsky, O. The Geometry of the master equation and topological quantum field theory. *Int. J. Mod. Phys. A* **1997**, *12*, 1405–1430.
76. Grigoriev, M.A.; Damgaard, P.H. Superfield BRST charge and the master action. *Phys. Lett. B* **2000**, *474*, 323–330.
77. Vasiliev, M.A. Holography, Unfolding and Higher-Spin Theory. *J. Phys. A* **2013**, *46*, 1558–1561.
78. Eastwood, M.G. Higher symmetries of the Laplacian. *Ann. Math.* **2005**, *161*, 1645–1665.
79. Bekaert, X.; Cnockaert, S.; Iazeolla, C.; Vasiliev, M.A. Nonlinear higher spin theories in various dimensions. *arXiv* **2005**, arXiv:hep-th/0503128.
80. Sullivan, D. Infinitesimal computations in topology. *Publ. Math. l'IHÉS* **1977**, *47*, 269–331.
81. D'Auria, R.; Fre, P.; Townsend, P.K.; van Nieuwenhuizen, P. Invariance of actions, rheonomy and the new minimal n = 1 supergravity in the group manifold approach. *Ann. Phys.* **1984**, *155*, 423–446.
82. Van Nieuwenhuizen, P. Free graded differential superalgebras. In Proceedings of the 11th International Colloquium on Group Theoretical Methods in Physics, Istanbul, Turkey, 23–28 August 1982.
83. Barnich, G.; Grigoriev, M. BRST extension of the non-linear unfolded formalism. In Proceedings of the International School/Seminar on Quantum Field Theory, Supersymmetry, High Spin Fields, Gravity Tomsk, Russia, 20–26 March 2005.
84. Sharapov, A.A.; Skvortsov, E.D. Hochschild cohomology of the weyl algebra and vasiliev's equations. *Lett. Math. Phys.* **2017**, *2*, 1–18.
85. Bekaert, X.; Joung, E.; Mourad, J. Comments on higher-spin holography. *arXiv* **2012**, arXiv:1202.0543.
86. Brust, C.; Hinterbichler, K. Partially Massless Higher-Spin Theory. *J. High Energy Phys.* **2017**, *2017*, 86.
87. Joung, E.; Mkrtchyan, K. Partially-massless higher-spin algebras and their finite-dimensional truncations. *J. High Energy Phys.* **2016**, *2016*, 3.
88. Bonezzi, R.; Corradini, O.; Latini, E.; Waldron, A. Quantum Mechanics and Hidden Superconformal Symmetry. *Phys. Rev. D* **2017**, *96*, 126005.
89. Bailey, T.; Eastwood, M.; Gover, A. Thomas's structure bundle for conformal, projective and related structures. *Rocky Mt. J. Math.* **1994**, *24*, 1191–1217.
90. Gover, A.R.; Waldron, A. Boundary calculus for conformally compact manifolds. *Indiana Univ. Math. J.* **2014**, *63*, 119–163.

universe

MDPI

Article

Induced Action for Conformal Higher Spins from Worldline Path Integrals

Roberto Bonezzi

Physique Théorique et Mathématique, Université de Mons– UMONS 20, Place du Parc, 7000 Mons, Belgium; roberto.bonezzi@umons.ac.be

Received: 7 August 2017; Accepted: 29 August 2017; Published: 4 September 2017

Abstract: Conformal higher spin (CHS) fields, yet being non unitary, provide a remarkable example of a consistent interacting higher spin theory in flat space background, that is local to all orders. The non-linear action is defined as the logarithmically UV divergent part of a one-loop scalar effective action. In this paper we take a particle model, that describes the interaction of a scalar particle to the CHS background, and compute its path integral on the circle. We thus provide a worldline representation for the CHS action, and rederive its quadratic part. We plan to come back to the subject, to compute cubic and higher vertices, in a future work.

Keywords: higher spin fields; conformal symmetry; worldline formalism

1. Introduction

Four dimensional Maxwell ($s = 1$) theory is the first known example of a conformally invariant physical system. Similarly, massless matter lagrangians ($s = 0,\ 1/2$) have conformal symmetry in flat space, that can be enhanced to general covariance plus local Weyl symmetry when coupled to a curved spacetime metric. On the other hand, for spin greater than one ordinary two-derivative theories, such as (super)gravities ($s = 2,\ 3/2$) and massless higher spin theories ($s > 2$), are not conformal. Weyl squared gravity, with higher derivative lagrangian $\mathcal{L} = \sqrt{g}(W_{\mu\nu\rho\sigma})^2 \approx h^{\mu\nu}\Box^2 h_{\mu\nu} + \ldots$, and its supersymmetric extensions [1–3] are alternative models for $s \leq 2$ possessing local Weyl symmetry besides diffeomorphism invariance, and hence rigid conformal symmetry around flat space. Conformal higher spin fields (CHS) [3–14] are the $s > 2$ generalization of the Weyl graviton and conformal gravitino. In four dimensional flat space they are described by the free lagrangian[1]

$$S[h] = \sum_s \int d^4x\, h_s\, P_s\, \Box^s\, h_s\ , \tag{1}$$

where $h_s = h_{\mu_1\ldots\mu_s}$ and P_s is the spin-s transverse-traceless projector built out of s powers of $P_1 := \delta^\mu_\nu - \frac{\partial^\mu\partial_\nu}{\Box}$. The above action thus describes pure spin s states (transverse and traceless) off-shell, and is invariant under differential and algebraic gauge transformations:

$$\delta h_s = \partial\epsilon_{s-1} + \eta\, \alpha_{s-2} \tag{2}$$

generalizing linearized diffeomorphisms and Weyl symmetry of conformal (super)gravity. The higher derivative[2] kinetic operator ensures locality of the action, at the price of formally loosing unitarity.

---

1 We will discuss only bosonic totally symmetric fields. In arbitrary even dimensions one has to add a power $\Box^{\frac{d-4}{2}}$ of the laplacian.

2 One can describe CHS dynamics with ordinary two-derivative lagrangians, at the expense of introducing auxiliary fields [15,16].

Contrary to the case of free massless higher spins, that propagate only on maximally symmetric backgrounds, there is evidence [17–19] that conformal higher spins can propagate consistently on Bach-flat backgrounds, i.e., on the equations of motion of Weyl squared gravity. In fact, quite remarkably, the above action and linear gauge symmetry admit a consistent fully non-linear completion [6,7,11], that is well defined around flat space and local to all orders in the fields. Indeed, unlike the case of massless higher spins, the absence of dimensionful parameters fixes the number of derivatives of each vertex uniquely.[3] The CHS theory is power-counting renormalizable but, since Weyl and higher order algebraic symmetries are gauged, it has to be free of conformal and higher spin anomalies in order to be consistent at the quantum level. In the low spin case $s \leq 2$, vanishing of the total Weyl anomaly can be achieved only by $\mathcal{N} = 4$ conformal supergravity coupled to four $\mathcal{N} = 4$ SYM multiplets [20,21]. In the case of four dimensional CHS with one field of each integer spin, the a-coefficient[4] of the Weyl anomaly vanishes upon a (regularized) summation over all spins [22,23], while cancellation of the c-coefficient [23–26], as well as the higher spin algebraic symmetries, seems less clear [19].

Besides being interesting on its own, as it gives a nontrivial example of an interacting higher spin theory in flat space, CHS fields are intimately related to massless higher spin theories in Anti de Sitter space [27–33] via the vectorial AdS/CFT correspondence [34–39]. Moreover, the non-linear CHS action naturally arises as an induced action [6,7,11,40] in the holographic context: The free CFT of N complex scalars admits an infinite number of conserved conformal currents of every spin in the $U(N)$ singlet sector, $J_s \sim \phi_i^* \partial^s \phi^i$. The dual fields to these conformal currents are identified with massless higher spin gauge fields in AdS space, whose boundary values h_s source the J_s currents and can in turn be seen as CHS fields on the boundary. The scalar path integral with sources $\sum_s J_s h_s$ yields the generating functional $\Gamma[h]$ of correlators of the conformal currents and, according to AdS/CFT correspondence, should be equal to the on-shell value of the, yet unknown[5], action of massless higher spins in AdS[6]. However, the same generating functional $\Gamma[h]$ can be interpreted[7], from a pure boundary viewpoint, as a one-loop effective action for the CHS fields h_s, that inherit the linearized gauge symmetry (2) thanks to conservation and tracelessness of the currents J_s. The logarithmically divergent part of $\Gamma[h]$ is local and gauge invariant and can be thus identified as the classical non-linear action $S_{\text{CHS}}[h]$ for conformal higher spins [7,11].

The aim of this paper is to construct a quantum mechanical path integral to represent the effective action $\Gamma[h]$. Since the coupling $\sum_s J_s h_s$ is quadratic in the scalar fields, $\Gamma[h]$ is given by the functional determinant

$$\Gamma[h] = N \log \text{Det}[-\Box + \hat{H}] \,, \tag{3}$$

---

3 In arbitrary even dimension d the conformal weight of h_s is $2 - s$. Given an nth order vertex with fields of spin $(s_1, ..., s_n)$, the number of derivatives is fixed to $N = d + \sum_{i=1}^{n} s_i - 2n$.

4 In four dimensions the Weyl anomaly contains only two relevant structures: the Euler density, whose coefficient is usually named a, and the square of the Weyl tensor, whose coefficient is c.

5 Vasiliev's equations lack a standard variational principle. Non-standard actions of covariant hamiltonian type have been proposed in [41–45]. From an holographic perspective, CFT correlators have been used to reconstruct AdS vertices in [46–48].

6 Direct matching of free gauge theory correlators with AdS Witten diagrams has been investigated in [49–51] in order to exploit open-closed string duality. In particular, in [49] one-loop open string diagrams in the field theory limit (hence worldline loops) were shown to reproduce tree level diagrams in AdS by direct change of variables in the moduli space.

7 In the standard AdS/CFT context [52–54] the boundary values of bulk fields are fixed, non dynamical sources for CFT correlators. From a pure boundary perspective, however, one can see the coupling $\sum_s J_s h_s$ as a Noether coupling that gauges the infinite dimensional symmetry algebra [31,55] generated by the charges associated to the currents J_s. Moreover, even in the AdS/CFT context one can give different, Neumann type, boundary conditions to bulk fields, allowing them to fluctuate on the boundary [22,56,57].

where $\hat{H}$ is a differential operator linear in the CHS fields. Such type of one-loop effective actions is the most suitable to be computed by using first-quantized worldline models [58–69]. For instance, free massless scalar particles are described in first quantization by the relativistic worldline action

$$S[x,p,e] = \int_0^1 d\tau \left[p_\mu \dot{x}^\mu - \tfrac{e}{2}\, p^2\right] \quad \Leftrightarrow \quad S[x,e] = \int_0^1 d\tau \, \frac{\dot{x}^2}{2e}\,, \tag{4}$$

where $e(\tau)$ is the einbein, that enforces the mass-shell constraint $p^2 \approx 0$ and, equivalently, ensures local τ-reparametrization invariance. Coupling to a background curved metric $g_{\mu\nu}(x)$ and $U(1)$ gauge field $A_\mu(x)$ can be readily achieved by

$$S_{g,A}[x,p,e] = \int_0^1 d\tau \left[p_\mu \dot{x}^\mu - \tfrac{e}{2}\, g^{\mu\nu}(p_\mu - A_\mu)(p_\nu - A_\nu)\right] \quad \Leftrightarrow \quad S_{g,A}[x,e] = \int_0^1 d\tau \left[\tfrac{1}{2e}\, g_{\mu\nu}\dot{x}^\mu \dot{x}^\nu + A_\mu \dot{x}^\mu\right]. \tag{5}$$

Quantization of the above actions on the circle gives the scalar loop contribution to the QFT one-loop effective action for gravitons and photons. To be precise, when the hamiltonian is *not* of the form $H = p^2 + V(x)$, as it is the case in curved spacetime [70], the naive classical action does not give the correct quantum amplitudes, due to ordering issues in the quantum hamiltonian, and a local counterterm has to be added to the action (5) before using it in the path integral. One can add spinning degrees of freedom to the quantum particle [71–77], in order to give contributions of fields with nonzero spin in the loop. In the present case, since we are interested in the effective action $\Gamma[h]$ generated by a scalar loop, the scalar particle example will suffice.

In order to describe the interaction of the relativistic particle to background CHS fields, we shall employ the action proposed in [78], i.e.,

$$S_h[x,p,e] = \int_0^1 d\tau \left[p_\mu \dot{x}^\mu - e\, G(x,p)\right], \tag{6}$$

where in the generalized hamiltonian $G(x,p) = p^2 + \mathcal{H}(x,p)$ the conformal higher spin fields, contained in the p-power series expansion of $\mathcal{H}(x,p)$, are treated as perturbations over the flat space background p^2. In the low spin example (5) we gave the expression for the action both in phase space and configuration space. In most worldline applications one employs the configuration space action, but in the case at hand it seems much more convenient to stay with the phase space action and to perform the path integral directly in phase space.[8] Indeed, the arbitrary dependence on momenta of $\mathcal{H}(x,p)$ makes the inversion $p = p(\dot{x})$ quite cumbersome, along with the appearence of inverse powers of $\dot{x}^2$ that would produce singularities in perturbation theory. The issue of quantum ordering of the operator $\hat{H}$ of (3) in relation to the classical interaction vertex $\mathcal{H}(x,p)$ will be discussed in the main text.

In the next section we start by reviewing the construction of [7,11], that allows to find the explicit form of the operator $\hat{H}$. We proceed by introducing the above worldline model and discuss its symmetries. Finally, we quantize the action (6) by computing explicitly the path integral on the circle. By doing so we end up with a Scwhinger proper time representation of the effective action $\Gamma[h]$ that allows to extract the logarithmic divergence defining the classical action $S_{\mathrm{CHS}}[h]$. For illustrative purpose we shall rederive the quadratic action [3,7,11], while we plan to address cubic and higher vertices in a future work. We conclude in Section 3 by pointing out some aspects of the present formalism that may be improved, and discussing some interesting directions for future investigations.

---

[8] For very similar reasons, phase space worldline path integrals have been used in [79,80] in the context of non-commutative field theory.

2. Induced Action for Conformal Higher Spins

Let us start by considering a massless complex scalar field in flat spacetime of even dimension d, with action

$$S_0[\phi] = \int d^d x\, \partial^\mu \phi^* \partial_\mu \phi\,. \tag{7}$$

Being a free theory, it possesses an infinite number of conserved currents $J_{\mu(s)} = \phi^* \partial_{\mu_1} ... \partial_{\mu_s} \phi + ...$[9] of arbitrary integer spin $s = 0, 1, 2, ...$ and conformal dimension $\Delta_{J_s} = d - 2 + s$, that can be made traceless thanks to conformal invariance [81,82]. Conservation $\partial^\nu J_{\nu\mu(s-1)} \approx 0$ and tracelessness $J^\alpha{}_{\alpha\mu(s-2)} \approx 0$ hold on the scalar mass-shell $\Box\phi \approx 0$. In this setting one can introduce conformal higher spin fields (CHS) via the Noether interactions

$$S_{\text{int}}[\phi, h] = \sum_{s=0}^{\infty} \frac{(i)^s}{s!} \int d^d x\, J^{\mu(s)}\, h_{\mu(s)}\,, \tag{8}$$

that are invariant, on the free field equations $\Box\phi \approx 0$, under the gauge transformations

$$\delta_{\text{lin}} h_{\mu(s)} = \partial_\mu \varepsilon_{\mu(s-1)} + \eta_{\mu\mu}\, \alpha_{\mu(s-2)}\,, \tag{9}$$

that are the linearized higher spin generalization of the gauge symmetries of conformal gravity. These on-shell symmetries can be deformed to full off-shell ones leaving invariant the total action

$$S[\phi, h] = S_0[\phi] + S_{\text{int}}[\phi, h]\,, \tag{10}$$

by supplementing both the gauge fields and the scalar with extra transformations of the form

$$\delta\phi = \mathcal{O}(\phi)\,, \quad \delta h_s = \delta_{\text{lin}} h_s + \mathcal{O}(h)\,. \tag{11}$$

The UV logarithmically divergent part[10] of the effective action $\Gamma[h]$, induced by the scalar path integral

$$e^{-\Gamma[h]} = \int \mathcal{D}\phi^* \mathcal{D}\phi\, e^{-S[\phi,h]} = \text{Det}^{-1}\left(-\Box + \hat{H}\right)\,, \tag{12}$$

is local[11] and invariant under the full transformation $\delta h_s = \partial\varepsilon_{s-1} + \eta\, \alpha_{s-2} + \mathcal{O}(h)$. It can thus be used to define a fully non-linear classical action for conformal higher spin fields, and at the quadratic level it has been shown to reproduce the free action of [3]. In (12) $\hat{H}$ is a differential operator linear in h_s, whose precise form will be now reviewed following [11].

2.1. Noether Interaction and Symmetries

The generating function of all the traceless conserved currents $J_{\mu(s)}$

$$J(x, u) := \sum_{s=0}^{\infty} \frac{1}{s!} J_{\mu_1 ... \mu_s}(x)\, u^{\mu_1} ... u^{\mu_s} = \sum_{s=0}^{\infty} J_s(x, u) \tag{13}$$

obeying $\partial_u \cdot \partial_x J(x, u) \approx 0$ and $\partial_u^2 J(x, u) \approx 0$ can be written as [11]

$$J(x, u) = \Pi_d\, \mathcal{J}(x, u)\,, \quad \mathcal{J}(x, u) := \phi^*(x + u/2)\phi(x - u/2)\,, \tag{14}$$

---

9 Indices denoted with the same letter and groups of indices $\mu(k)$ are intended as symmetrized with strength one, e.g., $J_{\mu(s)} := J_{(\mu_1 ... \mu_s)}$.

10 The logarithmic divergence is present only in even dimensions, that is the only case we will treat here.

11 The induced action contains vertices with arbitrary powers of higher spin fields but, due to the absence of dimensionful parameters, the number of derivatives is bounded by the number of fields and sum of the spins involved.

where $\mathcal{J}(x,u)$ generates traceful conserved currents, that are mapped to the traceless ones by the operator

$$\Pi_d := \sum_{n=0}^{\infty} \frac{1}{n!(-\hat{N} - \frac{d-5}{2})_n} \left[\frac{\partial^2 - g\,\Box}{16}\right]^n , \tag{15}$$

with the Pochhammer symbol defined by $(a)_n := \frac{\Gamma(a+n)}{\Gamma(a)}$, and

$$\hat{N} := u \cdot \partial_u , \quad \partial := u \cdot \partial_x , \quad g := u^2 . \tag{16}$$

In terms of the higher spin generating function

$$h(x,u) := \sum_{s=0}^{\infty} \frac{1}{s!} h_{\mu_1 \dots \mu_s}(x)\, u^{\mu_1} \dots u^{\mu_s} = \sum_{s=0}^{\infty} h_s(x,u) , \tag{17}$$

the Noether interaction (8) can be written as

$$\begin{aligned} S_{\text{int}}[\phi, h] &= \int d^d x\, J(x, i\partial_u)\, h(x,u)|_{u=0} = \int d^d x\, \mathcal{J}(x, i\partial_u)\, \mathcal{H}(x,u)|_{u=0} \\ &= \int d^d x\, e^{i\partial_u \cdot \partial_v} \mathcal{J}(x,v)\, \mathcal{H}(x,u)|_{u,v=0} , \end{aligned} \tag{18}$$

where the transformed generating function of the gauge fields $\mathcal{H}(x,u)$ is obtained upon integrating by parts the spacetime derivatives in Π_d, and reads

$$\mathcal{H}(x,u) = \mathcal{P}_d\, h(x,u) , \quad \mathcal{P}_d := \sum_{n=0}^{\infty} \frac{1}{n!(\hat{N} + n + \frac{d-3}{2})_n} \left[\frac{\partial^{*2} - \text{Tr}\,\Box}{16}\right]^n , \tag{19}$$

the inverse map being given by [11]

$$h(x,u) = \mathcal{P}_d^{-1} \mathcal{H}(x,u) , \quad \mathcal{P}_d^{-1} := \sum_{n=0}^{\infty} \frac{(-1)^n}{n!(\hat{N} + \frac{d-1}{2})_n} \left[\frac{\partial^{*2} - \text{Tr}\,\Box}{16}\right]^n , \tag{20}$$

where we defined the divergence $\partial^* := \partial_u \cdot \partial_x$ and trace $\text{Tr} := \partial_u^2$ operators. Despite the infinite series appearing in (19), each spin-s component of the conformal fields h_s produces a finite tail of traces and divergences, as it can be seen by rewriting

$$\mathcal{H}(x,u) = \sum_{s=0}^{\infty} \sum_{n=0}^{[s/2]} \frac{1}{n!(s - n + \frac{d-3}{2})_n} \left[\frac{\partial^{*2} - \text{Tr}\,\Box}{16}\right]^n h_s(x,u) . \tag{21}$$

By introducing the Fourier transform of $\mathcal{J}(x,v)$ in v-space:

$$\mathcal{J}(x,v) = \int \frac{d^d p}{(2\pi)^d}\, e^{-iv\cdot p} \rho(x,p) , \tag{22}$$

the interaction (18) can be further rewritten in the form

$$S_{\text{int}}[\phi, h] = \int \frac{d^d x d^d p}{(2\pi)^d}\, \rho(x,p)\, \mathcal{H}(x,p) . \tag{23}$$

In [83] it has been shown that, upon introducing the first quantized Hilbert space where x^μ and $-i\frac{\partial}{\partial x^\mu}$ realize the algebra $[\hat{X}^\mu, \hat{P}_\nu] = i\,\delta^\mu_\nu$ and whereby the field $\phi(x)$ can be written as the wave function $\langle x|\phi\rangle$, the "density matrix" $\rho(x,p)$ is the Weyl symbol of the operator $|\phi\rangle\langle\phi|$. This allows, using the standard tools of Weyl quantization [84–86], to finally cast the action (23) as the inner product

$$S_{\rm int}[\phi,h] = {\rm Tr}\left[|\phi\rangle\langle\phi|\hat{H}\right] = \langle\phi|\hat{H}|\phi\rangle\,, \tag{24}$$

where $\hat{H}(\hat{X},\hat{P})$ is the operator with Weyl symbol given by $\mathcal{H}(x,p)$, i.e.,

$$\hat{H}(\hat{X},\hat{P}) = \int\frac{d^dxd^dp}{(2\pi)^d}\mathcal{H}(x,p)\int\frac{d^dyd^dk}{(2\pi)^d}\,e^{ik\cdot(x-\hat{X})-iy\cdot(p-\hat{P})}\,. \tag{25}$$

The total action entering the path integral (12) can thus be written as

$$S[\phi,h] = \langle\phi|\hat{P}^2+\hat{H}(\hat{X},\hat{P})|\phi\rangle\,, \tag{26}$$

which is clearly invariant under

$$|\phi\rangle \to \hat{O}^{-1}|\phi\rangle\,,\quad (\hat{P}^2+\hat{H}) \to \hat{O}^{\dagger}(\hat{P}^2+\hat{H})\hat{O}\,. \tag{27}$$

In terms of the hermitian operators $\hat{E}$ and $\hat{A}$ defined by $\hat{O} = \exp(\hat{A}+i\,\hat{E})$, the infinitesimal gauge transformations of the CHS fields contained in $\hat{H}$ are given by

$$\delta\hat{H} = i\,[\hat{P}^2+\hat{H},\hat{E}] + \{\hat{P}^2+\hat{H},\hat{A}\} = i\,[\hat{P}^2,\hat{E}] + \{\hat{P}^2,\hat{A}\} + \mathcal{O}(h)\,, \tag{28}$$

the linearized transformations (9) descending from the action of the $\hat{P}^2$ part. The symmetries associated to $\hat{E}$ correspond to the differential ones $\delta_\epsilon h_s = \partial\epsilon_{s-1} + ...$ and are preserved at the quantum level, while the generator $\hat{A}$, corresponding to the generalized Weyl symmetry $\delta_\alpha h_s = \eta\,\alpha_{s-2} + ...$, develops a quantum anomaly due to the non invariant measure of the path integral (12). Nonetheless, it can be shown [11] that the UV logarithmically divergent part of the effective action preserves the full symmetry generated by $\hat{A}+i\hat{E}$, and can indeed be identified with the conformal higher spin action.

2.2. Effective Action and Worldline Path Integral

The effective action $\Gamma[h]$ is given by

$$\Gamma[h] = {\rm Tr}\log\left(\hat{P}^2+\hat{H}\right) = -\int_0^\infty\frac{dT}{T}\,{\rm Tr}\left[e^{-T(\hat{P}^2+\hat{H})}\right] \tag{29}$$

in Schwinger proper time representation. The trace of the heat kernel

$$K[T;h] := {\rm Tr}\left[e^{-T(\hat{P}^2+\hat{H})}\right] \tag{30}$$

admits a Laurent expansion in powers of T when the higher spins in $\hat{H}$ are treated as a perturbation over the flat spin two background $\hat{P}^2$. Accordingly, upon introducing a cut-off Λ in the small-T region, the UV-regulated effective action can be organized according to its divergencies as

$$\Gamma_\Lambda[h] := -\int_{\frac{1}{\Lambda^2}}^\infty\frac{dT}{T}\,K[T;h] = \sum_{n=1}^\infty\Lambda^{2n}\,\Gamma_n[h] + \log\!\Lambda\,S_{\rm CHS}[h] + \Gamma_{\rm fin}[h] + \mathcal{O}(\Lambda^{-2})\,, \tag{31}$$

where the local, gauge invariant coefficient of the logarithmic divergence defines the conformal higher spin action being looked for.

As the heat kernel $K[T;h]$ can be viewed as the trace of the (euclidean) evolution operator associated to the quantum mechanical Hamiltonian $\hat{P}^2+\hat{H}(\hat{X},\hat{P})$, it is natural to represent it via a first quantized path integral. More so, the entire effective action (29) arises from the quantization of a relativistic particle model [78] that we briefly discuss. Consider the point particle hamiltonian action

$$S[x,p;e] = \int_0^1 d\tau \left[p_\mu \dot{x}^\mu - e\, G(x,p)\right] \,, \tag{32}$$

where

$$G(x,p) = p^2 + \mathcal{H}(x,p) \approx 0 \tag{33}$$

is the generalized mass-shell constraint imposed by the Lagrange multiplier $e(\tau)$, that is associated with τ-reparametrization invariance under

$$\delta x^\mu = \xi\{x^\mu, G\}_{\text{P.B.}} \,, \quad \delta p_\mu = \xi\{p_\mu, G\}_{\text{P.B.}} \,, \quad \delta e = \dot{\xi} \,, \tag{34}$$

where $\{,\}_{\text{P.B.}}$ denotes the Poisson bracket and $\xi(\tau)$ is a worldline local parameter. The action (32) describes the propagation of a relativistic spinless particle in the background of the CHS fields contained in $\mathcal{H}$ according to (21). The Lagrange multiplier $e(\tau)$, called einbein, is the gauge field for τ-reparametrizations, and it can be viewed as an intrinsic frame field on the worldline. The infinitesimal gauge transformations of the background fields, generated by

$$\delta_\epsilon \mathcal{H}(x,p) = \{\epsilon(x,p), p^2 + \mathcal{H}(x,p)\}_{\text{P.B.}} \,, \tag{35}$$

leave the action invariant when accompanied by the phase space transformations

$$\delta_\epsilon x^\mu = \{x^\mu, \epsilon(x,p)\}_{\text{P.B.}} \,, \quad \delta_\epsilon p_\mu = \{p_\mu, \epsilon(x,p)\}_{\text{P.B.}} \,. \tag{36}$$

This is the first quantized realization of the $\hat{E}$-type symmetries discussed in the field theory language, while the counterpart of the generalized Weyl symmetries $\hat{A}$ can be viewed as the invariance of the constraint surface $G(x,p) \approx 0$ under $G(x,p) \;\to\; e^{\alpha(x,p)}\, G(x,p)$. The action is indeed invariant under the combined transformations

$$\delta_\alpha \mathcal{H}(x,p) = \alpha(x,p)\left(p^2 + \mathcal{H}(x,p)\right), \quad \delta_\alpha e = -\alpha(x,p)\, e \tag{37}$$

and, as we shall see next, the transformation of the einbein is responsible for breaking the α-symmetry at the quantum level.

As it is well known from the cases of interaction with scalar, vector and gravitational backgrounds, the effective action $\Gamma[h]$ can be obtained by quantizing the action (32) on the circle:

$$\Gamma[h] = \int_{S^1} \frac{DxDpDe}{\text{VolGauge}}\, e^{-S_E[x,p;e]} \,, \tag{38}$$

where division by the gauge group volume entails the gauge fixing procedure for the local τ-reparametrizations, and S_E denotes the euclidean version of the action (32), i.e.,

$$S_E[x,p;e] = \int_0^1 d\tau \left[-i\, p_\mu \dot{x}^\mu + e\, G(x,p)\right] \,. \tag{39}$$

As mentioned in the Introduction, when the quantum hamiltonian contains mixing of coordinates and momenta, the naive path integral fails in general to provide the correct quantization. For a given classical hamiltonian $H(x,p)$, the functional integral in phase space produces transition amplitudes corresponding to the quantum Hamiltonian $\hat{H}_W(\hat{x},\hat{p})$ obtained from the classical one by Weyl ordering [87]. The correctness of the choice of $\mathcal{H}$ as classical vertex is thus ensured by the relation (25), that greatly simplifies the model. In gauge fixing the local symmetry (34) on the circle, it is customary to fix the einbein to a constant: $e(\tau) = T$ that plays the role of Schwinger's proper time and breaks the Weyl symmetry (37). The ghost system associated to τ-reparametrizations has locally trivial action, but its Faddeev-Popov determinant contributes on S^1 topology with a factor of T^{-1}, yielding

$$\Gamma[h] = \int_0^\infty \frac{dT}{T} \int_{S^1} DxDp\, e^{-S_E[x,p;T]} \,, \tag{40}$$

where the integral over T is the finite dimensional remnant[12] of the functional e-integral, and the gauge fixed action is simply obtained by replacing $e(\tau) = T$. We shall now compute the trace of the heat kernel

$$K[T;h] = \int_{S^1} DxDp\, e^{-S_E[x,p;T]} \tag{41}$$

by treating the phase space vertex $\mathcal{H}(x,p)$ as a perturbation over the free action

$$S_2[x,p] = \int_0^1 d\tau \left[T\, p^2 - i\, p_\mu \dot{x}^\mu\right] . \tag{42}$$

First of all we shall extract the zero mode from the periodic trajectories $x^\mu(\tau)$:

$$x^\mu(\tau) = x^\mu + q^\mu(\tau) \,, \quad x^\mu := \int_0^1 d\tau\, x^\mu(\tau) \;\to\; \int_0^1 d\tau\, q^\mu(\tau) = 0 \,, \tag{43}$$

so that the functional measure splits as $\int_{S^1} Dx = \int d^dx \int_{\mathrm{PBC}'} Dq$, where PBC′ denotes periodic boundary condition with the zero mode removed. Expectation values w.r.t. the free action (42) are denoted by

$$\langle F(q,p)\rangle := \frac{\int_{\mathrm{PBC}'} Dq \int Dp\, F(q,p)\, e^{-S_2[q,p]}}{\int_{\mathrm{PBC}'} Dq \int Dp\, e^{-S_2[q,p]}} \tag{44}$$

and the trace of the heat kernel can be written as

$$K[T;h] = \int \frac{d^dx}{(4\pi T)^{d/2}} \left\langle e^{-T\int_0^1 d\tau\, \mathcal{H}(x+q,p)} \right\rangle = \int \frac{d^dx}{(4\pi T)^{d/2}} \sum_{n=0}^{\infty} T^n\, \mathcal{V}_n[T;h] \,, \tag{45}$$

where $(4\pi T)^{-d/2}$ is the value of the free path integral and the n-field effective vertex is given by

$$\begin{aligned} \mathcal{V}_n[T;h] &= \frac{(-1)^n}{n!} \int_0^1 d\tau_1 \cdots \int_0^1 d\tau_n \left\langle \prod_{i=1}^n \mathcal{H}(x+q(\tau_i), p(\tau_i)) \right\rangle \\ &= \frac{(-1)^n}{n!} \int_0^1 d\tau_1 \cdots \int_0^1 d\tau_n \left\langle e^{\sum_{i=1}^n q_i\cdot\partial_{x_i} + p_i\cdot\partial_{u_i}} \right\rangle \mathcal{H}(x_1,u_1)\cdots\mathcal{H}(x_n,u_n)\big|_{\substack{x_i=x\\ u_i=0}} \\ &=: \hat{\mathcal{V}}_n(T;\partial_{x_i},\partial_{u_i})\, \mathcal{H}(x_1,u_1)\cdots\mathcal{H}(x_n,u_n)\big|_{\substack{x_i=x\\ u_i=0}} \,, \end{aligned} \tag{46}$$

where $q_i := q(\tau_i)$, $p_i := p(\tau_i)$ and we expanded the generating functions $\mathcal{H}(x+q_i,p_i)$ around $(x,0)$. In terms of the currents

$$j_n(\tau) := \sum_{i=1}^n \delta(\tau-\tau_i)\partial_{x_i} \,, \quad k_n(\tau) := \sum_{i=1}^n \delta(\tau-\tau_i)\partial_{u_i} \,, \tag{47}$$

the quantum average above can be recast in the form of a generating functional:

$$\left\langle e^{\sum_{i=1}^n q_i\cdot\partial_{x_i} + p_i\cdot\partial_{u_i}} \right\rangle = \left\langle e^{\int_0^1 d\tau [q(\tau)\cdot j_n(\tau) + p(\tau)\cdot k_n(\tau)]} \right\rangle . \tag{48}$$

---

[12] The quantity $T = \int_0^1 d\tau\, e(\tau)$ is gauge invariant on the circle; hence it constitutes a modulus to be integrated over after gauge fixing.

This is a quadratic path integral and can be computed exactly, yielding

$$\left\langle e^{\int_0^1 d\tau[q(\tau)\cdot j_n(\tau)+p(\tau)\cdot k_n(\tau)]}\right\rangle = \exp\Big\{\frac{1}{2}\int_0^1 d\tau \int_0^1 d\sigma K_n^T(\tau)\, G(\tau,\sigma) K_n(\sigma)\Big\} \tag{49}$$

for the column vector $K_n(\tau) = \begin{pmatrix} k_n(\tau) \\ j_n(\tau) \end{pmatrix}$. Here $G(\tau,\sigma)$ is the matrix of the phase space Green's functions[13]

$$\langle p_\mu(\tau) p_\nu(\sigma)\rangle = \frac{1}{2T}\eta_{\mu\nu}\,, \quad \langle p_\mu(\tau) q^\nu(\sigma)\rangle = i\,\delta_\mu^\nu\, f(\sigma-\tau)\,, \quad \langle q^\mu(\tau) q^\nu(\sigma)\rangle = 2T\,\eta^{\mu\nu}\, g(\tau-\sigma) \tag{50}$$

with propagators

$$f(\tau) = -\tau + \tfrac{1}{2}\,\mathrm{sign}(\tau)\,, \quad g(\tau) = \tfrac{1}{2}\,(\tau^2 - |\tau| + \tfrac{1}{6})\,, \quad \tau \in [-1,1]\,. \tag{51}$$

By using the currents (47) in (49) one obtains

$$\begin{aligned} \hat{V}_n(T;\partial_{x_i},\partial_{u_i}) &= \frac{(-1)^n}{n!}\int_0^1 d\tau_1 \cdots \int_0^1 d\tau_n \\ &\times \exp\frac{1}{2}\sum_{i,j=1}^{n}\left[\frac{1}{2T}\partial_{u_i}\cdot\partial_{u_j} + 2i\, f(\tau_i-\tau_j)\,\partial_{x_i}\cdot\partial_{u_j} + 2T\, g(\tau_i-\tau_j)\,\partial_{x_i}\cdot\partial_{x_j}\right]. \end{aligned} \tag{52}$$

To manipulate it further, let us notice that $\sum_{i=1}^n \partial_{x_i} \sim 0$ is a total derivative, according to (46) and (45). This allows to consistently drop the constant part in every $g(\tau)$ propagator, leaving the effective propagator $\hat{g}(\tau) := \frac{1}{2}(\tau^2 - |\tau|)$. The rigid translation invariance under $\tau_i \to \tau_i + c$, together with the periodicity of the trajectories over S^1, allows to fix one τ variable[14], let us say $\tau_n = 0$ and, thanks to the symmetry under permutations of the τ_i, that is manifest from the first line of (46), one can also transform the τ-integral: $\int_0^1 d\tau_1 ... \int_0^1 d\tau_{n-1} \to (n-1)! \int_0^1 d\tau_1 \int_0^{\tau_1} d\tau_2 ... \int_0^{\tau_{n-2}} d\tau_{n-1}$, finally obtaining for the effective vertex

$$\begin{aligned} \mathcal{V}_n[T;h] &= \frac{(-1)^n}{n} e^{\frac{1}{4T}\partial_U^2}\int_0^1 d\tau_1 \int_0^{\tau_1} d\tau_2 \cdots \int_0^{\tau_{n-2}} d\tau_{n-1} \\ &\times \exp \sum_{i<j}^{(\tau_n=0)} \Big\{-i(\tau_{ij}-\tfrac{1}{2})(\partial_{x_i}\cdot\partial_{u_j} - \partial_{x_j}\cdot\partial_{u_i}) + T\,\tau_{ij}(\tau_{ij}-1)\partial_{x_i}\cdot\partial_{x_j}\Big\}\, \mathcal{H}(x_1,u_1)\cdots\mathcal{H}(x_n,u_n)|_{\substack{x_i=x \\ u_i=0}} \end{aligned} \tag{53}$$

where $\tau_{ij} := \tau_i - \tau_j$ and $\partial_U := \sum_{i=1}^n \partial_{u_i}$. For any given set of spins $\{s_1, ..., s_n\}$ of the CHS fields h_{s_i}, the maximal number of u-derivatives is bounded by $S := \sum_{i=1}^n s_i$, so that the exponential $e^{\partial_U^2/4T}$ contributes with the maximal negative power $T^{-[S/2]}$, making the Laurent expansion

$$\mathcal{V}_n[T;h] = \sum_{k=-\infty}^{\infty} T^k\, \mathcal{V}_n^{(k)}[h] \tag{54}$$

---

13 See Appendix A for details.

14 This can be seen by just changing variables in integrals of periodic and translation invariant functions. However, a more precise justification comes from the gauge fixing procedure on the circle: The einbein $e(\tau)$ possesses, on S^1 topology, a Killing vector that is not fixed by the gauge $e(\tau) = T$ and that generates global translations around the circle. A natural way to fix the leftover global symmetry is then to fix the position of one vertex on the circle, e.g., by setting $\tau_n = 0$, as it is customary in String Theory.

well defined. The coefficient giving rise to the logarithmic divergence is thus the one of order $k = \frac{d}{2} - n$, and the CHS action can be identified as[15]

$$S_{\mathrm{CHS}}[h] = \int d^d x \sum_{n=2}^{\infty} \mathcal{V}_n^{(d/2-n)}[h] \,. \tag{55}$$

All the vertices of (55) can be in principle computed using (53) but, since locality has to be manifest in the spin decomposition of the $h(x,u)$ basis, it would be desirable to develop a formalism that avoids the introduction of the $\mathcal{H}(x,u)$ generating function. In such a case, all the differential operators $\hat{V}_n(T;\partial_{x_i},\partial_{u_i})$ should reduce to finite polynomials of homogeneous degree in spacetime derivatives.

2.3. The Quadratic Action

The effective vertex (53) is an equivalent representation of the one obtained in [11], but for illustrative purposes we shall rederive the quadratic action by computing $\mathcal{V}_2^{(\frac{d}{2}-2)}[h]$. From (53) one has

$$\mathcal{V}_2[T;h] = \tfrac{1}{2}\, e^{\frac{1}{4T}\partial_U^2 + \frac{i}{2}\partial_x\cdot\partial_U}\, \mathcal{F}(-i\partial_x\cdot\partial_U, T\,\Box)\, \mathcal{H}(x,u)\mathcal{H}(x',u')|_{\substack{x'=x\\ u,u'=0}} \tag{56}$$

where we used $\partial_{x'} \sim -\partial_x$ and

$$\mathcal{F}(\alpha,\beta) := \int_0^1 d\tau\, e^{\alpha\tau+\beta\tau(1-\tau)} = \sum_{n=0}^{\infty}\frac{1}{n!}\,\mathcal{F}_n(\alpha)\,\beta^n \,. \tag{57}$$

The functions $\mathcal{F}_n(\alpha)$ can be computed as hypergeometric integrals (A16):

$$\mathcal{F}_n(\alpha) = \frac{(n!)^2}{(2n+1)!}\, {}_1F_1(n+1;2n+2;\alpha)\,, \tag{58}$$

and using Kummer's Formula (A17) can be recast in terms of Bessel functions (A9), giving

$$\mathcal{V}_2[T;h] = \tfrac{\sqrt{\pi}}{2}\, e^{\frac{1}{4T}\partial_U^2} \sum_{n=0}^{\infty} (\partial_x\cdot\partial_U)^{-n-\frac{1}{2}}\, J_{n+\frac{1}{2}}\left(\frac{\partial_x\cdot\partial_U}{2}\right) (T\,\Box)^n\, \mathcal{H}(x,u)\mathcal{H}(x',u')|_{\substack{x'=x\\ u,u'=0}} \,. \tag{59}$$

It is now possible to extract the contribution of order $T^{\frac{d}{2}-2}$, yielding

$$\mathcal{V}_2^{(\frac{d-4}{2})}[h] = \sqrt{\tfrac{\pi}{8}}\left[\tfrac{1}{2}\sqrt{-\partial_{U\perp}^2\partial_x^2}\right]^{-\frac{d-3}{2}} J_{\frac{d-3}{2}}\left(\tfrac{1}{2}\sqrt{-\partial_{U\perp}^2\partial_x^2}\right)\left(\frac{\Box}{2}\right)^{\frac{d-4}{2}} \mathcal{H}(x,u)\mathcal{H}(x',u')|_{\substack{x'=x\\ u,u'=0}} \,, \tag{60}$$

where we used Lommel's expansion Formula (A11), and the transverse projection is defined as

$$v_\perp^\mu := v^\mu - \frac{v\cdot\partial_x\partial_x^\mu}{\Box} \,, \tag{61}$$

so that $\partial_{U\perp}^2\Box = \partial_U^2\Box - (\partial_x\cdot\partial_U)^2$. The above result coincides with the one of [11], and it can be seen that the Bessel function "undresses" the $\mathcal{H}$ fields, leaving a finite degree polynomial acting on the

[15] The field independent $\mathcal{V}_0(T) = 1$ cannot contribute to the logarithmic divergence and neither can the linear $\mathcal{V}_1[T;h]$ in $d \geq 4$.

two h fields. To do so one applies Gegenbauer addition theorem (A13) to the above Bessel function, with the triplet of variables $Z := \frac{1}{2}\sqrt{-\partial^2_{U\perp}\Box}$ and[16] $z_i := \frac{1}{2}\sqrt{-\partial^2_{u_i\perp}\Box_i}$ obeying

$$Z^2 = z^2 + z'^2 - 2\,zz'\,w \quad \text{for} \quad w = \frac{\partial_{u\perp}\cdot\partial_{u'\perp}}{\sqrt{\partial^2_{u\perp}\partial^2_{u'\perp}}}\,. \tag{62}$$

The resulting $J_{n+\frac{d-3}{2}}(z_i)$ that appear from the addition theorem produce the inverse maps $\mathcal{P}_d^{-1}$ when acting on the corresponding $\mathcal{H}(x_i, u_i)$ as[17]

$$\left[z_i^{-n-\frac{d-3}{2}}\, J_{n+\frac{d-3}{2}}(z_i)\,\mathcal{H}(x_i,u_i)\right]\Big|_n = h_n(x_i,u_i)\,, \tag{63}$$

and one is left with

$$\mathcal{V}_2^{(\frac{d-4}{2})}[h] = \sqrt{\tfrac{\pi}{8}}\sum_{n=0}^{\infty} c_n(d)\,\left(\frac{\Box}{2}\right)^{n+\frac{d-4}{2}}\left(\partial^2_{u\perp}\partial^2_{u'\perp}\right)^{n/2} C_n^{\frac{d-3}{2}}\left(\frac{\partial_{u\perp}\cdot\partial_{u'\perp}}{\sqrt{\partial^2_{u\perp}\partial^2_{u'\perp}}}\right) h(x,u)h(x',u')|_{\substack{x'=x\\u,u'=0}}\,, \tag{64}$$

where $C_n^{\nu}(w)$ is the Gegenbauer polynomial and $c_n(d) := \frac{2^{-3n-\frac{d-3}{2}}}{\Gamma(n+\frac{d-1}{2})(\frac{d-3}{2})_n}$. From the form of $\partial_{u_i\perp}$ one can see that the above expression is indeed local for each n and of homogeneous degree $2n+d-4$ in spacetime derivatives. It is also easy to view, from the definition of Gegenbauer polynomials (A14), that the sum over n is diagonal in contractions, being of homogeneous degree $(\partial_u\partial_{u'})^n$. The above expression is proportional for each n to the corresponding transverse and traceless projector[18] P_n, that can be displayed to write the quadratic action in the form (1) that is manifestly gauge invariant under the linearized transformations (9):

$$S_{2\,\mathrm{CHS}} = \int d^dx \sum_{s=0}^{\infty} c_s\, h_s(x,u)\, P_s\left(\overleftarrow{\partial_u}, \overrightarrow{\partial_v}\right)\Box^{s+\frac{d-4}{2}}\, h_s(x,v)\,, \tag{65}$$

where we discarded a spin-independent constant and $c_s = \frac{1}{2^{3s}\Gamma(s+\frac{d-1}{2})}$.

3. Discussion and Conclusions

In this paper we have provided a worldline path integral representation for the non-linear conformal higher spin action [7,11] in arbitrary even dimensions. We have rederived the quadratic part of the action, and we plan to come back in the future for the computation of cubic and higher vertices, some of which have been computed in [88,89], in transverse-traceless gauge, in the context of scattering amplitudes calculations.

The example of the quadratic action suggests that the "undressing" maps $\mathcal{P}_d^{-1}$ should appear at all orders in the differential operators $\hat{V}_n(T; \partial_{x_i}, \partial_{u_i})$, leaving finite degree polynomials acting on a string of fields $h_{s_1} \ldots h_{s_n}$. From the representation (53) it is not transparent how this should take place. For this reason, it would be interesting to find a way to avoid the introduction of the dressed generating function $\mathcal{H}(x,u)$, and work directly in the basis of CHS $h(x,u)$. To this goal, when restricting to four dimensions, a considerable advantage could come by working in terms of $sl(2,\mathbb{C})$ spinors instead of tensors. All the trace projections would become trivial and one could work directly in terms of conformal primary currents J_s.

16 In the variables z_i one can exchange x with x' for free.
17 For details see the original derivation [11].
18 See appendix C for the explicit form of the projectors.

Another issue of (non-linear) field redefinitions is apparent when looking at the low spin content of the Noether interaction (8): the linear coupling

$$\sum_{s=0}^{2} h_s J_s \sim h_0\,\phi^*\phi + h_1^{\mu}\,\phi^*\partial_\mu\phi + \tfrac{1}{2}\,h_2^{\mu\nu}\,\phi^*\partial_\mu\partial_\nu\phi \tag{66}$$

does not coincide with the standard Weyl and $U(1)$ invariant coupling of a complex scalar to a vector gauge field in curved spacetime, i.e.,

$$S = \int d^dx\sqrt{g}\left[g^{\mu\nu}D_\mu\phi^* D_\nu\phi + \tfrac{d-2}{4(d-1)}\,R\,\phi^*\phi\right], \tag{67}$$

and the basis (h_0, h_1, h_2) is related to the geometric $(A_\mu, g_{\mu\nu} = \eta_{\mu\nu} + h_{\mu\nu})$ by a non-linear field redefinition. This issue has been discussed, for instance, in [7,11,89]. In fact, the covariant description of CHS fields in curved (maybe Bach-flat) background is still an open problem [17–19], and it underpins the question of vanishing Weyl anomalies. To this end, it would be interesting to find a first quantized origin of CHS fields, since at the worldline (or worldsheet) level it could be easier to achieve a covariant description, and the sum over spins, that is crucial in proving the vanishing of anomalies as well as triviality of scattering amplitudes, would be accounted for by the worldline fields.

Acknowledgments: We would like to thank Arkady Tseytlin, Evgeny Skvortsov, Dmitry Ponomarev and David De Filippi for useful discussions. The author thanks the Imperial College London and the Ludwig Maximilian University of Munich for kind hospitality during the final stages of this work. The work of R.B. was supported by a PDR "Gravity and extensions" from the F.R.S.-FNRS (Belgium).

Conflicts of Interest: The authors declare no conflict of interest.

Appendix A. Worldline Phase Space Propagators

In this section we will derive the phase space propagators associated with the euclidean worldline action on the circle

$$S_2[x,p] = \int_0^1 d\tau\,\left[T\,p^2 - i\,p_\mu \dot{x}^\mu\right]. \tag{A1}$$

Upon extracting the zero mode from the periodic trajectories $x^\mu(\tau)$, one goes to Fourier space

$$\begin{aligned} x^\mu(\tau) &= x_0^\mu + q^\mu(\tau)\,, \quad q^\mu(\tau) = \sum_{n\in\mathbb{Z}\setminus\{0\}} q_n^\mu\,e^{2\pi i n\tau}\,, \\ p_\mu(\tau) &= \sum_{n\in\mathbb{Z}} p_{\mu\,n}\,e^{2\pi i n\tau}\,. \end{aligned} \tag{A2}$$

The Fourier modes obey the reality conditions

$$(x_0, p_0) \in \mathbb{R}\,, \quad q_n^* = q_{-n}\,, \quad p_n^* = p_{-n} \tag{A3}$$

and the action (A1) reads

$$S_2 = T\,p_0^2 + \sum_{n=1}^{\infty} Z_n^\dagger K_n\,Z_n \tag{A4}$$

in terms of the phase space vector Z_n and kinetic matrix K_n

$$Z_n := \begin{pmatrix} p_n \\ q_n \end{pmatrix}, \quad K_n := \begin{pmatrix} 2T & 2\pi n \\ -2\pi n & 0 \end{pmatrix}, \tag{A5}$$

where we suppressed all the spacetime indices. From the inverse matrix

$$K_n^{-1} = \begin{pmatrix} 0 & -\frac{1}{2\pi n} \\ \frac{1}{2\pi n} & \frac{2T}{4\pi^2 n^2} \end{pmatrix} \tag{A6}$$

it is immediate to extract the two-point functions

$$\langle p_\mu(\tau)\, p_\nu(\sigma)\rangle = \frac{\eta_{\mu\nu}}{2T}\,, \quad \langle p_\mu(\tau)\, q^\nu(\sigma)\rangle = i\,\delta^\nu_\mu\, f(\sigma-\tau)\,, \quad \langle q^\mu(\tau)\, q^\nu(\sigma)\rangle = 2T\,\eta_{\mu\nu}\, g(\tau-\sigma)\,. \tag{A7}$$

The above propagators are defined in terms of their Fourier series and read

$$\begin{aligned} f(\tau) &:= \sum_{n=1}^{\infty} \frac{1}{\pi n}\,\sin(2\pi n\,\tau) = -\tau + \tfrac{1}{2}\,\mathrm{sign}(\tau)\,, \\ g(\tau) &:= \sum_{n=1}^{\infty} \frac{1}{2\pi^2 n^2}\,\cos(2\pi n\,\tau) = \tfrac{1}{2}\,\tau^2 - \tfrac{1}{2}\,|\tau| + \tfrac{1}{12}\,, \end{aligned} \tag{A8}$$

where the latter expressions in terms of elementary functions hold in the interval $[-1,1]$.

Appendix B. Special Functions

We collect here the definitions and formulas that are relevant to the main text. The Bessel function of the first kind can be defined by the series

$$J_\nu(z) := \sum_{k=0}^{\infty} \frac{(-1)^k}{k!\,\Gamma(\nu+k+1)} \left(\frac{z}{2}\right)^{\nu+2k}\,, \tag{A9}$$

while the modified Bessel function $I_\nu(z)$ is given by

$$I_\nu(z) := \sum_{k=0}^{\infty} \frac{1}{k!\,\Gamma(\nu+k+1)} \left(\frac{z}{2}\right)^{\nu+2k} = i^{-\nu} J_\nu(iz)\,. \tag{A10}$$

The Lommel expansion formula reads

$$\sqrt{z+h}^{\,-\nu} J_\nu\left(\sqrt{z+h}\right) = \sum_{k=0}^{\infty} \tfrac{1}{k!} \left(-\tfrac{h}{2}\right)^k \sqrt{z}^{\,-\nu-k} J_{\nu+k}\left(\sqrt{z}\right) \tag{A11}$$

and, for a triplet (ω, x, y) obeying

$$\omega^2 = x^2 + y^2 - 2\,xy\,\cos\theta\,, \tag{A12}$$

one has the Gegenbauer addition theorem:

$$\omega^{-\nu} J_\nu(\omega) = 2^\nu \Gamma(\nu) \sum_{n=0}^{\infty} (\nu+n)\, x^{-\nu} J_{\nu+n}(x)\, y^{-\nu} J_{\nu+n}(y)\, C_n^\nu(\cos\theta)\,, \tag{A13}$$

where $C_n^\nu(z)$ is the Gegenbauer polynomial defined by

$$C_n^\nu(z) = \sum_{k=0}^{[n/2]} \frac{(-1)^k (\nu)_{n-k}}{k!\,(n-2k)!}\,(2z)^{n-2k}\,. \tag{A14}$$

The generalized hypergeometric function ${}_pF_q$ is defined by the series

$$ {}_pF_q(a_1,...,a_p;b_1,...,b_q;z) = \sum_{n=0}^{\infty} \frac{(a_1)_n \cdots (a_p)_n}{(b_1)_n \cdots (b_q)_n} \frac{z^n}{n!} . \tag{A15} $$

The confluent hypergeometric function ${}_1F_1(a;b;z)$ admits the integral representation

$$ {}_1F_1(a;b;z) = \frac{\Gamma(b)}{\Gamma(a)\Gamma(b-a)} \int_0^1 du\, e^{zu} u^{a-1}(1-u)^{b-a-1} \tag{A16} $$

and for $b = 2a$ it is related to the ${}_0F_1$ series and thus to the Bessel function via Kummer's formula:

$$ {}_1F_1(a;2a;z) = e^{z/2}{}_0F_1\left(;a+\tfrac{1}{2};\tfrac{z^2}{16}\right) = e^{z/2}\left(\frac{z}{4}\right)^{\frac{1}{2}-a} \Gamma(a+\tfrac{1}{2})\, I_{a-\frac{1}{2}}(z/2) . \tag{A17} $$

Appendix C. Transverse-Traceless Projectors

The transverse-traceless projectors ${P_{\mu(s)}}^{\nu(s)}$ of spin s, obeying

$$ \eta^{\alpha\beta}\, {P_{\alpha\beta\mu(s-2)}}^{\nu(s)} = 0 = {P_{\mu(s)}}^{\alpha\beta\nu(s-2)}\, \eta_{\alpha\beta} , \quad \partial^\alpha {P_{\alpha\mu(s-1)}}^{\nu(s)} = 0 = {P_{\mu(s)}}^{\alpha\nu(s-1)} \partial_\alpha \tag{A18} $$

with normalization

$$ {P_{\mu(s)}}^{\lambda(s)} {P_{\lambda(s)}}^{\nu(s)} = {P_{\mu(s)}}^{\nu(s)} , \tag{A19} $$

can be built from s powers of the corresponding spin one transverse projector

$$ {P_\mu}^\nu := \delta_\mu^\nu - \frac{\partial_\mu \partial^\nu}{\Box} \tag{A20} $$

as

$$ {P_{\mu(s)}}^{\nu(s)} = \sum_{k=0}^{[s/2]} \alpha_k(s)\, (P_{\mu\mu}\, P^{\nu\nu})^k\, ({P_\mu}^\nu)^{s-2k} , \tag{A21} $$

with coefficients $\alpha_k(s)$ being fixed by tracelessness as

$$ \alpha_k(s) = \frac{(-1)^k s!\, \Gamma(s-k+\frac{d-3}{2})}{4^k\, k!(s-2k)!\, \Gamma(s+\frac{d-3}{2})} . \tag{A22} $$

The generating function of the spin s projector:

$$ P_s(u,v) := \frac{1}{s!}\, u^{\mu_1} ... u^{\mu_s}\, {P_{\mu(s)}}^{\nu(s)}\, v_{\nu_1} ... v_{\nu_s} , \tag{A23} $$

acts on the generating function of a spin s field $h_s(x,u) = \frac{1}{s!} h_{\mu(s)} (u^\mu)^s$ as

$$ (P_s\, h_s)(x,u) := \tfrac{1}{s!}\, {P_{\mu(s)}}^{\nu(s)} h_{\nu(s)}(x) (u^\mu)^s = P_s(u, \partial_v)\, h_s(x,v) , \tag{A24} $$

and it can be written in terms of Gegenbauer polynomials as

$$ P_s(u,v) = \frac{\Gamma(\frac{d-3}{2})}{2^s\, \Gamma(s+\frac{d-3}{2})}\, (|u_\perp||v_\perp|)^s\, C_s^{\frac{d-3}{2}}\left(\frac{u_\perp \cdot v_\perp}{|u_\perp||v_\perp|}\right) , \tag{A25} $$

where transverse vectors are defined by

$$ u_\perp^\mu := u^\mu - \frac{u \cdot \partial \partial^\mu}{\Box} . \tag{A26} $$

References

1. Kaku, M.; Townsend, P.K.; van Nieuwenhuizen, P. Properties of Conformal Supergravity. *Phys. Rev. D* **1978**, *17*, 3179–3187.
2. Bergshoeff, E.; de Roo, M.; de Wit, B. Extended Conformal Supergravity. *Nucl. Phys. B* **1981**, *182*, 173–204.
3. Fradkin, E.S.; Tseytlin, A.A. Conformal Supergravity. *Phys. Rep.* **1985**, *119*, 233–362.
4. Siegel, W. All Free Conformal Representations in All Dimensions. *Int. J. Mod. Phys. A* **1989**, *4*, 2015–2020.
5. Fradkin, E.S.; Linetsky, V.Y. Cubic Interaction in Conformal Theory of Integer Higher Spin Fields in Four-dimensional Space-time. *Phys. Lett. B* **1989**, *231*, 97–106.
6. Tseytlin, A.A. On limits of superstring in $AdS_5 \times S^5$. *Theor. Math. Phys.* **2002**, *133*, 1376–1389.
7. Segal, A.Y. Conformal higher spin theory. *Nucl. Phys. B* **2003**, *664*, 59–130.
8. Shaynkman, O.V.; Tipunin, I.Y.; Vasiliev, M.A. Unfolded form of conformal equations in M dimensions and o(M + 2) modules. *Rev. Math. Phys.* **2006**, *18*, 823–886.
9. Marnelius, R. Lagrangian conformal higher spin theory. *arXiv* **2008**, arXiv:0805.4686.
10. Vasiliev, M.A. Bosonic conformal higher-spin fields of any symmetry. *Nucl. Phys. B* **2010**, *829*, 176–224.
11. Bekaert, X.; Joung, E.; Mourad, J. Effective action in a higher-spin background. *J. High Energy Phys.* **2011**, *2011*, 48.
12. Bandos, I.A.; de Azcarraga, J.A.; Meliveo, C. Extended supersymmetry in massless conformal higher spin theory. *Nucl. Phys. B* **2011**, *853*, 760–776.
13. Bekaert, X.; Grigoriev, M. Notes on the ambient approach to boundary values of AdS gauge fields. *J. Phys. A* **2013**, *46*, 214008.
14. Haehnel, P.; McLoughlin, T. Conformal Higher Spin Theory and Twistor Space Actions. *arXiv* **2016**, arXiv:1604.08209.
15. Metsaev, R.R. Ordinary-derivative formulation of conformal low spin fields. *J. High Energy Phys.* **2012**, *2012*, 64.
16. Metsaev, R.R. Ordinary-derivative formulation of conformal totally symmetric arbitrary spin bosonic fields. *J. High Energy Phys.* **2012**, *2012*, 62.
17. Nutma, T.; Taronna, M. On conformal higher spin wave operators. *J. High Energy Phys.* **2014**, *2014*, 66.
18. Grigoriev, M.; Tseytlin, A.A. On conformal higher spins in curved background. *J. Phys. A* **2017**, *50*, 125401.
19. Beccaria, M.; Tseytlin, A.A. On induced action for conformal higher spins in curved background. *Nucl. Phys. B* **2017**, *919*, 359–383.
20. Fradkin, E.S.; Tseytlin, A.A. One Loop Beta Function in Conformal Supergravities. *Nucl. Phys. B* **1982**, *203*, 157–178.
21. Fradkin, E.S.; Tseytlin, A.A. Conformal Anomaly in Weyl Theory and Anomaly Free Superconformal Theories. *Phys. Lett. B* **1984**, *134*, 187–193.
22. Giombi, S.; Klebanov, I.R.; Pufu, S.S.; Safdi, B.R.; Tarnopolsky, G. AdS Description of Induced Higher-Spin Gauge Theory. *J. High Energy Phys.* **2013**, *2013*, 16.
23. Tseytlin, A.A. On partition function and Weyl anomaly of conformal higher spin fields. *Nucl. Phys. B* **2013**, *877*, 598–631.
24. Giombi, S.; Klebanov, I.R.; Safdi, B.R. Higher Spin AdS_{d+1}/CFT_d at One Loop. *Phys. Rev. D* **2014**, *89*, 084004.
25. Beccaria, M.; Tseytlin, A.A. Higher spins in AdS_5 at one loop: vacuum energy, boundary conformal anomalies and AdS/CFT. *J. High Energy Phys.* **2014**, *2014*, 114.
26. Beccaria, M.; Tseytlin, A.A. On higher spin partition functions. *J. Phys. A* **2015**, *48*, 275401.
27. Vasiliev, M.A. Consistent equation for interacting gauge fields of all spins in (3+1)-dimensions. *Phys. Lett. B* **1990**, *243*, 378.
28. Vasiliev, M.A. Properties of equations of motion of interacting gauge fields of all spins in (3+1)-dimensions. *Class. Quant. Grav.* **1991**, *8*, 1387–1417.
29. Vasiliev, M.A. More on equations of motion for interacting massless fields of all spins in (3+1)-dimensions. *Phys. Lett. B* **1992**, *285*, 225–234.
30. Vasiliev, M.A. Higher spin gauge theories: Star product and AdS space. In *The Many Faces of the Superworld*; Shifman, M.A., Ed.; World Scientific: Singapore, 2000.
31. Vasiliev, M.A. Nonlinear equations for symmetric massless higher spin fields in (A)dS(d). *Phys. Lett. B* **2003**, *567*, 139–151.

32. Bekaert, X.; Cnockaert, S.; Iazeolla, C.; Vasiliev, M.A. Nonlinear higher spin theories in various dimensions. *arXiv* **2005**, arXiv:hep-th/0503128.
33. Didenko, V.E. Skvortsov, E.D. Elements of Vasiliev theory. *arXiv* **2014**, arXiv:1401.2975.
34. Sezgin, E.; Sundell, P. Massless higher spins and holography. *Nucl. Phys. B* **2002**, *644*, 303–370; Erratum in **2003**, *660*, 403.
35. Klebanov, I.R.; Polyakov, A.M. AdS dual of the critical O(N) vector model. *Phys. Lett. B* **2002**, *550*, 213–219.
36. Giombi, S.; Yin, X. Higher Spin Gauge Theory and Holography: The Three-Point Functions. *J. High Energy Phys.* **2010**, *2010*, 115.
37. Giombi, S.; Yin, X. Higher Spins in AdS and Twistorial Holography. *J. High Energy Phys.* **2011**, *2011*, 86.
38. Giombi, S.; Yin, X. The Higher Spin/Vector Model Duality. *J. Phys. A* **2013**, *46*, 214003.
39. Giombi, S. Higher Spin—CFT Duality. In *New Frontiers in Fields and Strings, Proceedings of the 2015 Theoretical Advanced Study Institute in Elementary Particle Physics, Boulder, Colorado, 1–26 June 2015*; World Scientific: Singapore, 2016; p. 137.
40. Liu, H.; Tseytlin, A.A. D = 4 superYang-Mills, D = 5 gauged supergravity, and D = 4 conformal supergravity. *Nucl. Phys. B* **1998**, *533*, 88–108.
41. Boulanger, N.; Sundell, P. An action principle for Vasiliev's four-dimensional higher-spin gravity. *J. Phys. A* **2011**, *44*, 495402.
42. Boulanger, N.; Colombo, N.; Sundell, P. A minimal BV action for Vasiliev's four-dimensional higher spin gravity. *J. High Energy Phys.* **2012**, *2012*, 43.
43. Boulanger, N.; Sezgin, E.; Sundell, P. 4D Higher Spin Gravity with Dynamical Two-Form as a Frobenius-Chern-Simons Gauge Theory. *arXiv* **2015**, arXiv:1505.04957.
44. Bonezzi, R.; Boulanger, N.; Sezgin, E.; Sundell, P. An Action for Matter Coupled Higher Spin Gravity in Three Dimensions. *J. High Energy Phys.* **2016**, *2016*, 3.
45. Bonezzi, R.; Boulanger, N.; Sezgin, E.; Sundell, P. Frobenius–Chern–Simons gauge theory. *J. Phys. A* **2017**, *50*, 055401.
46. Bekaert, X.; Erdmenger, J.; Ponomarev, D.; Sleight, C. Towards holographic higher-spin interactions: Four-point functions and higher-spin exchange. *J. High Energy Phys.* **2015**, *2015*, 170.
47. Bekaert, X.; Erdmenger, J.; Ponomarev, D.; Sleight, C. Quartic AdS Interactions in Higher-Spin Gravity from Conformal Field Theory. *J. High Energy Phys.* **2015**, *2015*, 149.
48. Sleight, C.; Taronna, M. Higher Spin Interactions from Conformal Field Theory: The Complete Cubic Couplings. *Phys. Rev. Lett.* **2016**, *116*, 181602.
49. Gopakumar, R. From free fields to AdS. *Phys. Rev. D* **2004**, *70*, 025009.
50. Gopakumar, R. From free fields to AdS. 2. *Phys. Rev. D* **2004**, *70*, 025010.
51. Gopakumar, R. From free fields to AdS: III. *Phys. Rev. D* **2005**, *72*, 066008.
52. Maldacena, J.M. The Large N limit of superconformal field theories and supergravity. *Int. J. Theor. Phys.* **1999**, *38*, 1113–1133.
53. Gubser, S.S.; Klebanov, I.R.; Polyakov, A.M. Gauge theory correlators from noncritical string theory. *Phys. Lett. B* **1998**, *428*, 105–114.
54. Witten, E. Anti-de Sitter space and holography. *Adv. Theor. Math. Phys.* **1998**, *2*, 253–291.
55. Eastwood, M.G. Higher symmetries of the Laplacian. *Ann. Math.* **2005**, *161*, 1645–1665.
56. Balasubramanian, V.; Gimon, E.G.; Minic, D.; Rahmfeld, J. Four-dimensional conformal supergravity from AdS space. *Phys. Rev. D* **2001**, *63*, 104009.
57. Compere, G.; Marolf, D. Setting the boundary free in AdS/CFT. *Class. Quant. Grav.* **2008**, *25*, 195014.
58. Schubert, C. Perturbative quantum field theory in the string inspired formalism. *Phys. Rept.* **2001**, *355*, 73–234.
59. Bern, Z.; Kosower, D.A. The Computation of loop amplitudes in gauge theories. *Nucl. Phys. B* **1992**, *379*, 451–561.
60. Strassler, M.J. Field theory without Feynman diagrams: One loop effective actions. *Nucl. Phys. B* **1992**, *385*, 145–184.
61. Bastianelli, F.; van Nieuwenhuizen, P. Trace anomalies from quantum mechanics. *Nucl. Phys. B* **1993**, *389*, 53–80.
62. D'Hoker, E.; Gagne, D.G. Worldline path integrals for fermions with general couplings. *Nucl. Phys. B* **1996**, *467*, 297–312.

63. Reuter, M.; Schmidt, M.G.; Schubert, C. Constant external fields in gauge theory and the spin 0, 1/2, 1 path integrals. *Ann. Phys.* **1997**, *259*, 313–365.
64. Bastianelli, F.; Zirotti, A. Worldline formalism in a gravitational background. *Nucl. Phys. B* **2002**, *642*, 372–388.
65. Bastianelli, F.; Corradini, O.; Zirotti, A. Dimensional regularization for N=1 supersymmetric sigma models and the worldline formalism. *Phys. Rev. D* **2003**, *67*, 104009.
66. Bastianelli, F.; Benincasa, P.; Giombi, S. Worldline approach to vector and antisymmetric tensor fields. *J. High Energy Phys.* **2005**, *2005*, 10.
67. Dai, P.; Huang, Y.T.; Siegel, W. Worldgraph Approach to Yang-Mills Amplitudes from N=2 Spinning Particle. *J. High Energy Phys.* **2008**, *2008*, 27.
68. Bastianelli, F.; Bonezzi, R. One-loop quantum gravity from a worldline viewpoint. *J. High Energy Phys.* **2013**, *2013*, 16.
69. Bastianelli, F.; Bonezzi, R.; Corradini, O.; Latini, E. Particles with non abelian charges. *J. High Energy Phys.* **2013**, *2013*, 98.
70. Bastianelli, F.; van Nieuwenhuizen, P. *Path Integrals and Anomalies in Curved Space*; Cambridge University Press: Cambridge, UK, 2006.
71. Gershun, V.D.; Tkach, V.I. Classical And Quantum Dynamics Of Particles With Arbitrary Spin. *J. Exp. Theor. Phys. Lett.* **1979**, *29*, 288–291.
72. Henneaux, M.; Teitelboim, C. First and second quantized point particles of any spin. In *Quantum Mechanics of Fundamental Systems 2*; Springer: Berlin, Germany, 1989; pp. 113–152.
73. Howe, P.S.; Penati, S.; Pernici, M.; Townsend, P.K. Wave Equations for Arbitrary Spin From Quantization of the Extended Supersymmetric Spinning Particle. *Phys. Lett. B* **1988**, *215*, 555–558.
74. Kuzenko, S.M.; Yarevskaya, Z.V. Conformal invariance, N extended supersymmetry and massless spinning particles in anti-de Sitter space. *Mod. Phys. Lett. A* **1996**, *11*, 1653–1664.
75. Bastianelli, F.; Corradini, O.; Latini, E. Spinning particles and higher spin fields on (A)dS backgrounds. *J. High Energy Phys.* **2008**, *2008*, 54.
76. Bastianelli, F.; Corradini, O.; Waldron, A. Detours and Paths: BRST Complexes and Worldline Formalism. *J. High Energy Phys.* **2009**, *2009*, 17.
77. Bastianelli, F.; Bonezzi, R.; Corradini, O.; Latini, E. Effective action for higher spin fields on (A)dS backgrounds. *J. High Energy Phys.* **2012**, *2012*, 113.
78. Segal, A.Y. Point particle in general background fields versus gauge theories of traceless symmetric tensors. *Int. J. Mod. Phys. A* **2003**, *18*, 4999–5021.
79. Bonezzi, R.; Corradini, O.; Franchino Vinas, S.A.; Pisani, P.A.G. Worldline approach to noncommutative field theory. *J. Phys. A* **2012**, *45*, 405401.
80. Ahmadiniaz, N.; Corradini, O.; D'Ascanio, D.; Estrada-Jiménez, S.; Pisani, P. Noncommutative U(1) gauge theory from a worldline perspective. *J. High Energy Phys.* **2015**, *2015*, 69.
81. Craigie, N.S.; Dobrev, V.K.; Todorov, I.T. Conformally Covariant Composite Operators in Quantum Chromodynamics. *Ann. Phys.* **1985**, *159*, 411–444.
82. Berends, F.A.; Burgers, G.J.H.; van Dam, H. Explicit Construction of Conserved Currents for Massless Fields of Arbitrary Spin. *Nucl. Phys. B* **1986**, *271*, 429–441.
83. Bekaert, X.; Joung, E.; Mourad, J. On higher spin interactions with matter. *J. High Energy Phys.* **2009**, *2009*, 126.
84. Weyl, H. Quantum mechanics and group theory. *Z. Phys.* **1927**, *46*, 1.
85. Wigner, E.P. On the quantum correction for thermodynamic equilibrium. *Phys. Rev.* **1932**, *40*, 749–759.
86. Moyal, J.E. Quantum mechanics as a statistical theory. *Proc. Camb. Philos. Soc.* **1949**, *45*, 99–124.
87. Sato, M.A. Operator Ordering and Perturbation Expansion in the Path Integration Formalism. *Prog. Theor. Phys.* **1977**, *58*, 1262–1270.
88. Joung, E.; Nakach, S.; Tseytlin, A.A. Scalar scattering via conformal higher spin exchange. *J. High Energy Phys.* **2016**, *2016*, 125.
89. Beccaria, M.; Nakach, S. Tseytlin, A.A. On triviality of S-matrix in conformal higher spin theory. *J. High Energy Phys.* **2016**, *2016*, 34.

universe

MDPI

Article

Higher Spin Superfield Interactions with the Chiral Supermultiplet: Conserved Supercurrents and Cubic Vertices

Ioseph L. Buchbinder [1,2,†], S. James Gates, Jr. [3,†] and Konstantinos Koutrolikos [4,*,†]

1 Department of Theoretical Physics, Tomsk State Pedagogical University, Tomsk 634041, Russia; joseph@tspu.edu.ru
2 National Research Tomsk State University, Tomsk 634050, Russia
3 Department of Physics, Brown University, Box 1843, 182 Hope Street, Barus & Holley 545, Providence, RI 02912, USA; sylvester_gates@brown.edu
4 Institute for Theoretical Physics and Astrophysics, Masaryk University, 61137 Brno, Czech
* Correspondence: kkoutrolikos@physics.muni.cz
† These authors contributed equally to this work.

Received: 30 October 2017; Accepted: 26 December 2017; Published: 10 January 2018

Abstract: We investigate cubic interactions between a chiral superfield and higher spin superfields corresponding to irreducible representations of the $4D$, $\mathcal{N}=1$ super-Poincaré algebra. We do this by demanding an invariance under the most general transformation, linear in the chiral superfield. Following Noether's method we construct an infinite tower of higher spin supercurrent multiplets which are quadratic in the chiral superfield and include higher derivatives. The results are that a single, massless, chiral superfield can couple only to the half-integer spin supermultiplets $(s+1, s+1/2)$ and for every value of spin there is an appropriate improvement term that reduces the supercurrent multiplet to a minimal multiplet which matches that of superconformal higher spins. On the other hand a single, massive, chiral superfield can couple only to higher spin supermultiplets of type $(2l+2, 2l+3/2)$ (only odd values of s, $s=2l+1$) and there is no minimal multiplet. Furthermore, for the massless case we discuss the component level higher spin currents and provide explicit expressions for the integer and half-integer spin conserved currents together with a R-symmetry current.

Keywords: supersymmetry; off-shell supermultiplets; higher spin

PACS: 11.30.Pb; 12.60.Jv

1. Introduction

Higher spin theories [1–10] have a considerable history and for a number of years drove the development of many ideas in theoretical physics. However, their role in fundamental interactions is still not clear. On the one hand, all the elementary particles observed in nature so far seem to be concentrated in a region of spin values (s) such that $s \leq 2$. Moreover, this observation appears to be supported by a substantial list of No-Go theorems [11–26] (for reviews look in [27,28]) suggesting that nature stops with spin 2. On the other hand, if we want to understand relativistic field theories and their quantum aspects in full generality, there is no a priori reason to exclude higher spin fields. In recent decades, this point was made undeniable due to the crucial part that massless and massive higher spin particles play in (*i*) the softness of string interactions at high energy scales, (*ii*) the possibilities to describe string effects in the framework of field theory, and (*iii*) investigations of some aspects of the holographic principle[1].

The construction of fully interacting higher spin theories is an extremely exciting topic but also very difficult, mostly due to the road blocks placed by the no-go results and maybe due to current lack of (still unknown) general principles. Also, one cannot exclude that higher spin field theory is an effective theory for an underlying, so far unknown, more fundamental theory. Nevertheless, there are few examples of successful approaches to higher spin theory such as Vasiliev's theory [10,29–31] (for reviews look in [32–34]) and the 3D Cherns-Simon higher spin-gravity formulation [35–37]. Despite their actual successes, these theories still appear very complicated. For example, Vasiliev's theory provides an infinite set of on-shell equations of motion and many conceptual questions about observables, Lagrangian formulation, locality[2] and quantization require continued study. In addition, the Chern-Simons description of interacting higher spins is restricted to 3D and has, in the massless case, no local degrees of freedom. Therefore, many important questions concerning higher spin field theory are still open[3].

In higher spin theories the structure of possible interaction vertices is essentially fixed by higher-spin symmetries. We will consider the construction of the simplest vertices in the supersymmetric higher-spin models. In this case, one can expect that the supersymmetry will impose the additional restrictions on the form of vertices and therefore one can hope to uncover clarifications and simplifications in comparison to non-supersymmetric higher-spin models.

The simplest higher spin interaction is described by the cubic vertex. Therefore, we will begin with the construction of a cubic vertex for supersymmetric field theory. It is well known that supersymmetric field models can be formulated on-shell in terms of component fields or off-shell in terms of appropriate superfields (see the text books [56,57]). Both these ways of constructing supersymmetric field models have their own advantages and disadvantages and complement each other. In this paper we will follow the superfield approach which allows us to keep manifest supersymmetry off-shell.

One kind of cubic interaction vertex for two types of fields can be written in the form jh, where j is a current constructed from fields of type ϕ (matter fields) and h is a field of another type (gauge fields). Because the gauge field h is defined up to gauge transformations, the current j must satisfy some conservation laws, i.e., it is conserved. Higher-spin interactions on the base of conserved current have been constructed and explored by many authors (see e.g., [58–67])[4].

In this work we will present the construction of the conserved $\mathcal{N} = 1$ higher superspin supercurrent and supertrace that generate the cubic interactions between super-Poincaré higher spin supermultiplets which play the role of gauge fields and the chiral supermultiplet which will play the role of matter. The higher spin supercurrent and higher spin supertrace together constitute the higher spin supercurrent multiplet and are the corresponding analogues to the low-spin supercurrent and supertrace of conventional supersymmetric theory (see [56,57]).

The strategy we follow is that of Noether's method, which is a perturbative procedure that allows one to constrain the allowed interactions by imposing invariance order by order in the number of (super)fields. Such a treatment of interactions will be very clear and useful for the cubic order. In our case the corresponding transformation for the matter superfield is the most general transformation, consistent with its chiral nature and up to linear order terms in the superfield and for the higher spin superfields is their gauge transformation.

---

1 For example, Fradkin-Vasiliev cubic interaction vertex of massless higher spin fields with gravity requires the *AdS* background.
2 See e.g., [38,39].
3 At present time, there is an extensive literature on different aspects of higher spin field theory. For example see the recent papers [40–55] and references therein.
4 A BRST approach to the construction of cubic vertex has been developed in [68].

The paper is organized as follows. Section 2 is devoted to discussing Noether's procedure and specific features of $4D$, $\mathcal{N} = 1$ super-Poincaré higher spin theories. In Section 3, we find the most general transformation of chiral superfield up to linear order and observe that the parameters of this transformation match the structure of the gauge transformations of specific higher spin supermultiplets. Sections 4–6 devoted to the construction of the higher spin supercurrent multiplet of a free massless chiral and generate the cubic interactions with higher spins. We find that the massless chiral can be coupled only to higher spin supermultiplets of type $(s + 1, s + 1/2)$. In Section 7, we show that for every value of integer s there are two types of higher spin supercurrent multiplets, the canonical and the minimal and one can go from the canonical to the minimal by an appropriate choice of improvement terms. Furthermore, we demonstrate that the minimal multiplet coincides with the supercurrent multiplet generated by superconformal higher spins. In Section 8, we discuss the on-shell superspace conservation equations for both supercurrent multiplets. For the case of minimal multiplet, we use the conservation equation alone to derive a simpler expression for the higher spin supercurrent. In Section 9, we project to components and find explicit expressions for the spacetime conserved integer spin, half-integer spin and R-symmetry currents. The integer spin current has two contributions, one of the boson - boson type that matches the known expressions for the integer spin currents of a complex scalar and the other is of the fermion - fermion type which agrees with the known expressions of integer spin currents of a spinor. The half-integer spin and R-symmetry currents, as far as we know, appear in the literature for the first time. Section 10, is devoted to the massive chiral superfield. We find that it can couple only to higher spin supermultiplets of type $(2l + 2\ ,\ 2l + 3)$ and we present new expressions for the higher spin supercurrent multiplet. For the massive chiral there is no minimal multiplet. In Section 11, we summarize and discuss the results.

2. Noether's Method

In general, finding *consistent* interactions is a very difficult problem if there is no guiding principle. For the cases of spin 2 (GR) and spin 1 (YM) there is a very well developed geometrical understanding (*Riemannian Manifolds* and *Principle Bundles* respectively) that plays the role of the guiding principle, but for higher spins we do not have this geometrical input. In some extent, the geometrical interpretation of higher spin fields is still mysterious. Therefore, we have to use alternative methods. The idea is to relax any geometrical prejudice and have only algebraic requirements. In this case the physical guiding principle is that of gauge invariance and *consistent* interactions are the ones that are in agreement with gauge symmetries. Keep in mind that this is a physical requirement in order for the interacting theory to have the same degrees of freedom as the free theory.

Noether's method is a systematic, perturbative, analysis of the invariance requirement. In this approach one expands the action $S[\phi, h]$ and the transformation of fields in a power series of a coupling constant g

$$S[\phi, h] = S_0[\phi] + gS_1[\phi, h] + g^2S_2[\phi, h] + \dots \ , \tag{1}$$

$$\delta\phi = \delta_0[\xi] + g\delta_1[\phi, \xi] + g^2\delta_2[\phi, \xi] + \dots \ , \tag{2}$$

$$\delta h = \delta_0[\zeta] + g\delta_1[h, \zeta] + g^2\delta_2[h, \zeta] + \dots \tag{3}$$

where $S_i[\phi, h]$ includes the interaction terms of order $i + 2$ in the number of fields and δ_i is the part of transformations with terms of order i in the number of fields. Hence, invariance can now be written iteratively up to the order we desire to investigate. For the free theory (g^0) and the cubic interactions (g^1), which is the first step beyond free theory, invariance demands:

$$g^0 : \ \frac{\delta S_0}{\delta\phi}\delta_0\phi + \frac{\delta S_0}{\delta h}\delta_0 h = 0 \ , \tag{4a}$$

$$g^1 : \ g\frac{\delta S_0}{\delta\phi}\delta_1\phi + g\frac{\delta S_1}{\delta\phi}\delta_0\phi + g\frac{\delta S_0}{\delta h}\delta_1 h + g\frac{\delta S_1}{\delta h}\delta_0 h = 0 \ . \tag{4b}$$

In our case, the role of matter will be played by the chiral supermultiplet, described by a chiral superfield Φ ($\bar{D}_{\dot{\alpha}}\Phi = 0$). At the free theory level the chiral superfield does not have any gauge transformation, $\delta_0\Phi = 0$.

For the role of gauge fields we consider the massless, higher spin, irreducible representations of the 4D, $\mathcal{N} = 1$, super-Poincaré algebra. In the pioneer papers [5,6,69], using a component formulation, free $\mathcal{N} = 1$ supersymmetric massless higher spin models in four dimensions have been constructed. A superfield formulation was proposed in [70–72] and further developed in subsequent papers [73,74] and generalized by different authors[5]. The results are[6]:

1. The integer superspin $Y = s$ supermultiplets $(s + 1/2, s)$ are described by a pair of superfields $\Psi_{\dot{\alpha}(s)\dot{\alpha}(s-1)}$ and $V_{\alpha(s-1)\dot{\alpha}(s-1)}$ with the following zero order gauge transformations

$$\delta_0\Psi_{\dot{\alpha}(s)\dot{\alpha}(s-1)} = -D^2 L_{\alpha(s)\dot{\alpha}(s-1)} + \frac{1}{(s-1)!}\bar{D}_{(\dot{\alpha}_{s-1}}\Lambda_{\alpha(s)\dot{\alpha}(s-2)}, \tag{5a}$$

$$\delta_0 V_{\alpha(s-1)\dot{\alpha}(s-1)} = D^{\alpha_s} L_{\alpha(s)\dot{\alpha}(s-1)} + \bar{D}^{\dot{\alpha}_s}\bar{L}_{\alpha(s-1)\dot{\alpha}(s)}. \tag{5b}$$

2. The half-integer superspin $Y = s + 1/2$ supermultiplets $(s + 1, s + 1/2)$ have two descriptions. One of them use the pair of superfields $H_{\alpha(s)\dot{\alpha}(s)}$, $\chi_{\alpha(s)\dot{\alpha}(s-1)}$ with the following zero order gauge transformations

$$\delta_0 H_{\alpha(s)\dot{\alpha}(s)} = \tfrac{1}{s!}D_{(\alpha_s}\bar{L}_{\alpha(s-1))\dot{\alpha}(s)} - \tfrac{1}{s!}\bar{D}_{(\dot{\alpha}_s}L_{\alpha(s)\dot{\alpha}(s-1))}, \tag{6a}$$

$$\delta_0\chi_{\alpha(s)\dot{\alpha}(s-1)} = \bar{D}^2 L_{\alpha(s)\dot{\alpha}(s-1)} + D^{\alpha_{s+1}}\Lambda_{\alpha(s+1)\dot{\alpha}(s-1)} \tag{6b}$$

and the other one use the superfields $H_{\alpha(s)\dot{\alpha}(s)}$, $\chi_{\alpha(s-1)\dot{\alpha}(s-2)}$ with

$$\delta_0 H_{\alpha(s)\dot{\alpha}(s)} = \tfrac{1}{s!}D_{(\alpha_s}\bar{L}_{\alpha(s-1))\dot{\alpha}(s)} - \tfrac{1}{s!}\bar{D}_{(\dot{\alpha}_s}L_{\alpha(s)\dot{\alpha}(s-1))}, \tag{7a}$$

$$\delta_0\chi_{\alpha(s-1)\dot{\alpha}(s-2)} = \bar{D}^{\dot{\alpha}_{s-1}}D^{\alpha_s}L_{\alpha(s)\dot{\alpha}(s-1)} + \tfrac{s-1}{s}D^{\alpha_s}\bar{D}^{\dot{\alpha}_{s-1}}L_{\alpha(s)\dot{\alpha}(s-1)} + \tfrac{1}{(s-2)!}\bar{D}_{(\dot{\alpha}_{s-2}}J_{\alpha(s-1)\dot{\alpha}(s-3))}. \tag{7b}$$

Consequently, the cubic interactions of the chiral superfield with the higher spin multiplets, according to (4) must satisfy:

$$\frac{\delta S_0}{\delta\Phi}\delta_1\Phi + \frac{\delta S_1}{\delta\mathcal{A}}\delta_0\mathcal{A} = 0 \tag{8}$$

where $\mathcal{A}$ is the set of superfields that participate in the description of higher spin supermultiplets for any value of s. In this language, the collection of *non-trivial* supercurrents that generate the cubic interaction terms correspond to the terms $\frac{\delta S_1}{\delta\mathcal{A}}$. The word *non-trivial* means that (*i*) the chiral superfield may not interact with all possible higher spin supermultiplets (trivially zero supercurrents) and (*ii*) for the ones that it interacts with, we must check that these interactions can not be adsorbed by redefinitions of the chiral superfield.

3. First Order Gauge Transformation for Chiral Superfield

In the previous section, we saw that the higher spin supercurrents of a chiral superfield are controlled by $\delta_1\Phi$. That is the part of the transformation of Φ which is linear in Φ. Examples of transformations of this type are generated by superdiffeomorphisms or the superconformal group and have been used in the past [76,77] in order to find the coupling of the chiral supermultiplet to supergravities.

---

5 See also a formulation of supersymmetric gauge theory in the framework of BRST approach [75].
6 This is the "economical" description according to [74].

In this section we present the higher spin version of this transformation. The most general ansatz one can write for such a transformation is[7]:

$$\delta_g\Phi = g\sum_{l=0}^{\infty}\sum_{k=0}^{\infty}\Big\{A_l^{\alpha(k+1)\dot\alpha(k)}\,\Box^l\,\mathrm{D}_{\alpha_{k+1}}\bar{\mathrm{D}}_{\dot\alpha_k}\mathrm{D}_{\alpha_k}\dots\bar{\mathrm{D}}_{\dot\alpha_1}\mathrm{D}_{\alpha_1}\Phi \tag{9}$$

$$+\Gamma_l^{\alpha(k)\dot\alpha(k+1)}\,\Box^l\,\bar{\mathrm{D}}_{\dot\alpha_{k+1}}\mathrm{D}^2\bar{\mathrm{D}}_{\dot\alpha_k}\mathrm{D}_{\alpha_k}\dots\bar{\mathrm{D}}_{\dot\alpha_1}\mathrm{D}_{\alpha_1}\Phi$$

$$+\Delta_l^{\alpha(k)\dot\alpha(k)}\,\Box^l\,\bar{\mathrm{D}}_{\dot\alpha_k}\mathrm{D}_{\alpha_k}\dots\bar{\mathrm{D}}_{\dot\alpha_1}\mathrm{D}_{\alpha_1}\Phi$$

$$+\,E_l^{\alpha(k)\dot\alpha(k)}\,\Box^l\,\mathrm{D}^2\bar{\mathrm{D}}_{\dot\alpha_k}\mathrm{D}_{\alpha_k}\dots\bar{\mathrm{D}}_{\dot\alpha_1}\mathrm{D}_{\alpha_1}\Phi\Big\}$$

and depends on four infinite families of coefficients $\{A^l_{\alpha(k+1)\dot\alpha(k)},\,\Gamma^l_{\alpha(k)\dot\alpha(k+1)},\,\Delta^l_{\alpha(k)\dot\alpha(k)},\,E^l_{\alpha(k)\dot\alpha(k)}\}$ with independently symmetrized dotted and undotted indices. To make this transformation consistent with the chiral nature of Φ we must have ($\bar{\mathrm{D}}_{\dot\beta}\delta_g\Phi = 0$):

$$A^l_{\alpha(k+1)\dot\alpha(k)} = -\tfrac{k+1}{k+2}\,\bar{\mathrm{D}}^{\dot\alpha_{k+1}}\Delta^l_{\alpha(k+1)\dot\alpha(k+1)}\,, \tag{10a}$$

$$\Gamma^l_{\alpha(k)\dot\alpha(k+1)} = \tfrac{1}{(k+1)!}\,\bar{\mathrm{D}}_{(\dot\alpha_{k+1}}\Delta^{l+1}_{\alpha(k)\dot\alpha(k))}\,, \tag{10b}$$

$$E^l_{\alpha(k)\dot\alpha(k)} = \bar{\mathrm{D}}^2\Delta^{l+1}_{\alpha(k)\dot\alpha(k)}\,, \tag{10c}$$

$$\bar{\mathrm{D}}_{(\dot\beta}\Delta^0_{\alpha(k)\dot\alpha(k))} = 0\,, \tag{10d}$$

$$\bar{\mathrm{D}}_{\dot\beta}\Delta^0 = 0\,. \tag{10e}$$

The conclusion is that parameters $A^l_{\alpha(k+1)\dot\alpha(k)},\,\Gamma^l_{\alpha(k)\dot\alpha(k+1)},\,E^l_{\alpha(k)\dot\alpha(k)}$ are not independent and furthermore

$$\Delta^0 = \bar{\mathrm{D}}^2\ell\,, \tag{11a}$$

$$\Delta^0_{\alpha(k)\dot\alpha(k)} = \tfrac{1}{k!}\bar{\mathrm{D}}_{(\dot\alpha_k}\ell_{\alpha(k)\dot\alpha(k-1))}\,, \tag{11b}$$

$$\Delta^l_{\alpha(k)\dot\alpha(k)}\ \text{is unconstrained for } l\geq 1 \tag{11c}$$

where ℓ, $\ell_{\alpha(k)\dot\alpha(k-1)}$ are arbitrary.

From Equation (8) it is evident that the parameters which appear in the transformation of Φ must also appear in the zeroth order gauge transformation of the higher spin superfields. Looking at the gauge parameters that appear in (5)–(7) we find that there is no unconstrained parameter with the structure of $\Delta^{l+1}_{\alpha(k)\dot\alpha(k)}$, but Equations (6) and (7) include unconstrained gauge parameters which match the structure of $\ell_{\alpha(k)\dot\alpha(k-1)}$. The conclusion is that in order to construct invariant theories where the chiral superfield couples to purely higher spin supermultiplets we have to consider the following transformation of Φ:

$$\delta_g\Phi = -g\sum_{k=0}^{\infty}\Big\{\bar{\mathrm{D}}^2\ell^{\alpha(k+1)\dot\alpha(k)}\,\mathrm{D}_{\alpha_{k+1}}\bar{\mathrm{D}}_{\dot\alpha_k}\mathrm{D}_{\alpha_k}\dots\bar{\mathrm{D}}_{\dot\alpha_1}\mathrm{D}_{\alpha_1}\Phi \tag{12}$$

$$-\tfrac{1}{(k+1)!}\bar{\mathrm{D}}^{(\dot\alpha_{k+1}}\ell^{\alpha(k+1)\dot\alpha(k))}\,\bar{\mathrm{D}}_{\dot\alpha_{k+1}}\mathrm{D}_{\alpha_{k+1}}\dots\bar{\mathrm{D}}_{\dot\alpha_1}\mathrm{D}_{\alpha_1}\Phi\Big\}$$

$$+g\bar{\mathrm{D}}^2\ell\,\Phi\ .$$

---

7 We use the conventions of *Superspace* [56] which include $\{\mathrm{D}_\alpha,\bar{\mathrm{D}}_{\dot\alpha}\} = i\partial_{\alpha\dot\alpha}$, $\mathrm{D}^\alpha\mathrm{D}_\alpha = 2\mathrm{D}^2$ and $\bar{\mathrm{D}}^{\dot\alpha}\bar{\mathrm{D}}_{\dot\alpha} = 2\bar{\mathrm{D}}^2$.

The last term of (12) will generate coupling to the vector multiplet, thus in order to consider purely higher spin interactions we should ignore it. However, for the sake of completeness we will not do that.

The second conclusion we can already reach, is that a theory of a single chiral superfield can couple *only* to half-integer superspin $Y = s + 1/2$ supermultiplets. This is a consequence of the constraint (10d) whose solution matches the structure of the transformation of bosonic superfields of half-integer superspin theories but crucially not that of integer superspin.

4. Constructing the Higher Spin Supercurrents I: Varying the Action

Having found the appropriate first order transformation for the chiral superfield, we use it to perform Noether's procedure for the cubic order terms, as described in Section 2 and construct the higher spin supercurrents of the chiral supermultiplet. We consider a free massless chiral superfield, so we start from the free action

$$S_o = \int d^8z\ \Phi\bar{\Phi} \tag{13}$$

and calculate its variation under $\delta_g\Phi$ [8]:

$$\begin{aligned}\delta_g S_o = -g\int\sum_{k=0}^{\infty}\Big\{&\bar{\mathrm{D}}^2\ell^{\alpha(k+1)\dot{\alpha}(k)}\ \mathrm{D}_{\alpha_{k+1}}\bar{\mathrm{D}}_{\dot{\alpha}_k}\mathrm{D}_{\alpha_k}\dots\bar{\mathrm{D}}_{\dot{\alpha}_1}\mathrm{D}_{\alpha_1}\Phi\,\bar{\Phi}\ + c.c.\\ &-\tfrac{1}{(k+1)!}\ \bar{\mathrm{D}}^{(\dot{\alpha}_{k+1}}\ell^{\alpha(k+1)\dot{\alpha}(k))}\ \bar{\mathrm{D}}_{\dot{\alpha}_{k+1}}\mathrm{D}_{\alpha_{k+1}}\dots\bar{\mathrm{D}}_{\dot{\alpha}_1}\mathrm{D}_{\alpha_1}\Phi\,\bar{\Phi}\ + c.c.\Big\}\\ +g\int\Big\{\bar{\mathrm{D}}^2\ell + \mathrm{D}^2\bar{\ell}\Big\}\,\Phi\,\bar{\Phi}\ .\end{aligned} \tag{14}$$

However, in the above expression we can freely add any pair of terms $A_{\alpha(k+1)\dot{\alpha}(k)}$, $B_{\alpha(k+1)\dot{\alpha}(k+1)}$ such that they identically satisfy the equation

$$\bar{\mathrm{D}}^2 A_{\alpha(k+1)\dot{\alpha}(k)} = \bar{\mathrm{D}}^{\dot{\alpha}_{k+1}} B_{\alpha(k+1)\dot{\alpha}(k+1)}\ . \tag{15}$$

These terms play the role of improvement terms. We can prove that there are at least two pairs of them

1. $A_{\alpha(k+1)\dot{\alpha}(k)} = W_{\alpha(k+1)\dot{\alpha}(k)}$, $B_{\alpha(k+1)\dot{\alpha}(k+1)} = \frac{k+1}{(k+2)(k+1)!}\bar{\mathrm{D}}_{(\dot{\alpha}_{k+1}}W_{\alpha(k+1)\dot{\alpha}(k))}$,
2. $A_{\alpha(k+1)\dot{\alpha}(k)} = \frac{1}{(k+1)!}\mathrm{D}_{(\alpha_{k+1}}\bar{\mathrm{D}}^{\dot{\alpha}_{k+1}}\bar{\mathcal{U}}_{\alpha(k))\dot{\alpha}(k+1)}$, $B_{\alpha(k+1)\dot{\alpha}(k+1)} = \frac{1}{(k+1)!}\mathrm{D}_{(\alpha_{k+1}}\bar{\mathrm{D}}^2\bar{\mathcal{U}}_{\alpha(k))\dot{\alpha}(k+1)}$

which will be relevant for our discussion. Hence, we can write for the variation of the S_o action:

$$\begin{aligned}\delta_g S_o = -g\int\sum_{k=0}^{\infty}\Big\{&\bar{\mathrm{D}}^2\ell^{\alpha(k+1)\dot{\alpha}(k)}\ \mathcal{T}_{\alpha(k+1)\dot{\alpha}(k)}\ + c.c.\\ &-\tfrac{1}{(k+1)!}\ \bar{\mathrm{D}}^{(\dot{\alpha}_{k+1}}\ell^{\alpha(k+1)\dot{\alpha}(k))}\ \mathcal{J}_{\alpha(k+1)\dot{\alpha}(k+1)}\ + c.c.\Big\}\\ +g\int\Big\{\bar{\mathrm{D}}^2\ell + \mathrm{D}^2\bar{\ell}\Big\}\ \mathcal{J}\end{aligned} \tag{16}$$

[8] From this point forward, when the integration is over the entire superspace the measure d^8z will not be explicitly written but it will be implied.

where

$$\mathcal{T}_{\alpha(k+1)\dot{\alpha}(k)} = \tfrac{1}{(k+1)!k!}\mathrm{D}_{(\alpha_{k+1}}\bar{\mathrm{D}}_{(\dot{\alpha}_k}\mathrm{D}_{\alpha_k}\dots\bar{\mathrm{D}}_{\dot{\alpha}_1)}\mathrm{D}_{\alpha_1)}\Phi\,\bar{\Phi} + W_{\alpha(k+1)\dot{\alpha}(k)} + \tfrac{1}{(k+1)!}\mathrm{D}_{(\alpha_{k+1}}\bar{\mathrm{D}}^{\dot{\alpha}_{k+1}}\bar{U}_{\alpha(k))\dot{\alpha}(k+1)}\,, \tag{17a}$$

$$\mathcal{J}_{\alpha(k+1)\dot{\alpha}(k+1)} = \tfrac{1}{(k+1)!(k+1)!}\bar{\mathrm{D}}_{(\dot{\alpha}_{k+1}}\mathrm{D}_{(\alpha_{k+1}}\dots\bar{\mathrm{D}}_{\dot{\alpha}_1)}\mathrm{D}_{\alpha_1)}\Phi\,\bar{\Phi} + \tfrac{1}{(k+1)!}\mathrm{D}_{(\alpha_{k+1}}\bar{\mathrm{D}}^2\bar{U}_{\alpha(k))\dot{\alpha}(k+1)} + \tfrac{k+1}{(k+2)(k+1)!}\bar{\mathrm{D}}_{(\dot{\alpha}_{k+1}}W_{\alpha(k+1)\dot{\alpha}(k))}\,, \tag{17b}$$

$$\mathcal{J} = \Phi\bar{\Phi} \quad . \tag{17c}$$

It is important to observe that these objects are not uniquely defined, but there is some freedom. For example $\mathcal{J}$ is defined up to terms $\mathrm{D}^{\alpha}\bar{\mathrm{D}}^2\lambda_{\alpha} + \bar{\mathrm{D}}^{\dot{\alpha}}\mathrm{D}^2\bar{\lambda}_{\dot{\alpha}}$ for an arbitrary λ_{α}[9], whereas $\mathcal{J}_{\alpha(k+1)\dot{\alpha}(k+1)}$ is defined up to terms $\bar{\mathrm{D}}^{\dot{\alpha}_{k+2}}\Xi_{\alpha(k+1)\dot{\alpha}(k+2)}$. Also $\mathcal{T}_{\alpha(k+1)\dot{\alpha}(k)}$ has the freedom

$$\mathcal{T}_{\alpha(k+1)\dot{\alpha}(k)} \sim \mathcal{T}_{\alpha(k+1)\dot{\alpha}(k)} + \bar{\mathrm{D}}_{(\dot{\alpha}_k}P^{(1)}_{\alpha(k+1)\dot{\alpha}(k-1))} + \bar{\mathrm{D}}^{\dot{\alpha}_{k+1}}P^{(2)}_{\alpha(k+1)\dot{\alpha}(k+1)} + \mathrm{D}_{(\alpha_{k+1}}\bar{\mathrm{D}}^2R^{(1)}_{\alpha(k))\dot{\alpha}(k)} + \mathrm{D}^{\alpha_{k+2}}\bar{\mathrm{D}}^2R^{(2)}_{\alpha(k+2)\dot{\alpha}(k)}\,. \tag{18}$$

Furthermore, Equation (16) points towards a coupling of the chiral with the first formulation (6) of $(s+1, s+1/2)$ supermultiplets, but for that to happen we must have $\mathcal{J}_{\alpha(k+1)\dot{\alpha}(k+1)}$ to be real. This is a consequence of the reality of superfield $H_{\alpha(s)\dot{\alpha}(s)}$ and transformation (6a). Thus, in order to couple the theory purely to half-integer superspin multiplet, we must make sure that we can select the improvement terms such that $\mathcal{J}_{\alpha(k+1)\dot{\alpha}(k+1)}=\bar{\mathcal{J}}_{\alpha(k+1)\dot{\alpha}(k+1)}$. This will depend on the detailed structure of the real and imaginary part of the term $\frac{1}{(k+1)!(k+1)!}\bar{\mathrm{D}}_{(\dot{\alpha}_{k+1}}\mathrm{D}_{(\alpha_{k+1}}\dots\bar{\mathrm{D}}_{\dot{\alpha}_1)}\mathrm{D}_{\alpha_1)}\Phi\,\bar{\Phi}$. The investigation of these structures is the purpose of the following section. Due to the chiral nature of Φ, this term can be simply written as $i^{k+1}\partial^{(k+1)}\Phi\,\bar{\Phi}$, where for simpicity we omit the uncontracted indices and complete symmetrization of them with appropriate symmetrization factors is understood. The symbol $\partial^{(k)}$ denotes a string of k spacetime derivatives.

5. The Combinatorics of the Imaginary Part

First of all, we decompose the quantity $i^{k+1}\partial^{(k+1)}\Phi\,\bar{\Phi}$ to a real and an imaginary part

$$i^{k+1}\partial^{(k+1)}\Phi\,\bar{\Phi} = \tfrac{i^{k+1}}{2}\left[\partial^{(k+1)}\Phi\,\bar{\Phi} + (-1)^{k+1}\Phi\,\partial^{(k+1)}\bar{\Phi}\right] + \tfrac{i^{k+1}}{2}\left[\partial^{(k+1)}\Phi\,\bar{\Phi} - (-1)^{k+1}\Phi\,\partial^{(k+1)}\bar{\Phi}\right] \tag{19}$$

and then we focus at the imaginary part with the goal to clarify whether the various improvement terms ($W_{\alpha(k+1)\dot{\alpha}(k)}$, $U_{\alpha(k+1)\dot{\alpha}(k)}$) can modify it in order to make $\mathcal{J}_{\alpha(k+1)\dot{\alpha}(k+1)}$ real. Notice the difference between even and odd values of $k+1$

$$\mathcal{I}^{(k+1)} \equiv i\mathrm{Im}[\,i^{k+1}\partial^{(k+1)}\Phi\,\bar{\Phi}\,] = \begin{cases} \frac{i}{2}(-1)^l\left(\partial^{(2l+1)}\Phi\,\bar{\Phi} + \Phi\,\partial^{(2l+1)}\bar{\Phi}\right), & \text{for } k+1 = 2l+1,\ l = 0,1,\dots \\ \frac{1}{2}(-1)^l\left(\partial^{(2l)}\Phi\,\bar{\Phi} - \Phi\,\partial^{(2l)}\bar{\Phi}\right), & \text{for } k+1 = 2l,\ l = 1,2,\dots \end{cases} \tag{20}$$

---

[9] λ_{α} has its own redundancy $\lambda_{\alpha} \sim \lambda_{\alpha} + \bar{\mathrm{D}}^{\dot{\alpha}}\zeta_{\alpha\dot{\alpha}} + i\mathrm{D}_{\alpha}\varrho$ with $\varrho = \bar{\varrho}$.

The type of terms that appear above are a special case to the more general type $\partial^{(m)}\Phi\,\partial^{(n)}\bar{\Phi}$ terms. It is easy to prove that this type of terms satisfy the following recursion relations:

$$\partial^{(m)}\Phi\,\partial^{(n)}\bar{\Phi} = \partial\left(\partial^{(m-1)}\Phi\,\partial^{(n)}\bar{\Phi}\right) - \partial^{(m-1)}\Phi\,\partial^{(n+1)}\bar{\Phi}\ , \tag{21}$$

$$\partial^{(m)}\Phi\,\partial^{(n)}\bar{\Phi} = \partial\left(\partial^{(m)}\Phi\,\partial^{(n-1)}\bar{\Phi}\right) - \partial^{(m+1)}\Phi\,\partial^{(n-1)}\bar{\Phi}\ . \tag{22}$$

Using these recursion formulas, one can prove that

$$\partial^{(2l+1)}\Phi\,\bar{\Phi} + \Phi\,\partial^{(2l+1)}\bar{\Phi} = \sum_{n=0}^{l} c_n\,\partial^{(2n+1)}\left\{\partial^{(l-n)}\Phi\,\partial^{(l-n)}\bar{\Phi}\right\}\ , \tag{23}$$

$$\partial^{(2l)}\Phi\,\bar{\Phi} - \Phi\,\partial^{(2l)}\bar{\Phi} = \sum_{n=0}^{l-1} d_n\,\partial^{(2n+1)}\left\{\partial^{(l-n)}\Phi\,\partial^{(l-n-1)}\bar{\Phi} - \partial^{(l-n-1)}\Phi\,\partial^{(l-n)}\bar{\Phi}\right\} \tag{24}$$

with

$$c_n = (-1)^{l-n}\left[\binom{l+n+1}{l-n} + \binom{l+n}{l-n-1}\right]\ ,\ d_n = (-1)^{l-n-1}\binom{l+n}{l-n-1}\ . \tag{25}$$

These identities hold in general, not just for the chiral but for any (super)function Φ. An alternative proof of them can be found by expanding the right hand side using the identity

$$\partial^{(m)}\left(A\,B\right) = \sum_{i=0}^{m}\binom{m}{i}\,\partial^{(m-i)}A\,\partial^{i}B \tag{26}$$

and matching the coefficients of the various terms with those of the left hand side. Doing that, one will find the following consistency conditions

$$\sum_{i=0}^{l} c_i\binom{2i+1}{l+i-p+1} = \begin{cases}1 \text{ for } p=0\\ 0 \text{ for } p=1,2,\ldots,l\end{cases}\ , \tag{27}$$

$$\sum_{i=0}^{l-1} d_i\left[\binom{2i+1}{l+i-p+1} - \binom{2i+1}{l+i-p}\right] = \begin{cases}-1 \text{ for } p=0\\ 0 \text{ for } p=1,2,\ldots,l-1\end{cases} \tag{28}$$

which define the coefficients c_n, d_n recursively and have (25) as solutions. Furthermore, due to (27) and (28) the coefficients c_i and d_i also satisfy

$$\sum_{i=0}^{l} c_i\binom{2i}{l-p+i} = (-1)^p\ ,\ \sum_{i=0}^{l-1} d_i\left[\binom{2i}{l-p+i} - \binom{2i}{l-p+i-1}\right] = (-1)^{p+1}\ . \tag{29}$$

5.1. Odd Values of $K+1$

With the above in mind, for the general odd case we get:

$$\mathcal{I}^{2l+1} = \sum_{n=0}^{l}\tfrac{(-1)^l}{2}c_n\,\partial^{(2n)}\left\{\mathrm{D},\bar{\mathrm{D}}\right\}\left[\partial^{(l-n)}\Phi\,\partial^{(l-n)}\bar{\Phi}\right]\ ,\ l=0,1,\ldots \tag{30}$$

where using the supersymmetry algebra we have converted $i\partial$ to the anticommutator of the spinorial covariant derivatives. Notice that with the exception of this part of the expression, everything else is real. So it will be beneficial if we convert the anticommutator of spinorial derivatives to a commutator of spinorial derivatives using the following identity,

$$\left\{\mathrm{D},\bar{\mathrm{D}}\right\} = \left[\mathrm{D},\bar{\mathrm{D}}\right] + 2\bar{\mathrm{D}}\mathrm{D}\ . \tag{31}$$

The part with the commutator will be a real contribution and the left over term has the structure $\bar{\mathrm{D}}\mathrm{D}\,(\dots)$. According to (17b) these terms can always be removed by an appropriate choice of the improvement term $W_{\alpha(2l+1)\dot{\alpha}(2l)}$, thus the reality of $\mathcal{J}_{\alpha(2l+1)\dot{\alpha}(2l+1)}$ can always be guaranteed. Specifically we get:

$$\begin{aligned} \mathcal{I}^{2l+1} = &(-1)^l \sum_{n=0}^{l} c_n\, \partial^{(2n)} \left[\partial^{(l-n)}\mathrm{D}\Phi\, \partial^{(l-n)}\bar{\mathrm{D}}\bar{\Phi}\right] \\ &- \tfrac{i}{2}(-1)^l \sum_{n=0}^{l} c_n\, \partial^{(2n)} \left[\partial^{(l-n+1)}\Phi\, \partial^{(l-n)}\bar{\Phi} - \partial^{(l-n)}\Phi\, \partial^{(l-n+1)}\bar{\Phi}\right] \\ &+ (-1)^l \sum_{n=0}^{l} c_n\, \partial^{(2n)}\bar{\mathrm{D}}\mathrm{D} \left[\partial^{(l-n)}\Phi\, \partial^{(l-n)}\bar{\Phi}\right] . \end{aligned} \tag{32}$$

The conclusion of this analysis is that the term $i^{2l+1}\partial^{(2l+1)}\Phi\,\bar{\Phi}$ which appears in the expression of $\mathcal{J}_{\alpha(2l+1)\dot{\alpha}(2l+1)}$ can be written as:

$$i^{2l+1}\partial^{(2l+1)}\Phi\,\bar{\Phi} = \mathrm{X}^{(2l+1)}_{\alpha(2l+1)\dot{\alpha}(2l+1)} + \frac{1}{[(2l+1)!]^2}\bar{\mathrm{D}}_{(\dot{\alpha}_{2l+1}}\mathrm{D}_{(\alpha_{2l+1}}Z^{(2l+1)}_{\alpha(2l))\dot{\alpha}(2l))} \tag{33}$$

where

$$\begin{aligned} \mathrm{X}^{(2l+1)}_{\alpha(2l+1)\dot{\alpha}(2l+1)} = &\tfrac{i}{2}(-1)^l \left[\partial^{(2l+1)}\Phi\,\bar{\Phi} - \Phi\,\partial^{(2l+1)}\bar{\Phi}\right] \\ &- \tfrac{i}{2}(-1)^l \sum_{n=0}^{l} c_n\, \partial^{(2n)} \left[\partial^{(l-n+1)}\Phi\, \partial^{(l-n)}\bar{\Phi} - \partial^{(l-n)}\Phi\, \partial^{(l-n+1)}\bar{\Phi}\right] \\ &+ (-1)^l \sum_{n=0}^{l} c_n\, \partial^{(2n)} \left[\partial^{(l-n)}\mathrm{D}\Phi\, \partial^{(l-n)}\bar{\mathrm{D}}\bar{\Phi}\right] , \end{aligned} \tag{34}$$

$$Z^{(2l+1)}_{\alpha(2L)\dot{\alpha}(2l)} = (-1)^l \sum_{n=0}^{l} c_n\, \partial^{(2n)} \left[\partial^{(l-n)}\Phi\, \partial^{(l-n)}\bar{\Phi}\right] \tag{35}$$

and both these quantities are real. These expressions can be further simplified using (26) and (29) to

$$\mathrm{X}^{(2l+1)}_{\alpha(2l+1)\dot{\alpha}(2l+1)} = i(-1)^l \sum_{p=1}^{2l} (-1)^p\, \partial^{(p)}\Phi\, \partial^{(2l+1-p)}\bar{\Phi} + (-1)^l \sum_{p=0}^{2l} (-1)^p\, \partial^{(p)}\mathrm{D}\Phi\, \partial^{(2l-p)}\bar{\mathrm{D}}\bar{\Phi}\,, \tag{36}$$

$$Z^{(2l+1)}_{\alpha(2l)\dot{\alpha}(2l)} = (-1)^l \sum_{p=0}^{2l} (-1)^p\, \partial^{(p)}\Phi\, \partial^{(2l-p)}\bar{\Phi}\ . \tag{37}$$

5.2. Even Values of $K+1$

The same analysis can be done for the general even case. For that situation we get

$$\begin{aligned} \mathcal{I}^{(2l)} = &\tfrac{1}{2}(-1)^{(l-1)} \sum_{n=0}^{l-1} d_n\, \partial^{(2n)} \left[\partial^{(l-n+1)}\Phi\, \partial^{(l-n-1)}\bar{\Phi} - 2\partial^{(l-n)}\Phi\, \partial^{(l-n)}\bar{\Phi} + \partial^{(l-n-1)}\Phi\, \partial^{(l-n+1)}\bar{\Phi}\right] \\ &+ i(-1)^{(l-1)} \sum_{n=0}^{l-1} d_n\, \partial^{(2n)} \left[\partial^{(l-n)}\mathrm{D}\Phi\, \partial^{(l-n-1)}\bar{\mathrm{D}}\bar{\Phi} - \partial^{(l-n-1)}\mathrm{D}\Phi\, \partial^{(l-n)}\bar{\mathrm{D}}\bar{\Phi}\right] \\ &+ i(-1)^{(l-1)} \sum_{n=0}^{l-1} d_n\, \partial^{(2n)}\bar{\mathrm{D}}\mathrm{D} \left[\partial^{(l-n)}\Phi\, \partial^{(l-n-1)}\bar{\Phi} - \partial^{(l-n-1)}\Phi\, \partial^{(l-n)}\bar{\Phi}\right] . \end{aligned} \tag{38}$$

Hence, the term $i^{2l}\partial^{(2l)}\Phi\,\bar{\Phi}$ can be expressed in the following way:

$$i^{2l}\partial^{(2l)}\Phi\,\bar{\Phi} = \mathrm{X}^{(2l)}_{\alpha(2l)\dot{\alpha}(2l)} + \frac{1}{[(2l)!]^2}\bar{\mathrm{D}}_{(\dot{\alpha}_{2l}}\mathrm{D}_{(\alpha_{2l}}Z^{(2l)}_{\alpha(2l-1))\dot{\alpha}(2l-1))} \tag{39}$$

where

$$X^{(2l)}_{\alpha(2l)\dot\alpha(2l)} = \tfrac{1}{2}(-1)^l\left[\partial^{(2l)}\Phi\,\bar\Phi + \Phi\,\partial^{(2l)}\bar\Phi\right] \tag{40}$$
$$+\tfrac{1}{2}(-1)^{(l-1)}\sum_{n=0}^{l-1} d_n\,\partial^{(2n)}\left[\partial^{(l-n+1)}\Phi\,\partial^{(l-n-1)}\bar\Phi - 2\partial^{(l-n)}\Phi\,\partial^{(l-n)}\bar\Phi + \partial^{(l-n-1)}\Phi\,\partial^{(l-n+1)}\bar\Phi\right]$$
$$+\,i(-1)^{(l-1)}\sum_{n=0}^{l-1} d_n\,\partial^{(2n)}\left[\partial^{(l-n)}\mathrm{D}\Phi\,\partial^{(l-n-1)}\bar{\mathrm{D}}\bar\Phi - \partial^{(l-n-1)}\mathrm{D}\Phi\,\partial^{(l-n)}\bar{\mathrm{D}}\bar\Phi\right]\ ,$$
$$Z^{(2l)}_{\alpha(2l-1)\dot\alpha(2l-1)} = i(-1)^{(l-1)}\sum_{n=0}^{l-1} d_n\,\partial^{(2n)}\left[\partial^{(l-n)}\Phi\,\partial^{(l-n-1)}\bar\Phi - \partial^{(l-n-1)}\Phi\,\partial^{(l-n)}\bar\Phi\right]\ .$$

As in the previous case, both $X^{(2l)}_{\alpha(2l)\dot\alpha(2l)}$ and $Z^{(2l)}_{\alpha(2l-1)\dot\alpha(2l-1)}$ are real. Using (26) and (29) we can simplify these expressions further

$$X^{(2l)}_{\alpha(2l)\dot\alpha(2l)} = (-1)^{(l-1)}\sum_{p=1}^{2l-1}(-1)^p\,\partial^{(p)}\Phi\,\partial^{(2l-p)}\bar\Phi + i(-1)^l\sum_{p=0}^{2l-1}(-1)^p\,\partial^{(p)}\mathrm{D}\Phi\,\partial^{(2l-1-p)}\bar{\mathrm{D}}\bar\Phi\ , \tag{41}$$

$$Z^{(2l)}_{\alpha(2l-1)\dot\alpha(2l-1)} = i(-1)^l\sum_{p=0}^{2l-1}(-1)^p\,\partial^{(p)}\Phi\,\partial^{(2l-1-p)}\bar\Phi\ . \tag{42}$$

6. Constructing the Higher Spin Supercurrents II: Gauge Invariance and Cubic Interactions

The main point of the previous section is to prove that for every value of integer m we can write

$$i^{(k+1)}\partial^{(k+1)}\Phi\,\bar\Phi = X^{(k+1)}_{\alpha(k+1)\dot\alpha(k+1)} + \tfrac{1}{[(k+1)!]^2}\bar{\mathrm{D}}_{(\dot\alpha_{k+1}}\mathrm{D}_{(\alpha_{k+1}}Z^{(k+1)}_{\alpha(k))\dot\alpha(k))} \tag{43}$$

where $X^{(k+1)}_{\alpha(k+1)\dot\alpha(k+1)}$ and $Z^{(k+1)}_{\alpha(k)\dot\alpha(k)}$ are:

$$X^{(k+1)}_{\alpha(k+1)\dot\alpha(k+1)} = (-i)^{k-1}\sum_{p=1}^{k}(-1)^p\,\partial^{(p)}\Phi\,\partial^{(k+1-p)}\bar\Phi + (-i)^k\sum_{p=0}^{k}(-1)^p\,\partial^{(p)}\mathrm{D}\Phi\,\partial^{(k-p)}\bar{\mathrm{D}}\bar\Phi\ , \tag{44}$$

$$Z^{(k+1)}_{\alpha(k)\dot\alpha(k)} = (-i)^k\sum_{p=0}^{k}(-1)^p\,\partial^{(p)}\Phi\,\partial^{(k-p)}\bar\Phi\ . \tag{45}$$

Thus the expression for $\mathcal{J}_{\alpha(k+1)\dot\alpha(k+1)}$ (17b) becomes:

$$\mathcal{J}_{\alpha(k+1)\dot\alpha(k+1)} = X^{(k+1)}_{\alpha(k+1)\dot\alpha(k+1)} + \tfrac{1}{(k+1)!(k+1)!}\bar{\mathrm{D}}_{(\dot\alpha_{k+1}}\mathrm{D}_{(\alpha_{k+1}}Z^{(k+1)}_{\alpha(k))\dot\alpha(k))} \tag{46}$$
$$+\tfrac{1}{(k+1)!}\mathrm{D}_{(\alpha_{k+1}}\bar{\mathrm{D}}^2\bar U_{\alpha(k))\dot\alpha(k+1)} + \tfrac{k+1}{(k+2)(k+1)!}\bar{\mathrm{D}}_{(\dot\alpha_{k+1}}W_{\alpha(k+1)\dot\alpha(k))}\ .$$

This is useful because it makes obvious that we can always make $\mathcal{J}_{\alpha(k+1)\dot\alpha(k+1)}$ real by choosing

$$W_{\alpha(k+1)\dot\alpha(k)} = -\tfrac{k+2}{k+1}\mathrm{D}^2 U_{\alpha(k+1)\dot\alpha(k)} - \tfrac{k+2}{k+1}\tfrac{1}{(k+1)!}\mathrm{D}_{(\alpha_{k+1}}Z^{(k+1)}_{\alpha(k))\dot\alpha(k)}\ . \tag{47}$$

With this choice we get

$$\mathcal{J}_{\alpha(k+1)\dot\alpha(k+1)} = X^{(k+1)}_{\alpha(k+1)\dot\alpha(k+1)} + \tfrac{1}{(k+1)!}\mathrm{D}_{(\alpha_{k+1}}\bar{\mathrm{D}}^2\bar U_{\alpha(k))\dot\alpha(k+1)} - \tfrac{1}{(k+1)!}\bar{\mathrm{D}}_{(\dot\alpha_{k+1}}\mathrm{D}^2 U_{\alpha(k+1)\dot\alpha(k))}\ , \tag{48}$$

$$\mathcal{T}_{\alpha(k+1)\dot\alpha(k)} = \tfrac{1}{(k+1)!}\mathrm{D}_{(\alpha_{k+1}}\mathcal{T}_{\alpha(k))\dot\alpha(k)}\ , \tag{49}$$

$$\mathcal{T}_{\alpha(k)\dot\alpha(k)} = i^k\partial^{(k)}\Phi\,\bar\Phi - \tfrac{k+2}{k+1}Z^{(k+1)}_{\alpha(k)\dot\alpha(k)} + \tfrac{k+2}{k+1}\mathrm{D}^{\alpha_{k+1}}U_{\alpha(k+1)\dot\alpha(k)} + \bar{\mathrm{D}}^{\dot\alpha_{k+1}}\bar U_{\alpha(k)\dot\alpha(k+1)}\ . \tag{50}$$

Due to Equation (49), the variation of the action can be enhanced from (16) to the following, with the addition of the $\lambda_{\alpha(k+2)\dot{\alpha}(k)}$ term:

$$\begin{aligned}\delta_g S_o =& -g \int \sum_{k=0}^{\infty} \Big\{ \Big[\bar{\mathrm{D}}^2 \ell^{\alpha(k+1)\dot{\alpha}(k)} - \mathrm{D}_{\alpha_{k+2}} \lambda^{\alpha(k+2)\dot{\alpha}(k)}\Big] \mathrm{D}_{\alpha_{k+1}} \mathcal{T}_{\alpha(k)\dot{\alpha}(k)} + c.c. \\ & - \tfrac{1}{(k+1)!} \bar{\mathrm{D}}^{(\dot{\alpha}_{k+1}} \ell^{\alpha(k+1)\dot{\alpha}(k))} \mathcal{J}_{\alpha(k+1)\dot{\alpha}(k+1)} + c.c. \Big\} \\ & + g \int \left\{\bar{\mathrm{D}}^2 \ell + \mathrm{D}^2 \bar{\ell}\right\} \mathcal{J} \end{aligned} \tag{51}$$

In order to complete Noether's procedure and get an invariant theory we have to add to the starting action S_o the following higher spin, cubic interaction terms

$$\begin{aligned} S_{\text{HS-}\Phi\text{ cubic interactions}} = & \, g \int \sum_{k=0}^{\infty} \Big\{ H^{\alpha(k+1)\dot{\alpha}(k+1)} \mathcal{J}_{\alpha(k+1)\dot{\alpha}(k+1)} \\ & + \chi^{\alpha(k+1)\dot{\alpha}(k)} \mathrm{D}_{\alpha_{k+1}} \mathcal{T}_{\alpha(k)\dot{\alpha}(k)} + c.c. \Big\} \\ & - g \int V \mathcal{J} \end{aligned} \tag{52}$$

where V is the real scalar superfield that describes the vector supermultiplet and has the gauge transformation $\delta_0 V = \bar{\mathrm{D}}^2 \ell + \mathrm{D}^2 \bar{\ell}$ and $H_{\alpha(k+1)\dot{\alpha}(k+1)}$, $\chi_{\alpha(k+1)\dot{\alpha}(k)}$ are the superfields that describe the super-Poincaré higher spin $(k+2, k+3/2)$ supermultiplet with the gauge transformations of (6). These cubic interaction terms generate the higher spin supercurrent $\mathcal{J}_{\alpha(k+1)\dot{\alpha}(k+1)}$ and the higher spin supertrace $\mathcal{T}_{\alpha(k)\dot{\alpha}(k)}$.

As expected, the supercurrent $\mathcal{J}_{\alpha(k+1)\dot{\alpha}(k+1)}$ and supertrace $\mathcal{T}_{\alpha(k)\dot{\alpha}(k)}$ include higher derivative terms. This is a corollary of the *Metsaev bounds* [78], where the number of derivatives that appear in a non-trivial cubic vertex is bounded from below by the highest spin involved and from above by the sum of the spins involved. In our case, there is no upper bound on the spins involved, which is consistent with the higher spin algebra structure[10] [79,80] thus making the number of derivatives that appear in (52) unbounded (as in string field theory).

Due to the higher derivative terms and the fixed engineering dimensions of $H_{\alpha(k+1)\dot{\alpha}(k+1)}$, $\chi_{\alpha(k+1)\dot{\alpha}(k)}$ from the free theory of massless, super-Poincaré higher spins [73,74], we need to have an appropriate dimensionful parameter M in order to balance the engineering dimensions of (52)[11], but since this effect can be easily tracked, for the sake of simplicity we will not explicitly include it. However, it is important to remember its presence since it introduces a scale into the theory. Also the parameter M gives the connection between the gauge parameters $\ell_{\alpha(k+1)\dot{\alpha}(k)}$, $\lambda_{\alpha(k+2)\dot{\alpha}(k)}$ that appear in (51) with the gauge parameters $L_{\alpha(k+1)\dot{\alpha}(k)}$, $\Lambda_{\alpha(k+2)\dot{\alpha}(k)}$ that appear in (6).

The conclusion of this section is that a single chiral superfield can have cubic interactions with only the half-integer superspin supermultiplets $(s+1, s+1/2)$ through the higher spin supercurrent and supertrace that have been constructed above, but more importantly although there are two possible descriptions of the $(s+1, s+1/2)$ supermultiplet, the chiral superfield has a preference to only one of them. The one that it chooses to interact with, is the one that appears in the higher spin, $\mathcal{N}$=2 theories as presented in [81].

---

10 The Jacobi identity requires an infinite tower of fields with unbounded spin.

11 Multiply the terms inside the curly bracket with $\left(\frac{1}{M}\right)^{k+1}$.

7. Minimal Multiplet of Noether Higher Spin Supercurrents

In the previous section, we found explicit expressions for the higher spin supercurrent and supertrace of the chiral superfield. Using the terminology of [77] these define the *canonical* multiplet of Noether higher spin supercurrents $\left\{\mathcal{J}_{\alpha(k+1)\dot{\alpha}(k+1)}, \mathcal{T}_{\alpha(k)\dot{\alpha}(k)}\right\}$. In this section we will show that for any value of the non-negative integer parameter k, there is another higher spin supercurrent multiplet, called the *minimal* multiplet $\left\{\mathcal{J}^{min}_{\alpha(k+1)\dot{\alpha}(k+1)}, \mathcal{T}^{min}_{\alpha(k)\dot{\alpha}(k)}\right\}$ and we arrive at it by an appropriate choice of the improvement terms such that $\mathcal{T}^{min}_{\alpha(k)\dot{\alpha}(k)} = 0$. In order to get some intuition about this process, it will be useful to examine first a simple example.

7.1. Coupling to Supergravity

For the case of $k = 0$ the canonical multiplet of supercurrents we obtain is

$$\begin{aligned} \mathcal{J}_{\alpha\dot{\alpha}} &= \mathrm{D}_\alpha\Phi\,\bar{\mathrm{D}}_{\dot{\alpha}}\bar{\Phi} + \mathrm{D}_\alpha\bar{\mathrm{D}}^2\bar{U}_{\dot{\alpha}} - \bar{\mathrm{D}}_{\dot{\alpha}}\mathrm{D}^2 U_\alpha\,, \\ \mathcal{T} &= -\Phi\bar{\Phi} + 2\mathrm{D}^\beta U_\beta + \bar{\mathrm{D}}^{\dot{\beta}}\bar{U}_{\dot{\beta}} \end{aligned} \tag{53}$$

and they generate the cubic interactions between the chiral and non-minimal supergravity supermultiplet. To investigate whether U_α has the potential to completely eliminate one of these supercurrents or reduce it to the point of being zero up to redefinitions of Φ, we consider the following ansatz

$$U_\alpha = f_1\,\mathrm{D}_\alpha\Phi\,\bar{\Lambda} + f_2\,\Phi\,\mathrm{D}_\alpha\bar{\Lambda} \tag{54}$$

where Λ is the prepotential of the chiral field (i.e., $\Phi = \bar{\mathrm{D}}^2\Lambda$). It is straight forward to find that:

$$\begin{aligned} \mathcal{J}_{\alpha\dot{\alpha}} = &\,[1+2f_1-2f_2]\,\mathrm{D}_\alpha\Phi\,\bar{\mathrm{D}}_{\dot{\alpha}}\bar{\Phi} - i[f_1-f_2]\,\partial_{\alpha\dot{\alpha}}\Phi\,\bar{\Phi} + i[f_1-f_2]\,\Phi\partial_{\alpha\dot{\alpha}}\bar{\Phi} \\ &+ [f_1-f_2]\,\bar{\mathrm{D}}_{\dot{\alpha}}\mathrm{D}_\alpha\left[\mathrm{D}^2\Phi\,\bar{\Lambda}\right] - [f_1-f_2]\,\mathrm{D}_\alpha\bar{\mathrm{D}}_{\dot{\alpha}}\left[\bar{\mathrm{D}}^2\bar{\Phi}\,\Lambda\right], \end{aligned} \tag{55}$$

$$\begin{aligned} \mathcal{T} = &\,[-1+3f_2-3f_1]\,\Phi\bar{\Phi} + 2[f_1-f_2]\,\mathrm{D}^2\Phi\,\bar{\Lambda} + [f_1-f_2]\,\bar{\mathrm{D}}^2\bar{\Phi}\,\Lambda \\ &+ 2[f_1+f_2]\,\mathrm{D}^2[\Phi\,\bar{\Lambda}] + [f_1+f_2]\bar{\mathrm{D}}^2[\bar{\Phi}\,\Lambda]\,. \end{aligned} \tag{56}$$

It is obvious that there is no choice of coefficients, f_1 and f_2 that can make $\mathcal{T}$ vanish. However, there is a choice that makes $\mathcal{T}$ proportional to the zeroth order (free theory) equation of motion of Φ. This is important because terms of this type can be absorbed by field redefinitions. If we choose $-f_1 = f_2 = 1/6$ we find

$$\begin{aligned} \mathcal{J}_{\alpha\dot{\alpha}} &= \tfrac{1}{3}\left\{\mathrm{D}_\alpha\Phi\,\bar{\mathrm{D}}_{\dot{\alpha}}\bar{\Phi} + i\partial_{\alpha\dot{\alpha}}\Phi\,\bar{\Phi} - i\Phi\partial_{\alpha\dot{\alpha}}\bar{\Phi}\right\} + \tfrac{1}{3}\left[\mathrm{D}_\alpha\bar{\mathrm{D}}_{\dot{\alpha}}(\Lambda\bar{\mathrm{D}}^2\bar{\Phi}) + c.c.\right] \\ \mathcal{T} &= -\tfrac{2}{3}\mathrm{D}^2\Phi\bar{\Lambda} - \tfrac{1}{3}\bar{\mathrm{D}}^2\bar{\Phi}\Lambda \end{aligned} \tag{57}$$

and therefore by redefining Φ in the following manner

$$\Phi \to \Phi + \tfrac{1}{3}g\bar{\mathrm{D}}^2(\Lambda\,\bar{\mathrm{D}}^{\dot{\alpha}}\mathrm{D}^\alpha H_{\alpha\dot{\alpha}}) - \tfrac{1}{3}g\bar{\mathrm{D}}^2(\Lambda\,\mathrm{D}^\alpha\chi_\alpha) - \tfrac{2}{3}g\bar{\mathrm{D}}^2(\Lambda\,\bar{\mathrm{D}}^{\dot{\alpha}}\bar{\chi}_{\dot{\alpha}}) \tag{58}$$

the S_o term will cancel the parts of the supercurrent and supertrace that have a $D^2\Phi$, $\bar{D}^2\bar{\Phi}$ dependence. The outcome of this procedure is the minimal multiplet of Noether supercurrent for the case of supergravity $\{\mathcal{J}^{min}_{\alpha\dot{\alpha}}, \mathcal{T}^{min}\}$, which is in agreement with the well known results in [77,82][12]

$$\mathcal{J}^{min}_{\alpha\dot{\alpha}} = \tfrac{1}{3}\left\{D_\alpha\Phi\bar{D}_{\dot{\alpha}}\bar{\Phi} + i\left(\partial_{\alpha\dot{\alpha}}\Phi\right)\bar{\Phi} - i\Phi\left(\partial_{\alpha\dot{\alpha}}\bar{\Phi}\right)\right\} , \tag{59a}$$

$$\mathcal{T}^{min} = 0 . \tag{59b}$$

Furthermore, the cubic interaction of the chiral superfield with supergravity becomes

$$S_{\text{SG-}\Phi\text{ cubic interactions}} = g\int H^{\alpha\dot{\alpha}}\,\mathcal{J}^{min}_{\alpha\dot{\alpha}} . \tag{60}$$

Nevertheless, we must keep in mind that Φ's redefinition (58) will generate order g^2 terms which we ignore because we focus on the cubic interaction terms. However, an interesting observation is that part of these g^2 terms modify our starting action S_o in the following way

$$\int\Phi\bar{\Phi} \rightarrow \int\left\{1 - \tfrac{1}{9}g^2\left[\bar{D}^{\dot{\alpha}}D^\alpha H_{\alpha\dot{\alpha}} - D^\alpha\chi_\alpha - 2\bar{D}^{\dot{\alpha}}\bar{\chi}_{\dot{\alpha}}\right]\left[D^\alpha\bar{D}^{\dot{\alpha}}H_{\alpha\dot{\alpha}} + 2D^\alpha\chi_\alpha + \bar{D}^{\dot{\alpha}}\bar{\chi}_{\dot{\alpha}}\right]\right\}\Phi\bar{\Phi} . \tag{61}$$

Of course this is nothing else than the perturbative construction of the volume element as one should expect for a theory that couples to supergravity.

7.2. Coupling to Higher Superspin Supermultiplets

Based on the previous example, we should check whether the minimal multiplet exists for the general case or not. According to (50), $\mathcal{T}_{\alpha(k)\dot{\alpha}(k)}$ is a linear combination of terms $\partial^{(p)}\Phi\,\partial^{(k-p)}\bar{\Phi}$ for various values of the non-negative integer p. Therefore a relevant ansatz for the improvement term is:

$$\mathcal{U}^{(p)}_{\alpha(k+1)\dot{\alpha}(k)} = f^{(p)}_1\partial^{(p)}D\Phi\,\partial^{(k-p)}\bar{\Lambda} + f^{(p)}_2\partial^{(p)}\Phi\,\partial^{(k-p)}D\bar{\Lambda} . \tag{62}$$

Following the instructions of (50) we calculate $D^{\alpha_{k+1}}\mathcal{U}^{(p)}_{\alpha(k+1)\dot{\alpha}(k)}$

$$\begin{aligned} D^{\alpha_{k+1}}\mathcal{U}^{(p)}_{\alpha(k+1)\dot{\alpha}(k)} =& f^{(p)}_2\tfrac{k+2}{k+1}\,\partial^{(p)}\Phi\,\partial^{(k-p)}\bar{\Phi} + f^{(p)}_1\tfrac{k+2}{k+1}\,\partial^{(p)}D^2\Phi\,\partial^{(k-p)}\bar{\Lambda} \\ &+ f^{(p)}_2\,\partial^{(p)}D^{\alpha_{k+1}}\Phi\,\partial^{(k-p)}D\bar{\Lambda} - f^{(p)}_1\,\partial^{(p)}D\Phi\,\partial^{(k-p)}D^{\alpha_{k+1}}\bar{\Lambda} . \end{aligned} \tag{63}$$

To avoid potential confusion, the explicit expression of the term $\partial^{(p)}D^{\alpha_{k+1}}\Phi\,\partial^{(k-p)}D\bar{\Lambda}$ is

$$\tfrac{1}{(k+1)k!}\partial_{(\alpha_1(\dot{\alpha}_1}\dots\partial_{\alpha_p\dot{\alpha}_p}D^{\alpha_{k+1}}\Phi\,\partial_{\alpha_{p+1}\dot{\alpha}_{p+1}}\dots\partial_{\alpha_k\dot{\alpha}_k)}D_{\alpha_{k+1})}\bar{\Lambda}$$

and by expanding the symmetrization of the indices, one can show that

$$\begin{aligned} &\partial^{(p)}D^{\alpha_{k+1}}\Phi\,\partial^{(k-p)}D\bar{\Lambda} = \\ &\quad -\tfrac{k-p+1}{k+1}\,\partial^{(p)}\Phi\,\partial^{(k-p)}\bar{\Phi} + i\tfrac{k-p}{k+1}\,\partial^{(p)}D\Phi\,\partial^{(k-p-1)}\bar{D}\bar{\Phi} \\ &\quad + i\tfrac{p}{k+1}\,\partial^{(p-1)}\bar{D}D^2\Phi\,\partial^{(k-p)}D\bar{\Lambda} + i\tfrac{k-p}{k+1}\,\partial^{(p)}D^2\Phi\,\partial^{(k-p-1)}\bar{D}D\bar{\Lambda} - \tfrac{1}{k+1}\,\partial^{(p)}D^2\Phi\,\partial^{(k-p)}\bar{\Lambda} \\ &\quad - i\tfrac{k-p}{k+1}D^2\left[\partial^{(p)}\Phi\,\partial^{(k-p-1)}\bar{D}D\bar{\Lambda}\right] + \tfrac{1}{k+1}D^2\left[\partial^{(p)}\Phi\,\partial^{(k-p)}\bar{\Lambda}\right] . \end{aligned} \tag{64}$$

[12] Keep in mind the difference in conventions for the covariant spinorial derivatives.

Similarly for the term $\partial^{(p)}\mathrm{D}\Phi\,\partial^{(k-p)}\mathrm{D}^{\alpha_{k+1}}\bar{\Lambda}$ we get

$$\begin{aligned}
&\partial^{(p)}\mathrm{D}\Phi\,\partial^{(k-p)}\mathrm{D}^{\alpha_{k+1}}\bar{\Lambda} = \\
&\quad \tfrac{p+1}{k+1}\,\partial^{(p)}\Phi\,\partial^{(k-p)}\bar{\Phi} + i\tfrac{k-p}{k+1}\,\partial^{(p)}\mathrm{D}\Phi\,\partial^{(k-p-1)}\bar{\mathrm{D}}\bar{\Phi} \\
&\quad +i\tfrac{p}{k+1}\,\partial^{(p-1)}\bar{\mathrm{D}}\mathrm{D}^2\Phi\,\partial^{(k-p)}\mathrm{D}\bar{\Lambda} - i\tfrac{k-p}{k+1}\,\partial^{(p)}\mathrm{D}^2\Phi\,\partial^{(k-p-1)}\mathrm{D}\bar{\mathrm{D}}\bar{\Lambda} + \tfrac{p+1}{k+1}\,\partial^{(p)}\mathrm{D}^2\Phi\,\partial^{(k-p)}\bar{\Lambda} \\
&\quad -i\tfrac{k-p}{k+1}\mathrm{D}^2\left[\partial^{(p)}\mathrm{D}\Phi\,\partial^{(k-p-1)}\bar{\mathrm{D}}\bar{\Lambda}\right] - \tfrac{p+1}{k+1}\mathrm{D}^2\left[\partial^{(p)}\Phi\,\partial^{(k-p)}\bar{\Lambda}\right].
\end{aligned} \tag{65}$$

Putting together all the above, we get

$$\begin{aligned}
\mathrm{D}^{\alpha_{k+1}}U^{(p)}_{\alpha(k+1)\dot{\alpha}(k)} = {}& \tfrac{p+1}{k+1}\left(f_2^{(p)} - f_1^{(p)}\right)\partial^{(p)}\Phi\,\partial^{(k-p)}\bar{\Phi} + i\tfrac{k-p}{k+1}\left(f_2^{(p)} - f_1^{(p)}\right)\partial^{(p)}\mathrm{D}\Phi\,\partial^{(k-p-1)}\bar{\mathrm{D}}\bar{\Phi} \\
&+\mathrm{D}^2\left[\vartheta\right] + \mathcal{O}(\mathrm{D}^2\Phi)
\end{aligned} \tag{66}$$

where $\mathrm{D}^2\left[\vartheta\right]$ is the sum of the terms that have the structure $\mathrm{D}^2\left[\ldots\right]$ and $\mathcal{O}(\mathrm{D}^2\Phi)$ is the sum of the terms that depend on the combination $\mathrm{D}^2\Phi$. Therefore the contribution of $U^{(p)}_{\alpha(k+1)\dot{\alpha}(k)}$ to $\mathcal{T}_{\alpha(k)\dot{\alpha}(k)}$ is

$$\begin{aligned}
&\tfrac{k+2}{k+1}\mathrm{D}^{\alpha_{k+1}}U^{(p)}_{\alpha(k+1)\dot{\alpha}(k)} + \bar{\mathrm{D}}^{\dot{\alpha}_{k+1}}\bar{U}^{(p)}_{\alpha(k)\dot{\alpha}(k+1)} = \\
&\tfrac{k+2}{k+1}\left(f_2^{(p)} - f_1^{(p)}\right)\partial^{(p)}\Phi\,\partial^{(k-p)}\bar{\Phi} + \tfrac{p+1}{k+1}\left(f_2^{(p)} - f_1^{(p)}\right)^{*}\partial^{(k-p)}\Phi\,\partial^{(p)}\bar{\Phi} - \tfrac{k-p}{k+1}\left(f_2^{(p)} - f_1^{(p)}\right)^{*}\partial^{(k-p-1)}\Phi\,\partial^{(p+1)}\bar{\Phi} \\
&+\tfrac{k+2}{k+1}\mathrm{D}^2\left[\vartheta\right] + \bar{\mathrm{D}}^2\left[\bar{\vartheta}\right] + \tfrac{k+2}{k+1}\mathcal{O}(\mathrm{D}^2\Phi) + \bar{\mathcal{O}}(\bar{\mathrm{D}}^2\bar{\Phi}) + \mathrm{D}\zeta
\end{aligned} \tag{67}$$

where we used $\partial^{(m)}\mathrm{D}\Phi\,\partial^{(n)}\bar{\mathrm{D}}\bar{\Phi} = \mathrm{D}\left(\partial^{(m)}\Phi\,\partial^{(n)}\bar{\mathrm{D}}\bar{\Phi}\right) - i\partial^{(m)}\Phi\,\partial^{(n+1)}\bar{\Phi}$ and $\mathrm{D}\zeta$ is the sum of terms that have the structure $\mathrm{D}(\ldots)$. It is important to observe that due to (i) Equation (49), (ii) the freedom in the definition of $\mathcal{T}_{\alpha(k+1)\dot{\alpha}(k)}$ (18) and (iii) the freedom to redefine the chiral superfield in a manner similar to Section 7.1, all the terms in the last line of (67) can be ignored. Furthermore, the terms in the first line contribute to the appropriate terms of $\mathcal{T}_{\alpha(k)\dot{\alpha}(k)}$. Hence, if we consider

$$U_{\alpha(k+1)\dot{\alpha}(k)} = \sum_{p=0}^{k} U^{(p)}_{\alpha(k+1)\dot{\alpha}(k)} \tag{68}$$

we have enough freedom to completely cancel $\mathcal{T}_{\alpha(k)\dot{\alpha}(k)}$. To illustrate this let us do this cancellation for $k=1$ and $k=2$ and then for the general case.

1. $\underline{k=1}$: The canonical supertrace is $i\partial\Phi\,\bar{\Phi} - \frac{3}{2}Z^{(2)} = -\frac{i}{2}\,\partial\Phi\,\bar{\Phi} + i\frac{3}{2}\,\Phi\,\partial\bar{\Phi}$.
   The contribution of $U^{(1)}$ is $\frac{3}{2}f^{(1)}\,\partial\Phi\,\bar{\Phi} + f^{(1)*}\,\Phi\,\partial\bar{\Phi}$, where $f^{(1)} = f_2^{(1)} - f_1^{(1)}$.
   The contribution of $U^{(0)}$ is $\frac{1}{2}f^{(0)*}\partial\Phi\,\bar{\Phi} + \left[\frac{3}{2}f^{(0)} - \frac{1}{2}f^{(0)*}\right]\Phi\,\partial\bar{\Phi}$, where $f^{(0)} = f_2^{(0)} - f_1^{(0)}$.
   We can cancel the supertrace competely if we select

$$\begin{aligned}
\tfrac{3}{2}f^{(1)} + \tfrac{1}{2}f^{(0)*} &= \tfrac{i}{2} \\
f^{(1)*} + \tfrac{3}{2}f^{(0)} - \tfrac{1}{2}f^{(0)*} &= -\tfrac{3i}{2}
\end{aligned} \quad \Rightarrow f^{(1)} = \tfrac{i}{10}\,,\ f^{(0)} = -\tfrac{7i}{10}\,. \tag{69}$$

2. $\underline{k=2}$: The canonical supertrace is $\frac{1}{3}\partial^2\Phi\,\bar{\Phi} - \frac{4}{3}\partial\Phi\,\partial\bar{\Phi} + \frac{4}{3}\Phi\,\partial^2\bar{\Phi}$.
   The contribution of $U^{(2)}$ is $\frac{4}{3}f^{(2)}\,\partial^2\Phi\,\bar{\Phi} + f^{(2)*}\,\Phi\,\partial^2\bar{\Phi}$, where $f^{(2)} = f_2^{(2)} - f_1^{(2)}$.
   The contribution of $U^{(1)}$ is $\left[\frac{4}{3}f^{(1)} + \frac{2}{3}f^{(1)*}\right]\partial\Phi\,\partial\bar{\Phi} - \frac{1}{3}f^{(1)*}\,\Phi\,\partial^2\bar{\Phi}$, where $f^{(1)} = f_2^{(1)} - f_1^{(1)}$.
   The contribution of $U^{(0)}$ is $\frac{1}{3}f^{(0)*}\,\partial^2\Phi\,\bar{\Phi} - \frac{2}{3}f^{(0)*}\,\partial\Phi\,\partial\bar{\Phi} + \frac{4}{3}f^{(0)}\,\Phi\,\partial^2\bar{\Phi}$, where $f^{(0)} = f_2^{(0)} - f_1^{(0)}$.
   If we select

$$\begin{aligned}
\tfrac{4}{3}f^{(2)} + \tfrac{1}{3}f^{(0)*} &= -\tfrac{1}{3} \\
\tfrac{4}{3}f^{(1)} + \tfrac{2}{3}f^{(1)*} - \tfrac{2}{3}f^{(0)*} &= \tfrac{4}{3} \\
f^{(2)*} - \tfrac{1}{3}f^{(1)*} + \tfrac{4}{3}f^{(0)} &= -\tfrac{4}{3}
\end{aligned} \quad \Rightarrow f^{(2)} = -\tfrac{1}{35}\,,\ f^{(1)} = \tfrac{13}{35}\,,\ f^{(0)} = -\tfrac{31}{35} \tag{70}$$

then we completely cancel the supertrace.

3. General k: For the general case, using (68) we can show that up to terms that can be ignored due to chiral redefinition and the freedom in the definitions of the supertrace (18) and (49) we get:

$$\tfrac{k+2}{k+1}\mathrm{D}^{\alpha_{k+1}}\mathcal{U}_{\alpha(k+1)\dot{\alpha}(k)}+\bar{\mathrm{D}}^{\dot{\alpha}_{k+1}}\bar{\mathcal{U}}_{\alpha(k)\dot{\alpha}(k+1)}=$$

$$\left\{\tfrac{k+2}{k+1}f^{(k)}+\tfrac{1}{k+1}f^{(0)*}\right\}\partial^{(k)}\Phi\,\bar{\Phi}+\sum_{p=0}^{k-1}\left\{\tfrac{k+2}{k+1}f^{(p)}+\tfrac{k+1-p}{k+1}f^{(k-p)*}-\tfrac{p+1}{k+1}f^{(k-1-p)*}\right\}\partial^{(p)}\Phi\,\partial^{(k-p)}\bar{\Phi}$$

where $f^{(p)}=f_2^{(p)}-f_1^{(p)}$. Then in order to cancel the supertrace, according to (45) and (50) we must have

$$(k+2)f^{(k)}+f^{(0)*}=(i)^k\,, \tag{71a}$$
$$(k+2)f^{(p)}+(k+1-p)f^{(k-p)*}-(p+1)f^{(k-1-p)*}=(-1)^{k+p}\,(i)^k\,(k+2)\,, \tag{71b}$$
$$p=0,1,\ldots,k-1\,.$$

This is a system of $k+1$ linear equations for the $k+1$ parameters $f^{(p)},\ p=0,1,...,k$. The solution is

$$f^{(p)}=(-1)^{k+p}\,(i)^k\,\frac{\sum\limits_{j=0}^{k-p}\binom{k+j+1}{p+j+1}\binom{k+1-j}{p+1}}{\binom{2k+3}{k+2}}\,,\quad p=0,1,...,k\,. \tag{72}$$

The result is that for any value of k, we can find an improvement term in order to go to the minimal multiplet of higher spin supercurrents $\{\,\mathcal{J}^{min}_{\alpha(k+1)\dot{\alpha}(k+1)}\,,\,\mathcal{T}^{min}_{\alpha(k)\dot{\alpha}(k)}\}$ where

$$\begin{aligned}\mathcal{J}^{min}_{\alpha(k+1)\dot{\alpha}(k+1)}=&\ if^{(k)}\,\partial^{(k+1)}\Phi\,\bar{\Phi}-if^{(k)*}\,\Phi\,\partial^{(k+1)}\bar{\Phi}\\&+i\sum_{p=1}^{k}\left\{(-1)^{k+p}\,(i)^k+f^{(p-1)}-f^{(k-p)*}\right\}\partial^{(p)}\Phi\,\partial^{(k+1-p)}\bar{\Phi}\\&+\sum_{p=0}^{k}\left\{(-1)^{k+p}\,(i)^k-f^{(p)}-f^{(k-p)*}\right\}\partial^{(p)}\mathrm{D}\Phi\,\partial^{(k-p)}\bar{\mathrm{D}}\bar{\Phi}\,,\end{aligned} \tag{73}$$

$$\mathcal{T}^{min}_{\alpha(k)\dot{\alpha}(k)}=0\,. \tag{74}$$

For $k=1$ and $k=2$ we get

$$\mathcal{J}^{min}_{\alpha\beta\dot{\alpha}\dot{\beta}}=-\tfrac{1}{10}\,\partial^{(2)}\Phi\,\bar{\Phi}-\tfrac{1}{10}\,\Phi\,\partial^{(2)}\bar{\Phi}+\tfrac{2}{5}\,\partial\Phi\,\partial\bar{\Phi}-\tfrac{1}{5}i\,\mathrm{D}\Phi\,\partial\bar{\mathrm{D}}\bar{\Phi}+\tfrac{1}{5}i\,\partial\mathrm{D}\Phi\,\bar{\mathrm{D}}\bar{\Phi}\,, \tag{75}$$

$$\begin{aligned}\mathcal{J}^{min}_{\alpha\beta\gamma\dot{\alpha}\dot{\beta}\dot{\gamma}}=&-\tfrac{i}{35}\,\partial^{(3)}\Phi\,\bar{\Phi}+\tfrac{i}{35}\,\Phi\,\partial^{(3)}\bar{\Phi}+i\tfrac{9}{35}\,\partial^{(2)}\Phi\,\partial\bar{\Phi}-i\tfrac{9}{35}\,\partial\Phi\,\partial^{(2)}\bar{\Phi}\\&-\tfrac{3}{35}\,\partial^{(2)}\mathrm{D}\Phi\,\bar{\mathrm{D}}\bar{\Phi}-\tfrac{3}{35}\,\mathrm{D}\Phi\,\partial^{(2)}\bar{\mathrm{D}}\bar{\Phi}+\tfrac{9}{35}\,\partial\mathrm{D}\Phi\,\partial\bar{\mathrm{D}}\bar{\Phi}\,.\end{aligned} \tag{76}$$

These expressions match the results of [41] which give the superconformal higher spin supercurrent. In the minimal supercurrent multiplet, the cubic interactions of the chiral supermultiplet with the higher spin supermultiplets are

$$S_{\text{HS-}\Phi\text{ cubic interactions}}=g\int\sum_{k=0}^{\infty}\,H^{\alpha(k+1)\dot{\alpha}(k+1)}\,\mathcal{J}^{min}_{\alpha(k+1)\dot{\alpha}(k+1)}\,. \tag{77}$$

8. On-Shell Conservation Equations

Using Noether's method, we have constructed an invariant action up to order g. Hence, for every unconstrained parameter $\ell_{\alpha(k+1)\dot{\alpha}(k)}$ and ℓ we generate a Bianchi identity, which express the invariance of the action. Once we go on-shell and take into account the equation of motion of Φ, the Bianchi identities reduce to the following on-shell conservation equations for the *canonical* multiplet of the higher spin supercurrents.

$$\bar{D}^{\dot{\alpha}_{k+1}}\mathcal{J}_{\alpha(k+1)\dot{\alpha}(k+1)} = \frac{1}{(k+1)!}\bar{D}^2 D_{(\alpha_{k+1}}\mathcal{T}_{\alpha(k))\dot{\alpha}(k)}\ ,\ k=0,1,2,\dots\ , \tag{78}$$

$$\bar{D}^2\mathcal{J} = 0\ . \tag{79}$$

It is straightforward to verify the validity of these on-shell equations using the expressions (48)–(50). For the *minimal* multiplet, the conservation equation takes the much simpler form

$$\bar{D}^{\dot{\alpha}_{k+1}}\mathcal{J}^{min}_{\alpha(k+1)\dot{\alpha}(k+1)} = 0\ ,\ k=0,1,2,\dots \tag{80}$$

After a bit of work, one can verify that Equation (73) satisfies this conservation equation. However, instead of using (73) we can get a simpler expression for the minimal higher spin supercurrent by using the conservation equation to define the coefficients of the various terms. From the previous section we know that the general ansatz for the minimal, higher spin supercurrent is

$$\mathcal{J}^{min}_{\alpha(s)\dot{\alpha}(s)} = \sum_{p=0}^{s} a_p\,\partial^{(p)}\Phi\,\partial^{(s-p)}\bar{\Phi} + \sum_{p=0}^{s-1} b_p\,\partial^{(p)}D\Phi\,\partial^{(s-p-1)}\bar{D}\bar{\Phi}\ . \tag{81}$$

We also know that $\mathcal{J}^{min}_{\alpha(s)\dot{\alpha}(s)}$ must be real, hence

$$a_p = a^*_{s-p}\ ,\ p=0,1,\dots,s\ , \tag{82}$$

$$b_p = b^*_{s-p-1}\ ,\ p=0,1,\dots,s-1 \tag{83}$$

and the on-shell conservation ($\bar{D}^{\dot{\alpha}_s}\mathcal{J}^{min}_{\alpha(s)\dot{\alpha}(s)} = 0$), also gives the constraint

$$i\,a_{p+1}\left[\tfrac{p+1}{s}\right] + b_p\left[\tfrac{s-p}{s}\right] = 0\ ,\ p=0,1,\dots,s-1\ . \tag{84}$$

The constraints (82)–(84) fix a_p and b_p to be (up to a real proportionality constant)

$$a_p = (-1)^p (i)^s \binom{s}{p}^2\ , \tag{85}$$

$$b_p = (-1)^p (i)^{s+1}\left(\frac{s-p}{p+1}\right)\binom{s}{p}^2 \tag{86}$$

and $\mathcal{J}^{min}_{\alpha(s)\dot{\alpha}(s)}$ is proportional to

$$\mathcal{J}^{min}_{\alpha(s)\dot{\alpha}(s)} \sim (i)^s\sum_{p=0}^{s}(-1)^p\binom{s}{p}^2\left\{\partial^{(p)}\Phi\,\partial^{(s-p)}\bar{\Phi} + i\left(\frac{s-p}{p+1}\right)\partial^{(p)}D\Phi\,\partial^{(s-p-1)}\bar{D}\bar{\Phi}\right\}. \tag{87}$$

We can fix the overall constant of proportionality by comparing this expression to (73), thus we get

$$\mathcal{J}^{min}_{\alpha(s)\dot{\alpha}(s)} = \frac{(-i)^s}{\binom{2s+1}{s+1}}\sum_{p=0}^{s}(-1)^p\binom{s}{p}^2\left\{\partial^{(p)}\Phi\,\partial^{(s-p)}\bar{\Phi} + i\left(\frac{s-p}{p+1}\right)\partial^{(p)}D\Phi\,\partial^{(s-p-1)}\bar{D}\bar{\Phi}\right\}. \tag{88}$$

It is easy to check that this expression agrees with Equations (59a), (73), (75) and (76) and up to an overall coefficient it also agrees with the results in [41].

9. Component Discussion

In the literature there are various sets of conserved currents that generate the cubic interactions of a complex scalar (two spin 0) and a spinor (one spin 1/2) with higher spins [58–63]. It is important to find how the results of previous sections translate at the component description.

In principle, we can start with Equation (77) and switch to the component formulation by evaluating the θ integrals in order to find the component analogue. However, for the purpose of identifying the higher spin, conserved currents, a conceptual cleaner approach would be to start with the superspace conservation Equation (80) and project it down to the component level, in order to derive the spacetime conservation equation of the currents. The latter is the approach that we will follow and the definition of components we will use is:

$$\begin{aligned}
\Phi^{(0,0)}_{\alpha(n)\dot{\alpha}(m)} &= \Phi_{\alpha(n)\dot{\alpha}(m)}|_{\theta=0} , & \Phi^{(1,0)}_{\beta\alpha(n)\dot{\alpha}(m)} &= D_\beta\Phi_{\alpha(n)\dot{\alpha}(m)}|_{\theta=0} , \\
\Phi^{(0,1)}_{\alpha(n)\dot{\beta}\dot{\alpha}(m)} &= \bar{D}_{\dot{\beta}}\Phi_{\alpha(n)\dot{\alpha}(m)}|_{\theta=0} , & \Phi^{(1,1)}_{\beta\alpha(n)\dot{\beta}\dot{\alpha}(m)} &= -\tfrac{1}{2}\left[D_\beta, \bar{D}_{\dot{\beta}}\right]\Phi_{\alpha(n)\dot{\alpha}(m)}|_{\theta=0} , \\
\Phi^{(2,0)}_{\alpha(n)\dot{\alpha}(m)} &= -D^2\Phi_{\alpha(n)\dot{\alpha}(m)}|_{\theta=0} , & \Phi^{(0,2)}_{\alpha(n)\dot{\alpha}(m)} &= -\bar{D}^2\Phi_{\alpha(n)\dot{\alpha}(m)}|_{\theta=0} , \\
\Phi^{(2,1)}_{\alpha(n)\dot{\beta}\dot{\alpha}(m)} &= -\tfrac{1}{2}\left\{D^2, \bar{D}_{\dot{\beta}}\right\}\Phi_{\alpha(n)\dot{\alpha}(m)}|_{\theta=0} , & \Phi^{(1,2)}_{\beta\alpha(n)\dot{\alpha}(m)} &= -\tfrac{1}{2}\left\{\bar{D}^2, D_\beta\right\}\Phi_{\alpha(n)\dot{\alpha}(m)}|_{\theta=0} , \\
\Phi^{(2,2)}_{\alpha(n)\dot{\alpha}(m)} &= \tfrac{1}{2}\{D^2, \bar{D}^2\}\Phi_{\alpha(n)\dot{\alpha}(m)}| - \tfrac{1}{4}\Box\Phi_{\alpha(n)\dot{\alpha}(m)}|_{\theta=0} .
\end{aligned} \tag{89}$$

The various components are labeled by the name of the superfield they come from and their position (n, m) in its θ expansion

$$\begin{aligned}
\Phi_{\alpha(n)\dot{\alpha}(m)} = \Phi_{\alpha(n)\dot{\alpha}(m)} &+ \theta^\beta\Phi^{(1,0)}_{\beta\alpha(n)\dot{\alpha}(m)} + \bar{\theta}^{\dot{\beta}}\Phi^{(0,1)}_{\alpha(n)\dot{\beta}\dot{\alpha}(m)} + \theta^2\Phi^{(2,0)}_{\alpha(n)\dot{\alpha}(m)} + \bar{\theta}^2\Phi^{(0,2)}_{\alpha(n)\dot{\alpha}(m)} \\
+\theta^\beta\bar{\theta}^{\dot{\beta}}\Phi^{(1,1)}_{\beta\alpha(n)\dot{\beta}\dot{\alpha}(m)} &+ \theta^\beta\bar{\theta}^2\Phi^{(1,2)}_{\beta\alpha(n)\dot{\alpha}(m)} + \theta^2\bar{\theta}^{\dot{\beta}}\Phi^{(2,1)}_{\alpha(n)\dot{\beta}\dot{\alpha}(m)} + \theta^2\bar{\theta}^2\Phi^{(2,2)}_{\alpha(n)\dot{\alpha}(m)} .
\end{aligned} \tag{90}$$

Furthermore, components with more than one index of the same type can be decomposed into symmetric (S) and anti-symmetric (A) pieces as follows

$$\begin{aligned}
F_{\beta\alpha(n)\dot{\alpha}(m)} &= F^{(S)}_{\beta\alpha(n)\dot{\alpha}(m)} + \tfrac{n}{(n+1)!}C_{\beta(\alpha_n}F^{(A)}_{\alpha(n-1))\dot{\alpha}(m)} , \\
F^{(S)}_{\beta\alpha(n)\dot{\alpha}(m)} &= \tfrac{1}{(n+1)!}F_{(\beta\alpha(n))\dot{\alpha}(m)} , \quad F^{(A)}_{\alpha(n-1)\dot{\alpha}(m)} = C^{\beta\alpha_n}F_{\beta\alpha(n)\dot{\alpha}(m)} .
\end{aligned} \tag{91}$$

Using the above, it is straightforward to project Equation (80) and the results we find for the bosonic components are:

$$\partial^{\alpha_s\dot{\alpha}_s}\mathcal{J}^{min\ (0,0)}_{\alpha(s)\dot{\alpha}(s)} = 0 , \tag{92a}$$

$$\mathcal{J}^{min\ (0,2)}_{\alpha(s)\dot{\alpha}(s)} = 0 , \tag{92b}$$

$$\mathcal{J}^{min\ (1,1)(S,A)}_{\alpha(s+1)\dot{\alpha}(s-1)} = -\frac{i}{2(s+1)!}\partial_{(\alpha_{s+1}}{}^{\dot{\alpha}_s}\mathcal{J}^{min\ (0,0)}_{\alpha(s))\dot{\alpha}(s)} , \tag{92c}$$

$$\mathcal{J}^{min\ (1,1)(A,A)}_{\alpha(s-1)\dot{\alpha}(s-1)} = 0 , \tag{92d}$$

$$\partial^{\alpha_{s+1}\dot{\alpha}_{s+1}}\mathcal{J}^{min\ (1,1)(S,S)}_{\alpha(s+1)\dot{\alpha}(s+1)} = 0 , \tag{92e}$$

$$\mathcal{J}^{min\ (2,2)}_{\alpha(k+1)\dot{\alpha}(k+1)} = -\tfrac{1}{4}\Box\mathcal{J}^{(0,0)}_{\alpha(k+1)\dot{\alpha}(k+1)} \tag{92f}$$

and for the fermionic components we get:

$$\mathcal{J}^{min\ (0,1)(A)}_{\alpha(s)\dot{\alpha}(s-1)} = 0 , \tag{93a}$$

$$\mathcal{J}^{min\ (1,2)(S)}_{\alpha(s+1)\dot{\alpha}(s)} = \frac{i}{2(s+1)!}\partial_{(\alpha_{s+1})}{}^{\dot{\alpha}_{s+1}}\mathcal{J}^{min\ (0,1)(S)}_{\alpha(s)\dot{\alpha}(s+1)} \ , \tag{93b}$$

$$\mathcal{J}^{min\ (1,2)(A)}_{\alpha(s-1)\dot{\alpha}(s)} = 0 \ , \tag{93c}$$

$$\partial^{\alpha_{s+1}\dot{\alpha}_s}\mathcal{J}^{min\ (1,0)(S)}_{\alpha(s+1)\dot{\alpha}(s)} = 0 \ . \tag{93d}$$

The lesson is that the component $\mathcal{J}^{min\ (1,1)(S,S)}_{\alpha(s+1)\dot{\alpha}(s+1)}$ is the *minimal* integer spin current and Equation (92e) is its conservation equation. The cubic interactions it generates are of the type

$$\int d^4x \sum_{s=0}^{\infty} h^{\alpha(s+1)\dot{\alpha}(s+1)}\mathcal{J}^{min\ (1,1)(S,S)}_{\alpha(s+1)\dot{\alpha}(s+1)} \tag{94}$$

where the field $h_{\alpha(s+1)\dot{\alpha}(s+1)}$ is the symmetric, traceless part of the free, massless, integer spin $j = s+1$ ($h_{\alpha(s+1)\dot{\alpha}(s+1)} \sim \left[\mathrm{D}_{(\alpha_{s+1}}, \bar{\mathrm{D}}_{(\dot{\alpha}_{s+1}}\right] H_{\alpha(s))\dot{\alpha}(s))}|$). From Equation (88) we get

$$\begin{aligned}\mathcal{J}^{min\ (1,1)(S,S)}_{\alpha(s+1)\dot{\alpha}(s+1)} \sim (-i)^s \sum_{p=0}^{s}(-1)^p \binom{s}{p}^2 \Big\{ & i\,\partial^{(p)}\phi\,\partial^{(s+1-p)}\bar{\phi} - i\left[\tfrac{2s+1-p}{p+1}\right]\partial^{(p+1)}\phi\,\partial^{(s-p)}\bar{\phi} \\ & + \left[\tfrac{s+p+2}{p+1}\right]\partial^{(p)}\chi\,\partial^{(s-p)}\bar{\chi} - \left[\tfrac{s-p}{p+1}\right]\partial^{(p+1)}\chi\,\partial^{(s-p-1)}\bar{\chi} \Big\} \ .\end{aligned} \tag{95}$$

Observe, that there are two contributions into these integer spin currents. The first one is the boson—boson contribution and includes the two terms of the first line, where $\phi = \Phi|$. This corresponds to the bosonic integer spin current that appears in [59] and also the traceless part of the currents in [58,63]. The second contribution is the fermion—fermion one and includes the two terms of the second line, where $\chi_\alpha = \mathrm{D}_\alpha\Phi|$. This corresponds to the fermionic integer spin current that appears in [59].

Furthermore, Equation (93d) gives the conservation of the half-integer spin current $\mathcal{J}^{min\ (1,0)(S)}_{\alpha(s+1)\dot{\alpha}(s)}$. The cubic interactions we get are:

$$\int d^4x \sum_{s=0}^{\infty} \psi^{\alpha(s+1)\dot{\alpha}(s)}\mathcal{J}^{min\ (1,0)(S)}_{\alpha(s+1)\dot{\alpha}(s)} + c.c. \tag{96}$$

where $\psi_{\alpha(s+1)\dot{\alpha}(s)}$ is the symmetric, traceless and γ-traceless part of the free, massless, half-integer spin $j = s+1/2$ ($\psi_{\alpha(s+1)\dot{\alpha}(s)} \sim \left\{\mathrm{D}_{(\alpha_{s+1}}, \bar{\mathrm{D}}^2\right\} H_{\alpha(s))\dot{\alpha}(s)}|$). Again using (88) we get

$$\mathcal{J}^{min\ (1,0)(S)}_{\alpha(s+1)\dot{\alpha}(s)} \sim (-i)^s \sum_{p=0}^{s}(-1)^p\binom{s}{p}^2\left(\frac{s+1}{p+1}\right)\partial^{(p)}\chi\,\partial^{(s-p)}\bar{\phi} \ . \tag{97}$$

This is the half-integer spin current and appears for the first time in the literature and it has only one contribution of the fermion—boson type.

Finally, we notice that Equation (92a) is the conservation of another current. This corresponds to the $\mathcal{R}$-symmetry current and it has the form

$$\mathcal{J}^{min\ (0,0)}_{\alpha(s)\dot{\alpha}(s)} \sim (-i)^s \sum_{p=0}^{s}(-1)^p\binom{s}{p}^2\left\{\partial^{(p)}\phi\,\partial^{(s-p)}\bar{\phi} + i\left(\frac{s-p}{p+1}\right)\partial^{(p)}\chi\,\partial^{(s-p-1)}\bar{\chi}\right\} \ . \tag{98}$$

10. Massive Chiral Superfield

10.1. Higher Spin Supercurrent and Supertrace

So far we have discussed the higher spin supercurrent multiplet of a free, massless chiral superfield. In this section, we repeat the analysis for a massive chiral superfield, with a starting action $S_o + S_m$ where S_o is given by (13) and S_m is the mass term:

$$S_m = \tfrac{m}{2} \int d^6z\, \Phi^2 \ + c.c.\, . \tag{99}$$

The variation of this extra term under (12) is

$$\begin{aligned}\delta_g S_m = -gm \sum_{k=0}^{\infty} \int d^6z \Big\{ & \bar{\mathrm{D}}^2 \ell^{\alpha(k+1)\dot\alpha(k)}\ i^k \partial^{(k)} \mathrm{D}\Phi\, \Phi \ + c.c. \\ & - \bar{\mathrm{D}}^{(\dot\alpha_{k+1}} \ell^{\alpha(k+1)\dot\alpha(k))}\ i^{k+1} \partial^{(k+1)} \Phi\, \Phi \ + c.c. \Big\} \\ + gm \int d^6z\, \bar{\mathrm{D}}^2 \ell\, \Phi\, \Phi \ + c.c. \end{aligned} \tag{100}$$

It is straight forward to show that:

$$\begin{aligned}\bar{\mathrm{D}}^2 \ell^{\alpha(k+1)\dot\alpha(k)}\ i^k \partial^{(k)} \mathrm{D}\Phi\, \Phi \ - \ \bar{\mathrm{D}}^{(\dot\alpha_{k+1}} \ell^{\alpha(k+1)\dot\alpha(k))}\ i^{k+1} \partial^{(k+1)} \Phi\, \Phi = & \\ \tfrac{1}{2} \bar{\mathrm{D}}^2 \Big[\bar{\mathrm{D}}^2 \ell^{\alpha(k+1)\dot\alpha(k)} \ \big\{ i^k \partial^{(k)} \mathrm{D}\Lambda\, \Phi \ + \ i^k \partial^{(k)} \mathrm{D}\Phi\, \Lambda \big\} & \\ - \bar{\mathrm{D}}^{(\dot\alpha_{k+1}} \ell^{\alpha(k+1)\dot\alpha(k))} \ \big\{ i^{k+1} \partial^{(k+1)} \Lambda\, \Phi \ + \ i^{k+1} \partial^{(k+1)} \Phi\, \Lambda \big\} \Big]\, , & \end{aligned} \tag{101}$$

$$\bar{\mathrm{D}}^2 \ell \Phi\, \Phi = \bar{\mathrm{D}}^2 \left[\bar{\mathrm{D}}^2 \ell\, \Lambda\, \Phi \right] \tag{102}$$

and by absorbing the overall $\bar{\mathrm{D}}^2$ factor, we can convert the integration over the entire superspace:

$$\begin{aligned}\delta_g S_m = \tfrac{g}{2} m \int \sum_{k=0}^{\infty} \Big\{ & \bar{\mathrm{D}}^2 \ell^{\alpha(k+1)\dot\alpha(k)} \ \left[i^k \partial^{(k)} \mathrm{D}\Lambda\, \Phi \ + \ i^k \partial^{(k)} \mathrm{D}\Phi\, \Lambda \right] \ + c.c. \\ & - \bar{\mathrm{D}}^{(\dot\alpha_{k+1}} \ell^{\alpha(k+1)\dot\alpha(k))} \ \left[i^{k+1} \partial^{(k+1)} \Lambda\, \Phi \ + \ i^{k+1} \partial^{(k+1)} \Phi\, \Lambda \right] \ + c.c. \Big\} \\ - gm \int \ \bar{\mathrm{D}}^2 \ell\, \Lambda\, \Phi \ + c.c. \end{aligned} \tag{103}$$

From this expression we can extract the contribution of the mass term to Equations (17a) and (17b). However, in order to couple the theory purely to higher spin supermultiplets the coefficient of $\bar{\mathrm{D}}^{(\dot\alpha_{k+1}} \ell^{\alpha(k+1)\dot\alpha(k))}$ must be written as a real term plus total spinorial or spacetime derivative terms. For the massless theory, we have proven this property via Equation (43) and it holds for any value of k. The story for a massive chiral is different as we will show that only the even values of $k = 2l$ can satisfy such a requirement.

The relevant quantity for the mass term is $i^{k+1} \partial^{(k+1)} \Lambda\, \Phi + i^{k+1} \partial^{(k+1)} \Phi\, \Lambda$. It is easy to show that this combination can be written in the following manner:

$$i^{k+1}\partial^{(k+1)}\Lambda\,\Phi\;+\;i^{k+1}\partial^{(k+1)}\Phi\,\Lambda = \tag{104}$$
$$= \begin{cases} i\partial\left[\sum\limits_{n=0}^{2l}(-1)^{l+n}\,\partial^{(n)}\Lambda\,\partial^{(2l-n)}\Phi\right], \; for\ k=2l\,,\ l=0,1,2,\dots \\ \partial\left[\sum\limits_{n=0}^{l}(-1)^{l+n+1}\,\partial^{(n)}\Lambda\,\partial^{(2l+1-n)}\Phi+\sum\limits_{n=l+1}^{2l+1}(-1)^{l+n}\,\partial^{(n)}\Lambda\,\partial^{(2l+1-n)}\Phi\right]+2\,\partial^{(l+1)}\Lambda\,\partial^{(l+1)}\Phi\,, \\ \qquad\qquad\qquad for\ k=2l+1\,,\ l=0,1,2,\dots \end{cases}$$

therefore, for odd values of k and due to the presence of the term $\partial^{(l+1)}\Lambda\,\partial^{(l+1)}\Phi$, there is no improvement term $W_{\alpha(2l+2)\dot{\alpha}(2l+1)}$ to eliminate the imaginary part of $\mathcal{J}_{\alpha(2l+2)\dot{\alpha}(2l+2)}$. Hence, in order to construct an invariant theory of a massive chiral interacting with irreducible higher spin supermultiplets, all terms in $\delta_g\,(S_o+S_m)$ that correspond to an odd value of k must be set to zero. For that reason the parameters ℓ and $\ell_{\alpha(2l+2)\dot{\alpha}(2l+1)}$ for $l=0,1,2,\dots$ must vanish and the transformation of Φ we must consider in this massive case is reduced to:

$$\delta_g\Phi = -g\sum_{l=0}^{\infty}\Big\{\;\bar{\mathrm{D}}^2\ell^{\alpha(2l+1)\dot{\alpha}(2l)}\,\mathrm{D}_{\alpha_{2l+1}}\bar{\mathrm{D}}_{\dot{\alpha}_{2l}}\mathrm{D}_{\alpha_{2l}}\dots\bar{\mathrm{D}}_{\dot{\alpha}_1}\mathrm{D}_{\alpha_1}\Phi \tag{105}$$
$$-\tfrac{1}{(2l+1)!}\bar{\mathrm{D}}^{(\dot{\alpha}_{2l+1}}\ell^{\alpha(2l+1)\dot{\alpha}(2l))}\,\bar{\mathrm{D}}_{\dot{\alpha}_{2l+1}}\mathrm{D}_{\alpha_{2l+1}}\dots\bar{\mathrm{D}}_{\dot{\alpha}_1}\mathrm{D}_{\alpha_1}\Phi\Big\}\;.$$

Moreover, we can show that for the case of $k=2l$ the quantity $i^{2l}\partial^{(2l)}\mathrm{D}\Lambda\,\Phi+i^{2l}\partial^{(2l)}\mathrm{D}\Phi\,\Lambda$ which appears in (103) as the coefficient of $\bar{\mathrm{D}}^2\ell^{\alpha(2l+1)\dot{\alpha}(2l)}$ can be expressed in the following way:

$$i^{2l}\partial^{(2l)}\mathrm{D}\Lambda\,\Phi+i^{2l}\partial^{(2l)}\mathrm{D}\Phi\,\Lambda=\mathrm{D}\Big[(-1)^l\,\Lambda\,\partial^{(2l)}\Phi\Big]+\partial\left[\sum_{n=0}^{2l-1}(-1)^{l+n+1}\,\partial^{(n)}\mathrm{D}\Lambda\,\partial^{(2l-1-n)}\Phi\right]\;. \tag{106}$$

With all the above into account, we get that

$$\mathcal{J}_{\alpha(2l+1)\dot{\alpha}(2l+1)}= X^{(2l+1)}_{\alpha(2l+1)\dot{\alpha}(2l+1)}+\tfrac{1}{(2l+1)!^2}\bar{\mathrm{D}}_{(\dot{\alpha}_{2l+1}}\mathrm{D}_{(\alpha_{2l+1}}Z^{(2l+1)}_{\alpha(2l))\dot{\alpha}(2l))}-\tfrac{im}{2(2l+1)!^2}\partial_{(\alpha_{2l+1}(\dot{\alpha}_{2l+1}}Y_{\alpha(2l))\dot{\alpha}(2l))} \tag{107}$$
$$+\tfrac{1}{(2l+1)!}\mathrm{D}_{(\alpha_{2l+1}}\bar{\mathrm{D}}^2\bar{U}_{\alpha(2l))\dot{\alpha}(2l+1)}+\tfrac{2l+1}{(2l+2)(2l+1)!}\bar{\mathrm{D}}_{(\dot{\alpha}_{2l+1}}W_{\alpha(2l+1)\dot{\alpha}(2l))}$$

with

$$Y_{\alpha(2l)\dot{\alpha}(2l)}=\sum_{n=0}^{2l}(-1)^{l+n}\,\partial^{(n)}\Lambda\,\partial^{(2l-n)}\Phi\;. \tag{108}$$

Now it is obvious that we can always make $\mathcal{J}_{\alpha(2l+1)\dot{\alpha}(2l+1)}$ real by selecting $W_{\alpha(2l+1)\dot{\alpha}(2l)}$ as follows:

$$W_{\alpha(2l+1)\dot{\alpha}(2l)}=-\tfrac{2l+2}{2l+1}\mathrm{D}^2U_{\alpha(2l+1)\dot{\alpha}(2l)}-\tfrac{2l+2}{(2l+1)(2l+1)!}\mathrm{D}_{(\alpha_{2l+1}}\Big[Z^{(2l+1)}_{\alpha(2l))\dot{\alpha}(2l)}-\tfrac{m}{2}\Big(Y_{\alpha(2l))\dot{\alpha}(2l)}+\bar{Y}_{\alpha(2l))\dot{\alpha}(2l)}\Big)\Big] \tag{109}$$

and the expressions for $\mathcal{J}_{\alpha(2l+1)\dot{\alpha}(2l+1)}$ and $\mathcal{T}_{\alpha(2l+1)\dot{\alpha}(2l)}$ become

$$\mathcal{J}_{\alpha(2l+1)\dot{\alpha}(2l+1)}=X^{(2l+1)}_{\alpha(2l+1)\dot{\alpha}(2l+1)}+\tfrac{m}{2(2l+1)!^2}\Big[\bar{\mathrm{D}}_{(\dot{\alpha}_{2l+1}}\mathrm{D}_{(\alpha_{2l+1}}\bar{Y}_{\alpha(2l))\dot{\alpha}(2l))}-\mathrm{D}_{(\alpha_{2l+1}}\bar{\mathrm{D}}_{(\dot{\alpha}_{2l+1}}Y_{\alpha(2l))\dot{\alpha}(2l))}\Big] \tag{110}$$
$$+\tfrac{1}{(2l+1)!}\Big[\mathrm{D}_{(\alpha_{2l+1}}\bar{\mathrm{D}}^2\bar{U}_{\alpha(2l))\dot{\alpha}(2l+1)}-\bar{\mathrm{D}}_{(\dot{\alpha}_{2l+1}}\mathrm{D}^2U_{\alpha(2l+1)\dot{\alpha}(2l))}\Big]\;,$$

$$\mathcal{T}_{\alpha(2l+1)\dot{\alpha}(2l)}=\tfrac{1}{(2l+1)!}\mathrm{D}_{(\alpha_{2l+1}}\mathcal{T}_{\alpha(2l))\dot{\alpha}(2l)}\;, \tag{111}$$

$$\mathcal{T}_{\alpha(2l)\dot{\alpha}(2l)}=(-1)^l\partial^{(2l)}\Phi\,\bar{\Phi}-\tfrac{2(l+1)}{2l+1}Z^{(2l+1)}_{\alpha(2l)\dot{\alpha}(2l)}+\tfrac{m(l+1)}{2l+1}\Big(Y_{\alpha(2l)\dot{\alpha}(2l)}+\bar{Y}_{\alpha(2l)\dot{\alpha}(2l)}\Big)+\tfrac{m}{2}\Omega_{\alpha(2l)\dot{\alpha}(2l)} \tag{112}$$
$$+\tfrac{2(l+1)}{2l+1}\mathrm{D}^{\alpha_{2l+1}}U_{\alpha(2l+1)\dot{\alpha}(2l)}+\bar{\mathrm{D}}^{\dot{\alpha}_{2l+1}}\bar{U}_{\alpha(2l)\dot{\alpha}(2l+1)}$$

where

$$\Omega_{\alpha(2l)\dot{\alpha}(2l)} = (-1)^{l+1}\,\Lambda\,\partial^{(2l)}\Phi + i\sum_{n=0}^{2l-1}\,(-1)^{l+1+n}\,\partial^{(n)}\bar{\mathrm{D}}\mathrm{D}\Lambda\,\partial^{(2l-1-n)}\Phi\ . \tag{113}$$

The result for the variation of the $S_o + S_m$ theory is

$$\begin{aligned}\delta_g\left(S_o + S_m\right) =& -g\int\sum_{l=0}^{\infty}\Big\{\left[\bar{\mathrm{D}}^2\ell^{\alpha(2l+1)\dot{\alpha}(2l)} - \mathrm{D}_{\alpha_{2l+2}}\lambda^{\alpha(2l+2)\dot{\alpha}(2l)}\right]\mathrm{D}_{\alpha_{2l+1}}\mathcal{T}_{\alpha(2l)\dot{\alpha}(2l)} + c.c. \\ & - \tfrac{1}{(2l+1)!}\bar{\mathrm{D}}^{(\dot{\alpha}_{2l+1}}\ell^{\alpha(2l+1)\dot{\alpha}(2l))}\,\mathcal{J}_{\alpha(2l+1)\dot{\alpha}(2l+1)} + c.c.\Big\}\end{aligned} \tag{114}$$

where $\mathcal{J}_{\alpha(2l+1)\dot{\alpha}(2l+1)}$ and $\mathcal{T}_{\alpha(2l+1)\dot{\alpha}(2l)}$ are given by (110) and (112). Therefore to get the invariant theory we have to add the following higher spin, cubic interaction terms

$$\begin{aligned}S_{\text{HS-massive chiral}} = g\int\sum_{l=0}^{\infty}\Big\{ & H^{\alpha(2l+1)\dot{\alpha}(2l+1)}\,\mathcal{J}_{\alpha(2l+1)\dot{\alpha}(2l+1)} \\ & + \chi^{\alpha(2l+1)\dot{\alpha}(2l)}\,\mathrm{D}_{\alpha_{2l+1}}\mathcal{T}_{\alpha(2l)\dot{\alpha}(2l)} + c.c.\Big\}\ .\end{aligned} \tag{115}$$

Apart from the various mass terms that deform the expressions for the higher spin supercurrent and supertrace, the biggest difference from the massless chiral story is that the massive chiral superfields has cubic interactions only with $(2l+2\ ,\ 2l+3/2)$ supermultiplets that correspond to superspin $Y = 2l + 3/2$. This includes supergravity ($l = 0$) but not the vector supermultiplet.

10.2. Minimal Multiplet of Higher Spin Supercurrents

Similar to the massless case, expressions (110) and (112) include an arbitrary improvement term $U_{\alpha(2l+1)\dot{\alpha}(2l)}$, hence we have to check whether this freedom can be used to completely eliminate the supertrace. For the case of supergravity the *canonical* supercurrent multiplet we get is:

$$\mathcal{J}_{\alpha\dot{\alpha}} = \mathrm{D}_\alpha\Phi\,\bar{\mathrm{D}}_{\dot{\alpha}}\bar{\Phi} + \tfrac{m}{2}\bar{\mathrm{D}}_{\dot{\alpha}}\mathrm{D}_\alpha\left(\bar{\Lambda}\bar{\Phi}\right) - \tfrac{m}{2}\mathrm{D}_\alpha\bar{\mathrm{D}}_{\dot{\alpha}}\left(\Lambda\Phi\right) + \mathrm{D}_\alpha\bar{\mathrm{D}}^2\bar{U}_{\dot{\alpha}} - \bar{\mathrm{D}}_{\dot{\alpha}}\mathrm{D}^2 U_\alpha\ , \tag{116}$$

$$\mathcal{T} = -\Phi\bar{\Phi} + \tfrac{m}{2}\Lambda\Phi + m\bar{\Lambda}\bar{\Phi} + 2\mathrm{D}^\alpha U_\alpha + \bar{\mathrm{D}}^{\dot{\alpha}}\bar{U}_{\dot{\alpha}}\ . \tag{117}$$

It is easy to see that there is no choice of U_α that can cancel the terms of $\mathcal{T}$ proportional to the mass. This is true not just for the case of supergravity, but for the higher spin supermultiplets as well. The higher spin supertrace $\mathcal{T}_{\alpha(2l)\dot{\alpha}(2l)}$ can not be eliminated and there is no *minimal* supercurrent multiplet for massive chirals.

However, we can use the procedure of Section 7 in order to absorb all the m independent terms of the supertrace and make it proportional to the mass. In this configuration the supercurrent will be the same as the minimal supercurrent of massless chiral (73) plus terms proportional to mass. For the case of supergravity this will give

$$\mathcal{J}_{\alpha\dot{\alpha}} = \mathcal{J}^{min}_{\alpha\dot{\alpha}} - \tfrac{m}{6}\mathrm{D}_\alpha\bar{\mathrm{D}}_{\dot{\alpha}}\left(\Lambda\Phi\right) + \tfrac{m}{6}\bar{\mathrm{D}}_{\dot{\alpha}}\mathrm{D}_\alpha\left(\bar{\Lambda}\bar{\Phi}\right)\ , \tag{118}$$

$$\mathcal{T} = \tfrac{m}{6}\Lambda\Phi + \tfrac{m}{3}\bar{\Lambda}\bar{\Phi} \tag{119}$$

where $\mathcal{J}^{min}_{\alpha\dot{\alpha}}$ is given in (59a).

10.3. Conservation Equation

The conservation equation that the $\mathcal{J}_{\alpha(2l+1)\dot{\alpha}(2l+1)}$ and $\mathcal{T}_{\alpha(2l)\dot{\alpha}(2l)}$ satisfy on-shell is

$$\bar{\mathrm{D}}^{\dot{\alpha}_{2l+1}}\mathcal{J}_{\alpha(2l+1)\dot{\alpha}(2l+1)} = \tfrac{1}{(2l+1)!}\bar{\mathrm{D}}^2\mathrm{D}_{(\alpha_{2l+1}}\mathcal{T}_{\alpha(2l))\dot{\alpha}(2l)}\ ,\ l = 0,1,2,\ldots \tag{120}$$

and it is straight forward to show that expressions (110) and (112) do that on-shell [13]. As we did for the massless chiral, we will use this conservation equation to derive a closed form expression for the higher spin supercurrent and supertrace. Based on the previous results the general ansatz for the higher spin supercurrent and supertrace is

$$\begin{aligned}\mathcal{J}_{\alpha(s)\dot{\alpha}(s)} = \mathcal{J}^{min}_{\alpha(s)\dot{\alpha}(s)} &+ m\sum_{p=0}^{s-1}\gamma_p\,\partial^{(p)}\mathrm{D}\bar{\mathrm{D}}\Lambda\,\partial^{(s-1-p)}\Phi + m\sum_{p=0}^{s-1}\delta_p\,\partial^{(p)}\bar{\mathrm{D}}\Lambda\,\partial^{(s-1-p)}\mathrm{D}\Phi \\ &- m\sum_{p=0}^{s-1}\gamma^*_p\,\partial^{(p)}\bar{\mathrm{D}}\mathrm{D}\bar{\Lambda}\,\partial^{(s-1-p)}\bar{\Phi} - m\sum_{p=0}^{s-1}\delta^*_p\,\partial^{(p)}\mathrm{D}\bar{\Lambda}\,\partial^{(s-1-p)}\bar{\mathrm{D}}\bar{\Phi}\,,\end{aligned} \tag{121}$$

$$\begin{aligned}\mathcal{T}_{\alpha(s-1)\dot{\alpha}(s-1)} = &\; m\sum_{p=0}^{s-1}\zeta_p\,\partial^{(p)}\Lambda\,\partial^{(s-1-p)}\Phi + m\sum_{p=0}^{s-1}\xi_p\,\partial^{(p)}\bar{\Lambda}\,\partial^{(s-1-p)}\bar{\Phi} \\ &+ m\sum_{p=0}^{s-2}\sigma_p\,\partial^{(p)}\bar{\mathrm{D}}\mathrm{D}\Lambda\,\partial^{(s-2-p)}\Phi\end{aligned} \tag{122}$$

with $\mathcal{J}^{min}_{\alpha(s)\dot{\alpha}(s)}$ given by (88). The conservation Equation (120) fixes the coefficients δ_p, ξ_p, ζ_p, σ_p:

$$\delta_p = -\gamma_p\,, \qquad p = 0,1,\ldots,s-1\,, \tag{123a}$$
$$\xi_p = -\tfrac{s+1}{s}\,\gamma^*_p\,, \qquad p = 0,1,\ldots,s-1\,, \tag{123b}$$
$$\zeta_0 = -\tfrac{1}{s}\,\gamma_0\,, \tag{123c}$$
$$\zeta_p = -\tfrac{p+1}{s}\,\gamma_p + \tfrac{s-p}{s}\,\gamma_{p-1}\,, \qquad p = 1,2,\ldots,s-1\,, \tag{123d}$$
$$\sigma_0 = -\tfrac{i}{s}\,\gamma_1 + i\,\tfrac{s-1}{s}\,\gamma_0\,, \tag{123e}$$
$$\sigma_p = (-1)^{p+1}\,\tfrac{i}{s}\,\gamma_1 + (-1)^p\,i\,\tfrac{s-1}{s}\,\gamma_0 + i\sum_{n=1}^{p}(-1)^{p+n+1}\left[\tfrac{n+1}{s}\,\gamma_{n+1} - \tfrac{s-2n-1}{s}\,\gamma_n - \tfrac{s-n}{s}\,\gamma_{n-1}\right] \qquad p = 1,2,\ldots,s-2 \tag{123f}$$

and the coefficients γ_p satisfy the constraints:

$$\gamma_p + \gamma_{s-p-1} = \frac{(-1)^{s+p}\,(i)^{s+1}}{\binom{2s+1}{s+1}}\sum_{n=0}^{p}\binom{s}{n}^2\left[\tfrac{s+1}{s+1-n} + (-1)^s\,\tfrac{s+1}{n+1}\right]\,,\; p = 0,1,\ldots,s-1\,, \tag{124a}$$
$$\sigma_{s-2} = -i\,\tfrac{s-1}{s}\,\gamma_{s-1} + \tfrac{i}{s}\,\gamma_{s-2}\,. \tag{124b}$$

Notice that the left hand side of (124a) is invariant under $p \to s-1-p$, therefore we get a consistency condition

$$\left[1 + (-1)^s\right]\sum_{n=0}^{s}\binom{s}{n}^2\frac{s+1}{n+1} = 0 \tag{125}$$

which selects only the odd values of s, in agreement with (115). For $s = 2l+1$, Equation (124a) fixes γ_l

$$\gamma_l = \frac{l+1}{\binom{4l+3}{2l+2}}\sum_{n=0}^{l}\binom{2l+1}{n}^2\left[\tfrac{1}{2l+2-n} - \tfrac{1}{n+1}\right]\,. \tag{126}$$

A consequence of that is $\xi_l \neq 0$ due to (123b). Therefore the supertrace can not be zero as in the massless case. Moreover, the constraints (124a) and (124b) provide a system of $l+2$ linear equations for the $2l+1$, γ_p coefficients, so there is a freedom of choice for $l-1$ of these coefficients. This freedom

[13] Keep in mind that the on-shell equation of motion for a free massive chiral is $\bar{\mathrm{D}}^2\bar{\Phi} = m\Phi$.

corresponds to the fact that there is no unique canonical supercurrent multiplet, in contrast with the massless case where the minimal multiplet is unique. An example of a choice is to have

$$\gamma_{l+2} = \gamma_{l+3} = \cdots = \gamma_{2l} = 0\,. \tag{127}$$

11. Summary and Discussion

Let us briefly summarize and discuss the results obtained. In Section 3 we presented the most general ansatz for the transformation of a 4D, $\mathcal{N} = 1$ chiral superfield with linear terms (9). The consistence with chirality, constrained the parameters (10) and revealed structures similar to the gauge transformations of free, massless, higher-superspin theories. This was a hint that chiral superfields can have cubic interactions with higher spin superfields. Therefore, using (12) and Noether's method we:

(*i*) Proved that a single, massless, chiral superfield can have cubic interactions (52) only with the half-integer superspin $(s+1, s+1/2)$ irreducible representations of the super-Poincaré group. Moreover, despite the fact that there are two different formulations of the half-integer superspin supermultiplets, the chiral superfield has a clear preference to couple only to one of them, the one that can be lifted to $\mathcal{N} = 2$ higher spin supermultiplets.

(*ii*) Generated the *canonical* multiplet of higher spin supercurrents $\left\{\mathcal{J}_{\alpha(k+1)\dot{\alpha}(k+1)}, \mathcal{T}_{\alpha(k)\dot{\alpha}(k)}\right\}$ (48) and (50) which satisfy conservation Equation (78) and leads to the cubic interactions

$$g\int\sum_{k=0}^{\infty}\left\{H^{\alpha(k+1)\dot{\alpha}(k+1)}\mathcal{J}_{\alpha(k+1)\dot{\alpha}(k+1)} + \chi^{\alpha(k+1)\dot{\alpha}(k)}\mathrm{D}_{\alpha_{k+1}}\mathcal{T}_{\alpha(k)\dot{\alpha}(k)} + \bar{\chi}^{\alpha(k)\dot{\alpha}(k+1)}\bar{\mathrm{D}}_{\dot{\alpha}_{k+1}}\bar{\mathcal{T}}_{\alpha(k)\dot{\alpha}(k)}\right\}\,. \tag{128}$$

The objects $\mathcal{J}_{\alpha(k+1)\dot{\alpha}(k+1)}$ and $\mathcal{T}_{\alpha(k)\dot{\alpha}(k)}$ are the higher spin supercurrent and higher spin supertrace respectively and are the higher spin analogues of the supercurrent and supertrace that appear in supergravity.

(*iii*) Proved that for every k, there is a unique alternative multiplet of higher spin supercurrents, called *minimal* $\left\{\mathcal{J}^{min}_{\alpha(k+1)\dot{\alpha}(k+1)}\,,\ 0\right\}$ (73) and (88) with conservation Equation (80). The cubic interactions for the minimal multiplet have the simpler form

$$g\int\sum_{k=0}^{\infty} H^{\alpha(k+1)\dot{\alpha}(k+1)}\mathcal{J}^{min}_{\alpha(k+1)\dot{\alpha}(k+1)}\,. \tag{129}$$

Furthermore, we presented the construction of the appropriate improvement term that will take us from the *canonical* to the *minimal* multiplet. The supercurrent $\mathcal{J}^{min}_{\alpha(k+1)\dot{\alpha}(k+1)}$ matches exactly the supercurrent generated by superconformal higher spins presented in [41].

The identification of the *minimal* multiplet with the results in [41] was expected because superconformal higher spin description does not include a compensator like $\chi_{\alpha(k+1)\dot{\alpha}(k)}$, hence the cubic interaction terms of the chiral with the superconformal higher spin supermultiplets can only take the form of (129). However, the superfield $H_{\alpha(k+1)\dot{\alpha}(k+1)}$ that appears in [41] is not the same because its dynamics involve higher derivative terms and also has different engineering dimensions.

In Section 9, we discuss the component structure of the theory and specifically we searched for the higher spin currents generated by the supercurrents. Starting from the superspace conservation equation we project down to the component level and we find:

(*iv*) An expression for the integer spin current $\mathcal{J}^{min\ (1,1)(S,S)}_{\alpha(s+1)\dot{\alpha}(s+1)}$ (95). There are two contributions to this current. The first is of the boson - boson type constructed out of a complex scalar ϕ which is defined as the the θ independent term of Φ ($\phi = \Phi|$). The second contribution is of the fermion—fermion type and is constructed out of a spinor χ_α defined as the θ term of Φ ($\chi_\alpha = \mathrm{D}_\alpha\Phi$). Both of these contributions agree with known results.

(v) An expression for the half-integer spin current $\mathcal{J}^{min\ (1,0)(S)}_{\alpha(s+1)\dot{\alpha}(s)}$ (97). This current appears for the first time in the literature because it requires both the complex scalar and the spinor, therefore non-supersymmetric theories can not be used to construct it.
(vi) An expression for an $\mathcal{R}$-symmetry current $\mathcal{J}^{min\ (0,0)}_{\alpha(s)\dot{\alpha}(s)}$ (98). This current also appears for the first time.

It is important to emphasize that in general the higher spin supercurrent and higher spin supertrace are independent quantities and the *minimal* multiplet can not always be reached. It depends on the peculiarities of the starting action and its symmetries, such as superconformal, to decide whether this can be done or not. In this work, we present a method of constructing the higher spin supercurrent and supertrace which is not restricted by these considerations. In Section 10, we discuss the higher spin supercurrent multiplet of a massive chiral superfield. Our results are:

(vii) A massive chiral can have cubic interactions only with the odd s $[s = 2l + 1]$ half-integer superspin supermultiplets $(2l + 2\,,\ 2l + 3/2)$.
(viii) The expressions for the higher spin supercurrent $\mathcal{J}_{\alpha(2l+1)\dot{\alpha}(2l+1)}$ (110) and (121) and supertrace $\mathcal{T}_{\alpha(2l)\dot{\alpha}(2l)}$ (112) and (122) of the *canonical* multiplet. These expressions have not been obtained before.
(ix) There is no *minimal* multiplet of supercurrents for this case since the supertrace can not be adsorbed by improvement terms. However, it can be arranged to be proportional to the mass parameter, so at the massless limit we land at the *minimal* multiplet of the massless chiral superfield.

There are several directions for the further development and generalization of the superfield interaction vertices studied in the paper. Firstly, the approach under consideration can directly be applied to derivation of the cubic interaction of the higher-superspin superfield with chiral superfield on the AdS superspace background. Secondly, it would be extremely interesting to construct the supercurrent and corresponding cubic interaction vertex for 4D, $\mathcal{N} = 2$ massless higher-superspin gauge superfield. In this case the supercurrent should apparently be built from hypermultiplet superfields on the framework of harmonic superspace [83] which provides unconstrained superfiled description for 4D, $\mathcal{N} = 2$ supermultiplets. Thirdly, it would be interesting to apply this approach to other matter supermultiplets such as the complex linear.

Acknowledgments: The authors are thankful to G. Tartaglino-Mazzucchelli for participation in the early stages of this work. Also the authors want to thank S. M. Kuzenko for extremely useful discussions and M. Taronna for correspondence. I.L.B. is grateful to the RFBR grant, project No. 18-02-00153 and to Russian Ministry of Education and Science, project No. 3.1386.2017 for partial support. The work of K.K. was supported by the grant P201/12/G028 of the Grant agency of Czech Republic.

Author Contributions: All authors contributed equally to all aspects of this work.

Conflicts of Interest: The authors declare no conflict of interest.

References

1. Singh, L.P.S.; Hagen, C.R. Lagrangian formulation for arbitrary spin. I. The boson case. *Phys. Rev. D* **1974**, *9*, 898.
2. Fronsdal, C. Massless fields with integer spin. *Phys. Rev. D* **1978**, *18*, 3624.
3. Fang, J.; Fronsdal, C. Massless fields with half-integral spin. *Phys. Rev. D* **1978**, *18*, 3630.
4. Fronsdal, C. Singletons and massless, integral-spin fields on de Sitter space. *Phys. Rev. D* **1979**, *20*, 848.
5. Curtright, T. Massless Field Supermultiplets With Arbitrary Spin. *Phys. Lett. B* **1979**, *85*, 219–224.
6. Curtright, T. High Spin Fields. *AIP Conf. Proc.* **1980**, *68*, 985.
7. Fang, J.; Fronsdal, C. Massless, half-integer-spin fields in de Sitter space. *Phys. Rev. D* **1980**, *22*, 1361.
8. De Wit, B.; Freedman, D.Z. Systematics of higher spin gauge fields. *Phys. Rev. D* **1980**, *21*, 358.
9. Fradkin, E.S.; Vasiliev, M.A. On the gravitational interaction of massless higher spin fields. *Phys. Lett. B* **1987**, *189*, 89–95.
10. Vasiliev, M.A. Consistent equation for interacting gauge fields of all spins in (3+1)-dimensions. *Phys. Lett. B* **1990**, *243*, 378–282.

11. Weinberg, S. Photons and Gravitons in S Matrix Theory: Derivation of Charge Conservation and Equality of Gravitational and Inertial Mass. *Phys. Rev. B* **1964**, *135*, 1049.
12. Weinberg, S. *The Quantum Theory of Fields. Volume I: Foundations*; Cambridge University Press: Cambridge, UK, 1995; Section 13.1.
13. Grisaru, M.T.; Pendleton, H.N.; van Nieuwenhuizen, P. Supergravity and the *S*-Matrix. *Phys. Rev. D* **1977**, *15*, 996.
14. Coleman, S.R.; Mandula, J. All possible symmetries of the *S*-matrix. *Phys. Rev.* **1967**, *159*, 1251–1256.
15. Haag, R.; Lopuszanski, J.T.; Sohnius, M. All possible generators of supersymmetries of the *S*-matrix. *Nucl. Phys. B* **1975**, *88*, 257–274.
16. Weinberg, S. *The Quantum Theory of Fields. Volume III: Supersymmetry*; Cambridge University Press: Cambridge, UK, 2000; Chapter 24.
17. Aragone, C.; Deser, S. Consistency Problems of Hypergravity. *Phys. Lett. B* **1979**, *86*, 161–163.
18. Berends, F.A.; van Holten, J.W.; de Wit, B.; van Nieuwenhuizen, P. On Spin 5/2 Gauge Fields. *J. Phys. A* **1980**, *13*, 1643.
19. Aragone, C.; La Roche, H. Massless Second Order Tetradic Spin 3 Fields And Higher Helicity Bosons. *Nuovo Cim. A* **1982**, *72*, 149–163.
20. Deser, S.; Yang, Z. Inconsistency Of Spin 4-Spin-2 Gauge Field Couplings. *Class. Quant. Grav.* **1990**, *7*, 1491.
21. Porrati, M. Universal Limits on Massless High-Spin Particles. *Phys. Rev. D* **2008**, *78*, 065016.
22. Taronna, M. On the Non-Local Obstruction to Interacting Higher Spins in Flat Space. *J. High Energy Phys.* **2017**, *2017*, 26.
23. Roiban, R.; Tseytlin, A.A. On four-point interactions in massless higher spin theory in flat space. *J. High Energy Phys.* **2017**, *2017*, 139.
24. Sleight, C.; Taronna, M. Higher spin gauge theories and bulk locality: A no-go result. *arXiv* **2017**, arXiv:1704.07859.
25. Boulanger, N.; Leclercq, S. Consistent couplings between spin-2 and spin-3 massless fields. *J. High Energy Phys.* **2006**, *2006*, 34.
26. Boulanger, N.; Leclercq, S.; Sundell, P. On the uniqueness of minimal coupling in higher-spin gauge theory. *J. High Energy Phys.* **2008**, *2008*, 56.
27. Bekaert, X.; Boulanger, N.; Sundell, P. How higher-spin gravity surpasses the spin two barrier. *Rev. Mod. Phys.* **2012**, *84*, 987.
28. Porrati, M. Old and New No Go Theorems on Interacting Massless Particles in Flat Space. *arXiv* **2012**, arXiv:1209.4876.
29. Vasiliev, M.A. Properties of equations of motion of interacting gauge fields of all spins in (3+1)-dimensions. *Class. Quant. Grav.* **1991**, *8*, 1387.
30. Vasiliev, M.A. More on Equations of Motion for Interacting Massless Fields of All Spins In (3+1)-Dimensions. *Phys. Lett. B* **1992**, *285*, 225–234.
31. Vasiliev, M.A. Nonlinear equations for symmetric massless higher spin fields in (A)dS(d). *Phys. Lett. B* **2003**, *567*, 139–151.
32. Vasiliev, M.A. Higher spin gauge theories in various dimensions. *Fortsch. Phys.* **2004**, *52*, 702–717.
33. Vasiliev, M.A. Higher spin gauge theories in any dimension. *Comptes Rendus Phys.* **2004**, *5*, 1101–1109.
34. Bekaert, X.; Cnockaert, S.; Iazeolla, C.; Vasiliev, M.A. Nonlinear higher spin theories in various dimensions. In Proceedings of the First Solvay Workshop on Higher-Spin Gauge Theories, Brussels, Belgium, 12–14 May 2004.
35. Gaberdiel, M.R.; Gopakumar, R. Minimal Model Holography. *J. Phys. A* **2013**, *46*, 214002.
36. Ammon, M.; Gutperle, M.; Kraus, P.; Perlmutter, E. Black holes in three dimensional higher spin gravity: A review. *J. Phys. A* **2013**, *46*, 214001.
37. Perez, A.; Tempo, D.; Troncoso, R. Higher Spin Black Holes. In *Lecture Notes in Physics*; Springer: Berlin/Heidelberg, Germany, 2015; Volume 892, pp. 265–288.
38. Taronna, M. Higher-spin interactions: four-point functions and beyond. *J. High Energy Phys.* **2012**, *2012*, 29.
39. Skvortsov, E.D.; Taronna, M. On locality, holographi and unfolding. *J. High Energy Phys.* **2015**, *2015*, 44.
40. Giombi, S.; Klebanov, I.R.; Safdi, B.R. Higher Spin AdS_{d+1}/CFT_d at One Loop. *Phys. Rev. D* **2014**, *89*, 84004.
41. Kuzenko, S.M.; Manvelyan, R.; Theisen, S. Off-shell superconformal higher spin multiplets in four dimensions. *J. High Energy Phys.* **2017**, *2017*, 34.

42. Kuzenko, S.M.; Tsulaia, M. Off-shell massive $\mathcal{N} = 1$ supermultiplets in three dimensions. *Nucl. Phys.* **2017**, *914*, 160–200.
43. Buchbinder, I.L.; Snegirev, T.V.; Zinoviev, Y.M. Lagrangian description of massive higher spin supermultiplets in AdS_3 space. *J. High Energy Phys.* **2017**, *2017*, 021,
44. Metsaev, R.R. Fermionic continous spin gauge fields in $(A)dS$ space. *Phys. Rev. B* **2017**, *773*, 135–141.
45. Didenko, V.E.; Vasiliev, M.A. Test of local form of higher-spin equations via AdS/CFT. *Phys. Rev. B* **2017**, *775*, 352–360.
46. Vasiliev, M.A. On the local framein nonlinear higher-spin equations. *arXiv* **2017**, arXiv:1707.03735.
47. Zinoviev, Y.M. Infinite spin fields in $d = 3$ and beyond. *Universe* **2017**, *3*, 63.
48. Basilie, T.; Bonezzi, R.; Boulanger, N. The Schouten tensor as a connection in the unfolding $3D$ conformal higher-spin fields. *J. High Energy Phys.* **2017**, *2017*, 54.
49. Bonezzi, R.; Boulanger, N.; De Filippi, D. Noncommutative Wilson lines in higher-spin theory and correlation functions for concerved currents for free conformal fields. *J. Phys. A* **2017**, *50*, 475401.
50. Beccaria, M.; Tseytlin, A.A. On induced action for conformal higher-spins in curved background. *Nucl. Phys. B* **2017**, *919*, 359–383.
51. Beccaria, M.; Tseytlin, A.A. C_T for conformal higher spin fields from partition function on conically deformed sphere. *J. High Energy Phys.* **2017**, *2017*, 123.
52. Sezgin, E.; Skvortsov, E.D.; Zhu, Y. Chern-Simons matter theories and higher-spin gravity. *J. High Energy Phys.* **2017**, *2017*, 133.
53. Skvortsov, E.D.; Tran, T. AdS/CFT in Fractional Dimension and Higher-Spins at One Loop. *Universe* **2017**, *3*, 61.
54. Taronna, M. A note on field redefinitions and higher-spin equations. *J. Phys. A* **2017**, *50*, 75401.
55. Sleight, C.; Taronna, M. Higher-Spin Algebras, Holography and Flat Space. *J. High Energy Phys.* **2017**, *2017*, 95.
56. Gates, S.J., Jr.; Grisaru, M.T.; Rocek, M.; Siegel, W. Superspace or One Thousand and One Lessons in Supersymmetry. *Front. Phys.* **1983**, *58*, 1–58.
57. Buchbinder, I.L.S.; Kuzenko, M. *Ideas and Methods of Supersymmetry and Supergravity Or the Way Through Supespace*; IOP Publishing: Bristol, UK; Philadelphia, PA, USA, 1998.
58. Berends, F.A.; Burgers, G.J.H.; van Dam, H. Explicit Construction of Conserved Currents for Massless Fields of Arbitrary Spin. *Nucl. Phys. B* **1986**, *271*, 429–441.
59. Anselmi, D. Higher spin current multiplets in operator product expansions. *Class. Quant. Grav.* **2000**, *17*, 1383.
60. Vasiliev, M.A. Higher spin gauge theories: Star-product and AdS space. In *The Many Faces of The Superworld*; Shifman, M., Ed.; World Scientific: Singapore, 2000.
61. Konstein, S.E.; Vasiliev, M.A.; Zaikin, V.N. Conformal higher spin currents in any dimension and AdS/CFT correspondence. *J. High Energy Phys.* **2000**, *2000*, 18.
62. Vasiliev, M.A.; Gelfond, O.G.A.; Skvortsov, E.D. Higher spin conformal currents in Minkowski space. *Theor. Math. Phys.* **2008**, *154*, 294–302.
63. Bekaert, X.; Joung, E.; Mourad, J. On higher spin interactions with matter. *J. High Energy Phys.* **2009**, *2009*, 126.
64. Bekaert, X.; Meunier, E. Higher spin interactions with scalar matter on a constant curvature spacetimes: Conserved current and cubic coupling generating functions. *J. High Energy Phys.* **2010**, *2010*, 116.
65. Bekaert, X.; Joung, E.; Mourad, J. Effective action in a higher spin background. *J. High Energy Phys.* **2011**, *2011*, 48.
66. Sleight, C.; Taronna, M. Higher Spin Interactions from Conformal Field Theory: The Complete Cubic Couplings. *Phys. Rev. Lett.* **2016**, *116*, 181602.
67. Gelfond, O.A.; Vasiliev, M.A. Current interactions from the one-form sector of nonlinear higher-spin equations. *arXiv* **2012**, arXiv:1706.03718.
68. Buchbinder, I.L.; Fotopoulos, A.; Petkou, A.C.; Tsulaia, M. Constructing the cubic interaction vertex of higher spin gauge fields. *Phys. Rev. D* **2006**, *74*, 105018.
69. Vasiliev, M.A. Gauge form of description of massless fields with arbitrary spin. *Phys. Atom. Nuclei* **1980**, *32*, 855–861.
70. Kuzenko, S.M.; Postnikov, V.V.; Sibiryakov, A.G. Massless gauge superfields of higher half integer superspins. *J. Exp. Theor. Phys. Lett.* **1993**, *57*, 534–538.
71. Kuzenko, S.M.; Sibiryakov, A.G. Massless gauge superfields of higher integer superspins. *J. Exp. Theor. Phys. Lett.* **1993**, *57*, 539–542.
72. Kuzenko, S.M.; Sibiryakov, A.G. Free massless higher-superspin superfields on the anti-de Sitter superspace. *Phys. Atom. Nucl* **1994**, *57*, 1257–1267.

73. Gates, S.J., Jr.; Koutrolikos, K. On 4D, $\mathcal{N} = 1$ massless gauge superfields of arbitrary superhelicity. *J. High Energy Phys.* **2014**, *2014*, 98.
74. Gates, S.J.; Koutrolikos, K. From Diophantus to Supergravity and massless higher spin multiplets. *J. High Energy Phys.* **2017**, *2017*, 63.
75. Buchbinder, I.L.; Koutrolikos, K. BRST Analysis of the Supersymmetric Higher Spin Field Models. *J. High Energy Phys.* **2015**, *2015*, 1–27.
76. Osborn, H. $\mathcal{N} = 1$ superconformal symmetry in four-dimensional quantum field theory. *Ann. Phys.* **1999**, *272*, 243–294.
77. Magro, M.; Sachs, I.; Wolf, S. Superfield Noether procedure. *Ann. Phys.* **2002**, *298*, 123–126.
78. Metsaev, R.R. Cubic interaction vertices of massive and massless higher spin fields. *Nucl. Phys. B* **2006**, *759*, 147–201.
79. Fradkin, E.S.; Vasiliev, M.A. Candidate to the Role of Higher Spin Symmetry. *Ann. Phys.* **1987**, *177*, 63–112.
80. Boulanger, N.; Ponomarev, D.; Skvortsov, E.; Taronna, M. On the uniqueness of higher-spin symmetries in AdS and CFT. *Int. J. Mod. Phys. A* **2013**, *28*, 1350162.
81. Gates, S.J.; Kuzenko, S.M.; Sibiryakov, A.G. $\mathcal{N} = 2$ supersymmetry of higher superspin massless theories. *Phys. Lett. B* **1997**, *412*, 59–68.
82. Ferrara, S.; Zumino, B. Transformation Properties of the Supercurrent. *Nucl. Phys. B* **1975**, *87*, 207–220.
83. Galperin, A.S.; Ivanov, E.A.; Ogievetsky, V.I.; Sokatchev, E.S. *Harmonic Superspace*; Cambridge University Press: Cambridge, UK, 2001.

universe

MDPI

Article

Asymptotic Charges at Null Infinity in Any Dimension

Andrea Campoleoni [1], Dario Francia [2,*] and Carlo Heissenberg [2]

[1] Institut für Theoretische Physik, ETH Zurich, Wolfgang-Pauli-Strasse, 27 8093 Zürich, Switzerland; campoleoni@itp.phys.ethz.ch

[2] Scuola Normale Superiore and INFN, Piazza dei Cavalieri, 7 I-56126 Pisa, Italy; carlo.heissenberg@sns.it

* Correspondence: dario.francia@sns.it

Received: 28 December 2017; Accepted: 6 February 2018; Published: 2 March 2018

Abstract: We analyse the conservation laws associated with large gauge transformations of massless fields in Minkowski space. Our aim is to highlight the interplay between boundary conditions and finiteness of the asymptotically conserved charges in any space-time dimension, both even and odd, greater than or equal to three. After discussing nonlinear Yang–Mills theory and revisiting linearised gravity, our investigation extends to cover the infrared behaviour of bosonic massless quanta of any spin.

Keywords: asymptotic symmetries; field theories in higher dimensions; Yang–Mills theory; BMS symmetry; higher spin symmetry

1. Introduction

In a previous work [1], we investigated the asymptotic symmetries of massless bosons of spin greater than two in four-dimensional Minkowski spacetime. We found that, upon assigning suitable boundary conditions on the rank$-s$ symmetric tensors $\varphi_{\mu_1 \cdots \mu_s}$, the asymptotic Killing equations for the rank$-(s-1)$ gauge parameters $\xi_{\mu_1 \cdots \mu_{s-1}}$ admit an infinite-dimensional set of solutions providing counterparts of the supertranslations emerging for spin-two fields in asymptotically flat spaces [2,3]. In particular, in strict analogy with the spin-two case [4,5][1], we found that Weinberg's soft theorem for any spin [7,8] could be derived as a consequence of the higher-spin supertranslation Ward identities. In addition, we studied the full asymptotic Killing tensor equation in any space-time dimension D for spin-three fields, finding in particular proper counterparts of superrotations in four dimensions [9,10].

In the present work, our goal is twofold: (1) to extend the analysis of asymptotic symmetries for all spins to arbitrary values of the space-time dimension; and (2) to compute the resulting charges and check their finiteness, thus proving the consistency of our choice of falloffs.

In order for our treatment to be as homogeneous as possible for any D, here we shall not make use of the notion of conformal null infinity. Indeed, its construction was shown to be obstructed in odd-dimensional spacetimes containing radiation because of singularities appearing in the components of the Weyl tensor of the unphysical space [11,12]. Differently, focussing on the falloffs of the solutions to the relevant equations of motion results in an exploration of null infinity that is devoid of such issues and thus allows for the same type of analysis in all dimensions [13].

Our general procedure can be summarised as follows: for all spins, we assume a power-like behaviour for the radial dependence of the field components keeping track of all possible subleading contributions (with some subtleties for the case of Yang–Mills theory in $D = 3$, where logarithmic

[1] See also [6] for a general review and more references.

dependence is also taken into account). In addition, we fix our boundary data through a set of Bondi-like conditions that can be interpreted as resulting from on-shell local gauge fixings. In this framework, a difference emerges between even and odd dimensionalities: whenever D is odd and greater than four, in order for both the radiation part and the Coulomb part of the solution to be accounted for, one finds that the expansion in powers of r requires both integer and half-integer exponents to be considered. Differently, only integer powers of r are needed whenever D is even. Moreover, radiation and Coulombic contributions behave as $r^{-(D-2)/2}$ and r^{3-D} respectively, and thus actually coincide in $D = 4$, thus justifying a separate analysis for dimensionalities higher than four on the one hand, and lower or equal to four on the other. Leading and subleading falloffs are determined by solving the equations of motion, while their consistency relies on checking that the energy flowing to null infinity per unit of retarded time is indeed finite. Once the falloffs are determined, we proceed to compute the asymptotic symmetries and the corresponding charges, while also checking finiteness of the latter.

Proceeding along these lines, in Section 2 we provide a full analysis of the nonlinear Yang–Mills theory in any dimension, starting from $D = 3$. The investigation of asymptotic symmetries and related aspects in dimensions other than four (both for spin one and spin two) has been performed in a number of works [14–19]. With respect to previous explorations of the Yang–Mills case in any D [14,17], here we also add the explicit computation of the charges, while, for the three-dimensional case already discussed in [16], we include the contribution of radiation. For the four-dimensional analysis of the spin-one case, see also [20–27].

In Section 3, we revisit the case of asymptotically flat gravity, for which an analysis in any dimension, both even and odd, can be found in [13]. Our review focusses on the linearised theory, which is useful for us in order to set the stage for the ensuing generalisation to higher spins that we first illustrate in Section 4 for the spin-three case. In particular, we complete the analysis in arbitrary dimension presented in [1] by computing the charges corresponding to the asymptotic symmetries. In Section 5, we pursue our exploration of the general spin$-s$ case initiated in [1]. In this respect, besides extending the study of large gauge transformations to higher space-time dimensions, we determine explicitly the proper counterpart of superrotations for any spin in $D = 4$. In addition, upon solving the equations of motion, we are led to a proposal for the boundary conditions eventually leading to finite asymptotic charges, which is explicitly tested in examples where we illustrate the on-shell cancellations of otherwise divergent terms.

The asymptotic symmetries that result from our analysis for all spins in $D > 4$ correspond to the solution to the global Killing tensor equations, and thus do not display the infinite-dimensional enhancement observed in $D \leq 4$. While this result is in agreement with similar conclusions drawn for spin two in previous works [28,29], it still leaves a number of questions unanswered, starting from the ultimate origin of Weinberg's soft theorem in $D > 4$.

While this work was in preparation, however, Ref. [19] appeared, with an alternative treatment of boundary conditions allowing for infinite-dimensional symmetries for linearised gravity in any even dimension, identified both as the origin of Weinberg's result for $D = 2k$ and as the sources of even-dimensional counterparts of the memory effect. (see also [15,17,18,30] for earlier discussions on the matter.)

The exploration of asymptotic symmetries for arbitrary-spin massless fields in any D, which we started in [1] and in the present work, presents a number of open challenges on which we plan to focus our attention in the future. Among the main ones, it ought to be stressed that our linearised analysis does not allow one to get a concrete grasp on the properties of the putative non-Abelian algebra underlying our findings, crucial to the issue of uncovering the physical meaning of such symmetries. This is relevant in particular in order to assess the role of higher-spin asymptotic symmetries in the high-energy regime of string scattering amplitudes (see e.g. [31–35]). In particular, in the latter respect, although once again of general interest in itself, the investigation on the possible infinite-dimensional enhancement of global asymptotic symmetries for all spins in $D > 4$ manifests special relevance.

2. Yang–Mills Theory

In this section, we analyse the equations of motion for Yang–Mills theory in D-dimensional Minkowski spacetime expanding their solutions in powers of $1/r$, thereby identifying the data that contribute to colour charge and to colour or energy flux at null infinity. In particular, we complement the related discussions in [16,20–24,26,27] by providing a unified treatment of all spacetime dimensions, and that in [14,17] by checking the finiteness of asymptotic charges in any dimension while also including radiation for $D = 3$.

We adopt the retarded Bondi coordinates $(x^\mu) = (u, r, x^i)$, where x^i, for $i = 1, 2, \ldots, n$, denotes the $n := D - 2$ angular coordinates on the sphere at null infinity. In these coordinates, the Minkowski metric reads

$$ds^2 = -du^2 - 2dudr + r^2\gamma_{ij}\, dx^i dx^j\,, \tag{1}$$

where γ_{ij} is the metric of the Euclidean n-sphere. The corresponding (flat) spacetime connection is denoted by ∇_μ, whose nonzero Christoffel symbols read

$$\Gamma^i{}_{jr} = r^{-1}\delta^i{}_j\,, \qquad \Gamma^u{}_{ij} = r\,\gamma_{ij} = -\,\Gamma^r{}_{ij}\,, \qquad \Gamma^k{}_{ij} = \frac{1}{2}\gamma^{kl}\left(\partial_i\gamma_{jl} + \partial_j\gamma_{il} - \partial_l\gamma_{ij}\right). \tag{2}$$

The Yang–Mills connection is denoted by $\mathcal{A}_\mu := \mathcal{A}_\mu^A T^A$, where the T^A are the generators of a compact Lie algebra $\mathfrak{g}$, whose gauge transformation is $\delta_\epsilon \mathcal{A}_\mu = \nabla_\mu \epsilon + [\mathcal{A}_\mu, \epsilon]$. The corresponding field strength is given by

$$\mathcal{F}_{\mu\nu} = \nabla_\mu \mathcal{A}_\nu - \nabla_\nu \mathcal{A}_\mu + [\mathcal{A}_\mu, \mathcal{A}_\nu]\,, \tag{3}$$

while the field equations are

$$\mathcal{G}_\nu := \nabla^\mu \mathcal{F}_{\mu\nu} + [\mathcal{A}^\mu, \mathcal{F}_{\mu\nu}] = 0\,. \tag{4}$$

Furthermore, we enforce the radial gauge

$$\mathcal{A}_r = 0\,, \tag{5}$$

which completely fixes the gauge in the bulk.

2.1. Boundary Conditions

For $D > 3$, we consider field configurations $\mathcal{A}_\mu$ whose asymptotic null behaviour is captured by an expansion[2] in powers of $1/r$, for $r \to \infty$. More explicitly, we parameterise their leading-order terms as follows:

$$\mathcal{A}_u(u, r, x^i) = r^a\, A_u(u, x^i) + \mathcal{O}(r^{a-1})\,, \qquad \mathcal{A}_i(u, r, x^j) = r^b\, A_i(u, x^j) + \mathcal{O}(r^{b-1})\,. \tag{6}$$

In order to determine the leading falloffs, we begin our analysis by substituting the conditions (6) into the u−component of the equations of motion, $\mathcal{G}_u = 0$. To leading order:

$$\begin{aligned} &-a\,\partial_u A_u\, r^{a-1} - \partial_u D^i A_i\, r^{b-2} + [\Delta A_u + a(a+1)A_u]\, r^{a-2} \\ &-\gamma^{ij}\,[A_i, \partial_u A_j]\, r^{2b-2} + \left(-D^i\,[A_u, A_i] + \gamma^{ij}\,[\partial_i A_j, A_u]\right) r^{a+b-2} + \gamma^{ij}\,[A_i, [A_j, A_u]]\, r^{a+2b-2} = 0\,, \end{aligned} \tag{7}$$

where D_i denotes the covariant derivative on the Euclidean n-sphere, while $\Delta := D^i D_i$.

Let us notice that, while the three linear terms in the first line are in principle independent, we can combine them in pairs upon imposing either $b = a + 1$, or $b = a$. As we shall see, the different types of solutions arising from these two options retain relevant physical meaning. Indeed, the first one

---

2 In the three-dimensional case ($n = 1$), to be discussed in Section 2.1.3, we shall also consider a logarithmic dependence in r.

corresponds to *radiation*, with the familiar falloff[3] $\mathcal{A}_u \sim r^{-n/2}$ of a spherical wave, which also carries a finite amount of energy per unit time through null infinity. The latter, on the other hand, leads to *Coulomb-type* solutions with the characteristic falloff $\mathcal{A}_u \sim r^{1-n}$ of the Coulomb potential, hence giving rise to a finite contribution to the colour charge.

Let us now discuss the asymptotic behaviour of colour and energy flux integrals. The definition of conserved charges associated with gauge symmetries is a subtle issue and we shall provide more details on the colour charge at null infinity in Section 2.2. Denoting the surface element of the n-sphere with unit radius by $d\Omega_n$, the $A-$th component of the colour charge at a given retarded time u is expressed as the following integral over the sphere S_u at a given value of u,

$$\mathcal{Q}^A(u) = \lim_{r\to\infty} \int_{S_u} \mathrm{tr}\left(\mathcal{F}_{ur}T^A\right) r^n d\Omega_n \,. \tag{8}$$

The energy flowing across S_u per unit time, on the other hand, can be cast as[4]

$$\mathcal{P}(u) = \lim_{r\to\infty} \int_{S_u} \gamma^{ij}\, \mathrm{tr}\left(\mathcal{F}_{ui}(\mathcal{F}_{rj} - \mathcal{F}_{uj})\right) r^{n-2} d\Omega_n \,. \tag{9}$$

The request that this quantity be finite imposes that the fields must go to zero at infinity in order to compensate for the factor of r^{n-2}, namely

$$a < 0\,, \qquad b < 0\,, \tag{10}$$

whenever $D > 4$. Due to this simplification, we restrict the present analysis to $D > 4$ and defer the discussion of the special cases $D = 3$ and $D = 4$ to a dedicated section.

In order to stress the relevant piece of physical information following from our choices of the falloffs, we first consider the leading-order terms in the equations of motion $\mathcal{G}_\mu = 0$ and analyse the outcome for the two options $b = a + 1$ and $b = a$. As a result, in particular, a radiation solution $(b = a + 1)$ is characterised by

$$\mathcal{A}_u = A_u\, r^{-n/2}\,, \qquad \mathcal{A}_i = A_i\, r^{1-n/2}\,, \tag{11}$$

where the $r-$independent components of the potential satisfy

$$A_u = \frac{2}{n}\, D^i A_i\,, \tag{12}$$

while, on the other hand, a Coulomb-type solution $(b = a)$ is such that

$$\mathcal{A}_u = \tilde{A}_u\, r^{1-n}\,, \qquad \mathcal{A}_i = \tilde{A}_i\, r^{1-n}\,, \tag{13}$$

and obeys

$$\partial_u \tilde{A}_u = 0\,, \qquad \partial_u \tilde{A}_i = \frac{1}{n}\, D_i \tilde{A}_u\,. \tag{14}$$

Let us stress that the presence of two distinct "branches" of solutions, radiation and Coulombic, is apparent only for $D > 4$, while in the four-dimensional case they effectively coincide. Notice also that, thanks to the condition (10), the nonlinear terms do not appear in these leading-order equations

---

3 The D-dimensional wave equation $-\partial_t^2 f + r^{-n}\partial_r(r^n\partial_r f) + \Delta f = 0$, where $t = u + r$, admits spherically symmetric solutions whose large-r behaviour is $r^{-n/2}\exp(iku)$.

4 The Yang–Mills Lagrangian for anti-Hermitian fields is $\mathcal{L} = \frac{1}{4}\mathrm{tr}(\mathcal{F}_{\mu\nu}\mathcal{F}^{\mu\nu})$, while the stress-energy tensor has the form $T_{\mu\nu} = -\,\mathrm{tr}\left(\mathcal{F}_{\mu\alpha}\mathcal{F}_\nu{}^\alpha\right) + \frac{1}{4}g_{\mu\nu}\mathrm{tr}\left(\mathcal{F}_{\alpha\beta}\mathcal{F}^{\alpha\beta}\right)$. The energy flux across S_u is then given by $-\int_{S_u} T_u{}^r r^n d\Omega_n = \int_{S_u}(T_{uu} - T_{ur})r^n d\Omega_n$ as $r \to \infty$.

for $D > 4$. We are now in the position to further justify the names we gave to these two kinds of solutions: a Coulomb solution has a generically non-zero colour charge at retarded time u,

$$Q^A(u) = (n-1)\int_{S_u} \operatorname{tr}\left(\tilde{A}_u T^A\right) d\Omega_n\,, \tag{15}$$

whereas its energy flux across S_u goes to zero, due to $\mathcal{F}_{ui} \sim r^{1-n}$. On the other hand, a radiation solution emits nonzero power at null infinity,

$$\mathcal{P}(u) = -\int_{S_u} \gamma^{ij} \operatorname{tr}\left(\partial_u A_i \partial_u A_j\right) d\Omega_n\,. \tag{16}$$

At this point, two general issues are in order. To begin with, it should be stressed that the colour charge of a radiation solution diverges off-shell like $r^{n/2-1}$ as $r \to \infty$. However, employing the relation (12) and recalling that the integral of any n-divergence $D_i v^i$ on S_u is zero by Stokes' theorem, we see that, at least to leading order, this potentially dangerous contribution vanishes on-shell. Performing a more detailed analysis, in the next section, we will prove that these kinds of cancellations ensure the finiteness of the colour charge to all orders.

In addition, one ought to study the behaviour of the colour flux for large r, namely the interplay occurring between radiation and Coulomb solutions due to the nonlinear nature of the theory. To do so, since the information on the colour charge is stored at order r^{1-n} in the $\mathcal{A}_u$ component, whereas radiation contributes at order $r^{-n/2}$ in the same component, we need to consider an expansion in $1/r$ that bridges the gap between these asymptotic behaviours. Due to the appearance of a half-integer exponent, the situation changes depending on the parity of the spacetime dimension, thus justifying to differentiate the discussion into two sections.

2.1.1. Even Space-Time Dimension

When $D > 4$ is even, we can consider an expansion of the following type:

$$\mathcal{A}_u = \sum_{J=1}^{\infty} a^{(J)} r^{1-n/2-J}\,, \qquad \mathcal{A}_i = \sum_{K=0}^{\infty} C_i^{(K)} r^{1-n/2-K}\,, \tag{17}$$

where $a^{(J)}$ and $C_i^{(K)}$ are r-independent functions. On the basis of the previous discussion, we expect

$$a^{(1)} = A_u\,, \qquad C_i^{(0)} = A_i\,, \tag{18}$$

to play the role of radiation terms, and

$$a^{(n/2)} = \tilde{A}_u\,, \qquad C_i^{(n/2)} = \tilde{A}_i\,, \tag{19}$$

to represent the Coulomb part of the solution. The components of the field strength are then given by:

$$\begin{aligned}
\mathcal{F}_{ur} &= \sum_{J=1}^{\infty}\left(\frac{n}{2}-1+J\right) a^{(J)} r^{-n/2-J}\,,\\
\mathcal{F}_{ir} &= \sum_{K=0}^{\infty}\left(\frac{n}{2}-J+K\right) C_i^{(K)} r^{-n/2-K}\,,\\
\mathcal{F}_{ui} &= \partial_u A_i\, r^{1-n/2} + \sum_{J=1}^{\infty}\left(\partial_u C_i^{(J)} - D_i a^{(J)}\right) r^{1-n/2-J} + \sum_{J=1}^{\infty} B_{ui}^{(J)} r^{2-n-J}\,,\\
\mathcal{F}_{ij} &= \sum_{K=0}^{\infty}\left(D_i C_j^{(K)} - D_j C_i^{(K)}\right) r^{1-n/2-K} + \sum_{K=0}^{\infty} B_{ij}^{(K)} r^{2-n-K}\,,
\end{aligned} \tag{20}$$

where $B_{ui}^{(J)}$ and $B_{ij}^{(K)}$ contain the nonlinear terms,

$$B_{ui}^{(J)} := \sum_{L=1}^{J} \left[a^{(L)}, C_i^{(J-L)}\right], \qquad B_{ij}^{(K)} := \sum_{L=0}^{K} \left[C_i^{(L)}, C_j^{(K-L)}\right]. \tag{21}$$

We first substitute this expansion into the equation $\mathcal{G}_r = 0$: denoting

$$\mathcal{G}_r^{(J)} := \left(\frac{n}{2} - J\right)\left(\frac{n}{2} + J - 1\right) a^{(J)} - \left(\frac{n}{2} + J - 2\right) D^i C_i^{(J-1)}, \tag{22}$$

this yields

$$\mathcal{G}_r^{(J)} = 0 \qquad \text{for } J = 1, 2, \dots, \frac{n}{2} - 1, \tag{23}$$

$$\mathcal{G}_r^{(J)} - \sum_{L=0}^{J-n/2} \left(\frac{n}{2} - 1 + L\right) \gamma^{ij} \left[C_i^{(J+n/2-L)}, C_j^{(L)}\right] = 0 \qquad \text{for } J = \frac{n}{2}, \frac{n}{2} + 1 \dots . \tag{24}$$

Then, we insert our expansion into the equation $\mathcal{G}_u = 0$: setting

$$\mathcal{G}_u^{(J)} := \left(\frac{n}{2} + J\right) \partial_u a^{J+1} + \left[\left(J - \frac{n}{2}\right)\left(\frac{n}{2} - 1 + J\right) + \Delta\right] a^{(J)} - D^i C_i^{(J)}, \tag{25}$$

$$\begin{aligned} \widehat{\mathcal{G}}_u^{(J)} := {} & D^i B_{iu}^{(J)} - \gamma^{ij} \left[C_i^{(J)}, \partial_u A_j\right] \\ & - \sum_{L=1}^{J} \left\{ \left(\frac{n}{2} - 1 + L\right) \left[a^{(J+1-L)}, a^{(L)}\right] + \gamma^{ij} \left[C_i^{(J-L)}, \partial_u C_j^{(L)} - D_j a^{(L)}\right] \right\}, \end{aligned} \tag{26}$$

we obtain

$$\frac{n}{2} \partial_u A_u - D^i A_i = 0, \tag{27}$$

$$\mathcal{G}_u^{(J)} = 0, \qquad \text{for } J = 1, 2, \dots, \frac{n}{2} - 2, \tag{28}$$

$$\mathcal{G}_u^{(n/2-1)} - \gamma^{ij}[A_i, \partial_u A_j] = 0, \tag{29}$$

$$\mathcal{G}_u^{(J)} + \widehat{\mathcal{G}}_u^{(J-n/2+1)} = 0, \qquad \text{for } J = \frac{n}{2}, \frac{n}{2} + 1, \dots, n-2, \tag{30}$$

$$\mathcal{G}_u^{(J)} + \widehat{\mathcal{G}}_{u,1}^{(J-n/2+1)} - \gamma^{ij} \sum_{L=1}^{J-n+2} \left[C_i^{(J-n+2-L)}, C_j^{(L)}\right] = 0, \qquad \text{for } J = n-1, n, \dots, \tag{31}$$

(when $D = 6$, Equation (28) reduces to (27)). It should be emphasised that the decoupling of the nonlinear terms, namely the linearity of Equations (23), (27) and (28) is a direct consequence of the assumptions (10) and only holds for $D > 4$. This asymptotic linearisation tells us that it is consistent to choose as boundary conditions near null infinity the falloffs (17), constrained by the linear Equations (23), (27) and (28). Indeed, this set of equations will allow us to then discuss the behaviour of the charges, and its main features are the following. First, from Equation (23), we obtain the constraints

$$a^{(J)} = \frac{2(n + 2J - 4)}{(n - 2J)(n + 2J - 2)} D^i C_i^{(J)} \qquad \text{for } J = 1, 2, \dots, \frac{n}{2} - 1, \tag{32}$$

namely that $a^{(1)} (= A_u), a^{(2)}, \dots, a^{(n/2-1)}$ are functions of the type $D_i v^i$, i.e., n-divergences, whereas, from Equation (24) evaluated for $J = n/2$, we note that $\tilde{A}_u = a^{(n/2)}$ does not bear the same form. Furthermore, Equations (27) and (28) together with (32) establish that $\partial_u A_u, \partial_u a^{(2)}, \dots, \partial_u a^{(n/2-1)}$ are n-divergences as well. On the other hand, by (29) and (32),

$$\begin{aligned}\partial_u \tilde{A}_u &= \frac{1}{n-1}\gamma^{ij}\left[A_i, \partial_u A_j\right] + \frac{1}{n-1}\left(D^i \tilde{A}_i + (n-2)a^{(n/2-1)} - \Delta a^{(n/2-1)}\right) \\ &= \frac{1}{n-1}\gamma^{ij}\left[A_i, \partial_u A_j\right] + (n\text{-divergence})\,.\end{aligned} \tag{33}$$

This equation allows one to compute the evolution of the leading Coulomb term $\tilde{A}_u$ along the $u-$direction in terms of the leading radiation terms A_i, and will therefore be at the basis of our colour flux formula across S_u.

2.1.2. Odd Space-Time Dimension

In the case of odd dimensions $D > 4$, we have to include two distinct expansions in $1/r$ in order to capture both radiation and Coulombic terms:

$$\begin{aligned}\mathcal{A}_u &= \sum_{J=1}^{\infty} a^{(J)} r^{1-n/2-J} + \sum_{K=0}^{\infty} \tilde{a}^{(K)} r^{1-n-K}\,, \\ \mathcal{A}_i &= \sum_{K=0}^{\infty} C_i^{(K)} r^{1-n/2-K} + \sum_{K=0}^{\infty} \tilde{C}_i^{(K)} r^{1-n-K}\,,\end{aligned} \tag{34}$$

where we identify

$$a^{(1)} = A_u\,, \qquad C_i^{(0)} = A_i\,, \tag{35}$$

and

$$\tilde{a}^{(0)} = \tilde{A}_u\,, \qquad \tilde{C}_i^{(0)} = \tilde{A}_i\,. \tag{36}$$

The relevant components of the field strength are then

$$\begin{aligned}\mathcal{F}_{ur} &= \sum_{J=1}^{\infty}\left(\frac{n}{2} - 1 + J\right) a^{(J)} r^{-n/2-J} + \sum_{K=0}^{\infty}(n-1+K)\,\tilde{a}^{(K)} r^{-n-K}\,, \\ \mathcal{F}_{ir} &= \sum_{K=0}^{\infty}\left(\frac{n}{2} - 1 + K\right) C_i^{(K)} r^{-n/2-K} + \sum_{K=0}^{\infty}(n-1+K)\,\tilde{C}_i^{(K)} r^{-n-K}\,.\end{aligned} \tag{37}$$

Likewise, the equations of motion will also contain two expansions in $1/r$: one in integer powers and one in half-integer powers. Expanding the equation $\mathcal{G}_r = 0$, we see that

$$\left(\frac{n}{2} - J\right)\left(\frac{n}{2} - 1 + J\right) a^{(J)} - \left(\frac{n}{2} - 2 + J\right) D^i C_i^{(J)} = 0 \qquad \text{for } J = 1, 2, \ldots, n-1\,, \tag{38}$$

while the terms containing $\tilde{A}_u = \tilde{a}^{(0)}$ cancel out identically.

Thus, in particular, the functions $a^{(1)} (= A_u), a^{(2)}, \ldots, a^{(n-1)/2}$ are n-divergences. Finally, the r^{-n} order of the equation $\mathcal{G}_u = 0$ provides us with the evolution of $\tilde{A}_u$ along the u direction for large r, namely

$$(n-1)\,\partial_u \tilde{A}_u = \gamma^{ij}\left[A_i, \partial_u A_j\right]. \tag{39}$$

Thus, we see that the phenomenon of asymptotic linearisation of the equations of motion, emphasised in the previous section for even-dimensional spacetimes, also occurs for odd dimensions and allowed us to derive the relevant set of boundary conditions for the definition of charge and energy flux integrals.

2.1.3. Three and Four Space-Time Dimensions

In $D = 4$, i.e., when $n = 2$, the leading radiation term and the Coulombic term coincide: indeed, finiteness of the energy flux (9) requires that, to leading order, a radiation solution behave like

$$\mathcal{A}_i(u, r, x^1, x^2) = A_i(u, x^1, x^2) + \mathcal{O}(r^{-1})\,, \tag{40}$$

while, using $b = a + 1$, we see that

$$\mathcal{A}_u(u, r, x^1, x^2) = \frac{A_u}{r}(u, x^1, x^2) + \mathcal{O}(r^{-2}). \tag{41}$$

This also gives generically a non-vanishing colour charge on the surface S_u via Equation (8). Using the leading terms in Equations (40) and (41), the only relevant dynamical information arises from $\mathcal{G}_u = 0$, which gives

$$\partial_u A_u = \partial_u D^i A_i + \gamma^{ij} \left[A_i, \partial_u A_j\right]. \tag{42}$$

The situation for $D = 3$ ($n = 1$) is rather different with respect to the previous cases, mainly because of two features. First, the factor of r^{-1} in Equation (9) tells us that, in order to produce a finite energy flux across S_u, the field components need not necessarily decay at infinity; consequently, one expects no clear distinction between radiation and Coulomb terms in the solution because no asymptotic linearisation occurs in the equations of motion. Second, the expression (15), and more specifically the factor of r, suggests that $\mathcal{A}_u$ should behave as $\log \frac{1}{r}$ in order to give a non-vanishing colour charge. These considerations motivate the following leading-order ansatz in three dimensions:

$$\mathcal{A}_u(u, r, \phi) \sim q \log \frac{1}{r} + p, \qquad \mathcal{A}_\phi(u, r, \phi) \sim \frac{\sqrt{r}}{\log r} C, \tag{43}$$

where q, p and C are r-independent functions. Indeed, with this choice, the colour flux and the energy flux read

$$Q^A(u) = \int_{S_u} q^A d\phi, \qquad \mathcal{P}(u) = -\int_{S_u} \text{tr}([q, C][q, C]) d\phi. \tag{44}$$

Using this ansatz, we find that the equation $\mathcal{G}_r = 0$ is identically satisfied at the leading order r^{-2}, whereas the equation $\mathcal{G}_u = 0$ gives

$$\partial_u q = -\left[q, p\right], \tag{45}$$

at order r^{-1}. This equation describes the u-evolution of q at null infinity and, hence, together with the first formula in (44), will lead to a formula for the colour flux.

2.2. Asymptotic Symmetries and Charges

In this section, we would like to discuss the form (8) of the colour charge at null infinity in the various dimensions. For related analyses, see [26,36–38]. To begin with, let us discuss which large gauge symmetries are admissible at null infinity. The residual gauge symmetry within the radial gauge is parameterised by an r-independent gauge parameter, since

$$0 = \delta_\epsilon A_r = \nabla_r \epsilon + \left[\mathcal{A}_r, \epsilon\right] \tag{46}$$

but $\mathcal{A}_r = 0$, hence $\nabla_r \epsilon = 0$. Then, we look for those parameters ϵ that preserve the leading falloff conditions imposed on the field $\mathcal{A}_\mu$. In the spirit of our previous illustration, we proceed by distinguishing the case of $D > 4$ from those of $D = 4$ and $D = 3$.

When $D > 4$, where radiation gives the dominant behaviour at infinity, we find, to leading order

$$r^{-n/2} \delta_\epsilon A_u = \partial_u \epsilon + r^{-n/2}[A_u, \epsilon], \tag{47}$$

which requires $\partial_u \epsilon = 0$. Furthermore,

$$r^{1-n/2} \delta_\epsilon A_i = \partial_i \epsilon + r^{1-n/2}[A_i, \epsilon], \tag{48}$$

but, since $1 - n/2 < 0$, this implies $\partial_i \epsilon = 0$. This means that ϵ is simply a constant. Hence, in $D > 4$, asymptotic symmetries coincide with the global part of the gauge group and the asymptotic charge is the ordinary colour charge computed via Equation (8). For even space-time dimensions, using Equation (20)

$$\begin{aligned} Q^A(u) &= \lim_{r\to\infty} \int_{S_u} \mathrm{tr}(\mathcal{F}_{ur} T^A)\, r^n d\Omega_n \\ &= \lim_{r\to\infty} \sum_{J=1}^{\infty} r^{n/2-J} \left(\frac{n}{2} - 1 + J\right) \int_{S_u} \mathrm{tr}(a^{(J)} T^A)\, d\Omega_n\,, \end{aligned} \tag{49}$$

where for $J < n/2$ all terms are integrals of n-divergences thanks to the relation (32), while the terms with $J > n/2$ go to zero as $r \to \infty$, thus

$$Q^A(u) = (n-1) \int_{S_u} \mathrm{tr}(\tilde{A}_u T^A)\, d\Omega_n\,. \tag{50}$$

For odd space-time dimensions,

$$\begin{aligned} Q^A(u) &= \lim_{r\to\infty} \sum_{J=1}^{\infty} r^{n/2-J} \left(\frac{n}{2} - 1 + J\right) \int_{S_u} \mathrm{tr}(a^{(J)} T^A)\, d\Omega_n \\ &+ \lim_{r\to\infty} \sum_{K=0}^{\infty} r^{-K} (n-1+K) \int_{S_u} \mathrm{tr}(\tilde{a}^{(K)} T^A)\, d\Omega_n\,, \end{aligned} \tag{51}$$

and, by the relation (38), the only nonzero contribution comes from the $K = 0$ term of the second series, giving again the result (50). For $D > 4$, we thus obtained that the colour charge is indeed expressed as an integral of the leading Coulombic component on S_u. Furthermore, on account of Equations (33) and (39), the colour flux is written as

$$\frac{d}{du} Q^A(u) = \int_{S_u} \gamma^{ij} \left[A_i, \partial_u A_j\right]^A d\Omega_n\,. \tag{52}$$

This is indeed consistent with the interpretation of A_i as the leading radiation term: this formula describes how Yang–Mills radiation across null infinity induces a change in the total colour of the space-time at successive retarded times u.

In $D = 4$, the gauge parameter must satisfy:

$$\begin{aligned} r^{-1} \delta_\epsilon A_u &= \partial_u \epsilon + r^{-1}[A_u, \epsilon]\,, \\ \delta_\epsilon A_i &= \partial_i \epsilon + [A_i, \epsilon]\,. \end{aligned} \tag{53}$$

The first equation again enforces $\partial_u \epsilon = 0$, whereas the second allows for an $\epsilon(x^1, x^2)$ with arbitrary dependence on the angles on the celestial sphere. The corresponding asymptotic charge is therefore

$$Q_\epsilon(u) = \lim_{r\to\infty} \int_{S_u} \mathrm{tr}(\mathcal{F}_{ur}\epsilon)\, r^2 d\Omega_2 = \int_{S_u} \mathrm{tr}(A_u \epsilon)\, d\Omega_2\,. \tag{54}$$

Taking into account Equation (42),

$$\frac{d}{du} Q_\epsilon(u) = \int_{S_u} \mathrm{tr}\left[\left(\partial_u D^i A_i + \gamma^{ij}[A_i, \partial_u A_j]\right)\epsilon\right] d\Omega_2\,. \tag{55}$$

To complete the picture, let us now turn to the situation in $D = 3$. There, neither $\mathcal{A}_u$ nor $\mathcal{A}_\phi$ fall off at infinity, and hence any $\epsilon(u, \phi)$ generates an allowed gauge transformation (the same result, in a slightly different setting, was already obtained in [16]). Thus, using the notation of the previous section,

$$\begin{aligned} Q_\epsilon(u) &= \int_{S_u} \mathrm{tr}(q\epsilon) d\phi\,, \\ \frac{d}{du} Q_\epsilon(u) &= \int_{S_u} \mathrm{tr}(q \partial_u \epsilon) d\phi - \int_{S_u} \mathrm{tr}([q, p]\epsilon) d\phi\,. \end{aligned} \tag{56}$$

Let us observe that these charges indeed form a representation of the underlying algebra: for $D \geq 4$, since $\delta_\epsilon A_u = [A_u, \epsilon]$,

$$[Q_{\epsilon_1}, Q_{\epsilon_2}] = \delta_{\epsilon_1} Q_{\epsilon_2} = \int_{S_u} \mathrm{tr}([A_u, \epsilon_1]\epsilon_2) d\Omega_n = \int_{S_u} \mathrm{tr}(A_u[\epsilon_1, \epsilon_2]) d\Omega_n = Q_{[\epsilon_1, \epsilon_2]}\,; \tag{57}$$

the same result holds for $D = 3$, noting that $\delta_\epsilon q = [q, \epsilon]$ and $\delta_\epsilon p = \partial_u \epsilon$, but p does not enter the charge formula. While the identity (57) holds in any dimension, it should be stressed that, when $D > 4$, the corresponding charge algebra coincides with $\mathfrak{g}$, whereas in $D = 4$ and $D = 3$, it is in fact an infinite-dimensional Kac–Moody algebra, owing to the arbitrary gauge parameters $\epsilon(x^1, x^2)$ and $\epsilon(u, \phi)$. In particular, we note the absence of a central charge, which could however emerge by performing the analysis for the linearised theory around a nontrivial background, as pointed out in [38].

Let us conclude this section by presenting some general observations that, although of basic nature, we found useful in order to frame the correctness of our procedure. For Yang–Mills theory, the following quantity

$$Q_\epsilon = \int_{\partial\Sigma} dx_{\mu\nu}\, \mathrm{tr}\,(F^{\mu\nu}\epsilon)\,, \tag{58}$$

where Σ is a generic Cauchy surface, provides both the conserved charge, as obtained by the Noether algorithm, and the Hamiltonian generator of the gauge symmetry parameterised by ϵ on the space tangent to the surface of solutions, as calculated via covariant phase space methods. Indeed, a generic variation of the Yang–Mills Lagrangian, after integrating by parts, reads

$$\delta\mathcal{L} = -\,\mathrm{tr}\,(\mathcal{G}^\mu \delta A_\mu) + \partial_\mu \mathrm{tr}\,(F^{\mu\nu}\delta A_\nu) =: -\mathrm{tr}\,(\mathcal{G}^\mu \delta A_\mu) + \partial_\mu \theta^\mu(\delta A)\,, \tag{59}$$

where we defined the symplectic potential $\theta^\mu(\delta A) = \mathrm{tr}\,(F^{\mu\nu}\delta A_\nu)$, while $\mathcal{G}^\mu$ denotes the Euler–Lagrange derivatives of $\mathcal{L}$, given in Equation (4). The presymplectic form is then given by

$$\omega^\mu(\delta_1 A, \delta_2 A) = \delta_{[1}\theta^\mu(\delta_{2]} A)\,, \tag{60}$$

with square brackets denoting antisymmetrisation, and correspondingly the formal variation of the Hamiltonian generator of the gauge symmetry H_ϵ is

$$\not{\delta} H_\epsilon = \int_\Sigma dx_\mu \omega^\mu(\delta A, \delta_\epsilon A) = \delta \int_{\partial\Sigma} dx_{\mu\nu} \mathrm{tr}(F^{\mu\nu}\epsilon) - \int_\Sigma dx_\mu \mathrm{tr}(\delta\mathcal{G}^\mu\epsilon)\,. \tag{61}$$

Noting that the last term is proportional to the linearised equations of motion, i.e., that it vanishes on the space tangent to the surface of solutions, we can write

$$\not{\delta} H_\epsilon \approx \delta Q_\epsilon\,,$$

which explicitly shows that $\not{\delta} H_\epsilon$ is integrable and that we may choose to set $H_\epsilon = Q_\epsilon$ by requiring a flat connection to have zero colour charge. Furthermore, the Noether charge is simply

$$\int_\Sigma dx_\mu\, \theta^\mu(\delta_\epsilon A) = Q_\epsilon - \int_\Sigma dx_\mu\, \mathrm{tr}(\mathcal{G}^\mu\epsilon) \approx Q_\epsilon\,, \tag{62}$$

so that the two approaches agree in this case. The definition of Q_ϵ is in principle subject to ambiguities stemming from $\theta^\mu \mapsto \theta^\mu + \partial_\nu\lambda^{\mu\nu}$, where $\lambda^{\mu\nu} = -\lambda^{\nu\mu}$, which does not alter the variation (59). In the spirit of [39], we may choose to set to zero the corresponding additional terms, precisely because this choice defines an integrable Hamiltonian, as shown above. Further motivation for the absence of these terms is provided by the agreement with the general analysis of [37] and by the fact that they play no role in the generation of Ward identities for residual gauge freedom [38].

In order to finally make contact with Equation (8), we may then apply Equation (58) choosing as a Cauchy surface

$$\Sigma = \Sigma_u \cup \mathscr{I}^+_{<u} \,,$$

where Σ_u is any space-like hypersurface such that $\partial\Sigma_u = S_u$ while $\mathscr{I}^+_{<u}$ is the portion of null infinity up to the retarded time u. Then, using the general expression for the charge (58) and Stokes' theorem, we see that Q_ϵ can be expressed as a sum of the total charge at the retarded time u, as in Equation (8), and the charge flown across $\mathscr{I}^+_{<u}$ due to radiation.

3. Linearised Gravity

Boundary conditions giving finite energy and angular momentum at null infinity have been first proposed for spacetimes of any even dimensions in [11] (see also [19,28,40]). The proposal has been extended to encompass also odd space-time dimensions in [13,29]. We refer to these works for a detailed analysis of asymptotic charges and fluxes at null infinity in nonlinear Einstein gravity. Here, we revisit instead the problem within the linearised theory. In particular, we point out that the boundary conditions discussed in previous works can be inferred by demanding finiteness of the linearised asymptotic charges. In analogy with the Yang–Mills example, the fluxes of energy and angular-momentum at null infinity are instead affected by interactions, so that they will be excluded from our analysis. Besides its intrinsic interest, the ensuing discussion is also instrumental for us in order to better frame the results that we will present for higher-spin fields in Sections 4 and 5.

3.1. Boundary Conditions

We parameterise the Minkowski background with the retarded Bondi coordinates (1), and we analyse the linearised metric fluctuations in the "Bondi gauge"

$$h_{r\mu} = 0\,, \quad \gamma^{ij}h_{ij} = 0 \quad \Rightarrow \quad g^{\mu\nu}h_{\mu\nu} = 0\,. \tag{63}$$

Differently from the spin-one radial gauge (5), these conditions cannot be reached by means of an off-shell gauge fixing, but the number of constraints is the same as in the transverse-traceless gauge. We therefore assume that they can be imposed on shell.[5]

When Equation (63) holds, the linearised vacuum Einstein equations reduce to

$$R_{\mu\nu} = \Box h_{\mu\nu} - \nabla_{(\mu}\nabla\cdot h_{\nu)} = 0\,. \tag{64}$$

In the following, we will solve these equations assuming that the metric fluctuations admit an expansion in powers of $1/r$ around null infinity. The main idea, suggested by the Yang–Mills example, is that asymptotically the interactions deform the linearised solutions only starting from a subleading order in their expansion in powers of $1/r$. The conditions (63) imply $R_{rr} = 0$ identically, while the other equations of motion read[6]

$$R_{ru} = \frac{1}{r^2}\left\{\left(r^2\partial_r^2 + n\,r\partial_r\right)h_{uu} - \partial_r D\cdot h_u\right\} = 0\,, \tag{65}$$

$$R_{ri} = \frac{1}{r^2}\left(r^2\partial_r^2 + (n-2)\,r\partial_r - 2(n-1)\right)h_{ui} - \frac{1}{r^3}\left(r\partial_r - 2\right)D\cdot h_i = 0\,, \tag{66}$$

---

[5] With hindsight, our choice is legitimated, e.g., by the agreement between the charges and asymptotic symmetries derived in this framework and those obtained by assuming only suitable falloff conditions on the components of the metric that cannot be set to zero with an off-shell gauge fixing (compare e.g. the conditions (63) with Equation (8) of [11]).

[6] From now on, we shall often denote a derivative with respect to u with a dot, i.e., $\partial_u f = \dot{f}$.

$$
\begin{aligned}
R_{ij} = & -\tfrac{1}{r}\left(2\,r\partial_r + n - 4\right)\dot{h}_{ij} + \tfrac{1}{r}\left\{\left(r\partial_r + n - 2\right) D_{(i}\dot{h}_{j)u} + 2\,\gamma_{ij} D\cdot \dot{h}_u\right\} \\
& + \tfrac{1}{r^2}\left\{\left(\Delta + r^2\partial_r^2 + (n-4)\,r\partial_r - 2(n-2)\right) h_{ij} - D_{(i}D\cdot h_{j)}\right\} \\
& - 2\,\gamma_{ij}\left(r\partial_r + n - 1\right) h_{uu} = 0\,.
\end{aligned} \tag{67}
$$

When the previous equations are satisfied, the following ones are satisfied as well at almost all orders in an expansion in powers of $1/r$ (see Section 5 for more details):

$$
R_{uu} = \frac{n}{r}\,\dot{h}_{uu} - \frac{2}{r^2}\,D\cdot \dot{h}_u + \frac{1}{r^2}\left(\Delta + r^2\partial_r^2 + n\,r\partial_r\right) h_{uu} = 0\,, \tag{68}
$$

$$
\begin{aligned}
R_{ui} = & -\tfrac{1}{r}\left(r\partial_r - 2\right)\dot{h}_{ui} - \tfrac{1}{r^2}\,D\cdot \dot{h}_i + \tfrac{1}{r}\left(r\partial_r + n - 2\right)\partial_i h_{uu} \\
& + \tfrac{1}{r^2}\left\{\left(\Delta + r^2\partial_r^2 + (n-2)\,r\partial_r - n + 1\right) h_{ui} - D_i D\cdot h_u\right\} = 0\,.
\end{aligned} \tag{69}
$$

As for the Yang–Mills case, the only exception is given by the leading order of a stationary solution. By substituting a power-law ansatz,

$$
h_{uu} = r^a B(u,x^k) + \mathcal{O}(r^{a-1})\,, \quad h_{ui} = r^b U_i(u,x^k) + \mathcal{O}(r^{b-1})\,, \quad h_{ij} = r^c C_{ij}(u,x^k) + \mathcal{O}(r^{c-1})\,, \tag{70}
$$

Equations (65)–(67) turn into

$$
R_{ru} = r^{a-2} a(a+n-1) B - r^{b-3} b\, D\cdot U + \cdots = 0\,, \tag{71}
$$

$$
R_{ri} = r^{b-2}(b-2)(b+n-1)\,U_i - r^{c-3}(c-2) D\cdot C_i + \cdots = 0\,, \tag{72}
$$

$$
\begin{aligned}
R_{ij} = & -r^{c-1}(2c+n-4)\,\dot{C}_{ij} + r^{b-1}\left\{(b+n-2) D_{(i}\dot{U}_{j)} + 2\,\gamma_{ij} D\cdot \dot{U}\right\} \\
& - 2\,r^a(a+n-1)\gamma_{ij}\dot{B} + \cdots = 0\,,
\end{aligned} \tag{73}
$$

where the dots stand for subleading terms. Imposing $b = a+1$ and $c = b+1$ allows one to mutually cancel the addenda in Equations (71) and (72), while Equation (73) is solved to leading order provided that the coefficient of r^{c-1} vanishes. This is the analogue of the choice that gives the radiation solution in the Yang–Mills case: it does not impose any constraint on C_{ij} while, for $D > 3$, it fixes the leading exponents as follows:

$$
a = -\frac{n}{2}\,, \qquad b = -\frac{n}{2} + 1\,, \qquad c = -\frac{n}{2} + 2\,. \tag{74}
$$

Besides this formal analogy, one can verify that a solution of this type carries a finite amount of energy per unit of retarded time through null infinity, [7]

$$
\mathcal{P}(u) = \lim_{r\to\infty}\int_{S_u}\left(T_{uu} - T_{ur}\right) r^n d\Omega_n = \int_{S_u} \gamma^{i_1 j_1}\gamma^{i_2 j_2}\,\dot{C}_{i_1 i_2}\dot{C}_{j_1 j_2} d\Omega_n\,, \tag{75}
$$

and can therefore be interpreted as a gravitational wave propagating on the Minkowski background, thus providing a convincing justification for the falloffs (74). In Section 3.2, we shall also show that a solution of the Einstein equations with these leading falloffs is endowed with finite energy and angular momentum charges at null infinity.

---

7 The massless Fierz–Pauli Lagrangian $\mathcal{L} = \frac{1}{2}h^{\mu\nu}(\Box h_{\mu\nu} - \nabla_{(\mu}\nabla\cdot h_{\nu)} + \nabla_\mu\nabla_\nu h^\alpha{}_\alpha - \eta_{\mu\nu}(\Box h^\alpha{}_\alpha - \nabla\cdot\nabla\cdot h))$ gives rise, in Bondi gauge, to the canonical stress-energy tensor

$$
T_{\alpha\beta} = \nabla_\alpha h_{\mu\nu}\nabla_\beta h^{\mu\nu} - 2\nabla\cdot h^\mu \nabla_\beta h_{\alpha\mu} + \eta_{\alpha\beta}\mathcal{L}\,.
$$

While in our linearised setup one cannot capture the flux of energy associated with the self-interactions of the gravitational field, it still makes sense to evaluate the flux pertaining to an *eternal* radiating source in the interior, which is constant over u. Indeed, this is a quantity that is well defined also in the linearised theory and is given by (75).

The charges actually depend on the subleading (for $D > 3$) terms in the expansion in powers of r with exponents

$$a = b = c = 1 - n\,. \tag{76}$$

When n is even, these contributions are actually "integration constants" (they anyway admit a dependence on x^i) in the radiation solution with leading falloffs (74), while when n is odd they appear as the leading order of a companion solution with its own expansion in powers of $1/r$. For this reason, we treat separately the two cases, while discussing the peculiarities of the $n = 1$ and $n = 2$ instances in Section 3.1.3.

3.1.1. Even Space-Time Dimension

When $D = n - 2$ is even, we consider the following ansatz for the linearised fluctuations in Bondi gauge (63):

$$h_{uu} = \sum_{k=0}^{\infty} r^{-\frac{n}{2}-k} B^{(k)}(u,x^m)\,, \quad h_{ui} = \sum_{k=0}^{\infty} r^{-\frac{n}{2}-k+1} U_i^{(k)}(u,x^m)\,, \quad h_{ij} = \sum_{k=0}^{\infty} r^{-\frac{n}{2}-k+2} C_{ij}^{(k)}(u,x^m)\,, \tag{77}$$

with $\gamma^{ij}C_{ij}^{(k)} = 0$. As discussed above, the leading falloffs have been chosen such that the linearised solution carries a finite amount of energy per unit time at null infinity. From Equation (65), one can then compute the coefficients of h_{uu},

$$B^{(k)} = \begin{cases} \frac{2(n+2k-2)}{(n+2k)(n-2k-2)}\, D\cdot U^{(k)} & \text{for } k \neq \frac{n-2}{2} \\ 2\, m_B & \text{for } k = \frac{n-2}{2} \end{cases}\,, \tag{78}$$

while Equation (66) fixes h_{ui} as

$$U_i^{(k)} = \begin{cases} \frac{2(n+2k)}{(n+2k+2)(n-2k)}\, D\cdot C_i^{(k)} & \text{for } k \neq \frac{n}{2} \\ N_i & \text{for } k = \frac{n}{2} \end{cases}\,. \tag{79}$$

The quantities m_B and N_i do not contribute to Equations (65) and (66) thanks to the cancellation of the coefficients in front of the corresponding $B^{(k)}$ and $U_i^{(k)}$. Their dependence on the retarded time u is however fixed by Equations (68) and (69) that, when $n > 2$, read

$$\dot{m}_B = \frac{3-n}{4n(n-1)}\,(\Delta - n + 2)\, D\cdot D\cdot C^{(\frac{n-4}{2})}\,, \tag{80a}$$

$$\dot{N}_i = \frac{1}{n+1}\left\{2\,\partial_i m_B - \frac{n-1}{n}\,(\Delta-1)\, D\cdot C_i^{(\frac{n-2}{2})}\right\}, \tag{80b}$$

where $\Delta = D_iD^i$. These equations are solved by

$$m_B(u,x^j) = \mathcal{M}(x^j) - \frac{n-3}{4(n-1)n}\int_{-\infty}^{u} du'\,(\Delta - n + 2)\, D\cdot D\cdot C^{(\frac{n-4}{2})}(u',x^j)\,, \tag{81a}$$

$$\begin{aligned} N_i(u,x^j) = \mathcal{N}_i(x^j) &+ \frac{2\,u}{n+1}\,\partial_i\mathcal{M}(x^j) - \frac{n-1}{n(n+1)}\int_{-\infty}^{u} du'\,(\Delta-1)\, D\cdot C_i^{(\frac{n-2}{2})}(u',x^j) \\ &- \frac{n-3}{2(n-1)n(n+1)}\int_{-\infty}^{u} du' \int_{-\infty}^{u'} du''\, D_i\,(\Delta - n + 2)\, D\cdot D\cdot C^{(\frac{n-4}{2})}(u'',x^j)\,. \end{aligned} \tag{81b}$$

Note that the expressions for m_B and N_i (which are the linearised counterparts of the Bondi mass and angular momentum aspects) contain two types of contributions: one depends on the "integration constants" $\mathcal{M}$ and $\mathcal{N}_i$, which enter in combinations with a fixed dependence on u, while the other depends on the integrals over u of certain combinations of the tensors $C_{ij}^{(k)}$. We anticipate that in Section 3.2 we shall show that, for $n \neq 2$, the integration constants $\mathcal{M}$ and $\mathcal{N}_i$ completely specify the asymptotic linearised charges, while the integral terms, which are not even present when the dimension of spacetime is odd, do not contribute to them.

The tensors in the expansion of h_{ij} are instead fixed recursively in terms of $C_{ij}{}^{(0)}$—whose u-derivative is the linearised analogue of the Bondi news—up to an arbitrary function of x^i for each term of the expansion. Equation (67) indeed implies

$$\begin{aligned} \dot{C}_{ij}^{(k+1)} &= -\frac{1}{2(k+1)}\left\{\left[\Delta - \frac{n(n-2)}{4} + k(k+1) - 2\right] C_{ij}^{(k)} \right.\\ &\left. - \frac{4}{(n+2k+2)(n-2k)}\left[n\, D_{(i}D\cdot C_{j)}{}^{(k)} - 2\,\gamma_{ij}D\cdot D\cdot C^{(k)}\right]\right\} \quad \forall\, k \neq \frac{n}{2}. \end{aligned} \tag{82}$$

The value of k excluded from this expression shows that from $k = \frac{n+2}{2}$ onwards the tensors also depend on N_i:

$$\dot{C}_{ij}^{(\frac{n+2}{2})} = \frac{1}{n+2}\left\{D_{(i}N_{j)} - \frac{2}{n}\gamma_{ij}D\cdot N - (\Delta + n - 2)\,C_{ij}^{(\frac{n}{2})} + D_{(i}\,D\cdot C_{j)}^{(\frac{n}{2})}\right\}. \tag{83}$$

These terms in the expansion (77), anyway, do not contribute to the linearised charges and, generically, they receive nonlinear corrections in Einstein gravity [13,29].

For the values of k that do not impose any constraints on $B^{(k)}$ and $U_i{}^{(k)}$, Equations (65) and (66) imply instead

$$(n-2)(n-1)\,D\cdot D\cdot C^{(\frac{n-2}{2})} = 0\,, \qquad D\cdot C_i^{(\frac{n}{2})} = 0\,. \tag{84}$$

These conditions do not constrain $C^{(0)}$ because they are compatible with the divergences of Equation (82). For instance, it implies

$$D\cdot D\cdot \dot{C}^{(k+1)} = -\frac{(n+2k-2)(n-2k-4)}{2(k+1)(n+2k+2)(n-2k)}\left[\Delta - \frac{(n+2k)(n-2k-2)}{4}\right] D\cdot D\cdot C^{(k)} \tag{85}$$

and, for $n > 2$, the r.h.s. vanishes for $k+1 = \frac{n-2}{2}$. Similarly, Equations (68) and (69) reduce to divergences of Equation (82) for all values of k aside from those that fix the u−dependence (81) of the Bondi mass and angular momentum aspects.

3.1.2. Odd Space-Time Dimension

When n is odd and greater than one, in order to obtain non-zero asymptotic charges at null infinity, one has to complement the ansatz (77), that in this case contains half-integer powers of r, with a companion expansion including integer powers of the radial coordinate. We therefore consider the ansatz

$$h_{uu} = \sum_{k=0}^{\infty} r^{-\frac{n}{2}-k}B^{(k)}(u,x^m) + \sum_{k=0}^{\infty} r^{1-n-k}\tilde{B}^{(k)}(u,x^m)\,, \tag{86a}$$

$$h_{ui} = \sum_{k=0}^{\infty} r^{-\frac{n}{2}-k+1}U_i^{(k)}(u,x^m) + \sum_{k=0}^{\infty} r^{1-n-k}\tilde{U}_i^{(k)}(u,x^m)\,, \tag{86b}$$

$$h_{ij} = \sum_{k=0}^{\infty} r^{-\frac{n}{2}-k+2}C_{ij}^{(k)}(u,x^m) + \sum_{k=0}^{\infty} r^{1-n-k}\tilde{C}_{ij}^{(k)}(u,x^m)\,, \tag{86c}$$

with $\gamma^{ij}C_{ij}^{(k)} = \gamma^{ij}\tilde{C}_{ij}^{(k)} = 0$. Since n is odd, the factors entering the expansion of Equations (65) and (66) in powers of $\sqrt{r}$ are always different from zero. As a result, for any k one has again

$$B^{(k)} = \frac{2(n+2k-2)}{(n+2k)(n-2k-2)}\, D\cdot U^{(k)}\,, \qquad U_i^{(k)} = \frac{2(n+2k)}{(n+2k+2)(n-2k)}\, D\cdot C_i^{(k)}\,, \tag{87}$$

while the tensors $C^{(k)}$ satisfy (82). These conditions imply that the Equations (68) and (69) are identically satisfied. In Section 3.2, we shall see that the relations (87) guarantee that the radiation solution does not contribute to the asymptotic charges. Equations (68) and (69) fix instead the u-evolution of the leading terms in the Coulomb-type solution as

$$m_B(u,x^j) := \frac{\tilde{B}^{(0)}}{2} = \mathcal{M}(x^j)\,, \tag{88a}$$

$$N_i(u,x^j) := \tilde{U}_i^{(0)} = \mathcal{N}_i(x^j) + \frac{2u}{n+1}\,\partial_i\mathcal{M}(x^j)\,. \tag{88b}$$

The subleading terms in the expansion in powers of r are fixed by the analogues of the relations (82) and (87). For instance,

$$\tilde{B}^{(l+1)} = -\frac{n+l-1}{(l+1)(n+l)}\, D\cdot\tilde{U}^{(l)}\,, \qquad \tilde{U}_i^{(l)} = -\frac{n+l+1}{(l+1)(n+l+2)}\, D\cdot\tilde{C}_i^{(l)}\,, \tag{89}$$

while the $\tilde{C}^{(l)}$ are fixed recursively by an equation with the same form as (82) with shifted coefficients $k \to k - n/2 - 1$ (see also the general analysis in Section 5 for more details). At any rate, these relations will be irrelevant for the computation of the charges in Section 3.2 and, in general, will receive nonlinear corrections in Einstein gravity as shown by the comparison of our analysis with [13,29].

3.1.3. Three and Four Space-Time Dimensions

In three and four space-time dimensions, i.e., when $n = 1$ or $n = 2$, the previous analysis has to be amended for some details, which however introduce significant physical consequences. We begin by considering the peculiarities that emerge in four dimensions: in this case, inserting our ansatz (77) in the equations of motion (65) and (66) leads to

$$h_{uu} = \frac{2}{r}\, m_B + \mathcal{O}(r^{-2})\,, \qquad h_{ui} = \frac{1}{2}\, D\cdot C_i + \frac{1}{r}\, N_i + \mathcal{O}(r^{-2})\,, \qquad h_{ij} = r\, C_{ij} + \mathcal{O}(1)\,, \tag{90}$$

where, for brevity, we defined $C_{ij} := C_{ij}{}^{(0)}$, while we used the same notation as in the previous subsections for the leading terms of the Coulomb-like solution. Note that in the component h_{uu} the leading order of the radiation and Coulomb-like solutions coincide. Moreover, Equation (65) does not impose any constraint on the double divergence of C_{ij} (the factor in front of it vanishes when $n = 2$ as we recalled in Equation (84)), while $D\cdot C_i{}^{(1)} = 0$ as for generic n. Since C_{ij} has now a non-vanishing double divergence, the equations fixing the dependence on u of m_B and N_i have to be modified as follows (cf. (80)):

$$\dot{m}_B = \frac{1}{4}\,\partial_u D\cdot D\cdot C\,, \tag{91a}$$

$$\dot{N}_i = \frac{2}{3}\,\partial_i m_B - \frac{1}{6}\left\{(\Delta-1)\, D\cdot C_i - D_i D\cdot D\cdot C\right\}. \tag{91b}$$

Consequently, the leading terms of the Coulomb-like branch depend on the radiation solution and on the usual set of integration constants as

$$m_B(u,x^j) = \mathcal{M}(x^j) + \frac{1}{4} D\cdot D\cdot C(u,x^j)\,, \tag{92a}$$

$$N_i(u,x^j) = \mathcal{N}_i(x^j) + \frac{2u}{3}\,\partial_i\mathcal{M}(x^j) - \frac{1}{6}\int_{-\infty}^{u} du'\,[(\Delta-1)\,D\cdot C_i - 2\,D_iD\cdot D\cdot C](u',x^j)\,. \tag{92b}$$

The dependence on C_{ij} in m_B and N_i will be crucial in Section 3.2, where we shall compute the asymptotic charges.

When $n = 1$, in the component h_{uu} the radiation branch becomes subleading with respect to the Coulomb-type one. Given that in three space-time dimensions fields of spin two do not propagate any local degrees of freedom, it is therefore natural to ignore the radiation branch altogether and work with boundary conditions that only encompass Coulomb-type solutions of the equations of motion:

$$h_{uu} = 2\mathcal{M}(\phi) + \mathcal{O}(r^{-1})\,, \qquad h_{u\phi} = \mathcal{N}(\phi) + u\,\partial_\phi\mathcal{M}(\phi) + \mathcal{O}(r^{-1})\,, \qquad h_{\phi\phi} = 0\,, \tag{93}$$

where ϕ denotes the angular coordinate on the circle at null infinity while we already displayed the constraints on the leading terms imposed by the equations of motion. Notice that we set to zero the component $h_{\phi\phi}$, consistently with our choice of boundary conditions in any D according to which the tensor h_{ij} is traceless (and thus identically zero if $n = 1$). Alternatively, one can consider $h_{\phi\phi} = rC(\phi) + \mathcal{O}(1)$ [41]. Our choice is not restrictive, however, as it still allows for an enhancement of the asymptotic symmetry algebra from Poincaré to $\mathfrak{bms}_3$.

3.2. Asymptotic Symmetries and Charges

We can now identify the gauge transformations preserving the form of the linearised solutions, which are the asymptotic symmetries of the system. These determine the asymptotic charges, which play the dual role of being conserved quantities labelled by the parameters of asymptotic symmetries and of generating the latter via the Poisson bracket derived from the action. In the Bondi gauge (63), the asymptotic symmetries must satisfy

$$\delta h_{r\mu} = 0\,, \quad \delta h_{uu} = \mathcal{O}(r^{-\frac{n}{2}})\,, \quad \delta h_{ui} = \mathcal{O}(r^{-\frac{n}{2}+1})\,, \quad \delta h_{ij} = \mathcal{O}(r^{-\frac{n}{2}+2})\,, \tag{94}$$

where the variations are given by linearised diffeomorphisms $\delta h_{\mu\nu} = \nabla_{(\mu}\xi_{\nu)}$ that leave (64) invariant. These conditions are appropriate for spacetimes of both even and odd dimensions, since the differences highlighted in Sections 3.1.1 and 3.1.2 only affect subleading terms in the expansion of the solutions.[8] Equations (94) are solved by

$$\xi_r = -\left(T + \frac{u}{n}\,D\cdot v\right), \tag{95a}$$

$$\xi_u = \frac{r}{n}\,D\cdot v - \frac{1}{n}\,(\Delta+n)\,T\,, \tag{95b}$$

$$\xi_i = r^2\,v_i + r\,D_i\xi_r\,, \tag{95c}$$

where T and v_i only depend on the angular coordinates x^i and, when $n > 2$, are constrained by the following differential conditions:

8 The exception is given by $n = 1$. Accordingly with the discussion in Section 3.1.3, in three dimensions, the conditions (94) are substituted by

$$\delta h_{uu} = \mathcal{O}(1)\,, \quad \delta h_{u\phi} = \mathcal{O}(1)\,, \quad \delta h_{\phi\phi} = 0\,, \quad \delta h_{r\mu} = 0\,.$$

$$D_{(i}v_{j)} - \frac{2}{n}\,\gamma_{ij} D \cdot v = 0\,, \tag{96}$$

$$D_{(i}D_{j)}T - \frac{2}{n}\,\gamma_{ij} \Delta T = 0\,. \tag{97}$$

Equation (96), which states that v_i is a conformal Killing vector on the sphere at null infinity, actually holds in any space-time dimension, while T has to satisfy (97) only when $n \neq 2$ (when $n = 2$ the variations of the field components which are proportional to the combination (97) are of the same order as the falloffs (90)). When $n > 2$, these conditions imply that the ξ_μ of the form (95) are Killing vectors of the Minkowski background, while for $n \leq 2$ (i.e. $D \leq 4$) a larger residual symmetry is allowed. When $n = 2$, the constraint (97) is indeed absent, so that supertranslations generated by an arbitrary $T(x^i)$ are allowed. Moreover, in this case, Equation (96) admits locally infinitely many independent solutions, which generate superrotations. To analyse the $n = 1$ case, note that both (96) and (97) are traceless combinations: as a result, they vanish identically when $n = 1$, so that both $v_\phi(\phi)$ and $T(\phi)$ are arbitrary functions.

The asymptotically conserved charges corresponding to the previous residual symmetries are given by

$$Q(u) = \lim_{r\to\infty} r^{n-1} \int_{S_u} d\Omega_n \left\{ h_{uu}\,(r\partial_r + n)\,\xi_r + \frac{1}{r}\,\gamma^{ij}\left[\, \xi_i\,\partial_r h_{uj} - h_{ui}\,\partial_r \xi_j - \xi_r D_i h_{uj} \right] \right\}, \tag{98}$$

where the parameters ξ_μ are understood to satisfy Equations (95) and (96) (together with (97) when $n > 2$). The integral is evaluated over the sphere S_u at constant retarded time on null infinity as, e.g., in Equation (8) (see Appendix A for more details). When n is even, thanks to the limit $r \to \infty$, for fields satisfying the ansatz (77) $Q(u)$ reduces to

$$\begin{aligned} Q(u) = &- \textstyle\sum_{k=0}^{\frac{n-2}{2}} r^{\frac{n-2k}{2}} \int_{S_u} d\Omega_n \left\{ \frac{n+2k+2}{2}\, v^i U_i^{(k)} + \left(T + \frac{u}{n}\, D\cdot v\right) \left[n B^{(k-1)} + \frac{n+2k-2}{2\,r^2}\, D \cdot U^{(k)} \right] \right\} \\ &- \textstyle\int_{S_u} d\Omega_n \left\{ 2n\, m_B \left(T + \frac{u}{n}\, D\cdot v\right) + (n+1)\,\gamma^{ij} v_i N_j \right\}. \end{aligned} \tag{99}$$

When n is odd, the only difference is that the extremum of the sum becomes $\frac{n-3}{2}$, so that the following considerations apply verbatim also to this case. As we shall see, at the linearised level, the $u-$dependence of the charges will turn out to be fictitious when $n \neq 2$. This is actually expected on general grounds whenever the charges are computed on exact Killing vectors of the background [37], as recalled in Appendix A.

The integrals in the first line must vanish in order to have finite asymptotic charges and this is indeed the case if one considers the relations imposed by the equations of motion. From the form (79) of the solutions, after integration by parts, one finds that, for each value of k, the first term in the charge formula (99) contains a contribution proportional to

$$C_{ij}^{(k)} D^{(i} v^{j)} = C_{ij}^{(k)} \left(D^{(i} v^{j)} - \frac{2}{n}\,\gamma^{ij} D \cdot v \right), \tag{100}$$

where the trace condition $\gamma^{ij} C_{ij}^{(k)} = 0$ was also used. Similarly, by the relation (78), the second term in (99), absent when $n = 2$, gives contributions than can be cast in the form

$$C_{ij}^{(k)} D^{(i} D^{j)} \left(T + \frac{u}{n}\, D \cdot v \right) = C_{ij}^{(k)} \left(D^{(i} D^{j)} T - \frac{2}{n}\,\gamma^{ij} \Delta T \right) + \frac{u}{n}\, C_{ij}^{(k)} D^{(i} D^{j)} D \cdot v\,. \tag{101}$$

This implies that the combination (100) vanishes in any space-time dimension (including $n = 2$). The contribution in T in (101) vanishes instead only when $n > 2$, but this does not cause any problem since the whole expression is actually absent when $n = 2$. The last term in Equation (101) vanishes as well when $n > 2$, since the divergence of the conformal Killing Equation (96) implies

$$\Delta v_i = \frac{2-n}{n} D_i D \cdot v - (n-1)\, v_i \,. \tag{102}$$

Acting with the Laplacian operator on Equation (96) and substituting this identity, one eventually obtains

$$(n-2)\, D_{(i} D_{j)} D \cdot v + 2\,\gamma_{ij}\,(\Delta + 2)\, D \cdot v = 0\,, \tag{103}$$

which implies $C^{(k)}_{ij} D^{(i} D^{j)} D \cdot v = 0$ for a traceless $C^{(k)}_{ij}$ (when $n > 2$).

Similar arguments allow one to prove that the integral terms in Equation (81) (which are absent when n is odd—cf. (88)) do not contribute as well to the charges when $n > 2$. In evaluating the last line of the charge formula (99), one eventually has to take into account the precise u dependence of N_i dictated by (81b). For $n \neq 2$, this gives

$$Q = -\int_{S^n} d\Omega_n \left\{ 2n\, T \mathcal{M} + (n+1)\, v^i \mathcal{N}_i \right\}, \tag{104}$$

where the dependence on u in the gauge parameter and in the field precisely cancels. The linearised Poincaré charges (104) depend on the "integration constants" that specify the Coulomb-type branch of the solutions of the equations of motion.[9] Each integration constant, in its turn, is conjugated in (104) to one of generators of the asymptotic symmetries. Let us also notice that these asymptotic charges, due to their constancy in u, should correspond in particular to the charges that one can measure at spatial infinity. A dependence on the retarded time, reflecting the changes in the total energy of the system induced by the flux of energy carried by the gravitational radiation, is reinstated when considering interactions [11,13,28,29,40].

The four-dimensional ($n = 2$) case requires instead a separate analysis. As we have seen, the linear divergence in r that appears in the charge (99) in this case vanishes on account of the identity (101) as for generic n. By substituting the expansions (92) into the second line of Equation (99), one obtains instead

$$Q(u) = -\int_{S_u} d\Omega_n \left\{ T\,(4\mathcal{M} + D \cdot D \cdot C) + v^i \left(3\mathcal{N}_i - \frac{u}{2} D_i D \cdot D \cdot C + D_i \int_{-\infty}^{u} du' D \cdot D \cdot C \right) \right\}. \tag{105}$$

The terms that appear for generic n are reproduced, cf. (104), but there is a sharp difference with respect to $n = 2$: the charges now depend also on the boundary data of the radiation part of the solution, and this brings back a dependence on the retarded time. The charges associated with ordinary Poincaré transformations, however, still take the same form as in Equation (104). For a translation, Equation (97) indeed holds also in four dimensions. Similarly, for a Lorentz transformation, one actually has $D_{(i} D_{j)} D \cdot v = 0$ also when $n = 2$. These conditions are instead not satisfied by the T and v^i generating supertranslations and superrotations, respectively (see e.g. [43]). In the latter case, in particular, in order for the charges to be well defined, one should impose in addition suitable boundary conditions on $D_i D \cdot D \cdot C$ at the past boundary of null infinity. The corresponding dependence on C_{ij} in the asymptotic charges is instrumental in deriving Weinberg's soft graviton theorem from the Ward identities of the supertranslation symmetry [5].[10]

---

9 The same expression for the charges holds also when the dimension of spacetime is equal to three, and it corresponds to the natural presentation of Q that one obtains in the Chern–Simons formulation of three-dimensional gravity (see e.g. Section 4.2 of [42]). The only difference is that when $n = 1$ T and v^i are arbitrary functions of the angular coordinate ϕ on the circle at null infinity.

10 While this work was under completion, analogous, u-dependent asymptotic charges, associated with infinite-dimensional asymptotic symmetries in any even space-time dimension, were presented in [19]. In addition, in this case, the arbitrary function on the sphere generating the residual symmetry is conjugated to the boundary data of the radiation branch.

4. Spin 3

We now consider a single spin-3 field on a Minkowski background and we prove that the boundary conditions at null infinity proposed in [1] give finite higher-spin charges in any space-time dimension. We work in the linearised theory, postponing to future work an analysis of the possible effects of the known cubic vertices[11] on the asymptotic charges and their canonical algebra.

4.1. Boundary Conditions

Following [1], we bound the spin-three field to satisfy the "Bondi-like gauge"

$$\varphi_{r\mu\nu} = 0\,, \quad \gamma^{ij}\varphi_{ij\mu} = 0\,, \quad \Rightarrow \quad g^{\mu\nu}\varphi_{\mu\nu\rho} = 0\,, \tag{106}$$

which is a natural generalisation of the Bondi gauge (63) that we used in the analysis of linearised gravity. As in the latter case, the conditions (106) cannot be reached with an off-shell gauge fixing, but the number of constraints is the same as in the transverse-traceless gauge. We therefore assume that they can be imposed in a neighbourhood of null infinity and consider the reduced Fronsdal equations[12]

$$\mathcal{F}_{\mu\nu\rho} = \Box\varphi_{\mu\nu\rho} - \nabla_{(\mu}\nabla\cdot\varphi_{\nu\rho)} = 0\,. \tag{107}$$

The expansion of (107) in Bondi coordinates has been presented in Appendix A of [1], and can be extracted from the general spin-s expressions presented below in Equations (140) and (141). Following the same logic as in the previous section, one can substitute a power-law ansatz in the equations of motion and realise that they are satisfied at the leading order provided that

$$\varphi_{uuu} = \mathcal{O}(r^{-\frac{n}{2}})\,, \quad \varphi_{uui} = \mathcal{O}(r^{-\frac{n}{2}+1})\,, \quad \varphi_{uij} = \mathcal{O}(r^{-\frac{n}{2}+2})\,, \quad \varphi_{ijk} = \mathcal{O}(r^{-\frac{n}{2}+3})\,. \tag{108}$$

These are the fall-off conditions that have been proposed in [1]; in the following, we shall exhibit the full solution of the linear equations of motion with these falloffs and we prove that it carries finite and non-trivial conserved spin-three charges at null infinity. As for gravity, the charges actually depend on the subleading terms at order r^{1-n}. We shall analyse separately even and odd space-time dimensions also in this case, while discussing in a dedicated subsection the main peculiarities emerging in three and four dimensions.

4.1.1. Even Space-Time Dimension

When n is even, we then consider the following ansatz for the fields in the Bondi gauge (106):

$$\varphi_{uuu} = \sum_{l=0}^{\infty} r^{-\frac{n}{2}-l}B^{(l)}(u,x^m)\,, \qquad \varphi_{uui} = \sum_{l=0}^{\infty} r^{-\frac{n}{2}-l+1}U_i^{(l)}(u,x^m)\,, \tag{109a}$$

$$\varphi_{uij} = \sum_{l=0}^{\infty} r^{-\frac{n}{2}-l+2}V_{ij}^{(l)}(u,x^m)\,, \qquad \varphi_{ijk} = \sum_{l=0}^{\infty} r^{-\frac{n}{2}-l+3}C_{ijk}^{(l)}(u,x^m)\,, \tag{109b}$$

with $\gamma^{ij}V_{ij}^{(l)} = \gamma^{ij}C_{ijk}^{(l)} = 0$. This choice is further motivated by the observation that, in complete analogy with the lower-spin cases, a solution of this type gives rise to a generically non-zero energy flux through null infinity:

$$\mathcal{P}(u) = \lim_{r\to\infty}\int_{S_u}(T_{uu}-T_{ur})\,r^n d\Omega_n = \int_{S_u}\gamma^{i_1j_1}\gamma^{i_2j_2}\gamma^{i_3j_3}\,\dot{C}^{(0)}_{i_1i_2i_3}\dot{C}^{(0)}_{j_1j_2j_3}d\Omega_n\,, \tag{110}$$

---

[11] For recent works and extensive references, see [44–49].

[12] Actually Equation (107), as well as its counterpart (139) for spin s, follows from the Lagrangian equations of an alternative formulation for massless particles of any spin [50–52].

where $T_{\mu\nu}$ denotes the energy-momentum tensor of the solution (see Section 5 for more details). It can be therefore interpreted as a spin-three wave reaching null infinity.

Substituting the ansatz (109) in the equations $\mathcal{F}_{r\mu\nu} = 0$, one obtains

$$B^{(k)} = \frac{2(n+2k-2)}{(n+2k)(n-2k-2)}\, D\cdot U^{(k)} \qquad \text{for } k \neq \frac{n-2}{2}, \tag{111a}$$

$$U_i^{(k)} = \frac{2(n+2k)}{(n+2k+2)(n-2k)}\, D\cdot V_i^{(k)} \qquad \text{for } k \neq \frac{n}{2}, \tag{111b}$$

$$V_{ij}^{(k)} = \frac{2(n+2k+2)}{(n+2k+4)(n-2k+2)}\, D\cdot C_{ij}^{(k)} \qquad \text{for } k \neq \frac{n+2}{2}. \tag{111c}$$

In the cases excluded from the previous formulae, the coefficients in front of, respectively, $B^{(\frac{n-2}{2})}$, $U^{(\frac{n}{2})}$ and $V^{(\frac{n+2}{2})}$ vanish and the equations of motion imply instead

$$(n-2)\, D\cdot D\cdot D\cdot C^{(\frac{n-2}{2})} = 0, \qquad D\cdot D\cdot C_i^{(\frac{n}{2})} = 0, \qquad D\cdot C_{ij}^{(\frac{n+2}{2})} = 0. \tag{112}$$

The factor $(n-2)$ in the first constraint shows that $C^{(0)}$ remains arbitrary even when $n = 2$. Substituting the same ansatz in the equation $\mathcal{F}_{ijk} = 0$, one obtains, for $l \neq \frac{n+2}{2}$,

$$\begin{aligned} \dot{C}_{ijk}^{(l+1)} = -\frac{1}{2(l+1)} \Big\{ & \Big[\Delta - \frac{n(n-2)}{4} + l(l+1) - 3 \Big] C_{ijk}^{(l)} \\ & - \frac{4}{(n+2l+4)(n-2l+2)} \Big[(n+2)\, D_{(i} D\cdot C_{jk)}{}^{(l)} - 2\,\gamma_{(ij} D\cdot D\cdot C_{k)}{}^{(l)} \Big] \Big\}. \end{aligned} \tag{113}$$

The value of l excluded from this expression shows that from there on the tensors $C^{(l)}$ also depend on $V^{(\frac{n+2}{2})}$,

$$\dot{C}_{ijk}^{(\frac{n+4}{2})} = \frac{1}{n+4} \Big\{ D_{(i} V_{jk)}^{(\frac{n+2}{2})} - \frac{2}{n+2}\,\gamma_{(ij} D\cdot V_{k)}^{(\frac{n+2}{2})} - (\Delta + 2n - 1)\, C_{ijk}^{(\frac{n+2}{2})} \Big\}, \tag{114}$$

although—as for gravity—these terms will not play any role in the analysis of the linearised charges. The u-evolution of $B^{(\frac{n-2}{2})}$, $U^{(\frac{n}{2})}$ and $V^{(\frac{n+2}{2})}$ is fixed instead by the equations $\mathcal{F}_{u\mu\nu} = 0$ (with $\mu, \nu \neq r$) as

$$B^{(\frac{n-2}{2})} = \mathcal{M} - \frac{n-3}{6(n+1)n} \int_{-\infty}^{u} du' (\Delta - n + 2)\, D\cdot D\cdot D\cdot C^{(\frac{n-4}{2})}, \tag{115a}$$

$$\begin{aligned} U_i^{(\frac{n}{2})} = \mathcal{N}_i + \frac{u}{n+2}\,\partial_i \mathcal{M} - \frac{n-1}{2(n+1)(n+2)} \int_{-\infty}^{u} du' (\Delta - 1)\, D\cdot D\cdot C_i^{(\frac{n-2}{2})} \\ - \frac{n-3}{6n(n+1)(n+2)} \int_{-\infty}^{u} du' \int_{-\infty}^{u'} du''\, D_i\, (\Delta - n + 2)\, D\cdot D\cdot D\cdot C^{(\frac{n-4}{2})}, \end{aligned} \tag{115b}$$

$$\begin{aligned} V_{ij}^{(\frac{n+2}{2})} = \mathcal{P}_{ij} + \frac{u}{n+3} \Big(D_{(i}\mathcal{N}_{j)} - \frac{2}{n}\,\gamma_{ij} D\cdot\mathcal{N} \Big) + \frac{u^2}{2(n+2)(n+3)} \Big(D_{(i}D_{j)}\mathcal{M} - \frac{2}{n}\,\gamma_{ij}\,\Delta\mathcal{M} \Big) \\ - \frac{n+1}{(n+2)(n+3)} \int_{-\infty}^{u} du' (\Delta + n - 2)\, D\cdot C_{ij}^{(\frac{n}{2})} + \cdots. \end{aligned} \tag{115c}$$

The omitted terms in the last equation correspond to the multiple integrals in the retarded time that one obtains by integrating the differential equation

$$\dot{V}_{ij}^{(\frac{n+2}{2})} = \frac{1}{n+3} \Big\{ D_{(i} U_{j)}^{(\frac{n}{2})} - \frac{2}{n}\,\gamma_{ij}\, D\cdot U^{(\frac{n}{2})} - \frac{n+1}{n+2} (\Delta + n - 2)\, D\cdot C_{ij}^{(\frac{n}{2})} \Big\} \tag{116}$$

given by the equation $\mathcal{F}_{uij} = 0$. At any rate, in Section 4.2, we shall show that all integrals in these expressions do not contribute to the linearised charges, provided that one impose suitable regularity conditions that make them finite.

The relevant terms to determine the charges are therefore those depending on the "integration constants" $\mathcal{M}$, $\mathcal{N}_i$ and $\mathcal{P}_{ij}$, which actually admit an arbitrary dependence on the coordinates x^m on the sphere at null infinity. They all appear at order r^{n-1} in the expansions (109) and enter (115) in combinations with a fixed polynomial dependence on the retarded time u. For all other powers of $1/r$, the equations $\mathcal{F}_{u\mu\nu} = 0$ (with $\mu, \nu \neq r$) reduce to divergences of Equation (113) and are therefore identically satisfied (see (151) below for more details). As in the examples with lower spin, the divergences of Equation (113) also imply the constraints (112) (to be more precise their derivative in u). As a result, the latter do not impose any further condition on the $C^{(l)}$ with lower values of l. Let us stress that some of the considerations above are valid only for $n > 2$; see Section 4.1.3 for a discussion of the four- and three-dimensional cases.

4.1.2. Odd Space-Time Dimension

In complete analogy with linearised gravity, when n is odd, one has to add further terms to the ansatz (109) in order to obtain non-trivial asymptotic charges at null infinity. We therefore consider the ansatz

$$\varphi_{uuu} = \varphi_{uuu}[B] + \sum_{l=0}^{\infty} r^{1-n-l}\tilde{B}^{(l)}(u,x^m)\,, \qquad \varphi_{uui} = \varphi_{uui}[U] + \sum_{l=0}^{\infty} r^{1-n-l}\tilde{U}_i^{(l)}(u,x^m)\,, \tag{117a}$$

$$\varphi_{uij} = \varphi_{uij}[V] + \sum_{l=0}^{\infty} r^{1-n-l}\tilde{V}_{ij}^{(l)}(u,x^m)\,, \qquad \varphi_{ijk} = \varphi_{ijk}[C] + \sum_{l=0}^{\infty} r^{1-n-l}\tilde{C}_{ijk}^{(l)}(u,x^m)\,, \tag{117b}$$

where $\varphi_{uuu}[B]$, etc. denote the terms introduced in the expansions (109), which are still necessary if one desires to describe radiation, that is if one wishes to have a non-vanishing energy flux through null infinity (which is still given by Equation (110)). The new contributions to the expansion of the field components satisfy $\gamma^{ij}\tilde{V}_{ij}^{(l)} = \gamma^{ij}\tilde{C}_{ijk}^{(l)} = 0$. Since n is odd, all factors entering the expansion of the equations of motion in powers of $\sqrt{r}$ are different from zero. As a result, the tensors $B^{(l)}$, $U^{(l)}$ and $V^{(l)}$ satisfy the same conditions as in (111), but without any constraint on the allowed values of l. Similarly, the tensors $C^{(l)}$ satisfy Equation (113) for any l. The tensors appearing at the leading order of the new, Coulomb-like branch of our ansatz must satisfy

$$\tilde{B}^{(0)} = \mathcal{M}\,, \qquad \tilde{U}_i^{(0)} = \mathcal{N}_i + \frac{u}{n+2}\,\partial_i\mathcal{M}\,, \tag{118a}$$

$$\tilde{V}_{ij}^{(0)} = \mathcal{P}_{ij} + \frac{u}{n+3}\left(D_{(i}\mathcal{N}_{j)} - \frac{2}{n}\,\gamma_{ij}D\cdot\mathcal{N}\right) + \frac{u^2}{2(n+2)(n+3)}\left(D_{(i}D_{j)}\mathcal{M} - \frac{2}{n}\,\gamma_{ij}\,\Delta\mathcal{M}\right), \tag{118b}$$

on account of the equations of motion $\mathcal{F}_{u\mu\nu} = 0$ (with $\mu, \nu \neq r$). Notice the similarity with Equation (115): the only difference is that, when n is odd, there is no contribution from the data of the solution that encode radiation (here stored in the $\sqrt{r}$ branch). As we shall see below, the latter terms anyway do not contribute to the linearised charges. The subleading $\mathcal{O}(r^{-n})$ terms in our ansatz do not contribute as well to the linearised charges. For this reason, we refrain from displaying here the relations that they have to satisfy in order to solve the equations of motion. The interested reader can extract them from the general expression (156) presented in Section 5.

4.1.3. Three and Four Space-Time Dimensions

When $n = 1$ or $n = 2$, the previous analysis has to be modified, in complete analogy with what we have seen for spin two. We begin by revisiting the four-dimensional case, where, in the component φ_{uuu}, the leading order of the radiation and of the Coulomb-like solutions now coincide. Moreover, the equation $\mathcal{F}_{ruu} = 0$ does not impose any constraint on the triple divergence of $C^{(0)}$. As a result, the equations fixing the dependence on u of $B^{(0)}$, $U^{(1)}$ and $V^{(2)}$ are slightly modified as follows:

$$B^{(0)} = \mathcal{M} + \frac{1}{6}\, D \cdot D \cdot D \cdot C^{(0)}\,, \tag{119a}$$

$$U_i{}^{(1)} = \mathcal{N}_i + \frac{u}{4}\,\partial_i \mathcal{M} - \frac{1}{24}\int_{-\infty}^{u} du' \left[(\Delta - 1)\, D \cdot D \cdot C_i{}^{(0)} - 2\, D_i D \cdot D \cdot D \cdot C^{(0)} \right], \tag{119b}$$

$$V_{ij}{}^{(2)} = \mathcal{P}_{ij} + \frac{u}{5}\left(D_{(i}\mathcal{N}_{j)} - \gamma_{ij} D \cdot \mathcal{N} \right) + \frac{u^2}{40}\left(D_{(i}D_{j)}\mathcal{M} - \gamma_{ij}\,\Delta\mathcal{M} \right) + \cdots . \tag{119c}$$

$\mathcal{M}, \mathcal{N}_i$ and $\mathcal{P}_{ij}$ are arbitrary functions of x^i as in Equation (115) while in (119c) we omitted the integrals that are obtained by the substitution of the previous formulae in the differential Equation (116), which is not modified even when $n = 2$. In analogy with the spin-two case, the additional terms in $D \cdot D \cdot D \cdot C^{(0)}$ will be instrumental in building the charges associated with the spin-three generalisations of supertranslations and superrotations identified in [1].

When $n = 1$, the radiation branch becomes again subleading with respect to the Coulomb-type one in φ_{uuu}. Moreover, fields of spin greater than one do not propagate local degrees of freedom in three dimensions. It is therefore natural to ignore the radiation branch and work with boundary conditions that only encompass Coulomb-type solutions of the equations of motion. The only non-vanishing components of the field in the Bondi gauge (106) are

$$\varphi_{uuu} = \mathcal{M}(\phi) + \mathcal{O}(r^{-1})\,, \qquad \varphi_{uu\phi} = \mathcal{N}(\phi) + \frac{u}{3}\,\partial_\phi \mathcal{M}(\phi) + \mathcal{O}(r^{-1})\,, \tag{120}$$

where ϕ denotes again the angular coordinate on the circle at null infinity and we already included the constraints on the leading terms imposed by the equations of motion. The same conditions on the field $\varphi_{\mu\nu\rho}$ have been previously obtained in [53,54] by translating boundary conditions proposed in the Chern–Simons formulation of three-dimensional spin-three gravity. The latter were designed to obtain asymptotic symmetries given by a contraction of the $\mathcal{W}_3 \oplus \mathcal{W}_3$ algebra of asymptotic symmetries of spin-three gravity in AdS_3 [55]. In analogy with the metric-like analysis performed in AdS_3 in [56], in the following we will recover the same infinite dimensional symmetries within our setup.

4.2. Asymptotic Symmetries and Charges

We now recall the key features of gauge transformations preserving the fall-off conditions (108), which have been identified in [1]. Our current goal is to prove that the linearised charges associated with these asymptotic symmetries are finite. In Appendix A, we show that, in the Bondi gauge (106), the charges are expressed in terms of the fields and the parameters of asymptotic symmetries as

$$\begin{aligned} Q(u) = -\lim_{r\to\infty} \frac{r^{n-1}}{2} \int_{S_u} d\Omega_n \Big\{ & r\,\varphi_{uuu}\,\partial_r\,\xi_{rr} + \xi_{rr}\,(r\partial_r + 2n)\,\varphi_{uuu} - \frac{2}{r}\,\xi_{rr} D \cdot \varphi_{uu} \\ & - \frac{2}{r^2}\,\gamma^{ij}\left[\varphi_{uui}\,(r\partial_r + n)\,\xi_{rj} - \frac{1}{r}\,\xi_{ri} D \cdot \varphi_{uj} \right] + \frac{1}{r^3}\,\gamma^{ik}\gamma^{jl}\left[\varphi_{uij}\,\partial_r\,\xi_{kl} - \xi_{ij}\,\partial_r\,\varphi_{ukl} \right] \Big\} . \end{aligned} \tag{121}$$

As we shall see, also in this case at the linearised level the $u-$dependence of the charges will turn out to be fictitious when $n \neq 2$. (For $n > 2$, this is due to the fact that the only symmetries are exact symmetries of the background.) As far as the computation of charges is concerned, the only relevant components of the gauge parameters generating the residual gauge symmetry are

$$\xi_{rr} = T - \frac{2u}{n+1}\, D\cdot\rho + \frac{u^2}{(n+1)(n+2)}\, D\cdot D\cdot K\,, \tag{122a}$$

$$\xi_{ri} = r^2\left(\rho_i - \frac{u}{n+2}\, D\cdot K_i\right) + \frac{r}{2}\, D_i\xi_{rr}\,, \tag{122b}$$

$$\begin{aligned}\xi_{ij} &= r^4\, K_{ij} + r\left(D_{(i}\xi_{j)r} - \frac{2}{n+1}\,\gamma_{ij}\, D\cdot\xi_r\right)\\ &\quad - \frac{r^2}{4}\left(D_{(i}D_{j)}\xi_{rr} - \frac{2(n+3)}{(n+1)(n+2)}\,\gamma_{ij}\left(\Delta + \frac{2(n+1)}{n+3}\right)\xi_{rr}\right).\end{aligned} \tag{122c}$$

Here, K_{ij}, ρ_i and T are tensors defined on the sphere at null infinity, which generalise the vector v_i and the function T that characterised the asymptotic symmetries of linearised gravity in Equation (95). They thus depend only on the coordinates x^m, while all dependence on u is explicit in the expressions above, which have been determined assuming an expansion in powers of r and u. The remaining components ξ_{uu}, ξ_{ur} and ξ_{ui} are also non-vanishing and depend on K_{ij}, ρ_i and T. We refer to [1] for their explicit expressions.

The tensors K_{ij}, ρ_i and T that characterise the asymptotic symmetries are not arbitrary: for $n > 2$, they are bound to satisfy the differential equations

$$\mathcal{K}_{ijk} \equiv D_{(i}K_{jk)} - \frac{2}{n+2}\,\gamma_{(ij}D\cdot K_{k)} = 0\,, \qquad \gamma^{ij}K_{ij} = 0\,, \tag{123}$$

$$\mathcal{R}_{ijk} \equiv D_{(i}D_j\rho_{k)} - \frac{2}{n+2}\left(\gamma_{(ij}\Delta\rho_{k)} + \gamma_{(ij}\{D_{k)}, D_l\}\rho^l\right) = 0\,, \tag{124}$$

$$\mathcal{T}_{ijk} \equiv D_{(i}D_jD_{k)}T - \frac{2}{n+2}\left(\gamma_{(ij}\Delta D_{k)}T + \gamma_{(ij}\{D_{k)}, D_l\}D^lT\right) = 0\,. \tag{125}$$

When the dimension of spacetime is equal to four, i.e., when $n = 2$, the last condition does not apply and the function $T(x^m)$ is instead arbitrary [1]. In this case, the corresponding symmetry is the analogue of gravitational supertranslations. Equation (123) generalises the conformal Killing Equation (96) and states that K_{ij} is a conformal (traceless) Killing tensor of rank-two on the celestial sphere (see e.g. [57]). For $n > 2$, this equation admits a finite number of solutions, while, when $n = 2$, locally there are infinitely many solutions, generalising gravitational superrotations. The same is true for the less familiar Equation (124) satisfied by ρ_i: when $n > 2$, it admits a finite number of solutions, while, for $n = 2$, locally one can build infinitely many independent solutions. For the details, we refer again to [1]. All combinations above are traceless: as a result, they vanish identically when the dimension of spacetime is equal to three, i.e., when $n = 1$. This implies that, in three dimensions, $T(\phi)$ and $\rho(\phi)$ are arbitrary functions, while the symmetry generated by the traceless K_{ij} is actually absent.

Substituting the gauge parameters (122) into the expression for the charges (121), one obtains

$$\begin{aligned}Q(u) = \lim_{r\to\infty}\frac{r^{n-1}}{2}\int_{S_u} d\Omega_n\, \Bigg\{ &\chi\left(T - \frac{2u}{n+1}\, D\cdot\rho + \frac{u^2}{(n+1)(n+2)}\, D\cdot D\cdot K\right)\\ &+ \chi^i\left(\rho_i - \frac{u}{n+2}\, D\cdot K_i\right) + \chi^{ij}K_{ij}\Bigg\},\end{aligned} \tag{126}$$

with

$$\chi = -\left(r\partial_r + 2n\right)\varphi_{uuu} - \frac{n-1}{r}\,D\cdot\varphi_{uu} + \frac{1}{2r}\,\partial_r D\cdot D\cdot\varphi_u\,, \tag{127a}$$

$$\chi_i = 2(n+2)\,\varphi_{uui} - \frac{2}{r}\left(r\partial_r - 2\right)D\cdot\varphi_{ui}\,, \tag{127b}$$

$$\chi_{ij} = \left(r\partial_r - 4\right)\varphi_{uij}\,. \tag{127c}$$

The next task is to evaluate the charge (126) on the solutions of the equations of motion discussed above. For n even and greater than two Equations (111), (112) and (115) lead to

$$\begin{aligned}\chi = &-\sum_{k=0}^{\frac{n-4}{2}} r^{-\frac{n}{2}-k}\,\frac{n+2k-2}{2(n-2k-2)}\,D\cdot D\cdot D\cdot C^{(k)} - r^{1-n}(n+1)\mathcal{M}\\ &+ r^{1-n}\,\frac{n-3}{6n}\int_{-\infty}^{u} du'(\Delta-n+2)\,D\cdot D\cdot D\cdot C^{(\frac{n-4}{2})} + o(r^{1-n})\,,\end{aligned} \tag{128a}$$

$$\begin{aligned}\chi_i = &\sum_{k=0}^{\frac{n-2}{2}} r^{-\frac{n}{2}-k+1}\,\frac{2(n+2k)}{n-2k}\,D\cdot D\cdot C_i^{(k)} + 2\,r^{1-n}\Big\{(n+2)\,\mathcal{N}_i + u\,\partial_i\mathcal{M}\Big\} + o(r^{1-n})\\ &- \frac{r^{1-n}}{n+1}\int_{-\infty}^{u} du'\Big\{(n-1)\,(\Delta-1)\,D\cdot D\cdot C_i^{(\frac{n-2}{2})} + \frac{n-3}{3n}\int_{-\infty}^{u'} du''D_i(\Delta-n+2)\,(D\cdot)^3C^{(\frac{n-4}{2})}\Big\}\,,\end{aligned} \tag{128b}$$

$$\begin{aligned}\chi_{ij} = &-\sum_{k=0}^{\frac{n}{2}} r^{-\frac{n}{2}-k+2}\,\frac{n+2k+2}{n-2k+2}\,D\cdot C_{ij}{}^{(k)}\\ &- r^{1-n}\left\{(n+3)\,\mathcal{P}_{ij} + u\left(D_{(i}\mathcal{N}_{j)} - \frac{2}{n}\,\gamma_{ij}D\cdot\mathcal{N}\right) + \frac{u^2}{2(n+2)}\left(D_{(i}D_{j)}\mathcal{M} - \frac{2}{n}\,\gamma_{ij}\,\Delta\mathcal{M}\right)\right\} + \cdots\,,\end{aligned} \tag{128c}$$

where, in (128c), besides the terms $o(r^{1-n})$, we also omitted the integrals in the retarded time that one obtains by substituting (115c). For $n=2$, the first two expressions are modified as follows:

$$\chi = -\frac{1}{r}\left\{3\,\mathcal{M} + \frac{1}{2}\,D\cdot D\cdot D\cdot C^{(0)}\right\} + \mathcal{O}(r^{-2})\,, \tag{129a}$$

$$\begin{aligned}\chi_i = &\,2\,D\cdot D\cdot C_i^{(0)} + \frac{2}{r}\left\{4\mathcal{N}_i + u\,\partial_i\mathcal{M} + \frac{1}{6}\int_{-\infty}^{u} du'\left[(\Delta-1)\,D\cdot D\cdot C_i^{(0)} - 2\,D_iD\cdot D\cdot D\cdot C^{(0)}\right]\right\}\\ &+ \mathcal{O}(r^{-2})\,.\end{aligned} \tag{129b}$$

The correct χ_{ij} is instead obtained by setting $n=2$ in (128c) and by correcting the integral terms according to (119c). For n odd, the extrema of the sums become, respectively, $\frac{n-3}{2}$, $\frac{n-1}{2}$ and $\frac{n+1}{2}$, while the terms in the second lines of Equations (128) are absent.

Looking only at the r-dependence, the sums in the previous formulae would give a divergent contribution to the charges. These vanish, however, thanks to the differential constraints on the parameters in (123), (124) and, when, $n>2$, (125). Let us begin by exhibiting this mechanism in the simplest case: the term $\chi^{ij}K_{ij}$ in the charge (126) contains divergent contributions that, integrating by parts, can be cast in the form

$$C^{(l)}_{ijk}\,D^{(i}K^{jk)} = C^{(l)}_{ijk}\left\{\frac{2}{n+2}\,\gamma^{(ij}D\cdot K^{k)} - \mathcal{K}^{ijk}\right\} = 0\,, \tag{130}$$

where we recall that $\mathcal{K}_{ijk}$ is the shortcut introduced in (123) to denote the differential equation satisfied by K_{ij}. This cancellation is the analogue of the one involving the conformal Killing equation in linearised gravity: it holds because the conformal Killing tensor equation allows for substituting the symmetrised gradient with a term in γ_{ij} and the tensors $C^{(l)}$ are traceless. The next cancellation is slightly more involved: integrating by parts one obtains

$$\int_{S_u} d\Omega_n\, \chi^i \left(\rho_i - \frac{u}{n+2} D\cdot K_i\right) \sim \sum_{l=0}^{[\frac{n-1}{2}]} r^{-\frac{n}{2}-l} \int_{S_u} d\Omega_n\, C^{(l)}_{ijk} \left(D^{(i}D^j\rho^{k)} - \frac{u}{n+2} D^{(i}D^j D\cdot K^{k)}\right) + \cdots . \tag{131}$$

To cancel the contribution in ρ_i, one can use Equation (124), which again allows one to substitute the symmetrised gradient with a term in γ_{ij}. To cancel the contribution in K_{ij}, one can instead use the following consequence of the conformal Killing tensor Equation (123):

$$\begin{aligned} D_{(i}D_j D\cdot K_{k)} =& -2\,\gamma_{(ij}D\cdot K_{k)} + \frac{3}{n+1}\gamma_{(ij}D_{k)}D\cdot D\cdot K \\ & - \frac{n+2}{n}\left\{\Delta\mathcal{K}_{ijk} - D_{(i}D\cdot\mathcal{K}_{jk)} + \frac{1}{n+1}\gamma_{(ij}D\cdot D\cdot\mathcal{K}_{k)} + (n-3)\,\mathcal{K}_{ijk}\right\}. \end{aligned} \tag{132}$$

Similar considerations apply to the integral terms in the second line of (128b). The remaining contribution in the charge formula (126) contains three addenda whose divergent parts can be cast in the following form by integrating by parts:

$$C^{(l)}_{ijk} D^{(i}D^j D^{k)} T\,, \qquad C^{(l)}_{ijk} D^{(i}D^j D^{k)} D\cdot\rho\,, \qquad C^{(l)}_{ijk} D^{(i}D^j D^{k)} D\cdot D\cdot K\,. \tag{133}$$

These terms are actually absent when $n = 2$. For $n > 2$, the first contribution vanishes thanks to the differential constraint (125). The other two types of terms vanish thanks to the following consequences of the equations satisfied by ρ_i and K_{ij}:

$$D_{(i}D_j D_{k)} D\cdot\rho = \frac{2}{n+2}\gamma_{(ij}D_{k)}\,(3\Delta + 2(n-1))\,D\cdot\rho + \text{terms in } \mathcal{R}_{ijk}\,, \tag{134a}$$

$$D_{(i}D_j D_{k)} D\cdot D\cdot K = -8\,\gamma_{(ij}D_{k)} D\cdot D\cdot K + \text{terms in } \mathcal{K}_{ijk}\,. \tag{134b}$$

The precise form of the omitted terms is displayed in Equations (B.8) and (B.9) of [1].

We have therefore proven that, in the Bondi gauge (106), a spin-three field with the falloffs (108) at null infinity given in [1] and satisfying the Fronsdal equations up to the contributions of order r^{n-1} to its components admits finite asymptotic linearised charges. For any $n > 2$, these depend on the "integration constants" specifying the solution as

$$Q = -\frac{1}{2}\int_{S^n} d\Omega_n \left\{(n+1)\,T\mathcal{M} - 2(n+2)\,\rho^i\mathcal{N}_i + (n+3)\,K^{ij}\mathcal{P}_{ij}\right\}, \tag{135}$$

where $\mathcal{M}, \mathcal{N}_i$ and $\mathcal{P}_{ij}$ are the tensors on the sphere at null infinity introduced in Equation (115) (cf. (118) for n odd). As anticipated, the charges are constant along null infinity when the dimension of spacetime is greater than four. The same is true also in three dimensions: in this case, both K^{ij} and $\mathcal{P}_{ij}$ are not present and the asymptotic charges take the form

$$Q = -\int d\phi \left\{T(\phi)\mathcal{M}(\phi) - 3\,\rho(\phi)\mathcal{N}(\phi)\right\}, \tag{136}$$

in agreement with the result derived in the Chern–Simons formulation [53,54]. When the dimension of spacetime is equal to three or bigger than four, with our boundary conditions, the spin-three charges thus display a structure very similar to that of the corresponding charges computed on anti de Sitter backgrounds in [58]. The latter, indeed, in the limit of vanishing cosmological constant should reproduce the flat-space charges at spatial infinity.

When $n = 2$, the modifications in the dependence on u of the leading terms in the Coulomb-type solution recalled in Equation (119) (and (129)) lead to the following expression for the asymptotic charges:

$$Q(u) = -\frac{1}{2}\int_{S_u} d\Omega_n \left\{ 3\,T\left(\mathcal{M} + \frac{1}{6}\,D\cdot D\cdot D\cdot C^{(0)}\right) - 8\,\rho^i\mathcal{N}_i + 5\,K^{ij}\mathcal{P}_{ij} + \cdots \right\}. \tag{137}$$

In this formula, we omitted other u–dependent terms in $C^{(0)}$, whose form is not particularly illuminating and can be readily obtained by substituting (129) in (126). The main information is anyway that in four dimensions a dependence on the retarded time appears already in the linearised theory, thanks to the contribution to the charges of the radiation solution. As shown in [1], where the terms in T in the charge (137) have been actually already presented, the dependence on radiation data is instrumental in deriving Weinberg's theorem for spin-three soft quanta from the Ward identities of the supertranslation symmetry generated by the arbitrary function $T(x^i)$.

5. Arbitrary Spin

In this section, we first extend to arbitrary values of the spin the analysis of the linearised equations of motion displayed in Sections 3 and 4 for fields of spin two and three. We then perform a preliminary study of the residual gauge symmetry preserving the form of the solutions. In particular, we fix the structure of asymptotic symmetries to leading order in an expansion in powers of r. This allows us both to exhibit some examples of on-shell cancellations of divergent contributions to the charges and to propose a general expression for their finite part in terms of the integration constants, arising after integration over u, which specify the solutions. This allows us to motivate a proposal for boundary conditions giving finite asymptotic spin-s charges in Minkowski backgrounds of any dimension.

5.1. Boundary Conditions

In the retarded Bondi coordinates (1), we study Fronsdal's equations in the "Bondi gauge" defined by

$$\varphi_{r\mu_1\cdots\mu_{s-1}} = 0\,, \qquad \gamma^{ij}\varphi_{ij\mu_1\cdots\mu_{s-2}} = 0\,. \tag{138}$$

These constraints imply that the fields are traceless, so that Fronsdal's tensor take the Maxwell-like form [51]

$$\mathcal{F}_{\mu_s} = \Box\varphi_{\mu_s} - \nabla_\mu\nabla\cdot\varphi_{\mu_{s-1}}\,. \tag{139}$$

Here and in the following, groups of symmetrised indices are denoted by a single Greek letter with a label indicating the total number of indices, so that, e.g., $\varphi_{\mu_1\cdots\mu_s} \to \varphi_{\mu_s}$; repeated indices denote instead a symmetrisation involving the minimum number of terms needed and without any overall factor, so that, e.g., $A_\mu B_\mu \equiv A_{\mu_1}B_{\mu_2} + A_{\mu_2}B_{\mu_1}$. For more details, see [59]. As before, we also assume that their solutions can be expanded in (half-integer) powers of r^{-1} in a neighbourhood of null infinity.

Equation (138) also implies that all components of the Fronsdal tensor with at least two radial indices vanish identically. The components with a single radial index read instead

$$\mathcal{F}_{r\,u_{s-k-1}i_k} = \frac{1}{r^2}\left\{r^2\partial_r^2 + (n-2k)\,r\partial_r - 2k(n-1)\right\}\varphi_{u_{s-k}i_k} - \frac{1}{r^3}\,(r\partial_r - 2k)\,D\cdot\varphi_{u_{s-k-1}i_k}\,. \tag{140}$$

The remaining components, without any radial index, are

$$\begin{aligned}\mathcal{F}_{u_{s-k}i_k} &= \frac{1}{r}\left\{(s-k-2)\,r\partial_r + n(s-k-1) + 2k\right\}\dot{\varphi}_{u_{s-k}i_k} - \frac{s-k}{r^2}\,D\cdot\dot{\varphi}_{u_{s-k-1}i_k} \\ &+ \frac{1}{r^2}\left\{\left[\Delta + r^2\partial_r^2 + (n-2k)\,r\partial_r - k(n-k)\right]\varphi_{u_{s-k}i_k} - D_iD\cdot\varphi_{u_{s-k}i_{k-1}}\right\} \\ &+ \frac{1}{r}\left\{(r\partial_r + n-2)\,D_i\varphi_{u_{s-k+1}i_{k-1}} + 2\,\gamma_{ii}D\cdot\varphi_{u_{s-k+1}i_{k-2}}\right\} - 2\,(r\partial_r + n-1)\,\gamma_{ii}\varphi_{u_{s-k+2}i_{k-2}}\,.\end{aligned} \tag{141}$$

We begin by studying the previous equations for n even. In this case, we employ the ansatz

$$\varphi_{u_{s-k}i_k} = \sum_{l=0}^{\infty} r^{-\frac{n}{2}+k-l} C_{i_k}^{(k,l)}(u,x^m)\,. \tag{142}$$

As in the previous examples, the leading behaviour of our ansatz is designed to give a finite flux of energy per unit time across the sphere S_u at fixed u, a feature that we interpret as radiation crossing null infinity. The canonical energy-momentum tensor of the Fronsdal Lagrangian in Bondi gauge,

$$\mathcal{L} = \frac{1}{2}\,\varphi^{\mu_s}\left(\Box\varphi_{\mu_s} - \nabla_\mu\nabla\cdot\varphi_{\mu_{s-1}} + \eta_{\mu\mu}\nabla\cdot\nabla\cdot\varphi_{\mu_{s-2}}\right), \tag{143}$$

reads indeed

$$T_{\alpha\beta} = \nabla_\alpha\varphi_{\mu_s}\nabla_\beta\varphi^{\mu_s} - s\,\nabla\cdot\varphi^{\mu_{s-1}}\nabla_\beta\varphi_{\alpha\mu_{s-1}} + \eta_{\alpha\beta}\mathcal{L}\,. \tag{144}$$

The corresponding power flowing through null infinity is then

$$\mathcal{P}(u) = \lim_{r\to\infty}\int_{S_u}(T_{uu}-T_{ur})\,r^n d\Omega_n = \int_{S_u}\gamma^{i_1j_1}\cdots\gamma^{i_sj_s}\,\dot{C}^{(s,0)}_{i_1\cdots i_s}\dot{C}^{(s,0)}_{j_1\cdots j_s}d\Omega_n\,. \tag{145}$$

When substituting the ansatz (142), the components $\mathcal{F}_{r\mu_{s-1}}$ of the Fronsdal tensor vanish provided that

$$C_{i_k}^{(k,l)} = \frac{2\,[n+2(k+l-1)]}{[n+2(k+l)]\,[n+2(k-l-1)]}\,D\cdot C_{i_k}^{(k+1,l)} \qquad \text{for } l \neq \frac{n}{2}+k-1\,, \tag{146}$$

while, for $l = n/2+k-1$, the equations of motion imply

$$(n+2k-2)D\cdot C_{i_k}^{(k+1,\frac{n}{2}+k-1)} = 0\,. \tag{147}$$

In the formula above, we exhibited the factor emerging in the computation that vanishes when $n=2$ and $k=0$. This anticipates that the peculiarities of the four-dimensional case that we encountered before persist for arbitrary values of the spin. At any rate, Equation (146) shows that, for all even values of n, the tensors entering the ansatz (142) are fixed in terms of the $C^{(s,l)}$, with the exception of

$$C_{i_k}^{(k,\frac{n}{2}+k-1)} \equiv \mathcal{Q}_{i_k}^{(k)} \qquad \text{for } k<s\,. \tag{148}$$

The tensors $C^{(s,l)}$ are then determined (up to integrations constants) in terms of an arbitrary tensor $C^{(s,0)}(u,x^m)$ via the equation $\mathcal{F}_{i_s}=0$, which gives

$$\begin{aligned}\dot{C}_{i_s}^{(s,l+1)} &= -\frac{1}{2(l+1)}\left\{\left[\Delta - \frac{n(n-2)}{4} + l(l+1) - s\right]C_{i_s}^{(s,l)}\right.\\ &\left.- \frac{4(n+2s-4)}{[n+2(s+l-1)][n+2(s-l-2)]}\left[D_iD\cdot C_{i_{s-1}}^{(s,l)} - \frac{2}{n+2s-4}\gamma_{ii}D\cdot D\cdot C_{i_{s-2}}^{(s,l)}\right]\right\}.\end{aligned} \tag{149}$$

The remaining components of the equations of motion fix the u-evolution of the tensors $\mathcal{Q}^{(k)}$ defined in Equation (148). To this end, it is convenient to expand the Fronsdal tensor in powers of r^{-1}. When the ansatz (142) holds, one has

$$\mathcal{F}_{u_{s-k}i_k}[C] = \sum_{l=0}^{\infty} r^{-\frac{n}{2}+k-l-1}\mathfrak{F}_{i_k}^{(k,l)}(u,x^m) \tag{150}$$

and, when the equations $\mathcal{F}_{r\mu_{s-1}}=0$ are satisfied, one can recast the expansion in the following form:

$$\begin{aligned}\mathcal{F}_{u_{s-k}i_k} = & \sum_{\substack{l=0 \\ l\neq\frac{n}{2}+k-2}}^{\infty} \frac{2\,r^{-\frac{n}{2}+k-l-2}}{n+2(k-l-2)}\, D\cdot\mathfrak{F}_{i_k}{}^{(k+1,l+1)} \\ & + r^{-n}\Big\{(n+s+k-2)\dot{\mathcal{Q}}_{i_k}{}^{(k)} - D_i\mathcal{Q}_{i_{k-1}}{}^{(k-1)} + \frac{2}{n+2k-4}\gamma_{ii}D\cdot\mathcal{Q}_{i_{k-2}}{}^{(k-1)} \\ & + \frac{n+2k-3}{(s-k)!(n+s+k-3)}\left[\Delta+k(k-4)+n(k-1)+2\right](D\cdot)^{s-k}\mathcal{C}^{(s,\frac{n}{2}+k-2)}\Big\}.\end{aligned} \tag{151}$$

This implies that, on shell, the $\mathcal{Q}^{(k)}$ are n-divergences as the other $C^{(k,l)}$, however up to a set of integrations constants $\mathcal{M}^{(k)}(x^m)$. The second line of Equation (151) actually dictates that $\mathcal{Q}^{(k)}$ depends on the integrations constants of all $\mathcal{Q}^{(l)}$ with $l<k$ with a precise polynomial dependence on u. As we have seen in the previous examples, this is instrumental in making the independence on the retarded time of the asymptotic charges explicit. Concretely, the $\mathcal{Q}^{(k)}$ depend on the integrations constants as

$$\mathcal{Q}_{i_k}{}^{(k)} = \sum_{l=0}^{k}\frac{(n+s+k-l-2)!}{l!(n+s+k-2)!}\,u^l\,\underbrace{D_i\cdots D_i}_{l\ \text{terms}}\mathcal{M}_{i_{k-l}}{}^{(k-l)}+\cdots. \tag{152}$$

In this formula, we omitted both the terms in $\mathcal{M}^{(k-l)}$ that make the expression above traceless and the terms in the $C^{(k,l)}$ resulting from the integration of Equation (151). Both types of contributions anyway will not contribute to the asymptotic charges when $n>2$. When $n=2$, Equation (151) has to be modified in analogy with the discussions in Sections 3.1.3 and 4.1.3, since $(D\cdot)^sC^{(0)}$ does not vanish anymore.

When n is odd, one has to consider an ansatz containing both integer and half-integer powers of r. For $n>1$, we set

$$\varphi_{u_{s-k}i_k} = \sum_{l=0}^{\infty} r^{-\frac{n}{2}-l+k}C^{(k,l)}(u,x^m) + \sum_{l=0}^{\infty} r^{1-n-l}\tilde{C}^{(k,l)}(u,x^m), \tag{153}$$

so that the leading order has the same form as in the ansatz (142). Equation (145) thus guarantees that we have a finite flux of energy per unit time across S_u in this case too. When $n=1$, consistently with the absence of propagating degrees of freedom for fields of spin $s>1$ in three dimensions, the radiation branch of the solution becomes subleading in the field component with only u indices and we will ignore it as in the examples with spin two and three. Notice that, due to the trace constraint in (138), in this case, the only non-vanishing components of the field are

$$\varphi_{u_s} = \mathcal{M}^{(s)}(\phi)+\mathcal{O}(r^{-1}), \qquad \varphi_{u_{s-1}\phi} = \mathcal{N}^{(s)}(\phi)+\frac{u}{s}\partial_\phi\mathcal{M}^{(s)}(\phi)+\mathcal{O}(r^{-1}). \tag{154}$$

The boundary conditions therefore contain only two arbitrary functions for each value of the spin, in analogy with what happens in AdS_3 [55,56,60–62].

The equations $\mathcal{F}_{r\mu_{s-1}}=0$ imply the relations (146) also when n is odd and greater than one, but without any limitation on the allowed values of k. The leading order of the Coulomb branch is instead not constrained by these equations, so that we can define

$$\tilde{C}^{(k,0)} \equiv \mathcal{Q}^{(k)} \qquad \text{for } k<s. \tag{155}$$

The remaining $\tilde{C}^{(k,l)}$ are again fixed in terms of the $\tilde{C}^{(s,l)}$ by Equation (146), provided that one identifies $\tilde{C}^{(k,l)} = C^{(k,\frac{n}{2}+k+l-1)}$, which is

$$\tilde{C}^{(k,l+1)} = -\frac{n+2k+l-1}{(l+1)(n+2k+l)}\,D\cdot\tilde{C}^{(k+1,l)}. \tag{156}$$

Performing the same substitution in Equation (149) gives the relation fixing all $\tilde{C}^{(s,l)}$ in terms of an arbitrary $\tilde{C}^{(s,0)}(u,x^m)$ (again up to integration constants).

Expanding the contributions to the Fronsdal tensor of the terms with integer powers of r as

$$\mathcal{F}_{u_{s-k}i_k}[\tilde{C}] = \sum_{l=0}^{\infty} r^{-n-l}\tilde{\mathfrak{F}}^{(k,l)}(u,x^m)\,, \tag{157}$$

one eventually obtains

$$\begin{aligned}\mathcal{F}_{u_{s-k}i_k} &= \sum_{l=0}^{\infty} \frac{2\,r^{-\frac{n}{2}+k-l-2}}{n+2(k-l-2)}\,D\cdot\mathfrak{F}_{i_k}^{(k+1,l+1)} - \sum_{l=0}^{\infty}\frac{r^{-n-l-1}}{l+1}\,D\cdot\tilde{\mathfrak{F}}_{i_k}^{(k+1,l)} \\ &+ r^{-n}\left\{(n+s+k-2)\dot{\mathcal{Q}}_{i_k}^{(k)} - D_i\mathcal{Q}_{i_{k-1}}^{(k-1)} + \frac{2}{n+2k-4}\,\gamma_{ii}D\cdot\mathcal{Q}_{i_{k-2}}^{(k-1)}\right\}.\end{aligned} \tag{158}$$

As a result, when n is odd, the $\mathcal{Q}^{(k)}$ satisfy a relation analogous to Equation (152), where the omitted terms, which anyway do not contribute to the charges, are actually absent.

By analogy with the examples of spin two and three, one is led to conclude that, in the Bondi gauge (138), the boundary conditions to be imposed on a spin-s field in order to obtain finite asymptotic charges for any D, even or odd, are the following:

$$\begin{aligned}\varphi_{u_{s-k}i_k} &= \sum_{l=0}^{\left[\frac{n+1}{2}\right]+k-2} r^{-\frac{n}{2}+k-l}\,\frac{2^{s-k}(n+2(k-l-2))!!(n+2(k+l-1))}{(n+2(s-l-2))!!(n+2(s+l-1))}\,(D\cdot)^{s-k}C_{i_k}^{(s,l)} \\ &+ r^{1-n}\mathcal{Q}_{i_k}^{(k)} + \mathcal{O}(r^{-\frac{n}{2}-\left[\frac{n}{2}\right]}),\end{aligned} \tag{159}$$

where the $C^{(k,l)}$ satisfy (146), while the $\mathcal{Q}^{(k)}$ satisfy (152). In this work, we do not perform a complete analysis of the asymptotic symmetries for fields of arbitrary spin. Still, we can provide support to the correctness of the boundary conditions (159) by showing that they lead to the cancellation of some of the potentially divergent contributions to the linearised charges, while also proving that the $\mathcal{Q}^{(k)}$ give a finite contribution to them. This is the goal of the next section.

5.2. Asymptotic Symmetries and Charges

In order to preserve the boundary conditions (159), the variations of the field components in Bondi coordinates,

$$\begin{aligned}\delta\varphi_{r_{s-k-l}u_l i_k} &= l\,\dot{\xi}_{r_{s-k-l}u_{l-1}i_k} + \frac{s-k-l}{r}\,(r\partial_r - 2k)\,\xi_{r_{s-k-l-1}u_l i_k} \\ &+ D_i\xi_{r_{s-k-l}u_l i_{k-1}} - 2r\,\gamma_{ii}\xi_{r_{s-k-l}u_{l+1}i_{k-2}} + 2r\,\gamma_{ii}\xi_{r_{s-k-l+1}u_l i_{k-2}}\,,\end{aligned} \tag{160}$$

must satisfy (for $n \geq 2$)

$$\delta\varphi_{r\mu_{s-1}} = 0\,, \qquad \delta\varphi_{u_{s-k}i_k} = \mathcal{O}(r^{-\frac{n}{2}+k})\,. \tag{161}$$

When $n=1$, one should have instead $\delta\varphi_{u_s} = \mathcal{O}(1)$ and $\delta\varphi_{u_{s-1}\phi} = \mathcal{O}(1)$, while all other variations must vanish.

From the examples discussed in the previous sections, we are led to consider the ansatz

$$\xi_{r_{s-k-l-1}u_l i_k} = r^{2k+l}\lambda_{i_k}^{(k,l)}(u,x^m) + \mathcal{O}(r^{2k+l-1})\,. \tag{162}$$

Under these conditions, the terms in the second line of the variations (160) become subleading for $l>0$, while the first line gives a first-order differential equation in u which fixes the u-dependence of the $\lambda^{(k,l)}$. To proceed, one can notice that the gauge parameters generating the residual symmetry must be both divergenceless and traceless. These constraints are indeed necessary to leave the gauge-fixed version (139) of the equations of motion invariant. They imply

$$\begin{aligned}\nabla\cdot\xi_{r_{s-k-l-1}u_{l-1}i_k} &= -\dot{\xi}_{r_{s-k-l}u_{l-1}i_k} - \frac{1}{r}\left(r\partial_r + n\right)\xi_{r_{s-k-l-1}u_l i_k} + \frac{1}{r^2}\,D\cdot\xi_{r_{s-k-l-1}u_{l-1}i_k} \\ &- \frac{s-k-l-1}{r^3}\,\xi'_{r_{s-k-l-2}u_{l-1}i_k} + \frac{1}{r}\left(r\partial_r + n\right)\xi_{r_{s-k-l}u_{l-1}i_k} = 0\,,\end{aligned} \tag{163}$$

and

$$\xi_{r_{s-k-l}u_{l-1}i_k} - 2\,\xi_{r_{s-k-l-1}u_l i_k} + \frac{1}{r^2}\,\xi'_{r_{s-k-l-2}u_{l-1}i_k} = 0\,, \tag{164}$$

where we denoted with a prime a contraction with the n-dimensional metric γ_{ij}. Combining this information with the requirement that the variations (160) vanish at leading order, one obtains

$$\dot{\lambda}_{i_k}^{(k,l)} + \frac{s-k-l-1}{n+s+k-2}\,D\cdot\lambda_{i_k}^{(k+1,l)} = 0\,, \tag{165}$$

which, as anticipated, fixes the u-dependence of the leading order in the expansion in powers of r.

From now on, we focus on the components of the gauge parameter that are relevant to the computation of asymptotic charges. In Appendix A, we shall show that, in the Bondi gauge (138), they can be expressed in terms of the non-vanishing field components as

$$\begin{aligned}Q(u) = &- \lim_{r\to\infty}\int_{S_u}\frac{r^{n-1}d\Omega_n}{(s-1)!}\sum_{p=0}^{s-1}\binom{s-1}{p}\Big\{\varphi_{u_{s-p}i_p}\left(r\partial_r + n + 2p\right)\xi^{u_{s-p-1}i_p} \\ &+ \xi^{u_{s-p-1}i_p}\left[(s-p-2)\left(r\partial_r + n\right)\varphi_{u_{s-p}i_p} - \frac{s-p-1}{r}\,D\cdot\varphi_{u_{s-p-1}i_p}\right]\Big\}\,.\end{aligned} \tag{166}$$

Therefore, they only depend on the components $\xi_{r_{s-k-1}i_k} = (-1)^{s-k-1}\xi^{u_{s-k-1}}{}_{i_k}$ of the gauge parameters of asymptotic symmetries. According to Equations (162) and (165), these satisfy

$$\xi_{r_{s-k-1}i_k} = r^{2k}\left(K_{i_k}^{(k)} + \sum_{m=1}^{s-k-1}\frac{(-1)^m u^m}{(n+s+k-2)_m}\binom{s-k-1}{m}(D\cdot)^m K_{i_k}^{(k+m)}\right) + \mathcal{O}(r^{2k-1})\,, \tag{167}$$

where the tensors $K^{(k)}(x^m)$ only depend on the coordinates on the n-dimensional sphere at null infinity, while $(a)_n \equiv a(a+1)\cdots(a+n-1)$ is the Pochhammer symbol. Moreover, at the leading order in r, Equation (164) implies that the $K^{(k)}$ are traceless (with respect to contractions with γ_{ij}). This implies that, in three dimensions, only $K^{(0)}$ and $K^{(1)}$ are actually present, in analogy with the corresponding reduction in the number of integration constants.

The tensors $K^{(k)}$ must also satisfy suitable differential constraints, which generalise those displayed in Equations (123) and (124) for the spin-three case. To identify them, it is convenient to focus on the variations of the field components without any u index. The absence of u-derivatives indeed allows one to study the resulting equations order by order in an expansion in powers of u. We can thus introduce the following expansion of the relevant components of the gauge parameter:[13]

$$\xi_{r_{s-k-1}i_k} = \sum_{l=0}^{k} r^{2k-l}A_{i_k}^{(k,l)}(x^m) + \text{terms in } u\,, \tag{168}$$

$$\xi_{r_{s-k-2}u i_k} = \sum_{l=0}^{k+1} r^{2k-l+1}B_{i_k}^{(k,l)}(x^m) + \text{terms in } u\,, \tag{169}$$

13 We already encoded the information on the minimum power of r entering the decomposition that can be extracted starting from the inspection of the equation $\delta\varphi_{r_s} = 0$ and substituting the result in the other variations of the form $\delta\varphi_{r_{s-k}i_k}$.

where $A^{(k,0)} \equiv K^{(k)}$ as dictated by Equation (167). Substituting this decomposition in the variations (160) gives

$$\begin{aligned}\delta\varphi_{r_{s-k-1}i_{k+1}} = \sum_{l=0}^{k} r^{2k-l}\Big\{ & -(l+1)(s-k-1)\,A_{i_{k+1}}^{(k+1,l+1)} + D_i A_{i_k}^{(k,l)} \\ & - 2\,\gamma_{ii} B_{i_{k-1}}^{(k-1,l)} + 2\,\gamma_{ii} A_{i_{k-1}}^{(k-1,l-1)} \Big\} + \text{terms in } u\,. \end{aligned} \tag{170}$$

Preserving the boundary conditions (161) thus requires, for all $k < s-1$,

$$A_{i_k}^{(k,l)} = \frac{1}{l(s-k)}\, D_i A_{i_{k-1}}^{(k-1,l-1)} + \gamma_{ii}\,(\cdots) \quad \Rightarrow \quad A_{i_{s-1}}^{(s-1,l)} = \frac{1}{(l!)^2}\,(D_i)^l K_{i_{s-l-1}}^{(s-l-1)} + \gamma_{ii}\,(\cdots)\,. \tag{171}$$

Notice that, for any value of l, the omitted combination in the $A^{(s-1,l)}$ tensor contains at least one new tensor of the $B^{(k,l)}$ family with respect to those appearing for lower values of l. As a result, by substituting back in Equation (160) (with $k = s-1$), one obtains the condition

$$\delta\varphi_{i_s} = \sum_{l=0}^{s-1} \frac{r^{2s-l-2}}{(l!)^2}\left\{ (D_i)^{l+1} K_{i_{s-l-1}}^{(s-l-1)} + \gamma_{ii}\,\Xi_{i_{s-2}}^{(s-l-1)} \right\} + \text{terms in } u = \mathcal{O}(r^{-\frac{n}{2}+s})\,, \tag{172}$$

where the tensors $\Xi^{(k)}$ can be considered as independent.[14]

If the space-time dimension is greater than four, i.e., if $n > 2$, the traceless tensors $K^{(k)}$ defined in Equation (167) must therefore satisfy the differential constraints

$$\underbrace{D_i \cdots D_i}_{s-k} K_{i_k}^{(k)} + \gamma_{ii}\,\Xi_{i_{s-2}}^{(k)} = 0\,, \qquad \gamma^{mn} K_{i_{k-2}mn}^{(k)}\,, \qquad 0 \le k \le s-1\,, \tag{173}$$

which, in particular, imply that the $(s-k)$th trace of $\Xi^{(k)}$ vanishes. Actually, these tensors can be eliminated by computing successive traces of Equations (173). For instance, for $k = s-1$, we obtained the conformal Killing tensor equation on the sphere (see e.g. [57]); one can eliminate the tensor $\Xi^{(s-1)}$ by computing a trace of (173) to obtain

$$D_i K_{i_{s-1}}^{(s-1)} - \frac{2}{n+2s-4}\,\gamma_{ii} D\cdot K_{i_{s-2}}^{(s-1)} = 0\,, \qquad \gamma^{kl} K_{i_{s-3}kl}^{(s-1)} = 0\,, \tag{174}$$

which is the formulation that we used in Equations (96) and (123) for fields of spin two and three, respectively.

When $n = 2$, the last term in the sum (172) does not have to vanish in order to preserve the boundary conditions. As a result, the function on the sphere denoted by $K^{(0)}(x^m)$ is completely arbitrary, as already pointed out in [1]. This infinite-dimensional enhancement of the asymptotic symmetries, generalising BMS supertranslations, is accompanied by a local infinite-dimensional enhancement of the symmetries generated by the tensors $K^{(s-1)}(x^m)$. The conformal Killing tensor Equation (174) admits in general $\frac{1}{s+1}\binom{n+s+1}{n+1}\binom{n+s}{n}$ independent globally defined solutions for $n \ge 2$ [57]. When $n = 2$, in addition, one also finds the further, local solutions

$$K_{z\cdots z}^{(s-1)} = K(z)\,, \qquad K_{\bar{z}\cdots\bar{z}}^{(s-1)} = \tilde{K}(\bar{z})\,, \qquad K_{z\cdots z\,\bar{z}\cdots\bar{z}}^{(s-1)} = 0\,, \tag{175}$$

where $z = e^{ix^1}\cot\frac{x^2}{2}$ together with its conjugate provide complex coordinates on the celestial sphere. These local solutions provide an extension to arbitrary values of the spin of superrotations [9,10].

---

[14] For instance, the dependence of the $\Xi^{(k)}$ on the $K^{(k)}$ tensors can be eliminated by redefining $B^{(k,l)} \to B^{(k,l)} + A^{(k,l-1)}$ in (169).

Moreover, for $s = 3$, in [1], we have shown that the equation for the tensor $K^{(s-2)}$ also admits locally infinitely many solutions. We defer to future work a complete analysis of Equation (173), but it is tempting to conjecture that all these equations admit infinitely-many solutions when the dimension of space-time is equal to four, that is when $n = 2$.

When $n = 1$, the equations for the surviving tensors $K^{(0)}(\phi)$ and $K^{(1)}(\phi)$ trivialise. As a result, for each spin-s field, the asymptotic symmetries are generated by two arbitrary functions of the angular coordinate on the circle at null infinity, in analogy with the spin-three case [53,54]. The number of free functions in the asymptotic symmetries is also the same as in AdS_3 higher-spin theories [55,56,60–62].

Before moving to the actual evaluation of the linearised charges, let us mention that one should complete our preliminary analysis of the asymptotic symmetries by studying the cancellation of the subleading orders in the gauge transformations (160), as we did for $s = 3$ in [1]. Nevertheless, we stress that the resulting (naively overdetermined) system of equations admits at least the solutions corresponding to rank-$(s-1)$ traceless and divergenceless Killing tensors of the Minkowski background. They indeed generate spin-s gauge transformations leaving the Minkowski background invariant, so that they obviously satisfy the weaker conditions (161). Consequently, in analogy with the spin-three example, besides the differential constraints (173), we do not expect any additional constraint on the tensors $K^{(k)}$ that fully characterises the asymptotic symmetries.

We now return to the expression (166) for the asymptotic charges, aiming to make manifest their $u-$independence for $n \neq 2$. While performing this analysis, one first has to keep track of the cancellation of all potentially divergent terms in the limit $r \to \infty$. As we have seen in the previous sections, these usually occur after integrations by parts; for instance, the coefficient of the leading power in r is of the form

$$\int_{S_u} d\Omega_n\, K^{(s-1)}_{i_{s-1}} D \cdot \mathcal{C}^{(s,0)\, i_{s-1}} = \int_{S_u} d\Omega_n\, D_i K^{(s-1)}_{i_{s-1}} \mathcal{C}^{(s,0)\, i_s}\,, \tag{176}$$

and hence vanishes because $K^{(s-1)}$ satisfies the conformal Killing tensor Equation (174) and $\mathcal{C}^{(s,0)}$ is traceless. Although we do not perform here an exhaustive inspection of these cancellations, the systematics suggested by the examples of spin one, two and three leads us to expect that they hold in general. Assuming that this is indeed the case, the finite contribution to the charges is determined by the Coulomb-like terms $\mathcal{Q}^{(k)}$ in the boundary conditions (159). Substituting them into the charge formula (166), while taking into account their dependence on the integration constants in (152), one obtains after integration by parts

$$\begin{aligned} \mathcal{Q}(u) = & \int_{S_u} \frac{d\Omega_n}{(s-1)!} \sum_{p=0}^{s-1} \sum_{m=0}^{s-p-1} \sum_{l=0}^{p} \binom{s-1}{p} \binom{s-p-1}{m} \binom{p}{l} \frac{(s+p+n-2)(s+n+p-l-2)!}{(s+n+p-2)_m (s+n+p-2)!} \\ & \times (-1)^{s+m+p+l}\, u^{l+m} \mathcal{M}^{(p-l)}_{i_{p-l}} (D\cdot)^{l+m} K^{(m+p)\, i_{p-l}} \end{aligned} \tag{177}$$

$$= \int_{S_u} \sum_{k,q} \frac{d\Omega_n\, (-1)^{s+k} (s+n+q-2)!}{(s-1)!(s+n+k+q-3)!} \left[\sum_{p=0}^{s-1} (-1)^p \binom{s-1}{p} \binom{s-p-1}{k+q-p} \binom{p}{q} \right] u^k \mathcal{M}^{(q)}_{i_q} (D\cdot)^k K^{(k+q)\, i_q}\,.$$

The final expression has been obtained by introducing the new labels $k = l + m$ and $q = p - l$ and by swapping the sums, whose ranges precisely correspond to the values of the labels for which the integrand does not vanish (with the convention $\binom{N}{n} = 0$ for $n < 0$ or $n > N$). One can eventually verify that the sum within square brackets in the second line of Equation (177) vanishes for any $k > 0$, thus providing, via the disappearance of the $u-$dependence, a good consistency check of our formulae for any value of the spin. The u-independent contribution then reads

$$\mathcal{Q} = \int_{S^n} \frac{d\Omega_n}{(s-1)!} \sum_{q=0}^{s-1} (-1)^{s+q} (s+n+q-2) \binom{s-1}{q} K^{(q)}_{i_q} \mathcal{M}^{(q)\, i_q}\,, \tag{178}$$

in full analogy with the result that we presented for $s = 3$ in Equation (135).[15]

When the dimension of spacetime is equal to three, similar considerations apply, with the additional simplification that the tensors $K^{(k)}$ with $k \geq 2$ and the integration constants $\mathcal{M}^{(k)}$ with $k \geq 2$ are actually absent. By defining $K^{(0)} \equiv T$, $\mathcal{M}^{(0)} = \mathcal{M}$ and $K^{(1)}_{\phi} \equiv v(\phi)$, $\mathcal{M}^{(1)}_{\phi} \equiv \mathcal{N}(\phi)$, in this case the charge (178) becomes

$$Q = \frac{(-1)^s}{(s-2)!} \int d\phi \left\{ T(\phi)\mathcal{M}(\phi) - s\,\rho(\phi)\mathcal{N}(\phi) \right\} \tag{179}$$

for any value of spin. In four dimensions, additional terms in u, depending on the data of the radiation solution, do appear. In particular, in agreement with the discussion in sect. 3.2 of [1], the charge formula (178) receives the following additional contribution in the supertranslation sector:

$$Q(u) = \frac{(-1)^s s}{(s-1)!} \int_{S_u} d\Omega_n\, K^{(0)} \left(\mathcal{M}^{(0)} + \frac{1}{s!}\,(D\cdot)^s C^{(0)} \right) + \cdots . \tag{180}$$

The contributions of the radiation data to the terms involving (generalised) superrotations can be determined following the same steps as in Section 4.2.

Acknowledgments: A.C. acknowledges the support of the Université libre de Bruxelles, where part of this work has been done. His work has been partially supported by the ERC Advanced Grant High-Spin-Grav, by FNRS-Belgium (convention FRFC PDR T.1025.14 and convention IISN 4.4503.15) and by the NCCR SwissMAP, funded by the Swiss National Science Foundation. A.C., D.F. and C.H. are grateful to the Service de Mécanique et Gravitation, Université de Mons and to Kyung Hee University, Seoul for the kind hospitality extended to some or all of us when part of this work was being done. The work of D.F. and of C.H. has been supported in part by Scuola Normale Superiore and by INFN Pisa.

Author Contributions: All authors contributed equally to this work.

Conflicts of Interest: The authors declare no conflict of interest.

Appendix A. Spin-s Charges in Bondi Gauge

In this appendix, we consider the linearised charges at null infinity associated with bosonic gauge fields of arbitrary spin. We work in generic space-time dimension and we express the charges in terms of the non-vanishing components of the fields in the Bondi gauge (138).

Appendix A.1. On-Shell Closed Two-Form for Arbitrary Spin

We begin from the following on-shell closed two-form, which gives the spin-s linearised charges upon integration on a codimension-two surface [63]:

$$\begin{aligned} k^{[\alpha\beta]} = \frac{\sqrt{-g}}{(s-1)!} \Big\{ & \nabla^{[\alpha} \varphi^{\beta]}{}_{\mu_{s-1}} \xi^{\mu_{s-1}} + \varphi_{\mu_{s-1}}{}^{[\alpha} \nabla^{\beta]} \xi^{\mu_{s-1}} + (s-1)\, \nabla \cdot \varphi_{\mu_{s-2}}{}^{[\alpha} \xi^{\beta]\mu_{s-2}} \\ & + (s-1)\, \xi^{\mu_{s-2}[\alpha} \nabla^{\beta]} \varphi_{\mu_{s-2}} + \frac{s-1}{2} \left(\varphi_{\mu_{s-2}} \nabla^{[\alpha} \xi^{\beta]\mu_{s-2}} - \nabla_{\mu} \varphi_{\mu_{s-3}}{}^{[\alpha} \xi^{\beta]\mu_{s-2}} \right) \Big\} . \end{aligned} \tag{A1}$$

As in Section 5, groups of symmetrised indices have been denoted by a single Greek letter with a label denoting the total number of indices, while repeated indices denote a symmetrisation involving the minimum number of terms needed and without any overall factor. Furthermore, omitted indices denote a trace, that is $\varphi_{\mu_{s-2}} = g^{\alpha\beta} \varphi_{\alpha\beta\mu_{s-2}}$. Equation (A1) has been obtained by eliminating the triplet auxiliary fields from Equation (48) of [63] and it applies to any space-time dimension D. If the field

[15] In order to compare with the spin-two charge (104), one should take into account the factor of 2 introduced by the definition of the Bondi mass aspect (see e.g. (78)) and that $T = -K^{(0)}$ as dictated by Equation (95).

satisfies Fronsdal's equations of motion and the gauge parameter satisfies the Killing tensor equation $\nabla_{\mu}\xi_{\mu_{s-1}} = 0$, then $\nabla_{\alpha}k^{[\alpha\beta]} = 0$.

We are interested in the charges at null infinity, which are defined as an integral over the sphere of dimension $n = D - 2$ at each point u, which we denote by S_u. As a result, in the Bondi coordinates (1), they involve only a specific component of the two-form (A1) and they are defined as

$$Q(u) = \lim_{r\to\infty} \int_{S_u} k^{ur}[\varphi, \xi]\, dx^1 \cdots dx^n \,. \tag{A2}$$

Appendix A.2. Rewriting in Bondi Gauge

We now wish to manifest the simplifications of Equation (A1) that are induced by the conditions (138) defining the Bondi gauge. First of all, the terms in its second line involve the trace of the field and therefore vanish in Bondi gauge. Since the only non-vanishing Christoffel symbols are those displayed in Equation (2) and the non-vanishing components of the inverse metric are

$$g^{ur} = -1\,, \qquad g^{rr} = 1\,, \qquad g^{ij} = r^{-2}\gamma^{ij}\,, \tag{A3}$$

these conditions also imply

$$\nabla_u\, \varphi_{r\mu_{s-1}} = \nabla_r\, \varphi_{r\mu_{s-1}} = \gamma^{ij}\nabla_i\, \varphi_{rj\mu_{s-2}} = 0\,. \tag{A4}$$

As a result, in the "Bondi gauge", the relevant component of the two-form (A1) reads

$$k^{ur} = \frac{r^n\sqrt{\gamma}}{(s-1)!}\left\{ \xi^{\rho_{s-1}}\nabla_r\, \varphi_{u\rho_{s-1}} - \varphi_{u\rho_{s-1}}\nabla_r\, \xi^{\rho_{s-1}} + (s-1)\, \xi^{u\rho_{s-2}}\nabla\cdot\varphi_{u\rho_{s-2}} \right\}. \tag{A5}$$

Using Equations (2) and (138), one can see that the covariant derivatives that are relevant for Equation (A5) are

$$\nabla_r\varphi_{u_{s-p}i_p} = \frac{1}{r}\,(r\partial_r - p)\,\varphi_{u_{s-p}i_p}\,, \tag{A6a}$$

$$\nabla_r\xi^{u_{s-p-1}i_p} = \frac{1}{r}\,(r\partial_r + p)\,\xi^{u_{s-p-1}i_p}\,, \tag{A6b}$$

$$\nabla\cdot\varphi_{u_{s-p-1}i_p} = -\frac{1}{r}\,(r\partial_r + n)\,\varphi_{u_{s-p}i_p} + \frac{1}{r^2}\,D\cdot\varphi_{u_{s-p-1}i_p}\,, \tag{A6c}$$

where we recall that D_i denotes the Levi–Civita connection for the metric γ_{ij} on the sphere at null infinity. All in all, by expanding Equation (A5) in components, one eventually gets:

$$\begin{aligned} k^{ur} = \frac{r^{n-1}\sqrt{\gamma}}{(s-1)!}\Bigg\{ &\xi^{i_{s-1}}\,(r\partial_r - s + 1)\,\varphi_{u\,i_{s-1}} - \varphi_{u\,i_{s-1}}\,(r\partial_r + s - 1)\,\xi^{i_{s-1}} \\ &- \sum_{p=0}^{s-2}\binom{s-1}{p}\Big[(s-p-2)\,\xi^{u_{s-p-1}i_p}\,(r\partial_r + n)\,\varphi_{u_{s-p}i_p} \\ &+ \varphi_{u_{s-p}i_p}\,(r\partial_r + n + 2p)\,\xi^{u_{s-p-1}i_p} - \frac{s-p-1}{r}\,\xi^{u_{s-p-1}i_p}\,D\cdot\varphi_{u_{s-p-1}i_p}\Big]\Bigg\}. \end{aligned} \tag{A7}$$

References

1. Campoleoni, A.; Francia, D.; Heissenberg, C. On higher-spin supertranslations and superrotations. *J. High Energy Phys.* **2017**, *2017*, 120.
2. Bondi, H.; van der Burg, M.G.J.; Metzner, A.W.K. Gravitational waves in general relativity. VII. Waves from axisymmetric isolated systems. *Proc. R. Soc. Lond. A* **1962**, *269*, 21–52.
3. Sachs, R. Asymptotic symmetries in gravitational theory. *Phys. Rev.* **1962**, *128*, 2851–2864.

4. Strominger, A. On BMS Invariance of Gravitational Scattering. *J. High Energy Phys.* **2014**, *2014*, 152.
5. He, T.; Lysov, V.; Mitra, P.; Strominger, A. BMS supertranslations and Weinberg's soft graviton theorem. *J. High Energy Phys.* **2015**, *2015*, 151.
6. Strominger, A. Lectures on the Infrared Structure of Gravity and Gauge Theory. *arXiv* **2017**, arXiv:1703.05448.
7. Weinberg, S. Photons and Gravitons in *S*-Matrix Theory: Derivation of Charge Conservation and Equality of Gravitational and Inertial Mass. *Phys. Rev. B* **1964**, *135*, 1049–1056.
8. Weinberg, S. Infrared photons and gravitons. *Phys. Rev. B* **1965**, *140*, 516–524.
9. Barnich, G.; Troessaert, C. Symmetries of asymptotically flat 4 dimensional spacetimes at null infinity revisited. *Phys. Rev. Lett.* **2010**, *105*, 111103.
10. Barnich, G.; Troessaert, C. Aspects of the BMS/CFT correspondence. *J. High Energy Phys.* **2010**, *2010*, 62.
11. Hollands, S.; Ishibashi, A. Asymptotic flatness and Bondi energy in higher dimensional gravity. *J. Math. Phys.* **2005**, *46*, 022503.
12. Hollands, S.; Wald, R.M. Conformal null infinity does not exist for radiating solutions in odd spacetime dimensions. *Class. Quantum Gravity* **2004**, *21*, 5139–5145.
13. Tanabe, K.; Kinoshita, S.; Shiromizu, T. Asymptotic flatness at null infinity in arbitrary dimensions. *Phys. Rev. D* **2011**, *84*, 044055.
14. Barnich, G.; Lambert, P.H. Einstein-Yang–Mills theory: Asymptotic symmetries. *Phys. Rev. D* **2013** , *88*, 103006.
15. Kapec, D.; Lysov, V.; Pasterski, S.; Strominger, A. Higher-Dimensional Supertranslations and Weinberg's Soft Graviton Theorem. *Ann. Math. Sci. Appl.* **2017**, *2*, 69–94.
16. Barnich, G.; Lambert, P.H.; Mao, P. Three-dimensional asymptotically flat Einstein-Maxwell theory. *Class. Quantum Gravity* **2015**, *32*, 245001.
17. Mao, P.; Ouyang, H. Note on soft theorems and memories in even dimensions. *Phys. Lett. B* **2017**, *774*, 715–722.
18. Campiglia, M.; Coito, L. Asymptotic charges from soft scalars in even dimensions. *arXiv* **2017**, arXiv:1711.05773.
19. Pate, M.; Raclariu, A.M.; Strominger, A. Gravitational Memory in Higher Dimensions. *arXiv* **2017**, arXiv:1712.01204.
20. Strominger, A. Asymptotic Symmetries of Yang–Mills Theory. *J. High Energy Phys.* **2014**, *2014*, 151.
21. He, T.; Mitra, P.; Porfyriadis, A.P.; Strominger, A. New Symmetries of Massless QED. *J. High Energy Phys.* **2014**, *2014*, 112.
22. He, T.; Mitra, P.; Strominger, A. 2D Kac–Moody Symmetry of 4D Yang–Mills Theory. *J. High Energy Phys.* **2016**, *2016*, 137.
23. Adamo, T.; Casali, E. Perturbative gauge theory at null infinity. *Phys. Rev. D* **2015**, *91*, 125022.
24. Campiglia, M.; Laddha, A. Asymptotic symmetries of QED and Weinberg's soft photon theorem. *J. High Energy Phys.* **2015** , *2015*, 115.
25. Mao, P.; Ouyang, H.; Wu, J.B.; Wu, X. New electromagnetic memories and soft photon theorems. *Phys. Rev. D* **2017**, *95*, 125011.
26. Mao, P.; Wu, J.-B. Note on asymptotic symmetries and soft gluon theorems. *Phys. Rev. D* **2017**, *96*, 065023.
27. Pate, M.; Raclariu, A.M.; Strominger, A. Color Memory: A Yang-Mills Analog of Gravitational Wave Memory. *Phys. Rev. Lett.* **2017**, *119*, 261602.
28. Hollands, S.; Ishibashi, A.; Wald, R.M. BMS Supertranslations and Memory in Four and Higher Dimensions. *Class. Quantum Gravity* **2017**, *34*, 155005.
29. Tanabe, K.; Shiromizu, T.; Kinoshita, S. Angular momentum at null infinity in higher dimensions. *Phys. Rev. D* **2012**, *85*, 124058.
30. Garfinkle, D.; Hollands, S.; Ishibashi, A.; Tolish, A.; Wald, R.M. The Memory Effect for Particle Scattering in Even Spacetime Dimensions. *Class. Quantum Gravity* **2017**, *34*, 145015.
31. Gross, D.J. High-Energy Symmetries of String Theory. *Phys. Rev. Lett.* **1988**, *60*, 1229–1232.
32. Moeller, N.; West, P.C. Arbitrary four string scattering at high energy and fixed angle. *Nucl. Phys. B* **2005**, *729*, 1–48.
33. Sagnotti, A.; Taronna, M. String Lessons for Higher-Spin Interactions. *Nucl. Phys. B* **2011**, *842*, 299–361.
34. Sagnotti, A. Notes on Strings and Higher Spins. *J. Phys. A* **2013**, *46*, 214006.
35. Casali, E.; Tourkine, P. On the null origin of the ambitwistor string. *J. High Energy Phys.* **2016**, *2016*, 036.
36. Abbott, L.F.; Deser, S. Charge Definition in Nonabelian Gauge Theories. *Phys. Lett. B* **1982**, *116*, 259–263.

37. Barnich, G.; Brandt, F. Covariant theory of asymptotic symmetries, conservation laws and central charges. *Nucl. Phys. B* **2002**, *633*, 3–82
38. Avery, S.G.; Schwab, B.U.W. Noether's second theorem and Ward identities for gauge symmetries. *J. High Energy Phys.* **2016**, *2016*, 31.
39. Wald, R.M.; Zoupas, A. A General definition of 'conserved quantities' in general relativity and other theories of gravity. *Phys. Rev. D* **2000**, *61*, 084027.
40. Hollands, S.; Thorne, A. Bondi mass cannot become negative in higher dimensions. *Commun. Math. Phys.* **2015**, *333*, 1037–1059.
41. Barnich, G.; Compere, G. Classical central extension for asymptotic symmetries at null infinity in three spacetime dimensions. *Class. Quantum Gravity* **2007**, *24*, 5.
42. Riegler, M.; Zwikel, C. Canonical Charges in Flatland. *arXiv*, **2017**, arXiv:1709.09871.
43. Barnich, G.; Troessaert, C. BMS charge algebra. *J. High Energy Phys.* **2011**, *2011*, 105.
44. Metsaev, R.R. BRST-BV approach to cubic interaction vertices for massive and massless higher-spin fields. *Phys. Lett. B* **2013**, *720*, 237–243.
45. Joung, E.; Lopez, L.; Taronna, M. Solving the Noether procedure for cubic interactions of higher spins in (A)dS. *J. Phys. A* **2013**, *46*, 214020.
46. Boulanger, N.; Ponomarev, D.; Skvortsov, E.D. Non-abelian cubic vertices for higher-spin fields in anti-de Sitter space. *J. High Energy Phys.* **2013**, *2013*, 8.
47. Sleight, C.; Taronna, M. Higher Spin Interactions from Conformal Field Theory: The Complete Cubic Couplings. *Phys. Rev. Lett.* **2016**, *116*, 181602.
48. Conde, E.; Joung, E.; Mkrtchyan, K. Spinor-Helicity Three-Point Amplitudes from Local Cubic Interactions. *J. High Energy Phys.* **2016**, *2016*, 40.
49. Francia, D.; Monaco, G.L.; Mkrtchyan, K. Cubic interactions of Maxwell-like higher spins. *J. High Energy Phys.* **2017**, *2017*, 68.
50. Francia, D. Low-spin models for higher-spin Lagrangians. *Prog. Theor. Phys. Suppl.* **2011**, *188*, 94–105.
51. Campoleoni, A.; Francia, D. Maxwell-like Lagrangians for higher spins. *J. High Energy Phys.* **2013**, *2013*, 168.
52. Francia, D. Generalised connections and higher-spin equations. *Class. Quantum Gravity* **2012**, *29*, 245003.
53. Afshar, H.; Bagchi, A.; Fareghbal, R.; Grumiller, D.; Rosseel, J. Spin-3 Gravity in Three-Dimensional Flat Space. *Phys. Rev. Lett.* **2013**, *111*, 121603.
54. Gonzalez, H.A.; Matulich, J.; Pino, M.; Troncoso, R. Asymptotically flat spacetimes in three-dimensional higher spin gravity. *J. High Energy Phys.* **2013**, *2013*, 16.
55. Campoleoni, A.; Fredenhagen, S.; Pfenninger, S.; Theisen, S. Asymptotic symmetries of three-dimensional gravity coupled to higher-spin fields. *J. High Energy Phys.* **2010**, *2010*, 7.
56. Campoleoni, A.; Henneaux, M. Asymptotic symmetries of three-dimensional higher-spin gravity: The metric approach. *J. High Energy Phys.* **2015**, *2015*, 143.
57. Eastwood, M.G. Higher symmetries of the Laplacian. *Ann. Math.* **2005**, *161*, 1645–1665.
58. Campoleoni, A.; Henneaux, M.; Hörtner, S.; Leonard, A. Higher-spin charges in Hamiltonian form. I. Bose fields. *J. High Energy Phys.* **2016**, *2016*, 146.
59. Francia, D. Geometric Lagrangians for massive higher-spin fields. *Nucl. Phys. B* **2008**, *796*, 77–122.
60. Henneaux, M.; Rey, S.-J. Nonlinear W_∞ as Asymptotic Symmetry of Three-Dimensional Higher Spin Anti-de Sitter Gravity. *J. High Energy Phys.* **2010**, *2010*, 7.
61. Gaberdiel, M.R.; Hartman, T. Symmetries of Holographic Minimal Models. *J. High Energy Phys.* **2011**, *2011*, 31.
62. Campoleoni, A.; Fredenhagen, S.; Pfenninger, S. Asymptotic W-symmetries in three-dimensional higher-spin gauge theories. *J. High Energy Phys.* **2011**, *2011*, 113.
63. Barnich, G.; Bouatta, N.; Grigoriev, M. Surface charges and dynamical Killing tensors for higher spin gauge fields in constant curvature spaces. *J. High Energy Phys.* **2005**, *2005*, 10.

universe

MDPI

Article

The ABC of Higher-Spin AdS/CFT

Simone Giombi [1,*], Igor R. Klebanov [1,2] and Zhong Ming Tan [1,3]

1 Department of Physics, Princeton University, Princeton, NJ 08544, USA; klebanov@princeton.edu (I.R.K.); tanzhongm@gmail.com (Z.M.T.)
2 Princeton Center for Theoretical Science, Princeton University, Princeton, NJ 08544, USA
3 Département de Physique, École Normale Supérieure, 45 rue d'Ulm, 75005 Paris, France
* Correspondence: sgiombi@princeton.edu

Received: 27 October 2017; Accepted: 11 December 2017; Published: 19 January 2018

Abstract: In recent literature, one-loop tests of the higher-spin AdS_{d+1}/CFT_d correspondences were carried out. Here, we extend these results to a more general set of theories in $d > 2$. First, we consider the Type B higher spin theories, which have been conjectured to be dual to CFTs consisting of the singlet sector of N free fermion fields. In addition to the case of N Dirac fermions, we carefully study the projections to Weyl, Majorana, symplectic and Majorana–Weyl fermions in the dimensions where they exist. Second, we explore theories involving elements of both Type A and Type B theories, which we call Type AB. Their spectrum includes fields of every half-integer spin, and they are expected to be related to the $U(N)/O(N)$ singlet sector of the CFT of N free complex/real scalar and fermionic fields. Finally, we explore the Type C theories, which have been conjectured to be dual to the CFTs of p-form gauge fields, where $p = \frac{d}{2} - 1$. In most cases, we find that the free energies at $O(N^0)$ either vanish or give contributions proportional to the free-energy of a single free field in the conjectured dual CFT. Interpreting these non-vanishing values as shifts of the bulk coupling constant $G_N \sim 1/(N-k)$, we find the values $k = -1, -1/2, 0, 1/2, 1, 2$. Exceptions to this rule are the Type B and AB theories in odd d; for them, we find a mismatch between the bulk and boundary free energies that has a simple structure, but does not follow from a simple shift of the bulk coupling constant.

Keywords: AdS/CFT correspondence; Higher-Spin Symmetry; 1/N expansion

---

1. Introduction

Extensions of the original AdS/CFT correspondence [1–3] to relations between the "vectorial" d-dimensional CFTs and the Vasiliev higher-spin theories in $(d+1)$-dimensional AdS space [4–8] have attracted considerable attention (for recent reviews of the higher-spin AdS_{d+1}/CFT_d correspondence, see [9,10]). The CFTs in question are quite well understood; their examples include the singlet sector of the free $U(N)/O(N)$ symmetric theories where the dynamical fields are in the vectorial representation (rather than in the adjoint representation) or of the vectorial interacting CFTs such as the $d = 3$ Wilson–Fisher and Gross–Neveu models [11–13]. Some years ago, the singlet sectors of $U(N)/O(N)$ symmetric d-dimensional CFTs of scalar fields were conjectured to be dual to the Type A Vasiliev theory in AdS_{d+1} [11], while the CFTs of fermionic fields to the Type B Vasiliev theory [12,13]. In $d = 3$, the $U(N)/O(N)$ singlet constraint is naturally imposed by coupling the massless matter fields to the Chern–Simons gauge field [14,15]. While the latter is in the adjoint representation, it carries no local degrees of freedom so that the CFT remains vectorial. More recently, a new similar set of dualities was proposed in even d and called Type C [16–18]; it involves the CFTs consisting of some number N of $\left(\frac{d}{2}-1\right)$-form gauge fields projected onto the $U(N)/O(N)$ singlet sector.

The higher-spin AdS/CFT conjectures were tested through matching of three-point correlation functions of operators at order N, corresponding to tree level in the bulk [9,19]; further work on the correlation functions includes [20–25]. Another class of tests, which involves calculations at order N^0, corresponding to the one-loop effects in the bulk, was carried out in [16–18,26–29]. It concerned the calculation of one-loop vacuum energy in Euclidean AdS_{d+1}, corresponding to the sphere free energy $F = -\log Z_{S^d}$ in CFT_d; in even/odd d, this quantity enters the a/F theorems [30–36]. Similar tests using the thermal AdS_{d+1}, where the Vasiliev theory is dual to the vectorial CFT on $S^{d-1} \times S^1$, have also been conducted [16–18,28,37]. Such calculations serve as a compact way of checking the agreement of the spectra in the two dual theories. The quantities of interest are the formula for the thermal free energies at arbitrary temperature β, as well as the temperature-independent Casimir energy E_c.

In this paper, we continue and extend the earlier work [16–18,26–29] on the one-loop tests of higher-spin AdS/CFT. In particular, we will compare the Type B theories in various dimensions d and their dual CFTs consisting of the Dirac fermionic fields (we also consider the theories with Majorana, symplectic, Weyl or Majorana–Weyl fermions in the dimensions where they are admissible). Let us also comment on the Sachdev–Ye–Kitaev (SYK) model [38,39], which is a quantum mechanical theory of a large number N of Majorana fermions with random interactions; it has been attracting a great deal of attention recently [40–44]. After the use of the replica trick, this model has manifest $O(N)$ symmetry [40], and it is tempting to look for its gravity dual using some variant of Type B higher spin theory. Following [45], one may speculate that the SYK model provides an effective IRdescription of a background of a Type B Vasiliev theory asymptotic to AdS_4, which is dual to a theory of Majorana fermions; this background should describe RGflow from AdS_4 to AdS_2 (one could also search for RG flow from HStheory in AdS_{d+1} to AdS_2 with $d = 2, 4, \ldots$).

Two other types of theories with no explicitly-constructed Vasiliev equations are also explored. First, we consider the theories whose CFT duals are expected to consist of both scalar and fermionic fields, with a subsequent projection onto the singlet sector. These theories, which we call Type AB, are then expected to have half-integral spin gauge fields in addition to the integral spin gauge fields of Type A and Type B theories. Depending on the precise scalar and fermion field content, the Type AB theories may be supersymmetric in some specific dimension d. For example, the $U(N)$ singlet sector of one fundamental Dirac fermion and one fundamental complex scalar is supersymmetric in $d = 3$, and a similar theory with one fundamental Dirac fermion and two fundamental complex scalars is supersymmetric in $d = 5$ [46].[1] Second, we study the Type C theories, where the CFT dual consists of some number of p-form gauge fields, with $p = \frac{d}{2} - 1$; the self-duality condition on the field strength may also be imposed. Such theories were studied in [16–18] for $d = 4$ and 6, and we extend them to more general dimensions.

The organization of the paper is as follows. In Section 2, we review how the comparison of the partition functions of the higher-spin theory and the corresponding CFT allows us to draw useful conclusions about their duality. We will also go through the various HS theories that will be examined in this paper. This will allow us to summarize our results in Tables 1, 2 and 6. In Section 3, we present our results for the free energy of Vasiliev theory in Euclidean AdS_{d+1} space asymptotic to the round sphere S^d. In addition, in Appendix A.1, we detail the calculations for the free energy of Vasiliev theory in the thermal AdS_{d+1} space, which is asymptotic to $S^{d-1} \times S^1$.

Note: Shortly before completion of this paper, we became aware of independent forthcoming work on related topics by M.Gunaydin, E. Skvortsov and T. Tran [47]. After the original submission of this paper, we also noticed a related paper by Y. Pang, E. Sezgin and Y. Zhu [48].

2. Review and Summary of Results

2.1. Higher Spin Partition Functions in Euclidean AdS Spaces

According to the AdS/CFT dictionary, the CFT partition function Z_{CFT} on the round sphere S^d has to match the partition function of the bulk theory on the Euclidean AdS_{d+1} asymptotic to S^d.

1 This theory may be coupled to the $U(N)$ 5dChern–Simons gauge theory to impose the singlet constraint.

This is the hyperbolic space $\mathbb{H}_{d+1}$ with the metric, $ds^2 = d\rho^2 + \sinh^2\rho\, d\Omega_d$, where $d\Omega_d$ is the metric of a unit d-sphere. After defining the free energy $F = -\log Z$, the AdS/CFT correspondence implies $F_{\rm CFT} = F_{\rm bulk}$.

For a vectorial CFT with $U(N)$, $O(N)$ or $USp(N)$ symmetry, the large N expansion is:

$$F_{\rm CFT} = N f^{(0)} + f^{(1)} + \frac{1}{N} f^{(2)} + \dots . \tag{1}$$

For a CFT consisting of N free fields, one obviously has $f^{(n)} = 0$ for all $n \geq 1$.

For the bulk gravitational theory with Newton constant G_N the perturbative expansion of the free energy assumes the form:

$$F_{\rm bulk} = \frac{1}{G_N} F^{(0)} + F^{(1)} + G_N F^{(2)} + \dots \tag{2}$$

The leading contribution is the on-shell classical action of the theory; it should match the leading term in the CFT answer which is of order N. Such a matching seems impossible at present due to the lack of a conventional action for the higher spin theories.[2] However, as first noted in [26], the one-loop correction $F^{(1)}$ requires the knowledge of only the free quadratic actions for the higher-spin fields in AdS$_{d+1}$; it can be obtained by summing the logarithms of functional determinants of the relevant kinetic operators. The latter were calculated by Camporesi and Higuchi [49–52], who derived the spectral zeta function for fields of arbitrary spin in (A)dS. What remains is to carry out the appropriately regularized sum over all spins present in a particular version of the higher spin theory.

The corresponding sphere free energy in a free CFT is given by $F_{\rm CFT} = NF$, where F may be extracted from the determinant for a single conformal field (see, for example, [35]); the examples of the latter are conformally coupled scalars, massless fermions, or p-form gauge fields. For vectorial theories with double-trace interactions, such as the Wilson–Fisher and Gross–Neveu models, the CFT itself has a non-trivial $\frac{1}{N}$ expansion, and so $F_{\rm CFT} = NF + \mathcal{O}(N^0)$. To match the large N scaling, the Newton constant of the bulk theory must behave as:

$$\frac{1}{G_N} \propto N, \tag{3}$$

in the large N limit. If one assumes that $\frac{1}{G_N} F^{(0)} = F_{\rm CFT}$, then all the higher-loop corrections to $F_{\rm bulk}$ must vanish for $F_{\rm CFT} = F_{\rm bulk}$ to hold. In [26,27], it was found that for the Vasiliev Type A theories in all dimensions d, the non-minimal theories containing each integer spin indeed have a vanishing one-loop correction to F. However, the minimal theories with even spins only were found to have a non-vanishing one-loop contribution that matched exactly the value of the sphere free-energy of a single conformal real scalar. This surprising result was then interpreted as a one-loop shift:

$$\frac{1}{G_N} \sim N - 1, \tag{4}$$

where the one-loop contribution cancels exactly the shift in the coupling constant. Such an integer shift is consistent with the quantization condition for $\frac{1}{G_N}$ established in [20,21]. The rule $N \to N-1$ does not apply to all the variants of the HS theory. In [16,17] it was shown that the one-loop calculations in Type C higher spin theories dual to free $U(N)/O(N)$ Maxwell fields in $d = 4$ required that $\frac{1}{G_N} \sim N - 1$ or $N - 2$ respectively. If the Maxwell fields are taken to be self-dual then $\frac{1}{G_N} \sim N - 1/2$; in view of this half-integer shift it is not clear if such a theory is consistent.

---

2 In the collective field approach to the bulk theory the action does exist, and the matching of free energies works by construction [29]. However, the precise connection of this formalism with the Vasiliev equations remains an open problem.

2.2. Variants of Higher Spin Theories and Key Results

The simplest and best understood HS theory is the Type A Vasiliev theory in AdS_{d+1}, which is known at non-linear level for any d [7]. The spectrum consists of a scalar with $m^2 = -2(d-2)$ and a tower of totally symmetric HS gauge fields (in the minimal theory, only the even spins are present). This is in one to one correspondence with the spectrum of $O(N)/U(N)$ invariant "single trace" operators on the CFT side, which consists of the $\Delta = d-2$ scalar:

$$J_0 = \phi_i^* \phi^i \tag{5}$$

and the tower of conserved currents:

$$J_{\mu_1 \cdots \mu_s} = \phi_i^* \partial_{(\mu_1} \cdots \partial_{\mu_s)} \phi^i + \cdots , \qquad s \geq 1. \tag{6}$$

This spectrum can be confirmed for instance by computing the tensor product of two free scalar representations, which yields the result [8,53,54]

$$\left(\frac{d}{2} - 1; 0\right) \otimes \left(\frac{d}{2} - 1; 0\right) = (d-2; 0, \ldots, 0) + \sum_{s=1}^{\infty} (d-2+s; s, 0, \ldots, 0) \tag{7}$$

where the notation $(\Delta; m_1, m_2, \ldots)$ indicates a representation of the conformal algebra with conformal dimension Δ and $SO(d)$ representation labeled by $[m_1, m_2, \ldots]$ (on the left-hand side, $(d/2-1; 0)$ is a shorthand for $(d/2-1; 0, \ldots, 0)$). Equivalently, one may obtain the same result by computing the "thermal" partition function of the free CFT on $S^1 \times S^{d-1}$, using a flat connection to impose the $U(N)$ singlet constraint [28,37]. Similarly one can consider real scalars and $O(N)$ singlet constraint, where one obtains the same spectrum but with odd spins removed (this corresponds to symmetrizing the product in (7)).

Another version of the HS theory is the so-called "Type B" theory, which is defined to be the HS gauge theory in AdS_{d+1} dual to the free fermionic CFT_d restricted to its singlet sector. The field content of such theories can be deduced from CFT considerations, by deriving the spectrum of singlet operators which are bilinears in the fermionic fields. In the case of Dirac fermions, one has the following results for the tensor product of two free fermion representations [8,54]: in even d:

$$\begin{aligned}\left(\tfrac{d-1}{2}; \tfrac{1}{2}\right) \otimes \left(\tfrac{d-1}{2}; \tfrac{1}{2}\right) = 2(d-1; 0, \ldots, 0) + 2\textstyle\sum_{s=1}^{\infty} [(d-2+s; s, 0, \ldots, 0) + (d-2+s; s, 1, 0, \ldots, 0) \\ + (d-2+s; s, 1, 1, 0, \ldots, 0) + \ldots + (d-2+s; s, 1, 1, 1, \ldots, 1, 0) \\ + (d-2+s; s, 1, 1, \ldots, 1, 1) + (d-2+s; s, 1, 1, \ldots, 1, -1)]\end{aligned} \tag{8}$$

and in odd d:

$$\begin{aligned}\left(\frac{d-1}{2}; \frac{1}{2}\right) \otimes \left(\frac{d-1}{2}; \frac{1}{2}\right) = (d-1; 0, \ldots, 0) + \sum_{s=1}^{\infty} [(d-2+s; s, 0, \ldots, 0) + \\ +(d-2+s; s, 1, 0, \ldots, 0) + \ldots + (d-2+s; s, 1, 1, \ldots, 1, 0) + (d-2+s; s, 1, 1, \ldots, 1, 1)] \, .\end{aligned} \tag{9}$$

Note that in the case $d = 3$, the spectra of the Type A and Type B theory are the same, except for the fact that the $m^2 = -2$ scalar is parity even in the former and parity odd in the latter (and also quantized with conjugate boundary conditions, $\Delta = 1$ versus $\Delta = 2$). In this special case, the fully non-linear equations for the Type B HS gauge theory in AdS_4 are known and closely related to those of the Type A theory [6]. For all $d > 3$, however, the spectra of Type B theories differ considerably from Type A theories, since they contain towers of spins with various mixed symmetries, see (8) and (9), and the corresponding non-linear equations are not known. As an example, and to clarify the meaning of (8) and (9), let us consider $d = 4$ [28,55–57]. In this case, on the CFT side one can construct the two scalar operators:

$$J_0 = \bar{\psi}_i \psi^i, \qquad \tilde{J}_0 = \bar{\psi}_i \gamma_5 \psi^i, \tag{10}$$

as well as (schematically) the totally symmetric and traceless bilinear currents:

$$J_{\mu_1 \cdots \mu_s} = \bar{\psi}_i \gamma_{(\mu_1} \partial_{\mu_2} \cdots \partial_{\mu_s)} \psi^i + \cdots, \qquad \tilde{J}_{\mu_1 \cdots \mu_s} = \bar{\psi}_i \gamma_5 \gamma_{(\mu_1} \partial_{\mu_2} \cdots \partial_{\mu_s)} \psi^i + \cdots, \qquad s \geq 1, \tag{11}$$

and a tower of mixed higher symmetry bilinear current,

$$M_{\mu_1 \cdots \mu_s, \nu} = \bar{\psi}_i \gamma_{\nu(\mu_1} \partial_{\mu_2} \cdots \partial_{\mu_s)} \psi^i + \cdots, \qquad s \geq 1, \tag{12}$$

where $\gamma_{\nu\mu_1} = \gamma_{[\nu}\gamma_{\mu_1]}$ is the antisymmetrized product of the gamma matrices. These operators are dual to corresponding HS fields in AdS_5. In particular, in addition to two towers of Fronsdal fields and a tower of mixed symmetry gauge fields, there are two bulk scalar fields and a massive antisymmetric tensor dual to $\bar{\psi}_i \gamma_{\mu\nu} \psi^i$. Similarly, in higher dimensions one can construct the tower of mixed symmetry operators appearing in (8) and (9) by using the antisymmetrized product of several gamma matrices. In the Young tableaux notation, these operators correspond to the hook type diagrams:

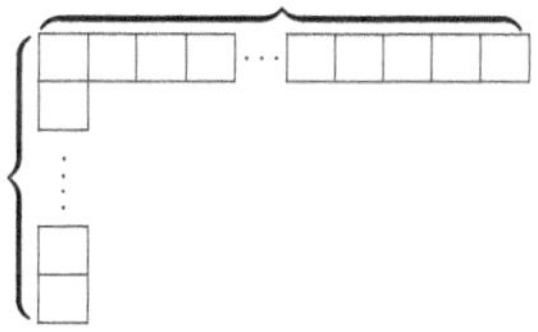

(13)

where $1 < j \leq p$, with $p = d/2$ for even d and $p = (d-1)/2$ for odd d. For $s > 1$, these operators are conserved currents and are dual to massless gauge fields in the bulk, while for $s = 1$ they are dual to massive antisymmetric fields.

For even d, we find evidence that the non-minimal Type B theory is exactly dual to the singlet sector of the $U(N)$ free fermionic CFT. The one-loop free energy of the Vasiliev theory vanishes exactly. This generalizes the result given in [16] for the non-minimal Type B theory in AdS_5; namely, there is no shift to the coupling constant in the non-minimal Type B theory dual to the singlet sector of Dirac fermions.

However, for all odd d, the one-loop free energy does not vanish. Instead, it follows a surprising formula:

$$F^{(1)}_{\text{type B}} = -\frac{1}{\Gamma(d+1)} \int_0^{1/2} du\, u \sin(\pi u) \Gamma\left(\frac{d}{2} + u\right) \Gamma\left(\frac{d}{2} - u\right), \tag{14}$$

which has an equivalent form for integer d:

$$F^{(1)}_{\text{type B}} = \frac{1}{2\Gamma(d+1)} \int_0^{1} du \cos(\pi u) \Gamma\left(\frac{d+1}{2} + u\right) \Gamma\left(\frac{d+1}{2} - u\right), \tag{15}$$

For example, for $d = 3,5,7$, one finds:

$$\begin{aligned} F^{(1)}_{\text{type B}} &= -\frac{\zeta(3)}{8\pi^2}, & d=3, \\ F^{(1)}_{\text{type B}} &= -\frac{\zeta(3)}{96\pi^2} - \frac{\zeta(5)}{32\pi^2}, & d=5, \\ F^{(1)}_{\text{type B}} &= -\frac{\zeta(3)}{720\pi^2} - \frac{\zeta(5)}{192\pi^2} - \frac{\zeta(7)}{128\pi^2}, & d=7. \end{aligned} \tag{16}$$

and similarly for higher d. Obviously, these complicated shifts cannot be accommodated by an integer shift of N. While the reason for this is not fully clear to us, it may be related to the fact that the imposition of the singlet constraint requires introduction of other terms in F. For example, in $d = 3$ the theory also contains a Chern–Simons sector, whose leading contribution to F is of order N^2. Perhaps a detailed understanding of these additional terms holds the key to resolving the puzzle for the fermionic theories in odd d.

We note that (14) always produces only linear combinations of $\zeta(2k+1)/\pi^2$ with rational coefficients. Interestingly, these formulas are related to the change in F due to certain double-trace deformations [58]. In particular, the first formula gives (up to sign) the change in free energy due to the double-trace deformation $\sim \int d^d x O_\Delta^2$, where O_Δ is a scalar operator of dimension $\Delta = \frac{d-1}{2}$, and the second formula is proportional to the change in free energy due to the deformation $\sim \int d^d x \bar{\Psi}_\Delta \Psi_\Delta$, where Ψ_Δ is a fermionic operator of dimension $\Delta = \frac{d-2}{2}$. The reason for this formal relation to the double-trace flows is unclear to us.

We also consider bulk Type B theories where various truncations have been imposed on the non-minimal Type B theory and we provide evidence that they are dual to the singlet sectors of various free fermionic CFTs. In $d = 2,3,4,8,9 \bmod 8$ we study the CFT of N Majorana fermions with the $O(N)$ singlet constraint, while in $d = 5,6,7 \bmod 8$ we study the theory of N symplectic Majorana fermions with the $USp(N)$ singlet constraint.[3] We also study the CFT of Weyl fermions in even d, and of Majorana–Weyl fermions when $d = 2 \bmod 8$. We will discuss these truncations in more detail in Section 3.2.1. For even d, we find that under the Weyl truncation, the Type B theories have vanishing F at the one-loop level. Under the Majorana/symplectic Majorana condition, the free energy of the truncated Type B theory gives (up to sign) the free energy of one free conformal fermionic field on S^d. This is logarithmically divergent due to the CFT a-anomaly, $F_f^{S^d} = a_f \log(\mu R)$, where the anomaly coefficient a_f is given by [58]:

$$a_f = 2^{\frac{d}{2}} \frac{(-1)^{\frac{d}{2}}}{\pi\Gamma(1+d)} \int_0^1 du \cos\left(\frac{\pi u}{2}\right) \Gamma\left(\frac{1+d+u}{2}\right) \Gamma\left(\frac{1+d-u}{2}\right) \tag{17}$$

$$= \left\{-\frac{1}{6}, \frac{11}{180}, -\frac{191}{7560}, \frac{2497}{226,800}, -\frac{14,797}{2,993,760}, \frac{92,427,157}{40,864,824,000}, -\frac{36,740,617}{35,026,992,000}, \ldots\right\} \tag{18}$$

for $d = \{2,4,6,8,\ldots\}$. Finally, under the Majorana–Weyl condition, the free energy of the corresponding truncated Type B theory reproduces half of the anomaly coefficients given in (18), corresponding to a single Majorana–Weyl fermion.

For the odd d case, the minimal Type B theories dual to the Majorana (or symplectic Majorana) projections again have unexpected values of their one-loop free energies. They are listed in Table 6.

[3] Let us note that one can also consider a "non-minimal" $USp(N)$ type-B model by starting with the free theory of N complex fermions (with N even), and imposing a $USp(N)$ singlet constraint on the spectrum (but not the symplectic Majorana condition on the fermions). This can be done in any d, and the resulting higher-spin theories are the fermionic analog of the symplectic type-A theories discussed in [26]. In AdS_4, one obtains this way a spectrum containing a $\Delta = 2$ parity odd scalar, one tower of higher-spin fields of all even spins $s \geq 2$, and three towers of odd-spin fields, as in the scalar case in [26]. One can analogously work out the spectra of such non-minimal $USp(N)$ type-B theories in higher dimensions, following similar steps as outlined in Section 3.2.1 for minimal $USp(N)$ theories.

We did not find a simple analytic formula that reproduces these numbers, but we note that, as in the non-minimal Type B result (14), these values are always linear combinations of $\zeta(2k+1)/\pi^2$ with rational coefficients. It would be very interesting to understand the origin of these "anomalous" results in the Type B theories.

One may also consider free CFTs which involve both the conformal scalars and fermions in the fundamental of $U(N)$ (or $O(N)$), with action:

$$S = \int d^d x \sum_{i=1}^{N} \left[(\partial_\mu \phi_i^*)(\partial^\mu \phi^i) + \bar{\psi}_i (\partial\!\!\!/) \psi^i \right]. \tag{19}$$

When we impose the $U(N)$ singlet constraint, the spectrum of single trace operators contains not only the bilinears in ϕ and ψ, which are the same as discussed above, but also fermionic operators of the form:

$$\Psi_{\mu_1 \cdots \mu_s} = \bar{\psi}^i \partial_{(\mu_1} \cdots \partial_{\mu_{s-\frac{1}{2}})} \phi^i + \ldots, \qquad \text{where } s = \frac{1}{2}, \frac{3}{2}, \cdots, \tag{20}$$

The dual HS theory in AdS should then include, in addition to the bosonic fields that appear in Type A and Type B theories, a tower of massless half integer spin particles with $s = 3/2, 5/2, \ldots$, plus a $s = 1/2$ matter field. We will call the resulting HS theory the "Type AB" theory. Note that in $d = 3$ this leads to a supersymmetric theory, but in general d the action (19) is not supersymmetric. One may also truncate the model to the $O(N)/USp(N)$ by imposing suitable reality conditions. There is no qualitative difference in the spectrum of the half-integer operators in the truncated models, with the only quantitative difference being a doubling of the degrees of freedom of each half-integer spin particle when going from $O(N)/USp(N)$ to $U(N)$ in the dual CFT.

The partition function for the Type AB theory is,

$$Z = e^{-F} = e^{-\frac{1}{G_N} F^{(0)} + F^{(1)} + G_N F^{(2)} + \cdots}, \qquad \text{where } F^{(1)} = F_f^{(1)} + F_b^{(1)}, \tag{21}$$

with F_b being for the contributions from bosonic higher-spin fields, which arise from purely Type A and purely Type B contributions, and $F_f^{(1)}$ is the contribution of the HS fermions dual to (20). Up to one-loop level, the bosonic and fermionic contributions are decoupled, as indicated in (21). A similar decoupling of the Casimir energy occurs at the one-loop level, i.e., $E_c^{(1)} = E_{c,f}^{(1)} + E_{c,b}^{(1)}$.

Our calculations for the Euclidean-AdS higher spin theory shows that $F_f^{(1)} = 0$ at the one-loop level for both $U(N)/O(N)$ theories for all d. Similarly, the Casimir energies are found to vanish: $E_{c,f} = 0$. In even d, from our results on the Type B theories and the earlier results on Vasiliev Type A theories, we see that $F_b^{(1)} = 0$ for the non-minimal Type AB theory, and this suggests that Type AB theories at one-loop have vanishing $F^{(1)}$. For odd d, $F^{(1)}$ is non-vanishing with the non-zero contribution coming from the Type B theory's free energy, as discussed above.

Finally, we consider the Type C higher-spin theories, which are conjectured to be dual to the singlet sector of massless p-forms, where $p = (\frac{d}{2} - 1)$.[4] The first two examples of these theories are the $d = 4$ case discussed in [16,17], where the dynamical fields are the N Maxwell fields, and the $d = 6$ case [18] where the dynamical fields are N 2-form gauge fields with field strength $H_{\mu\nu\rho}$. In these theories, there are also an infinite number of totally symmetric conserved higher-spin currents, in addition to various fields of mixed symmetry. We will extend these calculations to even $d > 6$.

As for Type B theories in $d > 3$, there are no known equations of motion for type C theories, but we can still infer their free field spectrum from CFT considerations, using the results of [54].

4 The choice of the p-form is made to ensure that the current operators satisfy the unitary bound, as well as conformal invariance.

The non-minimal theory is obtained by taking N complex $(d/2-1)$-form gauge fields A, and imposing a $U(N)$ singlet constraint. One may further truncate these models by taking real fields and $O(N)$ singlet constraint, which results in the "minimal type C" theory. In addition, one can further impose a self-duality condition on the $d/2$-form field strength $F = dA$. Since $*^2 = +1$ in $d = 4m+2$ and $*^2 = -1$ in $d = 4m$, where $*$ is the Hodge-dual operator, one can impose the self-duality condition $F = *F$ only in $d = 4m+2$ (for m integer); this can be done both for real ($O(N)$) and complex ($U(N)$) fields. In $d = 4m$, and only in the non-minimal case with N complex fields, one can impose the self-duality condition $F = i*F$. Decomposing $F = F_1 + iF_2$ into its real and imaginary parts, this condition implies $F_1 = -*F_2$, and self-dual and anti-self-dual parts of F are complex conjugate of each other.

As an example, let us consider $d = 4$ and take N complex Maxwell fields with a $U(N)$ singlet constraint. The spectrum of the the single trace operators arising from the tensor product $\bar{F}^i_{\mu\nu} \otimes F^{\rho\sigma}_i$ can be found to be [17,54]:

$$\begin{aligned}(2;1,1)_c \otimes (2;1,1)_c &= 2(4;0,0) + (4;1,1)_c + (4;2,2)_c \\ &\quad + 2\sum_{s=2}^{\infty}(s+2;s,0) + \sum_{s=3}^{\infty}(s+2;s,2)_c\end{aligned} \tag{22}$$

where we use the notation $(2;1,1)_c = (2;1,1) + (2;1,-1)$, corresponding to the sum of the self-dual and anti self-dual 2-form field strength with $\Delta = 2$, and similarly for the representations appearing on the right-hand side. Note that we use $SO(4)$ notations $[m_1, m_2]$ to specify the representation. The operators in the first line are dual to matter fields in AdS_5 in the corresponding representations, while the second line corresponds to massless HS gauge fields. Note that a novel feature compared to Type A and Type B is the presence of mixed symmetry representations with two boxes in the second row:

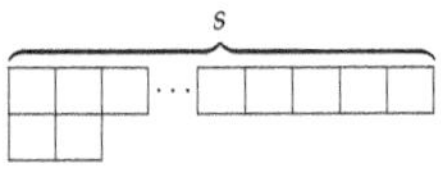

(23)

Imposing a reality condition and $O(N)$ singlet constraint, one obtains the minimal spectrum [17]:

$$\begin{aligned}[(2;1,1)_c \otimes (2;1,1)_c]_{\text{symm}} &= 2(4;0,0) + (4;2,2)_c \\ &\quad + \textstyle\sum_{s=2}^{\infty}(s+2;s,0) + \sum_{s=4,6,\ldots}^{\infty}(s+2;s,2)_c\,.\end{aligned} \tag{24}$$

Similarly, one may obtain the spectrum in all higher dimensions $d = 4m$ and $d = 4m+2$, as will be explained in detail in Section 3.2.3. As an example, in the $d = 8$ type C theory we find the representations:

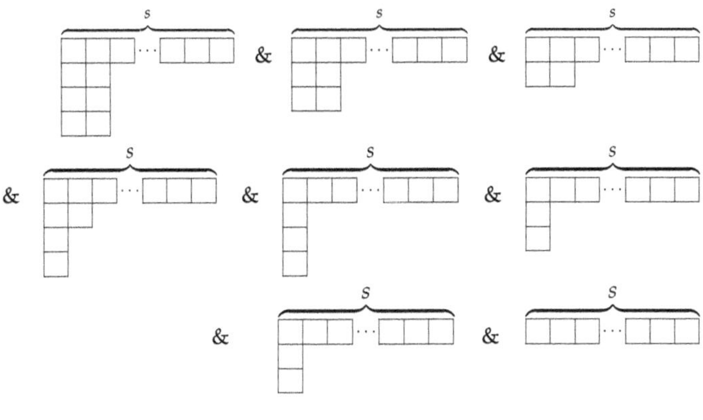

(25)

Our results for the one-loop calculations in type C theories are summarized in Table 1. We find that the non-minimal $U(N)$ theories have non-zero one-loop contributions, unlike the Type A and type B theories (in even d). The results can be grouped into two subclasses depending on the spacetime dimension, namely those in $d = 4m$ or in $d = 4m + 2$, where m is an integer. In the minimal type C theories with $O(N)$ singlet constraint, we find that for all $d = 4m$ the identification of the bulk coupling constant is $1/G_N \sim N - 2$, while in $d = 4m + 2$, the bulk one-loop free energy vanishes, and therefore no shift is required. In the self-dual $U(N)/O(N)$ theories, the one-loop free energy does not vanish, but can be accounted for by half-integer shifts $1/G_N \sim N \pm 1/2$, as mentioned earlier. We find that all of these results are consistent with calculations of Casimir energies in thermal AdS space, which are collected in the Appendix.

Table 1. Summary of results of one-loop calculations for even $d > 0$. By no shift, we mean that there are no shifts to the relation $G_N \sim 1/N$ due to one-loop free energy of the particular theory. Results for Type A theories taken from [27].

Type of Theory		Shift to $\frac{1}{G_N} \sim N$
Type A Theories		
Non-Minimal $U(N)$:		No shift
Minimal $O(N)$:		$N \to N-1$
Type B Theories		
Non-Minimal $U(N)$:		No shift
Minimal	$O(N)$ in $d = 2,4,8$ (mod 8):	$N \to N-1$
	$USp(N)$ in $d = 6$ (mod 8):	$N \to N+1$
Weyl Projection:		No shift
Majorana–Weyl:	$d = 2$ (mod 8):	$N \to N-1$
Type C Theories (p-Forms)		
Non-minimal $U(N)$	$d = 4, 8, 12, \ldots$:	$N \to N-1$
	$d = 6, 10, 14, \ldots$:	$N \to N+1$
Minimal $O(N)$	$d = 4, 8, 12, \ldots$:	$N \to N-2$
	$d = 6, 10, 14, \ldots$:	No shift
Self-dual $U(N)$	$d = 4, 8, 12, \ldots$:	$N \to N-\frac{1}{2}$
	$d = 6, 10, 14, \ldots$:	$N \to N+\frac{1}{2}$
Self-dual $O(N)$	$d = 4, 8, 12, \ldots$:	Not defined
	$d = 6, 10, 14, \ldots$:	$N \to N-\frac{1}{2}$

Table 2. Summary of results of one-loop calculations for odd $d > 0$. Again, by no shift, we mean that there are no shifts to the coupling constant coming from the spectrum of the particular theory. Results for Type A theories taken from [27].

	Type of Theory	Shift to $\frac{1}{G_N} \sim N$
	Type A Theories	
	Non-Minimal $U(N)$:	No shift
	Minimal $O(N)$:	$N \to N-1$
	Type B Theories	
	Non-Minimal $U(N)$:	Shifted by (14)
Minimal	$O(N)$ in $d = 3,9$ (mod 8):	See Section 3.3.3
Minimal	$USp(N)$ in $d = 5,7$ (mod 8):	See Section 3.3.3

3. Matching the Sphere Free Energy

3.1. The AdS Spectral Zeta Function

Let us first review the calculation of the one-loop partition function on the hyperbolic space in the case of the totally symmetric HS fields [26,27]. After gauge fixing of the linearized gauge invariance, the contribution of a spin s ($s \geq 1$) totally symmetric gauge field to the bulk partition function is obtained as [59–61]:

$$Z_s = \frac{\left[\det_{s-1}^{STT}\left(-\nabla^2 + (s+d-2)(s-1)\right)\right]^{\frac{1}{2}}}{\left[\det_{s}^{STT}\left(-\nabla^2 + (s+d-2)(s-2) - s\right)\right]^{\frac{1}{2}}} \tag{26}$$

where the label STT stands for symmetric traceless transverse tensors, and the numerator corresponds to the contributions of the spin $s-1$ ghosts. The mass-like terms in the above kinetic operators are related to the conformal dimension of the dual fields. For a totally symmetric field with kinetic operator $-\nabla^2 + \kappa^2$, the dual conformal dimension is given by:

$$\Delta(\Delta - d) - s = \kappa^2 \,. \tag{27}$$

For the values of κ in (26), one finds for the physical spin s field in the denominator[5]

$$\Delta^{\mathrm{ph}} = s + d - 2 \tag{28}$$

which corresponds to the scaling dimension of the dual conserved current in the CFT. Similarly the conformal dimension obtained from the ghost kinetic operator in (26) is:

$$\Delta^{\mathrm{ph}} = s + d - 1 \,. \tag{29}$$

From CFT point of view, this is the dimension of the divergence $\partial \cdot J_s$, which is a null state that one has to subtract to obtain the short representation of the conformal algebra corresponding to a conserved current.

---

5 We choose the root Δ_+ above the unitarity bound. The alternate root corresponds to gauging the HS symmetry at the boundary [62].

The determinants in (26) can be computed using the heat kernel, or equivalently spectral zeta functions techniques.[6] The spectral zeta function for a differential operator on a compact space with discrete eigenvalues λ_n and degeneracy d_n is defined as

$$\zeta(z) = \sum_n d_n \lambda_n^{-z}. \tag{30}$$

In our case, the differential operators in hyperbolic space have continuous spectrum, and the sum over eigenvalues is replaced by an integral. Let us consider a field labeled by the representation $\alpha_s = [s, m_2, m_3, \ldots]$ of $SO(d)$[7], where we have denoted by $m_1 = s$ the length of the first row in the corresponding Young diagram, which we may call the spin of the particle (for example, for a totally symmetric field, we have $\alpha_s = [s, 0, 0, \ldots, 0]$). For a given representation α_s, the spectral zeta function takes the form:

$$\zeta_{(\Delta;\alpha_s)}(z) = \frac{\mathrm{vol}\,(\mathrm{AdS}_{d+1})}{\mathrm{vol}(S^d)} \frac{2^{d-1}}{\pi} g_{\alpha_s} \int_0^\infty du \frac{\mu_{\alpha_s}(u)}{\left[u^2 + \left(\Delta - \frac{d}{2}\right)^2\right]^z}, \tag{31}$$

where $\mu_{\alpha_s}(u)$ is the spectral density of the eigenvalues, which will be given shortly, and g_{α_s} is the dimension of the representation α_s (see Equations (43) and (44) below). The denominator corresponds to the eigenvalues of the kinetic operator, and Δ is the dimension of the dual CFT operator.[8] The regularized volume of AdS is given explicitly by [63–65]:

$$\mathrm{vol}(\mathrm{AdS}_{d+1}) = \begin{cases} \pi^{d/2}\Gamma(-\frac{d}{2}), & d \text{ odd}, \\ \frac{2(-\pi)^{d/2}}{\Gamma(1+\frac{d}{2})} \log R, & d \text{ even}, \end{cases} \tag{32}$$

where R is the radius of the boundary sphere. The logarithmic dependence on R in even d is related to the presence of the Weyl anomaly in even dimensional CFTs. Finally, the volume of the round sphere of unit radius is:

$$\mathrm{vol}(S^d) = \frac{2\pi^{(d+1)/2}}{\Gamma[(d+1)/2]}. \tag{33}$$

Once the spectral zeta function is known, the contribution of the field labeled by $(\Delta; \alpha_s)$ to the bulk free energy is obtained as:

$$F^{(1)}_{(\Delta;\alpha_s)} = \sigma \left[-\frac{1}{2}\zeta'_{(\Delta;\alpha_s)}(0) - \zeta_{(\Delta;\alpha_s)}(0) \log(\ell\,\Lambda) \right], \tag{34}$$

where $\sigma = +1$ or -1 depending on whether the field is bosonic or fermionic. Here ℓ is the AdS curvature, which we will set to one henceforth, and Λ is a UVcut-off. In general, the coefficient of the logarithmic divergence $\zeta_{(\Delta;\alpha_s)}(0)$ vanishes for each α_s in even dimension d, but it is non-zero for odd d.

When the dimension $\Delta = s + d - 2$, the field labeled by α_s is a gauge field and one has to subtract the contribution of the corresponding ghosts in the α_{s-1} representation.[9] We find it convenient to introduce the notation:

6 The heat-kernel is related to the spectral zeta-function by a Mellin transformation.
7 This can be thought as the representation that specifies the dual CFT operator. From AdS point of view, one may view $SO(d)$ as the little group for a massive particle in $d+1$ dimensions.
8 For the case of totally symmetric fields, this form of the eigenvalues can be deduced from the results of [51]. See for example the Appendix of [16,18] for an explicit derivation in AdS_5 and AdS_7 for arbitrary representations.
9 As in the case of totally symmetric fields, the representation labeling the ghosts can be understood from CFT point of view from the structure of the character of the short representations of the conformal algebra and the corresponding null states, see [54].

$$\mathfrak{Z}_{(\Delta^{\rm ph}=s+d-2;\alpha_s)}(z) \equiv \zeta_{(\Delta^{\rm ph};\alpha_s)}(z) - \zeta_{(\Delta^{\rm ph}+1;\alpha_{s-1})}(z) \tag{35}$$

to indicate the spectral zeta function of the HS gauge fields in the α_s representation, with ghost contribution subtracted. The full one-loop free energy may be then obtained by summing over all representations α_s appearing in the spectrum. For instance, in the case of the non-minimal Type A theory, we may define the "total" spectral zeta function:

$$\zeta^{\rm HS}_{\text{type A}}(z) = \zeta_{(d-2;[0,\dots,0])}(z) + \sum_{s=1}^{\infty} \mathfrak{Z}_{(s+d-2;[s,0,\dots,0])}(z) \tag{36}$$

from which we can obtain the full one-loop free energy:

$$F^{(1)}_{\text{type A}} = \left[-\frac{1}{2}(\zeta^{\rm HS}_{\text{type A}})'(0) - \frac{1}{2}\zeta^{\rm HS}_{\text{type A}}(0)\log(\ell^2\Lambda^2)\right]. \tag{37}$$

Similarly, one can obtain $\zeta^{\rm HS}_{\rm total}(z)$ and the one-loop free energy in the other higher spin theories we discuss. As these calculations requires summing over infinite towers of fields, one has of course to suitably regularized the sums, as discussed in [26,27] and reviewed in the explicit calculations below.

3.1.1. The Spectral Density for Arbitrary Representation

A general formula for the spectral density for a field labeled by the representation $\alpha = [m_1, m_2, \dots]$ was given in [52], and we summarize their result below.

In AdS$_{d+1}$, arranging the weights for the irreps of $SO(d)$ as $m_1 \geq m_2 \geq \cdots \geq |m_p|$, where $p = \frac{d-1}{2}$ for odd d and $p = \frac{d}{2}$ for even d, we may define:

$$\ell_j = m_{p-j+1} + j - 1, \qquad \text{for } d = \text{even}, \tag{38}$$

$$\ell_j = m_{p-j+1} + j - \frac{1}{2}, \qquad \text{for } d = \text{odd}. \tag{39}$$

In terms of these, the spectral density takes the form of:

$$\mu_\alpha(u) = \frac{\pi}{\left(2^{d-1}\Gamma\left(\frac{d+1}{2}\right)\right)^2}\prod_{j=1}^{p}(u^2+\ell_j^2), \qquad \text{for } d = \text{even}, \tag{40}$$

$$\mu_\alpha(u) = \frac{\pi}{\left(2^{d-1}\Gamma\left(\frac{d+1}{2}\right)\right)^2} f(u)\, u \prod_{j=1}^{p}(u^2+\ell_j^2), \qquad \text{for } d = \text{odd}, \tag{41}$$

where:

$$f(u) = \begin{cases} \tanh(\pi u), & \ell_j = \text{half-integer}, \\ \coth(\pi u), & \ell_j = \text{integer}. \end{cases} \tag{42}$$

The pre-factor of $\frac{\pi}{\left(2^{d-1}\Gamma\left(\frac{d+1}{2}\right)\right)^2}$ arises as a normalization constant found by imposing the condition that as we approach flat space from hyperbolic space, the spectral density should approach that of flat space.

The number of degrees of freedom g_α is equal to the dimension of the corresponding representation of $SO(d)$, and is given by [66]:

$$g_{\alpha_s} = \prod_{1\leq i<j\leq p} \frac{m_i - m_j + j - i}{j-i} \prod_{1\leq i<j\leq p} \frac{m_i + m_j + 2p - i - j}{2p-i-j}, \quad \text{for } d = 2p, \tag{43}$$

and:

$$g_{\alpha_s} = \prod_{1\le i\le p} \frac{2m_i+2p-2i+1}{2p-2i+1} \prod_{1\le i<j\le p} \frac{m_i-m_j+j-i}{j-i} \times \prod_{1\le i<j\le p} \frac{m_i+m_j+2p-i-j+1}{2p-i-j+1}, \qquad \text{for } d = 2p+1, \tag{44}$$

where $\alpha = [m_1, \ldots, m_p]$. As an example, in the Type A case in AdS_{d+1}, the only representation we need to consider is $m_1 = s$, and for all $j \neq 1$, $m_j = 0$. This gives us:

$$\mu_{[s,0,\ldots,0]}(u) = \frac{\pi}{(2^{d-1}\Gamma(\frac{d+1}{2}))^2}\left[u^2 + \left(s+\frac{d-2}{2}\right)^2\right]\left|\frac{\Gamma(iu+\frac{d-2}{2})}{\Gamma(iu)}\right|^2$$
$$= \begin{cases} \dfrac{\pi}{\left(2^{d-1}\Gamma\left(\frac{d+1}{2}\right)\right)^2}\left[u^2+\left(s+\dfrac{d-2}{2}\right)^2\right]\displaystyle\prod_{j=0}^{(d-4)/2}(u^2+j^2), & d = \text{even}, \\ \dfrac{\pi}{\left(2^{d-1}\Gamma\left(\frac{d+1}{2}\right)\right)^2} u\tanh(\pi u)\left[u^2+\left(s+\dfrac{d-2}{2}\right)^2\right]\displaystyle\prod_{j=0}^{(d-5)/2}\left[u^2+(j+\frac{1}{2})^2\right], & d = \text{odd}. \end{cases} \tag{45}$$

and:

$$g_{[s,0,\ldots,0]} = \frac{(2s+d-2)(s+d-3)!}{(d-2)!s!}, \qquad d \geq 3. \tag{46}$$

The results agree with the formulas derived in [49] and used in [27].

In Type AB theories, we need the spectral density for fermion fields in the $\alpha = [s, 1/2, 1/2, \ldots, 1/2]$ representation. We find that the above general formulas for even and odd d can be expressed in the compact form valid for all d:

$$\mu_{[s,\frac{1}{2},\ldots,\frac{1}{2}]}(u) = \frac{\pi}{\left(2^{d-1}\Gamma\left(\frac{d+1}{2}\right)\right)^2}\left[u^2+\left(s+\frac{d-2}{2}\right)^2\right]\left|\frac{\Gamma\left(iu+\frac{d-1}{2}\right)}{\Gamma(iu+\frac{1}{2})}\right|^2, \tag{47}$$

and:

$$g_{[s,\frac{1}{2},\ldots,\frac{1}{2}]} = \frac{(s-\frac{5}{2}+d)!}{(s-\frac{1}{2})!(d-2)!}n_F(d), \qquad n_F(d) = \begin{cases} 2^{\frac{d-2}{2}}, & \text{if } d = \text{even}, \\ 2^{\frac{d-1}{2}}, & \text{if } d = \text{odd}. \end{cases} \tag{48}$$

The spectral densities for the mixed symmetry fields appearing in Type B and C theories can be obtained in a straightforward way from the above general formulas, and we present the explicit results in the next sections.

3.2. *Calculations in Even d*

3.2.1. Type B Theories

Spectrum

The non-minimal Type B higher spin theory, which is conjectured to be dual to the $U(N)$ singlet sector of the free Dirac fermion theory, contains towers of mixed symmetry gauge fields of all integer spins. From the spectrum given in (8), we obtain the total spectral zeta function:

$$\begin{aligned}\zeta^{\text{HS}}_{\text{type B}}(z) &= 2\zeta_{(\Delta=d-1;[0,0,\ldots,0])}(z) \\ &+2\sum_{s=1}^{\infty}\Big[\mathfrak{Z}_{(\Delta^{\text{ph}};[s,1,1,\ldots,1,0])}(z)+\mathfrak{Z}_{(\Delta^{\text{ph}};[s,1,1,\ldots,1,0,0])}(z)+\ldots+\mathfrak{Z}_{(\Delta^{\text{ph}};[s,1,0,\ldots,0])}(z) \\ &\qquad +\mathfrak{Z}_{(\Delta^{\text{ph}};[s,0,0,\ldots,0])}(z)\Big] \\ &+\sum_{s=1}^{\infty}\Big[\mathfrak{Z}_{(\Delta^{\text{ph}};[s,1,1,\ldots,1,1])}(z)+\mathfrak{Z}_{(\Delta^{\text{ph}};[s,1,1,\ldots,1,1,-1])}(z)\Big].\end{aligned} \tag{49}$$

In the third line of (49), the representations $[s,1,1,\ldots,1,1]$ and $[s,1,1,\ldots,1,-1]$ give the self-dual and anti-self-dual parts of the corresponding fields. At the level of the spectral ζ functions, they yield equal contributions.[10] Using the spectral zeta function formulas listed in Section 3.1.1 and summing over all representations given above, we find that for all even d:

$$\zeta^{\text{HS}}_{\text{type B}}(z) = \mathcal{O}(z^2)\,, \tag{50}$$

and consequently the one-loop free energy in the non-minimal Type B theory in even d exactly vanishes:

$$F^{(1)}_{\text{type B}} = 0\,. \tag{51}$$

There are various truncations to the non-minimal Type B theory that results in the Weyl, Majorana and Majorana–Weyl projections on the free fermionic CFT. While the Weyl projection can be applied in all even dimensions d, the Majorana projection can be applied in dimensions $d = 2,3,4,8,9 \pmod 8$, and the Majorana–Weyl projection only in dimensions $d = 2 \pmod 8$. An interesting example is $d = 10$ (AdS_{11}), where we can consider all four types of Type B theories.

Weyl projection The projection from the non-minimal Type B theory described above is slightly different when the theory is in $d = 4m$ or $d = 4m+2$. Using the results of [54] for the product of chiral fermion representations, we find[11]

$$\begin{aligned}\zeta^{\text{HS}}_{\text{type B Weyl}}(z) &= \begin{cases}\sum\limits_{s=1}^{\infty}\left[\zeta_{(\Delta^{\text{ph}};[s,0,0,\ldots,0])}(z) + \zeta_{(\Delta^{\text{ph}};[s,1,1,0,\ldots,0])}(z) + \ldots + \zeta_{(\Delta^{\text{ph}};[s,1,1,\ldots,1])}(z)\right], & \text{for } d = 4m+2,\\ \sum\limits_{s=1}^{\infty}\left[\zeta_{(\Delta^{\text{ph}};[s,0,0,\ldots,0])}(z) + \zeta_{(\Delta^{\text{ph}};[s,1,1,0,\ldots,0])}(z) + \ldots + \zeta_{(\Delta^{\text{ph}};[s,1,1,\ldots,1,0])}(z)\right], & \text{for } d = 4m,\end{cases}\\ &= \begin{cases}\sum\limits_{s=1}^{\infty}\sum\limits_{\substack{t_i\geq 0\\ t_i\geq t_{i+1}}}^{1}\zeta_{(\Delta^{\text{ph}};[s,t_1,t_1,\ldots,t_m,t_m])}(z), & \text{for } d = 4m+2,\\ \sum\limits_{s=1}^{\infty}\sum\limits_{\substack{t_i\geq 0\\ t_i\geq t_{i+1}}}^{1}\zeta_{(\Delta^{\text{ph}};[s,t_1,t_1,\ldots,t_{m-1},t_{m-1},0])}(z), & \text{for } d = 4m,\end{cases}\end{aligned} \tag{52}$$

Note that under this projection, there are no scalars in the spectrum. The case $d = 4$ (AdS_5) was already discussed in [28]. Summing over all representations, we find that for all even d:

$$\zeta^{\text{HS}}_{\text{type B Weyl}}(z) = \mathcal{O}(z^2)\,, \tag{53}$$

and so

$$F^{(1)}_{\text{type B Weyl}} = 0\,. \tag{54}$$

Minimal Theory (Majorana projection) The Majorana condition $\bar{\psi} = \psi^T C$, where C is the charge conjugation matrix, can be imposed in $d = 2,3,4,8,9 \pmod 8$, see for instance [67]. In these dimensions, we can consider the theory of N free Majorana fermions and impose an $O(N)$ singlet constraint. In $d = 6 \pmod 8$, provided one has an even number N of fermions, one can impose instead a symplectic Majorana condition $\bar{\psi}^i = \psi^T_j C\Omega^{ij}$, where C is the charge conjugation matrix and Ω^{ij} the antisymmetric symplectic metric. In this case, we consider the theory of N free symplectic Majoranas with a $USp(N)$ singlet constraint.

---

[10] Note that, technically, for all Type B theories the field of spin $s = 1$ in the tower of spins of representation $[s,1,\ldots]$ is not a gauge field. However, for conciseness we still use the symbol $\zeta_{(\Delta^{\text{ph}};[s,1,\ldots])}$ for these fields; the corresponding ghost contribution is zero, so it does not make a practical difference.

[11] To obtain this result, we note that in $d = 4m$, complex conjugation flips the chirality of a Weyl spinor, while in $d = 4m+2$ the Weyl representation is self-conjugate. Therefore, in order to obtain $U(N)$ invariant operators, we should use Equation (4.20) of [54] for $d = 4m$, and Equation (4.23) of the same reference for $d = 4m+2$.

The operator spectrum in the minimal theory can be deduced by working out which operators of the non-minimal theory are projected out by the Majorana constraint. The bilinear operators in the non-minimal theory are of the schematic form:

$$J_{\mu_1\cdots\mu_s,\nu_1\cdots\nu_{n-1}} \sim \bar{\psi}_i(\Gamma^{(n)})_{\nu_1\cdots\nu_{n-1}(\mu_1}\partial_{\mu_2}\partial_{\mu_3}\cdots\partial_{\mu_s)}\psi^i+\ldots \tag{55}$$

where $n=0,\ldots,\frac{d}{2}-1$, and $\Gamma^{(n)}$ is the antisymmetrized product of n gamma matrices. For Majorana fermions, we have $\bar{\psi}=\psi^T C$, and so the operators are projected out or kept depending on whether $C\Gamma^{(n)}$ is symmetric or antisymmetric. If $C\Gamma^{(n)}$ is symmetric, then the operators with an even number of derivatives (i.e., odd spin) are projected out; if it is antisymmetric, then the operators with an odd number of derivatives (i.e., even spin) are projected out. In addition to (55), the non-minimal Type B theories in even d include two scalars $J_0=\bar{\psi}_i\psi^i$ and $\tilde{J}_0=\bar{\psi}_i\gamma_*\psi^i$, where $\gamma_*\sim\Gamma^{(d)}$ is the chirality matrix. When C is symmetric, J_0 is projected out, and when $C\gamma_*$ is symmetric, $\tilde{J}_0$ is projected out.[12]

For instance, in $d=4$, the non-minimal theory contains the operators given in (10)–(12). In $d=4$, one has that both C and $C\gamma_5$ are antisymmetric, so both scalars in (10) are retained. Then, one has that $C\gamma^\mu$ is symmetric while $C\gamma^\mu\gamma^5$ antisymmetric, and so we keep the first tower in (11) for even s and the other tower for odd s: together, they make up a single tower in the $[s,0]$ representation with all integer spins. Finally, $C\Gamma_{\mu\nu}$ is symmetric, so we keep the mixed symmetry fields (12) with an odd number of derivatives, i.e., the spectrum contains the representations $[s,1]_c=[s,1]+[s,-1]$ for all even s.

Higher dimensions can be analyzed similarly, using the symmetry/antisymmetry properties of $C\Gamma^{(n)}$ in various d [67]. The results are summarized in Table 3. One finds that under the Majorana projections the operators with the "heaviest" weight $[s,1,1,\ldots,1]_c$ always form a tower containing all even s. The next representation $[s,1,\ldots,1,0]$ form a tower of all integer s. Then, $[s,1,\ldots,1,0,0]$ appears in two towers of all odd s. Finally, $[s,1,\ldots,1,0,0,0]$ form a tower of all integer spins, after which this cycle repeats. The number of scalars with $\Delta=d-1$ to be included also changes in a cycle of 4. In AdS_5, we have 2 scalars; in AdS_7, we have 1 (this case, though, should be discussed separately, see below); in AdS_9, we have 0; in AdS_{11}, we have 1, and the cycle repeats. In a more compact notation, the total spectral zeta function in the minimal Type B theories dual to the $O(N)$ Majorana theories is:

$$\begin{aligned}\zeta^{\rm HS}_{\text{type B Maj.}}(z) &= \chi(d)\zeta_{(d-1;[0,0,\ldots,0])}(z)\\
&+\sum_{s=2,4,6,\ldots}^{\infty}\ \sum^{1}_{\substack{t_i\geq t_{i+1}\\ t_i\geq 0\\ \sum_w t_w=w\,(\mathrm{mod}\ 4)}}\left(\mathfrak{Z}_{(\Delta;[s,t_1,t_2,\ldots,t_{w-1},t_w])}(z)+\mathfrak{Z}_{(\Delta;[s,t_1,t_2,\ldots,t_{w-1},-t_w])}(z)\right)\\
&+\sum_{s=1,2,3,\ldots}\ \sum^{1}_{\substack{t_i\geq t_{i+1}\\ t_i\geq 0\\ \sum_w t_w=(w-1)\,(\mathrm{mod}\ 4)}}\mathfrak{Z}_{(\Delta;[s,t_1,t_2,\ldots,t_{w-1},t_w])}(z)\\
&+\sum_{s=1,2,3,\ldots}\ \sum^{1}_{\substack{t_i\geq t_{i+1}\\ t_i\geq 0\\ \sum_w t_w=(w-3)\,(\mathrm{mod}\ 4)}}\mathfrak{Z}_{(\Delta^{\rm ph};[s,t_1,t_2,\ldots,t_{w-1},t_w])}(z)\\
&+\sum_{s=1,3,5,\ldots}\ \sum^{1}_{\substack{t_i\geq t_{i+1}\\ t_i\geq 0\\ \sum_w t_w=(w-2)\,(\mathrm{mod}\ 4)}}\left(\mathfrak{Z}_{(\Delta^{\rm ph};[s,t_1,t_2,\ldots,t_{w-1},t_w])}(z)+\mathfrak{Z}_{(\Delta^{\rm ph};[s,t_1,t_2,\ldots,t_{w-1},-t_w])}(z)\right)\end{aligned} \tag{56}$$

where $\chi(d)=1,2,0$ when $d=0,2,4\,(\mathrm{mod}\ 8)$ respectively. Explicit illustrations of this formula are given in Table 3. Using these spectra we find, in all even d where the Majorana condition is possible:

---

[12] As an example, consider the bilinear $\psi^T M\psi$. If M is symmetric, this operator clearly vanishes. On the other hand, consider $\psi^T M\partial_\mu\psi$. In this case, if M is an antisymmetric matrix, then this is equal to $+\partial_\mu\psi^T M\psi$. In turn, this means that $\psi^T M\partial_\mu\psi=\frac{1}{2}\partial_\mu(\psi^T M\psi)$, and so this operator is a total derivative and is not included in the spectrum of primaries.

$$F^{(1)}_{\text{type B Maj.}} = a_f \log R \tag{57}$$

where R is the radius of the boundary sphere, and a_f is the a-anomaly coefficient of a single Majorana fermion in dimension d, given in (18). As explained earlier, this is consistent with the duality, provided $G_N^{\text{type B Maj.}} \sim 1/(N-1)$. As mentioned above, in $d = 6$ (mod 8), i.e., $\text{AdS}_{7(\text{mod } 8)}$, we should impose a symplectic Majorana condition and consider the $USp(N)$ invariant operators. In terms of the operators (55), since $\bar{\psi} = \psi^T C\Omega$ with Ω antisymmetric, all this means is that now odd spins are projected out when $C\Gamma^{(n)}$ is antisymmetric, and even spins are projected out when $C\Gamma^{(n)}$ is symmetric. Similarly, the scalar operators $\bar{\psi}_i\psi^i$ and $\psi_i\gamma_*\psi^i$ are now projected out when C and $C\gamma_*$ are antisymmetric, respectively. In $d = 6$ (mod 8), one has that C is symmetric and $C\gamma_*$ is antisymmetric, so we retain a single scalar field. On the other hand, $C\gamma_\mu$ and $C\gamma_\mu\gamma_*$ are both antisymmetric, and so we have two towers of totally symmetric representations of all even s.[13] The projection of the mixed symmetry representations can be deduced similarly. The total spectral zeta function is given by the formula:

$$\begin{aligned}
\zeta^{\text{HS}}_{\text{type B Symp.Maj.}}(z) &= \zeta_{(d-1;[0,0,\ldots,0])}(z) \\
&+ \sum_{s=1,3,5,\ldots}^{\infty} \sum_{\substack{t_i \geq t_{i+1} \\ t_i \geq 0 \\ \sum_w t_w = w \,(\text{mod } 4)}}^{1} \left(3_{(\Delta^{\text{ph}};[s,t_1,t_2,\ldots,t_{w-1},t_w])}(z) + 3_{(\Delta^{\text{ph}};[s,t_1,t_2,\ldots,t_{w-1},-t_w])}(z)\right) \\
&+ \sum_{s=1,2,3,\ldots} \sum_{\substack{t_i \geq t_{i+1} \\ t_i \geq 0 \\ \sum_w t_w = (w-1) \,(\text{mod } 4)}}^{1} 3_{(\Delta^{\text{ph}};[s,t_1,t_2,\ldots,t_{w-1},t_w])}(z) \\
&+ \sum_{s=1,2,3,\ldots} \sum_{\substack{t_i \geq t_{i+1} \\ t_i \geq 0 \\ \sum_w t_w = (w-3) \,(\text{mod } 4)}}^{1} 3_{(\Delta^{\text{ph}};[s,t_1,t_2,\ldots,t_{w-1},t_w])}(z) \\
&+ \sum_{s=2,4,6,\ldots} \sum_{\substack{t_i \geq t_{i+1} \\ t_i \geq 0 \\ \sum_w t_w = (w-2) \,(\text{mod } 4)}}^{1} \left(3_{(\Delta^{\text{ph}};[s,t_1,t_2,\ldots,t_{w-1},t_w])}(z) + 3_{(\Delta^{\text{ph}};[s,t_1,t_2,\ldots,t_{w-1},-t_w])}(z)\right)
\end{aligned} \tag{58}$$

An illustration of the formula is given in Table 3 for the AdS_7 and AdS_{15} cases. Using these spectra, we find that the one loop free energy of the minimal Type B theory corresponding to the symplectic Majorana projection is given by:

$$F^{(1)}_{\text{type B sympl.Maj.}} = -a_f \log R\,, \tag{59}$$

i.e., the opposite sign compared to (57). This is consistent with the duality, provided $G_N^{\text{type B sympl.Maj.}} \sim 1/(N+1)$.

Majorana–Weyl Projection

Finally the spectra arising from the Majorana–Weyl projection, which can be imposed in dimensions $d = 2(\text{mod } 8)$, is the overlap of the individual Majorana and Weyl projection. The resulting spectrum yields the total zeta function:

[13] Note that, had we tried to impose the standard Majorana condition, we would have retained the totally symmetric fields of all odd spins. Then, the spectrum would not include a graviton, i.e., the dual CFT would not have a stress tensor.

$$\zeta^{\rm HS}_{\text{Type B MW}}(z) = \sum_{s=2,4,6,\ldots}^{\infty} \sum^{1}_{\substack{t_i \geq t_{i+1} \\ t_i \geq 0 \\ \sum_w t_w = w \, (\text{mod } 4)}} \mathfrak{Z}_{(\Delta;[s,t_1,t_2,\ldots,t_{w-1},t_w])}(z) + \sum_{s=1,3,5,\ldots} \sum^{1}_{\substack{t_i \geq t_{i+1} \\ t_i \geq 0 \\ \sum_w t_w = (w-2) \, (\text{mod } 4)}} \mathfrak{Z}_{(\Delta^{\rm ph};[s,t_1,t_2,\ldots,t_{w-1},t_w])}(z) \,. \tag{60}$$

Table 3. Projection of the non-minimal Type B theory to the Majorana/symplectic Majorana minimal Type B theory in even d. Notice that in AdS_7 and AdS_{15}, where we impose a symplectic Majorana projection, the pattern does not exactly follow the one seen in the other dimensions, as explained in the text. Instead, they are 'inverted', with the swapping of the towers for each weight from being only even integer spins to only odd integer spins. Their shift is highlighted in cyan. As defined earlier, the subscript 'c' indicates that both self-dual and anti-self-dual parts are included, corresponding to the weights $[t_1, \ldots, t_{k-1}, t_k]$ and $[t_1, \ldots, t_{k-1}, -t_k]$.

AdS_3 $O(N)$

α	$s=$		
	$1,2,3,\ldots$	$2,4,6,\ldots$	$1,3,5,\ldots$
$[s]$		2	
Scalar ($\Delta = 1$)		1	
$F^{(1)}$		$-\frac{1}{6}\log R$	

AdS_5 $O(N)$

α	$s=$		
	$1,2,3,\ldots$	$2,4,6,\ldots$	$1,3,5,\ldots$
$[s,1]_c$		1	
$[s,0]$	1		
Scalar ($\Delta = 3$)		2	
$F^{(1)}$		$\frac{11}{180}\log R$	

AdS_7 $USp(N)$

α	$s=$		
	$1,2,3,\ldots$	$2,4,6,\ldots$	$1,3,5,\ldots$
$[s,1,1]_c$			1
$[s,1,0]$	1		
$[s,0,0]$		2	
Scalar ($\Delta = 5$)		1	
$F^{(1)}$		$\frac{191}{7560}\log R$	

AdS_9 $O(N)$

α	$s=$		
	$1,2,3,\ldots$	$2,4,6,\ldots$	$1,3,5,\ldots$
$[s,1,1,1]_c$		1	
$[s,1,1,0]$	1		
$[s,1,0,0]$			2
$[s,0,0,0]$	1		
Scalar ($\Delta = 7$)		0	
$F^{(1)}$		$\frac{2497}{226{,}800}\log R$	

AdS_{11} $O(N)$

α	$s=$		
	$1,2,3,\ldots$	$2,4,6,\ldots$	$1,3,5,\ldots$
$[s,1,1,1,1]_c$		1	
$[s,1,1,1,0]$	1		
$[s,1,1,0,0]$			2
$[s,1,0,0,0]$	1		
$[s,0,0,0,0]$		2	
Scalar ($\Delta = 9$)		1	
$F^{(1)}$		$-\frac{14{,}797}{2{,}993{,}760}\log R$	

AdS_{13} $O(N)$

α	$s=$		
	$1,2,3,\ldots$	$2,4,6,\ldots$	$1,3,5,\ldots$
$[s,1,1,1,1,1]_c$		1	
$[s,1,1,1,1,0]$	1		
$[s,1,1,1,0,0]$			2
$[s,1,1,0,0,0]$	1		
$[s,1,0,0,0,0]$		2	
$[s,0,0,0,0,0]$	1		
Scalar ($\Delta = 11$)		2	
$F^{(1)}$		$\frac{92{,}427{,}157}{40{,}864{,}824{,}000}\log R$	

AdS_{15} $USp(N)$

α	$s=$		
	$1,2,3,\ldots$	$2,4,6,\ldots$	$1,3,5,\ldots$
$[s,1,1,1,1,1,1]_c$			1
$[s,1,1,1,1,1,0]$	1		
$[s,1,1,1,1,0,0]$		2	
$[s,1,1,1,0,0,0]$	1		
$[s,1,1,0,0,0,0]$			2
$[s,1,0,0,0,0,0]$	1		
$[s,0,0,0,0,0,0]$		2	
Scalar ($\Delta = 13$)		1	
$F^{(1)}$		$-\frac{36{,}740{,}617}{35{,}026{,}992{,}000}\log R$	

AdS_{17} $O(N)$

α	$s=$		
	$1,2,3,\ldots$	$2,4,6,\ldots$	$1,3,5,\ldots$
$[s,1,1,1,1,1,1,1]_c$		1	
$[s,1,1,1,1,1,1,0]$	1		
$[s,1,1,1,1,1,0,0]$			2
$[s,1,1,1,1,0,0,0]$	1		
$[s,1,1,1,0,0,0,0]$		2	
$[s,1,1,0,0,0,0,0]$	1		
$[s,1,0,0,0,0,0,0]$			2
$[s,0,0,0,0,0,0,0]$	1		
Scalar ($\Delta = 15$)		0	
$F^{(1)}$		$\frac{61{,}430{,}943{,}169}{125{,}046{,}361{,}440{,}000}\log R$	

AdS_{19} $O(N)$

α	$s=$		
	$1,2,3,\ldots$	$2,4,6,\ldots$	$1,3,5,\ldots$
$[s,1,1,1,1,1,1,1,1]_c$		1	
$[s,1,1,1,1,1,1,1,0]$	1		
$[s,1,1,1,1,1,1,0,0]$			2
$[s,1,1,1,1,1,0,0,0]$	1		
$[s,1,1,1,1,0,0,0,0]$		2	
$[s,1,1,1,0,0,0,0,0]$	1		
$[s,1,1,0,0,0,0,0,0]$			2
$[s,1,0,0,0,0,0,0,0]$	1		
$[s,0,0,0,0,0,0,0,0]$		2	
Scalar ($\Delta = 15$)		1	
$F^{(1)}$		$-\frac{23{,}133{,}945{,}892{,}303}{99{,}786{,}996{,}429{,}120{,}000}\log R$	

AdS_{21} $O(N)$

α	$s=$		
	$1,2,3,\ldots$	$2,4,6,\ldots$	$1,3,5,\ldots$
$[s,1,1,1,1,1,1,1,1,1]_c$		2	
$[s,1,1,1,1,1,1,1,1,0]$	1		
$[s,1,1,1,1,1,1,1,0,0]$			2
$[s,1,1,1,1,1,1,0,0,0]$	1		
$[s,1,1,1,1,1,0,0,0,0]$		2	
$[s,1,1,1,1,0,0,0,0,0]$	1		
$[s,1,1,1,0,0,0,0,0,0]$			2
$[s,1,1,0,0,0,0,0,0,0]$	1		
$[s,1,0,0,0,0,0,0,0,0]$		2	
$[s,0,0,0,0,0,0,0,0,0]$	1		
Scalar ($\Delta = 17$)		2	
$F^{(1)}$		$\frac{16{,}399{,}688{,}681{,}447}{149{,}003{,}207{,}337{,}600{,}000}\log R$	

An illustration of this can be seen in Table 4, where we list the spectra of AdS_{11} and AdS_{19}. Summing up over these spectra, we find the result:

$$F^{(1)}_{\text{type B MW}} = \frac{1}{2} a_f \log R \tag{61}$$

which is the a-anomaly coefficient of a single Majorana–Weyl fermion at the boundary.

In $d = 6$ (mod 8), one may impose a symplectic Majorana–Weyl projection. The resulting spectra are the overlap between the symplectic Majorana and Weyl projections. For instance, in $d = 6$ we find a minimal theory with a totally symmetric tower $[s, 0, 0]$ of all even spins, and a tower of the mixed symmetry fields $[s, 1, 1]$ of all odd spins (see Table 4). In this case (and similarly for higher $d = 14, 22, \ldots$), we find:

$$F^{(1)}_{\text{type B SMW}} = -\frac{1}{2} a_f \log R\,. \tag{62}$$

Since the a-anomaly of the boundary free theory of N symplectic Majorana–Weyl fermions is $a_{N\ \text{SMW}} = \frac{N}{2} a_f$, this result is consistent with a shift $G_N^{\text{type B SMW}} \sim 1/(N+1)$.

Table 4. Table of weights and their towers of spins for (top left) AdS_{11} and (top right) AdS_{19} under Majorana–Weyl projection, and for (bottom left) AdS_7 and (bottom right) AdS_{15} under the Symplectic Majorana–Weyl projection. There are no subscripts $_c$ for the $[s, 1, \ldots, 1]$ representations because the dual representations $[s, 1, \ldots, 1, -1]$ are not included.

AdS_{11} (Majorana–Weyl)

α	$s =$ 1,2,3,...	2,4,6,...	1,3,5,...
$[s,1,1,1,1]$		1	
$[s,1,1,0,0]$			1
$[s,0,0,0,0]$		1	
$F^{(1)}$	$\frac{-14{,}797}{5{,}987{,}520} \log R$		

AdS_{19} (Majorana–Weyl)

α	$s =$ 1,2,3,...	2,4,6,...	1,3,5,...
$[s,1,1,1,1,1,1,1,1]$		1	
$[s,1,1,1,1,1,1,0,0]$			1
$[s,1,1,1,1,0,0,0,0]$		1	
$[s,1,1,0,0,0,0,0,0]$			1
$[s,0,0,0,0,0,0,0,0]$		1	
$F^{(1)}$	$-\frac{23{,}133{,}945{,}892{,}303}{199{,}573{,}992{,}858{,}240{,}000} \log R$		

AdS_7 (Symplectic Majorana–Weyl)

α	$s =$ 1,2,3,...	2,4,6,...	1,3,5,...
$[s,1,1]$			1
$[s,0,0]$		1	
$F^{(1)}$	$\frac{191}{15{,}120} \log R$		

AdS_{15} (Symplectic Majorana–Weyl)

α	$s =$ 1,2,3,...	2,4,6,...	1,3,5,...
$[s,1,1,1,1,1,1]$			1
$[s,1,1,1,1,0,0]$		1	
$[s,1,1,0,0,0,0]$			1
$[s,0,0,0,0,0,0]$		1	
$F^{(1)}$	$-\frac{36{,}740{,}617}{70{,}053{,}984{,}000} \log R$		

Sample Calculations

AdS_5 Following (49) for the non-minimal Type B theory,

$$\zeta^{\text{HS}}_{\text{type B}}(z) = 2\zeta_{(3;[0,0])}(z) + \sum_{s=1}^{\infty} \left(3_{(\Delta^{\text{ph}};[s,1])}(z) + 3_{(\Delta^{\text{ph}};[s,-1])}(z) \right) + 2\sum_{s=1}^{\infty} 3_{(\Delta^{\text{ph}};[s,0])}(z). \tag{63}$$

We see that there are two weights to consider in AdS_5, corresponding to $[s, 0]$ and $[s, \pm 1]$ representation. Using (31) and (40), we have:

$$\begin{aligned}\frac{\zeta_{(\Delta;[s,1])}(z)}{\log R} &= \pi^2 \int_0^\infty du \frac{(u^2+1)\left[(s+1)^2+u^2\right]}{12\pi^3} \frac{s(2+s)}{\left[u^2+(\Delta-2)^2\right]^z} \\ &= \frac{s(s+2)}{8\sqrt{\pi}\Gamma(z)}\left[\frac{(s+1)^2(\Delta-2)^{1-2z}\Gamma(z-\frac{1}{2})}{3} + \frac{\left((s+1)^2+1\right)(\Delta-2)^{3-2z}\Gamma(z-\frac{3}{2})}{6}\right. \\ &\qquad \left. + \frac{(\Delta-2)^{5-2z}\Gamma(z-\frac{5}{2})}{4}\right]\end{aligned} \tag{64}$$

In the above, we made use of the formula,

$$\int_0^\infty du \frac{u^{2p}}{\left[u^2+\nu^2\right]^z} = \nu^{2p+1-2z}\int_0^\infty du \frac{u^{2p}}{\left[u^2+1\right]^z} = \nu^{2p+1-2z}\frac{\Gamma(p+\frac{1}{2})}{2}\frac{\Gamma(z-p-\frac{1}{2})}{\Gamma(z)}, \tag{65}$$

to go from the first to the second line.

In our regularization scheme, we sum over the physical modes separately from the ghost modes. We introduce $\zeta(k,\nu)$, the Hurwitz zeta function (analytically extended to the entire complex plane), which is given by:

$$\zeta(k,\nu) = \sum_{s=0}^{\infty} \frac{1}{(s+\nu)^k}. \tag{66}$$

Then, using $\Delta^{\mathrm{gh}} = s+3$,

$$\begin{aligned}&\frac{1}{\log R}\sum_{s=1}^\infty \zeta_{(\Delta^{\mathrm{gh}};[s-1,1])}(z) \\ &= \frac{1}{96\sqrt{\pi}\Gamma(z)}\Big[2\zeta(2z-7)\Gamma\left(z-\tfrac{3}{2}\right)+3\zeta(2z-7)\Gamma\left(z-\tfrac{5}{2}\right)+8\zeta(2z-6)\Gamma\left(z-\tfrac{3}{2}\right) \\ &\quad +6\zeta(2z-6)\Gamma\left(z-\tfrac{5}{2}\right)+4\zeta(2z-5)\Gamma\left(z-\tfrac{1}{2}\right)+12\zeta(2z-5)\Gamma\left(z-\tfrac{3}{2}\right) \\ &\quad +16\zeta(2z-4)\Gamma\left(z-\tfrac{1}{2}\right)+8\zeta(2z-4)\Gamma\left(z-\tfrac{3}{2}\right)+20\zeta(2z-3)\Gamma\left(z-\tfrac{1}{2}\right) \\ &\quad +8\zeta(2z-2)\Gamma\left(z-\tfrac{1}{2}\right)\Big].\end{aligned} \tag{67}$$

Similarly, using $\Delta^{\mathrm{ph}} = s+2$,

$$\begin{aligned}&\frac{1}{\log R}\sum_{s=1}^\infty \zeta_{(\Delta^{\mathrm{ph}};[s,1])}(z) \\ &= \frac{1}{96\sqrt{\pi}\Gamma(z)}\Big\{\Gamma\left(z-\tfrac{5}{2},1\right)\left[6\zeta(2z-6,1)+3\zeta(2z-7,1)\right] \\ &\quad +\Gamma\left(z-\tfrac{3}{2}\right)\left[12\zeta(2z-5,1)+2\zeta(2z-7,1)+8\zeta(2z-6,1)\Gamma+8\zeta(2z-4,1)\right] \\ &\quad +\left(z-\tfrac{1}{2}\right)\left[16\zeta(2z-4,1)\Gamma+8\zeta(2z-2,1)+20\zeta(2z-3,1)+4\zeta(2z-5,1)\right]\Big\}\end{aligned} \tag{68}$$

Putting (67) and (68) together,

$$\begin{aligned}&\frac{1}{\log R}\sum_{s=1}^\infty \left[\zeta_{(\Delta^{\mathrm{ph}};[s,1])}(z) - \zeta_{(\Delta^{\mathrm{gh}};[s-1,1])}(z)\right] \\ &= -\frac{1}{90}z + \frac{1}{180}\left[56\zeta'(-6)-160\zeta'(-4)-120\zeta'(-2)-2\gamma+3\psi\left(-\tfrac{5}{2}\right)-5\psi\left(-\tfrac{3}{2}\right)\right]z^2+\mathcal{O}(z^3)\end{aligned} \tag{69}$$

where $\psi(x)$ is the digamma function, γ the Euler–Mascheroni constant, and ζ' the derivative of the Riemann Zeta function $\zeta(z)$ (which is related to the Hurwitz Zeta function $\zeta(z) = \zeta(z,1)$). Similarly, for the totally symmetric representation $[s,0]$, we have:

$$\begin{aligned}
&\tfrac{1}{\log R}\textstyle\sum_{s=1}^{\infty}\mathfrak{Z}_{(\Delta^{\mathrm{ph}};[s,0])}(z)\\
&=\tfrac{1}{\log R}\textstyle\sum_{s=1}^{\infty}\left[\zeta_{(\Delta^{\mathrm{ph}};[s,0])}(z)-\zeta_{(\Delta^{\mathrm{gh}};[s-1,0])}(z)\right]\\
&=\tfrac{1}{24\sqrt{\pi}\Gamma(z)}\left[3\zeta(2(z-3))\Gamma\left(z-\tfrac{5}{2}\right)+4(\zeta(2(z-3))+\zeta(2(z-2)))\Gamma\left(z-\tfrac{3}{2}\right)\right]\\
&=\left(\tfrac{14}{45}\zeta'(-6)+\tfrac{4}{9}\zeta'(-4)\right)z^2+\mathcal{O}(z^3)
\end{aligned}\tag{70}$$

Finally, for the massive scalar with $\Delta=3$, we have:

$$\begin{aligned}
\frac{\zeta_{(3;[0,0])}(z)}{\log R}&=\int_0^{\infty}du\frac{(s+1)^2u^2\left[(s+1)^2+u^2\right]}{12\pi[(\Delta-2)^2+u^2]^z}\Bigg|_{\substack{\Delta=3,\\ s=0}}\\
&=\left[\frac{(s+1)^4(\Delta-2)^{3-2z}\Gamma\left(z-\frac{3}{2}\right)}{48\sqrt{\pi}\Gamma(z)}+\frac{(s+1)^2(\Delta-2)^{5-2z}\Gamma\left(z-\frac{5}{2}\right)}{32\sqrt{\pi}\Gamma(z)}\right]_{\substack{\Delta=3,\\ s=0}}\\
&=\frac{\Gamma\left(z-\frac{3}{2}\right)}{48\sqrt{\pi}\Gamma(z)}+\frac{\Gamma\left(z-\frac{5}{2}\right)}{32\sqrt{\pi}\Gamma(z)}\\
&=\tfrac{1}{90}z+\tfrac{1}{180}\left[2\gamma-3\psi\left(-\tfrac{5}{2}\right)+5\psi\left(-\tfrac{3}{2}\right)\right]z^2+\mathcal{O}(z^3).
\end{aligned}\tag{71}$$

When summing (69)–(71) together, there are no terms of order $\mathcal{O}(z^0)$ or $\mathcal{O}(z^1)$ in the sum, and hence, taking $z\to 0$, we obtain $F^{(1)}=0$ for the non-minimal Type B theory.

For the Type B minimal theory, we should evaluate, according to (56), the following sum:

$$\zeta^{\mathrm{HS}}_{\mathrm{Total-Type\ B}}(z)=2\zeta_{(3;[0,0])}(z)+\sum_{s=2,4,6,\ldots}\left(\mathfrak{Z}_{(\Delta^{\mathrm{ph}};[s,1])}(z)+\mathfrak{Z}_{(\Delta^{\mathrm{ph}};[s,-1])}(z)\right)+\sum_{s=1}^{\infty}\mathfrak{Z}_{(\Delta^{\mathrm{ph}};[s,0])}(z).\tag{72}$$

The first and third term of the sum have already been evaluated for in the non-minimal theory in (70) and (71) respectively. For the second term,

$$\begin{aligned}
&\textstyle\sum_{s=2,4,6,\ldots}\left(\mathfrak{Z}_{(\Delta^{\mathrm{ph}};[s,1])}+\mathfrak{Z}_{(\Delta^{\mathrm{ph}};[s,-1])}\right)\\
&=2\textstyle\sum_{s=2,4,6,\ldots}\left[\frac{(s+1)^2(s+2)s^{2-2z}\Gamma\left(z-\frac{1}{2}\right)}{24\sqrt{\pi}\Gamma(z)}+\frac{(s+2)\left((s+1)^2+1\right)s^{4-2z}\Gamma\left(z-\frac{3}{2}\right)}{48\sqrt{\pi}\Gamma(z)}\right.\\
&\qquad\left.+\frac{(s+2)s^{6-2z}\Gamma\left(z-\frac{5}{2}\right)}{32\sqrt{\pi}\Gamma(z)}\right]\log R\\
&\quad-2\textstyle\sum_{s=2,4,6,\ldots}\left[\frac{(s-1)s^2(s+1)^{2-2z}\Gamma\left(z-\frac{1}{2}\right)}{24\sqrt{\pi}\Gamma(z)}+\frac{(s-1)\left(s^2+1\right)(s+1)^{4-2z}\Gamma\left(z-\frac{3}{2}\right)}{48\sqrt{\pi}\Gamma(z)}\right.\\
&\qquad\left.+\frac{(s-1)(s+1)^{6-2z}\Gamma\left(z-\frac{5}{2}\right)}{32\sqrt{\pi}\Gamma(z)}\right]\log R
\end{aligned}\tag{73}$$

To illustrate the zeta-regularization, let us consider the last term,

$$\begin{aligned}
&\textstyle\sum_{s=2,4,6,\ldots}\frac{(s-1)(s+1)^{6-2z}\Gamma\left(z-\frac{5}{2}\right)}{32\sqrt{\pi}\Gamma(z)}\\
&=\textstyle\sum_{s=2,4,6,\ldots}\left[\frac{(s+1)^{7-2z}\Gamma\left(z-\frac{5}{2}\right)}{32\sqrt{\pi}\Gamma(z)}-2\frac{(s+1)^{6-2z}\Gamma\left(z-\frac{5}{2}\right)}{32\sqrt{\pi}\Gamma(z)}\right]\\
&=\textstyle\sum_{s=1,2,3,\ldots}\left[\frac{2^{7-2z}\left(s+\frac{1}{2}\right)^{7-2z}\Gamma\left(z-\frac{5}{2}\right)}{32\sqrt{\pi}\Gamma(z)}-2\frac{2^{6-2z}\left(s+\frac{1}{2}\right)^{6-2z}\Gamma\left(z-\frac{5}{2}\right)}{32\sqrt{\pi}\Gamma(z)}\right]\\
&=\frac{2^{2-2z}\zeta\left(2z-7,\frac{3}{2}\right)\Gamma\left(z-\frac{5}{2}\right)}{\sqrt{\pi}\Gamma(z)}-\frac{2^{2-2z}\zeta\left(2z-6,\frac{3}{2}\right)\Gamma\left(z-\frac{5}{2}\right)}{\sqrt{\pi}\Gamma(z)}
\end{aligned}\tag{74}$$

where on the second line we used the substitution $s \to 2s$, followed by rewriting $2s+1 = 2(s+\frac{1}{2})$.[14] The partial results coming from summing each tower are given in Table A3. Putting everything together, we obtain $F^{(1)}_{\text{type B Maj.}} = \frac{11}{180}\log R = a_f^{d=4}\log R$, which agrees with the results of [16].

Finally, for the Weyl truncated theory,

$$\zeta^{\text{HS}}_{\text{type B Weyl}}(z) = \sum_{s=1}^{\infty} \mathfrak{Z}_{(\Delta^{\text{ph}};[s,0])}(z) = \mathcal{O}(z^2), \tag{75}$$

which gives us $F^{(1)} = 0$.

AdS$_{11}$ We skip the $d = 7,9$ case, whose spectrum for the various theories follow from the discussion in Section 3.3.3. For reference, the calculated free energy of each weight $F^{(1)}$ is given in Tables A4 and A5.

Instead, let us consider the $d = 11$ case, where we can compare the four different types of fermions: non-minimal ($U(N)$), Weyl, minimal ($O(N)$), and Majorana–Weyl. The calculations of $F^{(1)}$ for each the various weights and their spectra are given in Table A6. In the non-minimal and Weyl projected theories, the bulk $F^{(1)}$ contributions sum to zero, whereas in the minimal and Majorana–Weyl theories, the bulk $F^{(1)}$ contributions are $-14,797/2,993,760\log R$ and $-14,797/5,987,520\log R$ respectively. The numerical parts of these free energies correspond exactly to the values of the free energy of one real fermion, $-14,797/2,993,760$ and one real Weyl fermion on S^{10}, $-14,797/5,987,520$.

3.2.2. Fermionic Higher Spins in Type AB Theories

Spectrum

We described earlier that there is only one irrep of $SO(d)$ of interest here that describes the tower of spins corresponding to the fermionic bilinears in Type AB theories, namely $\alpha_s = [s,\frac{1}{2},\frac{1}{2},\ldots,\frac{1}{2}]$. Therefore, in the non-minimal theories dual to complex scalars and fermions in the $U(N)$ singlet sector, the purely fermionic contribution to the total zeta function is:

$$\zeta^{\text{HS}}_{\text{type AB ferm}}(z) = 2\zeta^{\text{HS}}_{(\Delta_{1/2};[\frac{1}{2},\frac{1}{2},\frac{1}{2},\ldots,\frac{1}{2}])}(z) + 2\sum_{s=\frac{3}{2},\frac{5}{2},\ldots}^{\infty} \mathfrak{Z}_{(\Delta^{\text{ph}};[s,\frac{1}{2},\frac{1}{2},\ldots,\frac{1}{2}])}(z), \tag{76}$$

where $\Delta_{1/2} = \frac{1}{2} + d - 2 = d - \frac{3}{2}$. Thus, the spectrum of spins gives us a massive Dirac fermion[15], and a tower of complex massless higher-spin fermionic fields.[16]

Sample Calculation: AdS$_5$

After collecting our equations following (76), we have,

$$\zeta^{\text{HS}}_{\text{type AB ferm}}(z) = 2\zeta^{\text{HS}}_{(\Delta_{1/2};[\frac{1}{2},\frac{1}{2}])}(z) + 2\sum_{s=\frac{3}{2},\frac{5}{2},\ldots}^{\infty} \mathfrak{Z}_{(\Delta^{\text{ph}};[s,\frac{1}{2}])}(z), \tag{77}$$

with $\Delta^{\text{ph}} = 2 + s$. For the massive fermion contribution,

$$\zeta_{(\Delta_{1/2};[s,\frac{1}{2}])}(z) = \frac{2^{2z-10}\left(36\Gamma\left(z-\frac{1}{2}\right) + 20\Gamma\left(z-\frac{3}{2}\right) + 3\Gamma\left(z-\frac{5}{2}\right)\right)}{3\sqrt{\pi}\Gamma(z)}. \tag{78}$$

---

14 Similar shifts and scaling will be applied in the higher dimensional Type B cases, as well as the Type AB and C cases, and details of transformations to the Hurwitz-zeta function can be found in Appendix B.1.

15 With mass $|m| = (\Delta_{1/2} - d/2)/2 = (d-3)/4$.

16 The factor of two in (76) just accounts for the fact that the representations are complex.

Then,

$$\frac{\zeta_{(\Delta;[s,\frac{1}{2}])}(z)}{\log R} = \frac{(s+\frac{1}{2})(s+\frac{3}{2})}{48\pi(\Delta-2)^{2z-1}\Gamma(z)} \times \Big[(\Delta-2)^4\Gamma\left(\tfrac{5}{2}\right)\Gamma\left(z-\tfrac{5}{2}\right)+\Gamma\left(\tfrac{3}{2}\right)\Gamma\left(z-\tfrac{3}{2}\right)(\Delta-2)^2\left(\tfrac{1}{4}+(s+1)^2\right) +\Gamma\left(\tfrac{1}{2}\right)\Gamma\left(z-\tfrac{1}{2}\right)\tfrac{(s+1)^2}{4}\Big]. \tag{79}$$

This gives us,

$$\begin{aligned}&\textstyle\sum_{s=\frac{3}{2},\frac{5}{2},\ldots}\zeta_{(\Delta^{\text{ph}};[s,\frac{1}{2}])}(z)\\ &=\tfrac{1}{1536\sqrt{\pi}\Gamma(z)}\left\{6\left[4\zeta\left(2z-7,\tfrac{3}{2}\right)+8\zeta\left(2z-6,\tfrac{3}{2}\right)+3\zeta\left(2z-5,\tfrac{3}{2}\right)\right]\Gamma\left(z-\tfrac{5}{2}\right)\right.\\ &+\left[16\zeta\left(2z-7,\tfrac{3}{2}\right)+64\zeta\left(2z-6,\tfrac{3}{2}\right)+96\zeta\left(2z-5,\tfrac{3}{2}\right)+64\zeta\left(2z-4,\tfrac{3}{2}\right)\right.\\ &\left.+15\zeta\left(2z-3,\tfrac{3}{2}\right)\right]\Gamma\left(z-\tfrac{3}{2}\right)+2\left[4\zeta\left(2z-5,\tfrac{3}{2}\right)+16\zeta\left(2z-4,\tfrac{3}{2}\right)\right.\\ &\left.\left.+23\zeta\left(2z-3,\tfrac{3}{2}\right)+14\zeta\left(2z-2,\tfrac{3}{2}\right)+3\zeta\left(2z-1,\tfrac{3}{2}\right)\right]\Gamma\left(z-\tfrac{1}{2}\right)\right\}.\end{aligned} \tag{80}$$

The technicalities of the shift to the Hurwitz Zeta function in the sum above is similar to the case for the minimal Type B theory in AdS_5 which we worked out earlier. More details can be found in Appendix B.1. Similarly,

$$\begin{aligned}&\textstyle\sum_{s=\frac{3}{2},\frac{5}{2},\ldots}\zeta_{(\Delta^{\text{gh}};[s-1,\frac{1}{2}])}(z)\\ &=\tfrac{1}{1536\sqrt{\pi}\Gamma(z)}\left\{6\left[4\zeta\left(2z-7,\tfrac{5}{2}\right)-8\zeta\left(2z-6,\tfrac{5}{2}\right)+3\zeta\left(2z-5,\tfrac{5}{2}\right)\right]\Gamma\left(z-\tfrac{5}{2}\right)\right.\\ &+\left[16\zeta\left(2z-7,\tfrac{5}{2}\right)-64\zeta\left(2z-6,\tfrac{5}{2}\right)+96\zeta\left(2z-5,\tfrac{5}{2}\right)-64\zeta\left(2z-4,\tfrac{5}{2}\right)\right.\\ &\left.+15\zeta\left(2z-3,\tfrac{5}{2}\right)\right]\Gamma\left(z-\tfrac{3}{2}\right)+2\left[4\zeta\left(2z-5,\tfrac{5}{2}\right)-16\zeta\left(2z-4,\tfrac{5}{2}\right)\right.\\ &\left.\left.+23\zeta\left(2z-3,\tfrac{5}{2}\right)-14\zeta\left(2z-2,\tfrac{5}{2}\right)+3\zeta\left(2z-1,\tfrac{5}{2}\right)\right]\Gamma\left(z-\tfrac{1}{2}\right)\right\}.\end{aligned} \tag{81}$$

Quite clearly, the Hurwitz-zeta function shifts differently for the physical and ghost modes. Adding all three contributions and expanding near $z=0$,

$$\zeta^{\text{HS}}_{\text{type AB ferm}}(z)=\mathcal{O}(z^2) \tag{82}$$

which implies that $F^{(1)}_{\text{type AB ferm}}=0$, consistently with the duality.

For reference, we also report the expected expression of $\zeta^{\text{HS}}_{\text{type AB ferm}}(z)$ for AdS_7 and AdS_9, expanded in z up to the second order, in Appendix C.2.

3.2.3. Type C Theories

Calculations for Type C theories are similar to those described above and we will not go through all details explicitly. In the following sections, we list the spectrum of fields in these theories, including their various possible truncations. The free energy contributions in a few explicit examples are collected for reference in Appendix C.3.

Spectrum

The spectrum of the non-minimal type C theories, dual to the free theory of N complex $d/2$-form gauge fields with $U(N)$ singlet constraint, can be obtained from the character formulas derived in [54]. While the resulting spectra may look complicated, they follow a clear pattern that can be rather easily identified if one refers to the tables given in Appendix C.3. The results are split into the cases $d = 4m$ and $d = 4m + 2$. For $d = 4m$, the total spectral zeta function is given by[17,18]

$$\begin{aligned}\zeta^{\text{HS}}_{\text{type C}}(z) &= 2 \sum_{\substack{k_i \geq 0 \\ k_i \geq k_{i+1}}}^{1} \zeta_{(4m;[k_1,k_1,\ldots,k_m,k_m])}(z) \\ &+ \sum_{s=2}^{\infty} \left[\sum_{\substack{t_i \geq 0 \\ t_i \geq t_{i+1}}}^{2} 2\mathfrak{Z}_{(\Delta^{\text{ph}};[s,t_1,t_1,t_2,t_2,\ldots,t_{m-1},t_{m-1},0])}(z) \right. \\ &+ \left. \sum_{\substack{j_i \geq 0 \\ j_i \geq j_{i+1}}}^{2} \left(\mathfrak{Z}_{(\Delta^{\text{ph}};[s,2,j_1,j_1,j_2,j_2,\ldots,j_{m-1},+j_{m-1}])}(z) + \mathfrak{Z}_{(\Delta^{\text{ph}};[s,2,j_1,j_1,j_2,j_2,\ldots,j_{m-1},-j_{m-1}])}(z) \right) \right]\end{aligned} \tag{83}$$

and for $d = 4m + 2$:

$$\begin{aligned}\zeta^{\text{HS}}_{\text{type C}}(z) &= 2 \sum_{\substack{k_i \geq 0 \\ k_i \geq k_{i+1}}}^{1} \zeta_{(4m+2;[k_1,k_1,\ldots,k_m,k_m,0])}(z) \\ &+ \sum_{s=2}^{\infty} \left[\sum_{\substack{t_i \geq 0 \\ t_i \geq t_{i+1}}}^{2} 2\mathfrak{Z}_{(\Delta^{\text{ph}};[s,2,t_1,t_1,t_2,t_2,\ldots,t_{m-1},t_{m-1},0])}(z) \right. \\ &+ \left. \sum_{\substack{j_i \geq 0 \\ j_i \geq j_{i+1}}}^{2} \left(\mathfrak{Z}_{(\Delta^{\text{ph}};[s,j_1,j_1,j_2,j_2,\ldots,j_{m-1},+j_{m-1}])}(z) + \mathfrak{Z}_{(\Delta^{\text{ph}};[s,j_1,j_1,j_2,j_2,\ldots,j_{m-1},-j_{m-1}])}(z) \right) \right]\end{aligned} \tag{84}$$

Using these spectra and (40) to compute the zeta functions, we find the results:

$$\begin{aligned}F^{(1)}_{\text{type C}} &= 2a_{d/2-\text{form}} \log R\,, \quad d = 4m \\ F^{(1)}_{\text{type C}} &= -2a_{d/2-\text{form}} \log R\,, \quad d = 4m + 2\end{aligned} \tag{85}$$

where $a_{d/2-\text{form}}$ is the a-anomaly coefficient of a single real $(d/2 - 1)$-form gauge field in dimension d. The first few values in $d = 4, 6, 8, \ldots$ read [69]:

$$a_{d/2-\text{form}} = \left\{ \frac{62}{90}, -\frac{221}{210}, \frac{8051}{5670}, -\frac{1,339,661}{748,440}, \frac{525,793,111}{243,243,000}, -\frac{3,698,905,481}{1,459,458,000}, \ldots \right\}. \tag{86}$$

Thus, we see that (85) is consistent with the duality provided $G_N^{\text{type C}} \sim 1/(N-1)$ in $d = 4m$, and $G_N^{\text{type C}} \sim 1/(N+1)$ in $d = 4m + 2$.

Minimal Type C $O(N)$ The "minimal type C" theory corresponds to the $O(N)$ singlet sector of the free theory of N $(d/2 - 1)$-form gauge fields. Its spectrum can be in principle obtained by appropriately "symmetrizing" the character formulas given in [54] and used above to obtain the non-minimal spectrum. The spectra in $d = 4$ and $d = 6$ were obtained in [16–18]. Generalizing those results for all d, we arrive at the following total spectral zeta functions. In $d = 4m$,

---

17 (83) and (84) correspond to Equations (4.20) and (4.21) and (4.22) and (4.23) of [54] respectively, and the tensorial decomposition in these quoted equations can be further simplified by the formulas on p. 104 of [68].

18 For all Type C theories, the field of spin $s = 2$ in the towers of spins of representation $[s, 2, \ldots]$ are not gauge fields, but we will still use the symbol $\mathfrak{Z}$ for conciseness. See footnote 10 for similar remarks.

$$
\begin{aligned}
\zeta^{\text{HS}}_{\text{min. type C}}(z) &= 2\sum_{\substack{k_i \geq 0 \\ k_i \geq k_{i+1}}}^{1} \zeta_{(4m;[k_1,k_1,k_1,k_1,\ldots,k_{\lfloor \frac{m}{2} \rfloor},k_{\lfloor \frac{m}{2} \rfloor},k_{\lfloor \frac{m}{2} \rfloor},k_{\lfloor \frac{m}{2} \rfloor},0])}(z) \\
&+ \sum_{s=2}^{\infty} \sum_{\substack{t_i \geq 0 \\ t_i \geq t_{i+1}}}^{2} \mathfrak{z}_{(\Delta^{\text{ph}};[s,t_1,t_1,t_2,t_2,\ldots,t_{m-1},t_{m-1},0])}(z) \\
&+ \sum_{s=2,4,6,\ldots} \sum_{\substack{j_i \geq 0 \\ j_i \geq j_{i+1} \\ \sum_i j_i = 0 \ (\text{mod } 2)}}^{2} \left(\mathfrak{z}_{(\Delta^{\text{ph}};[s,2,j_1,j_1,j_2,j_2,\ldots,j_{m-1},+j_{m-1}])}(z) + \mathfrak{z}_{(\Delta^{\text{ph}},s;[s,2,j_1,j_1,j_2,j_2,\ldots,j_{m-1},-j_{m-1}])}(z) \right) \\
&+ \sum_{s=3,5,7,\ldots} \sum_{\substack{j_i \geq 0 \\ j_i \geq j_{i+1} \\ \sum_i j_i = 1 \ (\text{mod } 2)}}^{2} \left(\mathfrak{z}_{(\Delta^{\text{ph}};[s,2,j_1,j_1,j_2,j_2,\ldots,j_{m-1},+j_{m-1}])}(z) + \mathfrak{z}_{(\Delta^{\text{ph}};[s,2,j_1,j_1,j_2,j_2,\ldots,j_{m-1},-j_{m-1}])}(z) \right)
\end{aligned}
\tag{87}
$$

and in $d = 4m + 2$,

$$
\begin{aligned}
\zeta^{\text{HS}}_{\text{min. type C}}(z) &= \sum_{\substack{k_i \geq 0 \\ k_i \geq k_{i+1}}}^{1} \zeta_{(4m+2;[k_1,k_1,\ldots,k_m,k_m,0])}(z) \\
&+ \sum_{s=2}^{\infty} \sum_{\substack{t_i \geq 0 \\ t_i \geq t_{i+1}}}^{2} \mathfrak{z}_{(\Delta^{\text{ph}};[s,2,t_1,t_1,t_2,t_2,\ldots,t_{m-1},t_{m-1},0])}(z) \\
&+ \sum_{s=2,4,6,\ldots} \sum_{\substack{j_i \geq 0 \\ j_i \geq j_{i+1} \\ \sum_i j_i = 0 \ (\text{mod } 2)}}^{2} \left(\mathfrak{z}_{(\Delta^{\text{ph}};[s,j_1,j_1,j_2,j_2,\ldots,j_m,+j_m])}(z) + \mathfrak{z}_{(\Delta^{\text{ph}};[s,j_1,j_1,j_2,j_2,\ldots,j_m,-j_m])}(z) \right) \\
&+ \sum_{s=3,5,7,\ldots} \sum_{\substack{j_i \geq 0 \\ j_i \geq j_{i+1} \\ \sum_i j_i = 1 \ (\text{mod } 2)}}^{2} \left(\mathfrak{z}_{(\Delta^{\text{ph}};[s,j_1,j_1,j_2,j_2,\ldots,j_m,+j_m])}(z) + \mathfrak{z}_{(\Delta^{\text{ph}};[s,j_1,j_1,j_2,j_2,\ldots,j_m,-j_m])}(z) \right),
\end{aligned}
\tag{88}
$$

where $\lfloor \frac{m}{2} \rfloor$ denotes the integer part of $\frac{m}{2}$.

As a consistency check of these spectra, in Appendix A.1 we computed the corresponding partition functions in thermal AdS with $S^1 \times S^{d-1}$ boundary. After summing up over all representations appearing in the zeta functions above, the result matches the (symmetrized) square of the one-particle partition function of a $(d/2-1)$-form gauge field, see Equation (A29).

Evaluating the spectral zeta functions with the help of the formulas in Section 3.1.1, we obtain the results:

$$
\begin{aligned}
F^{(1)}_{\text{min. type C SD}} &= 2a_{d/2-\text{form}} \log R\,, \qquad & d &= 4m \\
F^{(1)}_{\text{min. type C SD}} &= 0\,, \qquad & d &= 4m+2
\end{aligned}
\tag{89}
$$

These correspond to the shifts given in Table 1. Interestingly, in the minimal type C theory in $d = 6, 10, \ldots$ the bulk one-loop free energy vanishes and no shift of the coupling constant is required.

Self-dual $U(N)$ In $d = 4m$, we can impose a self-duality constraint $F^i = i * F^i$ in the theory of N complex p-forms. The resulting spectrum of $U(N)$ invariant bilinears leads to the following total zeta function in the bulk[19]

$$\zeta^{\rm HS}_{\text{type C SD}}(z) = \sum_{s=2}^{\infty} \sum_{\substack{t_i \geq 0 \\ t_i \geq t_{i+1}}}^{2} \mathfrak{Z}_{(\Delta^{\rm ph};[s,t_1,t_1,t_2,t_2,\ldots,t_{m-1},t_{m-1},0])}(z)\,. \tag{90}$$

In $d = 4m + 2$, we can impose the self-duality condition $F^i = *F^i$, and the resulting truncated spectrum gives the following total zeta function[20]

$$\zeta^{\rm HS}_{\text{type C SD}}(z) = \sum_{s=2}^{\infty} \sum_{\substack{j_i \geq 0 \\ j_i \geq j_{i+1}}}^{2} \mathfrak{Z}_{(\Delta^{\rm ph};[s,j_1,j_1,j_2,j_2,\ldots,j_{m-1},+j_{m-1}])}(z)\,. \tag{91}$$

Using these spectra, we find the results:

$$\begin{aligned} F^{(1)}_{\text{type C SD}} &= \frac{1}{2} a_{d/2-\text{form}} \log R\,, \qquad && d = 4m \\ F^{(1)}_{\text{type C SD}} &= -\frac{1}{2} a_{d/2-\text{form}} \log R\,, \qquad && d = 4m+2 \end{aligned} \tag{92}$$

which correspond to the shifts given in Table 1.

Self-dual $O(N)$ In $d = 4m + 2$, we can impose a self-duality condition on the theory of N real forms with $O(N)$ singlet constraint. The spectrum is given by the "overlap" of the minimal type C and self-dual $U(N)$ spectra given above. The resulting total zeta function is given by:

$$\begin{aligned} \zeta^{\rm HS}_{\text{min. type C SD}}(z) = & \sum_{s=2,4,6,\ldots} \sum_{\substack{j_i \geq 0 \\ j_i \geq j_{i+1} \\ \sum_i j_i = 0 \,(\text{mod } 2)}}^{2} \mathfrak{Z}_{(\Delta;[s,j_1,j_1,j_2,j_2,\ldots,j_m,+j_m])}(z) \\ & + \sum_{s=3,5,7,\ldots} \sum_{\substack{j_i \geq 0 \\ j_i \geq j_{i+1} \\ \sum_i j_i = 1 \,(\text{mod } 2)}}^{2} \mathfrak{Z}_{(\Delta;[s,j_1,j_1,j_2,j_2,\ldots,j_m,+j_m])}(z) \end{aligned} \tag{93}$$

from which we find the result:

$$F^{(1)}_{\text{min. type C SD}} = \frac{1}{4} a_{d/2-\text{form}} \log R\,, \quad d = 4m + 2\,. \tag{94}$$

3.3. Calculations in Odd d

3.3.1. Preliminaries

Alternate Regulators

In the calculations for even d discussed above, we chose to sum over the spins before sending the spectral parameter $z \to 0$. This analytic continuation in z is a natural way to regulate the sums. In practice, this is possible in the even d case because the spectral density is polynomial in the integrating variable u. In the case of odd d, summing before sending $z \to 0$ is not easy to do, and we

[19] This corresponds to Equation (4.20) in [54]. This is because in $d = 4m$ complex conjugation maps self-dual to anti self-dual forms.

[20] This corresponds to Equation (4.23) in [54].

will instead first send $z \to 0$ and then evaluate the regularized sums over spins. There are two equivalent ways to do this. The first involves using exponential factors to suppress the spins:

$$\sum_{\substack{\text{all spins}\\ \text{in } \alpha_s}} \left[\mathfrak{Z}_{(\Delta^{\text{ph}};\alpha_s)}(z)\Big|_{z=0} \right] = \lim_{\epsilon\to 0} \sum_{\substack{\text{all spins}\\ \text{in } \alpha_s}} e^{-\epsilon(\Delta^{\text{ph}}-\frac{d}{2})} (\zeta_{(\Delta^{\text{ph}};\alpha_s)})(0) - \lim_{\epsilon\to 0} \sum_{\substack{\text{all spins}\\ \text{in } \alpha_s}} e^{-\epsilon(\Delta^{\text{gh}}-\frac{d}{2})} (\zeta_{(\Delta^{\text{gh}};\alpha_{s-1})})(0), \tag{95}$$

and similarly:

$$\sum_{\substack{\text{all spins}\\ \text{in } \alpha_s}} \left[\frac{\partial}{\partial z}\mathfrak{Z}_{(\Delta^{\text{ph}};\alpha_s)}(z)\Big|_{z=0} \right] = \lim_{\epsilon\to 0} \sum_{\substack{\text{all spins}\\ \text{in } \alpha_s}} e^{-\epsilon(\Delta^{\text{ph}}-\frac{d}{2})} (\zeta_{(\Delta^{\text{ph}};\alpha_s)})'(0) - \lim_{\epsilon\to 0} \sum_{\substack{\text{all spins}\\ \text{in } \alpha_s}} e^{-\epsilon(\Delta^{\text{gh}}-\frac{d}{2})} (\zeta_{(\Delta^{\text{gh}};\alpha_{s-1})})'(0), \tag{96}$$

where we recall that $\Delta^{\text{ph}} = s + d - 2$ and $\Delta^{\text{gh}} = s + d - 1$. In even d, one can show that this procedure, with the shifted exponentials as above, gives the same result as first summing over all representations and then sending the spectral parameter $z \to 0$. Equivalently, instead of the exponential regulators, one can use the analytic continuation of the Hurwitz zeta function by evaluating:

$$\sum_{\substack{\text{all spins}\\ \text{in } \alpha_s}} \left[\mathfrak{Z}_{(\Delta^{\text{ph}};\alpha_s)}(z)\Big|_{z=0} \right] = \lim_{\epsilon\to 0} \sum_{\substack{\text{all spins}\\ \text{in } \alpha_s}} \left(\Delta^{\text{ph}} - \tfrac{d}{2}\right)^{-\epsilon} (\zeta^{\text{HS}}_{(\Delta^{\text{ph}},s;\alpha_s)})(0) - \lim_{\epsilon\to 0} \sum_{\substack{\text{all spins}\\ \text{in } \alpha_s}} \left(\Delta^{\text{gh}} - \tfrac{d}{2}\right)^{-\epsilon} (\zeta^{\text{HS}}_{(\Delta^{\text{gh}};\alpha_{s-1})})(0), \tag{97}$$

and:

$$\sum_{\substack{\text{all spins}\\ \text{in } \alpha_s}} \left[\frac{\partial}{\partial z}\mathfrak{Z}_{(\Delta^{\text{ph}};\alpha_s)}(z)\Big|_{z=0} \right] = \lim_{\epsilon\to 0} \sum_{\substack{\text{all spins}\\ \text{in } \alpha}} \left(\Delta^{\text{ph}} - \tfrac{d}{2}\right)^{-\epsilon} (\zeta_{(\Delta^{\text{ph}};\alpha_s)})'(0) - \lim_{\epsilon\to 0} \sum_{\substack{\text{all spins}\\ \text{in } \alpha_s}} \left(\Delta^{\text{gh}} - \tfrac{d}{2}\right)^{-\epsilon} (\zeta^{\text{HS}}_{(\Delta^{\text{gh}};\alpha_{s-1})})'(0), \tag{98}$$

This method, which is closely related to the one previously used in [27],[21] will be described in the next sections in greater detail.

Note that, while in even d $\zeta_{(\Delta_s;\alpha_s)}(0)$ vanishes identically for any representation, this is not true in odd d. Vanishing of the logarithmic divergence in the one-loop free energy requires in this case summing over all the bulk fields, as reviewed below.

Integrals

In all odd d calculations, we encounter the integrals of the type:

$$\int_0^\infty \frac{u^k}{e^{2\pi u} \pm 1} \log[u^2 + b^2] = \int_0^\infty \frac{u^k\, du}{e^{2\pi u} \pm 1} \left[\log(u^2) + \int_0^{b^2} \frac{1}{u^2 + x} dx \right]. \tag{99}$$

We define:

$$A_k^{\pm}(x) \equiv \int_0^\infty \frac{u^k}{e^{2\pi u} \pm 1} \frac{du}{u^2 + x}, \qquad B_k^{\pm} \equiv \int_0^\infty \frac{u^k}{e^{2\pi u} \pm 1}. \tag{100}$$

[21] In that paper, an "averaged" regulator of $(\frac{\Delta^{\text{ph}}+\Delta^{\text{gh}}}{2} - \frac{d}{2})^{-\epsilon}$ was preferred for the Type A theory calculations, and it can be shown to give the same result as the regulators (97) and (98) that we will use in our calculations. In Type AB theories, however, it appears that "averaged" regulator does not work, and we will use the shifts defined in (97) and (98) in all theories consistently.

There exists a recursive relation between the various A_k's and B_k's for any odd integer $2k+1$ (see Appendix B.2 for a proof):

$$A_{2k+1}^{\pm}(x) = (-x)^k A_1^{\pm}(x) + \sum_{j=1}^{k} (-x)^{k-j} B_{2j-1}^{\pm}. \tag{101}$$

As a consequence of this relation, we only need the explicit analytic expressions of the integrals $A_1^{\pm}$,[22] which is given by [49]:

$$A_1^{+}(x) = \frac{1}{2}\left[-\log(\sqrt{x}) + \psi\left(\sqrt{x}+\frac{1}{2}\right)\right] \tag{103}$$

$$A_1^{-}(x) = \frac{1}{2}\left[\log(\sqrt{x}) - \frac{1}{2\sqrt{x}} - \psi(\sqrt{x})\right], \tag{104}$$

where $\psi(x)$ is the digamma function $\psi(x) = \Gamma'(x)/\Gamma(x)$.

3.3.2. Calculational Method and Type A Example

To illustrate the method of calculation, we first review the calculation in the non-minimal Type A theory [26,27]. The calculations for the various Type B theories are similar and we will not give all details. Calculations for the Type AB theory are similar with slight differences that will be discussed below.

Unlike the even d case, the spectral function $\mu_\alpha(u)$ is no longer polynomial in u, but a polynomial in u multiplied by a hyperbolic function,

$$\mu_\alpha(u) = \mu_\alpha^{\text{poly}}(u) \times f_\pm(u), \qquad \text{where} \quad f_\pm(u) = \begin{cases} f_+(u) = \tanh(\pi u), & \text{for bosons,} \\ f_-(u) = \coth(\pi u), & \text{for fermions.} \end{cases} \tag{105}$$

Then, for a particular spectral weight α, the partition function can be written as:

$$\zeta_{(\Delta;\alpha)}(z) = \frac{\text{vol}\,(\text{AdS}_{d+1})}{\text{vol}(S^d)} \frac{2^{d-1}}{\pi} \int_0^\infty du \frac{g_\alpha \mu_\alpha^{\text{poly}}(u)}{\left[u^2 + \left(\Delta - \frac{d}{2}\right)^2\right]^z} f_\pm(u). \tag{106}$$

We will use the example of the Type A theory in AdS_4 to walk us through the calculations. In the non-minimal Type A theory in AdS_4, the only representations are the totally symmetric ones $\alpha = [s]$, $s \geq 0$, and the spectral zeta function for a given spin s is:

$$\begin{aligned} \zeta_{(\Delta;\alpha_s)}(z) &= \frac{\text{vol}(\text{AdS}_4)}{\text{vol}(S^3)} \frac{4}{\pi} \int_0^\infty du \frac{g_{[s]}(s)\mu_{[s]}^{\text{poly}}(u)}{\left[u^2+\left(\Delta^{\text{ph}}-\frac{3}{2}\right)^2\right]^z} f_+(u) \\ &= \int_0^\infty du \frac{(2s+1)u\left[\left(s+\frac{1}{2}\right)^2+u^2\right]}{6\left[\left(\Delta-\frac{3}{2}\right)^2+u^2\right]^z} \tanh(\pi u), \end{aligned} \tag{107}$$

where $\mu_{[s]}^{\text{poly}}(u) = \frac{u\left(u^2+\left(s+\frac{1}{2}\right)^2\right)}{8\pi^2}$ and $g_{[s]} = 2s+1$.

To calculate the one-loop free energy, we will need to evaluate $\sum \zeta_{(\Delta;\alpha)}(0)$ and $\sum \zeta'_{(\Delta;\alpha)}(0)$.

---

[22] While not needed, the integral results for $B_k^{\pm}$, can be identified with the Hurwitz-Lerch Phi function $\Phi(z,s,v)$,

$$\int_0^\infty du \frac{u^k}{e^{2\pi u} \pm 1} = \int_0^\infty du \frac{1}{2\pi} \frac{\left(\frac{u}{2\pi}\right)^k e^{-u}}{1 \pm e^{-u}} = \frac{\Gamma(k+1)}{(2\pi)^{k+1}} \Phi(\pm 1, k+1, 1) \tag{102}$$

Computing $\sum\zeta_{(\alpha;\Delta)}(0)$:

Setting $z = 0$ in (107), we find:

$$\zeta_{(\Delta;[s])}(0) = \int_0^\infty du \frac{(2s+1)u\left[\left(s+\frac{1}{2}\right)^2 + u^2\right]}{6} \tanh(\pi u). \tag{108}$$

Regulating this sum by inserting the prefactor $(\Delta - \frac{d}{2})^{-\epsilon}$ as in (97), we find[23]

$$\begin{aligned}
\textstyle\sum_{s=1}^\infty \zeta_{(\Delta^{\rm ph};[s])}(0) &= \textstyle\lim_{\epsilon\to 0}\sum_{s=1}^\infty \int_0^\infty du \frac{(2s+1)u\left[\left(s+\frac{1}{2}\right)^2+u^2\right]}{6}\left(s-\frac{1}{2}\right)^{-\epsilon}\tanh(\pi u) \\
&= \textstyle\int_0^\infty \frac{du}{6}\lim_{\epsilon\to 0}\left[2\zeta\left(-1+\epsilon,\frac{1}{2}\right)u^3 + 2\zeta\left(-3+\epsilon,\frac{1}{2}\right)u + 6\zeta\left(-2+\epsilon,\frac{1}{2}\right)u\right. \\
&\qquad \left. +6\zeta\left(-1+\epsilon,\frac{1}{2}\right)u + (2u^3+2u)\zeta\left(\epsilon,\frac{1}{2}\right)\right]\tanh(\pi u) \\
&= \textstyle\int_0^\infty du\left[\frac{u^3}{72} + \frac{113u}{2880}\right]\tanh(\pi u).
\end{aligned} \tag{109}$$

A similar calculation for the ghost modes using the prefactor $(\Delta^{\rm gh} - \frac{d}{2})^{-\epsilon}$ yields:

$$\sum_{s=1}^\infty \zeta_{(\Delta^{\rm gh};[s-1])}(0) = \int_0^\infty \frac{du}{e^{2\pi i u}+1}\left[\frac{233u}{2880} + \frac{13u^3}{72}\right]\tanh(\pi u). \tag{110}$$

For the bulk scalar, we simply set $s = 0$ in $\zeta_{(\Delta^{\rm ph};[s])}(0)$, and obtain $\zeta_{(1;[0])}(0) = \int_0^\infty du\left[\frac{u}{24} + \frac{u^3}{6}\right]\tanh(\pi u)$. Putting all contributions together, the coefficient of the logarithmic divergence in the one-loop free energy is:

$$\begin{aligned}
\zeta^{\rm HS}_{\rm type\ A}(0) &= \textstyle\zeta_{(1;[0])}(0) + \sum_{s=1}^\infty \zeta_{(\Delta^{\rm ph};[s])}(0) - \sum_{s=1}^\infty \zeta_{(\Delta^{\rm gh};[s-1])}(0) \\
&= \textstyle\int_0^\infty du\tanh(\pi u)\left[\frac{u^3}{72} + \frac{113u}{2880} - \frac{233u}{2880} - \frac{13u^3}{72} + \frac{u}{24} + \frac{u^3}{6}\right] \\
&= 0.
\end{aligned} \tag{111}$$

It is remarkable that when we sum over the entire spectrum of bulk fields, we get:

$$\zeta^{\rm HS}_{\rm Total}(0) = 0, \tag{112}$$

which indicates that the one-loop free energies have no logarithmic divergences. We find that this result holds not only in Type A theories [26,27], but also in all of the Type B and Type AB theories we discuss below.

Computing $\zeta'_{(\Delta;\alpha_s)}(0)$:

The evaluation of $\zeta'(0)$ in odd d is considerably more complicated. One may begin by splitting the $f_\pm(u)$ term as: $f_\pm(u) = 1 \mp \frac{2}{e^{2\pi i u}\pm 1}$ so that

$$\zeta_{(\Delta;\alpha)}(z) = \zeta^{\rm poly}_{(\Delta;\alpha)}(z) + \zeta^{\rm exp}_{(\Delta;\alpha)}(z) \tag{113}$$

[23] Alternatively, one could first write $\tanh(\pi u) = 1 - 2/(e^{2\pi i u}+1)$, evaluate the integral coming from the first term by analytic continuation in z, and the one coming from the second term directly at $z = 0$, since it converges.

where:

$$\zeta^{\text{poly}}_{(\Delta;\alpha)}(z) = \frac{\text{vol}\left(\text{AdS}_{d+1}\right)}{\text{vol}(S^d)}\frac{2^{d-1}}{\pi}\int_0^\infty du \frac{g_\alpha \mu^{\text{poly}}_\alpha(u)}{\left[u^2+\left(\Delta-\frac{d}{2}\right)^2\right]^z} \tag{114}$$

$$\zeta^{\text{exp}}_{(\Delta;\alpha)}(z) = \mp\frac{\text{vol}\left(\text{AdS}_{d+1}\right)}{\text{vol}(S^d)}\frac{2^{d-1}}{\pi}\int_0^\infty du \frac{g_\alpha \mu^{\text{poly}}_\alpha(u)}{\left[u^2+\left(\Delta-\frac{d}{2}\right)^2\right]^z}\frac{2}{e^{2\pi i u}\pm 1} \tag{115}$$

Additionally, by differentiating (113),

$$\frac{\partial}{\partial z}\zeta_{(\Delta;\alpha)}(z)\big|_{z=0} = \frac{\partial}{\partial z}\zeta^{\text{poly}}_{(\Delta;\alpha)}(z)\big|_{z=0} + \frac{\partial}{\partial z}\zeta^{\text{exp}}_{(\Delta;\alpha)}(z)\big|_{z=0} \tag{116}$$

The integral in $\zeta^{\text{poly}}_{(\Delta;\alpha)}(z)$ may be evaluated at arbitrary z, and after taking the derivative and summing over spins, one finds a zero contribution to the free energy. The evaluation of $\zeta^{\text{exp}}_{(\Delta;\alpha)}(z)$ is more involved, and we refer the reader to Appendix B.3 and [26,27] for more details. The final result is that, in the non-minimal theory [26]:

$$\zeta'_{(1;[0])}(0) + \sum_{s=1}^{\infty}\zeta'_{(s+1;[s])}(0) - \sum_{s=1}^{\infty}\zeta'_{(s+2;[s-1])}(0) = 0\,, \tag{117}$$

which implies that the one loop free energy vanishes. In the non-minimal theory, one finds instead:

$$-\frac{1}{2}\left[\zeta'_{(1;[0])}(0) + \sum_{s=2,4,6,\ldots}^{\infty}\zeta'_{(s+1;[s])}(0) - \sum_{s=2,4,6,\ldots}^{\infty}\zeta'_{(s+2;[s-1])}(0)\right] = \frac{\log 2}{8} - \frac{3\zeta(3)}{16\pi^2}\,, \tag{118}$$

which is the free energy of a single real conformal scalar on S^3. An analogous result is found for the Type A theory in AdS_{d+1} for all d [27].

3.3.3. Type B Theories

Non-Minimal Theory

The full spectral zeta function for the non-minimal Type B theory in odd d follows from Equation (9), and reads:

$$\begin{aligned}\zeta^{\text{HS}}_{\text{type B}}(z) &= \zeta_{(d-1;[0,0,\ldots,0])}(z)\\ &+\textstyle\sum_{s=1}^{\infty}\left(\mathfrak{z}_{(\Delta^{\text{ph}};[s,1,1,\ldots,1,1])}(z) + \mathfrak{z}_{(\Delta^{\text{ph}};[s,1,1,\ldots,1,1,0])}(z) + \ldots + \mathfrak{z}_{(\Delta^{\text{ph}};[s,1,0,\ldots,0])}(z) + \mathfrak{z}_{(\Delta^{\text{ph}};[s,0,0,\ldots,0])}(z)\right)\\ &= \zeta_{(d-1;[0,0,\ldots,0])}(z) + \textstyle\sum_{s=1}^{\infty}\sum^{1}_{\substack{t_i\geq t_{i+1}\\ t_i\geq 0}}\mathfrak{z}_{(\Delta^{\text{ph}};[s,t_1,t_2,\ldots,t_{w-1},t_w])}(z),\end{aligned} \tag{119}$$

Note that instead of two towers, there is only one tower for each representation, due to the lack of the chirality matrix. Using this spectrum and the procedure outlined above to regulate the sums, we find that the logarithmic divergence correctly cancels:

$$\zeta^{\text{HS}}_{\text{type B}}(0) = 0\,. \tag{120}$$

However, as summarized in Section 2.2, the evaluation of $(\zeta^{\text{HS}}_{\text{type B}})'(0)$ leads to a surprising result. The one-loop free energy of the non-minimal Type B theories in all odd d does not vanish, but is given by (14), or equivalently by (15). This apparent mismatch with the expected result $F^{(1)} = 0$ remains to be understood.

Minimal Theories

Majorana fermions in odd d can be defined for $d = 3, 9 \pmod 8$. When the Majorana condition is not possible, one can impose the symplectic Majorana (SM) condition and consider the $USp(N)$ singlet sector of N free SM fermions, as explained in the even d case above.

The spectra of the minimal theories can be again deduced from the symmetry/antisymmetry properties of the $C\Gamma^{(n)}$ matrices. In the Majorana case, if $C\Gamma^{(n)}$ is symmetric the operators of the form (55) are retained for even spins and projected out for odd spins, and vice-versa if $C\Gamma^{(n)}$ is antisymmetric. The scalar operator $\bar{\psi}_i\psi^i$ is projected out if C is symmetric. For instance, in $d = 3$ the C matrix is antisymmetric and $C\gamma_\mu$ is symmetric, and so the spectrum of the minimal theory includes the $\Delta = 2$ (pseudo)-scalar and the tower of totally symmetric fields of even spin. Higher dimensional cases can be worked out similarly, and the first few examples are listed in Table 5. In a compact notation, the total spectral zeta function of the minimal theories dual to the Majorana projected fermion model reads:

$$\begin{aligned}
\zeta^{\mathrm{HS}}_{\text{type B Maj.}}(z) &= \chi(d)\zeta_{(d-1;[0,0,\ldots,0])}(z) \\
&+ \sum_{s=2,4,6,\ldots}^{\infty} \sum_{\substack{t_i \geq t_{i+1} \\ t_i \geq 0 \\ \sum_i t_i = w \pmod 4}}^{1} \mathfrak{Z}_{(\Delta;[s,t_1,t_2,\ldots,t_{w-1},t_w])}(z) + \sum_{s=1,3,5,\ldots} \sum_{\substack{t_i \geq t_{i+1} \\ t_i \geq 0 \\ \sum_i t_i = (w-1) \pmod 4}}^{1} \mathfrak{Z}_{(\Delta;[s,t_1,t_2,\ldots,t_{w-1},t_w])}(z) \\
&+ \sum_{s=1,3,5,\ldots} \sum_{\substack{t_i \geq t_{i+1} \\ t_i \geq 0 \\ \sum_i t_i = (w-2) \pmod 4}}^{1} \mathfrak{Z}_{(\Delta^{\mathrm{ph}};[s,t_1,t_2,\ldots,t_{w-1},t_w])}(z) + \sum_{s=2,4,6,\ldots} \sum_{\substack{t_i \geq t_{i+1} \\ t_i \geq 0 \\ \sum_i t_i = (w-3) \pmod 4}}^{1} \mathfrak{Z}_{(\Delta^{\mathrm{ph}};[s,t_1,t_2,\ldots,t_{w-1},t_w])}(z)
\end{aligned} \tag{121}$$

where $\chi(d) = 1, 0$ when $d = 3, 9 \pmod 8$ respectively.

In $d = 5, 7 \pmod 8$ we can impose instead the symplectic Majorana projection. In this case, the condition for which spins are projected out is reversed compared to the Majorana case, in a way analogous to what discussed earlier in the even d case. For instance, in $d = 5$ (AdS_6) one has that C is antisymmetric, and so the scalar operator $\bar{\psi}_i\psi^i$ is now projected out. Then, $C\gamma_\mu$ is antisymmetric, and so the spectrum includes the totally symmetric $[s, 0]$ representations for even s only. Finally, $C\gamma_{\mu\nu}$ is symmetric, and so we keep the representations $[s, 1]$ with odd s only. Higher dimensional cases are worked out similarly, and the first few examples are listed in Table 5. The total spectral zeta function can be expressed as:

$$\begin{aligned}
\zeta^{\mathrm{HS}}_{\text{type B Symp.Maj.}}(z) &= \chi(d)\zeta_{(d-1;[0,0,\ldots,0])}(z) \\
&+ \sum_{s=1,3,5,\ldots}^{\infty} \sum_{\substack{t_i \geq t_{i+1} \\ t_i \geq 0 \\ \sum_i t_i = w \pmod 4}}^{1} \mathfrak{Z}_{(\Delta;[s,t_1,t_2,\ldots,t_{w-1},t_w])}(z) + \sum_{s=2,4,6,\ldots} \sum_{\substack{t_i \geq t_{i+1} \\ t_i \geq 0 \\ \sum_i t_i = (w-1) \pmod 4}}^{1} \mathfrak{Z}_{(\Delta;[s,t_1,t_2,\ldots,t_{w-1},t_w])}(z) \\
&+ \sum_{s=2,4,6,\ldots} \sum_{\substack{t_i \geq t_{i+1} \\ t_i \geq 0 \\ \sum_i t_i = (w-2) \pmod 4}}^{1} \mathfrak{Z}_{(\Delta^{\mathrm{ph}};[s,t_1,t_2,\ldots,t_{w-1},t_w])}(z) + \sum_{s=1,3,5,\ldots} \sum_{\substack{t_i \geq t_{i+1} \\ t_i \geq 0 \\ \sum_i t_i = (w-3) \pmod 4}}^{1} \mathfrak{Z}_{(\Delta^{\mathrm{ph}};[s,t_1,t_2,\ldots,t_{w-1},t_w])}(z)
\end{aligned} \tag{122}$$

where $\chi(d) = 0, 1$ when $d = 5, 7 \pmod 8$ respectively. In both versions of the minimal truncation, we find that the coefficient of the logarithmic divergence still vanishes after summing up over the full spectrum. However, similarly to the non-minimal case, the minimal Type B theories in odd d appear to have a non-zero one-loop free energy, which we report in Table 6.

Table 5. Spectra of the minimal Type B theory dual to the fermionic vector model with Majorana (or symplectic Majorana) projection. The corresponding values of $F^{(1)}$ can be found in Table 6.

AdS$_4$ $O(N)$

α	$s =$ 1,2,3,...	2,4,6,...	1,3,5,...
$[s]$		1	
Scalar ($\Delta = 2$)		1	

AdS$_6$ $USp(N)$

α	$s =$ 1,2,3,...	2,4,6,...	1,3,5,...
$[s,1]$			1
$[s,0]$		1	
Scalar ($\Delta = 4$)			

AdS$_8$ $USp(N)$

α	$s =$ 1,2,3,...	2,4,6,...	1,3,5,...
$[s,1,1]$			1
$[s,1,0]$		1	
$[s,0,0]$		1	
Scalar ($\Delta = 6$)		1	

AdS$_{10}$ $O(N)$

α	$s =$ 1,2,3,...	2,4,6,...	1,3,5,...
$[s,1,1,1]$		1	
$[s,1,1,0]$			1
$[s,1,0,0]$			1
$[s,0,0,0]$		1	
Scalar ($\Delta = 8$)			

AdS$_{12}$ $O(N)$

α	$s =$ 1,2,3,...	2,4,6,...	1,3,5,...
$[s,1,1,1,1]$		1	
$[s,1,1,1,0]$			1
$[s,1,1,0,0]$			1
$[s,1,0,0,0]$		1	
$[s,0,0,0,0]$		1	
Scalar ($\Delta = 10$)		1	

AdS$_{14}$ $USp(N)$

α	$s =$ 1,2,3,...	2,4,6,...	1,3,5,...
$[s,1,1,1,1,1]$			1
$[s,1,1,1,1,0]$		1	
$[s,1,1,1,0,0]$		1	
$[s,1,1,0,0,0]$			1
$[s,1,0,0,0,0]$			1
$[s,0,0,0,0,0]$		1	
Scalar ($\Delta = 12$)			

AdS$_{16}$ $USp(N)$

α	$s =$ 1,2,3,...	2,4,6,...	1,3,5,...
$[s,1,1,1,1,1,1]$			1
$[s,1,1,1,1,1,0]$		1	
$[s,1,1,1,1,0,0]$		1	
$[s,1,1,1,0,0,0]$			1
$[s,1,1,0,0,0,0]$			1
$[s,1,0,0,0,0,0]$		1	
$[s,0,0,0,0,0,0]$		1	
Scalar ($\Delta = 14$)		1	

We did not find an analytic formula for these results similar to (14). However, we note that all these "anomalous" values only involve the Riemann zeta functions $\zeta(2k+1)$ divided by π^2, and interestingly all other transcendental constants that appear in intermediate steps of the calculation cancel out.

Table 6. One-loop free energy of the minimal Type B HS theory in AdS_{d+1} for odd d.

d	F_{computed} (Minimal Type B)
3	$\frac{\log(2)}{8} - \frac{5\zeta(3)}{16\pi^2}$
5	$\frac{3\log(2)}{64} + \frac{7\zeta(3)}{192\pi^2} - \frac{49\zeta(5)}{128\pi^4}$
7	$\frac{5\log(2)}{128} + \frac{227\zeta(3)}{3840\pi^2} - \frac{5\zeta(5)}{256\pi^4} - \frac{441\zeta(7)}{512\pi^6}$
9	$-\frac{35\log(2)}{2048} + \frac{315\zeta(7)}{2048\pi^6} + \frac{3825\zeta(9)}{4096\pi^8} - \frac{617\zeta(3)}{21,504\pi^2} - \frac{85\zeta(5)}{2048\pi^4}$
11	$\frac{63\log(2)}{16,384} + \frac{68,843\zeta(3)}{10,321,920\pi^2} + \frac{31,033\zeta(5)}{2,211,840\pi^4} - \frac{29\zeta(7)}{98,304\pi^6} - \frac{13,579\zeta(9)}{98,304\pi^8} - \frac{31,745\zeta(11)}{65,536\pi^{10}}$
13	$\frac{231\log(2)}{131,072} + \frac{1,933,151\zeta(3)}{619,315,200\pi^2} + \frac{27,993,331\zeta(5)}{3,715,891,200\pi^4} + \frac{1,056,541\zeta(7)}{123,863,040\pi^6} - \frac{285,799\zeta(9)}{11,796,480\pi^8}$ $-\frac{150,541\zeta(11)}{786,432\pi^{10}} - \frac{258,049\zeta(13)}{524,288\pi^{12}}$
15	$\frac{429\log(2)}{524,288} + \frac{2,423,526,031\zeta(3)}{1,653,158,707,200\pi^2} + \frac{41,124,367\zeta(5)}{10,899,947,520\pi^4} + \frac{12,837\zeta(7)}{2,097,152\pi^6} + \frac{47,549\zeta(9)}{66,060,288\pi^8}$ $-\frac{104,687\zeta(11)}{2,097,152\pi^{10}} - \frac{503,685\zeta(13)}{2,097,152\pi^{12}} - \frac{2,080,641\zeta(15)}{4,194,304\pi^{14}}$

3.3.4. Type AB Theories

Spectrum and Results

As in the even d case, the only irrep of $SO(d)$ describing the tower of half-integer spins is $\alpha_s = [s, \frac{1}{2}, \frac{1}{2}, \ldots, \frac{1}{2}]$. Thus, the total spectral zeta function is given by the same equation as in (76).

The calculation is rather similar to the one we outlined for the Type A theory. The only difference is that the spectral density $\mu_\alpha(u)$ includes $\coth(\pi u)$ instead of $\tanh(\pi u)$. For example, in the Type AB theory in AdS_4, the higher-spin zeta-function is given by:

$$\zeta_{(\Delta;[s])}(z) = \int_0^\infty du \frac{(2s+1)u\left[\left(s+\frac{1}{2}\right)^2 + u^2\right]}{6\left[\left(\Delta - \frac{3}{2}\right)^2 + u^2\right]^z} \coth(\pi u). \tag{123}$$

The calculations for $\sum \zeta_{(\Delta;\alpha_s)}(0)$ are essentially identical to that of Type A theories, and in particular we find that the contribution to the logarithmic divergence due to the fermionic fields vanishes after summing over the whole tower. Heading straight to the calculation of $\sum \zeta'_{(\Delta;\alpha_s)}(0)$, if we follow the procedure outlined for the Type A case, we have:

$$\zeta'_{(\Delta;[s])}(0) = -\int_0^\infty du \frac{(2s+1)u}{3(e^{2\pi u}-1)}\left[u^2 + \left(s+\frac{1}{2}\right)^2\right] \log\left[\left(\Delta - \frac{3}{2}\right)^2 + u^2\right]. \tag{124}$$

Rewriting the exponential terms using (99), we should use $A_1^-(x)$ instead of $A_1^+(x)$. This introduces an extra $\frac{1}{2\sqrt{x}}$ term in $\zeta^{\text{HS}-\exp'}_{\Delta,s;[s]}(0)$, i.e.,

$$\begin{aligned}\zeta^{\exp'}_{(\Delta;[s])}(0) &= -\int_0^\infty du \frac{(2s+1)u}{3(e^{2\pi u}-1)}\left[u^2+\left(s+\frac{1}{2}\right)^2\right]\log(u^2) \\ &\quad -\int_0^{(s-\frac{1}{2})^2} dx \frac{1}{12}(2s+1)\left[(2s+1)^2+4u^2\right]\left(-\frac{1}{2\sqrt{x}}+\frac{\log(\sqrt{x})}{2}-\frac{\psi(\sqrt{x})}{2}\right)\end{aligned} \tag{125}$$

In any case, the terms involving $\frac{1}{2\sqrt{x}}$, which we can call $\zeta^{\exp-\text{sqrt}'}_{(\Delta;[s])}(0)$, will not contribute to the value of $\zeta^{\exp'}_{(\Delta;[s])}(0)$. Only the contributions from the terms involving $\psi(\sqrt{x})$, namely the third term inside the bracket of (125) will contribute. After putting all together, the end result is:

$$\zeta'_{(\frac{3}{2};[\frac{1}{2}])}(0) + \sum_{s=\frac{3}{2},\frac{5}{2},\ldots}\left(\zeta'_{(s+1;[s])}(0) - \zeta'_{(s+2;[s-1])}(0)\right) = 0\,, \tag{126}$$

i.e., the tower of fermionic fields in Type AB theories yields a vanishing contribution to the bulk one-loop free energy. This result extends to all higher d.

Acknowledgments: The work of SG was supported in part by the U.S. NSF under Grants No. PHY-1318681 and PHY-1620542. The work of IRK was supported in part by the U.S. NSF under Grants No. PHY-1314198 and PHY-1620059. The work of ZMT was supported in part by the École Normale Supérieure, France under the grant LabEx ENS-ICFP: ANR-10-LABX-0010/ANR-10-IDEX-0001-02 PSL, and the Public Service Commission, Singapore. Parts of this paper are based on ZMT's Princeton University 2014 undergraduate thesis. IRK thanks the Niels Bohr International Academy for hospitality during the final stages of his work on this paper.

Author Contributions: All authors contribute equally to all aspects of the paper.

Conflicts of Interest: The authors declare no conflict of interest.

Appendix A

Appendix A.1. $S^1 \times S^{d-1}$ Partition Functions and Casimir Energies

Besides testing the gauge/gravity duality by comparing the partition functions on the Euclidean AdS_{d+1} (hyperbolic space) and CFT_d on S^d, we can also compare thermal partition functions of higher spin theories on thermal AdS_{d+1} and boundary CFTs defined on $S^1 \times S^{d-1}$, where the inverse temperature β of the thermal AdS space is interpreted as the length of S^1. Calculations of the thermal free energy and Casimir energy serve as useful checks on our results in hyperbolic space with S^d boundary. The results below follow and generalize [28], which considered Type A theories in all d and Type B theories in $d = 2, 3, 4$, and [16–18], where Type B theories in $d = 6$ and type C theories in $d = 4, 6$ were discussed.

The free energy on $S^1 \times S^{d-1}$ takes the form:

$$F = F_\beta + \beta E_c \tag{A1}$$

where F_β depends non-trivially on the temperature and goes to zero at large β, and E_c is the Casimir energy. The latter is related to the "one-particle" partition function on $S^1 \times S^{d-1}$ by (see e.g., [28] for a review):

$$E_c = \sigma\frac{1}{2}\zeta_E(-1) = \sigma\frac{1}{2\Gamma(z)}\int_0^\infty d\beta\beta^{z-1}\mathcal{Z}(\beta)\bigg|_{z=-1}\,. \tag{A2}$$

where $\sigma = +1$ for bosonic fields, and $\sigma = -1$ for fermionic ones, and $\mathcal{Z}(\beta)$ denotes the one-particle partition function. This also determines F_β by:

$$\begin{aligned} F_\beta &= -\sum_{m=1}^{\infty} \frac{1}{m} \mathcal{Z}(m\beta) , \quad \text{boson} \\ F_\beta &= \sum_{m=1}^{\infty} \frac{(-1)^m}{m} \mathcal{Z}(m\beta) , \quad \text{fermion} \end{aligned} \tag{A3}$$

Note that E_c vanishes for a CFT$_d$ in odd d, but it is non-zero in even d.

In the vector models restricted to the singlet sector, one finds that $F_\beta = \mathcal{O}(N^0)$, due to the integration over the flat connection which enforces the gauge singlet constraint [28,37]. This term should then match the temperature dependent part of the bulk one-loop thermal free energy, obtained by summing over all fields in the AdS spectrum, and the agreement serves as a useful check on the bulk spectra. The Casimir term, on the other hand, is just given by N times the Casimir energy of a single conformal field. If no shift is expected in the map between the bulk coupling constant and N, then the CFT Casimir contribution should be reproduced just by a classical calculation in AdS (which we have no access to at present), and bulk loop corrections to the Casimir energy should vanish. However, when a shift $G_N \sim 1/(N-k)$ is expected, the one-loop correction to the Casimir energy should precisely be consistent with such a shift. We will see below that this is the case in all higher spin theories we considered in this paper.

On the CFT side, the one-particle partition functions of a conformal scalar and Majorana (or Weyl) fermion are given by:

$$\mathcal{Z}_0(q) = \frac{q^{\frac{d}{2}-1}(1+q)}{(1-q)^{d-1}} , \qquad \mathcal{Z}_{\frac{1}{2}}(q) = \frac{2^{\lfloor \frac{d}{2} \rfloor} q^{\frac{d-1}{2}}}{(1-q)^{d-1}} , \qquad q = e^{-\beta} \tag{A4}$$

Using (A2) and the identity $(1-q)^{-b} = \sum_{n=1}^{\infty} \binom{n+b-2}{b-1} q^{n-1}$, one then finds the Casimir energies:

$$\begin{aligned} E_{c,0} &= \sum_{n=0}^{\infty} \frac{(n+d-3)!}{(d-2)!n!} [n + \frac{1}{2}(d-2)]^{1-z}|_{z=-1} , \\ E_{c,1/2} &= -2^{\lfloor \frac{d}{2} \rfloor -1} \sum_{n=0}^{\infty} \frac{(n+d-2)!}{(d-2)!n!} [n + \frac{1}{2}(d-1)]^{1-z}|_{z=-1} . \end{aligned} \tag{A5}$$

Evaluating this with Hurwitz zeta regularization, one obtains the values in $d = 4, 6, 8, \ldots$:

$$\begin{aligned} E_{c,0} &= \left\{ \frac{1}{240}, -\frac{31}{60,480}, \frac{289}{3,628,800}, -\frac{317}{22,809,600}, \frac{6,803,477}{2,615,348,736,000}, \ldots \right\} \\ E_{c,1/2} &= \left\{ \frac{17}{960}, -\frac{367}{48,384}, \frac{27,859}{8,294,400}, -\frac{1,295,803}{851,558,400}, \frac{5,329,242,827}{7,608,287,232,000}, \ldots \right\} \end{aligned} \tag{A6}$$

For the real $(d/2-1)$-form gauge field, with no self-duality imposed on the $d/2$-form field strength, the one-particle partition function is given by (see for instance Appendix D of [18]):

$$\mathcal{Z}_{\frac{d}{2}\text{-form}}(q) = \frac{2q^{d/2}}{(1-q)^{d-1}} \left(\sum_{j=1}^{\frac{d}{2}} a_{d,j} (-q)^{\frac{d}{2}-j} \right) , \qquad a_{d,j} = \binom{d-1}{j-1} . \tag{A7}$$

Note that when we expand $\mathcal{Z}_{\frac{d}{2}\text{-form}}(q)$ around $q = 1$, the leading pole term is,

$$\mathcal{Z}_{\frac{d}{2}\text{-form}}(q) \sim \frac{2}{(1-q)^{d-1}} n(d), \qquad \text{where} \quad n(d) = \binom{d-2}{\frac{d}{2}-1}, \tag{A8}$$

which gives the correct number of propagating degrees of freedom of a $(d/2-1)$-form gauge field. Inserting (A7) into (A2), one finds the Casimir energies in $d = 4, 6, 8, \ldots$: [24]

$$E_{c,d/2-\text{form}} = \left\{ \frac{11}{120}, -\frac{191}{2016}, \frac{2497}{25{,}920}, -\frac{14{,}797}{152{,}064}, \frac{92{,}427{,}157}{943{,}488{,}000}, \ldots \right\}. \tag{A9}$$

On the AdS side, at the level of the one-particle partition functions, the contribution of a bulk field to the thermal free energy is given essentially by the character of the corresponding representation of the conformal group. For the representations α_s dual to massless gauge fields, we have:

$$\mathcal{Z}_{\alpha_s}(q) = \frac{q^{\Delta^{\text{ph}}}}{(1-q)^d} [g_{\alpha_s} - q g_{\alpha_{s-1}}], \tag{A10}$$

where $\Delta^{\text{ph}} = s + d - 2$ and g_{α_s} is the dimension of the representation α_s (the number of propagating degrees of freedom in the bulk is $g_{\alpha_s} - g_{\alpha_{s-1}}$). For the massive fields, the ghost contribution is not present, and one has:

$$\mathcal{Z}_{\alpha} = \frac{q^{\Delta}}{(1-q)^d} g_{\alpha}. \tag{A11}$$

One may obtain a "total" one-particle partition function $\mathcal{Z}(\beta)$ in the bulk by summing over all representations in the spectrum, and from it one may then find the bulk one-loop Casimir energy by (A2) and F_β by (A3). In the following we summarize the result of these calculations in the various higher spin theories considered in this paper.

Type A Theories

In [28], it was shown that:

$$\text{Non-Minimal Type A:} \qquad \mathcal{Z}(\beta) = \sum_{\alpha} \mathcal{Z}_{\alpha}(q) = [\mathcal{Z}_0(q)]^2, \tag{A12}$$

$$\text{Minimal Type A:} \qquad \mathcal{Z}(\beta) = \sum_{\gamma} \mathcal{Z}_{\gamma}(q) = \frac{1}{2}\left[[\mathcal{Z}_0(q)]^2 + \mathcal{Z}_0(q^2) \right] \tag{A13}$$

where α refers to the spectrum containing the weights $[s, 0, \ldots, 0]$ with each integer spin $s = 0, 1, 2, \ldots$, and γ refers to the spectra containing the weights $[s, 0, \ldots, 0]$ with each even integer spin $s = 0, 2, 4, \ldots$. The result on the right-hand side, where $\mathcal{Z}_0(\beta)$ is the scalar one-particle partition function given in (A4), precisely agrees with the singlet sector CFT calculation [28,37].

The bulk Casimir energy can be obtained by inserting the right-hand side of (A12) and (A13) into (A2) (alternatively, one may compute the Casimir contributions spin by spin, and sum up at the end). One finds that $[\mathcal{Z}_0(q)]^2$ gives zero contribution to the Casimir energy,[25] while $\mathcal{Z}_0(q^2)$ gives a contribution equal to $2E_{c,0}$. Then, $E_{c,\text{type A}} = 0$ and $E_{c,\text{min. type A}} = E_{c,0}$, consistently with the expected shift of G_N deduced from the S^d calculations.

---

[24] The values obtained for $d = 4, 6$ agree with those in the literature [16–18], while the values for other dimensions are new as far as we know.

[25] This is because of symmetry under $q \to 1/q$. Any function symmetric under this exchange gives a zero contribution under the integral in (A2).

Type B Theories

In the Type B theories and their various truncations, we find:

$$\text{Non-Minimal Type B:} \qquad \sum_{\alpha} \mathcal{Z}_{\alpha}(q) = [\mathcal{Z}_{\frac{1}{2}}(q)]^2, \tag{A14}$$

$$\text{Weyl-Projection:} \qquad \sum_{\gamma} \mathcal{Z}_{\gamma}(q) = \frac{1}{4}[\mathcal{Z}_{\frac{1}{2}}(q)]^2, \tag{A15}$$

$$\text{Minimal Type B:} \qquad \sum_{\delta} \mathcal{Z}_{\delta}(q) = \begin{cases} \frac{1}{2}\Big[[\mathcal{Z}_{\frac{1}{2}}(q)]^2 - \mathcal{Z}_{\frac{1}{2}}(q^2)\Big], & \text{for } O(N), \\ \frac{1}{2}\Big[[\mathcal{Z}_{\frac{1}{2}}(q)]^2 + \mathcal{Z}_{\frac{1}{2}}(q^2)\Big], & \text{for } USp(N), \end{cases} \tag{A16}$$

$$\text{Majorana–Weyl:} \qquad \sum_{\epsilon} \mathcal{Z}_{\epsilon}(q) = \frac{1}{2}\left(\frac{1}{4}[\mathcal{Z}_{\frac{1}{2}}(q)]^2 - \frac{1}{2}\mathcal{Z}_{\frac{1}{2}}(q^2)\right) \tag{A17}$$

$$\text{Symplectic Majorana–Weyl:} \qquad \sum_{\kappa} \mathcal{Z}_{\kappa}(q) = \frac{1}{2}\left(\frac{1}{4}[\mathcal{Z}_{\frac{1}{2}}(q)]^2 + \frac{1}{2}\mathcal{Z}_{\frac{1}{2}}(q^2)\right) \tag{A18}$$

where α, γ, δ are the spectra given by (49), (52), (56)–(58), and ϵ, κ the Majorana–Weyl truncations discussed in Section 3.2.1.The right-hand side of all the above equations, with $\mathcal{Z}_{\frac{1}{2}}(q)$ given in (A4), is again in precise agreement with the thermal calculations in the singlet sector of the fermionic CFT (with the relevant fermion projection and gauge group). As an explicit example, in AdS_{11}, we have:

$$\text{Non-Minimal Type B:} \qquad \sum_{\alpha} \mathcal{Z}_{\alpha}(q) = \frac{1024q^9}{(q-1)^{18}} = \left(\frac{32q^{9/2}}{(1-q)^9}\right)^2, \tag{A19}$$

$$\text{Weyl-Projection:} \qquad \sum_{\gamma} \mathcal{Z}_{\delta}(q) = \frac{256q^9}{(q-1)^{18}}, \tag{A20}$$

$$\text{Minimal Type B:} \qquad \sum_{\delta} \mathcal{Z}_{\gamma}(q) = \frac{1}{2}\left(\frac{1024q^9}{(1-q)^{18}} - \frac{32q^9}{(1-q^2)^9}\right), \tag{A21}$$

$$\text{Majorana–Weyl:} \qquad \sum_{\epsilon} \mathcal{Z}_{\epsilon}(q) = \frac{1}{2}\left(\frac{256q^9}{(1-q)^{18}} - \frac{16q^9}{(1-q^2)^9}\right), \tag{A22}$$

which all agree with the formulas in (A14)–(A18). For instance, using the spectrum found in Table 3, the explicit computations for the Majorana–Weyl case is as follows:

$$\begin{aligned}[s,1,1,1,1]: \quad & \textstyle\sum_{s=2,4,6,\ldots} \frac{q^{s+8}}{576(1-q)^{10}}\left[\frac{(s+8)!}{(s+4)(s-1)!} - q\frac{(s+7)!}{(s+3)(s-2)!}\right] \\ &= \tfrac{q^8}{(q-1)^{18}(q+1)^9}\big(-q^{13} - q^{12} + 8q^{11} + 134q^{10} + 98q^9 + 3914q^8 + 2948q^7 \\ &\quad +12,984q^6 + 4983q^5 + 8799q^4 + 924q^3 + 1050q^2\big)\end{aligned} \tag{A23}$$

$$\begin{aligned}[s,1,1,0,0]: \quad & \textstyle\sum_{s=1,3,5,\ldots} \frac{q^{s+8}}{720(1-q)^{10}}\left[\frac{(s+4)(s+8)!}{(s+2)(s+6)(s-1)!} - q\frac{(s+3)(s+7)!}{(s+1)(s+5)(s-2)!}\right] \\ &= \tfrac{q^7}{(q-1)^{18}(q+1)^9}\big(q^{18} + q^{17} - 8q^{16} - 8q^{15} + 29q^{14} + 29q^{13} - 64q^{12} - 64q^{11} + 1043q^{10} \\ &\quad +923q^9 + 6992q^8 + 3760q^7 + 10,039q^6 + 2407q^5 + 3352q^4 + 120q^3 + 120q^2\big)\end{aligned} \tag{A24}$$

$$[s,0,0,0,0]: \quad \sum_{s=2,4,6,\ldots} \frac{q^{s+8}}{20160(1-q)^{10}} \left[\frac{(s+4)(s+7)!}{s!} - q\frac{(s+3)(s+6)!}{(s-1)!} \right]$$
$$= \frac{q^8}{(q-1)^{18}(q+1)^9} \big(-q^{17} - q^{16} + 8q^{15} + 8q^{14} - 28q^{13} - 28q^{12} + 56q^{11} + 66q^{10} \tag{A25}$$
$$-61q^9 + 59q^8 + 140q^7 + 392q^6 + 98q^5 + 218q^4 + 44q^3 + 54q^2 \big)$$

Summing (A23)–(A25) up, we obtain (A22).

The bulk one-loop contribution to the Casimir energy in Type B theories can be obtained by inserting the right-hand sides of (A14)–(A18) into (A2). The only non-zero contribution comes from $\mathcal{Z}_{\frac{1}{2}}(q^2)$, which yields $2E_{c,1/2}$. Then, we see that the bulk one-loop Casimir energies in all variants of the Type B theories are in precise agreement with the shifts of the coupling constant summarized in Table 1. Note that in odd d we get zero Casimir energy on both CFT and bulk sides, as it should be, so this calculation does not shed light on the anomalous shifts we encountered in Type B theories in odd d. A few explicit values of the bulk one-loop Casimir energies are collected in Table A1.

Table A1. Type B Casimir energies. The grey boxes indicate that the particular type of fermion is not defined for the given dimension.

d	Non-Minimal	Weyl	Minimal ($O(N)/USp(N)$)	Majorana–Weyl
3	0		0	
4	0	0	$\frac{17}{960}$	
5	0		0	
6	0	0	$\frac{367}{48,384}$	
7	0		0	
8	0	0	$\frac{27,859}{8,294,400}$	$\frac{27,859}{16,588,800}$
9	0		0	
10	0	0	$-\frac{12,950,803}{851,558,400}$	$-\frac{12,950,803}{1,703,116,800}$
11	0		0	

Type AB Theories

In the purely fermionic sector of the Type AB theories, the only representations are given by the weights $[s, \frac{1}{2}, \ldots, \frac{1}{2}]$, which lead to a simple computation that gives for a generic d,

$$\begin{aligned} \mathcal{Z}_{\text{type AB ferm}}(q) &= \frac{q^{d-\frac{3}{2}}}{(1-q)^d} g_{[1/2,1/2,\ldots,1/2]} + \sum_{s=\frac{3}{2},\frac{5}{2},\ldots} \frac{q^{s+d-2}}{(1-q)^d} \left[g_{[s,1/2,\ldots,1/2]} - q g_{[s-1,1/2,\ldots,1/2]} \right] \\ &= \frac{2^{\lfloor \frac{d}{2} \rfloor} q^{d-\frac{3}{2}}(1+q)}{(1-q)^{2(d-1)}} = \mathcal{Z}_0(q) \mathcal{Z}_{\frac{1}{2}}(q). \end{aligned} \tag{A26}$$

A quick calculation gives us $E_c = 0$ for the contribution of the fermionic tower in the Type AB theories, which is nicely consistent with what we obtained in the S^d calculations, namely that there are no shifts due to the purely fermionic fields.

Type C Theories

In type C theories, summing up over the relevant bulk spectra given in Section 3.2.3, we find:

$$\text{Non-Minimal Type C:} \qquad \sum_{\alpha} \mathcal{Z}_{\alpha}(q) = [\mathcal{Z}_{\frac{d}{2}\text{-form}}(q)]^2, \tag{A27}$$

$$U(N)\ \text{Self-Dual:} \qquad \sum_{\gamma} \mathcal{Z}_{\delta}(q) = \frac{1}{4}[\mathcal{Z}_{\frac{d}{2}\text{-form}}(q)]^2, \tag{A28}$$

$$\text{Minimal Type C:} \qquad \sum_{\delta} \mathcal{Z}_{\gamma}(q) = \frac{1}{2}\left[[\mathcal{Z}_{\frac{d}{2}\text{-form}}(q)]^2 + \mathcal{Z}_{\frac{d}{2}\text{-form}}(q^2) \right], \tag{A29}$$

$$O(N)\ \text{Self-Dual:} \qquad \sum_{\epsilon} \mathcal{Z}_{\epsilon}(q) = \frac{1}{2}\left[\frac{1}{4}\left(\mathcal{Z}_{\frac{d}{2}\text{-form}}(q) \right)^2 + \frac{1}{2} \mathcal{Z}_{\frac{d}{2}\text{-form}}(q^2) \right], \tag{A30}$$

where $\mathcal{Z}_{\frac{d}{2}\text{-form}}(q)$ is the one-particle partition function (A7) of a single real $(d/2-1)$-form gauge field. The results on the right-hand side have the correct structure expected from the CFT thermal free energy in the $U(N)/O(N)$ singlet sector of the theory of N differential form gauge fields. This calculation was carried out explicitly in the $S^1 \times S^3$ case in [17], and we expect it to generalize to all d.

The one-loop Casimir energies of Type C theories can be obtained by plugging the right-hand side of (A27)–(A30) into (A2). The calculation can be simplified by noting that, due to the symmetry properties under $q \to 1/q$, the term $[\mathcal{Z}_{\frac{d}{2}\text{-form}}(q)]^2$ contributes $2(-1)^{d/2} E_{c,d/2-\text{form}}$ after the integration in (A2),[26] and $\mathcal{Z}_{\frac{d}{2}\text{-form}}(q^2)$ contributes $2E_{c,d/2-\text{form}}$. Then we see that in all cases the one-loop Casimir energies in the bulk are consistent with the shifts of the coupling constant summarized in Table 1. A few explicit values are reported in Table A2.

[26] See Appendix D of [18] for a discussion of this.

Table A2. Type C Casimir energies. The grey boxes indicate that the particular type of p-form is not defined for the dimension.

d	Non-Minimal $U(N)$	Self-Dual $U(N)$	Minimal $O(N)$	Self-dual $O(N)$
4	$\frac{11}{60}$	$\frac{11}{240}$	$\frac{11}{60}$	
6	$\frac{191}{1008}$	$\frac{191}{4032}$	0	$-\frac{191}{8064}$
8	$\frac{2497}{12,960}$	$\frac{2497}{51,840}$	$\frac{2497}{12,960}$	
10	$\frac{14,797}{76,032}$	$\frac{14,797}{304,128}$	0	$-\frac{14,797}{608,256}$
12	$\frac{92,427,157}{471,744,000}$	$\frac{92,427,157}{1,886,976,000}$	$\frac{92,427,157}{471,744,000}$	
14	$\frac{36,740,617}{186,624,000}$	$\frac{36,740,617}{746,496,000}$	0	$-\frac{36,740,617}{1,492,992,000}$

Appendix B. Some Technical Details on the One-Loop Calculations in Hyperbolic Space

Appendix B.1. Hurwitz Zeta Regularization

To implement ζ-function regularization, we identify the conventionally divergent term $\sum_{s=1}^{\infty} 1/(s+\nu)^k$ as $\sum_{s=0}^{\infty} 1/(s+\nu+1)^k$, and treating it as the Hurwitz zeta function,

$$\zeta(k,\beta) \equiv \sum_{n=0}^{\infty} \frac{1}{(n+\beta)^k}, \tag{A31}$$

where we then analytically extend to the full complex plane. This allows us to regulate systematically the sums to obtain their finite contributions.

Suppose we want to start summing all integer spins $s \geq \ell \geq 0$, then,

$$\sum_{s=\ell}^{\infty} \frac{1}{(s+\nu)^k} = \zeta(k, \ell+\nu). \tag{A32}$$

This is the convention we applied in this paper, and avoids potential inconsistencies that can occur with the Hurwitz zeta function. We might also consider sums that only incorporate a particular subset of spins, such as either all odd integer spins or all even integer spins. To do so, we can transform the summing variable of the original Hurwitz zeta function appropriately. We give two examples:

To sum over all even spins, consider:

$$\sum_{s=2,4,6,\ldots}^{\infty} \frac{1}{(s+\nu)^k} = \sum_{s=1}^{\infty} \frac{1}{(2s+\nu)^k} = \sum_{s=1}^{\infty} \frac{2^{-k}}{\left(s+\frac{\nu}{2}\right)^k} = \sum_{s=0}^{\infty} \frac{2^{-k}}{\left(s+\frac{\nu}{2}+1\right)^k} = 2^{-k}\zeta\left(k, \frac{\nu}{2}+1\right). \tag{A33}$$

A similar scheme for summing over all odd spins is:

$$\sum_{s=1,3,5,\ldots}^{\infty} \frac{1}{(s+\nu)^k} = \sum_{s=1}^{\infty} \frac{1}{(2s-1+\nu)^k} = \sum_{s=1}^{\infty} \frac{2^{-k}}{\left(s+\frac{\nu-1}{2}\right)^k} = 2^{-k}\zeta\left(k, \frac{\nu-1}{2}+1\right). \tag{A34}$$

Unlike conventional summation where rearrangement of terms may lead to problems, ζ-function regularization allows for rearrangement. In particular, the ζ-function satisfies,

$$\sum_{s=2,4,6,\ldots}^{\infty} \frac{1}{(s+\nu)^k} + \sum_{s=1,3,5,\ldots}^{\infty} \frac{1}{(s+\nu)^k} = \zeta(k, \nu+1), \tag{A35}$$

which allows us to obtain the regularization over both the odd or the even integer spins by just doing one of the two calculation.

Appendix B.2. Identity for Odd d Free Energy Calculations

The relationship described in (101) can be derived by:

$$\begin{aligned} A_k^{\pm}(x) &= \int_0^{\infty} \frac{u^k}{e^{2\pi u}\pm 1} \frac{du}{u^2+x} \\ &= \frac{\partial}{\partial a}\left[\int_0^{\infty} du\, \frac{u^{k-2}}{e^{2\pi u}\pm 1} \log[a u^2 + x] \right]_{a=1} \\ &= \frac{\partial}{\partial a}\left\{ \log(a) \int_0^{\infty} du\, \frac{u^{k-2}}{e^{2\pi u}\pm 1} + \int_0^{\infty} du\, \frac{u^{k-2}}{e^{2\pi u}\pm 1} \log\left[u^2 + \frac{x}{a}\right] \right\}_{a=1} \\ &= \int_0^{\infty} du\, \frac{u^{k-2}}{e^{2\pi u}\pm 1} - x\int_0^{\infty} \frac{u^{k-2}}{e^{2\pi u}\pm 1}\frac{du}{u^2+x} \\ &= B_{k-2}^{\pm}(x) - xA_{k-2}^{\pm}(x). \end{aligned} \tag{A36}$$

Appendix B.3. Evaluating $\zeta'_{\Delta,\alpha}(0)$

Here we collect some details on the evaluation of the term $\partial_z \zeta^{\exp}_{(\Delta;\alpha_s)}(z)|_{z=0}$ in (115), in the explicit example of the Type A theory in AdS$_4$. The calculations in the other theories studied in this paper go through in a similar way. After some integral identities and algebraic manipulations, we may write:

$$\frac{\partial}{\partial z}\zeta^{\exp}_{(\Delta;\alpha_s)}(z)\big|_{z=0} = \zeta^{\exp-\log-1'}_{(\Delta;\alpha_s)}(0) + \zeta^{\exp-\log-2'}_{(\Delta;\alpha_s)}(0) + \zeta^{\exp-\mathrm{const}'}_{(\Delta;\alpha_s)}(0) + \zeta^{\exp-\psi'}_{(\Delta;\alpha_s)}(0). \tag{A37}$$

The only overall non-zero contribution will come from the fourth term, $\zeta^{\exp-\psi'}_{(\Delta;\alpha_s)}(0)$, and the contributions of the first three will cancel out, after taking into account the ghost modes and all other particles in the entire spectra of the theory.

To understand what these three terms are, let's return to the Type A non-minimal theory, the l.h.s. of (A37) is now,

$$\frac{\partial}{\partial z}\zeta^{\exp}_{(\Delta;\alpha_s)}(z)\big|_{z=0} = \int_0^{\infty} du \frac{(2s+1)u^3\left((2s+1)^2+u^2\right)\log\left(\left(\Delta-\frac{3}{2}\right)^2+u^2\right)}{12\left(e^{2\pi u}+1\right)}. \tag{A38}$$

Using (99), we can rewrite the above term into:

$$\underbrace{\int_0^{\infty} du \frac{(2s+1)u^3\left((2s+1)^2+u^2\right)}{12\left(e^{2\pi u}+1\right)}\log(u^2)}_{=\zeta^{\exp-\log-1'}_{(\Delta;\alpha_s)}(0)} + \int_0^{\infty} du \int_0^{(s-\frac{1}{2})^2} dx \frac{(2s+1)u^3\left((2s+1)^2+u^2\right)}{12\left(e^{2\pi u}+1\right)} \frac{1}{u^2+x} \tag{A39}$$

where $\Delta^{\mathrm{ph}} - \frac{3}{2} = s - \frac{1}{2}$. The second term can then be explicitly integrated using the recursive relation for $\int_0^k dx \frac{u^k}{e^{2\pi i u}\pm 1}\frac{1}{u^2+x}$ found in Appendix B.2,

$$
\begin{aligned}
&\int_0^{(s-\frac{1}{2})^2} dx \int_0^\infty du \, \frac{(2s+1)u^3\left((2s+1)^2+u^2\right)}{12(e^{2\pi u}+1)} \frac{1}{u^2+x} \\
&= \underbrace{\frac{1}{2}\int_0^{(s-\frac{1}{2})^2} dx \log(x) \left(\frac{1}{6}x(2s+1) - \frac{1}{6}\left(s+\frac{1}{2}\right)^2 (2s+1)\right)}_{\equiv \zeta^{\text{exp}-\log-2'}_{(\Delta\text{ph};[s])}(0)} + \underbrace{\frac{1}{3}B_1^+(2s+1)}_{\equiv \zeta^{\text{exp}-\text{const}'}_{(\Delta\text{ph};[s])}(0)} \\
&\quad \underbrace{+ \int_0^{(s-\frac{1}{2})^2} dx \, \psi\left(\sqrt{x}+\frac{1}{2}\right)\left(\frac{1}{6}x(-2s-1) + \frac{1}{6}(2s+1)\left(s+\frac{1}{2}\right)^2\right)}_{\equiv \zeta^{\text{exp}-\psi'}_{(\Delta\text{ph};[s])}(0)},
\end{aligned}
\tag{A40}
$$

where $B_k^{\pm} := \int_0^\infty du \frac{u^k}{e^{2\pi u}\pm 1}$, and $\psi(x)$ is the digamma function $\psi(x) = \Gamma'(x)/\Gamma(x)$. We concentrate on the last term including the digamma function, since it is the only term that contributes to the final partition function. To integrate the digamma function, we make use of its integral representation:

$$
\psi(x) = \int_0^\infty \left(\frac{e^{-t}}{t} - \frac{e^{-xt}}{1-e^{-t}}\right) dt \tag{A41}
$$

so that we get:

$$
\begin{aligned}
\zeta^{\text{exp}-\psi'}_{(\Delta\text{ph};[s])}(0) &= \int_0^\infty dt \int_0^{(s-\frac{1}{2})^2} dx \left(\frac{e^{-t}}{t} - \frac{e^{-(\sqrt{x}+\frac{1}{2})}}{1-e^{-t}}\right)\left(-\tfrac{1}{6}x(2s+1) + \tfrac{1}{24}(2s+1)^3\right) \\
&= \int_0^\infty dt \, \frac{(2s+1)e^{-t}}{24(e^t-1)t}\left\{e^{-st+2t}\left[-\frac{4(4s^2-8s+1)}{t} + 16s^2 + \frac{24-48s}{t^2} - 8s - \frac{48}{t^3}\right]\right. \\
&\qquad \left. + \tfrac{1}{4}\left(1-4s^2\right)^2\left(e^t-1\right) - \frac{2e^{\frac{3t}{2}}}{t^3}\left[(2st+t)^2 - 24\right] - \tfrac{1}{8}(1-2s)^4\left(e^t-1\right)\right\}
\end{aligned}
\tag{A42}
$$

The terms in the integrand above split into those that include a prefactor of e^{-st}, and those that do not. For the terms with the prefactor, we can sum over the spins easily and without a regulator,

$$
\begin{aligned}
&\sum_{s=1}^\infty \frac{(2s+1)e^{-t}}{24(e^t-1)t}\left\{e^{-st+2t}\left[16s^2 - 8s + \frac{-16s^2+32s-4}{t} + \frac{24-48s}{t^2} - \frac{48}{t^3}\right]\right\} \\
&= \frac{e^t}{6(e^t-1)^5 t^4}\left[t^2 + 3e^{3t}\left(2t^3+3t^2-6t-12\right) + e^t\left(6t^3 - 17t^2 + 42t - 60\right)\right. \\
&\qquad \left. + e^{2t}\left(36t^3 - 41t^2 - 18t + 84\right) - 6t + 12\right].
\end{aligned}
\tag{A43}
$$

For those terms without the prefactor, we sum using the same regulator as in the previous segment,

$$
\begin{aligned}
&\sum_{s=1}^\infty \left(s-\frac{1}{2}\right)^{-\epsilon} \frac{(2s+1)e^{-t}}{24(e^t-1)t}\left\{\tfrac{1}{4}\left(1-4s^2\right)^2\left(e^t-1\right) - \frac{2e^{\frac{3t}{2}}}{t^3}\left[(2st+t)^2-24\right] - \tfrac{1}{8}(1-2s)^4\left(e^t-1\right)\right\} \\
&= \frac{e^{t/2}}{6(e^t-1)t^4} - \frac{113e^{t/2}}{1440(e^t-1)t^2} + \frac{1609e^{-t}}{241{,}920(e^t-1)t} - \frac{1609}{241{,}920(e^t-1)t}.
\end{aligned}
\tag{A44}
$$

Combining (A43) and (A44) under the integrand, we obtain the expression for $\zeta^{\text{exp}-\psi'}_{\Delta\text{ph};[s]}(0)$. Then, repeating the calculations for the ghost calculations, we obtain:

$$
\begin{aligned}
\zeta^{\exp-\psi'}_{(\Delta^{\rm gh};[s-1])}(0) &= \int_0^\infty dt\left[\frac{13e^{t/2}}{6(e^t-1)t^4}+\frac{2}{(e^t-1)^2t^4}-\frac{4e^t}{(e^t-1)^3t^4}-\frac{4e^t\left(e^t+1\right)}{(e^t-1)^4t^3}+\frac{1}{(e^t-1)^2t^3}\right.\\
&\quad-\frac{2e^t\left(e^t+1\right)}{(e^t-1)^4t^2}-\frac{233e^{t/2}}{1440(e^t-1)t^2}+\frac{1}{6(e^t-1)^2t^2}+\frac{e^t}{(e^t-1)^3t^2}\\
&\quad-\frac{4e^t\left(4e^t+e^{2t}+1\right)}{3(e^t-1)^5t^2}+\frac{349e^{-t}}{241{,}920(e^t-1)t}-\frac{349}{241{,}920(e^t-1)t}+\frac{e^t}{3(e^t-1)^3t}\\
&\quad\left.-\frac{4e^t\left(4e^t+e^{2t}+1\right)}{3(e^t-1)^5t}\right].
\end{aligned}
\tag{A45}
$$

After combining these above with the integral representation for the scalar term, we then make use of the integral representation of the Hurwitz-Lerch transcendental function,

$$
\Phi(z,s,\nu)=\frac{1}{\Gamma(s)}\int_0^\infty dt\frac{t^{s-1}e^{-\nu t}}{1-ze^{-t}}=\sum_0^\infty(n+\nu)^{-s}z^n,
\tag{A46}
$$

to transform the expressions into sums of derivatives of Hurwitz-Lerch transcendental functions [27]. Finally, the Type A non-minimal theory will give us an expression of:

$$
\zeta^{\exp-\psi'}_{(1;[0])}(0)+\sum_{s=1}^\infty\zeta^{\exp-\psi'}_{(\Delta^{\rm ph};[s])}(0)-\sum_{s=1}^\infty\zeta^{\exp-\psi'}_{(\Delta^{\rm gh};[s-1])}(0)=0.
\tag{A48}
$$

Appendix C. Spectra of Higher Spin Theories and Their Free Energy Contributions

Appendix C.1. Type B Theories

Table A3. Results for Type B theory in AdS_5.

AdS_5				
Towers of Spins		**Contribution to F from One Tower Summed Over:**		
$(\Delta^{\rm ph};\alpha)$		**$s=1,2,3,\ldots$**	**$s=2,4,6,\ldots$**	**$s=1,3,5,\ldots$**
$(\Delta^{\rm ph};[s,1])$	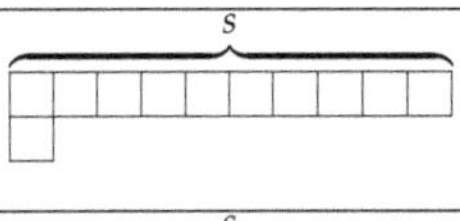	$\frac{1}{180}\log R$	$\frac{13}{360}\log R$	$-\frac{11}{360}\log R$
$(\Delta^{\rm ph};[s,0])$	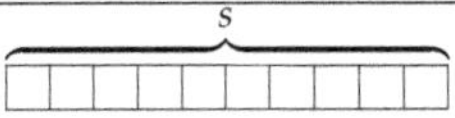	0	$\frac{1}{90}\log R$	$-\frac{1}{90}\log R$
Scalar		**Contribution to F by one scalar**		
$(3;[0,0])$		$-\frac{1}{180}\log R$		

[27] This makes use of the identity,

$$
\frac{1}{(1-e^{-t})^{n+1}(1+e^{-t})^{m+1}}=\frac{(-1)^n}{n!m!}\partial^n_{z_1}\partial^m_{z_2}\left[\frac{1}{z_1-z_2}\left(\frac{1}{z_1-e^{-t}}-\frac{1}{z_2-e^{-t}}\right)\right]\Bigg|_{z_1=1,z_2=-1}
\tag{A47}
$$

Table A4. Results for Type B theory in AdS_7.

AdS_7			
Towers of Spins	**Contribution to F from One Tower Summed Over:**		
$(\Delta^{ph}; \alpha)$	**$s = 1, 2, 3, \ldots$**	**$s = 2, 4, 6, \ldots$**	**$s = 1, 3, 5, \ldots$**
$(\Delta^{ph}; [s, 1, 1])$	$\frac{1}{1512} \log R$	$-\frac{211}{15,120} \log R$	$\frac{221}{15,120} \log R$
$(\Delta^{ph}; [s, 1, 0])$	$\frac{4}{945} \log R$	$-\frac{2}{315} \log R$	$\frac{2}{189} \log R$
$(\Delta^{ph}; [s, 0, 0])$	$-\frac{1}{1512} \log R$	$-\frac{1}{504} \log R$	$\frac{1}{756} \log R$
Scalar	**Contribution to F by one scalar**		
$(5; [0, 0, 0])$	$-\frac{4}{945} \log R$		

Table A5. Results for Type B theory in AdS_9.

AdS_9			
Towers of Spins	**Contribution to F from One Tower Summed Over:**		
$(\Delta^{ph}; \alpha)$	**$s = 1, 2, 3, \ldots$**	**$s = 2, 4, 6, \ldots$**	**$s = 1, 3, 5, \ldots$**
$(\Delta^{ph}; [s, 1, 1, 1])$	$\frac{23}{226,800} \log R$	$\frac{3463}{453,600} \log R$	$-\frac{1139}{151,200} \log R$
$(\Delta^{ph}; [s, 1, 1, 0])$	$\frac{13}{28,350} \log R$	$\frac{133}{16,200} \log R$	$-\frac{293}{37,800} \log R$
$(\Delta^{ph}; [s, 1, 0, 0])$	$\frac{353}{113,400} \log R$	$-\frac{1189}{226,800} \log R$	$-\frac{23}{10,800} \log R$
$(\Delta^{ph}; [s, 0, 0, 0])$	$-\frac{13}{28,350} \log R$	$-\frac{29}{113,400} \log R$	$-\frac{23}{113,400} \log R$
Scalar	**Contribution to F by one scalar**		
$(7; [0, 0, 0, 0])$	$-\frac{9}{2800} \log R$		

Table A6. Results for Type B theory in AdS_{11}.

$\mathbf{AdS_{11}}$				
Towers of Spins		**Contribution to F from One Tower Summed Over:**		
$(\Delta^{ph};\alpha)$		$s=1,2,3,\ldots$	$s=2,4,6,\ldots$	$s=1,3,5,\ldots$
$(\Delta^{ph};[s,1,1,1,1])$	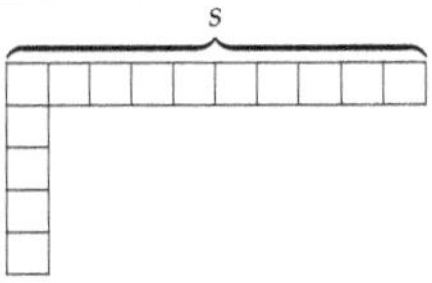	$\frac{263}{14,968,800}\log R$	$-\frac{19771}{4,276,800}\log R$	$\frac{138,923}{29,937,600}\log R$
$(\Delta^{ph};[s,1,1,1,0])$	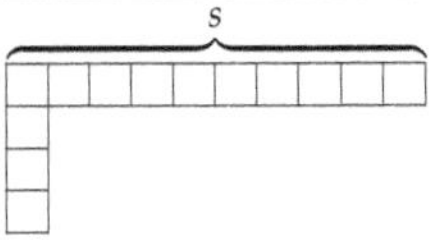	$\frac{31}{467,775}\log R$	$-\frac{2273}{374,220}\log R$	$\frac{11,489}{1,871,100}\log R$
$(\Delta^{ph};[s,1,1,0,0])$	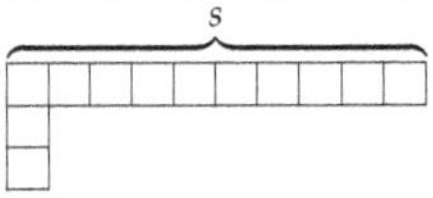	$\frac{311}{1,069,200}\log R$	$-\frac{6599}{2,993,760}\log R$	$\frac{37,349}{14,968,800}\log R$
$(\Delta^{ph};[s,1,0,0,0])$	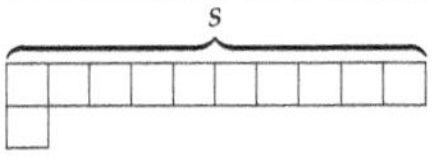	$\frac{1153}{467,775}\log R$	$-\frac{3947}{1,871,100}\log R$	$\frac{19}{53,460}\log R$
$(\Delta^{ph};[s,0,0,0,0])$	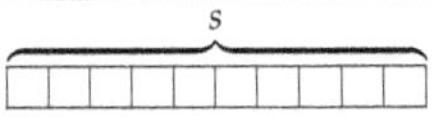	$-\frac{19}{61,600}\log R$	$-\frac{5143}{14,968,800}\log R$	$\frac{263}{7,484,400}\log R$
Scalar			**Contribution to F by one scalar**	
$(9;[0,0,0,0,0])$			$-\frac{1184}{467,775}\log R$	

Appendix C.2. Calculation of $\mathfrak{Z}^{\rm HS}_{\rm total}(z)$ in Type AB Theories

Appendix C.2.1. AdS_7

In this case,

$$\mathfrak{Z}^{\rm HS}_{\rm total}(z) = \frac{z^2\pi}{86,016}\Big(-11,253\zeta'(-10)+15,300\zeta'(-8)+119,658\zeta'(-6)-137,900\zeta'(-4)+21,735\zeta'(-2)\Big) + \mathcal{O}\left(z^3\right) \tag{A49}$$

which gives us $F^{(1)}_f = 0$, as we set $z\to 0$.

Appendix C.2.2. AdS_9

In this case,

$$\mathfrak{Z}^{\rm HS}_{\rm total}(z) = \frac{\pi}{16,647,192,576,000}\left[-136,525\zeta'(-14)+1,242,150\zeta'(-12)+2,651,957\zeta'(-10)\right.$$
$$\left.-42,097,100\zeta'(-8)+100,665,453\zeta'(-6)-71,501,850\zeta'(-4)+9,993,375\zeta'(-2)\right]z^2 + \mathcal{O}\left(z^3\right) \tag{A50}$$

which gives us $F_f^{(1)} = 0$, as we set $z = 0$.

Appendix C.3. Free Energy Values for Type C Theories in AdS_9

Table A7. Results for Type C theory in AdS_9.

AdS_9 Type C				
Towers of Spins		**Contribution to F from One Tower Summed Over:**		
$(\Delta^{ph}; \alpha)$		$s = 2, 3, 4, \ldots$	$s = 2, 4, 6, \ldots$	$s = 3, 5, 7, \ldots$
$(\Delta^{ph}; [s, 2, 2, 2])$	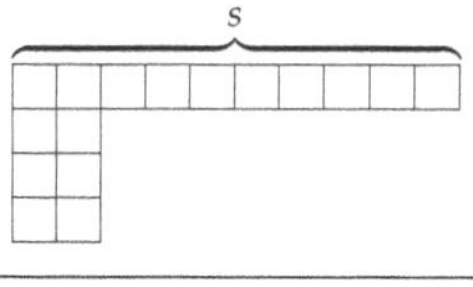	$\frac{23}{1800} \log R$	$\frac{2213}{3600} \log R$	$-\frac{2167}{3600} \log R$
$(\Delta^{ph}; [s, 2, 0, 0])$	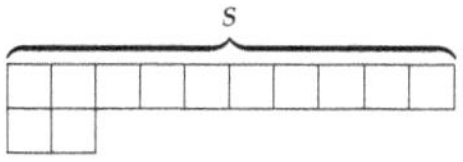	$\frac{3121}{6300} \log R$	$\frac{14,281}{37,800} \log R$	$\frac{127}{1080} \log R$
$(\Delta^{ph}; [s, 2, 1, 1])$	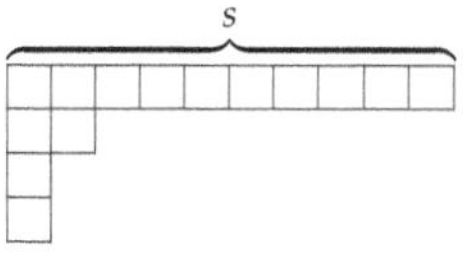	$\frac{19,409}{37,800} \log R$	$\frac{19,679}{75,600} \log R$	$-\frac{19,139}{75,600} \log R$
$(\Delta^{ph}; [s, 2, 2, 0])$	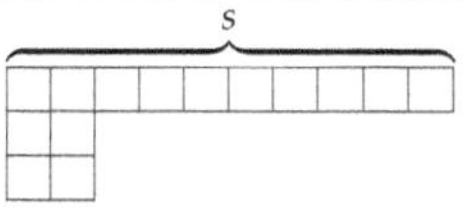	$\frac{329}{2700} \log R$	$-\frac{569}{5400} \log R$	$\frac{409}{1800} \log R$
$(\Delta^{ph}; [s, 1, 1, 0])$	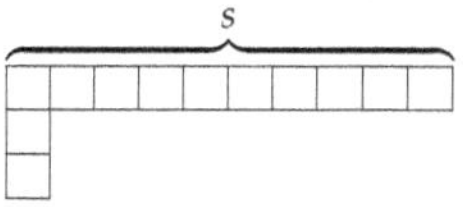	$\frac{31,399}{113,400} \log R$	$\frac{133}{16,200} \log R$	$\frac{2539}{9450} \log R$
$(\Delta^{ph}; [s, 0, 0, 0])$	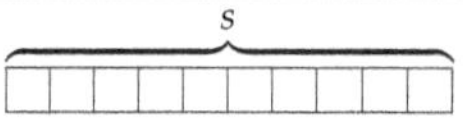	$\frac{35,293}{113,400} \log R$	$-\frac{29}{113,400} \log R$	$\frac{841}{2700} \log R$
Other Particles		**Contribution to F by one particle**		
$(8; [1, 1, 1, 1])$		$-\frac{908}{2835} \log R$		
$(8; [1, 1, 0, 0])$		$-\frac{1856}{14,175} \log R$		
$(8; [0, 0, 0, 0])$		$\frac{1978}{14,175} \log R$		

Appendix C.3.1. Spectra of Spins for Type C Theories

In these following results, $\alpha = [t_1, t_2, \ldots, t_{k-1}, t_k]_c$ denote two towers $\alpha = [t_1, t_2, \ldots, t_{k-1}, t_k]$ and $\alpha = [t_1, t_2, \ldots, t_{k-1}, -t_k]$, which at the level of computation of the spin factor $g_\alpha(s)$ and $\mu_\alpha(s)$ are indistinguishable. Hence, a single tower of $[t_1, t_2, \ldots, t_{k-1}, t_k]_c$ encompasses one of each of the towers, and correspondingly, a 1/2 tower encompasses only the $[t_1, t_2, \ldots, t_{k-1}, t_k]$ tower (the one with the positive t_k).

Table A8. Spectra of type C Theories.

AdS$_5$

	Non-Minimal $U(N)$			**Minimal $O(N)$**			**$U(N)$ Self-Dual**			**$O(N)$ Self-Dual**		
	$s=$			$s=$			$s=$			$s=$		
Towers $(\Delta^{\text{ph}};\alpha)$	$2,3,4,\ldots$	$2,4,6,\ldots$	$3,5,7,\ldots$	$2,3,4,\ldots$	$2,4,6,\ldots$	$3,5,7,\ldots$	$2,3,4,\ldots$	$2,4,6,\ldots$	$3,5,7,\ldots$	$2,3,4,\ldots$	$2,4,6,\ldots$	$3,5,7,\ldots$
$(\Delta^{\text{ph}};[s,0])$	2			2			1			$O(N)$ self-dual theory		
$(\Delta^{\text{ph}};[s,2]_c)$	1				1/2					not defined for CFT$_4$		
Other Particles												
$(4;[1,1])$	2											
$(4;[0,0])$	2			2								
$F^{(1)}$ Value	$+\frac{62}{45}\log R$			$+\frac{62}{45}\log R$			$+\frac{1}{4}\frac{62}{45}\log R$					

AdS$_7$

	Non-Minimal $U(N)$			**Minimal $O(N)$**			**$U(N)$ Self-Dual**			**$O(N)$ Self-Dual**		
	$s=$			$s=$			$s=$			$s=$		
Towers $(\Delta^{\text{ph}};\alpha)$	$2,3,4,\ldots$	$2,4,6,\ldots$	$3,5,7,\ldots$	$2,3,4,\ldots$	$2,4,6,\ldots$	$3,5,7,\ldots$	$2,3,4,\ldots$	$2,4,6,\ldots$	$3,5,7,\ldots$	$2,3,4,\ldots$	$2,4,6,\ldots$	$3,5,7,\ldots$
$(\Delta^{\text{ph}};[s,2,0])$	2			1								
$(\Delta^{\text{ph}};[s,1,1]_c)$	1					1/2	1/2					1/2
$(\Delta^{\text{ph}};[s,0,0])$	2				2		1				1	
$(\Delta^{\text{ph}};[s,2,2]_c)$	1				1/2		1/2				1/2	
Other Particles												
$(6;[1,1,0])$	2			1								
$(6;[0,0,0])$	2			1								
$F^{(1)}$ Value	$+\frac{221}{105}\log R$			0			$+\frac{1}{4}\frac{221}{105}\log R$			$-\frac{1}{8}\frac{221}{105}\log R$		

AdS$_9$	**Non-Minimal $U(N)$**			**Minimal $O(N)$**			**$U(N)$ Self-Dual**			**$O(N)$ Self-Dual**		
	$s=$			$s=$			$s=$			$s=$		
Towers $(\Delta^{\text{ph}};\alpha)$	$2,3,4,\ldots$	$2,4,6,\ldots$	$3,5,7,\ldots$	$2,3,4,\ldots$	$2,4,6,\ldots$	$3,5,7,\ldots$	$2,3,4,\ldots$	$2,4,6,\ldots$	$3,5,7,\ldots$	$2,3,4,\ldots$	$2,4,6,\ldots$	$3,5,7,\ldots$
$(\Delta^{\text{ph}};[s,0,0,0])$	2			1			1			$O(N)$ self-dual theory not defined for CFT$_8$		
$(\Delta^{\text{ph}};[s,1,1,0])$	2			1			1					
$(\Delta^{\text{ph}};[s,2,2,0])$	2			1			1					
$(\Delta^{\text{ph}};[s,2,1,1]_c)$	1					1/2						
$(\Delta^{\text{ph}};[s,2,0,0])$	2				2							
$(\Delta^{\text{ph}};[s,2,2,2]_c)$	1				1/2							
Other Particles												
$(8;[1,1,1,1])$		2			2							
$(8;[1,1,0,0])$		2										
$(8;[0,0,0,0])$		2			2							
$F^{(1)}$ Value		$+\frac{8051}{2835}\log R$			$+\frac{8051}{2835}\log R$			$+\frac{1}{4}\frac{8051}{2835}\log R$				

AdS$_{11}$

	Non-Minimal $U(N)$			Minimal $O(N)$			$U(N)$ Self-Dual			$O(N)$ Self-Dual		
	$s=$			$s=$			$s=$			$s=$		
Towers $(\Delta^{ph};\alpha)$	$2,3,4,\ldots$	$2,4,6,\ldots$	$3,5,7,\ldots$	$2,3,4,\ldots$	$2,4,6,\ldots$	$3,5,7,\ldots$	$2,3,4,\ldots$	$2,4,6,\ldots$	$3,5,7,\ldots$	$2,3,4,\ldots$	$2,4,6,\ldots$	$3,5,7,\ldots$
$(\Delta^{ph};[s,2,0,0,0])$	2			1								
$(\Delta^{ph};[s,2,1,1,0])$	2			1								
$(\Delta^{ph};[s,2,2,2,0])$	2			1								
$(\Delta^{ph};[s,1,1,0,0])$	2					2	1					1
$(\Delta^{ph};[s,2,2,1,1]_c)$	1					1/2	1/2					1/2
$(\Delta^{ph};[s,0,0,0,0])$	2				2		1				1	
$(\Delta^{ph};[s,1,1,1,1]_c)$	1				1/2		1/2				1/2	
$(\Delta^{ph};[s,2,2,0,0])$	2				2		1				1	
$(\Delta^{ph};[s,2,2,2,2]_c)$	1				1/2		1/2				1/2	
Other Particles												
$(10;[1,1,1,1,0])$	2			1								
$(10;[1,1,0,0,0])$	2			1								
$(10;[0,0,0,0,0])$	2			1								
$F^{(1)}$ Value	$+\frac{13,39,661}{374,220}\log R$			0			$+\frac{1}{4}\frac{1,339,661}{374,220}\log R$			$-\frac{1}{8}\frac{1,339,661}{374,220}\log R$		

AdS_{13}

	Non-Minimal $U(N)$			**Minimal $O(N)$**			**$U(N)$ Self-Dual**			**$O(N)$ Self-Dual**		
	$s=$			$s=$			$s=$			$s=$		
Towers $(\Delta^{ph};\alpha)$	$2,3,4,\ldots$	$2,4,6,\ldots$	$3,5,7,\ldots$	$2,3,4,\ldots$	$2,4,6,\ldots$	$3,5,7,\ldots$	$2,3,4,\ldots$	$2,4,6,\ldots$	$3,5,7,\ldots$	$2,3,4,\ldots$	$2,4,6,\ldots$	$3,5,7,\ldots$
$(\Delta^{ph};[s,0,0,0,0,0]$	2			1			1					
$(\Delta^{ph};[s,1,1,0,0,0]$	2			1			1					
$(\Delta^{ph};[s,1,1,1,1,0]$	2			1			1					
$(\Delta^{ph};[s,2,2,0,0,0]$	2			1			1					
$(\Delta^{ph};[s,2,2,1,1,0]$	2			1			1					
$(\Delta^{ph};[s,2,2,2,2,0]$	2			1			1				$O(N)$ self-dual theory not defined for CFT_{12}	
$(\Delta^{ph};[s,2,1,1,0,0]$	2					2						
$(\Delta^{ph};[s,2,2,2,1,1]_c$	1					1/2						
$(\Delta^{ph};[s,2,0,0,0,0]$	2				2							
$(\Delta^{ph};[s,2,1,1,1,1]_c$	1				1/2							
$(\Delta^{ph};[s,2,2,2,0,0]$	2				2							
$(\Delta^{ph};[s,2,2,2,2,2]_c$	1				1/2							
Other Particles												
$(12;[1,1,1,1,1,1])$		2										
$(12;[1,1,1,1,0,0])$		2			2							
$(12;[1,1,0,0,0,0])$		2										
$(12;[0,0,0,0,0,0])$		2			2							
$F^{(1)}$ Value	$+\frac{525,793,111}{121,621,500}\log R$			$+\frac{525,793,111}{121,621,500}\log R$			$+\frac{1}{4}\frac{525,793,111}{121,621,500}\log R$					

AdS$_{15}$

	Non-Minimal $U(N)$			Minimal $O(N)$			$U(N)$ Self-Dual			$O(N)$ Self-Dual		
	$s=$			$s=$			$s=$			$s=$		
Towers $(\Delta^{\text{ph}};\alpha)$	$2,3,4,\ldots$	$2,4,6,\ldots$	$3,5,7,\ldots$	$2,3,4,\ldots$	$2,4,6,\ldots$	$3,5,7,\ldots$	$2,3,4,\ldots$	$2,4,6,\ldots$	$3,5,7,\ldots$	$2,3,4,\ldots$	$2,4,6,\ldots$	$3,5,7,\ldots$
$(\Delta^{\text{ph}};[s,2,0,0,0,0,0]$	2			1								
$(\Delta^{\text{ph}};[s,2,1,1,0,0,0]$	2			1								
$(\Delta^{\text{ph}};[s,2,1,1,1,1,0]$	2			1								
$(\Delta^{\text{ph}};[s,2,2,2,0,0,0]$	2			1								
$(\Delta^{\text{ph}};[s,2,2,2,1,1,0]$	2			1								
$(\Delta^{\text{ph}};[s,2,2,2,2,2,0]$	2			1								
$(\Delta^{\text{ph}};[s,1,1,0,0,0,0]$	2					2	1					1
$(\Delta^{\text{ph}};[s,1,1,1,1,1,1]_c$	1					1/2	1/2					1/2
$(\Delta^{\text{ph}};[s,2,2,1,1,0,0]$	2					2	1					1
$(\Delta^{\text{ph}};[s,2,2,2,2,1,1]_c$	1					1/2	1/2					1/2
$(\Delta^{\text{ph}};[s,0,0,0,0,0,0]$	2				2		1				1	
$(\Delta^{\text{ph}};[s,1,1,1,1,0,0]$	2				2		1				1	
$(\Delta^{\text{ph}};[s,2,2,0,0,0,0]$	2				2		1				1	
$(\Delta^{\text{ph}};[s,2,2,1,1,1,1]_c$	1				1/2		1/2				1/2	
$(\Delta^{\text{ph}};[s,2,2,2,2,0,0]$	2				2		1				1	
$(\Delta^{\text{ph}};[s,2,2,2,2,2,2]_c$	1				1/2		1/2				1/2	
Other Particles												
$(14;[1,1,1,1,1,1,0])$	2			1								
$(14;[1,1,1,1,0,0,0])$	2			1								
$(14;[1,1,0,0,0,0,0])$	2			1								
$(14;[0,0,0,0,0,0,0])$	2			1								
$F^{(1)}$ Value	$+\frac{3,698,905,481}{729,729,000}\log R$			0			$-\frac{1}{4}\frac{3,698,905,481}{729,729,000}\log R$			$-\frac{1}{8}\frac{3,698,905,481}{729,729,000}\log R$		

References

1. Maldacena, J.M. The Large *N* Limit of Superconformal Field Theories and Supergravity. *Adv. Theor. Math. Phys.* **1998**, *2*, 231–252.
2. Gubser, S.S.; Klebanov, I.R.; Polyakov, A.M. Gauge Theory Correlators from Non-Critical String Theory. *Phys. Lett. B* **1998**, *428*, 105–114.
3. Witten, E. Anti-de Sitter Space and Holography. *Adv. Theor. Math. Phys.* **1998**, *2*, 253–291.
4. Fradkin, E.; Vasiliev, M.A. On the Gravitational Interaction of Massless Higher Spin Fields. *Phys. Lett. B* **1987**, *189*, 89–95.
5. Vasiliev, M.A. Consistent equation for interacting gauge fields of all spins in (3 + 1)-dimensions. *Phys. Lett. B* **1990**, *243*, 378–382.
6. Vasiliev, M.A. More on equations of motion for interacting massless fields of all spins in (3 + 1)-dimensions. *Phys. Lett. B* **1992**, *285*, 225–234.
7. Vasiliev, M. Nonlinear equations for symmetric massless higher spin fields in (A)dS(d). *Phys. Lett. B* **2003**, *567*, 139–151.
8. Vasiliev, M.A. Higher spin superalgebras in any dimension and their representations. *J. High Energy Phys.* **2004**, *2004*, 046.
9. Giombi, S.; Yin, X. The Higher Spin/Vector Model Duality. *J. Phys. A Math. Theor.* **2013**, *46*, 214003.
10. Giombi, S. TASI Lectures on the Higher Spin-CFT Duality. In Proceedings of the 2015 Theoretical Advanced Study Institute in Elementary Particle Physics, Boulder, CO, USA, 1–26 June 2015; pp. 137–214.
11. Klebanov, I.R.; Polyakov, A.M. AdS dual of the critical $O(N)$ vector model. *Phys. Lett. B* **2002**, *550*, 213–219.
12. Sezgin, E.; Sundell, P. Holography in 4D (super) higher spin theories and a test via cubic scalar couplings. *J. High Energy Phys.* **2005**, *2005*, 044.
13. Leigh, R.G.; Petkou, A.C. Holography of the N = 1 higher-spin theory on AdS(4). *J. High Energy Phys.* **2003**, *2003*, 011.
14. Aharony, O.; Gur-Ari, G.; Yacoby, R. d = 3 Bosonic Vector Models Coupled to Chern-Simons Gauge Theories. *J. High Energy Phys.* **2012**, *2012*, 037.
15. Giombi, S.; Minwalla, S.; Prakash, S.; Trivedi, S.P.; Wadia, S.R.; Yin, X. Chern-Simons Theory with Vector Fermion Matter. *Eur. Phys. J. C* **2012**, *72*, 2112.
16. Beccaria, M.; Tseytlin, A.A. Higher spins in AdS_5 at one loop: Vacuum energy, boundary conformal anomalies and AdS/CFT. *J. High Energy Phys.* **2014**, *2014*, 114.
17. Beccaria, M.; Tseytlin, A.A. Vectorial AdS_5/CFT_4 duality for spin-one boundary theory. *J. Phys. A Math. Theor.* **2014**, *47*, 492001.
18. Beccaria, M.; Macorini, G.; Tseytlin, A.A. Supergravity one-loop corrections on AdS_7 and AdS_3, higher spins and AdS/CFT. *Nucl. Phys. B* **2015**, *892*, 211–238.
19. Giombi, S.; Yin, X. Higher Spin Gauge Theory and Holography: The Three-Point Functions. *J. High Energy Phys.* **2010**, *2010*, 115.
20. Maldacena, J.; Zhiboedov, A. Constraining Conformal Field Theories with A Higher Spin Symmetry. *J. Phys. A Math. Theor.* **2013**, *46*, 214011.
21. Maldacena, J.; Zhiboedov, A. Constraining conformal field theories with a slightly broken higher spin symmetry. *Class. Quantum Gravity* **2013**, *30*, 104003.
22. Didenko, V.; Skvortsov, E. Towards higher-spin holography in ambient space of any dimension. *J. Phys. A Math. Theor.* **2013**, *46*, 214010.
23. Didenko, V.E.; Skvortsov, E.D. Exact higher-spin symmetry in CFT: All correlators in unbroken Vasiliev theory. *J. High Energy Phys.* **2013**, *2013*, 158.
24. Boulanger, N.; Kessel, P.; Skvortsov, E.D.; Taronna, M. Higher spin interactions in four-dimensions: Vasiliev versus Fronsdal. *J. Phys. A Math. Theor.* **2016**, *49*, 095402.
25. Bekaert, X.; Erdmenger, J.; Ponomarev, D.; Sleight, C. Quartic AdS Interactions in Higher-Spin Gravity from Conformal Field Theory. *J. High Energy Phys.* **2015**, *2015*, 149.
26. Giombi, S.; Klebanov, I.R. One Loop Tests of Higher Spin AdS/CFT. *J. High Energy Phys.* **2013**, *2013*, 068.
27. Giombi, S.; Klebanov, I.R.; Safdi, B.R. Higher Spin AdS_{d+1}/CFT_d at One Loop. *Phys. Rev. D* **2014**, *89*, 084004.
28. Giombi, S.; Klebanov, I.R.; Tseytlin, A.A. Partition Functions and Casimir Energies in Higher Spin AdS_{d+1}/CFT_d. *Phys. Rev. D* **2014**, *90*, 024048.

29. Jevicki, A.; Jin, K.; Yoon, J. 1/N and loop corrections in higher spin AdS_4/CFT_3 duality. *Phys. Rev. D* **2014**, *89*, 085039.
30. Zamolodchikov, A.B. Irreversibility of the Flux of the Renormalization Group in a 2D Field Theory. *J. Exp. Theor. Phys. Lett.* **1986**, *43*, 730–732.
31. Cardy, J.L. Is There a c Theorem in Four-Dimensions? *Phys. Lett. B* **1988**, *215*, 749–752.
32. Komargodski, Z.; Schwimmer, A. On Renormalization Group Flows in Four Dimensions. *J. High Energy Phys.* **2011**, *2011*, 099.
33. Myers, R.C.; Sinha, A. Seeing a C-Theorem with Holography. *Phys. Rev. D* **2010**, *82*, 046006.
34. Jafferis, D.L.; Klebanov, I.R.; Pufu, S.S.; Safdi, B.R. Towards the F-Theorem: $\mathcal{N} = 2$ Field Theories on the Three-Sphere. *J. High Energy Phys.* **2011**, *2011*, 102.
35. Klebanov, I.R.; Pufu, S.S.; Safdi, B.R. F-Theorem without Supersymmetry. *J. High Energy Phys.* **2011**, *2011*, 038.
36. Casini, H.; Huerta, M. On the RG running of the entanglement entropy of a circle. *Phys. Rev. D* **2012**, *85*, 125016.
37. Shenker, S.H.; Yin, X. Vector Models in the Singlet Sector at Finite Temperature. *arXiv* **2011**, arXiv:1109.3519.
38. Sachdev, S.; Ye, J.-W. Gapless spin fluid ground state in a random, quantum Heisenberg magnet. *Phys. Rev. Lett.* **1993**, *70*, 3339.
39. Kitaev, A. A Simple Model of Quantum Holography. Available online: http://online.kitp.ucsb.edu/online/entangled15/kitaev/ and http://online.kitp.ucsb.edu/online/entangled15/kitaev2/ (accessed on 11 January 2018)
40. Sachdev, S. Bekenstein-Hawking Entropy and Strange Metals. *Phys. Rev. X* **2015**, *5*, 041025.
41. Polchinski, J.; Rosenhaus, V. The Spectrum in the Sachdev-Ye-Kitaev Model. *J. High Energy Phys.* **2016**, *2016*, 1.
42. Jevicki, A.; Suzuki, K.; Yoon, J. Bi-Local Holography in the SYK Model. *J. High Energy Phys.* **2016**, *2016*, 007.
43. Maldacena, J.; Stanford, D. Remarks on the Sachdev-Ye-Kitaev model. *Phys. Rev. D* **2016**, *94*, 106002.
44. Maldacena, J.; Stanford, D.; Yang, Z. Conformal symmetry and its breaking in two dimensional Nearly Anti-de-Sitter space. *Progr. Theor. Exp. Phys.* **2016**, *2016*, 12C104.
45. Almheiri, A.; Polchinski, J. Models of AdS_2 backreaction and holography. *J. High Energy Phys.* **2015**, *2015*, 014.
46. Fernando, S.; Gunaydin, M. Minimal unitary representation of 5*d* superconformal algebra $F(4)$ and AdS_6/CFT_5 higher spin (super)-algebras. *Nucl. Phys. B* **2015**, *890*, 570–605.
47. Günaydin, M.; Skvortsov, E.D.; Tran, T. Exceptional $F(4)$ higher-spin theory in AdS_6 at one-loop and other tests of duality. *J. High Energy Phys.* **2016**, *2016*, 168.
48. Pang, Y.; Sezgin, E.; Zhu, Y. One Loop Tests of Supersymmetric Higher Spin AdS_4/CFT_3. *Phys. Rev. D* **2017**, *95*, 026008.
49. Camporesi, R. Zeta function regularization of one loop effective potentials in anti-de Sitter space-time. *Phys. Rev. D* **1991**, *43*, 3958–3965.
50. Camporesi, R.; Higuchi, A. Arbitrary spin effective potentials in anti-de Sitter space-time. *Phys. Rev. D* **1993**, *47*, 3339–3344.
51. Camporesi, R.; Higuchi, A. Spectral functions and zeta functions in hyperbolic spaces. *J. Math. Phys.* **1994**, *35*, 4217–4246.
52. Camporesi, R.; Higuchi, A. The Plancherel measure for p-forms in real hyperbolic spaces. *J. Geom. Phys.* **1994**, *15*, 57–94.
53. Flato, M.; Fronsdal, C. Quantum Field Theory of Singletons: The Rac. *J. Math. Phys.* **1981**, *22*, 1100.
54. Dolan, F. Character formulae and partition functions in higher dimensional conformal field theory. *J. Math. Phys.* **2006**, *47*, 062303.
55. Anselmi, D. Theory of higher spin tensor currents and central charges. *Nucl. Phys. B* **1999**, *541*, 323–368.
56. Anselmi, D. Higher spin current multiplets in operator product expansions. *Class. Quantum Gravity* **2000**, *17*, 1383–1400.
57. Alkalaev, K. Mixed-symmetry tensor conserved currents and AdS/CFT correspondence. *J. Phys. A Math. Theor.* **2013**, *46*, 214007.
58. Giombi, S.; Klebanov, I.R. Interpolating between *a* and *F*. *J. High Energy Phys.* **2015**, *2015*, 117.
59. Gaberdiel, M.R.; Gopakumar, R.; Saha, A. Quantum *W*-symmetry in AdS_3. *J. High Energy Phys.* **2011**, *2011*, 1–22.

60. Gaberdiel, M.R.; Grumiller, D.; Vassilevich, D. Graviton 1-loop partition function for 3-dimensional massive gravity. *J. High Energy Phys.* **2010**, *2010*, 094.
61. Gupta, R.K.; Lal, S. Partition Functions for Higher-Spin theories in AdS. *J. High Energy Phys.* **2012**, *2012*, 071.
62. Giombi, S.; Klebanov, I.R.; Pufu, S.S.; Safdi, B.R.; Tarnopolsky, G. AdS Description of Induced Higher-Spin Gauge Theory. *J. High Energy Phys.* **2013**, *2013*, 016.
63. Casini, H.; Huerta, M. Entanglement Entropy for the N-Sphere. *Phys. Lett. B* **2010**, *694*, 167–171.
64. Diaz, D.E.; Dorn, H. Partition functions and double-trace deformations in AdS/CFT. *J. High Energy Phys.* **2007**, *2007*, 046.
65. Casini, H.; Huerta, M.; Myers, R.C. Towards a Derivation of Holographic Entanglement Entropy. *J. High Energy Phys.* **2011**, *2011*, 036.
66. Frappat, L.; Sorba, P.; Sciarrino, A. Dictionary on Lie superalgebras. *arXiv* **1996**, arXiv:9607161.
67. Van Proeyen, A. Tools for supersymmetry. *Ann. Univer. Craiova Phys.* **1999**, *9*, 1–48.
68. Frappat, L.; Sciarrino, A.; Sorba, P. *Dictionary on Lie Algebras and Superalgebras;* Academic Press: San Diego, CA, USA, 2000.
69. Cappelli, A.; D'Appollonio, G. On the trace anomaly as a measure of degrees of freedom. *Phys. Lett. B* **2000**, *487*, 87–95.

universe

MDPI

Article

Three Point Functions in Higher Spin AdS_3 Holography with $1/N$ Corrections

Yasuaki Hikida [1,*] and Takahiro Uetoko [2]

1 Center for Gravitational Physics, Yukawa Institute for Theoretical Physics, Kyoto University, Kyoto 606-8502, Japan
2 Department of Physical Sciences, College of Science and Engineering, Ritsumeikan University, Shiga 525-8577, Japan; rp0019fr@ed.ritsumei.ac.jp
* Correspondence: yhikida@yukawa.kyoto-u.ac.jp

Received: 31 August 2017; Accepted: 25 September 2017; Published: 9 October 2017

Abstract: We examine three point functions with two scalar operators and a higher spin current in 2d W_N minimal model to the next non-trivial order in $1/N$ expansion. The minimal model was proposed to be dual to a 3d higher spin gauge theory, and $1/N$ corrections should be interpreted as quantum effects in the dual gravity theory. We develop a simple and systematic method to obtain three point functions by decomposing four point functions of scalar operators with Virasoro conformal blocks. Applying the method, we reproduce known results at the leading order in $1/N$ and obtain new ones at the next leading order. As confirmation, we check that our results satisfy relations among three point functions conjectured before.

Keywords: conformal field theory; W algebra; AdS/CFT correspondence; higher spin gauge theory

1. Introduction

Holography is expected to offer a way to learn quantum corrections of gravity theory from $1/N$ corrections in dual conformal field theory. In this paper, we address this issue by utilizing one of the simplest holographies proposed in [1][1], where 2d W_N minimal model is dual to Prokushkin-Vasiliev theory on AdS_3 given by [6]. We examine three point functions with two scalar operators and one higher spin current in the minimal model up to the next leading order in $1/N$ expansion. They should be interpreted as one-loop corrections to three point interactions between two bulk scalars and one higher spin gauge fields in the dual higher spin theory. We develop a simple and systematic method to compute the three point functions by decomposing four point functions of scalar operators with Virasoro conformal blocks. Among others, we expect that this way of computation makes the dual higher spin interpretation easier. Applying the method, we reproduce known results at the leading order in $1/N$ obtained by [7,8]. Exact results are available only up to correlators with spin 5 current [9–11], but a simple relation was conjectured for generic s in [11]. We obtain the $1/N$ corrections of correlators with spin $s \leq 8$ current, and the results for $s = 6, 7, 8$ should be new. We check that they satisfy the conjectured relation as confirmation of our results.

1 Recently, a different method to the issue has been adopted in [2–4] by analyzing the strongly coupled regime of conformal field theories in $1/N$ expansion. This becomes possible because of recent developments on conformal bootstrap technique, e.g., in [5].

We would like to examine the W_N minimal model in $1/N$ expansion, but we should specify the expansion in more details. The minimal model has a coset description

$$\frac{\mathrm{su}(N)_k \oplus \mathrm{su}(N)_1}{\mathrm{su}(N)_{k+1}} \tag{1}$$

whose central charge is given by

$$c = (N-1)\left(1 - \frac{N(N+1)}{(N+k)(N+k+1)}\right) \tag{2}$$

The model has two parameters N, k. For our purpose, it is convenient to define the 't Hooft coupling

$$\lambda = \frac{N}{N+k} \tag{3}$$

and label the model by N, λ instead of N, k. We then expand the model in $1/N$, where each order depends on the other parameter λ. The expansion is almost the same as $1/c$ expansion because of $c \sim N(1-\lambda^2) + \mathcal{O}(N^0)$, but details are different.

The minimal model is argued to be dual to the higher spin theory of [6], which includes higher spin gauge fields $\varphi^{(s)}$ $(s = 2, 3, 4, \ldots)$ and complex scalar fields $\phi_\pm$ with mass $m^2 = -1 + \lambda^2$. The large N limit of minimal model with λ in Equation (3) kept finite corresponds to the classical limit of higher spin theory, where λ is identified with the parameter in bulk scalar mass. The higher spin gauge fields $\varphi^{(s)}$ and bulk scalars $\phi_\pm$ are dual to higher spin currents $J^{(s)}$ and scalar operators $\mathcal{O}_\pm$, respectively. Here different boundary conditions are assigned to the bulk scalars $\phi_\pm$ and the dual conformal dimensions are given by $\Delta_\pm = 2h_\pm = 1 \pm \lambda$ at the tree level.

Basic data of conformal field theory may be given by spectrum and three point functions of primary operators. Since higher spin symmetry of the minimal model is exact, spectrum does not receive any corrections in $1/N$. Namely, there is no anomalous dimension for higher spin current $J^{(s)}$. Therefore, as simple but non-trivial examples, we examine three point functions and specifically focus on those with two scalar operators and one higher spin current as

$$\langle \mathcal{O}_\pm(z_1)\bar{\mathcal{O}}_\pm(z_2)J^{(s)}(z_3)\rangle \tag{4}$$

with $s = 2, 3, 4, \ldots$. Here $\bar{\mathcal{O}}_\pm$ are complex conjugate of $\mathcal{O}_\pm$. In [7,8], the three point functions in the large N limit of the minimal model have been computed from classical higher spin theory. They were reproduced with conformal field theory approach in [8,12,13][2], but these methods are applicable only to the leading order analysis in $1/N$. Since the W_N minimal model is solvable, for instance, by making use of the coset description in Equation (1), we can obtain the three point functions in Equation (4) with finite N, k in principle. However, in practice, the computation would be quite complicated, and only explicit expressions are available only with spin $3, 4, 5$ currents [9–11] (see also [15] for an alternative algebraic method).

In this paper, we develop a different way to compute the three point functions in Equation (4) from the decomposition of scalar four point functions by Virasoro conformal blocks. Our method may

[2] The analysis in [12,13] were made in the context of $\mathcal{N} = 2$ holographic duality in [14], but we can see that the analysis reduces to that for the bosonic case under a suitable truncation at the leading order in $1/N$.

be explained as follows; Let us consider a generic operator product expansion of scalar operators $\mathcal{O}_i$ with conformal weights (h_i, h_i) as

$$\mathcal{O}_1(z_1)\mathcal{O}_2(z_2) = \sum_p \frac{C_{12p}}{z_{12}^{h_1+h_2-h_p}\bar{z}_{12}^{h_1+h_2-\bar{h}_p}} \mathcal{A}_p(z_2) + \cdots \tag{5}$$

where the coefficient C_{12p} includes the information of three point function. Moreover, $\mathcal{A}_p$ has conformal weights $(h_p, \bar{h}_p)$, and dots denote contributions from descendants. Using the expansion, we can decompose scalar four point function as

$$\langle \mathcal{O}_1(\infty)\mathcal{O}_2(1)\mathcal{O}_3(z)\mathcal{O}_4(0)\rangle = \sum_p \frac{C_{12p}C_{34p}}{|z|^{2(h_3+h_4)}} \mathcal{F}(c, h_i, h_p, z)\bar{\mathcal{F}}(c, h_i, \bar{h}_p, \bar{z}) \tag{6}$$

Here $\mathcal{F}(c, h_i, h_p, z)$ is Virasoro conformal block, which can be fixed only from the symmetry in principle. Once we know scalar four point functions and Virasoro conformal blocks, we can read off coefficients as C_{12p} by solving constraint equations coming from Equation (6). For our case with $\mathcal{O}_i = \mathcal{O}_\pm$ or $\bar{\mathcal{O}}_\pm$, four point functions can be computed exactly with finite N, k, for instance, by applying Coulomb gas approach as in [16]. On the other hand, Virasoro conformal blocks are quite complicated, but explicit forms may be obtained by applying Zamolodchikov's recursion relation [17], see also [18,19]. We can find other works on the $1/c$ expansion of Virasoro conformal blocks in, e.g., [20–23]. Gathering these knowledges, we shall obtain the coefficients as C_{12p} up to the next leading order in $1/N$ expansion.

The paper is organized as follows; In order to study the decomposition in Equation (6), we need to examine scalar four point functions and Virasoro conformal blocks. In the next section we decompose scalar four point functions in terms of cross ratio z, and in Section 3 we give the explicit expressions of Virasoro conformal blocks in expansions both in $1/N$ and z. After these preparations, we compute three point functions in Equation (4) by solving constraint equations coming from Equation (6) in Section 4. In Section 4.1 we reproduce known results at the leading order in $1/N$. In Section 4.2 we obtain the $1/N$ corrections of three point functions for $s = 3, 4, \ldots, 8$, and check that they satisfy the relation conjectured in [11]. Section 5 is devoted to conclusion and discussions. In Appendix A we examine Virasoro conformal blocks in expansions of $1/c$ and z by analyzing Zamolodchikov's recursion relation. In Appendix B we compute three point functions with higher spin currents of double trace type.

2. Expansions of Four Point Functions

We would like to obtain the coefficients as C_{12p} by solving Equation (6). For the purpose, we need information on the both sides of the equation, i.e., scalar four point functions and Virasoro conformal blocks. In this section we examine scalar four point functions. We are interested in three point functions of two scalar operators $\mathcal{O}_\pm$ and a higher spin current $J^{(s)}$ as in Equation (4). We consider the following four point functions with scalar operators $\mathcal{O}_\pm$ as

$$G_{++}(z) \equiv \langle \mathcal{O}_+(\infty)\bar{\mathcal{O}}_+(1)\mathcal{O}_+(z)\bar{\mathcal{O}}_+(0)\rangle \tag{7}$$
$$G_{--}(z) \equiv \langle \mathcal{O}_-(\infty)\bar{\mathcal{O}}_-(1)\mathcal{O}_-(z)\bar{\mathcal{O}}_-(0)\rangle \tag{8}$$
$$G_{-+}(z) \equiv \langle \mathcal{O}_-(\infty)\mathcal{O}_+(1)\bar{\mathcal{O}}_+(z)\bar{\mathcal{O}}_-(0)\rangle \tag{9}$$

Exact expressions with finite N, k may be found in [16]. From the expansions in z, we can read off what kind of operators are involved in the decomposition by Virasoro conformal blocks. In the rest of this section, we obtain the explicit forms of four point functions in z expansion for parts relevant to later analysis.

Let us first examine the z expansion of $G_{++}(z)$ in Equation (7), and see generic properties of the four point functions. The expression with finite N, k is [16]

$$G_{++}(z) = |z(1-z)|^{-2\Delta_+} \left[\left| (1-z)^{1+\lambda} {}_2F_1\left(1+\frac{\lambda}{N}, -\frac{\lambda}{N}; -\lambda; z\right) \right|^2 + \mathcal{N}_1 \left| z^{1+\lambda} {}_2F_1\left(1+\frac{\lambda}{N}, -\frac{\lambda}{N}; 2+\lambda; z\right) \right|^2 \right] \tag{10}$$

with

$$\mathcal{N}_1 = -\frac{\Gamma(1+\lambda-\frac{\lambda}{N})\Gamma(-\lambda)^2\Gamma(2+\lambda+\frac{\lambda}{N})}{\Gamma(-1-\lambda-\frac{\lambda}{N})\Gamma(\frac{\lambda}{N}-\lambda)\Gamma(2+\lambda)^2} \tag{11}$$

Here the exact value of conformal dimensions $\Delta_+ = 2h_+$ is

$$\Delta_+ = \frac{(N-1)(2N+1+k)}{N(N+k)} = 1+\lambda-\frac{1}{N}-\frac{1}{N^2}\lambda+\mathcal{O}(N^{-3}) \tag{12}$$

which is expanded in $1/N$ up to the N^{-2} order.

In the expansion in z, we would like to pick up the terms corresponding to the three point function in Equation (4). The operator product of $\mathcal{O}_+$ may be expanded as

$$\begin{aligned} \mathcal{O}_+(z)\bar{\mathcal{O}}_+(0) =& \frac{1}{|z|^{2\Delta_+}} + \sum_s \frac{C_+^{(s)} z^s}{|z|^{2\Delta_+}} J^{(s)}(0) \\ &+ \sum_{(s_1,s_2;s')} \frac{C_+^{(s_1,s_2;s')} z^{s'}}{|z|^{2\Delta_+}} J^{(s_1,s_2;s')}(0) + \sum_{n,m} C_+^{(n,m)} z^n \bar{z}^m \mathcal{A}_{(n,m)}(0) \cdots \end{aligned} \tag{13}$$

Here $J^{(s_1,s_2;s')}(z)$ are higher spin currents of double trace type as

$$J^{(s_1,s_2;s')} = J^{(s_1)}\partial^{s'-s_1-s_2} J^{(s_2)} + \cdots \tag{14}$$

with $s' \geq 6$ as $s_1, s_2 \geq 3$ and $s' - s_1 - s_2 \geq 0$. If we use the normalization as $\langle J^{(s)} J^{(s)} \rangle \propto N$, then the two point function of this type of operator becomes $\langle J^{(s_1,s_2;s')} J^{(s_1,s_2;s')} \rangle \propto N^2$. This is related to the fact that $C_+^{(s)} \propto N^{-1/2}$, while $C_+^{(s_1,s_2;s)} \propto N^{-1}$. There could be currents of other multi-trace type, but the contributions are more suppressed in $1/N$. Furthermore, $\mathcal{A}_{(n,m)}(z)$ are double trace type operators of the form as

$$\mathcal{A}_{(n,m)} = \mathcal{O}_+ \partial^n \bar{\partial}^m \bar{\mathcal{O}}_+ + \cdots \tag{15}$$

and the conformal weights are $(h_{n,m}, \bar{h}_{n,m}) = (2h_+ + n, 2h_+ + m)$. The dots in Equation (13) include the operators dressed by higher spin currents $J^{(s)}(z), \bar{J}^{(s)}(\bar{z})$ for instance.

The operator product expansion in Equation (13) suggests that the contributions from $J^{(s)}$ or its descendants are included in terms like $z^{s+l}/|z|^{2\Delta_+}$, where $l = 0, 1, 2, \ldots$ corresponds to the level of descendant. In Equation (10), such terms appear as

$$G_{++}(z) = |z|^{-2\Delta_+}(1-z)^{-\Delta_+ +1+\lambda} {}_2F_1\left(1+\frac{\lambda}{N}, -\frac{\lambda}{N}; -\lambda; z\right) + \cdots \tag{16}$$

Note that they also include effects from higher spin currents of double trace type $J^{(s_1,s_2;s')}(z)$ among others. For the first term in Equation (10), the other contributions involve at least one anti-holomorphic current $\bar{J}^{(s)}(\bar{z})$. For the second term in Equation (10), the expansions become polynomials of z and $\bar{z}$ at the leading order in $1/N$, and this implies that double trace type operators $\mathcal{A}_{(n,m)}$ should appear as $\mathcal{A}_p$ in Equation (5). At the leading order in $1/N$, we can expand Equation (16) around $z \sim 0$ as

$$G_{++}(z) \sim |z|^{-2(\lambda+1)} \tag{17}$$

This corresponds to the expansion by the identity operator in Equation (13). Thus the non-trivial contributions to our three point functions come form the terms at least of order $1/N$.

At the next and next-to-next orders in $1/N$, there are two types of contributions in Equation (16). One comes from

$$(1-z)^{-\Delta_+ +1+\lambda} = (1-z)^{\frac{1}{N}+\frac{\lambda}{N^2}} + \mathcal{O}(N^{-3}) \tag{18}$$

which becomes

$$\begin{aligned}&(1-z)^{\frac{1}{N}}(1-z)^{\frac{\lambda}{N^2}}\\ &\quad = 1 - \frac{1}{N}\sum_{k=1}\frac{1}{k}z^k + \frac{1}{N^2}\left[\sum_{k=1}^{\infty}\left(-\frac{\lambda}{k}z^k\right) + \sum_{k=2}^{\infty}\frac{1}{k}H_{k-1}z^k\right] + \mathcal{O}(N^{-3})\end{aligned} \tag{19}$$

Here we have used for $k \geq 2$

$$\begin{aligned}\binom{\frac{1}{N}}{k} &= \frac{\Gamma(1+\frac{1}{N})}{k!\Gamma(1-k+\frac{1}{N})} = (-1)^{k-1}\frac{1}{Nk!}\left(1-\frac{1}{N}\right)\left(2-\frac{1}{N}\right)\cdots\left(k-1-\frac{1}{N}\right)\\ &= (-1)^{k-1}\frac{1}{Nk}\left(1-\frac{1}{N}\sum_{i=1}^{k-1}\frac{1}{i}\right) + \mathcal{O}(N^{-3})\end{aligned} \tag{20}$$

and the definition of harmonic number

$$H_n = \sum_{j=1}^{n}\frac{1}{j} \tag{21}$$

The other comes from the hypergeometric function, which can be similarly expanded in $1/N$ as

$$\begin{aligned}&{}_2F_1\left(1+\frac{\lambda}{N}, -\frac{\lambda}{N}, -\lambda; z\right) = \frac{\Gamma(-\lambda)}{\Gamma(1+\frac{\lambda}{N})\Gamma(-\frac{\lambda}{N})}\sum_{n=0}^{\infty}\frac{\Gamma(1+\frac{\lambda}{N}+n)\Gamma(-\frac{\lambda}{N}+n)}{\Gamma(-\lambda+n)}\frac{z^n}{n!}\\ &= 1 + \frac{\Gamma(1-\lambda)}{N}\sum_{n=1}^{\infty}\frac{\Gamma(n)}{\Gamma(n-\lambda)}z^n + \frac{1}{N^2}\left(\lambda z + \lambda\Gamma(1-\lambda)\sum_{n=2}^{\infty}\frac{\Gamma(n)}{n\Gamma(n-\lambda)}z^n\right) + \mathcal{O}(N^{-3})\end{aligned} \tag{22}$$

In total, we have

$$|z|^{2\Delta_+}G_{++}(z) \sim 1 + \frac{1}{N}\sum_{n=1}^{\infty}\left(-\frac{1}{n} + \frac{\Gamma(1-\lambda)\Gamma(n)}{\Gamma(n-\lambda)}\right)z^n + \frac{1}{N^2}\sum_{n=2}^{\infty}f_{++}^{(n)}z^n \tag{23}$$

where

$$f_{++}^{(n)} = -\sum_{l=1}^{n-1}\frac{\Gamma(1-\lambda)\Gamma(l)}{(n-l)\Gamma(l-\lambda)} - \frac{\lambda}{n} + \frac{H_{n-1}}{n} + \frac{\lambda\Gamma(1-\lambda)\Gamma(n)}{n\Gamma(n-\lambda)} \tag{24}$$

First few expressions are

$$f^{(2)}_{++} = \frac{1}{2}\left(-2 - \frac{1}{\lambda - 1} - \lambda\right) \tag{25}$$

$$f^{(3)}_{++} = \frac{1}{3}\left(\frac{4}{\lambda - 2} + \frac{1}{\lambda - 1} - \lambda\right) \tag{26}$$

$$f^{(4)}_{++} = \frac{1}{8}\left(1 - \frac{18}{\lambda - 3} + \frac{8}{\lambda - 2} + \frac{14}{\lambda - 1} - 2\lambda\right) \tag{27}$$

We would like to move to another four point function $G_{--}(z)$ in Equation (8), whose expression with finite N, k can be again found in [16]. We use the four point function in order to obtain the three point function in Equation (4) with the other type of scalar operator $\mathcal{O}_-$. As for $G_{++}(z)$, the relevant part is

$$G_{--}(z) = |z|^{-2\Delta_-}(1-z)^{\frac{1}{N} - \frac{\lambda}{N^2}} {}_2F_1\left(1 - \frac{\lambda}{N}, \frac{\lambda}{N} - \frac{\lambda^2}{N^2}; \lambda - \frac{\lambda^2}{N}; z\right) + \cdots \tag{28}$$

Here we may need

$$\begin{aligned} \Delta_- &= 2h_- = \tfrac{N-1}{N}\left(1 - \tfrac{N+1}{N+k+1}\right) \\ &= 1 - \lambda - \tfrac{1}{N}(1-\lambda^2) + \tfrac{1}{N^2}\lambda(1-\lambda^2) + \mathcal{O}(N^{-3}) \end{aligned} \tag{29}$$

Similarly to $G_{++}(z)$ we can expand $G_{--}(z)$ in z as

$$|z|^{2\Delta_-}G_{--}(z) \sim 1 + \frac{1}{N}\sum_{n=1}^{\infty}\left(-\frac{1}{n} + \frac{\Gamma(1+\lambda)\Gamma(n)}{\Gamma(n+\lambda)}\right)z^n + \frac{1}{N^2}\sum_{n=2}^{\infty} f^{(n)}_{--}z^n \tag{30}$$

where

$$f^{(n)}_{--} = -\sum_{l=1}^{n-1}\tfrac{\Gamma(1+\lambda)\Gamma(l)}{(n-l)\Gamma(l+\lambda)} + \tfrac{\lambda}{n} + \tfrac{H_{n-1}}{n} + \tfrac{\lambda\Gamma(1+\lambda)\Gamma(n)}{\Gamma(n+\lambda)}\left(\sum_{k=1}^{n-1}\tfrac{\lambda}{k+\lambda} - \tfrac{1}{n}\right) \tag{31}$$

First few expressions are

$$f^{(2)}_{--} = \frac{\lambda}{2} - \frac{3}{2(\lambda+1)} + \frac{1}{(\lambda+1)^2} \tag{32}$$

$$f^{(3)}_{--} = \frac{\lambda}{3} - \frac{13}{3(\lambda+1)} + \frac{2}{(\lambda+1)^2} + \frac{20}{3(\lambda+2)} - \frac{8}{(\lambda+2)^2} \tag{33}$$

$$f^{(4)}_{--} = \frac{\lambda}{4} - \frac{31}{4(\lambda+1)} + \frac{3}{(\lambda+1)^2} + \frac{23}{\lambda+2} - \frac{24}{(\lambda+2)^2} - \frac{63}{4(\lambda+3)} + \frac{27}{(\lambda+3)^2} + \frac{1}{8} \tag{34}$$

From the four point functions $G_{\pm\pm}(z)$, we can read off the square root of coefficients $(C^{(s)}_{\pm})^2$, but relative phase factor cannot be fixed. In order to determine it, we also need to examine $G_{-+}(z)$ in Equation (9), which can be computed as [16]

$$G_{-+}(z) = |1-z|^{-2\Delta_+}|z|^{\frac{2}{N}}\left|1 + \frac{1-z}{Nz}\right|^2 \tag{35}$$

with finite N, k. For later arguments, we need

$$|1-z|^{2\Delta_+}G_{-+}(z) \sim 1 + \frac{1}{N}\sum_{n=2}^{\infty}\frac{n-1}{n}(1-z)^n \tag{36}$$

which is expanded in $(1-z)$ up to the $1/N$ order.

3. Virasoro Conformal Blocks

In the previous section we analyzed the left hand side of Equation (6). In order to obtain three point functions by solving the equations in Equation (6), we further need information on $\mathcal{F}(c, h_i, h_p, z)$. In general, the forms of Virasoro conformal blocks are quite complicated. In practice, we actually do not need to know closed forms but expansions in z up to some orders. For this purpose, a standard approach may be solving Zamolodchikov's recursion relation in [17]. Following the algorithm developed in [18] (see also [19]), we obtain the expressions of Virasoro conformal blocks to several orders in z and $1/c$ in Appendix A. Related works may be found in [19–23], and in particular, some closed form expressions were given, e.g., in [20]. Our findings agree with their results after minor modifications.

Let us consider the four point function in the decomposition of Equation (6) with $h_1 = h_2$ and $h_3 = h_4$. In the decomposition, intermediate operator $\mathcal{A}_p$ can be the identity or other. As observed in the examples of previous section, only the Virasoro conformal block with the identity operator (called as vacuum block) survives at the leading order in $1/N$. This simply means that the four point functions are factorized into the products of two point ones at the leading order in $1/N$. Virasoro conformal block with $\mathcal{A}_p$ of single trace type would appear at the next leading order in $1/N$. We would like to examine $1/N$ corrections to three point functions, so we need $1/N$ corrections to the Virasoro block of $\mathcal{A}_p$. This also implies that we need the expression of vacuum block up to the next-to-next leading order in $1/N$.

Let us first examine the vacuum block with $h_1 = h_3 = h_\pm$. As was explained in Appendix A, the $1/c$-expansion of vacuum block is given by

$$\begin{aligned} \mathcal{V}_0(x) = \; & 1 + \tfrac{2h_1h_3}{c} z^2 {}_2F_1(2,2;4;z) \\ & + \tfrac{1}{c^2}\left[h_1^2h_3^2 k_a(z) + h_1h_3(h_1+h_3)k_b(z) + h_1h_3k_c(z)\right] + \mathcal{O}(c^{-3}) \end{aligned} \tag{37}$$

with

$$\begin{aligned} k_a(z) &= 2z^4 + 4z^5 + \frac{28z^6}{5} + \frac{34z^7}{5} + \frac{2687z^8}{350} + \mathcal{O}(z^9) \\ k_b(z) &= \frac{2z^4}{5} + \frac{4z^5}{5} + \frac{39z^6}{35} + \frac{47z^7}{35} + \frac{263z^8}{175} + \mathcal{O}(z^9) \\ k_c(z) &= \frac{2z^4}{25} + \frac{4z^5}{25} + \frac{109z^6}{490} + \frac{131z^7}{490} + \frac{1879z^8}{6300} + \mathcal{O}(z^9) \end{aligned} \tag{38}$$

The $1/c$ order term corresponds to the exchange of spin 2 current (energy momentum tensor) in terms of global block. We need to rewrite the expansion in $1/c$ by that in $1/N$ as

$$\mathcal{V}_0(z) = \mathcal{V}_0^{(0)}(z) + \mathcal{V}_0^{(1)}(z)\frac{1}{N} + \mathcal{V}_0^{(2)}(z)\frac{1}{N^2} + \mathcal{O}(N^{-3}) \tag{39}$$

The first two terms can be easily read off as

$$\mathcal{V}_0^{(0)}(z) = 1\,, \quad \mathcal{V}_0^{(1)}(z) = \frac{1}{2}\left(\frac{1\pm\lambda}{1\mp\lambda}\right) z^2 {}_2F_1(2,2;4;z) \tag{40}$$

Since there are two types of contributions to $\mathcal{V}_0^{(2)}$, we separate it into two parts as

$$\mathcal{V}_0^{(2)} = \mathcal{V}_0^{(2,1)}(z) + \mathcal{V}_0^{(2,2)}(z) \tag{41}$$

One comes from the $1/c$ order term with the next leading contribution from $h_\pm^2/c$ as

$$\begin{aligned}\mathcal{V}_0^{(2,1)}(z) &= f_{\pm\pm}^{(2)} z^2 {}_2F_1(2,2;4;z) \\ &= f_{\pm\pm}^{(2)} \left(z^2 + z^3 + \tfrac{9z^4}{10} + \tfrac{4z^5}{5} + \tfrac{5z^6}{7} + \tfrac{9z^7}{14} + \tfrac{7z^8}{12}\right) + \mathcal{O}(z^9)\end{aligned} \tag{42}$$

where $f_{\pm\pm}^{(2)}$ were given in Equations (25) and (32). Here we have used

$$\frac{2h_\pm^2}{c} = \frac{1}{2N}\left(\frac{1\pm\lambda}{1\mp\lambda}\right) + \frac{1}{N^2} f_{\pm\pm}^{(2)} + \mathcal{O}(N^{-3}) \tag{43}$$

which are obtained from the $1/N$ expansions of $h_\pm$ as in Equations (12) and (29) and c in Equation (2) as

$$c = N(1-\lambda^2)\left[1 - \frac{1}{N}\left(\lambda + \frac{1}{1+\lambda}\right)\right] + \mathcal{O}(N^{-1}) \tag{44}$$

The other comes from the $1/c^2$ order terms in Equation (37) as

$$\mathcal{V}_0^{(2,2)}(z) = \frac{(1\pm\lambda)^2}{16(1\mp\lambda)^2} k_a(z) + \frac{(1\pm\lambda)}{4(1\mp\lambda)^2} k_b(z) + \frac{1}{4(1\mp\lambda)^2} k_c(z) \tag{45}$$

with $k_a(z)$, $k_b(z)$, and $k_c(z)$ in Equation (38).

We also need Virasoro blocks of $\mathcal{A}_p$ up to the next non-trivial order in $1/N$. It is known that the Virasoro block is expanded in $1/c$ as (see, e.g., [23])

$$\mathcal{V}_p(z) = g(h_p, z) + \frac{1}{c}\left[h_1 h_3 f_a(h_p, z) + (h_1 + h_3) f_b(h_p, z) + f_c(h_p, z)\right] + \mathcal{O}(c^{-2}) \tag{46}$$

Here $g(h_p, z)$ is the global block of $\mathcal{A}_p$ and the expressions of $f_a(h_p, z)$, $f_b(h_p, z)$, and $f_c(h_p, z)$ were obtained in [23]. See also Appendix A. For our application, we set $h_1 = h_3 = h_\pm$ and $h_p = s$. We need the expansion in $1/N$ instead of $1/c$ as

$$\mathcal{V}_s(z) = \mathcal{V}_s^{(0)}(z) + \mathcal{V}_s^{(1)}(z)\frac{1}{N} + \mathcal{O}(N^{-2}) \tag{47}$$

The leading term $\mathcal{V}_p^{(0)}(z)$ is given by the global block as

$$\mathcal{V}_s^{(0)}(z) = g(s, z) = z^s {}_2F_1(s, s; 2s; x) \tag{48}$$

The next order contributions in $1/N$ are

$$\mathcal{V}_s^{(1)}(z) = \frac{1}{4}\frac{1\pm\lambda}{1\mp\lambda} f_a(s,z) + \frac{1}{(1\mp\lambda)} f_b(s,z) + \frac{1}{(1-\lambda^2)} f_c(s,z) \tag{49}$$

where the functions $f_a(s,z)$, $f_b(s,z)$, and $f_c(s,z)$ are given by

$$\begin{aligned}
f_a(s,z) &= z^s\Big[2z^2+(s+2)z^3+\tfrac{(s+3)(5s(s+3)+6)z^4}{20s+10}+\tfrac{(s+3)(s+4)(5s(s+4)+8)z^5}{60(2s+1)}\\
&\quad+\mathcal{O}(z^6)\Big]\\
f_b(s,z) &= s(s-1)z^s\Big[\tfrac{z^2}{2s+1}+\tfrac{(s+2)z^3}{4s+2}+\tfrac{(s+3)(5s(s+4)+18)z^4}{20(4s(s+2)+3)}\\
&\quad+\tfrac{(s+3)(s+4)(5s(s+5)+24)z^5}{120(4s(s+2)+3)}+\mathcal{O}(z^6)\Big]\\
f_c(s,z) &= \tfrac{s^2(s-1)^2}{2(2s+1)^2}z^s\Big[z^2+\tfrac{(s+2)z^3}{2}+\tfrac{(s+3)(s(10s(2s+11)+191)+108)z^4}{40(2s+3)^2}\\
&\quad+\tfrac{(s+3)(s+4)(s(10s(2s+13)+243)+144)z^5}{240(2s+3)^2}+\mathcal{O}(z^6)\Big]
\end{aligned} \tag{50}$$

4. Three Point Functions

After the preparations in previous sections, we now work on the decompositions of four point functions by Virasoro conformal blocks as in Equation (6). In the current case, the decompositions are

$$|z|^{2\Delta_\pm}G_{\pm\pm}(z)=\mathcal{V}_0(z)+\sum_{s=3}^{\infty}(C_\pm^{(s)})^2\mathcal{V}_s(z)+\sum_{(s_1,s_2;s')}(C_\pm^{(s_1,s_2;s')})^2\mathcal{V}_{s'}(z)+\cdots \tag{51}$$

Here $G_{\pm\pm}(z)$ are four point functions defined in Equations (7) and (8), and the expansions in z were obtained as in Equations (23) and (30). Moreover, $\mathcal{V}_0(z)$ is the vacuum block and $\mathcal{V}_s(z)$ is the Virasoro block of higher spin current $J^{(s)}$ (or $J^{(s_1,s_2;s)}$). Their expansions in z can be found in the previous section.

Solving constraint equations from Equation (51), we read off the coefficients $C_\pm^{(s)}$, which are proportional to the three point functions in Equation (4). It is convenient to expand the coefficients in $1/N$ as

$$C_\pm^{(s)}=\frac{1}{N^{1/2}}\left(C_{\pm,0}^{(s)}+\frac{1}{N}C_{\pm,1}^{(s)}+\mathcal{O}(N^{-2})\right) \tag{52}$$

Then we can see that the constraint equations from Equation (51) at the order N^0 is trivially satisfied as $1=1$. The non-trivial conditions arise from order $1/N$ terms, and they determine the leading order expressions $C_{\pm,0}^{(s)}$ as seen in the next subsection. The main purpose of this paper is to compute $C_{\pm,1}^{(s)}$, which are $1/N$ corrections to the leading order expressions. We derive them by solving order N^{-2} conditions up to $s=8$ in Section 4.2. Notice that we should take care of $C_\pm^{(s_1,s_2;s')}$ in Equation (51) for $s\geq 6$, which may be expanded as

$$C_\pm^{(s_1,s_2;s')}=\frac{1}{N}C_{\pm,0}^{(s_1,s_2;s')}+\mathcal{O}(N^{-2}) \tag{53}$$

The coefficients $C_{\pm,0}^{(s_1,s_2;s')}$ are analyzed in Appendix B.

4.1. Leading Order Expressions in $1/N$

We start from three point functions at the leading order in $1/N$. We examine the constraint equations from Equation (51) up to $1/N$ order. Up to this order, the vacuum block is given by (see Equation (37))

$$\mathcal{V}_0(z)=1+\frac{1}{N}(C_{\pm,0}^{(2)})^2z^2{}_2F_1(2,2;4;z)+\mathcal{O}(N^{-2}) \tag{54}$$

where we have defined

$$C^{(2)}_{\pm,0} = \sqrt{\frac{2(h_\pm)^2}{c}}\Bigg|_{\mathcal{O}(N^{-1/2})} = \sqrt{\frac{1}{2}\frac{1\pm\lambda}{1\mp\lambda}} \tag{55}$$

The Virasoro block of $J^{(s)}$ is

$$\mathcal{V}_s(z) = z^s {}_2F_1(s,s;2s;z) + \mathcal{O}(N^{-1}) \tag{56}$$

as in Equation (47) with Equation (48). Therefore, the expansion in Equation (51) can be written as

$$|z|^{2\Delta_\pm} G_{\pm\pm}(z) = 1 + \frac{1}{N}\sum_{s=2}^{\infty}(C^{(s)}_{\pm,0})^2 z^s {}_2F_1(s,s;2s;z) + \cdots \tag{57}$$

up to the order of $1/N$. The four point functions $G_{\pm\pm}(z)$ can be expanded as

$$|z|^{2\Delta_\pm} G_{\pm\pm}(z) \sim 1 + \frac{1}{N}\sum_{n=1}^{\infty} z^n\left(-\frac{1}{n} + \frac{\Gamma(1\mp\lambda)\Gamma(n)}{\Gamma(n\mp\lambda)}\right) + \cdots \tag{58}$$

as in Equations (23) and (30) up to the same order. On the other hand, the global blocks can be written as

$$z^s {}_2F_1(s,s;2s;z) = \frac{\Gamma(2s)}{(\Gamma(s))^2}\sum_{n=0}^{\infty}\frac{(\Gamma(s+n))^2}{\Gamma(2s+n)}\frac{z^{n+s}}{n!} \tag{59}$$

Comparing the coefficients in front of z^n, we obtain

$$-\frac{1}{n} + \frac{\Gamma(1\mp\lambda)\Gamma(n)}{\Gamma(n\mp\lambda)} = (\Gamma(n))^2\sum_{s=2}^{n}\frac{\Gamma(2s)(C^{(s)}_{\pm,0})^2}{(\Gamma(s))^2\Gamma(s+n)(n-s)!} \tag{60}$$

They are the constraint equations for $(C^{(s)}_{\pm,0})^2$ with $s = 3,4,\ldots$.

In order to fix relative phase factor, we examine $G_{-+}(z)$ in Equation (9) as well. The decomposition in Equation (6) become

$$|1-z|^{2\Delta_+} G_{-+}(z) \sim 1 + \frac{1}{N}\sum_{s=2}^{\infty}(-1)^s C^{(s)}_{-,0} C^{(s)}_{+,0}(1-z)^s {}_2F_1(s,s;2s;1-z) \tag{61}$$

in this case. The extra phase factor $(-1)^s$ may require explanation; Now we need to use a slightly different expression of operator product expansion as

$$\mathcal{O}_+(1)\bar{\mathcal{O}}_+(z) = C^{(s)}_+ \frac{(1-z)^s}{|1-z|^{2\Delta_+}} J^{(s)}(1) + \cdots \tag{62}$$

Then the coefficients in front of global blocks are given by

$$C^{(s)}_+\langle\mathcal{O}_-(\infty)J^{(s)}(1)\bar{\mathcal{O}}_-(0)\rangle = C^{(s)}_+(-1)^s\langle\mathcal{O}_-(\infty)\bar{\mathcal{O}}_-(1)J^{(s)}(0)\rangle \propto (-1)^s C^{(s)}_- C^{(s)}_+ \tag{63}$$

Here the factor $(-1)^s$ can be obtained from the coordinate dependence of three point function, which is completely fixed by conformal symmetry, see Equation (65) below. Therefore, we have constraint equations for three point functions as

$$\frac{n-1}{n} = (\Gamma(n))^2\sum_{s=2}^{n}\frac{(-1)^s\Gamma(2s)C^{(s)}_{-,0}C^{(s)}_{+,0}}{\Gamma(s)^2\Gamma(s+n)(n-s)!} \tag{64}$$

by comparing the coefficients in front of z^n.

Now we have three types of constraint equation as in Equations (60) and (64), and we would like to show that the known results satisfy these equations. At the leading order in $1/N$, the three point functions have been computed as [8]

$$\langle \mathcal{O}_\pm(z_1)\bar{\mathcal{O}}_\pm(z_2)J^{(s)}(z_3)\rangle = \frac{\eta_\pm^{(s)}}{2\pi}\frac{\Gamma(s)^2}{\Gamma(2s-1)}\frac{\Gamma(s\pm\lambda)}{\Gamma(1\pm\lambda)}\left(\frac{z_{12}}{z_{13}z_{23}}\right)^s \langle \mathcal{O}_\pm(z_1)\bar{\mathcal{O}}_\pm(z_2)\rangle \tag{65}$$

The phase factors $\eta_\pm^{(s)}$ depends on the convention of higher spin currents, but we may set $\eta_+^{(s)} = 1$ and $\eta_-^{(s)} = (-1)^s$. The two point function of higher spin current $J^{(s)}$ in Equation (65) is (see (6.1) of [8])

$$\langle J^{(s)}(z_1)J^{(s)}(z_2)\rangle = \frac{B^{(s)}}{z_{12}^{2s}}, \quad B^{(s)} = \frac{N}{2^{2s}\pi^{5/2}}\frac{\sin(\pi\lambda)}{\lambda}\frac{\Gamma(s)\Gamma(s-\lambda)\Gamma(s+\lambda)}{\Gamma(s-\frac{1}{2})} \tag{66}$$

at the leading order in $1/N$. The coefficients $C_{\pm,0}^{(s)}$ are given by normalization independent ratios as

$$C_{\pm,0}^{(s)} = \left.\frac{\langle \mathcal{O}_\pm\bar{\mathcal{O}}_\pm J^{(s)}\rangle}{\langle \mathcal{O}_\pm\bar{\mathcal{O}}_\pm\rangle\langle J^{(s)}J^{(s)}\rangle^{1/2}}\right|_{\mathcal{O}(N^{-1/2})} \tag{67}$$

which become

$$C_{\pm,0}^{(s)} = \eta_\pm^{(s)}\sqrt{\frac{\Gamma(s)^2}{\Gamma(2s-1)}\frac{\Gamma(1\mp\lambda)}{\Gamma(1\pm\lambda)}\frac{\Gamma(s\pm\lambda)}{\Gamma(s\mp\lambda)}} \tag{68}$$

The first few coefficients are

$$C_{\pm,0}^{(3)} = \eta_\pm^{(3)}\sqrt{\frac{1}{6}\frac{(2\pm\lambda)(1\pm\lambda)}{(2\mp\lambda)(1\mp\lambda)}}, \qquad C_{\pm,0}^{(4)} = \sqrt{\frac{1}{20}\frac{(3\pm\lambda)(2\pm\lambda)(1\pm\lambda)}{(3\mp\lambda)(2\mp\lambda)(1\mp\lambda)}} \tag{69}$$

along with Equation (55) for $s = 2$. Using these explicit expressions, we can check that the constraint equations in Equations (60) and (64) are indeed satisfied[3].

4.2. $1/N$ Corrections

We would like to move to the main part of this paper. In this subsection, we derive $1/N$ corrections to three point functions by examining the equations in Equation (51). With the help of analysis in previous sections, we have already ingredients necessary to the task. For examples, the expansions of $G_{\pm\pm}(z)$ were given in Equations (23) and (30) up to order $1/N^2$. Moreover, the vacuum block and the Virasoro block of $J^{(s)}$ are expanded as in Equations (39) and (47), respectively. Using these expansions, the equations in Equation (51) become

$$\begin{aligned}\textstyle\sum_{m=2}^\infty f_{\pm\pm}^{(m)} z^m = \;& \mathcal{V}_0^{(2)}(z) + \textstyle\sum_{s=3}^\infty 2C_{\pm,1}^{(s)}C_{\pm,0}^{(s)}\mathcal{V}_s^{(0)}(z) + \textstyle\sum_{s=3}^\infty (C_{\pm,0}^{(s)})^2\mathcal{V}_s^{(1)}(z) \\ & + \textstyle\sum_{(s_1,s_2;s')}(C_{\pm,0}^{(s_1,s_2;s')})^2\mathcal{V}_{s'}^{(0)}(z)\end{aligned} \tag{70}$$

at the order of $1/N^2$. Here $f_{\pm\pm}^{(m)}$ are defined in Equation (24) and Equation (31). At this order we should include the effects from higher spin currents of double trace type as $(C_{\pm,0}^{(s_1,s_2,s')})^2$ in Equation (70) with $s' \geq 6$.

---

3 We have confirmed this for Equation (60) with spin $s = 2, 3, \ldots, 70$ and for Equation (64) with all spin.

Let us examine the equations in Equation (70) from low order terms in z. There are no z^0 and z^1 order terms in the both sides. We can see that the equality in Equation (70) is satisfied at the order of z^2 from Equation (42). Non-trivial constraint equations appear at the z^3 order as

$$f^{(3)}_{\pm\pm} = f^{(2)}_{\pm\pm} + 2C^{(3)}_{\pm,0}C^{(3)}_{\pm,1} \tag{71}$$

where $f^{(2)}_{\pm\pm}$ comes from $\mathcal{V}_0^{(2,1)}$ in Equation (42). Solving them we find

$$\frac{C^{(3)}_{+,1}}{C^{(3)}_{+,0}} = -\tfrac{1}{2}\left(-\lambda + \tfrac{1}{1+\lambda} + \tfrac{4}{2+\lambda}\right), \quad \frac{C^{(3)}_{-,1}}{C^{(3)}_{-,0}} = \tfrac{1}{2}\left(-\lambda + \tfrac{1}{\lambda+1} + \tfrac{4}{\lambda+2} - 6\right) \tag{72}$$

The z^4 order constraints are

$$f^{(4)}_{\pm\pm} = f^{(2)}_{\pm\pm}\frac{9}{10} + \frac{(1\pm\lambda)^2}{8(1\mp\lambda)^2} + \frac{(1\pm\lambda)}{10(1\mp\lambda)^2} + \frac{1}{50(1\mp\lambda)^2} + 2C^{(4)}_{\pm,0}C^{(4)}_{\pm,1} + 2C^{(3)}_{\pm,0}C^{(3)}_{\pm,1}\frac{3}{2}$$

where the contribution from Equation (45) starts to enter. The constraints lead to

$$\begin{aligned}\frac{C^{(4)}_{+,1}}{C^{(4)}_{+,0}} &= \tfrac{1}{10}\left(5\lambda + \tfrac{6}{\lambda-1} - \tfrac{11}{\lambda+1} - \tfrac{20}{\lambda+2} - \tfrac{45}{\lambda+3}\right)\\ \frac{C^{(4)}_{-,1}}{C^{(4)}_{-,0}} &= \tfrac{1}{10}\left(-5\lambda + \tfrac{6}{\lambda-1} - \tfrac{1}{\lambda+1} + \tfrac{20}{\lambda+2} + \tfrac{45}{\lambda+3} - 60\right)\end{aligned} \tag{73}$$

We would like to keep going to the cases with $s \geq 5$, where $f_a(s,z)$, $f_b(s,z)$, and $f_c(s,z)$ in Equation (50) contribute. For $s = 5$, the conditions become

$$\begin{aligned}f^{(5)}_{\pm\pm} = {}& f^{(2)}_{\pm\pm}\frac{4}{5} + \frac{(1\pm\lambda)^2}{4(1\mp\lambda)^2} + \frac{(1\pm\lambda)}{5(1\mp\lambda)^2} + \frac{1}{25(1\mp\lambda)^2} + 2C^{(5)}_{\pm,0}C^{(5)}_{\pm,1} + 2C^{(4)}_{\pm,0}C^{(4)}_{\pm,1}\cdot 2\\ &+2C^{(3)}_{\pm,0}C^{(3)}_{\pm,1}\cdot\frac{12}{7} + (C^{(3)}_{\pm,0})^2\left[\frac{1}{2}\frac{1\pm\lambda}{1\mp\lambda} + \frac{6}{7(1\mp\lambda)} + \frac{18}{49(1-\lambda^2)}\right]\end{aligned} \tag{74}$$

We then find

$$\begin{aligned}\frac{C^{(5)}_{+,1}}{C^{(5)}_{+,0}} &= \tfrac{\lambda}{2} + \tfrac{25}{7(\lambda-1)} - \tfrac{57}{14(\lambda+1)} - \tfrac{2}{\lambda+2} - \tfrac{9}{2(\lambda+3)} - \tfrac{8}{\lambda+4}\\ \frac{C^{(5)}_{-,1}}{C^{(5)}_{-,0}} &= -\tfrac{\lambda}{2} + \tfrac{25}{7(\lambda-1)} - \tfrac{43}{14(\lambda+1)} + \tfrac{2}{\lambda+2} + \tfrac{9}{2(\lambda+3)} + \tfrac{8}{\lambda+4} - 10\end{aligned} \tag{75}$$

by solving the constraints.

For $s \geq 6$, the contributions from higher spin currents of double trace type should be considered. They are given by

$$J^{(3,3;6)} \sim: J^{(3)}J^{(3)} :, \quad J^{(3,4;7)} \sim: J^{(3)}J^{(4)} : \tag{76}$$

for $s = 6, 7$ and[4]

$$J^{(4,4;8)} \sim: J^{(4)}J^{(4)} :, \quad J^{(3,5;8)} \sim: J^{(3)}J^{(5)} :, \quad J^{(3,3;8)} \sim: J^{(3)}\partial^2 J^{(3)} : \tag{77}$$

---

4 There could be another current $J^{(3,4;8)} \sim: J^{(3)}\partial J^{(4)}$, but it does not give any contribution as shown in Appendix B.

for $s = 8$. Their precise forms are fixed such as to be primary in the sense of Virasoro algebra as derived in Appendix B[5]. Once we have the expressions of these currents, we can obtain the coefficients $(C^{(s_1,s_2;s')}_{\pm,0})^2$, which are defined as

$$(C^{(s_1,s_2;s')}_{\pm,0})^2 = \left. \frac{\langle \mathcal{O}_\pm \bar{\mathcal{O}}_\pm J^{(s_1,s_2;s')} \rangle^2}{\langle \mathcal{O}_\pm \bar{\mathcal{O}}_\pm \rangle^2 \langle J^{(s_1,s_2;s')} J^{(s_1,s_2;s')} \rangle} \right|_{\mathcal{O}(N^{-2})} \tag{78}$$

In Appendix B we also compute the three point functions $\langle \mathcal{O}_\pm \bar{\mathcal{O}}_\pm J^{(s_1,s_2;s')} \rangle$ and the two point functions $\langle J^{(s_1,s_2;s')} J^{(s_1,s_2;s')} \rangle$ for the currents in Equations (76) and (77) at the leading order in $1/N$.

Utilizing these results, we obtain $1/N$ corrections to three point functions with single trace currents of $s = 6, 7, 8$. The constraint equations for $s = 6$ are

$$\begin{aligned} f^{(6)}_{\pm\pm} &= f^{(2)}_{\pm\pm}\tfrac{5}{7} + \tfrac{7(1\pm\lambda)^2}{20(1\mp\lambda)^2} + \tfrac{39(1\pm\lambda)}{140(1\mp\lambda)^2} + \tfrac{109}{1960(1\mp\lambda)^2} + 2C^{(6)}_{\pm,0}C^{(6)}_{\pm,1} + 2C^{(5)}_{\pm,0}C^{(5)}_{\pm,1} \cdot \tfrac{5}{2} \\ &\quad + 2C^{(4)}_{\pm,0}C^{(4)}_{\pm,1} \cdot \tfrac{25}{9} + 2C^{(3)}_{\pm,0}C^{(3)}_{\pm,1} \cdot \tfrac{25}{14} + (C^{(3)}_{\pm,0})^2 \left[\tfrac{5}{4}\tfrac{1\pm\lambda}{1\mp\lambda} + \tfrac{15}{7(1\mp\lambda)} + \tfrac{90}{98(1-\lambda^2)}\right] \\ &\quad + (C^{(4)}_{\pm,0})^2 \left[\tfrac{1}{2}\tfrac{1\pm\lambda}{1\mp\lambda} + \tfrac{4}{3(1\mp\lambda)} + \tfrac{8}{9(1-\lambda^2)}\right] + (C^{(3,3;6)}_{\pm,0})^2 \end{aligned} \tag{79}$$

where the effect of $J^{(3,3;6)}$ in Equation (76) enters. Solving these equations we find

$$\begin{aligned} \frac{C^{(6)}_{+,1}}{C^{(6)}_{+,0}} &= \tfrac{\lambda}{2} - \tfrac{5}{3(\lambda-2)} + \tfrac{1315}{84(\lambda-1)} - \tfrac{1357}{84(\lambda+1)} - \tfrac{1}{3(\lambda+2)} - \tfrac{9}{2(\lambda+3)} - \tfrac{8}{\lambda+4} \\ &\quad - \tfrac{25}{2(\lambda+5)} \\ \frac{C^{(6)}_{-,1}}{C^{(6)}_{-,0}} &= -\tfrac{\lambda}{2} - \tfrac{5}{3(\lambda-2)} + \tfrac{1315}{84(\lambda-1)} - \tfrac{1273}{84(\lambda+1)} + \tfrac{11}{3(\lambda+2)} + \tfrac{9}{2(\lambda+3)} + \tfrac{8}{\lambda+4} \\ &\quad + \tfrac{25}{2(\lambda+5)} - 15 \end{aligned} \tag{80}$$

For spin 7, another double trace operator $J^{(3,4;7)}$ in Equation (76) should be considered as

$$\begin{aligned} f^{(7)}_{\pm\pm} &= f^{(2)}_{\pm\pm}\tfrac{9}{14} + \tfrac{17(1\pm\lambda)^2}{40(1\mp\lambda)^2} + \tfrac{846(1\pm\lambda)}{2520(1\mp\lambda)^2} + \tfrac{131}{1960(1\mp\lambda)^2} + 2C^{(7)}_{\pm,0}C^{(7)}_{\pm,1} + 2C^{(6)}_{\pm,0}C^{(6)}_{\pm,1} \cdot 3 \\ &\quad + 2C^{(5)}_{\pm,0}C^{(5)}_{\pm,1} \cdot \tfrac{45}{11} + 2C^{(4)}_{\pm,0}C^{(4)}_{\pm,1} \cdot \tfrac{10}{3} + 2C^{(3)}_{\pm,0}C^{(3)}_{\pm,1} \cdot \tfrac{25}{14} + (C^{(3)}_{\pm,0})^2 \tfrac{(3467\pm42\lambda(89\pm24\lambda))}{490(1-\lambda^2)} \\ &\quad + (C^{(4)}_{\pm,0})^2 \tfrac{(7\pm3\lambda)}{6(1-\lambda^2)} + (C^{(5)}_{\pm,0})^2 \tfrac{(31\pm11\lambda)}{242(1-\lambda^2)} + (C^{(3,3;6)}_{\pm,0})^2 \cdot 3 + (C^{(3,4;7)}_{\pm,0})^2 \end{aligned} \tag{81}$$

---

[5] From the decomposition by Virasoro conformal blocks as in Equation (6), we can read off three point functions among primary operators including intermediate one $\mathcal{A}_p$ by construction. For $s \le 5$, only $J^{(s)}$ starts to contribute as the intermediate operator at the z^s order, so we do not need to worry about if the operator is primary or not. However, for $s \ge 6$, there are degeneracies among $J^{(s)}$ and $J^{(s_1,s_2;s)}$, and the $1/N$ corrections $C^{(s)}_{\pm,1}$ could be read off once we have the information of $C^{(s_1,s_2;s)}_{\pm,0}$, see Equation (70). Since we compute $C^{(s_1,s_2;s)}_{\pm,0}$ by hand as explained in Appendix B, we have to explicitly construct primary operators of double trace type. We only need the leading order expressions, so it is enough to use commutation relations surviving in the large c limit as in Equation (A15) and higher spin charges at the leading order as in Equation (A18).

which lead to

$$\begin{aligned}\frac{C^{(7)}_{+,1}}{C^{(7)}_{+,0}} &= \frac{\lambda}{2} - \frac{490}{33(\lambda-2)} + \frac{8183}{132(\lambda-1)} - \frac{8249}{132(\lambda+1)} + \frac{424}{33(\lambda+2)} - \frac{9}{2(\lambda+3)} \\ &\quad - \frac{8}{\lambda+4} - \frac{25}{2(\lambda+5)} - \frac{18}{\lambda+6} \\ \frac{C^{(7)}_{-,1}}{C^{(7)}_{-,0}} &= -\frac{\lambda}{2} - \frac{490}{33(\lambda-2)} + \frac{8183}{132(\lambda-1)} - \frac{8117}{132(\lambda+1)} + \frac{556}{33(\lambda+2)} + \frac{9}{2(\lambda+3)} \\ &\quad + \frac{8}{\lambda+4} + \frac{25}{2(\lambda+5)} + \frac{18}{\lambda+6} - 21\end{aligned} \tag{82}$$

The constraint equations for $C^{(8)}_{\pm,1}$ are

$$\begin{aligned}&f^{(8)}_{\pm\pm} = f^{(2)}_{\pm\pm}\frac{7}{12} + \frac{(1\pm\lambda)^2}{(1\mp\lambda)^2}\frac{2687}{5600} + \frac{(1\pm\lambda)}{(1\mp\lambda)^2}\frac{263}{700} + \frac{1}{(1\mp\lambda)^2}\frac{1879}{25200} + 2C^{(8)}_{\pm,0}C^{(8)}_{\pm,1} \\ &+2C^{(7)}_{\pm,0}C^{(7)}_{\pm,1}\frac{7}{2} + 2C^{(6)}_{\pm,0}C^{(6)}_{\pm,1}\frac{147}{26} + 2C^{(5)}_{\pm,0}C^{(5)}_{\pm,1}\frac{245}{44} + +2C^{(4)}_{\pm,0}C^{(4)}_{\pm,1}\frac{245}{66} + 2C^{(3)}_{\pm,0}C^{(3)}_{\pm,1}\frac{7}{4} \\ &+(C^{(3)}_{\pm,0})^2\left(\frac{387\pm418\lambda+113\lambda^2}{40(1-\lambda^2)}\right) + (C^{(4)}_{\pm,0})^2\left(\frac{7(47977\pm41162\lambda+8833\lambda^2)}{21780(1-\lambda^2)}\right) \\ &+(C^{(5)}_{\pm,0})^2\left(\frac{7(31\pm11\lambda)^2}{484(1-\lambda^2)}\right) + (C^{(6)}_{\pm,0})^2\left(\frac{(43\pm13\lambda)^2}{338(1-\lambda^2)}\right) + (C^{(3,3;6)}_{\pm,0})^2\frac{147}{26} + (C^{(3,4;7)}_{\pm,0})^2\frac{7}{2} \\ &+(C^{(4,4;8)}_{\pm,0})^2 + (C^{(3,5;8)}_{\pm,0})^2 + (C^{(3,3;8)}_{\pm,0})^2\end{aligned} \tag{83}$$

Here we have taken care of double trace operators $J^{(4,4;8)}$, $J^{(3,5;8)}$, and $J^{(3,3;8)}$ in Equation (77). We then have

$$\begin{aligned}\frac{C^{(8)}_{+,1}}{C^{(8)}_{+,0}} &= \frac{\lambda}{2} + \frac{525}{143(\lambda-3)} - \frac{12572}{143(\lambda-2)} + \frac{101311}{429(\lambda-1)} - \frac{203051}{858(\lambda+1)} + \frac{12286}{143(\lambda+2)} \\ &\quad - \frac{2337}{286(\lambda+3)} - \frac{8}{\lambda+4} - \frac{25}{2(\lambda+5)} - \frac{18}{\lambda+6} - \frac{49}{2(\lambda+7)} \\ \frac{C^{(8)}_{-,1}}{C^{(8)}_{-,0}} &= -\frac{\lambda}{2} + \frac{525}{143(\lambda-3)} - \frac{12572}{143(\lambda-2)} + \frac{101311}{429(\lambda-1)} - \frac{202193}{858(\lambda+1)} + \frac{12858}{143(\lambda+2)} \\ &\quad + \frac{237}{286(\lambda+3)} + \frac{8}{\lambda+4} + \frac{25}{2(\lambda+5)} + \frac{18}{\lambda+6} + \frac{49}{2(\lambda+7)} - 28\end{aligned} \tag{84}$$

as solutions to the constraint equations.

Since the three point functions were already obtained with finite N, k in [9–11] for $s = 3, 4, 5$, they can be compared to our results in principle. Instead of doing so, we utilize a simpler relation, which is on the ratio of three point functions (see (4.52) of [11])

$$\frac{\langle \mathcal{O}_+ \bar{\mathcal{O}}_+ J^{(s)} \rangle}{\langle \mathcal{O}_- \bar{\mathcal{O}}_- J^{(s)} \rangle} = (-1)^s \frac{(k+N+1)}{(k+N)} \prod_{n=1}^{s-1} \left[\frac{nk + (n+1)N + n}{nk + (n-1)N} \right] \tag{85}$$

The relation was derived for $s = 2, 3, 4, 5$ by using the explicit results and conjectured for generic s based on them. The expression up to the $1/N$ order becomes

$$\begin{aligned}\frac{\langle \mathcal{O}_+ \bar{\mathcal{O}}_+ J^{(s)} \rangle}{\langle \mathcal{O}_- \bar{\mathcal{O}}_- J^{(s)} \rangle} &= \frac{C^{(s)}_{+,0} + \frac{1}{N} C^{(s)}_{+,1}}{C^{(s)}_{-,0} + \frac{1}{N} C^{(s)}_{-,1}} + \mathcal{O}(N^{-2}) \\ &= (-1)^s \prod_{n=1}^{s-1} \left(\frac{n+\lambda}{n-\lambda} \right) \left[1 + \frac{1}{N} \left(\lambda + \sum_{m=1}^{s-1} \frac{m\lambda}{m+\lambda} \right) + \mathcal{O}(N^{-2}) \right]\end{aligned} \tag{86}$$

Thus, at the leading order in $1/N$, we have

$$\frac{C^{(s)}_{+,0}}{C^{(s)}_{-,0}} = (-1)^s \prod_{n=1}^{s-1} \left(\frac{n+\lambda}{n-\lambda} \right) \tag{87}$$

We can easily check that Equation (68) satisfy this condition. The relation in Equation (86) at the next leading order in $1/N$ implies

$$\frac{C^{(s)}_{+,1}}{C^{(s)}_{+,0}} - \frac{C^{(s)}_{-,1}}{C^{(s)}_{-,0}} = \lambda + \sum_{m=1}^{s-1} \frac{m\lambda}{m+\lambda} \tag{88}$$

We have confirmed our results (and the conjectured relation in Equation (86)) by showing that our results on $C^{(s)}_{\pm,1}$ for $s = 3, \ldots, 8$ satisfy this equation.

Before ending this section, we would like to make comments on normalized three point functions

$$C^{(2)}_{\pm} = \frac{\langle \mathcal{O}_{\pm}\bar{\mathcal{O}}_{\pm} J^{(2)}\rangle}{\langle \mathcal{O}_{\pm}\bar{\mathcal{O}}_{\pm}\rangle \langle J^{(2)} J^{(2)}\rangle^{1/2}} \tag{89}$$

with the energy momentum tensor $T \propto J^{(2)}$. They do not appear in the decomposition of Virasoro conformal blocks but can be fixed by the conformal Ward identity as

$$C^{(2)}_{\pm} = \frac{1}{N^{1/2}} \left(C^{(2)}_{\pm,0} + \frac{1}{N} C^{(2)}_{\pm,1} + \mathcal{O}(N^{-2}) \right) = \sqrt{\frac{2h^2_{\pm}}{c}} \tag{90}$$

In particular, they lead to Equation (55) and

$$2C^{(2)}_{\pm,0} C^{(2)}_{\pm,1} = f^{(2)}_{\pm\pm} \tag{91}$$

with Equation (43), or equivalently

$$\frac{C^{(2)}_{+,1}}{C_{+,0}} = \frac{\lambda}{2} - \frac{1}{2(\lambda+1)}, \quad \frac{C^{(2)}_{-,1}}{C_{-,0}} = -\frac{\lambda}{2} + \frac{1}{2(\lambda+1)} - 1 \tag{92}$$

As a consistence check, we can show that they satisfy Equation (88) as well.

5. Conclusions and Open Problems

We have developed a new method to compute three point functions of two scalar operators and a higher spin current in Equation (4) in 2d W_N minimal model. This model can be described by the coset in Equation (1) with two parameters N, k, and we analyze it in $1/N$ expansion in terms of 't Hooft parameter $\lambda = N/(N+k)$ in Equation (3). We decompose scalar four point functions $G_{\pm\pm}(z)$ in Equations (7) and (8) and $G_{-+}(z)$ in Equation (9) by Virasoro conformal blocks. The four point functions were computed exactly with finite N, k in [16], and Virasoro conformal blocks can be obtained including $1/N$ corrections, say, by analyzing Zamolodchikov's recursion relation [17]. Solving the constraint equations from the decomposition, we can obtain three point functions including $1/N$ corrections. At the leading order in $1/N$, we can easily reproduce the known results in [8] because Virasoro conformal blocks reduce to global blocks in this case. At the next leading order, we have obtained $1/N$ corrections to the three point functions up to spin 8. Previously exact results were known for $s = 3, 4, 5$ in [9–11], and our findings for $s = 6, 7, 8$ are new. We have confirmed our results by checking that the conjectured relation in Equation (88) is satisfied.

We have evaluated $1/N$ corrections only up to spin 8 case because of the following two obstacles. One comes from $1/c$ corrections to Virasoro conformal blocks. Up to the required order in $1/c$, closed forms can be obtained, for instance, by following the method in [23] except for $f_c(s, z)$ in Equation (49). In Equation (50) (or in [23]), the function $f_c(s, z)$ is given up to the order z^{5+s}, but we need the term at order z^{6+s} with $s = 3$ for spin 9 computation. We have not tried to do so, but it should be possible to obtain the terms at higher orders in z without a lot of efforts. Another is related to the contributions from higher spin currents of double trace type as analyzed in Appendix B. In order to obtain primary operators of this type, we have used commutation relations in Equation (A15), which are borrowed from [24]. For spin 9, a current of the form $J^{(3,6;9)} \sim: J^{(3)}J^{(6)}:$ would give some contributions. However, in order to find its primary form, we need the commutation relation between W, Y, which is currently not available. At the order in $1/c$ which do not vanish at $c \to \infty$, we can derive the commutation relations involving more higher spin currents, for instance, from dual Chern-Simons description as in [25–28]. The computation is straightforward but might be tedious. In any case, it is definitely possible to obtain the $1/N$ corrections of three point functions for $s \geq 9$, and it is desired to have expressions for generic s.

There are many open problems we would like to think about. Because of the simplicity of our method, it is expected to be applicable to more generic cases. For example, it is worth generalizing the current analysis to supersymmetric cases. Recently, it becomes possible to discuss relations between 3d higher spin theory and superstrings by introducing extended supersymmetry to the duality by [1]. Higher spin holography with $\mathcal{N} = 3$ supersymmetry has been developed in a series of works [29–31], while large or small $\mathcal{N} = 4$ supersymmetry has been utilized through the well-studied holography with symmetric orbifold in [32,33]. Previous works on the subject may be found in [34–36]. As mentioned in introduction, the main motivation to examine $1/N$ corrections in 2d W_N minimal model is to learn quantum effects in dual higher spin theory. We would like to report on our recent progress in a separate publication [37].

Acknowledgments: We are grateful to Changhyun Ahn, Pawel Caputa, Takahiro Nishinaka, Volker Schomerus and Yuji Sugawara for useful discussion. Y.H. would like to thank the organizers of "Universität Hamburg—Kyoto University Symposium" and the workshop "New ideas on higher spin gravity and holography" at Kyung Hee University, Seoul for their hospitality. The work of Y.H. is supported by JSPS KAKENHI Grant Number 16H02182.

Author Contributions: Yasuaki Hikida and Takahiro Uetoko contributed equally to this work.

Conflicts of Interest: The authors declare no conflict of interest.

Appendix A. Recursion Relations and Virasoro Conformal Blocks

In this appendix we derive the expressions of Virasoro conformal blocks in expansions of $1/c$ and z by solving Zamolodchikov recursion relation in [17], and we compare our results to those previous obtained especially in [23]. We decompose a four point function by Virasoro conformal blocks $\mathcal{F}(c, h_i, h_p, z)$ as in Equation (6). In the following we set $h_1 = h_2$ and $h_3 = h_4$. The recursion relation for Virasoro conformal blocks is [17]

$$\begin{aligned}\mathcal{F}(c,h_i,h_p,z) = &\ z^{h_p}{}_2F_1(h_p,h_p;2h_p;z) \\ &+ \textstyle\sum_{m\geq 1,n\geq 2}^{\infty} \frac{R_{mn}(h_i,h_p)}{c-c_{mn}(h_p)}\mathcal{F}(c_{mn}(h_p),h_i,h_p+mn,z)\end{aligned} \tag{A1}$$

Here the poles for c are located at $c = c_{mn}(h_p)$ with

$$c_{mn}(h_p) = 13 - 6\left(t_{mn}(h_p)^{-1} + t_{mn}(h_p)\right) \tag{A2}$$

where

$$t_{mn}(h_p) = \left(2h_p + mn - 1 + \sqrt{4h_p(h_p + mn - 1) + (m-n)^2}\right)/(n^2 - 1) \tag{A3}$$

The residua are

$$R_{mn}(h_i,h_p) = A_{mn}(h_p)P_{mn}(h_i,h_p) \tag{A4}$$

where

$$
\begin{aligned}
P_{mn}(h_i,h_p) &= \prod_{j,k}\left(2l_1 - \tfrac{l_{jk}}{2}\right)\left(2l_3 - \tfrac{l_{jk}}{2}\right)\left(\tfrac{l_{jk}}{2}\right)^2 \\
A_{mn}(h_p) &= \frac{-12(t_{mn}^{-1} - t_{mn})}{(m^2-1)t_{mn}^{-1} - (n^2-1)t_{mn}} \prod_{a,b} \frac{1}{l_{ab}} \\
l_{jk}(m,n,h_p) &= (j - kt_{mn})t_{mn}^{-1/2}, \quad l_i(m,n,h_i,h_p) = (h_i + l_{11}^2/4)^{1/2}
\end{aligned} \tag{A5}
$$

The sum is taken over $j = -m+1, -m+3, \ldots, m-1$, $k = -n+1, -n+3, \ldots, n-1$, $a = -m+1, -m+2, \ldots, m$, $b = -n+1, -n+2, \ldots, n$ without $(a,b) = (0,0), (m,n)$.

For our purpose, it is enough to obtain first several terms of Virasoro blocks in z expansion, and we obtain them by following the strategy of [18], see also [19]. We decompose Virasoro conformal blocks by global blocks as

$$
\mathcal{F}(c,h_i,h_p,z) = z^{h_p} \sum_{q=0}^{\infty} \chi_q(c,h_i,h_p) z^q {}_2F_1(h_p+q, h_p+q; 2(h_p+q); z) \tag{A6}
$$

The generic expressions of χ_q are given in (2.28) of [18]. With $h_1 = h_2$ and $h_3 = h_4$, it can be shown that $\chi_q = 0$ for odd q. The explicit expressions for $q = 2,4,6$ can be found in (C.1) of the paper as

$$
\begin{aligned}
&\chi_2(c,h_p) = \gamma_{12}(c,h_p) \\
&\chi_4(c,h_p) = \gamma_{14}(c,h_p) + \gamma_{22}(c,h_p) + \gamma_{12}(c,h_p)\gamma_{12}(c_{12}(h_p, h_p+2)) \\
&\chi_6(c,h_p) = \gamma_{16}(c,h_p) + \gamma_{23}(c,h_p) + \gamma_{32}(c,h_p) + \gamma_{12}(c,h_p)\gamma_{14}(c_{12}(h_p), h_p+2) \\
&\quad + \gamma_{12}(c,h_p)\gamma_{22}(c_{12}(h_p), h_p+2) + \gamma_{14}(c,h_p)\gamma_{12}(c_{14}(h_p), h_p+4) \\
&\quad + \gamma_{22}(c,h_p)\gamma_{12}(c_{22}(h_p), h_p+4) + \gamma_{12}(c,h_p)\gamma_{12}(c_{12}(h_p), h_p+2)\gamma_{12}(c_{12}(h_p+2), h_p+4)
\end{aligned} \tag{A7}
$$

with

$$
\gamma_{mn}(c,h_p) = \frac{R_{mn}(h_i,h_p)}{c - c_{mn}(h_p)} \tag{A8}
$$

Inserting these expressions into Equation (A6), we can obtain the Virasoro conformal blocks up to the order of z^{h_p+7}.

Let us start from vacuum block. As discussed in the main context, we need its expression up to the $1/c^2$ order. For $h_p = 0$ the coefficients χ_q can be found in (2.15) of [18], and they are expended in $1/c$ as

$$
\begin{aligned}
\chi_2(c,h_i,0) &= \frac{2h_1h_3}{c} \\
\chi_4(c,h_i,0) &= \frac{2(5h_1^2+h_1)(5h_2^2+h_2)}{25c^2} + \mathcal{O}(c^{-3}) \\
\chi_6(c,h_i,0) &= \frac{(14h_1^2+h_1)(14h_3^2+h_3)}{4410c^2} + \mathcal{O}(c^{-3})
\end{aligned} \tag{A9}
$$

Note that there is no $1/c$-correction to $\chi_2(c,h_i,0)$. Using

$$
\begin{aligned}
{}_2F_1(4,4;8;z) &= 1 + 2z + \tfrac{25}{9}z^2 + \tfrac{10}{3}z^3 + \mathcal{O}(z^4) \\
{}_2F_1(6,6;12;z) &= 1 + 3z + \mathcal{O}(z^2)
\end{aligned} \tag{A10}
$$

and Equation (A6), we find Equation (37) with $k_a(z)$, $k_b(z)$, and $k_c(z)$ in Equation (38) but up to the order z^7.

We would like to compare the expressions to Equation (3.14) in [23]. Firstly, there is no contribution like $k_a(z)$[6]. Secondly, our $k_b(z)$ is twice of the corresponding one in [23]. Finally, we can see that $k_c(z)$ reproduces their expressions. In conclusion, we find very similar but different results. After carefully repeated the analysis, say, in [23], we obtain

$$\begin{aligned} k_a(z) &= 2z^4({}_2F_1(2,2;4;z))^2 \\ k_b(z) &= \frac{72}{z^2}((z-2)z\log(1-z) + 2(1-z)\log^2(1-z) - 4z^2) \\ k_c(z) &= \frac{12}{z^2}(12(z-2)z\mathrm{Li}_2(z) + 16z^2 + 6(z-1)^2\log^2(1-z) + (z-2)z\log(1-z)) \end{aligned} \tag{A11}$$

This version matches the above expressions in z expansion. Using these closed form results, we can go to more higher orders in z as in Equation (38).

Let us move to the case with non-trivial h_p, where expressions are needed up to the $1/c$ order. Using the expressions of χ_q in Equation (A7), we obtain

$$\begin{aligned} \chi_2(c,h_p) &= \frac{1}{c}\left(\frac{h_p^2(h_p-1)^2}{2(2h_p+1)^2} + (h_1+h_3)\frac{h_p(h_p-1)}{2h_p+1} + 2h_1h_3\right) + \mathcal{O}(c^{-2}) \\ \chi_4(c,h_p) &= \frac{1}{c}\left(\frac{(h_p-1)^2h_p^3(h_p+3)}{80(2h_p+1)(2h_p+3)^2(2h_p+5)}\right. \\ &\left. +(h_1+h_3)\frac{(h_p-1)h_p^2(h_p+3)}{20(2h_p+1)(2h_p+3)(2h_p+5)} + h_1h_3\frac{h_p(h_p+3)}{5(2h_p+1)(2h_p+5)}\right) + \mathcal{O}(c^{-2}) \end{aligned} \tag{A12}$$

With the expansions of hypergeometric function in z such as

$$\begin{aligned} {}_2F_1(h_p+2,h_p+2;2h_p+4;z) &= 1 + \frac{2+h_p}{2}z + \frac{(2+h_p)(3+h_p)^2}{4(5+2h_p)}z^2 \\ &+ \frac{(h_p+2)(h_p+3)(h_p+4)^2}{24(2h_p+5)}z^3 + \mathcal{O}(z^4) \\ {}_2F_1(h_p+4,h_p+4;2h_p+8;z) &= 1 + \frac{4+h_p}{2}z + \mathcal{O}(z^2) \end{aligned} \tag{A13}$$

we find Equation (46), where the functions $f_a(h_p,z)$, $f_b(h_p,z)$, and $f_c(h_p,z)$ are given by Equation (50) but with $s = h_p$. These were analyzed in [23] and, in particular, closed forms were obtained for $f_a(h_p,z)$ and $f_b(h_p,z)$ as

$$\begin{aligned} f_a(h_p,z) &= -12z^{h_p-1}{}_2F_1(h_p,h_p;2h_p;z)(2z + (2z + (2-z)\log(1-z))) \\ f_b(h_p,z) &= 12h_pz^{h_p}\left({}_2F_1(h_p,h_p;2h_p;z)\left(\log(1-z)\left(z^{-1}-1\right)+1\right)\right. \\ &\left. +\tfrac{1}{2}\log(1-z)\,{}_2F_1(h_p,h_p;2h_p+1;z)\right) \end{aligned} \tag{A14}$$

Our results match with their findings in this case.

Appendix B. Higher Spin Currents of Double Trace Type

In this appendix, we analyze higher spin currents of double trace type with $s' = 6,7,8$ in Equations (76) and (77). We first present basics on higher spin algebra, which are needed to obtain

6 It seems that the authors of [23] did not consider this type of contribution because it is not new but essentially given by the square of $1/c$ order term in Equation (37).

the precise expressions of these currents primary with respect to Virasoro algebra. We then derive the three and two point functions of these currents, which are used to obtain $(C^{(s_1,s_2;s')}_{\pm,0})^2$ in Equation (78).

Appendix B.1. Higher Spin Algebra

In order to find out higher spin currents of double trace type, which are primary to Virasoro algebra, we utilize commutation relations among higher spin currents given in [24] (see also [15,27,28]). The currents are denoted as W, U, X, Y, which are proportional to $J^{(s)}$ with $s = 3, 4, 5, 6$. In order to obtain the leading order expression $(C^{(s_1,s_2;s')}_{\pm,0})^2$, we only need commutation relations up to the terms vanishing at $c \to \infty$ as[7]

$$\begin{aligned}
&[L_m, L_n] = (m-n)L_{m+n} + \frac{c}{12}m(m^2-1)\delta_{m+n}\,, \quad [L_m, W_n] = (2m-n)W_{m+n} \\
&[L_m, U_n] = (3m-n)U_{m+n}\,, \quad [L_m, X_n] = (4m-n)X_{m+n}\,, \quad [L_m, Y_n] = (5m-n)Y_{m+n} \\
&[W_m, W_n] = 2(m-n)U_{m+n} - \frac{N_3}{12}(m-n)(2m^2+2n^2-mn-8)L_{m+n} \\
&-\frac{N_3 c}{144}m(m^2-1)(m^2-4)\delta_{m+n} \\
&[W_m, U_n] = (3m-2n)X_{m+n} - \frac{N_4}{15N_3}(n^3-5m^3-3mn^2+5m^2n-9n+17m)W_{m+n} \\
&[U_m, U_n] = 3(m-n)Y_{m+n} - n_{44}(m-n)(m^2-mn+n^2-7)U_{m+n} \\
&-\frac{N_4}{360}(m-n)(108-39m^2+3m^4+20mn-2m^3n-39n^2+4m^2n^2-2mn^2+3n^4)L_{m+n} \\
&-\frac{cN_4}{4320}m(m^2-1)(m^2-4)(m^2-9)\delta_{m+n} \\
&[W_m, X_n] = (4m-2n)Y_{m+n} + \frac{1}{56}\frac{N_5}{N_4}(28m^3-21m^2n+9mn^2-2n^3-88m+32n)U_{m+n}\,, \\
&[X_m, X_n] = -\frac{cN_5}{241920}m(m^2-1)(m^2-4)(m^2-9)(m^2-16)\delta_{m+n} + \cdots
\end{aligned} \tag{A15}$$

The constants are

$$\begin{aligned}
&N_3 = \tfrac{1}{5}(\lambda^2-4)\,, \quad N_4 = -\tfrac{3}{70}(\lambda^2-4)(\lambda^2-9) \\
&N_5 = \tfrac{1}{105}(\lambda^2-4)(\lambda^2-9)(\lambda^2-16)\,, \quad n_{44} = \tfrac{1}{30}(\lambda^2-19)
\end{aligned} \tag{A16}$$

in the current notation.

With the conventions, higher spin charges are given by

$$\begin{aligned}
&L_0|\mathcal{O}_\pm\rangle = h|\mathcal{O}_\pm\rangle\,, \quad W_0|\mathcal{O}_\pm\rangle = w|\mathcal{O}_\pm\rangle\,, \quad U_0|\mathcal{O}_\pm\rangle = u|\mathcal{O}_\pm\rangle \\
&X_0|\mathcal{O}_\pm\rangle = x|\mathcal{O}_\pm\rangle\,, \quad Y_0|\mathcal{O}_\pm\rangle = y|\mathcal{O}_\pm\rangle
\end{aligned} \tag{A17}$$

Here $|\mathcal{O}_\pm\rangle \equiv \mathcal{O}_\pm(0)|0\rangle$ and

$$\begin{aligned}
&h = \tfrac{1}{2}(1\pm\lambda)\,, \quad w = \pm\tfrac{1}{6}(2\pm\lambda)(1\pm\lambda)\,, \quad u = \tfrac{1}{20}(3\pm\lambda)(2\pm\lambda)(1\pm\lambda) \\
&x = \pm\tfrac{1}{70}(4\pm\lambda)(3\pm\lambda)(2\pm\lambda)(1\pm\lambda)\,, \quad y = \tfrac{1}{252}(5\pm\lambda)(4\pm\lambda)(3\pm\lambda)(2\pm\lambda)(1\pm\lambda)
\end{aligned} \tag{A18}$$

at the leading order in $1/c$.

[7] Here we have changed some signs, see, e.g., footnote 6 of [38]. The changes here are associated with redefinitions as $W \to iW$, $U \to -U$, $X \to -iX$, and Y untouched.

Appendix B.2. Three and Two Point Functions

We start from spin 6 current $J^{(3,3;6)} \sim: J^{(3)}J^{(3)}$: in Equation (76). Let us assume the form as

$$J^{(3,3;6)}(0)|0\rangle = J^{(3,3;6)}_{-6}|0\rangle = (W_{-3}W_{-3} + aU_{-6} + bL_{-6})|0\rangle \tag{A19}$$

Then the coefficients a, b are fixed by the condition $L_1 J^{(3,3;6)}_{-6}|0\rangle = 0$ as

$$a = -\frac{10}{9}, \quad b = \frac{5N_3}{7} \tag{A20}$$

We may rewrite

$$\begin{aligned} J^{(3,3;6)}(z) &= \sum_m \sum_n \frac{:W_m W_n:}{z^{m+n+6}} + \frac{a}{2}\sum_n \frac{(n+4)(n+5)U_n}{z^{n+6}} \\ &+ \frac{b}{24}\sum_n \frac{(n+2)(n+3)(n+4)(n+5)L_n}{z^{n+6}} \end{aligned} \tag{A21}$$

where the prescription of normal ordering is (see, e.g., (6.144) of [39])

$$: AB :_m = \sum_{n \le -h_A} A_n B_{m-n} + \sum_{n > -h_A} B_{m-n} A_n \tag{A22}$$

with h_A as the conformal weight of A. We then obtain $(C^{(3,3;6)}_{\pm,0})^2$ with the three point function

$$\begin{aligned} \langle \mathcal{O}_\pm | J^{(3,3;6)}_0 | \mathcal{O}_\pm \rangle &= \langle \mathcal{O}_\pm | (W_{+2}W_{-2} + W_{+1}W_{-1} + W_0 W_0 + 10aU_0 + 5bL_0) | \mathcal{O}_\pm \rangle \\ &= \left(\tfrac{8}{9}u + \tfrac{1}{14}N_3 h + w^2 \right) \langle \mathcal{O}_\pm | \mathcal{O}_\pm \rangle \end{aligned} \tag{A23}$$

and the normalization of higher spin current

$$\langle J^{(3,3;6)} J^{(3,3;6)} \rangle = 2\left(-\frac{5cN_3}{6} \right)^2 \tag{A24}$$

For spin 7 there is a double trace operator $J^{(3,4;7)} \sim: J^{(3)}J^{(4)}$: as in Equation (76). As above, we can show that

$$J^{(3,4;7)}(0)|0\rangle = J^{(3,4;7)}_{-7}|0\rangle = (W_{-3}U_{-4} + aX_{-7} + bW_{-7})\,|0\rangle \tag{A25}$$

with

$$a = -\frac{10}{11}, \quad b = -\frac{2}{9}\frac{N_4}{N_3} \tag{A26}$$

is primary. Rewriting

$$\begin{aligned} J^{(3,4;7)}(z) &= \sum_{m \ge 1, n} \frac{:W_m U_n:}{z^{m+n+7}} + \frac{a}{2}\sum_n \frac{(n+5)(n+6)X_n}{z^{n+7}} \\ &+ \frac{b}{24}\sum_n \frac{(n+3)(n+4)(n+5)(n+6)W_n}{z^{n+7}} \end{aligned} \tag{A27}$$

we find that

$$\begin{aligned} \langle \mathcal{O}_\pm | J^{(3,4;7)}_0 | \mathcal{O}_\pm \rangle &= \langle \mathcal{O}_\pm |\, (U_2 W_{-2} + U_1 W_{-1} + U_0 W_0 + 15aX_0 + 15bW_0)\, | \mathcal{O}_\pm \rangle \\ &= \left(\tfrac{15}{11}x - \tfrac{2}{15}\tfrac{N_4}{N_3}w + uw \right) \langle \mathcal{O}_\pm | \mathcal{O}_\pm \rangle \end{aligned} \tag{A28}$$

Normalization is given by

$$\langle J^{(3,4;7)} J^{(3,4;7)} \rangle = \left(-\frac{5cN_3}{6} \right) \left(-\frac{7cN_4}{6} \right) \tag{A29}$$

There are three types as in Equation (77) for $s = 8$, and we start from $J^{(4,4;8)} \sim: J^{(4)} J^{(4)} :$. We assume its form as

$$J^{(4,4;8)}(0)|0\rangle = J^{(4,4;8)}_{-8}|0\rangle = (U_{-4}U_{-4} + aY_{-8} + bU_{-8} + dL_{-8})|0\rangle \tag{A30}$$

The condition $L_1 J^{(4,4;8)}_{-8}|0\rangle = 0$ fixes the constants as

$$a = -\frac{21}{13}, \quad b = \frac{42}{11} n_{44}, \quad d = \frac{7N_4}{9} \tag{A31}$$

The operator $J^{(4,4;8)}$ is then obtained as

$$\begin{aligned} J^{(4,4;8)}(z) &= \sum_{m,n} \frac{:U_m U_n:}{z^{m+n+8}} + \frac{a}{2} \sum_n \frac{(n+6)(n+7)Y_n}{z^{n+8}} + \frac{b}{24} \sum_n \frac{(n+4)\cdots(n+7)U_n}{z^{n+8}} \\ &+ \frac{d}{6!} \sum_n \frac{(n+2)\cdots(n+7)L_n}{z^{n+8}} \end{aligned} \tag{A32}$$

Thus we find

$$\begin{aligned} &\langle \mathcal{O}_\pm | J^{(4,4;8)}_0 | \mathcal{O}_\pm \rangle \\ &= \langle \mathcal{O}_\pm | (U_3U_{-3} + U_2U_{-2} + U_1U_{-1} + U_0U_0 + 21aY_0 + 35bU_0 + 7dL_0 | \mathcal{O}_\pm \rangle \\ &= \left(-\frac{hN_4}{45} + \frac{18n_{44}u}{11} + u^2 + \frac{27y}{13} \right) \langle \mathcal{O}_\pm | \mathcal{O}_\pm \rangle \end{aligned} \tag{A33}$$

The normalization is

$$\langle J^{(4,4;8)} J^{(4,4;8)} \rangle = 2 \left(-\frac{7cN_4}{6} \right)^2 \tag{A34}$$

We then move to $J^{(3,5;8)} \sim: J^{(3)} J^{(5)} :$ in Equation (77). We find

$$J^{(3,5;8)}(0)|0\rangle = J^{(3,5;8)}_{-8}|0\rangle = (W_{-3}X_{-5} + aY_{-8} + bU_{-8})|0\rangle \tag{A35}$$

with

$$a = -\frac{10}{13}, \quad b = -\frac{15}{154} \frac{N_5}{N_4} \tag{A36}$$

is primary. With this expression, we compute

$$\langle \mathcal{O}_\pm | J^{(3,5;8)}_0 | \mathcal{O}_\pm \rangle = \langle \mathcal{O}_\pm | (X_2W_{-2} + X_1W_{-1} + X_0W_0 + 21aY_0 + 35bU_0 | \mathcal{O}_\pm \rangle \tag{A37}$$

$$= \left(-\frac{15N_5u}{77N_4} + wx + \frac{24y}{13} \right) \langle \mathcal{O}_\pm | \mathcal{O}_\pm \rangle \tag{A38}$$

and

$$\langle \Lambda^{(3,5)} \Lambda^{(3,5)} \rangle = \left(-\frac{5cN_3}{6} \right) \left(-\frac{9cN_5}{6} \right) \tag{A39}$$

For $J^{(3,3;8)} \sim: J^{(3)}\partial^2 J^{(3)}$: in Equation (77), we define

$$J^{(3,3;8)}_{-8}|0\rangle = (W_{-5}W_{-3} + aW_{-4}W_{-4} + bU_{-8} + dL_{-8})|0\rangle \tag{A40}$$

where the condition $L_1 J^{(3,3;8)}_{-8}|0\rangle = 0$ leads to

$$a = -\frac{7}{12}, \quad b = \frac{7}{11}, \quad d = -\frac{35N_3}{36} \tag{A41}$$

Using

$$\begin{aligned} J^{(3,3;8)}(z) &= \tfrac{1}{2}\sum_{m,n} \frac{:(m+3)(m+4)W_m W_n:}{z^{m+n+8}} + a\sum_{m,n} \frac{:(m+3)W_m(n+3)W_n:}{z^{m+n+8}} \\ &+ \tfrac{b}{24}\sum_n \frac{(n+4)\cdots(n+7)U_n}{z^{n+8}} + \tfrac{d}{6!}\sum_n \frac{(n+2)\cdots(n+7)L_n}{z^{n+8}} \end{aligned} \tag{A42}$$

we find

$$\begin{aligned} \langle \mathcal{O}_\pm | J^{(3,3;8)}_0 | \mathcal{O}_\pm \rangle &= \langle \mathcal{O}_\pm | \tfrac{1}{2}(2W_2W_{-2} + 6W_1W_{-1} + 12W_0W_0) \\ &+ a(5W_2W_{-2} + 8W_1W_{-1} + 9W_0W_0) + 21bY_0 + 35dU_0 | \mathcal{O}_\pm \rangle \\ &= \left(\frac{hN_3}{36} + \frac{3u}{11} + \frac{3w^2}{4} \right) \langle \mathcal{O}_\pm | \mathcal{O}_\pm \rangle \end{aligned} \tag{A43}$$

Here we have applied the normal ordering prescription as in Equation (A22). For instance, we may set

$$A_m = (m+3)(m+4)W_m, \quad B_n = W_n, \quad h_A = 5 \tag{A44}$$

The normalization is

$$\langle J^{(3,3;8)} J^{(3,3;8)} \rangle = \left(-\frac{5cN_3}{6}\right)\left(-\frac{35cN_3}{2}\right) + 2a^2(-5cN_3)^2 \tag{A45}$$

There could be another spin 8 current of double trace type as $J^{(3,4;8)} \sim: J^{(3)}\partial J^{(4)}$:. We can see that

$$J^{(3,4;8)}(0)|0\rangle = J^{(3,4;8)}_{-8}|0\rangle = (W_{-3}U_{-5} + aW_{-4}U_{-4} + bX_{-8} + dW_{-8})|0\rangle \tag{A46}$$

is primary for

$$a = -\frac{4}{3}, \quad b = -\frac{5}{3}, \quad d = -\frac{4N_4}{5N_3} \tag{A47}$$

Since $J^{(3,4;8)}(z)$ is given by

$$\begin{aligned} J^{(3,4;8)}(z) &= -\sum_{m,n} \frac{:W_m(n+4)U_n:}{z^{m+n+8}} - a\sum_{m,n} \frac{:(m+3)W_m U_n:}{z^{m+n+8}} \\ &- \tfrac{b}{6}\sum_n \frac{(n+5)(n+6)(n+7)X_n}{z^{n+8}} - \tfrac{d}{5!}\sum_n \frac{(n+3)\cdots(n+7)W_n}{z^{n+8}} \end{aligned} \tag{A48}$$

we find

$$\begin{aligned} \langle \mathcal{O}_\pm | J^{(3,4;8)}_0 | \mathcal{O}_\pm \rangle &= \langle \mathcal{O}_\pm | -(6U_2W_{-2} + 5U_1W_{-1} + 4U_0W_0) \\ &- a(U_2W_{-2} + 2U_1W_{-1} + 3U_0W_0) - 35bX_0 - 21dW_0 | \mathcal{O}_\pm \rangle = 0 \end{aligned} \tag{A49}$$

which means that there is no contribution from $J^{(3,4;8)}$.

References

1. Gaberdiel, M.R.; Gopakumar, R. An AdS_3 dual for minimal model CFTs. *Phys. Rev. D* **2011**, *83*, 066007.
2. Aharony, O.; Alday, L.F.; Bissi, A.; Perlmutter, E. Loops in AdS from conformal field theory. *J. High Energy Phys.* **2017**, *2017*, 36.
3. Alday, L.F.; Bissi, A. Loop corrections to supergravity on $AdS_5 \times S^5$. *arXiv* **2017**, arXiv:hep-th/1706.02388.
4. Aprile, F.; Drummond, J.M.; Heslop, P.; Paul, H. Quantum gravity from conformal field Theory. *arXiv* **2017**, arXiv:hep-th/1706.02822.
5. El-Showk, S.; Paulos, M.F.; Poland, D.; Rychkov, S.; Simmons-Duffin, D.; Vichi, A. Solving the 3D Ising model with the conformal bootstrap. *Phys. Rev. D* **2012**, *86*, 025022.
6. Prokushkin, S.; Vasiliev, M.A. Higher spin gauge interactions for massive matter fields in 3D AdS space-time. *Nucl. Phys. B* **1999**, *545*, 385–433.
7. Chang, C.M.; Yin, X. Higher spin gravity with matter in AdS_3 and its CFT dual. *J. High Energy Phys.* **2012**, *2012*, 24.
8. Ammon, M.; Kraus, P.; Perlmutter, E. Scalar fields and three-point functions in D=3 higher spin gravity. *J. High Energy Phys.* **2012**, *2012*, 113.
9. Bais, F.A.; Bouwknegt, P.; Surridge, M.; Schoutens, K. Coset construction for extended Virasoro algebras. *Nucl. Phys. B* **1988**, *304*, 371–391.
10. Ahn, C. The coset spin-4 Casimir operator and its three-point functions with scalars. *J. High Energy Phys.* **2012**, *2012*, 27.
11. Ahn, C.; Kim, H. Spin-5 Casimir operator its three-point functions with two scalars. *J. High Energy Phys.* **2014**, *2014*, 12.
12. Creutzig, T.; Hikida, Y.; Rønne, P.B. Three point functions in higher spin AdS_3 supergravity. *J. High Energy Phys.* **2013**, *2013*, 171.
13. Moradi, H.; Zoubos, K. Three-point functions in $\mathcal{N} = 2$ higher-spin holography. *J. High Energy Phys.* **2013**, *2013*, 18.
14. Creutzig, T.; Hikida, Y.; Rønne, P.B. Higher spin AdS_3 supergravity and its dual CFT. *J. High Energy Phys.* **2012**, *2012*, 109.
15. Gaberdiel, M.R.; Gopakumar, R. Triality in minimal model holography. *J. High Energy Phys.* **2012**, *2012*, 127.
16. Papadodimas, K.; Raju, S. Correlation functions in holographic minimal models. *Nucl. Phys. B* **2012**, *856*, 607–646.
17. Zamolodchikov, A.B. Conformal symmetry in two-dimensions: An explicit recurrence formula for the conformal partial wave amplitude. *Commun. Math. Phys.* **1984**, *96*, 419–422.
18. Perlmutter, E. Virasoro conformal blocks in closed form. *J. High Energy Phys.* **2015**, *2015*, 88.
19. Beccaria, M.; Fachechi, A.; Macorini, G. Virasoro vacuum block at next-to-leading order in the heavy-light limit. *J. High Energy Phys.* **2016**, *2016*, 72.
20. Fitzpatrick, A.L.; Kaplan, J.; Walters, M.T. Virasoro conformal blocks and thermality from classical background Fields. *J. High Energy Phys.* **2015**, *2015*, 200.
21. Fitzpatrick, A.L.; Kaplan, J. Conformal blocks beyond the semi-classical limit. *J. High Energy Phys.* **2016**, *2016*, 75.
22. Chen, H.; Fitzpatrick, A.L.; Kaplan, J.; Li, D.; Wang, J. Degenerate operators and the $1/c$ expansion: Lorentzian resummations, high order computations, and super-Virasoro blocks. *J. High Energy Phys.* **2017**, *2017*, 167.
23. Fitzpatrick, A.L.; Kaplan, J.; Li, D.; Wang, J. Exact Virasoro blocks from Wilson lines and background-independent operators. *J. High Energy Phys.* **2017**, *2017*, 92.
24. Gaberdiel, M.R.; Hartman, T.; Jin, K. Higher spin black holes from CFT. *J. High Energy Phys.* **2012**, *2012*, 103.
25. Henneaux, M.; Rey, S.J. Nonlinear W_∞ as asymptotic symmetry of three-dimensional higher spin anti-de Sitter gravity. *J. High Energy Phys.* **2010**, *2010*, 7.
26. Campoleoni, A.; Fredenhagen, S.; Pfenninger, S.; Theisen, S. Asymptotic symmetries of three-dimensional gravity coupled to higher-spin fields. *J. High Energy Phys.* **2010**, *2010*, 7.
27. Gaberdiel, M.R.; Hartman, T. Symmetries of holographic minimal models. *J. High Energy Phys.* **2011**, *2011*, 31.
28. Campoleoni, A.; Fredenhagen, S.; Pfenninger, S. Asymptotic W-symmetries in three-dimensional higher-spin gauge theories. *J. High Energy Phys.* **2011**, *2011*, 113.

29. Creutzig, T.; Hikida, Y.; Rønne, P.B. Extended higher spin holography and Grassmannian models. *J. High Energy Phys.* **2013**, *2013*, 38.
30. Creutzig, T.; Hikida, Y.; Rønne, P.B. Higher spin AdS_3 holography with extended supersymmetry. *J. High Energy Phys.* **2014**, *2014*, 163.
31. Hikida, Y.; Rønne, P.B. Marginal deformations and the Higgs phenomenon in higher spin AdS_3 holography. *J. High Energy Phys.* **2015**, *2015*, 125.
32. Gaberdiel, M.R.; Gopakumar, R. Large $\mathcal{N} = 4$ holography. *J. High Energy Phys.* **2013**, *2013*, 036.
33. Gaberdiel, M.R.; Gopakumar, R. Higher spins & strings. *J. High Energy Phys.* **2014**, *2014*, 44.
34. Ahn, C.; Kim, H. Three point functions in the large $\mathcal{N} = 4$ holography. *J. High Energy Phys.* **2015**, *2015*, 111.
35. Ahn, C.; Kim, H.; Paeng, J. Three-point functions in the $\mathcal{N} = 4$ orthogonal coset theory. *Int. J. Mod. Phys. A* **2016**, *31*, 1650090.
36. Ahn, C.; Kim, D.G.; Kim, M.H. The next 16 higher spin currents and three-point functions in the large $\mathcal{N} = 4$ holography. *Eur. Phys. J. C* **2017**, *77*, 523.
37. Hikida, Y.; Uetoko, T. Correlators in higher spin AdS_3 holography from Wilson lines with loop corrections. *arXiv* **2017**, arXiv:hep-th/1708.08657.
38. Creutzig, T.; Hikida, Y. Higgs phenomenon for higher spin fields on AdS_3. *J. High Energy Phys.* **2015**, *2015*, 164.
39. Di Francesco, P.; Mathieu, P.; Senechal, D. *Conformal Field Theory*; Springer: New York, NY, USA, 1997.

universe

MDPI

Article

A Note on (Non)-Locality in Holographic Higher Spin Theories

Dmitry Ponomarev

Theoretical Physics Group, Blackett Laboratory, Imperial College London, London SW7 2AZ, UK; d.ponomarev@imperial.ac.uk

Received: 23 October 2017; Accepted: 15 December 2017; Published: 1 January 2018

Abstract: It was argued recently that the holographic higher spin theory features non-local interactions. We further elaborate on these results using the Mellin representation. The main difficulty previously encountered in this method is that the Mellin amplitude for the free theory correlator is ill-defined. To resolve this problem, instead of literally applying the standard definition, we propose to define this amplitude by linearity using decompositions, where each term has the associated Mellin amplitude well-defined. Up to a sign, the resulting amplitude is equal to the Mellin amplitude for the singular part of the quartic vertex in the bulk theory and, hence, can be used to analyze bulk locality. From this analysis we find that the scalar quartic self-interaction vertex in the holographic higher spin theory has a singularity of a special form, which can be distinguished from generic bulk exchanges. We briefly discuss the physical interpretation of such singularities and their relation to the Noether procedure.

Keywords: higher spin theory; AdS/CFT correspondence; Mellin amplitudes; (non)-locality

1. Introduction

There is an overwhelming evidence that in the conventional sense, local higher spin theories do not exist in flat space. This evidence comes from numerous no-go results obtained within a wide range of approaches [1–7]. This problem was recently reconsidered using the light-cone [8] and the manifestly covariant [9,10] approaches. At the same time, recent analysis [11,12] indicates that at least in the self-dual sector there exist consistent local higher spin theories with properties very similar to those of their lower spin counterparts. One option could be to stop here and declare that higher spin theories in flat space cannot go beyond the self-dual sector. Alternatively, one can try to relax locality in some controllable way and push on with parity invariant higher spin theories.

So far in constructing higher spin theories locality was the main guiding principle and relaxing it will make the problem too ill-defined, see, for example, [13][1]. Also, usually non-locality has some undesirable physical consequences, such as superluminal propagation, Ostrogradsky instability, etc. This implies that the requirement of locality should be replaced with another guiding principle that would ensure both that the problem of higher spin interactions is well-posed and that physical pathologies are absent.

We do not have much to say on what these new guiding principles should be. However, what we can do instead is to look at higher spin theory in AdS and see how locality is violated there. This can give us a better idea of what locality violations to expect for putative parity invariant higher spin theories in flat space. The advantage of considering the AdS setting is that in this case the higher spin theory is known, at least at the level of scattering amplitudes. Indeed, via holography in the simplest

[1] Analogous statements can also be proven within the light-cone deformation procedure [14].

setup these can be identified with the correlators of the free $O(N)$ vector model [15,16][2]. The latter, in turn, can be easily computed. Thus, by studying the free theory correlators one can learn whether the bulk higher spin theory is local and if not, how exactly locality is violated. Another attractive feature of the holographic higher spin theory is that, being dual to a healthy theory on the boundary, it is unlikely to suffer from any serious problems.

Proceeding along these lines, recently it was argued [19] that holographic higher spin theory is non-local, see [20–22] for earlier discussions. This conclusion is based on the fact that via combining conformal block decompositions of bulk and boundary four-point functions in different channels one can show that the conformal block decomposition for the bulk theory contact interaction contains single trace conformal blocks. This, in turn, can be regarded as an indication that the quartic interaction vertex in the holographic higher spin theory is non-local.

While this argument, indeed, strongly suggests that the relevant vertex is non-local, it would be important to have an explicit formula clearly characterizing this non-locality. An explicit formula for the non-local part of the contact interaction may help us to understand whether the associated non-locality is general enough to trivialize the Noether procedure[3] along the lines of the argument [13]. This information may be then used to amend the standard Noether procedure in a way, that makes the problem of higher spin interactions a well-defined one. The resulting approach or, rather its flat space version, can then be employed to construct higher spin theories in the Minkowski space.

In this paper we analyze the Mellin amplitude for the four-point correlator in the free $O(N)$ vector model. Previously, this question was addressed in [21,22]. These attempts encountered problems that result into an ill-defined Mellin amplitude. Here we propose to *define* this amplitude using the superposition principle. To be more precise, one can present the boundary correlator in the form of a superposition of interfering processes and then define its Mellin amplitude as the sum of Mellin amplitudes for each individual process. We then use this Mellin amplitude and the fact that the boundary correlator up to a sign is equal to the singularity of the contact four-point interaction in the bulk higher spin theory to characterize non-locality of the latter.

This paper is organized as follows. In Section 2 we review how locality is defined in AdS. Then, in Section 3 we discuss how locality can be tested for holographic higher spin theories using the conformal block decomposition. In particular, we review [19] and discuss various related subtleties. Next, in Section 4 we propose a way to resolve a previously encountered problem with the definition of the Mellin amplitude for the boundary correlator. Finally, we conclude in Section 5.

2. Definitions of Locality

First, let us specify what we mean by locality in AdS. For any scattering process in AdS one can evaluate the Witten diagram, which results in some function $\mathcal{A}_n(x_i)$ of n points on the boundary x_i associated with the external lines of the scattering process. A particularly insightful representation for such amplitudes was recently proposed by Mack [24,25]

$$\mathcal{A}_n(x_i) = \prod_{1\le i<j\le n} \left(\int_{-i\infty}^{i\infty} \frac{d\delta_{ij}}{2\pi i} \Gamma(\delta_{ij})(x_{ij})^{-2\delta_{ij}} \right) \prod_{i=1}^{n} \delta\left(\Delta_i - \sum_{j=1}^{n} \delta_{ij} \right) \mathcal{M}_n(\delta_{ij}), \tag{1}$$

where $x_{ij}^2 = (x_i - x_j)^2$, Δ_i are dimensions of operators on external lines, δ_{ij} are variables dual to x_{ij} and $\mathcal{M}(\delta_{ij})$ is called the *Mellin amplitude*. The integration contours for independent δ_{ij} that remain after solving the delta-function constraints in (1) run parallel to the imaginary axes in a way that the series of poles produced by each Gamma-function as well as by $M(\delta_{ij})$ stay on one side of the

---

[2] Note that the idea of computing higher spin scattering amplitudes in AdS space from "the S-matrix" of singletons was discussed long before higher spin holography acquired its modern form [17,18].

[3] In the higher spin literature the procedure of perturbative construction of a gauge invariant action, see e.g., [23], is called the Noether procedure.

contours. The Mellin variables δ_{ij} can be thought of as the AdS counterparts of $q_i \times q_j$ for flat space scattering amplitudes, while operator dimensions Δ_i can be regarded as analogues of m_i^2. Then, one can see that the delta-functions in (1) impose constraints on δ_{ij}, which are equivalent to putting external momenta on-shell and imposing momentum conservation in flat space. Along these lines, the function $\mathcal{M}(\delta_{ij})$ plays the role of the AdS counterpart of the flat scattering amplitude. Moreover, it was shown on numerous examples that Mellin amplitudes for scattering processes in AdS have a clear analytic structure similar to the analytic structure of flat space scattering amplitudes for the same processes [26–28]. In particular, it was shown that contact interactions with a finite number of derivatives lead to polynomial Mellin amplitudes, while Mellin amplitudes for AdS exchanges are meromorphic functions featuring poles at locations, associated with dimensions of exchanged operators and their descendants. We refer the reader to [26–28] for more details.

This explains utility of the Mellin representation for studying locality in AdS. In flat space the theory is usually called local if its Lagrangian has a finite number of derivatives. This implies that the amplitudes associated with contact interactions should be polynomial. One can also consider a weaker notion of locality, which only demands that contact interactions produce amplitudes being entire functions. In this form, by employing Mellin amplitudes the notion of locality can be easily transferred from flat space to AdS. This is how (weak) locality was defined in [20] and we will adopt this definition here. To summarize, to verify whether the theory is local one would need to evaluate Mellin amplitudes for its contact diagrams and check whether they are given by entire functions.

First the issue of locality appears for quartic vertices and it is enough to consider self-interaction of scalar fields. Using holography, the associated amplitude can be defined by subtracting contributions of four-point exchanges from the boundary correlator. Cubic vertices needed to define exchanges are determined by matching three point Witten diagrams with the associated three-point boundary correlators. This is the approach that was undertaken in [20,29].

Unfortunately, due to computational difficulties with spinning exchanges, completion of this program directly within the Mellin representation remains technically prohibiting. Instead, quartic self-interaction vertex was implicitly constructed in [20] using a certain spectral representation for the conformal block decomposition, see [30] for a comprehensive review on the topic.

Luckily, locality can also be translated into the language of conformal blocks. In [31] it was shown that all consistent four-point functions featuring only double trace conformal blocks[4] in the conformal block decomposition are in one-to-one correspondence with contact four-point Witten diagrams in the bulk. On the other hand, conformal block decomposition of exchanges in the direct channel contains single trace conformal blocks with dimensions equal to dimensions of the exchanged operator. This suggests that absence/presence of single trace conformal blocks can be used as an alternative criterion of locality/non-locality of the associated contact vertex, see [20].

An important subtlety here is that even the exchange, while being non-local by definition, once expanded in the crossed channel, contains only double trace conformal blocks in the conformal block decomposition [32,33]. This phenomenon is similar to the one for flat space amplitudes, when the infinite series of local terms can hide a true singularity. Let us note, however, that in the same manner one can expect that an infinite series of single trace conformal blocks may, in principle, obscure the true locality nature of the vertex.

3. Locality in Holographic Higher Spin Theory

Let us now go into more details and see whether the quartic self-interaction vertex in the holographic higher spin theory is local by the single trace conformal block test. Below we will

4 Primary operators containing one/two trace contractions are called single/double trace operators. For vector models these are bilinear/quadrilinear operators in elementary fields. Conformal blocks with single/double trace operators exchanged are called single/double trace conformal blocks.

use the contact vertex in the crossing symmetric form, so it will be enough to check whether it has single trace conformal blocks in the s-channel conformal block decomposition. As explained above, the amplitude for the contact four-point vertex is defined as

$$\mathcal{A}_4^c(x_i) = G_4(x_i) - \sum_l \left(\mathcal{A}_4^{e|s,l}(x_i) + \mathcal{A}_4^{e|t,l}(x_i) + \mathcal{A}_4^{e|u,l}(x_i) \right), \tag{2}$$

where $G \equiv \langle \mathcal{O}(x_1)\mathcal{O}(x_2)\mathcal{O}(x_3)\mathcal{O}(x_4)\rangle_c$ is the connected part of the boundary correlator, and $\mathcal{A}_4^{e|s,l}$, $\mathcal{A}_4^{e|t,l}$ and $\mathcal{A}_4^{e|u,l}$ are s-, t- and u-channel exchanges respectively with l denoting spin of the exchange.

In [34] it was shown that once cubic vertices in the bulk theory agree with the CFT side at three-point level, then exchanges accommodate all direct channel single trace conformal block contributions of the boundary correlator. In other words, s-channel exchanges in (2) cancel all s-channel single trace contributions from the boundary correlator. On the other hand, as it was already mentioned, each of the exchanges in the crossed channels has only double trace conformal blocks in the s-channel conformal block decomposition. So, naively, one can conclude that the amplitude (2) for the contact vertex does not contain s-channel single trace conformal blocks and hence is local.

However, there is a flaw in this argument related to convergence of the spin sum for t- and u-exchanges. Unfortunately, the s-channel conformal block decomposition for this sum cannot be evaluated explicitly. However, one can see that it may hide s-channel singularities as follows.

First, we need to understand better the details of the conformal block decomposition of the boundary correlator. It reads

$$\langle \mathcal{O}(x_1)\mathcal{O}(x_2)\mathcal{O}(x_3)\mathcal{O}(x_4)\rangle_c = \frac{4}{N}\frac{1}{(x_{12}^2 x_{34}^2)^{\Delta}} \left[u^{\frac{\Delta}{2}} + \left(\frac{u}{v}\right)^{\frac{\Delta}{2}} + u^{\frac{\Delta}{2}}\left(\frac{u}{v}\right)^{\frac{\Delta}{2}} \right], \tag{3}$$

where u and v are conformally invariant cross-ratios

$$u = \frac{x_{12}^2 x_{34}^2}{x_{13}^2 x_{24}^2}, \qquad v = \frac{x_{14}^2 x_{23}^2}{x_{13}^2 x_{24}^2} \tag{4}$$

and $\Delta \equiv d-2$ is the dimension of the operator $\mathcal{O} = \phi^2$ of the free $O(N)$ vector model in d dimensions. For brevity, we denote the three terms on the right hand side of (3) as A, B and C.

It can be shown that A + B contains only single trace conformal blocks in the s-channel conformal block decomposition, while C contains only double trace blocks in the same channel. By doing cyclic permutations one can find analogous statements for conformal block decompositions in other channels.

Employing this information and the aforementioned result from [34], we can conclude that the t-channel single trace contributions from $\mathcal{A}_4^{e|t,l}$ add up to B + C, while the u-channel single trace contributions from $\mathcal{A}_4^{e|u,l}$ add up to A + C. In total this gives A + B + 2C, which, besides a double trace contribution 2C, contains a single trace piece A + B, when viewed from the point of view of the s-channel conformal block decomposition. This indirect argument allows to show that, in fact, the contact vertex $\mathcal{A}_4^c(x_i)$ has the s-channel singularity equal to minus that of the boundary correlator $G_4(x_i)$. Hence, the holographic higher spin theory is non-local by the single trace conformal block test, see [19].

Let us point out few subtleties related to this argument. First of all, as we already mentioned, while presence of a finite series of single trace conformal blocks in the conformal block decomposition does imply presence of poles in the Mellin amplitude and, therefore, non-locality, it is not clear what kind of singularity, if any, may be associated with an infinite series of single trace conformal blocks. For the case in question, the conformal block decomposition does contain an infinite series of single trace conformal blocks. Though, it is hard to expect that the singularity is absent at all, it would be interesting to have a more qualitative understanding of what kind of singularity we are dealing with. This is important at least to check whether this singularity is general enough to trivialise the deformation procedure along the lines of the argument in [13].

Secondly, the argument given above, strictly speaking, applies to the common domain of validity of the conformal block decompositions in all three channels, which is empty

$$(u < 1) \cap (v/u < 1) \cap (v^{-1} < 1) = \varnothing. \tag{5}$$

It was shown that the domains of convergence of conformal block decompositions are, in fact, much larger and can be applied to correlators analytically continued by these decompositions [35,36]. While we do not expect any difficulties with the analytic continuation of (3) in the coordinate space, for our purposes we rather need to make analytic continuations in the Mellin space, where neither we are aware of similar convergence theorems nor analytic continuations are straightforward if Mellin amplitudes involve distributions.

Finally, another subtlety is that exchanges besides single trace conformal blocks also inevitably contain double trace conformal blocks in the conformal block decomposition. This double trace contribution can be altered by field redefinitions, or, equivalently, by on-shell trivial cubic vertices. Nevertheless, it is not at all arbitrary. Singularities potentially can be generated from summation of these contributions over spin in the same way as it happens for single trace conformal blocks. This and some other subtleties were discussed in [19].

It is also interesting to confront the locality issue discussed here with its p-adic version. Holographic reconstruction of a quartic vertex in the p-adic case was performed in [37]. The striking difference with the Archimedean, that is the standard, analysis is that due to peculiar properties of the p-adics, one does not have any spinning operators and the sum (2) reduces to a single term with scalar exchanges. For this reason single trace contributions cancel out on both sides and the quartic vertex is local.

4. Mellin Amplitude for the Boundary Correlator

From the discussion in Section 2 it is clear that the conclusion about locality in the holographic higher spin theory depends exclusively on what the analytic structure of the Mellin amplitude for the boundary correlator is. Let us clarify what this amplitude is.

At four points (1) reads

$$\begin{aligned}\mathcal{A}_4(x_i) = \left(\frac{v}{x_{12}^2 x_{34}^2}\right)^{\Delta} &\int_{c_s-i\infty}^{c_s+i\infty} \frac{ds}{2\pi i} \int_{c_t-i\infty}^{c_t+i\infty} \frac{dt}{2\pi i} \\ &\times u^{s/2} v^{-(s+t)/2} \mathcal{M}_4(s,t) \Gamma^2\left[\frac{2\Delta - s}{2}\right] \Gamma^2\left[\frac{2\Delta - t}{2}\right] \Gamma^2\left[\frac{2\Delta - u}{2}\right],\end{aligned} \tag{6}$$

where we denoted

$$\begin{aligned} s &\equiv \Delta_1 + \Delta_2 - 2\delta_{12} = \Delta_3 + \Delta_4 - 2\delta_{34}, \\ t &\equiv \Delta_1 + \Delta_3 - 2\delta_{13} = \Delta_2 + \Delta_4 - 2\delta_{24}, \\ u &\equiv \Delta_1 + \Delta_4 - 2\delta_{14} = \Delta_2 + \Delta_3 - 2\delta_{23}, \end{aligned} \tag{7}$$

and then set $\Delta_i = \Delta$. The Mandelstam variables s, t and u are analogous to those in flat space and satisfy

$$s + t + u = 4\Delta. \tag{8}$$

The reason why the amplitude $\mathcal{M}$ is called the Mellin amplitude is because of its connection to the Mellin transform of $\mathcal{A}$. The *Mellin transform* of a function $f(u)$ is defined by

$$F(s) \equiv M[f(u)](s) \equiv \int_0^\infty du f(u) u^{s-1}. \tag{9}$$

This integral typically converges when s belongs to a strip in the complex plane defined by $a < \mathrm{Re}(s) < b$ with a and b real. The Mellin transform $F(s)$ is then analytic in this strip and this strip is called the *strip of analyticity* or *the analyticity domain*. The inverse transform is given by

$$f(u) = \int_{c-i\infty}^{c+i\infty} \frac{ds}{2\pi i} F(s) u^{-s}, \tag{10}$$

where $a < \mathrm{Re}(c) < b$. In other words, the integration contour in (10) runs parallel to the imaginary axis anywhere within the strip of analyticity.

By comparing (6) with the inverse Mellin transform formula (10), we can expect that the Mellin amplitude $\mathcal{M}$ can be obtained from the space-time amplitude $\mathcal{A}$ in the following two steps. First, one performs the Mellin transform of the amplitude $\mathcal{A}(u,v)$ expressed as a function of two independent cross-ratios u and v to find $M(s,t)$, called the *reduced Mellin amplitude*

$$\mathcal{A}_4(x_i) = \left(\frac{v}{x_{12}^2 x_{34}^2}\right)^\Delta \int_{-i\infty}^{i\infty} \frac{ds}{2\pi i} \int_{-i\infty}^{i\infty} \frac{dt}{2\pi i} u^{s/2} v^{-(s+t)/2} M_4(s,t). \tag{11}$$

Next one finds the Mellin amplitude $\mathcal{M}_4(s,t)$ by factoring out the combination of Gamma-functions from the reduced Mellin amplitude $M_4(s,t)$

$$M_4(s,t) = \mathcal{M}_4(s,t) \Gamma^2 \left[\frac{2\Delta - s}{2}\right] \Gamma^2 \left[\frac{2\Delta - t}{2}\right] \Gamma^2 \left[\frac{2\Delta - u}{2}\right]. \tag{12}$$

However, as it is not hard to see, for the correlator (3) this procedure leads to the ill-defined Mellin amplitude.

Indeed, first problem that one encounters is the necessity to make the Mellin transform of the power function, which according to the definition (9) leads to an integral that diverges for any s. Still, there is a consistent framework that allows to define it as a distribution, see [38]

$$M[u^\Delta](s) \equiv \int_0^\infty du\, u^\Delta u^{s-1} = \delta(\Delta + s). \tag{13}$$

Note that here the strip of analyticity is understood as consisting of a single line $\mathrm{Re}(s) = \Delta$. Then it is easy to verify that the inverse formula (10) does hold and the integration contour passes right through the singularity of a delta-function.

Applying this formula to (3) we find

$$M_4(s,t) = \frac{16}{N}(\delta(s-\Delta)\delta(t-\Delta) + \delta(s-\Delta)\delta(t-2\Delta) + \delta(s-2\Delta)\delta(t-\Delta)). \tag{14}$$

Now we plug this into (12) to find the associated Mellin amplitude. The reduced Mellin amplitude M_4 has support consisting of only three points and the same should be true for the Mellin amplitude $\mathcal{M}_4$ itself. Moreover, as it is not hard to see, for any of these three points the product of Gamma-functions in (12) is singular. Hence, the Mellin amplitude is the sum of terms of the form $x \times \delta(x)$, which is zero as a distribution. If we keep the inverse transform formula (6) intact, the vanishing Mellin amplitude implies that the correlator also vanishes in the coordinate representation, which is not the case. This problem was encountered in [21,22].

To summarize, the problem with a formal application of the rule that defines Mellin amplitudes for the case of free CFT's is as follows. The Mellin transform for each of the three terms in (3) is known, well-defined, invertible and given in terms of distributions. Performing the Mellin transform of the free correlator in the coordinate representation we find the associated reduced Mellin amplitude. As a final step, we are instructed to divide it by the combination of Gamma-functions as in (12). Usually, the reduced Mellin amplitude is a genuine function and this step does not cause any problems. However, for free CFT's the reduced Mellin amplitude is a distribution, which is, moreover, supported on points where the double trace Gamma-functions are singular. As a result, an algebraic operation of division by the product of double trace Gamma-functions is not invertible. One way to phrase this is to say that the Mellin amplitude for the free CFT correlator (3) is ill-defined.

It is worth to point out that the reduced Mellin amplitude (14) is also, strictly speaking, ill-defined, but for a different reason. As it was specified below (13), the Mellin transform of the power function is defined in a strip of analyticity consisting of a single line in the complex plane. For the three terms in (14) the strips of analyticity are

$$\{\mathrm{Re}(s) = \Delta,\ \mathrm{Re}(t) = \Delta\}, \quad \{\mathrm{Re}(s) = \Delta,\ \mathrm{Re}(t) = 2\Delta\}, \quad \{\mathrm{Re}(s) = 2\Delta,\ \mathrm{Re}(t) = \Delta\}. \tag{15}$$

They do not overlap, so the reduced Mellin amplitude (14) has the domain, which is the empty set.

4.1. *Inverse Mellin Transform vs. Superposition of Amplitudes*

In the previous discussion the Mellin amplitude was primarily understood as an alternative representation for the Witten diagrams and conformal correlators. In this respect, it was important that there is an unambiguous connection between amplitudes in the standard coordinate representation and in the Mellin form. So far this connection was realized via (1) and a prescription for the contour given below it. Similarly, the inverse relation is also known and involves the Mellin transform (9) as discussed in the previous section. For many relevant cases this dictionary is well-defined and is sufficient to establish an unambiguous connection between the Mellin and the coordinate representations. There are, however, cases, where a naive application of this dictionary leads to an ill-defined result. One of these cases we encountered in the previous section. Below we will try to answer the question of what should be our guiding principle for defining Mellin amplitudes if the standard dictionary with the coordinate representation breaks down and how the Mellin amplitudes can be extracted once the coordinate representation of the respective amplitudes is known.

As this guiding principle we suggest a natural requirement that the Mellin amplitude for a superposition of processes is the sum of Mellin amplitudes for each individual process. This *superposition property* is absolutely standard for probability amplitudes, so it is natural to require that Mellin amplitudes satisfy it as well. In particular, it holds for amplitudes in the standard coordinate representation. As we will see below, this property is in tension with the standard dictionary that relates the Mellin and the coordinate representations. More precisely, as it is not hard to see, the superposition property requires that the transform given by (1) is linear. This is only true if integration contours for all Mellin amplitudes for superposed processes coincide. As it will be illustrated below, the standard locations of the contours, as defined below (1), for different processes, relevant for the holographic higher spin theory, are incompatible with each other. This implies that the Mellin amplitude for a superposition of such processes can not be related in the standard way to the associated amplitude in the coordinate representation. For these problematic cases we propose to *define* the Mellin amplitude for a superposition of processes as a sum of constituent Mellin amplitudes irrespectively of the contour location constraints associated with these amplitudes. This enables us to extend the standard definition of the Mellin amplitude to cases for which it was previously inapplicable.

Below we will consider some natural examples of interfering processes in the holographic higher spin theories, for which the Mellin amplitudes defined in a standard way require incompatible integration contours. We will also discuss additional convergence subtleties that occur when the superposition involves an infinite set of processes. Our main goal is to understand how the Mellin amplitude defined by linearity as specified above can be related to the amplitude in the coordinate representation. Once this is clarified, we will propose the Mellin amplitude for the free theory correlator.

To start, we consider an example of the contact Witten diagram, for which the Mellin amplitude is free of singularities. As it was mentioned below (1), the integration contours should not break any of infinite series of poles generated by $\Gamma(\delta_{ij})$. For the four-point case (6), this implies that the contour for s integration should go between the poles generated by $\Gamma^2[(2\Delta - s)/2]$ and $\Gamma^2[(2\Delta - u)/2]$, while the contour for t integration should separate the poles of $\Gamma^2[(2\Delta - t)/2]$ and $\Gamma^2[(2\Delta - u)/2]$. To this end, one should require that

$$c_s < 2\Delta, \qquad c_t < 2\Delta, \qquad c_u < 2\Delta, \tag{16}$$

where $c_u \equiv 4\Delta - c_s - c_t$.

To reproduce the coordinate representation of the amplitude, we can first evaluate the s integral in (6). Closing the contour at $s \to \infty$ and arguing that the infinite arc integral vanishes, we reduce this integration to the sum of residues at $s = 2\Delta + n$ with $n \geq 0$. This gives a power series in the cross-ratio u/v with coefficients being functions of t. Then, for each term of the series we evaluate the t integral. Closing the contour at $t \to \infty$ we pick residues at $t = 2\Delta + m$ with $m \geq 0$, which produces an expansion in v^{-1}. Eventually, having evaluated both integrals, we find the amplitude presented as a power series in two cross-ratios u/v and v^{-1}. The poles of the reduced Mellin amplitude, which residues were evaluated to arrive to this representation are enclosed inside the red contour on Figure 1.

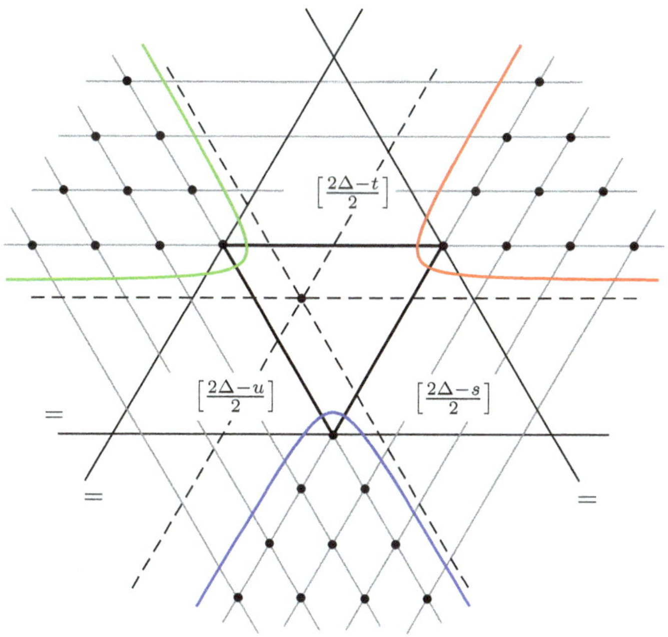

Figure 1. This figure represents the real (s, t, u) plane and locations of singularities of the reduced Mellin amplitude for the contact interaction. Singularities generated by $\Gamma(\delta_{ij})$ are shown as light-grey lines. The bold triangle is the real projection of the analyticity domain associated with the inverse Mellin transform of two variables, which defines admissible locations of the integration contour in the complex (s, t, u) space. Depending on the way how we choose to close this integration contour at infinity, we pick one of the three sets of poles, enclosed in the red, the blue and the green contours.

Alternatively, one can close the integration contour at large t and u, which produces an expansion of the amplitude in powers of the cross-ratios u^{-1} and v/u. The associated poles are enclosed inside the green contour on Figure 1. Finally, by closing the contour at large u and s we produce an expansion in powers of u and v. This expansion is generated by summing the residues of the reduced Mellin amplitude within the blue contour on Figure 1. The three expansions that we thus obtain are valid in the respective kinematics regimes. For domains where more than one expansion is valid, they produce the same result by virtue of various hypergeometric identities.

Let us now consider bulk exchanges with fields of dimension Δ. The associated Mellin amplitudes are well-known [26,27] and are given by hypergeometric functions. For generic dimensions of fields on external lines, the s-channel exchange Mellin amplitude features a series of poles at $s = \Delta + 2n$ with $n \geq 0$. The requirement that these poles are not separated by the contour puts an additional constraint $c_s < \Delta$ on its location. So, combining all constraints together, one finds that the domain of analyticity compared to the contact interaction case shrinks to, see Figure 2,

$$c_s < \Delta, \qquad c_t < 2\Delta, \qquad c_u < 2\Delta. \tag{17}$$

As in the case of a contact interaction, there are three different ways to close the integration contour of the Mellin integral, which results in three alternative representations of the Witten diagram as a series in the conformal cross-ratios.

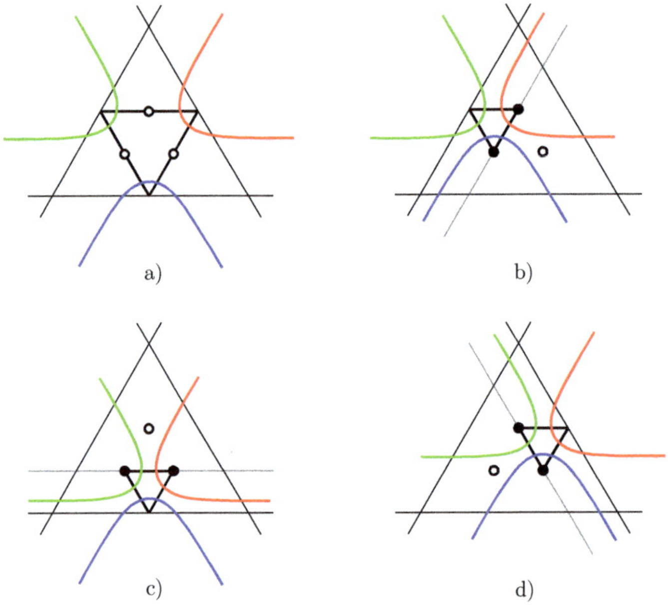

Figure 2. Here we illustrate the singularity structures of the reduced Mellin amplitudes for the contact Witten diagram and for exchanges with the field of dimension Δ in s, t and u channels. These are shown on figures (**a**–**d**) respectively. As before, bold triangles denote domains of analyticity of the reduced Mellin amplitude. Solid and empty circles represent locations of singularities associated with the three terms in the free theory correlator. Solid circles mean that the reduced Mellin amplitude has a given singularity, while empty circles mean that the reduced Mellin amplitude is regular at this point. The light-grey lines denote the leading single trace singularities of exchanges. For example, for the s-channel exchange it appears at $s = \Delta$.

A special attention should be payed to singularities appearing at

$$(s,t,u): \qquad (\Delta,\Delta,2\Delta), \quad (\Delta,2\Delta,\Delta), \quad (2\Delta,\Delta,\Delta) \tag{18}$$

as these produce contributions that remain in the complete boundary correlator (3). The reduced Mellin amplitude for the s-channel exchange has singularities at two of these locations. One is generated by

$$\frac{1}{(s-\Delta)(t-2\Delta)} \tag{19}$$

and contributes to the representation with the Mellin integration contour closed at $s,t \to \infty$ (the red contour) and does not contribute to the others. Another singularity is of the form

$$\frac{1}{(s-\Delta)(u-2\Delta)} \tag{20}$$

and its residue only contributes to the integral with the contour closed at $s,u \to \infty$ (the blue contour). Similar structures of poles and contour locations is exhibited by reduced Mellin amplitudes for other exchanges, see Figure 2.

It is now instructive to consider what happens if we add up three exchanges together. As we have just discussed, each of them has a well-defined Mellin amplitude, which allows to reproduce the associated amplitude in the coordinate representation using the standard dictionary. However, as it is not hard to see, admissible locations of the contours for different exchanges are incompatible with each other. For example, the analyticity domain for the t-channel exchange is

$$c_s < 2\Delta, \qquad c_t < \Delta, \qquad c_u < 2\Delta \tag{21}$$

and it has an empty overlap with the analyticity domain for the s-channel exchange (17). This manifests itself in a way that the singularity (20) at $(s,t,u) = (\Delta,\Delta,2\Delta)$ for the s-channel exchange is inside the blue contour, while the singularity

$$\frac{1}{(t-\Delta)(u-2\Delta)} \tag{22}$$

of the t-channel exchange is also located at the same point, but should be strictly outside the blue contour, as it is inside the green one. We would like to emphasize that contributions associated with these singularities are present in the free theory correlator, which means that we should expect similar inconsistencies with the contours for its Mellin amplitude as well.

How should we proceed in this situation? As we discussed previously, it seems natural to define the Mellin amplitude for the process that involves three exchanges in different channels as a sum of individual Mellin amplitudes associated with each exchange. This is what we are expected to do for consistency with the amplitude's superposition principle. In this case, however, the standard relation between the Mellin and the coordinate representations breaks down. Instead, to reconstruct the coordinate representation of the amplitude from its Mellin form one should first split the amplitude into pieces and then use different contours for each of them. Of course, amplitudes can be split into parts in different ways, which will result in different outcomes in the coordinate representation. This means that under some circumstances *Mellin amplitudes* do not define amplitudes in the usual coordinate form unambiguously, and, hence, *do not give a faithful representation* for scattering processes in AdS or, equivalently, correlators in the CFT.

It is also instructive to understand what happens with integration contours when we sum an infinite series of Witten diagrams. For example, one can expand the s-channel exchange in a series of contact diagrams and consider Mellin amplitudes for each of these diagrams. Then, as it is not hard to see from Figure 2, the constraint for the location of the contour for these diagrams is different from that for the s-channel exchange. For example, the blue contour for contact interactions can go either below or above the point $(s,t,u) = (\Delta, \Delta, 2\Delta)$. At the same time, for the s-channel exchange diagram, the only prescription that gives the desirable result is when the blue contour goes above this point. The reason why this happens is clear. Namely, additional poles in the exchange Mellin amplitude imply that the sum over contact Mellin amplitudes does not converge everywhere. As this expansion is closely related to the Taylor expansion in s at $s = 0$, one would rather expect that it converges when the blue contour goes above the pole at $(s,t,u) = (\Delta, \Delta, 2\Delta)$, because then it is closer to $s = 0$. From this example we learned that when dealing with series of Mellin amplitudes we may generate additional singularities and then the location of the contour should be chosen depending on where this series converges.

Another subtlety is related to the procedure that reduces integrals along the contours that go parallel to the imaginary axis to the sum of residues. For this procedure to commute with infinite summation, one has to ensure that the sum converges both in the strip of analyticity, where the initial contour runs, on the infinite arc contour that closes the initial contour and, in fact, everywhere inside the closed contour formed by joining these two contours together. To avoid these requirements one can deform the contour in the usual way before performing the summation. In this case it is enough to require that the series of Mellin amplitudes converges only in the vicinity of the real axes, where all their singularities are located. This, effectively, means that for the circumstances just described, one may reconstruct the Mellin amplitude from the coordinate representation by requirement that the reduced Mellin amplitude has correct residues irrespectively of what the standard Mellin integral gives.

4.2. Regularized Mellin Amplitude

Previous considerations motivate the following modification of the standard dictionary between the Mellin and the coordinate forms of the free correlator. First of all, we leave the possibility of using different integration contours for different terms in the reduced Mellin amplitude. To understand which contour should be chosen for each term, we will rely on how their singularities are located with respect to the contour in the reduced Mellin amplitudes for constituent Witten diagrams. Secondly, we replace the standard contour that runs parallel to the imaginary axis by a deformed one, which, essentially, replaces integration with the sum of residues of the reduced Mellin amplitude. As the standard integration contour can be deformed in different ways, we will require that sums of residues within each of the deformed contours produce the required coordinate representation.

With this said, let us adjust the reduced Mellin amplitude so that via the modified dictionary it translates into the free correlator (3) in the coordinate representation. First, we fix singularities located within the contour, which is closed at large values of s and t, that is the red contour, see Figures 1 and 2. It already encloses singularities of the form

$$\frac{1}{(s-\Delta)(t-2\Delta)}, \qquad \frac{1}{(s-2\Delta)(t-\Delta)}, \tag{23}$$

contributed by exchanges. Residues of these terms produce two out of three necessary contributions to the free theory correlator. As it is not hard to see, for constituent amplitudes the red contour never encloses singularities capable to produce the remaining term. Still, considering that this term is present in the correlator, the reduced Mellin amplitude should contain a singularity

$$\frac{1}{(s-\Delta)(t-\Delta)}. \tag{24}$$

As we discussed previously, this can happen as a result of summation of infinite series, e.g., exchanges over spin. Adding all contributions together, we find that the reduced Mellin amplitude, which is necessary to produce the boundary correlator from residues inside the red contour, reads[5]

$$M_4^r = \frac{16}{N}\left(\frac{1}{(s-\Delta)(t-2\Delta)} + \frac{1}{(s-2\Delta)(t-\Delta)} + \frac{1}{(s-\Delta)(t-\Delta)}\right). \tag{25}$$

Analogously, for the green and the blue contours we find

$$\begin{aligned} M_4^g &= \frac{16}{N}\left(\frac{1}{(t-\Delta)(u-2\Delta)} + \frac{1}{(t-2\Delta)(u-\Delta)} + \frac{1}{(t-\Delta)(u-\Delta)}\right), \\ M_4^b &= \frac{16}{N}\left(\frac{1}{(u-\Delta)(s-2\Delta)} + \frac{1}{(u-2\Delta)(s-\Delta)} + \frac{1}{(u-\Delta)(s-\Delta)}\right). \end{aligned} \tag{26}$$

Then, the total reduced Mellin amplitude is given by

$$M_4^w \equiv M_4^r + M_4^g + M_4^b = 0. \tag{27}$$

Accordingly, the Mellin amplitude for the free theory correlator (3) is

$$\mathcal{M}_4^w = \frac{M_4^r + M_4^g + M_4^b}{\Gamma^2\left[\frac{2\Delta-s}{2}\right]\Gamma^2\left[\frac{2\Delta-t}{2}\right]\Gamma^2\left[\frac{2\Delta-u}{2}\right]} \tag{28}$$

and also *formally* vanishes. Here "formally" additionally highlights the fact that cancellation occurs between Mellin amplitudes associated with incompatible integration contours.

In other words, we found that the total reduced Mellin amplitude for the free theory correlator is zero. Let us stress again, that this is not in contradiction with the correlator being non-zero in the coordinate representation—this happens, because to recover the coordinate representation of the amplitude, one has to use different integration contours for different terms in the reduced Mellin amplitude. To each term of the reduced Mellin amplitude (27) one can formally assign the following constraints on locations of integration contours

$$\begin{aligned} M^r: &\quad c_s < \Delta, \quad c_t < \Delta, \\ M^g: &\quad c_t < \Delta, \quad c_u < \Delta, \\ M^b: &\quad c_u < \Delta, \quad c_s < \Delta. \end{aligned} \tag{29}$$

However, let us emphasize again, that to recover the correlator in the coordinate representation, each of reduced Mellin amplitudes M^r, M^g and M^b should be integrated not along the standard contour that runs parallel to the imaginary axis, but rather along the deformed one, which runs in the vicinity of the real axis and encloses all singularities of the reduced Mellin amplitude located there.

One may try to avoid a seemingly unattractive feature of this proposal of not having a single integration contour for all components of the reduced Mellin amplitude by various regularizations. Focusing first on the three terms in (25), (26) with a singularity at $(s,t,u) = (\Delta,\Delta,2\Delta)$, one can infinitesimally change these contributions, so that locations of singularities generated by these terms shift one from another, thus developing a common domain of analyticity. Once this is done, the inverse Mellin transform can be performed by simply adding these three contributions and integrating them over a single integration contour inserted inside the common analyticity domain. Similar procedures

5 An analogous proposal for the Mellin transform for the power law function appeared in the context of the conformal bootstrap in Mellin space [39].

can be performed with singularities at other locations. An example of such regularization was recently considered in [40].

It is worth to note that these regularizations do change the total reduced Mellin amplitude. This can be illustrated by the following simple one-dimensional example

$$0 = \frac{1}{x} - \frac{1}{x} \quad \to \quad \lim_{\epsilon \to +0} \left(\frac{1}{x - i\epsilon} - \frac{1}{x + i\epsilon} \right) = 2\pi i \lim_{\epsilon \to +0} \left(\frac{1}{\pi} \frac{\epsilon}{x^2 + \epsilon^2} \right) = 2\pi i \delta(x), \tag{30}$$

where regularization replaces zero by a delta-function. By virtue of analogous manipulations one can replace the initial amplitude (25)-(26) by the one we started from in (14). As it was discussed above, the amplitude (14) still requires to use three different integration contours for each of the three terms in it. Hence, the necessity to use incompatible integration contours for different terms in the reduced Mellin amplitude cannot be avoided.

To summarize, in (28) we proposed an expression for the Mellin amplitude in the free CFT, which is defined not accordingly to a formal application of the Mellin transform as in [21,22], but rather as a sum of Mellin amplitudes for constituent bulk Witten diagrams, which are well-defined and known. Having studied how the integration contours of the inverse Mellin transform are located for each of these diagrams, we found that the total Mellin amplitude can be reconstructed from the boundary correlator in the coordinate representation by a certain modification of the standard procedure, which is described above. In particular, the standard integrals appearing in the inverse Mellin transform with integration contours going parallel to the imaginary axis were replaced by deformed ones going parallel to the real axis and enclosing singularities of the reduced Mellin amplitude located there. By doing that our goal was not to define a generalized version of the Mellin transform, but rather to reconstruct the Mellin amplitude, defined as a sum of constituent Mellin amplitudes, without actually evaluating this sum and imposing as little requirements on the convergence of this sum as possible. One can imagine scenarios where such a contour deformation is not necessary and by evaluating the standard Mellin integrals along imaginary axes one still reproduces the required power function in the coordinate representation[6]. It would be interesting to see what actually happens by evaluating the sum of the bulk Witten diagrams in the Mellin representation explicitly.

5. Conclusions

In conclusion, we briefly discuss what (28) implies for locality of the holographic higher spin theory. First of all, given that the singular part of the Mellin amplitude for the quartic scalar self-interaction in the higher spin theory differs from (28) just by the sign, we find that according to the definition of locality based on the analytic structure of Mellin amplitudes, the holographic higher spin theory is formally local. In all examples considered so far, this definition of locality appeared to be a successful counterpart of flat space locality defined in a standard way. In particular, in both flat and AdS spaces local interactions with a limited number of derivatives result into polynomial amplitudes, while amplitudes for processes that involve exchanges have singularities. In other words, we believe that our formal conclusion about locality of holographic higher spin theories has good reasons to be taken seriously.

On the other hand, it appears that Mellin amplitudes do not represent faithfully all scattering processes in AdS, in a sense that different processes may have identical Mellin amplitudes. Moreover,

[6] Instead of (25)–(27) one could split zero into parts, so that each of them has a vanishing arc integral at infinity, which, in turn, ensures that the standard Mellin integral along the imaginary axis can be replaced by the sum of residues. If integration contours for different terms are located so that all singularities except one stay on one side of the contours, then contributions from these singularities cancels out due to the fact that the total reduced Mellin amplitude is vanishing. Then, the only non-vanishing contribution to the amplitude in the coordinate representation is given in terms of residues of the singularity that for different terms appears on different sides of integration contours. Clearly, this contribution will be given by the power function. Some concrete examples of this mechanism at work can be found in [40].

this effect turns out to be crucial for amplitudes in the holographic higher spin theories. In particular, the Mellin amplitude for the conformal four-point correlator and, hence, the complete four-point higher spin scattering amplitude is formally zero. To be able to reconstruct a non-zero correlator in the coordinate representation, the Mellin amplitude should be first split into three parts and for each part one should use a different integration contour. The same effect is responsible for the formal vanishing of the singularity in the contact four-point interaction in the bulk higher spin theory. For this reason, one may argue that to make judgements about locality of the holographic higher spin theory, each of the three terms should rather be considered separately. As these terms contain singularities, one concludes that the associated non-localities are present in the bulk theory. In this way, our analysis can be reconciled with the conclusion of [19] that the holographic higher spin theory is non-local.

At the same time, we would not go as far as claiming that presence of the singularity (28) trivializes the Noether procedure as a tool to construct higher spin theories along the lines of the argument [13]. What trivialises the Noether procedure is generic non-localities associated with exchanges of fields present in the spectrum of the theory. For the holographic higher spin theory, a generic s-channel exchange has the Mellin amplitude featuring a sequence of singularities at $s = \Delta, \Delta + 2, \dots$. By inspecting explicit expressions in [26,27] one finds that for generic space-time dimensions this sequence is infinite. On the contrary, leaving aside that (28) formally vanishes, each of its terms has only a single singularity in each channel. In this respect it is more reminiscent of exchanges with singletons on the boundary [41–43], but with the dimension being Δ (or 2Δ), not $\Delta/2$. In fact, this similarity is not surprising given that the same singularity is produced by the exchange with an infinite tower of higher spin fields, which, in turn, via the Flato-Fronsdal theorem [44] can be related to a two-particle state of boundary singletons.

It would be interesting to turn these observations into more concrete proposals of how the functional class of admissible Lagrangians should be changed to make the Noether procedure for higher spin theories non-trivial. For example, one can propose that vertices that result into Mellin amplitudes with a finite number of poles are admissible[7]. Or, taking into account that the singularity of the four-point contact interaction formally vanishes in Mellin space, one may stick to the initial proposal of [20] and define locality as the requirement that the Mellin amplitude is an entire function. These proposals are based on a rather formal mathematical way to describe the difference between the known result (28) and a generic bulk exchange. Instead, it would be much more satisfactory if we had better understanding of how different types of singularities in Mellin amplitudes manifest themselves in bulk experiments. Then we would be able to motivate restrictions on functional classes of Lagrangians by the requirement that these restrictions rule out only theories that result into undesirable physical behavior. In this regard, let us stress again that being dual to physically healthy theories on the boundary, higher spin theories are not expected to have serious physical pathologies.

It is worth to note that the Noether procedure with locality defined as the requirement that Mellin amplitudes for contact interactions are free of poles has the following unattractive property: it treats a sum of the four-point exchanges with holographically fixed cubic couplings as a local quartic interaction. This means, in particular, that one can rescale all cubic couplings by a spin-independent prefactor and by an appropriate local change of the quartic vertex keep the total four-point Witten diagram intact. It is not hard to see that this procedure does not violate neither the consistency conditions of the Noether procedure nor locality, which, in turn, implies that cubic couplings may be fixed only up to an overall common factor. It is suggestive that the same pattern persists to all orders

[7] Four-point exchange diagrams with specially tuned dimensions of the exchanged field and fields on external lines may have a finite sequence of poles in the Mellin amplitude. This happens when the sequence of single trace poles of the reduced Mellin amplitude, e.g., $s = \Delta + 2n$ with $n \geq 0$ overlaps with poles of double trace Gamma-functions. In this case all except a finite number of single trace singularities are cancelled in the Mellin amplitude by zeros from $\Gamma^{-1}(\delta_{ij})$. Of course, generic exchanges with fields in the spectrum of the theory should not be regarded as local interactions. This means that for the special values of dimensions of fields as discussed above, locality cannot be defined as a requirement that contact interactions result into Mellin amplitudes with finite sequences of poles.

and the Noether procedure allows to define the theory up to one overall coupling constant at each order. In other words, it does not allow to reproduce the holographic higher spin theory unambiguously.

At the same time, such ambiguity may be used to achieve locality in the sense of the single trace conformal block test. Indeed, we can use the extra freedom to change the cubic couplings in a way that in (2) single trace conformal blocks cancel exactly. One can proceed in a similar manner to higher orders. A higher spin theory so defined is consistent in the sense of the bulk Noether procedure and local in the sense of the single trace conformal block test. Of course, via holography it is incompatible with the CFT consistency conditions, more precisely, with the OPE. It would be still interesting to see whether this inconsistency has manifestations in terms of the purely bulk physics.

One may argue that technical difficulties with evaluation of Mellin amplitudes for generic vertices make it hard to apply the amended Noether procedure in practice. In this respect we would like to stress that our goal in this paper was not to give a practical recipe for derivation of higher spin theories in AdS, but rather to explore at a formal level potential ways to define these theories perturbatively, taking into account a particular type of singularity present in holographic higher spin theories. After all, higher spin theories in AdS can already be reconstructed from holography and do not need another derivation. One still may hope that such analysis may lead to a better understanding of how locality should be relaxed and the Noether procedure be deformed for a more tractable problem of perturbative construction of higher spin theories in flat space.

In summary, as it was shown in [19], the holographic higher spin theory features non-local interactions. Possibly, the most lucid way to convey this statement without going into technical details is to say that the singularity of the contact four-point interaction is proportional to the singularity of the sum of exchanges in the theory. At the same time, this singularity takes a special form, which manifests itself, in particular, in the fact, that it vanishes in the appropriately defined Mellin representation. This information may be naturally used to define a non-trivial Noether procedure. It may also be instructive to view such a form of the singularity and of the total correlator in the Mellin representation as a consequence of the higher spin symmetry. In this respect, we would like to note that in the conformal higher spin theory the symmetry forces amplitudes to vanish everywhere except for points of zero measure in kinematic space [45,46]. It is also worth to note that in [47] the authors argued that already the Lorentz part of the higher spin symmetry implies that amplitudes in higher spin theories should be trivial. This conclusion is consistent with (28), if we add all contributions to the Mellin amplitude together. On the other hand, the correlator in the coordinate representation is non-vanishing and still covariant with respect to the higher spin symmetry. This once again highlights inequivalence between two languages—the Mellin and the coordinate representation—and may hint towards a way to go around no-go theorems for higher spin interactions in flat space.

Note added:

In the initial version of this paper we made a general observation that for constituent Witten diagrams the integration contour appearing in the Mellin transform encloses singularities, but does not go exactly through them as for the power law function. Based on that we suggested that the Mellin amplitude in the free CFT should be a rational function with simple poles at the required locations. The remaining ambiguities were fixed in a heuristic way. Then, in [40] the four-point correlator in IIB supergravity was computed. It was found that in Mellin representation the free part of the correlator arises as a regularization effect in the inverse Mellin transformation. This lead us to extend our previous analysis by carefully taking into account where the contour is located for each constituent Witten diagram. In agreement with [40] we found, that there is a consistent sense in which the Mellin amplitude can be defined to be zero.

Acknowledgments: I am grateful to S. Sarkar, E. Skvortsov and A. Tseytlin for fruitful discussions on different aspects of this work. I would also like to thank X. Bekaert, E. Skvortsov and A. Tseytlin for comments on the draft. This work was supported by the ERC Advanced grant No. 290456.

Conflicts of Interest: The author declares no conflict of interest. The founding sponsors had no role in the design of the study; in the collection, analyses, or interpretation of data; in the writing of the manuscript, and in the decision to publish the results.

References

1. Weinberg, S. Photons and Gravitons in S Matrix Theory: Derivation of charge conservation and equality of gravitational and inertial mass. *Phys. Rev. B* **1964**, *135*, 1049–1056.
2. Coleman, S.R.; Mandula, J. All possible symmetries of the S Matrix. *Phys. Rev.* **1967**, *159*, 1251–1256.
3. Aragone, C.; Deser, S. Consistency problems of hypergravity. *Phys. Lett. B* **1979**, *86*, 161–163.
4. Metsaev, R.R. Effective Action in String Theory. Ph.D. Thesis, Lebedev Physical Institute, Moscow, Russia, 1991.
5. Bekaert, X.; Boulanger, N.; Leclercq, S. Strong obstruction of the Berends-Burgers-van Dam spin-3 vertex. *J. Phys. A Math. Theor.* **2010**, *43*, 185401.
6. Dempster, P.; Tsulaia, M. On the structure of quartic vertices for massless higher spin fields on Minkowski Background. *Nucl. Phys.* **2012**, *865*, 353–375.
7. Joung, E.; Taronna, M. Cubic-interaction-induced deformations of higher-spin symmetries. *J. High Energy Phys.* **2014**, *2014*, 103.
8. Ponomarev, D.; Skvortsov, E.D. (Theoretical Physics Group, Blackett Laboratory, Imperial College London, London, UK). Local obstruction to the minimal gravitational coupling of higher-spin fields in flat space. Unpublished work, 2017.
9. Taronna, M. On the non-local obstruction to interacting higher spins in flat space. *J. High Energy Phys.* **2017**, *2017*, 26.
10. Roiban, R.; Tseytlin, A.A. On four-point interactions in massless higher spin theory in flat space. *J. High Energy Phys.* **2017**, *2017*, 139.
11. Ponomarev, D.; Skvortsov, E.D. Light-front higher-spin theories in flat space. *J. Phys. A Math. Theor.* **2017**, *50*, 095401.
12. Ponomarev, D. Chiral higher spin theories and self-duality. *arXiv* **2017**, arXiv:1710.00270.
13. Barnich, G.; Henneaux, M. Consistent couplings between fields with a gauge freedom and deformations of the master equation. *Phys. Lett. B* **1993**, *311*, 123–129.
14. Ponomarev, D. Off-shell spinor-helicity amplitudes from light-cone deformation procedure. *J. High Energy Phys.* **2016**, *2016*, 117.
15. Sezgin, E.; Sundell, P. Massless higher spins and holography. *Nucl. Phys. B* **2002**, *644*, 303–370.
16. Klebanov, I.R.; Polyakov, A.M. AdS dual of the critical O(N) vector model. *Phys. Lett. B* **2002**, *550*, 213–219.
17. Flato, M.; Fronsdal, C. On Dis and Racs. *Phys. Lett. B* **1980**, *97*, 236–240.
18. Fronsdal, C. Flat Space Singletons. *Phys. Rev. D* **1987**, *35*, 1262–1267.
19. Sleight, C.; Taronna, M. Higher spin gauge theories and bulk locality: A no-go result. *arXiv* **2017**, arXiv:1704.07859.
20. Bekaert, X.; Erdmenger, J.; Ponomarev, D.; Sleight, C. Quartic AdS interactions in higher-spin gravity from conformal field theory. *J. High Energy Phys.* **2015**, *2015*, 149.
21. Taronna, M. Pseudo-local Theories: A functional class proposal. In Proceedings of the International Workshop on Higher Spin Gauge Theories, Singapore, 4–6 November 2015; World Scientific: Singapore, 2017; pp. 59–84.
22. Bekaert, X.; Erdmenger, J.; Ponomarev, D.; Sleight, C. Bulk quartic vertices from boundary four-point correlators. In Proceedings of the International Workshop on Higher Spin Gauge Theories, Singapore, 4–6 November 2015; World Scientific: Singapore, 2017; pp. 291–303.
23. Berends, F.A.; Burgers, G.J.H.; van Dam, H. On the theoretical problems in constructing interactions involving higher spin massless particles. *Nucl. Phys. B* **1985**, *260*, 295–322.
24. Mack, G. D-independent representation of Conformal Field Theories in D dimensions via transformation to auxiliary Dual Resonance Models. Scalar amplitudes. *arXiv* **2009**, arXiv:0907.2407.
25. Mack, G. D-dimensional Conformal Field Theories with anomalous dimensions as Dual Resonance Models. *Bulg. J. Phys.* **2009**, *36*, 214–226.

26. Penedones, J. Writing CFT correlation functions as AdS scattering amplitudes. *J. High Energy Phys.* **2011**, *2011*, 25.
27. Paulos, M.F. Towards Feynman rules for Mellin amplitudes. *J. High Energy Phys.* **2011**, *2011*, 74.
28. Fitzpatrick, A.L.; Kaplan, J.; Penedones, J.; Raju, S.; van Rees, B.C. A natural language for AdS/CFT correlators. *J. High Energy Phys.* **2011**, *2011*, 95.
29. Bekaert, X.; Erdmenger, J.; Ponomarev, D.; Sleight, C. Towards holographic higher-spin interactions: Four-point functions and higher-spin exchange. *J. High Energy Phys.* **2015**, *2015*, 170.
30. Sleight, C. Interactions in higher-spin gravity: A holographic perspective. *J. Phys. A Math. Theor.* **2017**, *50*, 383001.
31. Heemskerk, I.; Penedones, J.; Polchinski, J.; Sully, J. Holography from conformal field Theory. *J. High Energy Phys.* **2009**, *2009*, 79.
32. Hoffmann, L.; Petkou, A.C.; Ruhl, W. Analyticity of AdS scalar exchange graphs in the crossed channel. *Phys. Lett. B* **2000**, *478*, 320–326.
33. El-Showk, S.; Papadodimas, K. Emergent spacetime and holographic CFTs. *J. High Energy Phys.* **2012**, *2012*, 106.
34. Liu, H. Scattering in anti-de Sitter space and operator product expansion. *Phys. Rev. D* **1999**, *60*, 106005.
35. Pappadopulo, D.; Rychkov, S.; Espin, J.; Rattazzi, R. Operator product expansion convergence in conformal field theory. *Phys. Rev. D* **2012**, *86*, 105043.
36. Rychkov, S.; Yvernay, P. Remarks on the convergence properties of the conformal block expansion. *Phys. Lett. B* **2016**, *753*, 682–686.
37. Gubser, S.S.; Parikh, S. Geodesic bulk diagrams on the Bruhat-Tits tree. *arXiv* **2017**, arXiv:1704.01149.
38. Bertrand, J.; Bertrand, P.; Ovarlez, J.P. *Transforms and Applications Handbook*; CRC Press: Boca Raton, FL, USA, 2000; Chapter 11.
39. Gopakumar, R.; Kaviraj, A.; Sen, K.; Sinha, A. Conformal bootstrap in Mellin space. *Phys. Rev. Lett.* **2017**, *118*, 081601.
40. Rastelli, L.; Zhou, X. How to succeed at holographic correlators without really trying. *arXiv* **2017**, arXiv:1710.05923.
41. Paulos, M.F.; Spradlin, M.; Volovich, A. Mellin amplitudes for dual conformal integrals. *J. High Energy Phys.* **2012**, *2012*, 72.
42. Nandan, D.; Paulos, M.F.; Spradlin, M.; Volovich, A. Star integrals, convolutions and simplices. *J. High Energy Phys.* **2013**, *2013*, 105.
43. Nizami, A.A.; Rudra, A.; Sarkar, S.; Verma, M. Exploring perturbative conformal field theory in Mellin space. *J. High Energy Phys.* **2017**, *2017*, 102.
44. Flato, M.; Fronsdal, C. One massless particle equals two dirac singletons: Elementary particles in a curved space. *Lett. Math. Phys.* **1978**, *2*, 421–426.
45. Joung, E.; Nakach, S.; Tseytlin, A.A. Scalar scattering via conformal higher spin exchange. *J. High Energy Phys.* **2016**, *2016*, 125.
46. Beccaria, M.; Nakach, S.; Tseytlin, A.A. On triviality of S-matrix in conformal higher spin theory. *J. High Energy Phys.* **2016**, *2016*, 34.
47. Sleight, C.; Taronna, M. Higher-spin algebras, holography and flat space. *J. High Energy Phys.* **2017**, *2017*, 95.

universe

MDPI

Article

AdS/CFT in Fractional Dimension and Higher-Spins at One Loop

Evgeny Skvortsov [1,2,*] **and Tung Tran** [1]

1 Arnold Sommerfeld Center for Theoretical Physics, Ludwig-Maximilians University Munich, Theresienstr. 37, D-80333 Munich, Germany; tung.tran@physik.uni-muenchen.de

2 Lebedev Institute of Physics, Leninsky ave. 53, 119991 Moscow, Russia

* Correspondence: eugene.d.skvortsov@gmail.com

Received: 6 July 2017; Accepted: 3 August 2017; Published: 14 August 2017

Abstract: Large-N, ϵ-expansion or the conformal bootstrap allow one to make sense of some of conformal field theories in non-integer dimension, which suggests that AdS/CFT may also extend to fractional dimensions. It was shown recently that the sphere free energy and the a-anomaly coefficient of the free scalar field can be reproduced as a one-loop effect in the dual higher-spin theory in a number of integer dimensions. We extend this result to all integer and also to fractional dimensions. Upon changing the boundary conditions in the higher-spin theory the sphere free energy of the large-N Wilson-Fisher CFT can also be reproduced from the higher-spin side.

Keywords: higher-spin theory; AdS/CFT; large-N

1. Introduction

There is some evidence that the AdS/CFT correspondence [1–3] may extend to fractional dimensions. Our goal is to support this idea by matching the sphere free energy of the free and critical vector models with the one-loop corrections to the vacuum partition function in the higher-spin gravity.

It was conjectured in [4] that the large-N $O(N)$ vector model, which describes the critical points of $O(N)$ magnets in three dimensions [5,6], should be dual to a theory with gauge fields of arbitrary high spin in the spectrum, which are known as higher-spin theories.[1] As it was mentioned already in [4], see also [11], the fact that the Wilson-Fisher critical point exists in $4-\epsilon$ expansion should allow one to make sense both of the dual higher-spin theory and of the duality itself in $AdS_{5-\epsilon}/CFT^{4-\epsilon}$. However, it is difficult at present to come up with an observable amenable to computation on the higher-spin side, especially in fractional dimension. In the paper we confirm the duality between free and critical large-N $O(N)$ vector models at the one-loop level with the AdS observable being the one-loop determinant of the Type-A higher-spin theory in Euclidean AdS_{d+1} and with the dual CFT observable being the sphere free energy $F = -\log Z_{S^d}$. The result holds true in all integer and non-integer dimensions, which extends and generalizes [12,13]. Upon changing the boundary conditions we reproduce the difference between the sphere free energy under a double trace deformation $(\phi^2)^2$ that drives the free model at UV to the critical model in IR.

The inspiration comes mostly from the CFT side, which is much better understood: there are different techniques available that allow one to make sense of at least some of the interacting conformal field theories in fractional dimensions. For example, the large-N expansion, see e.g., [14–17], can be used to compute the critical indices for any d, including non-integer ones. The large-N approximation is also important for the quasi-classical expansion on the AdS-side since the coupling constant

1 See also [7–10] for other developments of this and related conjectures that involve higher-spin theories.

in higher-spin theories G should be of order $1/N$. Another ubiquitous method is the ϵ-expansion [6]. Also, the conformal bootstrap can be set up in fractional space-time dimensions [18] and used to get predictions for the critical indices and to clearly show how the $2d$ Ising model smoothly turns into the $3d$ Ising model and ends up on the free theory in $4d$, the latter region can be accessed via the $4-\epsilon$ expansion too. The whole range $2 < d < 4$ is covered by the $1/N$-expansion whenever N is large. There are recent studies [19,20] pointing out that the critical vector model can be extended to a wider range of dimension $4 \leq d \leq 6$.

An interesting class of observables that can be computed on both sides of the duality comes from the sphere free energy $F = -\log Z_{S^d}$. It is also related to an important problem of how to define a measure for the number of effective degrees of freedom in a general QFT. Such an observable should decrease along RG flow and be stationary at fixed points which are described by conformal field theories. The $d=2$ case is solved by the c-theorem [21], while the $4d$ case by the a-theorem [22,23]. Both the central charge c and the a anomaly can be extracted from the sphere free energy: $F = a\log R$, where R is the radius of the sphere and $c=-3a$ in $2d$. In $d=3$ there is no conformal anomaly but it was first conjectured [24–26] and then proved [27,28] that F works in $3d$ as well. More generally, $\tilde{F} = (-)^{(d-1)/2}\log Z_{S^d}$ is expected [26] to work in odd d, in particular in $d=1$ it gives the g-theorem [29]. The last step [11] is to extend this definition to fractional dimensions by introducing generalized sphere free energy $\tilde{F} = \sin(\frac{\pi d}{2})\log Z_{S^d}$. This observable can be computed in non-integer dimension and the pole near even dimensions is resolved in such a way that the a-anomaly is captured, $\tilde{F} = (-1)^{d/2}\pi a/2$. For free CFT's or for the interactions induced by a double-trace deformation $\tilde{F}$ was computed in [26,30–33]. For the free scalar field it is

$$\tilde{F}^{\phi} = \frac{1}{\Gamma(d+1)}\int_0^1 u\sin(\pi u)\Gamma\left(\frac{d}{2}+u\right)\Gamma\left(\frac{d}{2}-u\right)du\,, \tag{1}$$

while for the change $\delta\tilde{F}$ induced by a double trace deformation due to an operator O_Δ of dimension Δ it is given by

$$\delta\tilde{F}_\Delta = \frac{1}{\Gamma(d+1)}\int_0^{\Delta-d/2} u\sin(\pi u)\Gamma\left(\frac{d}{2}+u\right)\Gamma\left(\frac{d}{2}-u\right)du\,. \tag{2}$$

and we are interested in the case $\Delta = d-2$ that corresponds to $O=\phi^2$.

On the dual AdS side the sphere free energy F should correspond to the partition function in the Euclidean AdS_{d+1}, whose boundary is the sphere S^d. Optimistically, one should be able to match all the terms in the two expansions

$$-\ln Z_{AdS} = F_{AdS} = \frac{1}{G}F^0_{AdS} + F^1_{AdS} + GF^2_{AdS} + \ldots\,, \tag{3}$$

$$-\ln Z_{CFT} = F_{CFT} = NF^0_{CFT} + F^1_{CFT} + \frac{1}{N}F^2_{CFT} + \ldots\,. \tag{4}$$

This idea was pursued in [12] and elaborated further in [13,34–48]. Generic duals of higher-spin theories are free CFT's as it is only in free CFT's one can have unbroken higher-spin symmetry [49–53]. For free CFT's only the leading term F^0_{CFT} is present, while it is not so for the large-N interacting vector model. However, F^0_{AdS} has not yet been computed since it should be equal to a regularized value of the classical action, which is not yet available. Still one can proceed to the one-loop term F^1_{AdS} that is equal to the determinant $|-\nabla^2+m^2|$ of the kinetic terms of free higher-spin fields summed over an appropriate spectrum determined by the symmetry (or by the spectrum of the currents in the free CFT dual). This one-loop vacuum partition function, i.e., the one-loop determinant, can be computed via ζ-function [54,55] as

$$F^1_{AdS} = -\frac{1}{2}\zeta'_{\Delta,s}(0) - \zeta_{\Delta,s}(0)\log l\Lambda \tag{5}$$

for each individual field of spin-s and dual operator dimension Δ and then summed up over a given spectrum (with the ghosts of the massless fields subtracted). We restrict ourselves to the simplest higher-spin theory, called Type-A, whose spectrum consists [56] of totally symmetric massless fields with all integer spins $s = 0, 1, 2, 3, 4, ...$ (non-minimal Type-A) or all even spins $s = 0, 2, 4, 6, ...$ (minimal Type-A). The minimal Type-A theory should be dual to free or critical $O(N)$ vector model while the non-minimal one should be dual to the $U(N)$ versions of the same CFT's. Whether the dual is free or interacting depends on the boundary conditions imposed on the scalar field, $s = 0$, of the higher-spin multiplet: $\Delta = d - 2$ for the free dual and $\Delta = 2$ for the (large-N) interacting one. Therefore, altogether we have four different cases:[2]

$$\begin{aligned} \zeta_{\text{HS,n.-m.}}(z) &= \zeta_{\Delta,0} + \sum_{s=1,2,...} \left[\zeta_{d+s-2,s} - \zeta_{d+s-1,s-1}\right], \\ \zeta_{\text{HS,min.}}(z) &= \zeta_{\Delta,0} + \sum_{s=2,4,...} \left[\zeta_{d+s-2,s} - \zeta_{d+s-1,s-1}\right], \end{aligned} \tag{6}$$

where Δ can be either $d-2$ or 2. The one-loop free energy in the higher-spin gravity is

$$F^1_{\text{HS}} = -\frac{1}{2}\zeta'_{\text{HS}}(0) - \zeta_{\text{HS}}(0)\log l\Lambda\,. \tag{7}$$

It was shown in a number of integer dimensions [12,13,34] that: (i) while each term in the sum may depend on the cutoff Λ, which makes the finite part ambiguous, the full one-loop vacuum energy does not depend on the cutoff Λ, i.e., $\zeta_{\text{HS}}(0) = 0$ for the (non)-minimal Type-A models; (ii) the finite part vanishes for the non-minimal Type-A, $\zeta'_{\text{HS,n.-m.}}(0) = 0$, and equals the sphere free energy F or the a-anomaly of the free scalar field, i.e., $a = -\frac{1}{2}\zeta'_{\text{HS,min.}}(0)$ for d even and $F = -\log Z_{S^d} = -\frac{1}{2}\zeta'_{\text{HS,min.}}(0)$ for d odd. This result can be consistent with the AdS/CFT duality provided the one-loop effect compensates for the integer shift in the relation between the bulk coupling constant G and the number of fields N on the CFT side, $G^{-1} \sim N-1$ (provided that F^0_{AdS} does match F^0_{CFT}).

The upshot of the one-loop computations in higher-spin theories is that the one-loop vacuum energy can reproduce the a-anomaly coefficient of the free scalar CFT in even dimensions and the sphere free energy in odd dimensions, which was shown for a number of dimensions. Upon changing the boundary conditions for the $s = 0$ member it was also shown that the difference $-\frac{1}{2}\delta\zeta'_{\text{HS}}(0)$, which is due to the scalar field, matches the sphere free energy of the large-N interacting vector model in $d = 3$ [12] and $d = 5$ [13].

It is worth stressing that the computations of ζ_{HS} are quite different for even and odd dimensions and the requirement for d to be an integer has been a crucial one. Another common feature of the one-loop computations in higher-spin theories is that the finite result for ζ_{HS} is obtained after an appropriate regularization of the sum over spins.

In the paper we revise the problem of one-loop computations in Type-A higher-spin theory aiming at the general proof for all integer dimensions and also for non-integer ones. We show that $\frac{1}{2}\sin(\frac{\pi d}{2})\zeta'_{\text{HS}}(0)$ for the minimal Type-A theory does reproduce the generalized sphere free energy (1) for all d. For $\Delta = 2$ boundary conditions on the $s = 0$ field the one-loop result matches the change in the sphere free energy (2) due to the double-trace deformation induced by operator $(\phi^2)^2$ on the CFT side [12,13,26]. Since the result is a technical one let us briefly discus the main steps. First of all, thanks to Camporesi and Higuchi [57] there is a representation of the spectral density that enters $\zeta_{\Delta,s}(z)$ such that it can be extended to non-integer dimensions. Next, we apply the Laplace transform to the spectral density, see also [42,43], which disentangles the integral over the spectral parameter and summation over spins. Then, we convert the integral into a sum over the residues. In order to handle the sum

2 The second term in the brackets is to subtract the ghosts.

we change the regularization prescription, see also [42], but it can be checked that this does not affect the result. Low and behold we arrive at the expression, which we refer to as *intermediate* form, whose regularized form gives (1). The intermediate form can also be obtained directly on the CFT side from the determinant on the sphere. The interacting large-N vector model requires taking into account the difference between the contributions of the $s=0$ fields for $\Delta=d-2$ and $\Delta=2$. Some other features and possible extensions are discussed in Conclusions.

2. Higher-Spin Partition Function in Fractional Dimensions

We first give in Section 2.1 the basic definitions related to the zeta-function and review the computation of one-loop determinants in even and odd dimensions, stressing the difference. Next, in Section 2.2, we proceed to non-integer dimensions and apply the main technical tools that allow us to handle fractional dimensions: Laplace transform, contour integration and modified regularization. In Section 2.3 we discuss the volume of the anti-de Sitter space that enters as an overall, but important, factor. The last steps on the AdS side—summation over spins and extraction of $\zeta_{\text{HS}}(0)$ and $\zeta'_{\text{HS}}(0)$ are done in Sections 2.4 and 2.5, where we arrive at certain *intermediate* form of the result that can be matched with the CFT side. The intermediate form is directly related to the free and critical vector models in Sections 2.6 and 2.7, which completes the proof.

2.1. Integer Dimensions

The general form of the spectral zeta-function in Euclidean anti-de Sitter space AdS_{d+1} for a field of any symmetry type is [57]:

$$\zeta_{\Delta,s} = \frac{\text{vol}(\mathbb{H}^{d+1})}{\text{vol}(S^d)} v_d g(s) \int_0^\infty d\lambda \, \frac{\mu(\lambda)}{\left[(\Delta-\frac{d}{2})^2+\lambda^2\right]^z}\,, \tag{8}$$

where $\mu(\lambda)$ is the most important factor—spectral density. It is normalized to its flat-space value:

$$\mu(\lambda)|_{\lambda\to\infty} = w_d \lambda^d\,, \qquad\qquad w_d = \frac{\pi}{[2^{d-1}\Gamma(\frac{d+1}{2})]^2}\,. \tag{9}$$

There are several overall factors that do not participate in the integral: $g(s)$ is the number of degrees of freedom, i.e., components of the irreducible transverse traceless tensor; $\text{vol}(\mathbb{H}^{d+1})$ is a regularized volume of the hyperbolic space [31]; and the leftover factor v_d can be combined with w_d:

$$v_d = \frac{2^{d-1}}{\pi}\,, \qquad\qquad u_d = v_d w_d = \frac{(\text{vol}(S^d))^2}{(2\pi)^{d+1}}\,. \tag{10}$$

We are interested in totally-symmetric massless spin-s fields that make the spectrum of the Type-A higher-spin theory, in which case the number of degrees of freedom is

$$g(s) = \frac{(d+2s-2)\Gamma(d+s-2)}{\Gamma(d-1)\Gamma(s+1)} = \dim^{so(d)} \mathbb{Y}(s)\,. \tag{11}$$

It counts the dimension of an irreducible $so(d)$ tensor of rank-s. The spectral density depends crucially on whether d is even or odd. For d even, the spectral density is a polynomial:

$$\mu(\lambda) = w_d \left(\left(\frac{d-2}{2}+s\right)^2 + \lambda^2 \right) \prod_{j=0}^{\frac{d-4}{2}} \left(j^2+\lambda^2\right)\,, \tag{12}$$

while for d odd it contains an additional *tanh* factor:

$$\mu(\lambda) = w_d \lambda \tanh(\pi\lambda) \left(\left(\frac{d-2}{2} + s \right)^2 + \lambda^2 \right) \prod_{j=1/2}^{\frac{d-4}{2}} \left(j^2 + \lambda^2 \right) . \tag{13}$$

The computation of $\zeta(z)$ in every even dimension d presents no difficulty, in principle, since the spectral density is a polynomial and the integral can be done with the help of

$$\int_0^\infty d\lambda \, \frac{\lambda^k}{(b^2+\lambda^2)^z} = \frac{\Gamma\left(\frac{k+1}{2}\right) b^{k-2z+1} \Gamma\left(-\frac{k}{2}+z-\frac{1}{2}\right)}{2\Gamma(z)} . \tag{14}$$

Here, the spectral parameter z serves also as a regulator. The only problem left is to evaluate the infinite sum over all spins (6) and there are several equivalent ways known how to do that [12,13,34]. Summation over spins can be done dimension by dimension with the complexity rapidly increasing with d.

On contrary, the computation in every odd dimension is a challenge. The coefficient of the log divergent piece, i.e., $\zeta_{\Delta,s}(0)$ can still be computed for any spin and weight Δ after splitting the integrand into the part that is simple and converges for z large enough and another part that is more complicated but converges for $z = 0$, which is done with the help of

$$\tanh x = 1 + \frac{-2}{1+e^x} . \tag{15}$$

Computation of $\zeta'_{\Delta,s}(0)$ leads to some integrals that cannot be evaluated analytically, but whose contribution cancels out after summing over the spectrum of the Type-A theory. Again the computation can be done dimension by dimension. We refer to [12,13] for more detail, see also [45,46].

Another issue that requires a separate treatment is the factor of the regularized volume of the Hyperbolic space $\mathbb{H}^{d+1}$, which via dimensional regularization can be found to be [31]:

$$\mathrm{vol}\, S^d = \frac{2\pi^{(d+1)/2}}{\Gamma\left(\frac{d+1}{2}\right)} , \qquad \mathrm{vol}\, \mathbb{H}^{d+1} = \begin{cases} \frac{2(-\pi)^{d/2}}{\Gamma(\frac{d}{2}+1)} \log R , & d = 2k , \\ \pi^{d/2}\Gamma\left(-\frac{d}{2}\right) , & d = 2k+1 . \end{cases} \tag{16}$$

The appearance of $\log R$ signals conformal anomaly. The sphere free energy also has the $\log R$ term, whose coefficient is the a-anomaly.

2.2. Fractional Dimensions

Coming to fractional dimensions we prefer to isolate all the factors, including the volume of the hyperbolic space, and denote the leftover as $\tilde{\mu}(\lambda)$

$$\zeta(z) = \mathcal{N} g(s) \int_0^\infty d\lambda \, \frac{\tilde{\mu}(\lambda)}{\left[\lambda^2 + \left(\Delta - \frac{d}{2}\right)^2\right]^z} , \qquad \mathcal{N} = \frac{v_d w_d \mathrm{vol}(\mathbb{H}^{d+1})}{\mathrm{vol}(S^d)} . \tag{17}$$

There is a representation of the spectral density that works in all dimensions [57]:

$$\tilde{\mu}(\lambda) = \left(\left(\frac{d-2}{2} + s \right)^2 + \lambda^2 \right) \left| \frac{\Gamma\left(\frac{d-2}{2} + i\lambda\right)}{\Gamma(i\lambda)} \right|^2 . \tag{18}$$

This is our starting point. We will show that the higher-spin zeta-function can be computed without having to make an assumption that d is an integer. The difference between even and odd dimensions can be observed in (18): a simple polynomial is obtained for d even, (12), and an additional non-polynomial factor is present for d odd, (13). In fact, the spectral density is not a polynomial in all dimensions, including fractional ones, except for the case of d even. Therefore, it is the case of d even that is special, all other dimensions being on equal footing. The computation we perform below is valid for all d except even (which is of measure zero on the real line). The result for d even is then obtained as a continuation from non-integer d.

Let us begin with the expression for the zeta-function that is obtained by collecting all the factors and expanding the gamma functions:

$$\zeta_{\nu,s}(z)=\mathcal{N}\frac{g(s)}{\pi}\int_0^\infty d\lambda\frac{\lambda\sinh(\pi\lambda)\left(\lambda^2+\left(\frac{d}{2}+s-1\right)^2\right)\Gamma\left(\frac{d}{2}+i\lambda-1\right)\Gamma\left(\frac{d}{2}-i\lambda-1\right)}{(\lambda^2+\nu^2)^z}\,, \tag{19}$$

where $\nu=\Delta-\frac{d}{2}$. The integrand is an even function of λ and therefore we can extend the range of integration to $(-\infty,\infty)$ at the price of $\frac{1}{2}$. It is convenient to perform the Laplace transform, see also [42,47],[3]

$$\frac{1}{(\lambda^2+\nu^2)^z}=\frac{\sqrt{\pi}}{\Gamma(z)}\int_0^\infty d\beta\, e^{-\beta\nu}\left(\frac{\beta}{2\lambda}\right)^{z-\frac{1}{2}}J_{z-\frac{1}{2}}(\lambda\beta)\,. \tag{20}$$

The main advantage is that the exponential $e^{-\beta\nu}$ times $g(s)$ can be summed over all spins in the spectrum directly. In other words, the sum over spins and the λ integral are now decoupled. This is one of the crucial steps that allows us to calculate the full zeta function ζ_{HS}, (6), in arbitrary dimension. Notice that in applying the Laplace transform we moved the branch point from $\pm i\nu$ in (19) to 0. Next, we split the Bessel function into

$$J_\alpha(x)=\frac{{}_1H_\alpha(x)+{}_2H_\alpha(x)}{2}\,, \tag{21}$$

where ${}_1H_\alpha(x)$ and ${}_2H_\alpha(x)$ are Hankel functions of the first kind and second kind.

Similarly to Green functions we close the contour for the part of ${}_1H_\alpha$ upward and the contour for the part of ${}_2H_\alpha$ downward. Let us show how to compute the contour integral of the part with ${}_1H_\alpha$ in (21) first. In order to evaluate the contribution coming from ${}_1H_\alpha$, we choose the contour as on Figure 1.

One needs to make sure that the upper arc of the contour does not cross any pole that comes from the $\Gamma\left(\frac{d}{2}+i\lambda-1\right)$. The residue theorem implies that

$$\oint_C f(\lambda)=2\pi i\sum_{l=0}^{\infty}\text{Res}_{\lambda\to i(\frac{d}{2}+l-1)}\left(\lambda-i\left(\frac{d}{2}+l-1\right)\right)f(\lambda)\,, \tag{22}$$

where we prefer to omit $g(s)/\pi$ for a moment:

$$f(\lambda)=\mathcal{N}\frac{\lambda\sinh(\pi\lambda)\left(\lambda^2+\left(\frac{d}{2}+s-1\right)^2\right)\Gamma\left(\frac{d}{2}+i\lambda-1\right)\Gamma\left(\frac{d}{2}-i\lambda-1\right)\beta^{z-\frac{1}{2}}\,{}_1H_{z-\frac{1}{2}}(\beta\lambda)}{2(2\lambda)^{z-\frac{1}{2}}} \tag{23}$$

We recall that the residues of Γ-function are

$$\text{Res}(\Gamma,-l)=\frac{(-1)^l}{\Gamma(l+1)}\,. \tag{24}$$

[3] One can represent the spectral zeta-function as a differential operator acting on some seed function that has enough parameters to produce $g(s)\mu(\lambda)$. Character is an example of such a function [42,47], which is also indispensable for taking tensor products. The characters are however difficult to define in non-integer dimension.

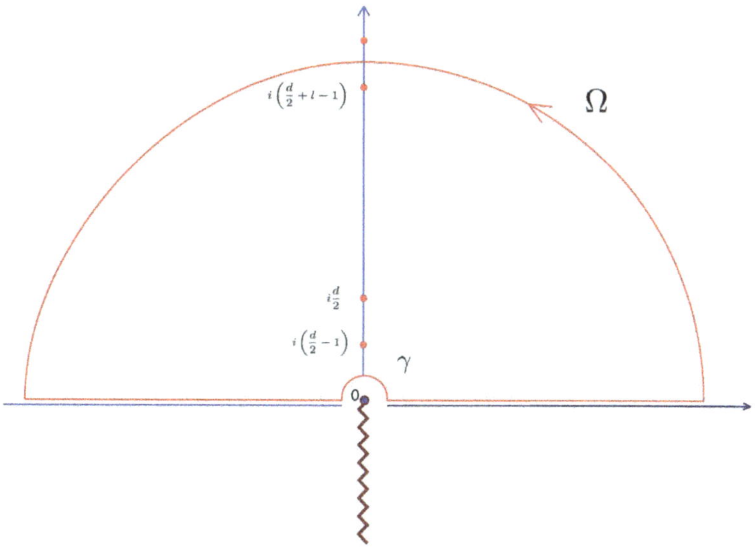

Figure 1. The contour for the part contains ${}_1H_\alpha$ lies in upper half plane where the poles are those of $\Gamma\left(\frac{d}{2}+i\lambda-1\right)$. As the λ integral approaches $(-\infty,\infty)$, the range of l also extends to infinity.

We, therefore, could change the integral over λ to an infinite sum over l. Before proceeding further, let us make sure that the upper arc and the contour around the branch point do not contribute to the whole contour integral. We make the change of variable $\lambda = Re^{i\theta}$:

$$\int_\Omega d\lambda f(\lambda) = \lim_{R\to\infty}\int_0^\pi d\theta f(Re^{i\theta}) \quad \text{and} \quad \int_\gamma d\lambda f(\lambda) = \lim_{R\to 0}\int_\pi^0 d\theta f(Re^{i\theta})\,. \tag{25}$$

Introducing z as a regulator [12,13,34,57] is useful in various ways. Let us consider the γ contour first, if we set z large enough then there is no contribution from the small contour

$$\lim_{\lambda\to 0}\lambda\sinh(\pi\lambda)\Gamma\left(\frac{d}{2}+i\lambda-1\right)\Gamma\left(\frac{d}{2}-i\lambda-1\right)\frac{\beta^{z-\frac{1}{2}}\,{}_1H_{z-\frac{1}{2}}(\beta\lambda)}{2(2\lambda)^{z-\frac{1}{2}}} = 0+\mathcal{O}(\lambda^2) \tag{26}$$

Therefore, (23) vanishes and the integral over the contour near the branch point in (25) also vanishes. Next, consider the large arc Ω, assuming that the contour goes in between the poles of the gamma function. The integrand (23) will also vanish as we make z large enough in the limit where the radius R goes to infinity.[4] Therefore, there is no contribution coming from γ and Ω arcs and (22) is equal to

$$\begin{aligned}\frac{1}{2}\int_{-\infty}^{\infty} d\lambda f(\lambda) = -\frac{\pi\mathcal{N}}{2}\sum_{l=0}^{\infty}\left(\frac{\beta^{z-\frac{1}{2}}\,{}_1H_{z-\frac{1}{2}}\left(i\left(\frac{d}{2}+l-1\right)\beta\right)}{\left(2i\left(\frac{d}{2}+l-1\right)\right)^{z-\frac{1}{2}}}\right)\left(\frac{d}{2}+l-1\right)\\ \times\left(\left(\frac{d}{2}+s-1\right)^2-\left(\frac{d}{2}+l-1\right)^2\right)\sin\left(\pi\left(\frac{d}{2}+l-1\right)\right)\frac{\Gamma(d+l-2)}{\Gamma(l+1)}(-1)^l\,.\end{aligned} \tag{27}$$

We notice that $\sin\left(\pi\left(\frac{d}{2}+l-1\right)\right) = -(-1)^l\sin\left(\frac{\pi d}{2}\right)$ and hence

$$\begin{aligned}\zeta(z) = &\frac{\mathcal{N}g(s)\sqrt{\pi}}{2\Gamma(z)}\int_0^{\infty} d\beta\sum_{l=0}^{\infty} e^{-\beta\nu}\left(\frac{\beta^{z-\frac{1}{2}}\,{}_1H_{z-\frac{1}{2}}\left(i\left(\frac{d}{2}+l-1\right)\beta\right)}{\left(2i\left(\frac{d}{2}+l-1\right)\right)^{z-\frac{1}{2}}}\right)\\ &\times\left(\frac{d}{2}+l-1\right)\left(\left(\frac{d}{2}+s-1\right)^2-\left(\frac{d}{2}+l-1\right)^2\right)\sin\left(\frac{\pi d}{2}\right)\frac{\Gamma(d+l-2)}{\Gamma(l+1)}\,.\end{aligned} \tag{28}$$

It is difficult to say anything about the sum in general, but eventually we are interested only in few terms around $z=0$. In [42], it was argued that one can change the regularization prescription so that the $z\to 0$ behaviour is not modified. Indeed, it is clear that to the leading order in z-expansion one can use

$$\lim_{z\to 0}\frac{\beta^{z-\frac{1}{2}}\,{}_1H_{z-\frac{1}{2}}(\beta\lambda)}{2\,(2\lambda)^{z-\frac{1}{2}}} = \frac{e^{i\beta\lambda}}{\sqrt{\pi\beta}}+\mathcal{O}(z)\,. \tag{29}$$

This way we obtain the following contribution coming from the ${}_1H_\alpha$ function with the contour in the upper half-plane, Figure 1:

$$\begin{aligned}\zeta_{{}_1H_\alpha}(z) = &\frac{\mathcal{N}g(s)\sqrt{\pi}}{\Gamma(z)}\int_0^{\infty} d\beta\sum_{l=0}^{\infty} e^{-\beta\nu}\frac{e^{-\beta\left(\frac{d}{2}+l-1\right)}}{\sqrt{\pi\beta}}\\ &\times\left(\frac{d}{2}+l-1\right)\left(\left(\frac{d}{2}+s-1\right)^2-\left(\frac{d}{2}+l-1\right)^2\right)\sin\left(\frac{\pi d}{2}\right)\frac{\Gamma(d+l-2)}{\Gamma(l+1)}\,.\end{aligned} \tag{30}$$

The presence of $1/\Gamma(z)\sim z$ factor in (20) implies that in order to get the right $\zeta(0)$ we can take only the constant term of (29) into account. However, there should be a discrepancy between $\zeta'(0)$ computed rigorously and the one after we drop the term $\mathcal{O}(z)$ in (29). The difference, which we call the *deficit*, originates from the term of order $\mathcal{O}(z)$ in (29). As was noted in [42] the deficit vanishes for representations that have even characters (even as a function of β, where $q=e^{-\beta}$ counts the energy via insertion of q^E). The deficit is discussed in Appendix B, where it is shown that it does not contribute to the full $\zeta'_{\rm HS}(0)$.

4 For a more detailed discussion see [57].

Next, we repeat the same steps for the contribution coming from ${}_2H_\alpha$ in (21) where we close the contour downwards. In this case, one has to use $-2\pi i$ when applying residue theorem[5] for the poles of $\Gamma\left(\frac{d}{2}-i\lambda-1\right)$. We obtain the same structure as in (30) since

$$\lim_{\lambda\to -i(\frac{d}{2}+l-1)}\lim_{z\to 0}\frac{\beta^{z-1/2}\,{}_2H_{z-\frac{1}{2}}(\beta\lambda)}{2\,(2\lambda))^{z-\frac{1}{2}}}=\frac{e^{-\beta(\frac{d}{2}+l-1)}}{\sqrt{\pi}\beta}+\mathcal{O}(z)\,. \tag{31}$$

Therefore, in order to compute the full one-loop free energy of the Type-A theory, (7), we can write the zeta-function in a modified form as

$$\begin{aligned}\tilde{\zeta}(z)=&\frac{\mathcal{N}g(s)}{\Gamma(2z)}\int_0^\infty d\beta\sum_{l=0}^{\infty}e^{-\beta\nu}\beta^{2z-1}e^{-\beta(\frac{d}{2}+l-1)}\left(\frac{d}{2}+l-1\right)\\&\times\left(\left(\frac{d}{2}+s-1\right)^2-\left(\frac{d}{2}+l-1\right)^2\right)\sin\left(\frac{\pi d}{2}\right)\frac{\Gamma(d+l-2)}{\Gamma(l+1)}\,.\end{aligned} \tag{32}$$

2.3. *Volume of Hyperbolic Space*

In integer dimensions, we can use the volume form of the sphere S^d and Hyperbolic space as in (16). This result arises [31] from the expansion of the formal volume $\pi^{D/2}\Gamma\left(-\frac{D}{2}\right)$ in $D=d-\epsilon$:

$$\mathrm{vol}\,\mathbb{H}^{d+1}=\frac{L_{d+1}}{\epsilon}+V_{d+1}+O(\epsilon)\,, \tag{33}$$

where ϵ-pole signals the $\log R$ divergence in $d=2k$ and V_{d+1} is the finite part that makes the leading contribution for $d=2k+1$. As it was already noted in [31], regularization of the volume IR divergences is not independent of regularization of the UV divergences that arise in one-loop determinants. Below, we propose an extension for the overall normalization factor which comes from the regularized volume to non-integer dimension. Note that one can write the general volume for Lobachevsky space as

$$\mathrm{vol}\,\mathbb{H}^{d+1}=-\frac{\pi^{\frac{d+2}{2}}}{\Gamma\left(\frac{d+2}{2}\right)\sin\left(\frac{\pi d}{2}\right)}\,, \tag{34}$$

which gives the right pole as in (33) and reduces to V_{d+1} for d odd. The $\sin\left(\frac{\pi d}{2}\right)$ factor inside the modified zeta function (32) will cancel with the one in (34) and gives us no poles for even dimensions. Together with the factor $\mathcal{N}$ in (17), one arrives at the overall normalization factor in general dimensions[6]

$$\widetilde{\mathcal{N}}=\mathcal{N}\sin\left(\frac{\pi d}{2}\right)=-\frac{1}{\Gamma(d+1)}\,. \tag{35}$$

This overall normalization factor is strikingly simple since we do not need to treat the cases of odd and even dimensions separately. Moreover, (35) can also be used in fractional dimension.

5 The contour is the reflection image of Figure 1 around the real axis.

6 Recall that $\mathrm{vol}\,S^d=\frac{2\pi^{(d+1)/2}}{\Gamma(\frac{d+1}{2})}$.

2.4. Non-Minimal Type-A

Using the regularized volume, we can now write the full modified zeta-function for the Type-A as

$$\begin{aligned}\tilde{\zeta}(z)_{\nu,s} = &-\frac{g(s)}{\Gamma(2z)\Gamma(d+1)}\int_0^\infty d\beta \sum_{l=0}^{\infty} e^{-\beta\nu}\beta^{2z-1}e^{-\beta(\frac{d}{2}+l-1)}\left(\frac{d}{2}+l-1\right)\\ &\times\left(\left(\frac{d}{2}+s-1\right)^2-\left(\frac{d}{2}+l-1\right)^2\right)\frac{\Gamma(d+l-2)}{\Gamma(l+1)}\end{aligned} \tag{36}$$

We first show that the modified zeta-function leads to $\zeta_{\rm HS}(0) = \zeta'_{\rm HS}(0)$ for the non-minimal Type-A theory. The total ζ-function for the non-minimal Type-A is

$$\tilde{\zeta}_{\text{n.-m.}}(z) = \tilde{\zeta}_{\frac{d}{2}-2,0}(z) + \sum_{s=1}^{\infty}\left(\tilde{\zeta}_{\frac{d}{2}+s-2,s}(z) - \tilde{\zeta}_{\frac{d}{2}+s-1,s-1}(z)\right), \tag{37}$$

where the labels of the zeta functions correspond to $\zeta_{\nu,s}$ as in (19). Using (36) and (11) we can perform the sum over all spins in (37) and get

$$\begin{aligned}\tilde{\zeta}_{\text{n.-m.}}(z) = &\sum_{l=0}^{\infty}\int_0^\infty \frac{d\beta\beta^{2z-1}}{\Gamma(d+1)\Gamma(2z)}\frac{e^{\frac{-\beta d}{2}}(-2+d+2l)\cosh\left(\frac{\beta}{2}\right)^2 e^{-\frac{\beta}{2}(-2+d+2l)}\Gamma(-2+d+l)}{(1-e^{-\beta})^d\Gamma(l+1)}\\ &\times\left(d^2+2(-2+l)l+d(-1+2l)-2l(-2+d+l)\cosh(\beta)\right)\\ &=0.\end{aligned} \tag{38}$$

It is the sum over l that makes the expression in (38) vanish. Next, we need to compute $\tilde{\zeta}'_{\text{n.-m.}}(0)$ using the modified zeta-function. Remember that

$$\lim_{z\to 0}\frac{\beta^{2z-1}}{\Gamma(2z)} \sim \frac{2z}{\beta} + \mathcal{O}(z^2). \tag{39}$$

In other words, the part of (38) without $1/\Gamma(2z)$ is $\tilde{\zeta}'(0)$. For the non-minimal Type-A we see that $\tilde{\zeta}'(0)$ vanishes. As a result we have proved that

$$\tilde{\zeta}_{\text{n.-m.}}(0) = \tilde{\zeta}'_{\text{n.-m.}}(0) = 0. \tag{40}$$

This extends the results of [12,13] to all odd dimensions as well as to fractional ones.

2.5. Minimal Type-A

The case of the minimal Type-A model is more interesting as we will not always find $0 = 0$-type of equality as in the non-minimal case. The ζ-function for the minimal Type-A is

$$\zeta_{\text{min.}}(z) = \zeta_{\frac{d}{2}-2,0}(z) + \sum_{s=2,4,\ldots}^{\infty}\left(\zeta_{\frac{d}{2}+s-2,s}(z) - \zeta_{\frac{d}{2}+s-1,s-1}(z)\right). \tag{41}$$

The final result after the summation is done has a very simple form:

$$\tilde{\zeta}_{\text{min.}}(z) = -\frac{1}{2\Gamma(2z)}\int_0^\infty d\beta\frac{\beta^{2z-1}e^{-\beta(2-d)}(1+e^{2\beta})^2}{(e^{2\beta}-1)^d}. \tag{42}$$

To obtain (42), it is suggestive to sum over s in (36) first. To do this we need to absorb all monomials in s into gamma functions. For example,

$$s\Gamma(d+s-2) = \Gamma(d+s-1) - (d-2)\Gamma(d+s-2) \tag{43}$$

After some algebra what we obtain are several terms of the form

$$\xi(\nu, p(s)) = e^{-\beta\nu}\frac{\Gamma(p(s))}{\Gamma(s+1)}. \tag{44}$$

Here $p(s)$ is of the form $s + \text{const}$ with different const. The sums are of the usual "statistical" form. Following (41) one should sum (44) according to

$$\xi\left(\frac{d}{2}-2, p(0)\right) + \sum_{s=2,4,\dots}^{\infty} \xi\left(\frac{d}{2}+s-2, p(s)\right) - \xi\left(\frac{d}{2}+s-1, p(s-1)\right), \tag{45}$$

where $\xi\left(\frac{d}{2}+s-1, p(s-1)\right)$ correspond to the ghosts. We, then, arrive at the sum over l:

$$\begin{aligned}
\tilde{\zeta}_{\text{min.}}(z) &= \sum_{l=0}^{\infty}\int_0^{\infty} d\beta \frac{e^{-\beta(\frac{d}{2}+l-1)}(-d+2l-2)\Gamma(d+l-2)}{\Gamma(d+1)\Gamma(l+1)} \frac{e^{\beta(1-\frac{d}{2})}(-1+\coth\beta)\sinh\frac{\beta}{2}}{(1-e^{-2\beta})^d} \\
&\times \frac{\beta^{2z-1}}{\Gamma(2z)}\Bigg[-2(1+e^{-\beta})^d\left(\cosh\frac{\beta}{2}\right)^3\Big((-1+d)d+2(-2+d)l+2l^2-2l(-2+d+l)\cosh\beta\Big) \\
&+\cosh\beta((-1+d)d+2(-2+d)l+2l^2+2l(-2+l+1)\cosh\beta)\sinh\frac{\beta}{2}\left(1-e^{-\beta}\right)^d\Bigg] \\
&= -\frac{1}{2\Gamma(2z)}\int_0^{\infty} d\beta \frac{\beta^{2z-1}e^{-\beta(2-d)}(1+e^{2\beta})^2}{(e^{2\beta}-1)^d} = (42).
\end{aligned}$$

Formula (42) is strikingly simple. Vanishing of $\tilde{\zeta}_{\text{min.}}(0)$ is due to the fact that $\lim_{z\to 0} 1/\Gamma(2z) = 0 + \mathcal{O}(z)$. For $\tilde{\zeta}'_{\text{min.}}(0)$, using (39), we arrive at

$$\tilde{\zeta}'_{\text{min.}}(0) = -\int_0^{\infty} d\beta \frac{e^{-\beta(2-d)}(1+e^{2\beta})^2}{\beta(e^{2\beta}-1)^d}. \tag{46}$$

The formula above is the *intermediate*[7] form. After a suitable regularization it will give the correct answer for the sphere free energy as we recall in the next Section. It is worth mentioning that some of the intermediate, usually divergent, expressions on the AdS side can be directly matched with their CFT cousins, see e.g., [34] for the Casimir Energy example. These facts accentuate the importance of careful adjustment of the regularization prescriptions on both sides of the duality.

2.6. Matching Free Vector Model

Having arrived at the intermediate form (46), we would like to show that exactly the same intermediate form emerges on the CFT side. It contains all the important information and can be directly used to derive the sphere free energy.

Let us review the main steps in [11,26,58] as to get the (generalized) sphere free energy $\tilde{F}$. The starting point is the expression for F for a free scalar field, which results from the sum over the eigen values of the Laplace operator on the sphere S^d [30,31]:

$$F^{\phi}_{\text{min.}} = \frac{1}{2}\sum_{l=0}^{\infty} d_l \log\frac{\Gamma(\frac{d}{2}+l-1)}{\Gamma(\frac{d}{2}+l+1)} = \frac{1}{2}\sum_{l=0}^{\infty} d_l \int_0^{\infty}\frac{d\beta}{\beta}\left(-2e^{-\beta}+e^{-\beta(l+\frac{d}{2})}+e^{-\beta(\frac{d}{2}+l-1)}\right), \tag{47}$$

7 We refer to it as intermediate as the integral is divergent and requires regularization.

where

$$d_l = \frac{(d+2l-1)\Gamma(d+l-1)}{\Gamma(d)\Gamma(l+1)} \tag{48}$$

is the degeneracy of eigen values. There is a clearly divergent part proportional to the total number of 'degrees of freedom', $\sum d_l$. This sum can be shown to vanish in a number of ways. For example, inserting cut-off $e^{-\epsilon l}$ we get

$$\sum_{l=0}^{\infty} d_l e^{-\epsilon l} \sim \epsilon^{-d} . \tag{49}$$

In order to regularize this divergence one can make d negative [31] and then continue d to the positive domain. In practice, this is equivalent to saying that the total number of degrees of freedom is zero:

$$\sum_{l=0}^{\infty} d_l = 0 . \tag{50}$$

Therefore, we successfully drop the first term in (47). In order to pass from $\log\Gamma$ to the intermediate form one needs to apply the integral representation of $\log\Gamma(x)$:

$$\log\frac{\Gamma(\mu+\nu+1)}{\Gamma(\mu+1)} = \int_0^\infty \frac{d\beta}{\beta}\left(\nu e^{-\beta} - \frac{e^{-\beta\mu}-e^{-\beta(\mu+\nu)}}{e^{\beta}-1}\right) . \tag{51}$$

As a result, (47) simplifies to

$$F^{\phi}_{min} = \frac{1}{2}\int_0^\infty \frac{d\beta}{\beta} e^{-\frac{\beta(2+d)}{2}} \frac{(1+e^{\beta})^2}{(1-e^{-\beta})^d} . \tag{52}$$

By making a change of variable, $\beta \to 2\beta$, we get exactly the intermediate form (46) obtained in AdS up to a factor of (-2). By definition, the AdS one-loop free energy is related to the sphere free energy as

$$F^{\phi}_{\text{min}} = -\frac{1}{2}\tilde{\zeta}'_{\text{min.}}(0) , \tag{53}$$

which explains the factor (-2) difference. We also note that (47) leads to

$$F^{\phi}_{\text{min.}} = \frac{1}{2}\sum_{l=0}^{\infty} d_l \log\frac{\Gamma(\frac{d}{2}+l-1)}{\Gamma(\frac{d}{2}+l+1)} = \frac{1}{\sin\left(\frac{\pi d}{2}\right)\Gamma(d+1)}\int_0^1 du u \sin(\pi u)\Gamma\left(\frac{d}{2}+u\right)\Gamma\left(\frac{d}{2}-u\right) . \tag{54}$$

In Appendix A we show that the same result can be obtained directly from the intermediate form, i.e., the AdS result suffices to reproduce (54) and there is no 'information loss' in going to the intermediate form. Then, the generalized sphere free energy $\tilde{F}^{\phi} = -\sin(\frac{\pi d}{2})F_{\phi}$ is [11,58]:

$$\tilde{F}^{\phi}_{\text{min.}} = \frac{1}{\Gamma(d+1)}\int_0^1 du \sin(\pi u)\Gamma\left(\frac{d}{2}-u\right)\Gamma\left(\frac{d}{2}+u\right) . \tag{55}$$

Finally, we have shown that the (generalized) sphere free energy of the free scalar field results from the one-loop determinant in the minimal Type-A higher-spin theory:

$$-\sin\left(\frac{\pi d}{2}\right)F^{\phi}_{\text{min.}} = \tilde{F}^{\phi}_{\text{min.}} = \frac{1}{2}\sin\left(\frac{\pi d}{2}\right)\tilde{\zeta}'_{\text{min.}}(0) , \tag{56}$$

which completes the proof. Despite the fact that our proof requires d not to be an even integer, the final result smoothly extrapolates to $d = 2k$, where there are poles that correspond to the a-anomaly. This extends the proof to even dimensions as well.

2.7. Matching Critical Vector Model

Let us consider the case of the duality between the critical $O(N)$ vector model and the (non)-minimal Type-A theory where the scalar field is quantized with $\Delta = 2$ ($\tilde{\nu}_\phi = 2 - \frac{d}{2}$) boundary condition. It is clear the we just need to add to $\zeta'_{\text{n.-m.}}(0)$ or $\zeta'_{\text{min.}}(0)$ the difference that is due to the change of boundary conditions for the scalar field. In this case, we see that $\tilde{\nu}_\phi = -\nu_\phi$. As we consider the modified zeta function (36), the exponential $\exp(-\beta\nu)$ will change sign. Also, it is clear, see Appendix B.2, that the deficit that can be missing from $\zeta'(0)$ due to the modified zeta-function, is absent thanks to $\tilde{\nu}_\phi = -\nu_\phi$. Repeating the procedure above, we obtain

$$\delta\tilde{\zeta}'_\phi(0) = \tilde{\zeta}'_{\frac{d}{2}-2,0}(0) - \tilde{\zeta}'_{2-\frac{d}{2},0}(0) = \int_0^\infty \frac{(1+e^\beta)\left(e^{2\beta} - e^{\beta(d-2)}\right)}{\beta(e^\beta - 1)^{d+1}}. \tag{57}$$

This is the intermediate form that after using the same regularization as on the CFT side will give the difference between the values of the generalized sphere free energy for the free and interacting $O(N)$ vector models:

$$\delta\tilde{F} = \tilde{F}_{IR} - \tilde{F}_{UV} = \frac{1}{\Gamma(d+1)} \int_0^{d/2-2} u \sin(\pi u) \Gamma\left(\frac{d}{2} - u\right) \Gamma\left(\frac{d}{2} + u\right) du\,. \tag{58}$$

Therefore, we come to the conclusion that

$$\delta F = -\frac{1}{2}\delta\tilde{\zeta}'_\phi(0) \tag{59}$$

Indeed, we can get (59) from the CFT side through an intermediate formula which is minus one half of (57). To be more explicit,

$$\begin{aligned}\delta F &= \frac{1}{2}\sum_{l=0}^{\infty} d_l \log\frac{\Gamma(l+2)}{\Gamma(l+d-2)} = \frac{1}{2}\sum d_l \int_0^\infty \frac{d\beta}{\beta}\left((4-d)e^{-\beta} - \frac{e^{-\beta(l+d-3)} - e^{-\beta(l+1)}}{e^\beta - 1}\right) \\ &= -\frac{1}{2}\int_0^\infty \frac{(1+e^\beta)\left(e^{2\beta} - e^{\beta(d-2)}\right)}{\beta(e^\beta - 1)^{d+1}}.\end{aligned} \tag{60}$$

The same procedure as in Appendix A allows one to relate the intermediate form to (58).

3. Discussion and Conclusions

Our main result is the derivation of the (generalized) sphere free energy $\tilde{F}$ of a free scalar field as a one-loop effect in the minimal Type-A higher-spin theory. Along the way we had to prove that the log-divergence, which is given by $\zeta_{\text{HS}}(0)$, vanishes identically both in the minimal and non-minimal Type-A theories, as well as $\zeta'_{\text{HS}}(0) = 0$ in the non-minimal Type-A theory. The main goal was to extend this result to all integer dimensions as well as to fractional ones. Also, we reproduced the $O(1)$ corrections to the (generalized) sphere free energy in the large-N critical $O(N)$ vector model, which should be dual to the minimal Type-A theory with $\Delta = 2$ boundary conditions for the scalar field. This supports the conjecture that the AdS/CFT duality may extend to fractional dimensions at least for some of the models and some of the observables that are well-defined in non-integer dimensions.

It would be interesting to extend the results of this paper to other models. For example, it should be possible to show directly in AdS_{d+1} that the generalized sphere free energy of higher-spin duals of $\Box^k\phi = 0$ free CFT's should follow

$$\tilde{F} = \frac{1}{\Gamma(d+1)} \int_0^{\Delta-\frac{d}{2}} u \sin(\pi u) \Gamma\left(\frac{d}{2} - u\right) \Gamma\left(\frac{d}{2} + u\right) du\,, \qquad \Delta = \frac{1}{2}(d-2k)\,, \quad k = 1, 2, \ldots\,, \tag{61}$$

which is in accordance with the values for integer d computed in [48,59]. Another motivation comes from the Type-B puzzle that has been observed in [12,44–46]. The Type-B is the higher-spin theory that is supposed to be dual to free fermion in generic dimension or to the Gross-Neveu model for $2 < d < 4$. The AdS one-loop determinant gives the a-anomaly coefficient of the free fermion for d even. However, the computations for d odd results in a sequence of numbers that do not match the sphere energy of the free fermion, but still can be deduced from the change of the F-energy under a double-trace deformation [45,46]. It would be interesting to see what happens in fractional dimensions with the Type-B theory.

Let us note that, as was noted in [57], there is some relation between the zeta-function of Laplace operator on the sphere S^{d+1} and on the hyperbolic space $\mathbb{H}^{d+1}$. This relation is obtained by choosing certain contour for the integral over the spectral parameter λ. As a result the sum over the residues gives the zeta-function on S^{d+1}, but there are other contributions as well. It would be advantageous to make the relation between the computations on sphere and hyperbolic spaces more direct in the higher-spin case. It is striking that the form for the zeta-function we obtained on the AdS side is very close to the one for the Casimir energies in [34]. In this regard let us mention that it is sometimes possible [34] to massage the AdS one-loop computations (the Casimir energy in the case of [34]) in such a way that the formally divergent sum over the spins is of the same form as on the CFT side. In this case, it is clear that one needs to use coherent regularizations on the two sides of the duality. Our computation follows the same strategy and ends up on the convenient intermediate form of the one-loop result that can be directly matched with the CFT one [11,26]. Mostly for technical reasons we used a modified regularization, see also [42,43], which leads to a certain deficit for representations whose character is not an even function of β. Fortunately, this issue does not affect the computation for the Type-A theories (after the summation over spins is performed).[8]

Let us also briefly mention other results that support the conjecture that at least the higher-spin/vector model duality should work in a continuous range of dimensions. The scalar cubic coupling in the higher-spin theory turns out to be extremal in AdS_4 [9,60] and vanishing of the coupling constant near AdS_4 for the dual of the bosonic vector-model [61] and near AdS_3 for the dual of the Gross-Neveu model [62] is properly compensated by the divergence of the bulk integral as a function of d. In principle, it should also be possible to compute the anomalous dimensions of the higher-spin currents in the critical vector and Gross-Neveu models that are well-defined for fractional dimensions. On the CFT side the anomalous dimensions of the higher-spin currents are known up to the $1/N^2$ order, [15,62–68] and up to ϵ^4 for the bosonic vector model in $4-\epsilon$ expansion [6,68–70].

Acknowledgments: We would like to thank Euihun Joung, Sebastian Konopka, Shailesh Lal and Tomas Prochazka for the very useful comments. We are grateful to Simone Giombi for a very useful discussion of the earlier draft. T.T. is grateful to Ivo Sachs, Igor Bertan and Katrin Hammer for useful discussions. The work of E.S. was supported in part by the Russian Science Foundation grant 14-42-00047 in association with Lebedev Physical Institute and by the DFG Transregional Collaborative Research Centre TRR 33 and the DFG cluster of excellence "Origin and Structure of the Universe".

Author Contributions: Both authors contributed equally to this work.

Conflicts of Interest: The authors declare no conflict of interest.

8 The deficit at order z is just the leftover of $P_{v,s}$ without the part including c_n^+ in [45]. This allows us to conclude that the modified zeta function works well for arbitrary dimension since $\mathcal{O}(z^2)$ is irrelevant at one loop.

Appendix A. From Intermediate to Final Form

As a result of the AdS computation we arrived at the intermediate form (52), which can easily be seen to arise in the computation of the determinant on the CFT side. Let us now show how to reach the (generalized) sphere free energy F_ϕ in its final form. In order to compute the β-integral we use

$$\frac{1}{\beta} = \frac{1}{2}\left(\frac{1}{1-e^{-\beta}}\int_0^1 due^{-u\beta} - \frac{1}{1-e^{\beta}}\int_0^1 due^{u\beta}\right). \tag{A1}$$

This allows for an analytic evaluation of the β integral. One obtains

$$F^{\phi}_{\text{min.}} = \frac{\Gamma(-d)}{4}\int_0^1 du(d+4(-1+u)u)\left(\frac{\Gamma\left(-1+\frac{d}{2}+u\right)}{\Gamma\left(1-\frac{d}{2}+u\right)} + \frac{\Gamma\left(\frac{d}{2}-u\right)}{\Gamma\left(2-\frac{d}{2}-u\right)}\right). \tag{A2}$$

After some straightforward algebra (A2) can be shown to split in two parts, the first one we can bring to the form of (for $\Delta = \frac{d}{2}-1$) [11,31]:

$$\begin{aligned} F_\Delta &= \Gamma(-d)\int_0^{\Delta-\frac{d}{2}} duu\left[\frac{\Gamma(\frac{d}{2}-u)}{\Gamma(1-u-\frac{d}{2})} - \frac{\Gamma(\frac{d}{2}+u)}{\Gamma(1+u-\frac{d}{2})}\right] \\ &= -\frac{1}{\sin(\frac{\pi d}{2})\Gamma(d+1)}\int_0^{\Delta-\frac{d}{2}} duu\sin(\pi u)\Gamma\left(\frac{d}{2}+u\right)\Gamma\left(\frac{d}{2}-u\right), \end{aligned} \tag{A3}$$

where the result for the free scalar field corresponds to $\Delta = \frac{d}{2}-1$. The second part has the form

$$\varpi = \frac{1}{4\Gamma(d)\sin\left(\frac{\pi d}{2}\right)}\int_0^1 (1-2u)\sin(\pi u)\Gamma\left(-1+\frac{d}{2}+u\right)\Gamma\left(\frac{d}{2}-u\right). \tag{A4}$$

However, this extra term vanishes due to the anti-symmetry of the integrand around $u = 1/2$. This shows that

$$F^{\phi}_{\text{min.}} = \frac{1}{2}\int_0^\infty \frac{d\beta}{\beta}\frac{e^{-\beta(2+d)/2}(1+e^\beta)^2}{(1-e^{-\beta})^d} = \frac{-1}{\Gamma(d+1)\sin\left(\frac{\pi d}{2}\right)}\int_0^1 du\,u\sin(\pi u)\Gamma\left(\frac{d}{2}-u\right)\Gamma\left(\frac{d}{2}+u\right).$$

Appendix B. Modified Zeta Function

In this Appendix we elaborate on the properties of the modified zeta-function we introduced in Section 2.2. It follows from the definition that the value of $\zeta_{\Delta,s}(0)$ is unaffected, which is illustrated in Appendix B.1. The value of $\zeta'_{\Delta,s}(0)$ differs in general from its true value. Fortunately, $\zeta'(0)$ is still the same for for the spectrum of (non)-minimal Type-A, which is studied in Appendix B.2. It is also shown there that there is no deficit for the difference between the scalars with $\Delta = d-2$ and $\Delta = 2$ boundary conditions.

Appendix B.1. Zeta

From (36), one can easily obtain the full zeta in various odd dimensions with the help of analytical continuation to the Lerch transcendent and then set $z \to 0$. For example,

$$d=3: \quad \tilde{\zeta}_{\nu,s} = \frac{(2s+1)(-17-40s-40s^2+240\nu^4-120(\nu+2s\nu)^2)}{5760}$$

$$d=5: \quad \tilde{\zeta}_{\nu,s} = -\frac{(1+s)(2+s)(3+2s)(-1835-2142s-714s^2-1260(3+2s)^2\nu^2+5040(5+2s(s+3))\nu^4-6720\nu^6)}{29030400}$$

It is easy to see that these polynomials in ν and s are exactly the zeta function for Type-A in [13], see also [45]. Therefore, there is no deficit at z^0 order, i.e.,

$$\tilde{\zeta}_{\nu,s}(0) - \zeta_{\nu,s}(0) = 0 + \mathcal{O}(z)\,. \tag{A5}$$

This explains how we can get all the correct $\tilde{\zeta}_{d,s}(0)$ for individual spins in general odd dimensions. There are many results on zeta-function at $d = 3$, see e.g., [12,13,71]. Let us illustrate that the modified zeta-function is solid enough to obtain these results. The spin factor in $d = 3$ is

$$g_3^A(s) = 2s + 1 \tag{A6}$$

Together with $\nu = s - \frac{1}{2}$, (36) becomes

$$\tilde{\zeta}_3^A(z) = -\frac{(2s+1)}{3!\Gamma(2z)} \int_0^\infty d\beta \sum_{l=0}^\infty e^{-\beta(s-\frac{1}{2})}\beta^{2z-1}e^{-\beta(\frac{1}{2}+l)}\left(\frac{1}{2}+l\right)\left(\left(\frac{1}{2}+s\right)^2 - \left(\frac{1}{2}+l\right)^2\right).$$

Now we can sum over l and obtain

$$\tilde{\zeta}_{3,s}^A(z) = \frac{1}{12\Gamma(2z)} \int_0^\infty d\beta \frac{\beta^{2z-1}e^{-\beta(s-1)}(1+e^\beta)(1+2s)(s(1+s)+e^{2\beta}s(1+s)-2e^\beta(3+s+s^2))}{(-1+e^\beta)^4}$$

In order to get to the actual numbers one needs to plug $s = 0, 1, 2, 3, ...$ then use the trick of analytical continuation via the Hurwitz-Lerch zeta function [12,13]. For example,

$$\tilde{\zeta}_{3,s}^A(0) = \left\{-\frac{1}{180}, -\frac{11}{60}, -\frac{181}{36}, -\frac{6097}{180}, ...\right\} \tag{A7}$$

Note that, after the continuation to the Hurwitz-Lerch transcendent, there will be another $\Gamma(2z)$ function in the nominator. This will cancel $1/\Gamma(2z)$ factor in the modified zeta function. Therefore, the modified zeta-function reproduces the correct result, which is expected.

Appendix B.2. Deficit

As we already explained in Section 2.2, we changed the regularization prescription. As a result the values of $\zeta'_{\Delta,s}(0)$ might be different from the correct ones for individual fields. It was noted in [42] that the deficit vanishes for certain representations (with even character). In particular, the deficit is absent for (non)-minimal Type-A theory. The purpose of this Section is to quantify the deficit for a number of cases.

For example, let us take the scalar field in $d = 3$. The zeta-prime can be derived by calculating $\zeta(z)$ at z order:

$$\zeta_{3,0}(z) = \tfrac{\zeta(-3+2z)}{6} + \tfrac{\zeta(-2+2z)}{4} + \tfrac{\zeta(-1+2z)}{12} = -\tfrac{1}{180} + \left(\tfrac{1}{72} - \tfrac{\log A}{6} + \tfrac{\zeta'(-3)}{3} + \tfrac{\zeta'(-2)}{2}\right)z + \mathcal{O}(z^2) \tag{A8}$$

One can already notice that there is a deficit between the value of $\tilde{\zeta}'^{A}_{3,0}(0)$ that is evaluated by the standard zeta function and (A8). This was also discussed in Appendix (B.1) of [42], when the authors use characters to evaluate $\tilde{\zeta}'(0)$ for different fields. Let us have a look at the deficit in $d = 3$ and $d = 5$ as to observe the general pattern.

Appendix B.2.1. d = 3

The result before sending z to 0 for the modified zeta function is

$$\begin{aligned}\tilde{\zeta}^{3}_{\nu,s}(z) =&\frac{(2s+1)}{24}\left[\nu\left((1+2s)^2-4\nu^2\right)\zeta(2z,\nu+\frac{1}{2})+4\zeta(-3+2z,\nu+\frac{1}{2})\right.\\&\left.-12\nu\zeta(-2+2z,\nu+\frac{1}{2})+(-1-4s(1+s)+12\nu^2)\zeta(-1+2z,\nu+\frac{1}{2})\right].\end{aligned} \tag{A9}$$

In order to compute the zeta-prime, one just needs to take the z derivative and set $z=0$:

$$\begin{aligned}\tilde{\zeta}'^{3}_{\nu,s}(0) =&\frac{(2s+1)}{12}\left[\nu\left((1+2s)^2-4\nu^2\right)\zeta'(0,\nu+\frac{1}{2})+4\zeta'(-3,\nu+\frac{1}{2})\right.\\&\left.-12\nu\zeta'(-2,\nu+\frac{1}{2})+(-1-4s(1+s)+12\nu^2)\zeta'(-1,\nu+\frac{1}{2})\right].\end{aligned} \tag{A10}$$

We then follow the procedure in [42] to find the deficit. First, we set $\nu=0$ and obtain

$$\tilde{\zeta}'^{3}_{0,s}(0) = \frac{(2s+1)}{12}\left(4\zeta'(-3,\frac{1}{2})-(2s+1)^2\zeta'(-1,\frac{1}{2})\right). \tag{A11}$$

Recall that for the standard zeta-prime in $d=3$, see [12,13], we have

$$\zeta'^{3}_{0,s}(0) = \frac{2s+1}{3}\left(c_3+\left(s+\frac{1}{2}\right)^2 c_1\right). \tag{A12}$$

We note that

$$\zeta'(-n,\frac{1}{2}) = (-)^{\frac{n+1}{2}}c_n, \qquad \text{where} \quad c_n=\int_0^\infty du\frac{2u^n\log u}{e^{2\pi u}+1}. \tag{A13}$$

Therefore, $\zeta'^{3}_{0,s}(0)$ and $\tilde{\zeta}'^{3}_{0,s}(0)$ do match. Then, we consider the ν derivatives for each of the zetas:

$$\begin{aligned}\partial_\nu\tilde{\zeta}'^{3}_{\nu,s}(0) =&\frac{(2s+1)}{12}\Bigg(\left((2s+1)^2-12\nu^2\right)\zeta'(0,\nu+\frac{1}{2})-12\zeta'(-2,\nu+\frac{1}{2})+24\nu\zeta'(-1,\nu+\frac{1}{2})\\&+\nu((2s+1)^2-4\nu^2)\partial_\nu\zeta'(0,\nu+\frac{1}{2})+4\partial_\nu\zeta'(-3,\nu+\frac{1}{2})-12\nu\partial_\nu\zeta'(-2,\nu+\frac{1}{2})\\&+(-1-4s(s+1)+12\nu^2)\partial_\nu\zeta'(-1,\nu+\frac{1}{2})+\nu((2s+1)^2-4\nu^2)\partial_\nu\zeta'(0,\nu+\frac{1}{2})\Bigg)\end{aligned} \tag{A14}$$

$$\partial_\nu\zeta'^{3}_{\nu,s}(0) = \frac{(2s+1)}{3}\left(\frac{\nu^3}{2}+\frac{\nu}{24}+\nu\left(\left(s+\frac{1}{2}\right)^2-\nu^2\right)\psi(\nu+\frac{1}{2})\right) \tag{A15}$$

Next, using the identities for Hurwitz zeta function

$$\partial_\nu\zeta(s,\nu)=-s\zeta(s+1,\nu), \qquad \partial_\nu\zeta'(0,\nu)=\psi(\nu) \tag{A16}$$

we can reduce the ν derivative of the modified zeta-prime to

$$\partial_\nu\tilde{\zeta}'^{3}_{\nu,s}(0) = \frac{(2s+1)\nu((2s+1)^2-4\nu^2)\psi(\nu+\frac{1}{2})}{12} \tag{A17}$$

Subtracting (A17) and (A15) together, then integrating over ν, one obtains the deficit for individual fields at order z:

$$\delta\zeta'_{\nu,s}(0) = \tilde{\zeta}'_{\nu,s}-\zeta'_{\nu,s} = -\frac{(2s+1)(\nu^2+6\nu^4)}{144} \tag{A18}$$

Since the deficit is an even function of ν, we can compute the difference between the scalars with $\Delta = d-2, 2$ boundary conditions using the modified zeta function thanks to $\delta\zeta'_{d-2,0} - \delta\zeta'_{2,0} = 0$. Using the cut-off $e^{-\epsilon(s+\frac{d-3}{2})}$, one can sum over either all spins or even spins and observe that the deficit does vanish:

$$\sum_s \delta\zeta'_{\nu,s}(0) = 0\,. \tag{A19}$$

Therefore, the deficit is absent both for the non-minimal and minimal Type-A theories at order z, which is what we need for $\zeta'_{\text{HS}}(0)$.

Appendix B.2.2. d = 5

In higher dimensions, there is another useful identity that we illustrate on the example of $d = 5$. Following the procedure outlined above, we obtain

$$\begin{aligned}\tilde{\zeta}'^{5}_{\nu,s}(0) =& \frac{(1+s)(2+s)(3+2s)}{5760}\Bigg[-16\zeta'(-5,\nu+\frac{3}{2}) + 8\nu(-3(5+2s(3+s))+20\nu^2)\zeta'(-2,\nu+\frac{3}{2})\\ &+80\nu\zeta'(-4,\nu+\frac{3}{2}) + (-(3+2s)^2+24(5+2s(3+s))\nu^2-80\nu^4)\zeta'(-1,\nu+\frac{3}{2})\\ &+\nu(-1+4\nu^2)(-9-4s(3+s)+4\nu^2)\zeta'(0,\nu+\frac{3}{2}) + 8(5+2s(3+s)-20\nu^2)\zeta'(-3,\nu+\frac{3}{2})\Bigg]\,.\end{aligned}$$

Setting $\nu = 0$ we arrive at

$$\tilde{\zeta}'^{5}_{0,s}(0) = \frac{(1+s)(2+s)(3+2s)}{5760}\Bigg[-16\zeta'(-5,\frac{3}{2}) + 8(5+2s(3+s))\zeta'(-3,\frac{3}{2}) - (3+2s)^2\zeta'(-1,\frac{3}{2})\Bigg]\,.$$

We massage the formula above as to be able to compare $\zeta'(-k,\frac{1}{2})$ with c_n, which can be done with the help of

$$\zeta(s,\nu) = \zeta(s,\nu+m) + \sum_{n=0}^{m-1}\frac{1}{(n+\nu)^s} \tag{A20}$$

We arrive at

$$\tilde{\zeta}'^{5}_{0,s}(0) = -\frac{(1+s)(2+s)(3+2s)}{5760}\Bigg[-16\zeta'(-5,\frac{1}{2}) + 8(5+2s(3+s))\zeta'(-3,\frac{1}{2}) - (3+2s)^2\zeta'(-1,\frac{1}{2})\Bigg]\,, \tag{A21}$$

which can be compared with the standard zeta-prime:

$$\zeta'^{5}_{0,s}(0) = -\frac{(1+s)(2+s)(3+2s)}{360}\left(c_5 + c_3\left(\frac{1}{4}+\left(s+\frac{3}{2}\right)^2\right) + \frac{c_1}{4}\left(s+\frac{3}{2}\right)^2\right)\,. \tag{A22}$$

Using the identity (A13), it is easy to realize that (A21) and (A22) are the same. Next, one can proceed as in the previous Section and get

$$\delta\zeta'^{5}_{\nu,s} = -\frac{(s+1)(s+2)(2s+3)\nu^2(107+580\nu^2-240\nu^4+120s(1+6\nu^2)+40s^2(1+6\nu^2))}{691200}\,. \tag{A23}$$

The sum over all (even) spins can be found to vanish, which guarantees that the deficit does not contribute to the zeta-prime of the (non)-minimal Type-A. Also the deficit is an even function of ν and therefore the difference due to $\Delta = d-2, 2$ boundary conditions for the scalar field is also free of any deficit.

Let us note that the deficit has already appeared in implicit form in the literature. It is the leftover of $P_{v,s}$ in [45] without the part including c_n^+, see also [13] where the same structures are present but in different notation.

References

1. Maldacena, J.M. The large N limit of superconformal field theories and supergravity. *Adv. Theor. Math. Phys.* **1998**, *2*, 231–252.
2. Witten, E. Anti-de Sitter space and holography. *Adv. Theor. Math. Phys.* **1998**, *2*, 253–291.
3. Gubser, S.S.; Klebanov, I.R.; Polyakov, A.M. Gauge theory correlators from non-critical string theory. *Phys. Lett. B* **1998**, *428*, 105–114.
4. Klebanov, I.R.; Polyakov, A.M. AdS dual of the critical O(N) vector model. *Phys. Lett. B* **2002**, *550*, 213–219.
5. Brezin, E.; Wallace, D.J.; Wilson, K. Feynman-graph expansion for the equation of state near the critical point. *Phys. Rev. B* **1973**, *7*, 232–239.
6. Wilson, K.G.; Kogut, J.B. The Renormalization group and the epsilon expansion. *Phys. Rep.* **1974**, *12*, 75–200.
7. Sundborg, B. Stringy gravity, interacting tensionless strings and massless higher spins. *Nucl. Phys. Proc. Suppl.* **2001**, *102*, 113–119.
8. Sezgin, E.; Sundell, P. Massless higher spins and holography. *Nucl. Phys.* **2002**, *B644*, 303–370.
9. Sezgin, E.; Sundell, P. Holography in 4D (super) higher spin theories and a test via cubic scalar couplings. *J. High Energy Phys.* **2005**, *2005*, 044.
10. Leigh, R.G.; Petkou, A.C. Holography of the N = 1 higher spin theory on AdS(4). *J. High Energy Phys.* **2003**, *2003*, 011.
11. Giombi, S.; Klebanov, I.R. Interpolating between *a* and *F*. *J. High Energy Phys.* **2015**, *2015*, 117.
12. Giombi, S.; Klebanov, I.R. One Loop Tests of Higher Spin AdS/CFT. *J. High Energy Phys.* **2013**, *2013*, 068.
13. Giombi, S.; Klebanov, I.R.; Safdi, B.R. Higher Spin AdS_{d+1}/CFT_d at One Loop. *Phys. Rev. D* **2014**, *89*, 084004.
14. Vasiliev, A.N.; Pismak, Y.M.; Khonkonen, Y.R. Simple Method of Calculating the Critical Indices in the $1/N$ Expansion. *Theor. Math. Phys.* **1981**, *46*, 104–113.
15. Lang, K.; Ruhl, W. The Critical O(N) sigma model at dimensions 2 < d < 4: Fusion coefficients and anomalous dimensions. *Nucl. Phys. B* **1993**, *400*, 597–623.
16. Petkou, A. Conserved currents, consistency relations and operator product expansions in the conformally invariant O(N) vector model. *Ann. Phys.* **1996**, *249*, 180–221.
17. Moshe, M.; Zinn-Justin, J. Quantum field theory in the large N limit: A Review. *Phys. Rep.* **2003**, *385*, 69–228.
18. El-Showk, S.; Paulos, M.; Poland, D.; Rychkov, S.; Simmons-Duffin, D.; Vichi, A. Conformal Field Theories in Fractional Dimensions. *Phys. Rev. Lett.* **2014**, *112*, 141601.
19. Fei, L.; Giombi, S.; Klebanov, I.R. Critical $O(N)$ models in $6-\epsilon$ dimensions. *Phys. Rev. D* **2014**, *90*, 025018.
20. Mati, P. Critical scaling in the large-N $O(N)$ model in higher dimensions and its possible connection to quantum gravity. *Phys. Rev. D* **2016**, *94*, 065025, doi:10.1103/PhysRevD.94.065025.
21. Zamolodchikov, A.B. Irreversibility of the Flux of the Renormalization Group in a 2D Field Theory. *JETP Lett.* **1986**, *43*, 730–732.
22. Cardy, J.L. Is There a c Theorem in Four-Dimensions? *Phys. Lett. B* **1988**, *215*, 749–752.
23. Komargodski, Z.; Schwimmer, A. On Renormalization Group Flows in Four Dimensions. *J. High Energy Phys.* **2011**, *2011*, 099.
24. Myers, R.C.; Sinha, A. Seeing a c-theorem with holography. *Phys. Rev. D* **2010**, *82*, 046006.
25. Jafferis, D.L.; Klebanov, I.R.; Pufu, S.S.; Safdi, B.R. Towards the F-Theorem: N = 2 Field Theories on the Three-Sphere. *J. High Energy Phys.* **2011**, *2011*, 102.
26. Klebanov, I.R.; Pufu, S.S.; Safdi, B.R. F-Theorem without Supersymmetry. *J. High Energy Phys.* **2011**, *2011*, 038.
27. Casini, H.; Huerta, M.; Myers, R.C. Towards a derivation of holographic entanglement entropy. *J. High Energy Phys.* **2011**, *2011*, 036.
28. Casini, H.; Huerta, M. On the RG running of the entanglement entropy of a circle. *Phys. Rev. D* **2012**, *85*, 125016.
29. Affleck, I.; Ludwig, A.W.W. Universal noninteger 'ground state degeneracy' in critical quantum systems. *Phys. Rev. Lett.* **1991**, *67*, 161–164.

30. Gubser, S.S.; Klebanov, I.R. A Universal result on central charges in the presence of double trace deformations. *Nucl. Phys. B* **2003**, *656*, 23–36.
31. Diaz, D.E.; Dorn, H. Partition functions and double-trace deformations in AdS/CFT. *J. High Energy Phys.* **2007**, *2007*, 046.
32. Allais, A. Double-trace deformations, holography and the c-conjecture. *J. High Energy Phys.* **2010**, *2010*, 040.
33. Aros, R.; Diaz, D.E. Determinant and Weyl anomaly of Dirac operator: A holographic derivation. *J. Phys. A* **2012**, *45*, 125401.
34. Giombi, S.; Klebanov, I.R.; Tseytlin, A.A. Partition Functions and Casimir Energies in Higher Spin AdS_{d+1}/CFT_d. *Phys. Rev. D* **2014**, *90*, 024048.
35. Beccaria, M.; Tseytlin, A.A. Higher spins in AdS_5 at one loop: Vacuum energy, boundary conformal anomalies and AdS/CFT. *J. High Energy Phys.* **2014**, *2014*, 114.
36. Beccaria, M.; Bekaert, X.; Tseytlin, A.A. Partition function of free conformal higher spin theory. *J. High Energy Phys.* **2014**, *2014*, 113.
37. Beccaria, M.; Tseytlin, A.A. Vectorial AdS_5/CFT_4 duality for spin-one boundary theory. *J. Phys. A* **2014**, *47*, 492001.
38. Beccaria, M.; Macorini, G.; Tseytlin, A.A. Supergravity one-loop corrections on AdS_7 and AdS_3, higher spins and AdS/CFT. *Nucl. Phys. B* **2015**, *892*, 211–238.
39. Basile, T.; Bekaert, X.; Boulanger, N. Flato-Fronsdal theorem for higher-order singletons. *J. High Energy Phys.* **2014**, *2014*, 131.
40. Beccaria, M.; Tseytlin, A.A. On higher spin partition functions. *J. Phys. A* **2015**, *48*, 275401.
41. Beccaria, M.; Tseytlin, A.A. Iterating free-field AdS/CFT: Higher spin partition function relations. *J. Phys. A* **2016**, *49*, 295401.
42. Bae, J.-B.; Joung, E.; Lal, S. One-loop test of free SU(N) adjoint model holography. *J. High Energy Phys.* **2016**, *2016*, 061.
43. Bae, J.-B.; Joung, E.; Lal, S. On the Holography of Free Yang-Mills. *J. High Energy Phys.* **2016**, *2016*, 074.
44. Pang, Y.; Sezgin, E.; Zhu, Y. One Loop Tests of Supersymmetric Higher Spin AdS_4/CFT_3. *Phys. Rev. D* **2017**, *95*, 026008.
45. Günaydin, M.; Skvortsov, E.D.; Tran, T. Exceptional $F(4)$ higher-spin theory in AdS_6 at one-loop and other tests of duality. *J. High Energy Phys.* **2016**, *2016*, 168.
46. Giombi, S.; Klebanov, I.R.; Tan, Z. M. The ABC of Higher-Spin AdS/CFT. *arXiv* **2006**, arXiv:1608.07611.
47. Bae, J.-B.; Joung, E.; Lal, S. A note on vectorial AdS_5/CFT_4 duality for spin-j boundary theory. *J. High Energy Phys.* **2016**, *2016*, 077.
48. Brust, C.; Hinterbichler, K. Partially Massless Higher-Spin Theory II: One-Loop Effective Actions. *J. High Energy Phys.* **2017**, *2017*, 126.
49. Maldacena, J.; Zhiboedov, A. Constraining Conformal Field Theories with A Higher Spin Symmetry. *arXiv* **2011**, arXiv:1112.1016.
50. Alba, V.; Diab, K. Constraining conformal field theories with a higher spin symmetry in d = 4. *arXiv* **2013**, arXiv:1307.8092.
51. Boulanger, N.; Ponomarev, D.; Skvortsov, E.; Taronna, M. On the uniqueness of higher-spin symmetries in AdS and CFT. *arXiv* **2013**, arXiv:1307.8092.
52. Stanev, Y.S. Constraining conformal field theory with higher spin symmetry in four dimensions. *Nucl. Phys.* **2013**, *B876*, 651–666.
53. Alba, V.; Diab, K. Constraining conformal field theories with a higher spin symmetry in $d > 3$ dimensions. *arXiv* **2015**, arXiv:1510.02535.
54. Dowker, J.S.; Critchley, R. Effective Lagrangian and Energy Momentum Tensor in de Sitter Space. *Phys. Rev. D* **1976**, *13*, 3224–3232.
55. Hawking, S.W. Zeta Function Regularization of Path Integrals in Curved Space-Time. *Commun. Math. Phys.* **1977**, *55*, 133–148.
56. Vasiliev, M.A. Nonlinear equations for symmetric massless higher spin fields in (A)dS(d). *Phys. Lett. B* **2003**, *567*, 139–151.
57. Camporesi, R.; Higuchi, A. Spectral functions and zeta functions in hyperbolic spaces. *J. Math. Phys.* **1994**, *35*, 4217–4246.

58. Giombi, S.; Klebanov, I.R.; Tarnopolsky, G. Conformal QED_d, *F*-Theorem and the ϵ Expansion. *J. Phys. A* **2016**, *49*, 135403.
59. Brust, C.; Hinterbichler, K. Free $\Box^k$ scalar conformal field theory. *J. High Energy Phys.* **2017**, *2017*, 066.
60. Giombi, S.; Yin, X. Higher Spin Gauge Theory and Holography: The Three-Point Functions. *J. High Energy Phys.* **2010**, *2010*, 115.
61. Bekaert, X.; Erdmenger, J.; Ponomarev, D.; Sleight, C. Towards holographic higher-spin interactions: Four-point functions and higher-spin exchange. *J. High Energy Phys.* **2015**, *2015*, 170.
62. Skvortsov, E.D. On (Un)Broken Higher-Spin Symmetry in Vector Models. In Proceedings of the International Workshop on Higher Spin Gauge Theories, Singapore, 4–6 November 2015; 2017; pp. 103–137.
63. Muta, T.; Popovic, D.S. Anomalous Dimensions of Composite Operators in the Gross-Neveu Model in Two + Epsilon Dimensions. *Prog. Theor. Phys.* **1977**, *57*, 1705–1719.
64. Giombi, S.; Gurucharan, V.; Kirilin, V.; Prakash, S.; Skvortsov, E. On the Higher-Spin Spectrum in Large N Chern-Simons Vector Models. *J. High Energy Phys.* **2017**, *2017*, 058.
65. Giombi, S.; Kirilin, V. Anomalous dimensions in CFT with weakly broken higher spin symmetry. *J. High Energy Phys.* **2016**, *2016*, 068.
66. Giombi, S.; Kirilin, V.; Skvortsov, E. Notes on Spinning Operators in Fermionic CFT. *J. High Energy Phys.* **2017**, *2017*, 041, doi:10.1007/JHEP05(2017)041.
67. Manashov, A.N.; Skvortsov, E.D. Higher-spin currents in the Gross-Neveu model at $1/n^2$. *J. High Energy Phys.* **2017**, *2017*, 132.
68. Manashov, A.N.; Skvortsov, E.D.; Strohmaier, M. Higher spin currents in the critical $O(N)$ vector model at $1/N^2$. *arXiv* **2017**, arXiv:1706.09256.
69. Braun, V.M.; Manashov, A.N. Evolution equations beyond one loop from conformal symmetry. *Eur. Phys. J. C* **2013**, *73*, 2544.
70. Derkachov, S.E.; Gracey, J.A.; Manashov, A.N. Four loop anomalous dimensions of gradient operators in ϕ^4 theory. *Eur. Phys. J. C* **1998**, *2*, 569–579.
71. Camporesi, R.; Higuchi, A. Arbitrary spin effective potentials in anti-de Sitter space-time. *Phys. Rev. D* **1993**, *47*, 3339–3344.

universe

MDPI

Article

Gauge Non-Invariant Higher-Spin Currents in AdS_4

Pavel Smirnov * and Mikhail Vasiliev

I.E.Tamm Department of Theoretical Physics, Lebedev Physical Institute of RAS, Leninsky prospect 53, 119991 Moscow, Russia; vasiliev@lpi.ru

* Correspondence: smirnov.mipt@gmail.com

Received: 10 October 2017; Accepted: 3 November 2017; Published: 16 November 2017

Abstract: Conserved currents of any spin $t > 0$ built from bosonic symmetric massless gauge fields of arbitrary integer spins $s_1 + s_2 > t$ in AdS_4 are found. Analogous to the case of $4d$ Minkowski space, currents considered in this paper are not gauge invariant, but generate gauge-invariant conserved charges.

Keywords: higher-spin theory; conserved currents; conserved charges

---

1. Introduction

Gauge-invariant conserved currents are well known and were deeply studied in the literature [1–9]. In the general case, a conserved current carries a set of three spins (t, s_1, s_2), where t is a spin of the current itself, and s_1 and s_2 are spins of the fields it is constructed from. For example, the so-called gravitational stress pseudo-tensor [10] ($s = t = 2$ conserved current) is not gauge invariant. The same fact is shown in [5] for the $t = 2$ current built from massless fields of spins $s > 2$. The spin-zero field has no gauge symmetry; thus, the currents with $s < 1$ are gauge invariant, while the spin-one current built from two massless spin-one fields is not.

The aim of this paper is to extend the Minkowski-space results of [11], presenting the full list of gauge non-invariant currents with integer spins in AdS_4 such that $t < s_1 + s_2$. Being gauge non-invariant, these currents give rise to the gauge-invariant conserved charges. Gauge non-invariant currents will be derived from the variation of the cubic action of [12,13], which is gauge invariant in the lowest order.

Conventions

In this paper, we consider AdS_4 space-time. Greek indices $\mu, \nu, \rho, \lambda, \sigma$ are the base and range from 0–3. Other Greek indices are spinorial and take values of one and two. The latter are raised and lowered by the $sp(2)$ antisymmetric forms: $\varepsilon_{\alpha\beta}, \varepsilon^{\alpha\beta}, \varepsilon_{\dot\alpha\dot\beta}, \varepsilon^{\dot\alpha\dot\beta}$

$$\varepsilon^{\alpha\beta}\varepsilon_{\alpha\gamma} = \delta^{\beta}_{\gamma}, \quad \varepsilon^{\dot\alpha\dot\beta}\varepsilon_{\dot\alpha\dot\gamma} = \delta^{\dot\beta}_{\dot\gamma}, \tag{1}$$

$$A_\alpha = A^\beta \varepsilon_{\beta\alpha}, \quad A^\alpha = A_\beta \varepsilon^{\alpha\beta}, \quad A_{\dot\alpha} = A^{\dot\beta}\varepsilon_{\dot\beta\dot\alpha}, \quad A^{\dot\alpha} = A_{\dot\beta}\varepsilon^{\dot\alpha\dot\beta}. \tag{2}$$

Complex conjugation $\bar{A}$ relates dotted and undotted spinors. Brackets ([...]) {...} imply complete (anti)symmetrization, i.e.,

$$A_{[\alpha}B_{\beta]} = \frac{1}{2}(A_\alpha B_\beta - A_\beta B_\alpha), \qquad A_{\{\alpha}B_{\beta\}} = \frac{1}{2}(A_\alpha B_\beta + A_\beta B_\alpha). \tag{3}$$

$A^{\alpha(m)}$ denotes a totally symmetric multispinor $A^{\{\alpha_1 \dots \alpha_m\}}$.

The wedge symbol $\wedge$ is implicit.

2. Fields, Equations, Actions

In the four-dimensional case considered in this paper, it is convenient to use the frame-like formalism in two-component spinor notation. In these terms, a bosonic spin-s Fronsdal field [14] is represented by multispinor one-forms [15]:

$$s \geq 1: \quad \varphi_{\mu_1 \dots \mu_s} \to \{\omega^{\alpha(m)}{}_{,}{}^{\dot\beta(n)} \mid m+n = 2(s-1)\}, \quad \omega^{\alpha(m)}{}_{,}{}^{\dot\beta(n)} = dx^\mu \omega_\mu{}^{\alpha(m)}{}_{,}{}^{\dot\beta(n)},$$

which are symmetric in all dotted and all undotted spinor indices and obey the reality condition [15]:

$$\omega^\dagger_{\alpha(m),\dot\beta(n)} = \omega_{\beta(n),\dot\alpha(m)}. \tag{4}$$

The frame-like field is a particular connection at $n = m = s-1$ (s is integer):

$$h_\mu{}^{\alpha(s-1)}{}_{,}{}^{\dot\beta(s-1)} dx^\mu := \omega_\mu{}^{\alpha(s-1)}{}_{,}{}^{\dot\beta(s-1)} dx^\mu. \tag{5}$$

By imposing appropriate constraints, the connections $\omega^{\alpha(m),\dot\beta(n)}$ can be expressed via $t = \frac{1}{2}|m-n|$ derivatives of the frame-like field [15].

Background gravity is described by the vierbein one-form $\tilde h^{\alpha}{}_{,}{}^{\dot\beta}$ and one-form connections $\tilde\omega^{\dot\alpha\dot\beta}$, $\tilde\omega^{\alpha\beta}$. Lorentz covariant derivative $\tilde D$ acts as usual:

$$\tilde D A^{\alpha(m)}{}_{,}{}^{\dot\beta(n)} = dA^{\alpha(m)}{}_{,}{}^{\dot\beta(n)} + m\tilde\omega^{\alpha}{}_{\gamma} A^{\alpha(m-1)\gamma}{}_{,}{}^{\dot\beta(n)} + n\tilde\omega^{\dot\beta}{}_{\dot\delta} A^{\alpha(m)}{}_{,}{}^{\dot\beta(n-1)\dot\delta} \tag{6}$$

for any multispinor $A^{\alpha(m),\dot\beta(n)}$. The torsion and curvature two-forms are:

$$\tilde R^{\alpha}{}_{,}{}^{\dot\beta} = d\tilde h^{\alpha}{}_{,}{}^{\dot\beta} + \tilde\omega^{\alpha}{}_{\gamma}\tilde h^{\gamma}{}_{,}{}^{\dot\beta} + \tilde\omega^{\dot\beta}{}_{\dot\delta}\tilde h^{\alpha}{}_{,}{}^{\dot\delta}, \tag{7}$$

$$\tilde R^{\alpha\alpha} = d\tilde\omega^{\alpha\alpha} + \tilde\omega^{\alpha}{}_{\gamma}\tilde\omega^{\alpha\gamma} + \lambda^2 \tilde h^{\alpha}{}_{,\dot\delta}\tilde h^{\alpha}{}_{,}{}^{\dot\delta}, \tag{8}$$

$$\tilde R^{\dot\beta\dot\beta} = d\tilde\omega^{\dot\beta\dot\beta} + \tilde\omega^{\dot\beta}{}_{\dot\gamma}\tilde\omega^{\dot\beta\dot\gamma} + \lambda^2 \tilde h_{\gamma,}{}^{\dot\beta}\tilde h^{\gamma}{}_{,}{}^{\dot\beta}, \tag{9}$$

where the parameter λ is proportional to the inverse AdS radius $\lambda \sim r^{-1}$. AdS_4 space is described by the vierbein and connections obeying the equations:

$$\tilde R^{\alpha}{}_{,}{}^{\dot\beta} = 0, \qquad \tilde R^{\alpha\alpha} = 0, \qquad \tilde R^{\dot\beta\dot\beta} = 0. \tag{10}$$

Linearized higher-spin (HS) curvatures are:

$$\begin{aligned} R_1{}^{\alpha(m)}{}_{,}{}^{\dot\beta(n)} = \tilde D\omega^{\alpha(m)}{}_{,}{}^{\dot\beta(n)} &+ n(\theta(m-n) + \lambda^2\theta(n-m-2))\tilde h_{\gamma,}{}^{\dot\beta}\omega^{\gamma\alpha(m)}{}_{,}{}^{\dot\beta(n-1)} \\ &+ m(\theta(n-m) + \lambda^2\theta(m-n-2))\tilde h^{\alpha}{}_{,\dot\delta}\omega^{\alpha(m-1)}{}_{,}{}^{\dot\beta(n)\dot\delta}, \end{aligned} \tag{11}$$

where $\theta(x)$ is the step-function:

$$\theta(x) = \begin{cases} 1 & \text{at } x \geq 0; \\ 0 & \text{at } x < 0. \end{cases} \tag{12}$$

Curvatures (11) obey the Bianchi identities [15]:

$$\begin{aligned} \tilde D R_1{}^{\alpha(m)}{}_{,}{}^{\dot\beta(n)} = -\lambda^{(|m-n|/2)+1}(m\lambda^{-|m-n-2|/2}\tilde h^{\alpha}{}_{,\dot\delta} R_1{}^{\alpha(m-1)}{}_{,}{}^{\dot\beta(n)\dot\delta} \\ + n\lambda^{-|m-n+2|/2}\tilde h_{\gamma,}{}^{\dot\beta} R_1{}^{\alpha(m)\gamma}{}_{,}{}^{\dot\beta(n-1)}). \end{aligned} \tag{13}$$

It is convenient to introduce two-forms $H_{\alpha\beta}$ and $\bar{H}_{\dot{\alpha}\dot{\beta}}$:

$$\tilde{h}_{\alpha,\dot{\beta}}\tilde{h}_{\gamma,\dot{\delta}} = \frac{1}{2}\epsilon_{\alpha\gamma}\bar{H}_{\dot{\beta}\dot{\delta}} + \frac{1}{2}\epsilon_{\dot{\beta}\dot{\delta}}H_{\alpha\gamma}\,, \tag{14}$$

$$H_{\alpha\beta} := \tilde{h}_{\alpha,\dot{\gamma}}\tilde{h}_{\beta,}{}^{\dot{\gamma}}\,, \quad \bar{H}_{\dot{\alpha}\dot{\beta}} := \tilde{h}_{\gamma,\dot{\alpha}}\tilde{h}^{\gamma}{}_{,\dot{\beta}}. \tag{15}$$

Free field equations for massless fields of spins $s \geq 2$ in Minkowski space can be written in the form [15]:

$$R_1{}^{\alpha(m)}{}_{,}{}^{\dot{\beta}(n)} = 0 \qquad \text{for} \qquad n > 0,\ m > 0,\ n+m = 2(s-1); \tag{16}$$

$$R_1{}^{\alpha(m)} = C^{\alpha(m)\gamma\delta}\ H_{\gamma\delta} \qquad \text{for} \qquad m = 2(s-1); \tag{17}$$

$$R_1{}^{\dot{\beta}(n)} = \bar{C}^{\dot{\beta}(n)\dot{\gamma}\dot{\delta}}\ \bar{H}_{\dot{\gamma}\dot{\delta}} \qquad \text{for} \qquad n = 2(s-1). \tag{18}$$

Equations (16)–(18) are equivalent to the equations of motion, which follow from the Fronsdal action [14] supplemented with certain algebraic constraints, which express connections $\omega_{\alpha(m),\dot{\beta}(n)}$ via $\frac{1}{2}|m-n|$ derivatives of the dynamical frame-like HS field. The multispinor zero-forms $C^{\alpha(2s)}$ and $\bar{C}^{\dot{\beta}(2s)}$, which remain non-zero on-shell, are spin-s analogues of the Weyl tensor in gravity.

HS gauge transformation is:

$$\delta\omega^{\alpha(m)}{}_{,}{}^{\dot{\beta}(n)} = \tilde{D}\epsilon^{\alpha(m)}{}_{,}{}^{\dot{\beta}(n)} + n(\theta(m-n) + \lambda^2\theta(n-m-2))\tilde{h}_{\gamma,}{}^{\dot{\beta}}\epsilon^{\gamma\alpha(m)}{}_{,}{}^{\dot{\beta}(n-1)}$$
$$+ m(\theta(n-m) + \lambda^2\theta(m-n-2))\tilde{h}^{\alpha}{}_{,\dot{\delta}}\epsilon^{\alpha(m-1)}{}_{,}{}^{\dot{\beta}(n)\dot{\delta}}, \tag{19}$$

where a gauge parameter $\epsilon^{\alpha(m)}{}_{,}{}^{\dot{\beta}(n)}(x)$ is an arbitrary function of x. Note that the limit $\lambda \to 0$ gives the proper description of HS fields in 4d Minkowski space.

As explained in [11], to obtain currents with odd and even spins, the connections $\omega^{i;\alpha(m),\dot{\beta}(n)}$ and curvatures $R^{i;\alpha(m),\dot{\beta}(n)}$ should be endowed with a color index $i = 1\ldots N$, which labels independent dynamical fields. To contract color indices, we introduce the real tensor c_{ijk}, which can be either symmetric or antisymmetric. Color indices are raised and lowered by the Euclidean metric g_{ij}. It is convenient to set $g_{ij} = \delta_{ij}$.

Free fields are described by the quadratic action [15]:

$$S_2 = \frac{1}{2}\int \sum_{m,n\geq 0} \frac{1}{m!n!}\,\varepsilon(m-n)\,\lambda^{-|m-n|}\,R_1{}^{i;\alpha(m)}{}_{,}{}^{\dot{\beta}(n)}R_{1\,i;\alpha(m),\dot{\beta}(n)}, \tag{20}$$

where $\varepsilon(x) = \theta(x) - \theta(-x)$ and $m+n = 2(s-1)$, s being a spin of the field.

Following [12,13], to obtain a cubic deformation of the quadratic action, the linear curvature R_1 in the action (20) has to be replaced by $R = R_1 + R_2$ where:

$$R_2{}^{i;\alpha(m)}{}_{,}{}^{\dot{\beta}(n)} = \sum_{p,q,k,l,u,v\geq 0} \lambda^{1+d_0-d_1-d_2}\,\frac{m!n!}{p!q!k!l!u!v!}\,c^{i}{}_{jk}\,\delta_{p+q,m}\delta_{u+v,n}$$
$$\times\,\omega^{j;\alpha(p)}{}_{\gamma(k),\dot{\delta}(l)}{}^{\dot{\beta}(u)}\,\omega^{k;\alpha(q)\gamma(k)}{}_{,}{}^{\dot{\delta}(l)\dot{\beta}(v)}, \tag{21}$$

$$d_0 = \frac{|m-n|}{2}, \quad d_1 = \frac{|p+k-l-u|}{2}, \quad d_2 = \frac{|q+k-l-v|}{2}.$$

The nonlinear action is:

$$S = \frac{1}{2}\int \sum_{m,n\geq 0} \frac{1}{m!n!}\,\varepsilon(m-n)\,\lambda^{-|m-n|}\,R^{i;\alpha(m)}{}_{,}{}^{\dot{\beta}(n)}R_{i;\alpha(m),\dot{\beta}(n)}. \tag{22}$$

3. Problem

It is convenient to describe currents as Hodge-dual differential forms. The on-shell closure condition for the latter is traded for the current conservation condition. In this paper, we consider spin-t currents in AdS_4 built from two connections of integer spins $s_1, s_2 > 0$ such that $t \leq s_1 + s_2 - 1$. Such currents contain the minimal possible number of derivatives of the dynamical fields. The analogous problem in $4d$ Minkowski space has been solved in [11] for the case of $s_1 = s_2$. The form of the currents will be derived from the nonlinear action (22).

An arbitrary variation of the action (22) can be represented in the form:

$$\delta S = \int \sum_{t,s_1,s_2} \sum_{m,n} \delta(m+n-2(t-1)) J_{t,s_1,s_2}{}^{i;\alpha(m)}{}_{,}{}^{\dot\beta(n)} \, \delta\omega_{i;\alpha(m),\dot\beta(n)} \,. \tag{23}$$

The current $J_{t,s_1,s_2}{}^{i;\alpha(m)}{}_{,}{}^{\dot\beta(n)}$ carries the color index i. Actually, there are N copies of a current, one for each value of i, and we can set $i = 1$ without loss of generality. In what follows, this index $i = 1$ will be omitted in all current forms. Furthermore, it is convenient to set $c_{jk} := c_{1jk}$ with c_{jk} being either symmetric or antisymmetric, i.e.,

$$c_{jk} = \eta c_{kj} \,, \qquad \eta^2 = 1 \,. \tag{24}$$

To define a nontrivial HS charge as an integral over a $3d$ space, one should find such a current three-form $J_{t,s_1,s_2}(x)$ built from dynamical HS fields that is closed by virtue of HS field Equations (16)–(18), but not exact. The closed current three-form is:

$$J_{t,s_1,s_2}(x) = \sum_{m,n} \frac{\lambda^{-|m-n|}}{m!n!} \delta(m+n-2(t-1)) \xi_{\alpha(m),\dot\beta(n)}(x) J_{t,s_1,s_2}{}^{\alpha(m)}{}_{,}{}^{\dot\beta(n)}(x), \tag{25}$$

where the factor of $\frac{\lambda^{-|m-n|}}{m!n!}$ is introduced for convenience and $\xi_{\alpha(m),\dot\beta(n)}$ are global symmetry parameters, which can be identified with those gauge symmetry parameters that leave the background gauge fields invariant. In accordance with (19), these parameters obey:

$$D\xi^{\alpha(m)}{}_{,}{}^{\dot\beta(n)} := \tilde{D}\xi^{\alpha(m)}{}_{,}{}^{\dot\beta(n)} + n(\theta(m-n) + \lambda^2\theta(n-m-2))\tilde{h}_{\gamma,}{}^{\dot\beta}\xi^{\gamma\alpha(m)}{}_{,}{}^{\dot\beta(n-1)} \\ + m(\theta(n-m) + \lambda^2\theta(m-n-2))\tilde{h}^{\alpha}{}_{,\dot\delta}\xi^{\alpha(m-1)}{}_{,}{}^{\dot\beta(n)\dot\delta} = 0. \tag{26}$$

One can see that:

$$dJ_{t,s_1,s_2} = \sum_{m,n} \frac{\lambda^{-|m-n|}}{m!n!} \left(D\xi_{\alpha(m),\dot\beta(n)} J_{t,s_1,s_2}{}^{\alpha(m)}{}_{,}{}^{\dot\beta(n)} + \xi_{\alpha(m),\dot\beta(n)} DJ_{t,s_1,s_2}{}^{\alpha(m)}{}_{,}{}^{\dot\beta(n)} \right). \tag{27}$$

Hence, for parameters obeying (26), the conservation condition amounts to equations:

$$DJ_{t,s_1,s_2}{}^{\alpha(m)}{}_{,}{}^{\dot\beta(n)} \simeq 0, \qquad m+n = 2(t-1). \tag{28}$$

For the currents defined via (23), the conservation condition (28) holds as a consequence of the gauge invariance of the action proven in [12].

Conserved currents generate conserved charges. By the Noether theorem, the latter are generators of global symmetries. Hence, one should expect as many conserved charges as global symmetry parameters. For a spin t, there are as many global symmetry parameters as the gauge parameters $\epsilon_{\alpha(m),\dot\beta(n)}$ with $m+n = 2(t-1)$.

In what follows, we will use notations:

$$D^{top}\omega^{\alpha(m)}{}_{,}{}^{\dot\beta(n)} := n\theta(m-n)\tilde{h}_{\gamma,}{}^{\dot\beta}\,\omega^{\gamma\alpha(m)}{}_{,}{}^{\dot\beta(n-1)} + m\theta(n-m)\tilde{h}^{\alpha}{}_{,\dot\delta}\,\omega^{\alpha(m-1)}{}_{,}{}^{\dot\beta(n)\dot\delta}, \tag{29}$$

$$D^{sub}\omega^{\alpha(m)}{}_{,}{}^{\dot\beta(n)} := n\theta(n-m-2)\tilde h_{\gamma,}{}^{\dot\beta}\,\omega^{\gamma\alpha(m)}{}_{,}{}^{\dot\beta(n-1)} + m\theta(m-n-2)\tilde h^{\alpha}{}_{,\dot\delta}\,\omega^{\alpha(m-1)}{}_{,}{}^{\dot\beta(n)\dot\delta}, \tag{30}$$

$$D^{cur}\omega^{\alpha(m)}{}_{,}{}^{\dot\beta(n)} := R_1{}^{\alpha(m)}{}_{,}{}^{\dot\beta(n)}. \tag{31}$$

As a consequence of (11):

$$D^{cur} = \tilde D + D^{top} + \lambda^2 D^{sub}. \tag{32}$$

Since the λ-dependent term vanishes in the Minkowski case, it is convenient to introduce the "flat" part of the covariant derivative:

$$D^{fl} := \tilde D + D^{top}. \tag{33}$$

It is also convenient to denote:

$$D^{h} := D^{top} + \lambda^2 D^{sub}. \tag{34}$$

Free field equations (16) imply that:

$$D^{cur}\omega^{\alpha(m)}{}_{,}{}^{\dot\beta(n)} \simeq \delta_{n,0}C^{\alpha(m)\gamma\delta}\,H_{\gamma\delta} + \delta_{m,0}\bar C^{\dot\beta(n)\dot\gamma\dot\delta}\,\bar H_{\dot\gamma\dot\delta}, \tag{35}$$

where $\simeq$ implies on-shell equality.

If the three-form J_{t,s_1,s_2} verifies (28) on-shell, the charge:

$$Q_\xi = \int_{M^3} J_{t,s_1,s_2} \tag{36}$$

is conserved by virtue of (26). As a result, there are as many conserved charges Q_ξ as independent global symmetry parameters ξ. Nontrivial charges are represented by the current cohomology, i.e., closed currents $J_{t,s_1,s_2}(x)$ modulo exact ones $J_{t,s_1,s_2} \simeq d\Psi_{t,s_1,s_2}$. Since the currents should be closed on-shell, i.e., by virtue of the free field Equations (16)–(18), analysis is greatly simplified by the fact that all linearized HS curvatures $R_1{}^{\alpha(m)}{}_{,}{}^{\dot\beta(n)}$ with $m>0$, $n>0$ are zero on-shell.

Conservation of currents does not imply that they are invariant under the gauge transformations (19). However, as shown below, the gauge variation of J_{t,s_1,s_2} is exact:

$$\delta J_{t,s_1,s_2}(x) \simeq dH_{t,s_1,s_2}(x) \tag{37}$$

so that the charge Q_ξ turns out to be gauge invariant.

Thus, the problem is:

- to find current three-forms (25) from the variation of action,
- to check that these forms obey the conservation condition (28),
- to check that in the flat limit $\lambda \to 0$, these forms give currents of [11],
- to check that the HS charges are gauge invariant.

4. Variation of the Action

Variation of the nonlinear curvature $R^{i;\alpha(m)}{}_{,}{}^{\dot\beta(n)}$ is:

$$\delta R^{i;\alpha(m)}{}_{,}{}^{\dot\beta(n)} = \delta R_1{}^{i;\alpha(m)}{}_{,}{}^{\dot\beta(n)} + \delta R_2{}^{i;\alpha(m)}{}_{,}{}^{\dot\beta(n)}, \tag{38}$$

where:

$$\begin{aligned}\delta R_1{}^{i;\alpha(m)}{}_{,}{}^{\dot\beta(n)} &= \tilde D\delta\omega^{i;\alpha(m)}{}_{,}{}^{\dot\beta(n)}\\ &+ n(\theta(m-n) + \lambda^2\theta(n-m-2))\,\tilde h_{\gamma,}{}^{\dot\beta}\,\delta\omega^{i;\gamma\alpha(m)}{}_{,}{}^{\dot\beta(n-1)}\\ &+ m(\theta(n-m) + \lambda^2\theta(m-n-2))\,\tilde h^{\alpha}{}_{,\dot\delta}\,\delta\omega^{i;\alpha(m-1)}{}_{,}{}^{\dot\beta(n)\dot\delta}\end{aligned} \tag{39}$$

and:

$$\delta R_2{}^{i;\alpha(m)}{}_{,}{}^{\dot\beta(n)} = \sum_{p,q,k,l,u,v} \lambda^{1+d_0-d_1-d_2} \frac{m!n!}{p!q!k!l!u!v!}(1+(-1)^{k+l}\eta)\, c^i{}_{jk}\, \delta_{p+q,m}\delta_{u+v,n} \times \delta\omega^{j;\alpha(p)}{}_{\gamma(k),\dot\delta(l)}{}^{\dot\beta(u)}\, \omega^{k;\alpha(q)\gamma(k)}{}_{,}{}^{\dot\delta(l)\dot\beta(v)} \quad (40)$$

with η defined in (24).

Variation of the action (22) is:

$$\delta S = \int \sum_{m,n} \varepsilon(m-n) \frac{\lambda^{-|m-n|}}{m!n!} (R_1{}^{i;\alpha(m)}{}_{,}{}^{\dot\beta(n)} \delta R_{1\; i;\alpha(m),\dot\beta(n)} + R_2{}^{i;\alpha(m)}{}_{,}{}^{\dot\beta(n)} \delta R_{1\; i;\alpha(m),\dot\beta(n)} + R_1{}^{i;\alpha(m)}{}_{,}{}^{\dot\beta(n)} \delta R_{2\; i;\alpha(m),\dot\beta(n)} + R_2{}^{i;\alpha(m)}{}_{,}{}^{\dot\beta(n)} \delta R_{2\; i;\alpha(m),\dot\beta(n)}). \quad (41)$$

The first term is the variation of the action S_2 (20), which vanishes on equations of motion (16)–(18). The last term is cubic in connections $\omega^{i;\alpha(m)}{}_{,}{}^{\dot\beta(n)}$, hence not contributing to bilinear currents. The second and third terms give rise to the currents. Using (11), (17), (18), (32), (39) and (40) and integrating by parts, we obtain:

$$\begin{aligned} \delta S \simeq \int \sum_{m,n} \varepsilon(m-n) \frac{\lambda^{-|m-n|}}{m!n!} & [-\tilde{D}R_2{}^{i;\alpha(m),\dot\beta(n)}\, \delta\omega_{i;\alpha(m),\dot\beta(n)} \\ & + n(\theta(m-n)+\lambda^2\theta(n-m-2))\, R_2{}^{i;\alpha(m),\dot\theta\dot\beta(n-1)}\, \tilde{h}^{\gamma}{}_{,\dot\theta}\, \delta\omega_{i;\gamma\alpha(m),\dot\beta(n-1)} \\ & + m(\theta(n-m)+\lambda^2\theta(m-n-2))\, R_2{}^{i;\alpha(m-1)\gamma,\dot\beta(n)}\, \tilde{h}_{\gamma,}{}^{\dot\delta}\, \delta\omega_{i;\alpha(m-1),\dot\beta(n)\dot\delta}] \\ & + \int \sum_{r>0} \frac{\lambda^{-r}}{r!} (C^{i;\alpha(r)\gamma\delta}\, H_{\gamma\delta}\, \delta R_{2\; i;\alpha(r)} - \bar{C}^{i;\dot\beta(r)\dot\gamma\dot\delta}\, \bar{H}_{\dot\gamma\dot\delta}\, \delta R_{2\; i;\dot\beta(r)}). \end{aligned} \quad (42)$$

Omitting the color index $i = 1$, this leads to the currents at $t > 1$ via:

$$J_{t,s_1,s_2} = \sum_{m,n} \delta(m+n-2(t-1))\xi_{\alpha(m),\dot\beta(n)} \frac{\delta S}{\delta\omega_{\alpha(m),\dot\beta(n)}}. \quad (43)$$

5. Examples

5.1. Spin-Two Current

To illustrate the structure of the current three-form and analyze the flat limit $\lambda \to 0$, consider a current with $t = 2$, $s_1 = s_2 = s > 1$:

$$J_{2,s} = \frac{\lambda^{-2}}{2}\xi_{\alpha\alpha} J_{2,s}{}^{\alpha\alpha} + \xi_{\alpha,\dot\beta} J_{2,s}{}^{\alpha}{}_{,}{}^{\dot\beta} + \frac{\lambda^{-2}}{2}\xi_{\dot\beta\dot\beta} J_{2,s}{}^{\dot\beta\dot\beta}, \quad (44)$$

where $J_{2,s} := J_{2,s,s}$. Using (21), (31), (35) and (40), we obtain:

$$\begin{aligned} J_{2,s}{}^{\alpha\alpha} = \sum_{m,n} \frac{4\lambda^{2-|m-n|}}{(m-1)!n!} c_{ij} [& n(\theta(m-n)+\lambda^2\theta(n-m-2))\omega^{i;\alpha\gamma(m-1)\varphi}{}_{,}{}^{\dot\delta(n-1)}\omega^{j;\alpha}{}_{\gamma(m-1),\dot\delta(n-1)\dot\theta}\tilde{h}_{\varphi,}{}^{\dot\theta} \\ & + (m-1)(\theta(n-m)+\lambda^2\theta(m-n-2))\omega^{i;\alpha\gamma(m-2)}{}_{,}{}^{\dot\delta(n)\dot\theta}\omega^{j;\alpha}{}_{\varphi\gamma(m-2),\dot\delta(n)}\tilde{h}^{\varphi}{}_{,\dot\theta} \\ & + (\theta(n-m)+\lambda^2\theta(m-n-2))\omega^{i;\gamma(m-1)}{}_{,}{}^{\dot\delta(n)\dot\theta}\omega^{j;\alpha}{}_{\gamma(m-1),\dot\delta(n)}\tilde{h}^{\alpha}{}_{,\dot\theta}], \end{aligned} \quad (45)$$

$$J_{2,s}{}^{\alpha}{}_{,}{}^{\dot\beta} = \sum_{m,n} 2\lambda^{2-|m-n|}\Big[\frac{1}{(m-1)!n!} c_{ij}\omega^{i;\alpha\gamma(m-1)}{}_{,}{}^{\dot\delta(n)}\omega^{j;\varphi}{}_{\gamma(m-1),\dot\delta(n)}\tilde h_{\varphi,}{}^{\dot\beta} - \frac{1}{m!(n-1)!} c_{ij}\omega^{i;\gamma(m)}{}_{,}{}^{\dot\delta(n-1)\dot\theta}\omega^{j;}{}_{\gamma(m),\dot\delta(n-1)}{}^{\dot\beta}\tilde h^{\alpha,}{}_{\dot\theta}\Big] + \frac{2\lambda^{4-2s}}{(2s-3)!}\big[c_{ij}C^{i;\alpha\gamma(2s-3)\varphi\rho}\, H_{\varphi\rho}\, \omega_{\gamma(2s-3)}{}^{\dot\beta} - c_{ij}\bar C^{i;\dot\delta(2s-3)\dot\beta\dot\psi\dot\theta}\, \bar H_{\dot\psi\dot\theta}\, \omega^{\alpha}{}_{\dot\delta(2s-3)}\big], \quad (46)$$

$$J_{2,s}{}^{\dot\beta\dot\beta} = \sum_{m,n} \frac{4\lambda^{2-|m-n|}}{m!(n-1)!} c_{ij}\big[m(\theta(n-m)+\lambda^2\theta(m-n-2))\omega^{i;\gamma(m-1)}{}_{,}{}^{\dot\delta(n-1)\dot\theta\dot\beta}\omega^{j;}{}_{\varphi\gamma(m-1),\dot\delta(n-1)}{}^{\dot\beta}\tilde h^{\varphi}{}_{,\dot\theta} + (n-1)(\theta(m-n)+\lambda^2\theta(n-m-2))\omega^{i;\varphi\gamma(m)}{}_{,}{}^{\dot\delta(n-2)\dot\beta}\omega^{j;}{}_{\gamma(m),\dot\delta(n-2)\dot\theta}{}^{\dot\beta}\tilde h_{\varphi,}{}^{\dot\theta} + (\theta(m-n)+\lambda^2\theta(n-m-2))\omega^{i;\varphi\gamma(m)}{}_{,}{}^{\dot\delta(n-1)}\omega^{j;}{}_{\gamma(m),\dot\delta(n-1)}{}^{\dot\beta}\tilde h_{\varphi,}{}^{\dot\beta}\big]. \quad (47)$$

Recall that $m+n=2(s-1)$, $m,n \geq 0$.

The terms in (45), (46) and (47) that contain inverse powers of λ contain higher derivatives. To obtain a proper $\lambda \to 0$ limit, such terms should be compensated by an exact form $d\Psi_{2,s}$ with:

$$\Psi_{2,s} = \sum_{m=0}^{s-3} \frac{2\lambda^{4-2(s-m)}}{m!(2s-3-m!)}\big[\xi_{\alpha\alpha} c_{ij}\omega^{i;\alpha\gamma(m)}{}_{,}{}^{\dot\delta(2s-3-m)}\omega^{i;\alpha}{}_{\gamma(m),\dot\delta(2s-3-m)} + \xi_{\alpha,\dot\beta}\big(c_{ij}\omega^{i;\alpha\gamma(2s-3-m)}{}_{,}{}^{\dot\delta(m)}\omega^{j;}{}_{\gamma(2s-3-m),\dot\delta(m)}{}^{\dot\beta} - c_{ij}\omega^{i;\alpha\gamma(m)}{}_{,}{}^{\dot\delta(2s-3-m)}\omega^{j;}{}_{\gamma(m),\dot\delta(2s-3-m)}{}^{\dot\beta}\big) - \xi_{\dot\beta\dot\beta} c_{ij}\omega^{i;\alpha\gamma(2s-3-m)}{}_{,}{}^{\dot\delta(m)\dot\beta}\omega^{i;}{}_{\gamma(2s-3-m),\dot\delta(m)}{}^{\dot\beta}\big]. \quad (48)$$

At $s=2$, it is not necessary to add this exact form since the current is regular in the flat limit.

The fact that complete antisymmetrization over any three two-component dotted or undotted indices gives zero yields the relation:

$$c_{ij}\omega^{i;\alpha\gamma(m-1)}{}_{,}{}^{\dot\delta(n)}\omega^{j;\varphi}{}_{\gamma(m-1),\dot\delta(n)}\,\tilde h_{\varphi,}{}^{\dot\beta} = -c_{ij}\omega^{i;\alpha\gamma(m-1)}{}_{,}{}^{\dot\delta(n-1)\dot\beta}\omega^{j;\varphi}{}_{\gamma(m-1),\dot\delta(n-1)}{}^{\dot\theta}\,\tilde h_{\varphi,\dot\theta} + c_{ij}\omega^{i;\alpha\gamma(m-1)}{}_{,}{}^{\dot\delta(n-1)\dot\theta}\omega^{j;\varphi}{}_{\gamma(m-1),\dot\delta(n-1)}{}^{\dot\beta}\,\tilde h_{\varphi,\dot\theta} \quad (49)$$

to be used in the sequel.

Straightforward calculation gives:

$$\hat J_{2,s} = J_{2,s} + d\Psi_{2,s} = \frac{\lambda^{-2}}{2}\xi_{\alpha\alpha}\hat J_{2,s}{}^{\alpha\alpha} + \xi_{\alpha,\dot\beta}\hat J_{2,s}{}^{\alpha}{}_{,}{}^{\dot\beta} + \frac{\lambda^{-2}}{2}\xi_{\dot\beta\dot\beta}\hat J_{2,s}{}^{\dot\beta\dot\beta}, \quad (50)$$

where

$$\hat J_{2,s}{}^{\alpha\alpha} = 2\lambda^2 c_{ij}\Big[\omega^{i;\alpha\varphi\gamma(s-2)}{}_{,}{}^{\dot\delta(s-2)}\omega^{j;\alpha}{}_{\gamma(s-2),\dot\delta(s-2)\dot\theta}\tilde h_{\varphi,}{}^{\dot\theta} + \frac{s-2}{s-1}\omega^{i;\alpha\gamma(s-3)}{}_{,}{}^{\dot\delta(s-1)\dot\theta}\omega^{j;\alpha}{}_{\varphi\gamma(s-3),\dot\delta(s-1)}\tilde h^{\varphi}{}_{,\dot\theta} + \frac{1}{s-1}\omega^{i;\alpha\gamma(s-2)}{}_{,}{}^{\dot\delta(s-1)}\omega^{j;}{}_{\gamma(s-2),\dot\delta(s-1)}{}^{\dot\theta}\tilde h^{\alpha}{}_{,\dot\theta}\Big], \quad (51)$$

$$\hat{J}_{2,s}{}^{\alpha}{}_{,}{}^{\dot\beta} = \frac{1}{s-1} c_{ij} [\omega^{i;\gamma(s-2)}{}_{,}{}^{\dot\delta(s-1)\dot\theta} \omega^{j;}{}_{\gamma(s-2),\dot\delta(s-1)}{}^{\dot\beta} \tilde{h}^{\alpha}{}_{,\dot\theta}$$
$$+ (s-2)\omega^{i;\alpha\gamma(s-3)}{}_{,}{}^{\dot\delta(s-1)\dot\theta} \omega^{j;}{}_{\varphi\gamma(s-3),\dot\delta(s-1)}{}^{\dot\beta} \tilde{h}^{\varphi}{}_{,\dot\theta} + (s-2)\omega^{i;\alpha\varphi\gamma(s-3)}{}_{,}{}^{\dot\delta(s-1)} \omega^{j;}{}_{\gamma(s-3),\dot\delta(s-1)}{}^{\dot\theta\dot\beta} \tilde{h}_{\varphi,\dot\theta}$$
$$+ (s-2)\omega^{i;\alpha\gamma(s-1)}{}_{,}{}^{\dot\delta(s-3)\dot\theta} \omega^{j;\varphi}{}_{\gamma(s-1),\dot\delta(s-3)}{}^{\dot\beta} \tilde{h}_{\varphi,\dot\theta} + (s-2)\omega^{i;\alpha\varphi\gamma(s-1)}{}_{,}{}^{\dot\delta(s-3)} \omega^{j;}{}_{\gamma(s-1),\dot\delta(s-3)\dot\theta}{}^{\dot\beta} \tilde{h}_{\varphi,}{}^{\dot\theta}$$
$$+ \omega^{i;\alpha\gamma(s-1)}{}_{,}{}^{\dot\delta(s-2)} \omega^{j;\varphi}{}_{\gamma(s-1),\dot\delta(s-2)} \tilde{h}_{\varphi,}{}^{\dot\beta}], \quad (52)$$

$$\hat{J}_{2,s}{}^{\dot\beta\dot\beta} = 2\lambda^2 c_{ij} [\omega^{i;\varphi\gamma(s-2)}{}_{,}{}^{\dot\delta(s-2)\dot\beta} \omega^{j;}{}_{\gamma(s-2),\dot\delta(s-2)\dot\theta}{}^{\dot\beta} \tilde{h}_{\varphi,}{}^{\dot\theta}$$
$$+ \frac{s-2}{s-1} \omega^{i;\varphi\gamma(s-1)}{}_{,}{}^{\dot\delta(s-3)\dot\beta} \omega^{j;}{}_{\gamma(s-1),\dot\delta(s-3)\dot\theta}{}^{\dot\beta} \tilde{h}_{\varphi}{}^{\dot\theta}$$
$$+ \frac{1}{s-1} \omega^{i;\varphi\gamma(s-1)}{}_{,}{}^{\dot\delta(s-2)} \omega^{j;}{}_{\gamma(s-1),\dot\delta(s-2)}{}^{\dot\beta} \tilde{h}_{\varphi,}{}^{\dot\beta}]. \quad (53)$$

Note that $\hat{J}_{2,s}$ does not contain λ explicitly. One can check, that (51), (52) and (53) obey (28). As a result, the form $\hat{J}_{2,s}$ (50) is closed by virtue of (26).

Since the AdS_4 current $\hat{J}_{2,s}$ (50) does not depend explicitly on λm it preserves its form in the flat limit $\lambda \to 0$. From (51)–(53), one can see that:

$$\hat{J}_{2,s} = J^{M}_{2,s} + D^{fl} \chi_{2,s}, \quad (54)$$

where $J^{M}_{2,s}$ at $\lambda = 0$ reproduces the spin-two current in Minkowski space and:

$$\chi_{2,s} = \frac{c_{ij}}{s-1} \Big(\xi_{\alpha\dot\beta} \big(\omega^{i;\alpha\gamma(s-2)}{}_{,}{}^{\dot\delta(s-1)} \omega^{j;}{}_{\gamma(s-2),\dot\delta(s-1)}{}^{\dot\beta} - \omega^{i;\alpha\gamma(s-1)}{}_{,}{}^{\dot\delta(s-2)} \omega^{j;}{}_{\gamma(s-1),\dot\delta(s-2)}{}^{\dot\beta} \big)$$
$$+ \lambda^2 \big(\xi_{\alpha\alpha} \omega^{i;\alpha\gamma(s-2)}{}_{,}{}^{\dot\delta(s-1)} \omega^{j;\alpha}{}_{\gamma(s-2),\dot\delta(s-1)} + \xi_{\dot\beta\dot\beta} \omega^{i;\gamma(s-1)}{}_{,}{}^{\dot\delta(s-2)\dot\beta} \omega^{j;}{}_{\gamma(s-1),\dot\delta(s-2)}{}^{\dot\beta} \big) \Big).$$

This proves that the flat limit of the current (50) reproduces the results of [11].

We observe that the current is Hermitian. It is nonzero if c_{ij} is symmetric.

5.2. Spin-One Current

Since the action (22) does not contain a kinetic term for spin-one field ω^i carrying no spinor indices, following [12], it should be added separately in a standard way:

$$S_{EM} = \int R_i{}^* R^i, \quad (55)$$

where * is the Hodge star operator and, in agreement with (11) and (21),

$$R^i = d\omega^i + \sum_{k,l \geq 0} \frac{\lambda^{1-|m-n|}}{k!l!} c^i{}_{jk}\, \omega^{j;}{}_{\gamma(k),\dot\delta(l)}\, \omega^{k;\gamma(k)}{}_{,}{}^{\dot\delta(l)}. \quad (56)$$

The full action is $S_{full} = S + S_{EM}$ with S (22). The spin-one part of the variation $\delta S_{t=1}$ of this action is:

$$\delta S_{t=1} = \int \sum_{r>0} \frac{\lambda^{-r}}{r!} \left(C^{i;\alpha(r)\gamma\delta} H_{\gamma\delta}\, \delta R_{2\; i;\alpha(r)} - \bar{C}^{i;\dot\beta(r)\dot\gamma\dot\delta} \bar{H}_{\dot\gamma\dot\delta}\, \delta R_{2\; i;\dot\beta(r)} \right) + \delta S_{EM}, \quad (57)$$

where:

$$\delta S_{EM} = \int [R_{1i}{}^* \delta R_2{}^i + R_{2i}{}^* \delta R_1{}^i]. \quad (58)$$

In the spin-one case equations, (16)–(18) amount to:

$$R_1{}^i = C^{i;\gamma\delta} H_{\gamma\delta} + \bar{C}^{i;\dot{\gamma}\dot{\delta}} \bar{H}_{\dot{\gamma}\dot{\delta}}, \tag{59}$$

where $C^{i;\gamma\delta}$ and $\bar{C}^{i;\dot{\gamma}\dot{\delta}}$ parametrize self-dual and anti-self-dual components of the spin-one field tensor. Using properties of the Pauli matrices, one can also see that:

$$R_1{}^{i*} = \mathrm{i}(C^{i;\gamma\delta} H_{\gamma\delta} - \bar{C}^{i;\dot{\gamma}\dot{\delta}} \bar{H}_{\dot{\gamma}\dot{\delta}}). \tag{60}$$

The sum of spin-one ($t = 1$) currents of fields of arbitrary spins $s_1 = s_2 \geq 1$ can be expressed as (the color index i is omitted):

$$\xi \frac{\delta S_{t=1}}{\delta \omega} = J_{1,1} + \sum_{s>1} J_{1,s}, \tag{61}$$

where ξ is a global symmetry parameter zero-form (26) with no spinor indices and:

$$J_{1,1} = 2\xi\lambda \mathrm{i} c_{ij}(C^{i;\gamma\delta} H_{\gamma\delta} - \bar{C}^{i;\dot{\gamma}\dot{\delta}} \bar{H}_{\dot{\gamma}\dot{\delta}})\omega^j - d(\xi R_2{}^{i*}), \tag{62}$$

$$J_{1,s} = 2\lambda^{3-2s}\xi c_{ij}(C^{i;\alpha(2s-2)\varphi\rho}\, H_{\varphi\rho}\, \omega^{j;}{}_{\alpha(2s-2)} - \bar{C}^{i;\dot{\beta}(2s-2)\dot{\gamma}\dot{\delta}}\, \bar{H}_{\dot{\gamma}\dot{\delta}}\, \omega^{j;}{}_{\dot{\beta}(2s-2)}\,). \tag{63}$$

In the case of $t = 1, s_1 = s_2 = s = 1$, one can transform $J_{1,1}$ into:

$$\hat{J}_{1,1} = \frac{1}{\lambda}(J_{1,1} + d(\xi R_2{}^{i*})) = 2\xi \mathrm{i} c_{ij}(C^{i;\gamma\delta} H_{\gamma\delta} - \bar{C}^{i;\dot{\gamma}\dot{\delta}} \bar{H}_{\dot{\gamma}\dot{\delta}})\omega^j. \tag{64}$$

It is not hard to see that the C-dependent terms are not exact provided that c_{ij} is antisymmetric. The current (64) coincides with Minkowski current $J^M_{1,1}$ from [11] modulo an overall factor of two.

In the case of $t = 1, s_1 = s_2 = s > 1$, the current results from the C-dependent terms of (42) by virtue of (40):

$$J_{1,s} = 2\lambda^{3-2s}\xi c_{ij}(C^{i;\alpha(2s-2)\varphi\rho}\, H_{\varphi\rho}\, \omega^{j;}{}_{\alpha(2s-2)} - \bar{C}^{i;\dot{\beta}(2s-2)\dot{\gamma}\dot{\delta}}\, \bar{H}_{\dot{\gamma}\dot{\delta}}\, \omega^{j;}{}_{\dot{\beta}(2s-2)}\,). \tag{65}$$

Note, that there are no currents with $t = 1, s_1 \neq s_2$. Furthermore, note that Equation (64) is a particular case of (65) at $s = 1$. The current three-form $J_{1,s}$ (65) is nontrivial if c_{ij} is antisymmetric.

For $s > 1$, $J_{1,s}$ (65) can be rewritten in the bilinear form in connections by adding an exact form:

$$\begin{aligned}\hat{J}_{1,s} = -\frac{1}{\lambda(-2)^{s-1}s(s-1)!}(J_{1,s} + d\Psi_{1,s}) = \\ \xi c_{ij}[\omega^{i;\varphi\gamma(s-2)}{}_{,}{}^{\dot{\delta}(s-1)}\omega^{j;}{}_{\gamma(s-2),\dot{\delta}(s-1)\dot{\theta}} + \omega^{i;\varphi\gamma(s-1)}{}_{,}{}^{\dot{\delta}(s-2)}\omega^{j;}{}_{\gamma(s-1),\dot{\delta}(s-2)\dot{\theta}}]\tilde{h}_{\varphi,}{}^{\dot{\theta}},\end{aligned} \tag{66}$$

with:

$$\begin{aligned}\Psi_{1,s} = 2\xi\lambda^{3-2s}\sum_{m=0}^{s-2}(-1)^{m+1}2^m\lambda^{2m}\frac{(s-1)!}{(s-m-1)!}c_{ij}(\omega^{i;\alpha(2s-2-m)}{}_{,}{}^{\dot{\beta}(m)}\omega^{j;}{}_{\alpha(2s-2-m),\dot{\beta}(m)} \\ - \omega^{i;\alpha(m)}{}_{,}{}^{\dot{\beta}(2s-2-m)}\omega^{j;}{}_{\alpha(m),\dot{\beta}(2s-2-m)}).\end{aligned} \tag{67}$$

This three-form is λ-independent, on-shell-closed, Hermitian and reproduces the result of [11]. Hence, it is non-trivial.

6. General Spins

The AdS_4 conserved currents J_{t,s_1,s_2} with $1 < t \leq s_1 + s_2 - 1$ (for definiteness, we set $s_1 \geq s_2$) result from the variation of action (42):

$$\begin{aligned} J_{t,s_1,s_2} = \sum_{m,n} \varepsilon(m-n) \, \frac{\lambda^{-|m-n|}}{m!n!} \, [-\xi_{\alpha(m),\dot\beta(n)} \; D^h \; R_2{}^{\alpha(m),\dot\beta(n)}|_{s_1,s_2} \\ - n(\theta(m-n) + \lambda^2\theta(n-m-2)) \, \xi_{\alpha(m+1),\dot\beta(n-1)} \; R_2{}^{\alpha(m),\dot\theta\dot\beta(n-1)}|_{s_1,s_2} \, \tilde{h}^{\alpha}{}_{,\dot\theta} \\ + m(\theta(n-m) + \lambda^2\theta(m-n-2)) \, \xi_{\alpha(m-1),\dot\beta(n+1)} \; R_2{}^{\alpha(m-1)\gamma,\dot\beta(n)}|_{s_1,s_2} \, \tilde{h}_{\gamma,}{}^{\dot\beta}] \\ + \sum_{p,q,k,v} \frac{2\lambda^{1-|q+k-p-v|}}{p!q!k!v!} \delta_{2p+q+v,2(t-1)} \delta_{p+k,2(s_1-1)} \delta_{p+q+k+v,2(s_2-1)} c_{ij} \xi_{\alpha(p+q),\dot\beta(p+v)} \\ \times [C^{i;\alpha(p)\gamma(k)\varphi\rho} \; H_{\varphi\rho} \; \omega^{j;\alpha(q)}{}_{\gamma(k),}{}^{\dot\beta(p+v)} - \bar{C}^{i;\dot\beta(p)\dot\delta(k)\dot\varphi\dot\rho} \; \bar{H}_{\dot\varphi\dot\rho} \; \omega^{j;\alpha(p+q)}{}_{,\dot\delta(k)}{}^{\dot\beta(v)}], \end{aligned} \quad (68)$$

where $R_2{}^{\alpha(m),\dot\beta(n)}|_{s_1,s_2}$ is the restriction of (21) to terms containing connections with spins s_1 and s_2. These currents contain $s_1 + s_2 - 2$ derivatives of the frame-like fields.

To check the non-exactness of the three-form (68), it suffices to add an exact form:

$$d\Psi_{t,s_1,s_2} = d(\sum_{m,n} \xi_{\alpha(m)\dot\beta(n)} \Psi_{t,s_1,s_2}{}^{\alpha(m),\dot\beta(n)}), \quad n + m = 2(t-1),$$

where:

$$\begin{aligned} \Psi_{t,s_1,s_2}{}^{\alpha(m),\dot\beta(n)} = \\ \sum_{p,q,k,l,u,v} \frac{2\lambda^{1-\frac{|p+k-l-u|-|q+k-l-v|}{2}}}{p!q!k!l!u!v!} \delta_{p+q,m} \delta_{u+v,n} \delta_{p+k+l+u,2(s_1-1)} \delta_{q+k+l+v,2(s_2-1)} \\ \times \theta(l+u-p-k-1) c_{ij} [\theta(m-n) \omega^{i;\alpha(p)\gamma(k),\dot\delta(l)\dot\beta(u)} \; \omega^{j;\alpha(q)}{}_{\gamma(k),\dot\delta(l)}{}^{\dot\beta(v)} \\ - \theta(n-m) \omega^{i;\alpha(u)\gamma(l),\dot\delta(k)\dot\beta(p)} \; \omega^{j;\alpha(v)}{}_{\gamma(l),\dot\delta(k)}{}^{\dot\beta(q)}], \end{aligned} \quad (69)$$

One can see that $\Psi_{t,s_1,s_2}{}^{\alpha(m),\dot\beta(n)}$ is adjusted to cancel the C-dependent terms.

The resulting current three-form:

$$\hat{J}_{t,s_1,s_2} := J_{t,s_1,s_2} + d\Psi_{t,s_1,s_2} \quad (70)$$

is

$$\begin{aligned} \hat{J}_{t,s_1,s_2}{}^{\alpha(m),\dot\beta(n)} = \\ \sum_{p,q,k,l,u,v} \frac{2\lambda^{1+2|m-n|+|p+k-l-u|-|q+k-l-v|}}{p!q!k!l!u!v!} \delta_{p+q,m} \delta_{u+v,n} \delta_{p+k+l+u,2(s_1-1)} \delta_{q+k+l+v,2(s_2-1)} c_{ij} \\ \times [\theta(p+k-l-u-1) D^h (\theta(m-n) \omega^{i;\alpha(p)\gamma(k),\dot\delta(l)\dot\beta(u)} \; \omega^{j;\alpha(q)}{}_{\gamma(k),\dot\delta(l)}{}^{\dot\beta(v)} \\ - \theta(n-m) \omega^{i;\alpha(u)\gamma(l),\dot\delta(k)\dot\beta(p)} \; \omega^{j;\alpha(v)}{}_{\gamma(l),\dot\delta(k)}{}^{\dot\beta(q)}) \\ - n(\theta(m-n) + \lambda^2\theta(n-m-2))((u+1)\omega^{i;\alpha(p-1)\gamma(k),\dot\delta(l)\dot\theta\dot\beta(u)} \; \omega^{j;\alpha(v)}{}_{\gamma(k),\dot\delta(l)}{}^{\dot\beta(v)} \tilde{h}^{\alpha}{}_{,\dot\theta} \\ + (v+1)\omega^{i;\alpha(p)\gamma(k),\dot\delta(l)\dot\beta(u)} \; \omega^{j;\alpha(q-1)}{}_{\gamma(k),\dot\delta(l)}{}^{\dot\theta\dot\beta(v)} \tilde{h}^{\alpha}{}_{,\dot\theta}) \\ + m(\theta(n-m) + \lambda^2\theta(m-n-2)) \, ((p+1)\omega^{i;\alpha(p)\varphi\gamma(k),\dot\delta(l)\dot\beta(u)} \; \omega^{j;\alpha(q)}{}_{\gamma(k),\dot\delta(l)}{}^{\dot\beta(v)} \tilde{h}_{\varphi,}{}^{\dot\beta} \\ + (q+1)\omega^{i;\alpha(p)\gamma(l),\dot\delta(k)\dot\beta(u-1)} \; \omega^{j;\alpha(q)\varphi}{}_{\gamma(l),\dot\delta(k)}{}^{\dot\beta(v-1)} \tilde{h}_{\varphi,}{}^{\dot\beta})]. \end{aligned} \quad (71)$$

This current contains $t - |s_1 - s_2|$ derivatives, which is the minimal possible number. The non-exactness of the current three-form $\hat{J}_{t,s_1,s_2}$ can be checked in the flat limit $\lambda \to 0$ just as in [11].

In the case of $s_1 = s_2 = s$:

$$\hat{J}_{t,s} = \sum_{n,m} \frac{\lambda^{-|m-n|}}{m!n!} \xi_{\alpha(m),\dot{\beta}(n)} \hat{J}_{t,s}{}^{\alpha(m)}{}_{,}{}^{\dot{\beta}(n)}, \quad n+m = 2(t-1),$$

where:

$$\begin{aligned}
\hat{J}_{t,s}{}^{\alpha(m)}{}_{,}{}^{\dot{\beta}(n)} = \\
\lambda^{|m-n|} m!n!(&\theta(n-m-4)\,\hat{g}(n)\, c_{ij}\omega^{i;\alpha(m)\varphi\gamma(s-2)}{}_{,}{}^{\dot{\delta}(s-t)\dot{\beta}(n-t+1)}\omega^{j;}{}_{\gamma(s-2),\dot{\delta}(s-t)\dot{\theta}}{}^{\dot{\beta}(t-1)}\tilde{h}_{\varphi,}{}^{\dot{\theta}} \\
&+\delta_{n,t}\, c_{ij}[2(t-1)\omega^{i;\alpha(m)\varphi\gamma(s-2)}{}_{,}{}^{\dot{\delta}(s-t)\dot{\beta}}\omega^{j;}{}_{\gamma(s-2),\dot{\delta}(s-t)\dot{\theta}}{}^{\dot{\beta}(n-1)}\tilde{h}_{\varphi,}{}^{\dot{\theta}} \\
&+\sum_{p=1}^{t-2}\hat{f}(p)\omega^{i;\alpha(m)\varphi\gamma(s-p-1)}{}_{,}{}^{\dot{\delta}(s-t+p)}\omega^{j;}{}_{\gamma(s-p-1),\dot{\delta}(s-t+p)}{}^{\dot{\beta}(n-1)}\tilde{h}_{\varphi,}{}^{\dot{\beta}}] \\
&+\delta_{m,t-1}\delta_{n,t-1}c_{ij}[\omega^{i;\alpha(t-1)\varphi\gamma(s-2)}{}_{,}{}^{\dot{\delta}(s-t)}\omega^{j;}{}_{\gamma(s-2),\dot{\delta}(s-t)\dot{\theta}}{}^{\dot{\beta}(t-1)}\tilde{h}_{\varphi,}{}^{\dot{\theta}} \\
&\quad+\omega^{i;\alpha(t-1)\varphi\gamma(s-t)}{}_{,}{}^{\dot{\delta}(s-2)}\omega^{j;}{}_{\gamma(s-t),\dot{\delta}(s-2)\dot{\theta}}{}^{\dot{\beta}(t-1)}\tilde{h}_{\varphi,}{}^{\dot{\theta}}] \\
&+\theta(m-n-4)\,\hat{g}(m)\, c_{ij}\omega^{i;\alpha(t-1)\varphi\gamma(s-t)}{}_{,}{}^{\dot{\delta}(s-2)}\omega^{j;\alpha(m-t+1)}{}_{\gamma(s-t),\dot{\delta}(s-2)\dot{\theta}}{}^{\dot{\beta}(n)}\tilde{h}_{\varphi,}{}^{\dot{\theta}} \\
&+\delta_{m,t}\, c_{ij}[2(t-1)\omega^{i;\alpha(m-1)\varphi\gamma(s-t)}{}_{,}{}^{\dot{\delta}(s-2)}\omega^{j;\alpha}{}_{\gamma(s-t),\dot{\delta}(s-2)\dot{\theta}}{}^{\dot{\beta}(n)}\tilde{h}_{\varphi,}{}^{\dot{\theta}} \\
&+\sum_{p=1}^{t-2}\hat{f}(p)\omega^{i;\alpha(m-1)\gamma(s-t+p)}{}_{,}{}^{\dot{\delta}(s-p-1)}\omega^{j;}{}_{\gamma(s-t+p),\dot{\delta}(s-p-1)}{}^{\dot{\theta}\dot{\beta}(n)}\tilde{h}^{\alpha}{}_{,\dot{\theta}}]), \qquad (72)
\end{aligned}$$

and:

$$\hat{g}(m) = \frac{2(t-1)!}{(2t-m-2)!(m-t+1)!}, \quad m \geq t+1, \qquad (73)$$

$$\hat{f}(1) = \frac{t-1}{s-t+1}, \quad \hat{f}(p) = (t-1)\frac{(s-t)!(s-p)!}{(s-3)!(s-t+p)!}, \quad p > 1. \qquad (74)$$

The second and last terms in (72) contribute to the special cases of $n = t$ and $m = t$, respectively.

One can check that the current (68) at $s_1 = s_2$ reproduces that of [11] up to a D^{fl}-exact form:

$$\chi_{t,s} = D^{fl}(\sum_{m,n} \xi_{\alpha(m)\dot{\beta}(n)}\chi_{t,s}{}^{\alpha(m),\dot{\beta}(n)}), \quad n+m = 2(t-1),$$

where:

$$\begin{aligned}
\chi_{t,s}{}^{\alpha(m)}{}_{,}{}^{\dot{\beta}(n)} = \\
&\theta(n-m-2)g(n)c_{ij}\sum_{p=1}^{m+1}\omega^{i;\alpha(m)\gamma(s-p),\dot{\delta}(s-t+p-1)\dot{\beta}(n-t+1)}\omega^{j;}{}_{\gamma(s-p),\dot{\delta}(s-t+p-1)}{}^{\dot{\beta}(t-1)} \\
&+\theta(m-n-2)g(m)c_{ij}\sum_{p=1}^{n+1}\omega^{i;\alpha(t-1)\gamma(s-t+p-1),\dot{\delta}(s-p)}\omega^{j;\alpha(m-t+1)}{}_{\gamma(s-t+p-1),\dot{\delta}(s-p)}{}^{\dot{\beta}(n)} \\
&+\delta_{m,t-1}\delta_{n,t-1}f c_{ij}\sum_{p=0}^{[\frac{t}{2}]}[\omega^{i;\alpha(t-p-1)\gamma(s-2),\dot{\delta}(s-t+1)\dot{\beta}(p)}\omega^{j;\alpha(p)}{}_{\gamma(s-2),\dot{\delta}(s-t+1)}{}^{\dot{\beta}(s-t-p)} \\
&\qquad+\omega^{i;\alpha(t-p-1)\gamma(s-1),\dot{\delta}(s-t)\dot{\beta}(p)}\omega^{j;\alpha(p)}{}_{\gamma(s-1),\dot{\delta}(s-t)}{}^{\dot{\beta}(s-t-p)}], \qquad (75)
\end{aligned}$$

where:

$$f = \frac{1}{s-t+1}, \qquad g(m) = \frac{t-m}{s-p}. \tag{76}$$

The conserved currents are nontrivial if c_{ij} is antisymmetric for odd $t + s_1 + s_2$ and symmetric for even.

Thus, the Hermitian current three-form $\hat{J}_{t,s_1,s_2}$ is on-shell closed, but not exact. It generates the corresponding real conserved charge $Q = \int \hat{J}_{t,s_1,s_2}$ that contains as many symmetry parameters as local HS gauge symmetries.

7. Gauge Transformations

Although the current three-form (68) is not invariant under the gauge transformations (19), its gauge variation is exact. Schematically, the proof consists of the following steps. By virtue of (32), the gauge variation of any term from (68) $\delta(\xi\omega_1\omega_2 h)$ can be written as:

$$\delta(\xi\omega_1\omega_2 h) = \xi\omega_1(\tilde{D}\varepsilon_2 + D^h\varepsilon_2)h + \xi(\tilde{D}\varepsilon_1 + D^h\varepsilon_1)\omega_2 h.$$

This gives:

$$\begin{aligned}\delta(\xi\omega_1\omega_2 h) = &-\tilde{D}(\xi\omega_1\varepsilon_2 h - \xi\varepsilon_1\omega_2 h) + \xi(D^{cur}\omega_1)\varepsilon_2 h + \xi\varepsilon_1(D^{cur}\omega_2)h \\ &+ (D^h\xi)\omega_1\varepsilon_2 h + \xi(D^h\omega_1)\varepsilon_2 h - \xi\omega_1(D^h\varepsilon_2)h + (D^h\xi)\varepsilon_1\omega_2 h + \xi\varepsilon_1(D^h\omega_2)h - \xi(D^h\varepsilon_1)\omega_2 h.\end{aligned}$$

All terms containing D^{cur} are canceled by $\delta(\xi C H\omega h)$. All terms with D^h cancel each other.

This gives:

$$\delta J_{t,s_1,s_2} \simeq dH_{t,s_1,s_2},$$

where:

$$\begin{aligned}
H_{t,s_1,s_2} = &\\
&-\sum_{m,n}\varepsilon(m-n)\sum_{p,q,k,l,u,v}\delta_{p+q,m}\delta_{u+v,n}\delta_{p+k+l+u,2(s_1-1)}\delta_{q+k+l+v,2(s_2-1)}c_{ij}\xi_{\alpha(p+q),\dot\beta(u+v)}\\
&\quad\times\frac{\lambda^{1-\frac{|m-n|}{2}-\frac{|p+k-l-u|}{2}-\frac{|q+k-l-v|}{2}}}{p!q!k!l!u!v!}[D^h\varepsilon^{i;\alpha(p)\gamma(k),\dot\delta(l)\dot\beta(u)}\,\omega^{j;\alpha(q)}{}_{\gamma(k),\dot\delta(l)}{}^{\dot\beta(v)}\\
&\quad-n(\theta(m-n)+\lambda^2\theta(n-m-2))\,\xi_{\alpha(p+q+1),\dot\beta(u+v-1)}\\
&\times(\varepsilon^{i;\alpha(p)\gamma(k),\dot\delta(l)\dot\theta\dot\beta(u)}\,\omega^{j;\alpha(q)}{}_{\gamma(k),\dot\delta(l)}{}^{\dot\beta(v)}\,\tilde{h}^{\alpha}{}_{,\dot\theta}+\varepsilon^{i;\alpha(p)\gamma(k),\dot\delta(l)\dot\beta(u)}\,\omega^{j;\alpha(q)}{}_{\gamma(k),\dot\delta(l)}{}^{\dot\theta\dot\beta(v)}\,\tilde{h}^{\alpha}{}_{,\dot\theta})\\
&\quad-m(\theta(n-m)+\lambda^2\theta(m-n-2))\,\xi_{\alpha(p+q-1),\dot\beta(u+v+1)}\\
&\times(\varepsilon^{i;\alpha(p)\gamma(k),\dot\delta(l)\dot\beta(u)}\,\omega^{j;\alpha(q)}{}_{\gamma(k),\dot\delta(l)}{}^{\dot\beta(v)}\,\tilde{h}^{\alpha}{}_{,\dot\theta}+\varepsilon^{i;\alpha(p)\varphi\gamma(k),\dot\delta(l)\dot\beta(u)}\,\omega^{j;\alpha(q)\varphi}{}_{\gamma(k),\dot\delta(l)}{}^{\dot\beta(v)}\,\tilde{h}_{\varphi,}{}^{\dot\beta})]\\
&+2\sum_{p,q,k,v}\frac{\lambda^{2-s_1-\frac{|k-p|}{2}}}{p!q!k!v!}\delta_{2p+q+v,2(t-1)}\delta_{p+k,2(s_1-1)}\delta_{p+q+k+v,2(s_2-1)}c_{ij}\xi_{\alpha(p+q),\dot\beta(p+v)}\\
&\quad\times[C^{i;\alpha(p)\gamma(k)\varphi\rho}\,H_{\varphi\rho}\,\varepsilon^{j;\alpha(q)}{}_{\gamma(k),}{}^{\dot\beta(p+v)}-\bar{C}^{i;\dot\beta(p)\dot\delta(k)\dot\varphi\dot\rho}\,\bar{H}_{\dot\varphi\dot\rho}\,\varepsilon^{j;\alpha(p+q)}{}_{,\dot\delta(k)}{}^{\dot\beta(v)}].
\end{aligned}\tag{77}$$

Details of the derivation of this formula are given in the Appendix for the case of $t = 2$, $s_1 = s_2 = s > 1$.

Thus, $\delta J_{t,s_1,s_2}$ is exact on-shell. The same is true for the spin-one current (65). Hence, though the current J_t is not gauge invariant, the corresponding charge is:

$$\delta Q_\xi \simeq \int dH_{t,s_1,s_2} = 0.$$

8. Conclusion

In this paper, spin-t HS currents J_{t,s_1,s_2} in AdS_4, built from boson fields of arbitrary spins obeying $t \leq s_1 + s_2 - 1$ are found from the variation principle. Being represented as three-forms, J_{t,s_1,s_2} are closed, but not exact, hence leading to nontrivial HS charges. These charges are gauge invariant because $\delta J_{t,s_1,s_2}$ is shown to be exact.

In the $4d$ Minkowski case, in addition to natural parity-even currents, we found "mysterious" parity-odd currents [11]. In agreement with the conjecture of [11], we were not able to extend parity-odd currents to AdS_4. The $\lambda \to 0$ limit of $\hat{J}_{t,s}$ (72) reproduces the parity-even currents of [11].

Currents constructed from fields of half-integer spins can be found analogously. How to operate with half-integer fields is shown in [15]. It is important to mention that for the currents built from fields of half-integer spin, the computations are essentially different, because Equation (11) for half-integer spins contains λ instead of λ^2 [15].

Let us stress that the derivation of the currents via the action applied in this paper leads to currents containing the non-minimal number of derivatives according to [16] with the higher-derivative terms corresponding to certain improvements. This is however anticipated since consistent cubic HS interactions are known [12] to contain higher-derivative terms allowing one to preserve HS gauge symmetries associated with gauge fields of different spins.

Acknowledgments: This research was supported in part by the Russian Foundation for Basic Research Grant No. 14-02-01172.

Author Contributions: Both authors contributed equally to this work.

Conflicts of Interest: The authors declare no conflict of interest.

Appendix Example of the Gauge Variation of $J^{t,s}$

Consider the case of $t = 2$, $s_1 = s_2 = s > 1$ (44):

$$J_{2,s} = \frac{\lambda^{-2}}{2}\xi_{\alpha\alpha} J_{2,s}{}^{\alpha\alpha} + \xi_{\alpha,\dot\beta} J_{2,s}{}^{\alpha}{}_{,}{}^{\dot\beta} + \frac{\lambda^{-2}}{2}\xi_{\dot\beta\dot\beta} J_{2,s}{}^{\dot\beta\dot\beta}.$$

The gauge variation of (44) under the gauge transformation (19) is:

$$\delta J_{2,s} = \frac{\lambda^{-2}}{2}\xi_{\alpha\alpha} \delta J_{2,s}{}^{\alpha\alpha} + \xi_{\alpha,\dot\beta} \delta J_{2,s}{}^{\alpha}{}_{,}{}^{\dot\beta} + \frac{\lambda^{-2}}{2}\xi_{\dot\beta\dot\beta} \delta J_{2,s}{}^{\dot\beta\dot\beta}, \tag{A1}$$

where:

$$\begin{aligned}\delta J_{2,s}{}^{\alpha\alpha} = -D^h\Big(\sum_{m,n}\frac{4\lambda^{2-|m-n|}}{(m-1)!n!}c_{ij}\big[\tilde{D}\varepsilon^{i;\alpha\gamma(m-1)}{}_{,}{}^{\dot\delta(n)}\omega^{j;\alpha}{}_{\gamma(m-1),\dot\delta(n)} \\ - D^{top}\varepsilon^{i;\alpha\gamma(m-1)}{}_{,}{}^{\dot\delta(n)}\omega^{j;\alpha}{}_{\gamma(m-1),\dot\delta(n)} - \lambda^2 D^{sub}\varepsilon^{i;\alpha\gamma(m-1)}{}_{,}{}^{\dot\delta(n)}\omega^{j;\alpha}{}_{\gamma(m-1),\dot\delta(n)}\big]\Big),\end{aligned} \tag{A2}$$

$$\delta J_{2,s}{}^{\alpha}{}_{,}{}^{\dot\beta} = \sum_{m,n} 2\lambda^{2-|m-n|}[\frac{1}{(m-1)!n!}c_{ij}\tilde{D}\epsilon^{i;\alpha\gamma(m-1)}{}_{,}{}^{\dot\delta(n)}\omega^{j;\varphi}{}_{\gamma(m-1),\dot\delta(n)}\tilde{h}_{\varphi,}{}^{\dot\beta}$$
$$- \frac{1}{(m-1)!n!}c_{ij}D^{top}\epsilon^{i;\alpha\gamma(m-1)}{}_{,}{}^{\dot\delta(n)}\omega^{j;\varphi}{}_{\gamma(m-1),\dot\delta(n)}\tilde{h}_{\varphi,}{}^{\dot\beta}$$
$$- \frac{\lambda^2}{(m-1)!n!}c_{ij}D^{sub}\epsilon^{i;\alpha\gamma(m-1)}{}_{,}{}^{\dot\delta(n)}\omega^{j;\varphi}{}_{\gamma(m-1),\dot\delta(n)}\tilde{h}_{\varphi,}{}^{\dot\beta}$$
$$- \frac{1}{m!(n-1)!}c_{ij}\omega^{i;\gamma(m)}{}_{,}{}^{\dot\delta(n-1)\dot\theta}\tilde{D}\epsilon^{j;}{}_{\gamma(m),\dot\delta(n-1)}{}^{\dot\beta}\tilde{h}^{\alpha,}{}_{\dot\theta}$$
$$+ \frac{1}{m!(n-1)!}c_{ij}\omega^{i;\gamma(m)}{}_{,}{}^{\dot\delta(n-1)\dot\theta}D^{top}\epsilon^{j;}{}_{\gamma(m),\dot\delta(n-1)}{}^{\dot\beta}\tilde{h}^{\alpha,}{}_{\dot\theta}]$$
$$+ \frac{\lambda^2}{m!(n-1)!}c_{ij}\omega^{i;\gamma(m)}{}_{,}{}^{\dot\delta(n-1)\dot\theta}D^{sub}\epsilon^{j;}{}_{\gamma(m),\dot\delta(n-1)}{}^{\dot\beta}\tilde{h}^{\alpha,}{}_{\dot\theta}]$$
$$+ \frac{2\lambda^{4-2s}}{(2s-3)!}c_{ij}[C^{i;\alpha\gamma(2s-3)\varphi\rho}\, H_{\varphi\rho}\, \tilde{D}\epsilon_{\gamma(2s-3)}{}^{\dot\beta} C^{i;\alpha\gamma(2s-3)\varphi\rho}\, H_{\varphi\rho}\, D^{top}\epsilon_{\gamma(2s-3)}{}^{\dot\beta}$$
$$- \bar{C}^{i;\dot\delta(2s-3)\dot\beta\dot\psi\dot\theta}\, \bar{H}_{\dot\psi\dot\theta}\, \tilde{D}\epsilon^{\alpha}{}_{\dot\delta(2s-3)} + \bar{C}^{i;\dot\delta(2s-3)\dot\beta\dot\psi\dot\theta}\, \bar{H}_{\dot\psi\dot\theta}\, D^{top}\epsilon^{\alpha}{}_{\dot\delta(2s-3)}], \quad \text{(A3)}$$

$$\delta J_{2,s}{}^{\dot\beta\dot\beta} = -D^h\Big(\sum_{m,n}\frac{4\lambda^{2-|m-n|}}{m!(n-1)!}c_{ij}[\tilde{D}\epsilon^{i;\gamma(m)}{}_{,}{}^{\dot\delta(n-1)\dot\beta}\omega^{j;}{}_{\gamma(m),\dot\delta(n-1)}{}^{\dot\beta}$$
$$- D^{top}\epsilon^{i;\gamma(m)}{}_{,}{}^{\dot\delta(n-1)\dot\beta}\omega^{j;}{}_{\gamma(m),\dot\delta(n-1)}{}^{\dot\beta} - \lambda^2 D^{sub}\epsilon^{i;\gamma(m)}{}_{,}{}^{\dot\delta(n-1)\dot\beta}\omega^{j;}{}_{\gamma(m),\dot\delta(n-1)}{}^{\dot\beta}]\Big). \quad \text{(A4)}$$

Rearranging terms in (A1), one can obtain:

$$\delta J_{t,s} = dH + \chi,$$

where $\chi \simeq 0$ (vanishes on-shell) and dH is exact with:

$$H = -\xi_{\alpha\alpha}D^h\big(\sum_{m,n}\frac{4\lambda^{2-|m-n|}}{(m-1)!n!}c_{ij}\epsilon^{i;\alpha\gamma(m-1)}{}_{,}{}^{\dot\delta(n)}\omega^{j;\alpha}{}_{\gamma(m-1),\dot\delta(n)}\big)$$
$$+ \sum_{m,n} 2\xi_{\alpha,\dot\beta}\frac{\lambda^{2-|m-n|}}{(m-1)!n!}c_{ij}(\epsilon^{i;\alpha\gamma(m-1)}{}_{,}{}^{\dot\delta(n)}\omega^{j;\varphi}{}_{\gamma(m-1),\dot\delta(n)}\tilde{h}_{\varphi,}{}^{\dot\beta} + \omega^{i;\alpha\gamma(m-1)}{}_{,}{}^{\dot\delta(n)}\epsilon^{j;\varphi}{}_{\gamma(m-1),\dot\delta(n)}\tilde{h}_{\varphi,}{}^{\dot\beta})$$
$$- \sum_{m,n} 2\xi_{\alpha,\dot\beta}\frac{\lambda^{2-|m-n|}}{m!(n-1)!}c_{ij}(\epsilon^{i;\gamma(m)}{}_{,}{}^{\dot\delta(n-1)\dot\theta}\omega^{j;}{}_{\gamma(m),\dot\delta(n-1)}{}^{\dot\beta}\tilde{h}^{\alpha,}{}_{\dot\theta} - \omega^{i;\gamma(m)}{}_{,}{}^{\dot\delta(n-1)\dot\theta}\epsilon^{j;}{}_{\gamma(m),\dot\delta(n-1)}{}^{\dot\beta}\tilde{h}^{\alpha,}{}_{\dot\theta})$$
$$- \xi_{\dot\beta\dot\beta}D^h\big(\sum_{m,n}\frac{4\lambda^{2-|m-n|}}{m!(n-1)!}c_{ij}\epsilon^{i;\gamma(m)}{}_{,}{}^{\dot\delta(n-1)\dot\beta}\omega^{j;}{}_{\gamma(m),\dot\delta(n-1)}{}^{\dot\beta}\big). \quad \text{(A5)}$$

Thus, the on-shell gauge variation of $J_{2,s}$ is exact.

References

1. Berends, F.A.; Burgers, G.J.H.; van Dam, H. Explicit construction of conserved currents for massless fields of arbitrary spin. *Nucl. Phys. B* **1986**, *271*, 429–441.
2. Vasiliev, M.A. Higher Spin Gauge Theories: Star-Product and AdS Space. In *The Many Faces of the Superworld*; World Scientific Pub Co Inc.: Singapore, 1999.
3. Anselmi, D. Higher-spin current multiplets in operator-product expansions. *Class. Quantum Gravity* **2000**, *17*, 1383–1400.
4. Konstein, S.E.; Vasiliev, M.A.; Zaikin, V.N. Conformal Higher Spin Currents in Any Dimension and AdS/CFT Correspondence. *J. High Energy Phys.* **2000**, *2000*, 018.
5. Deser, S.; Waldron, A. Stress and Strain: $T^{\mu\nu}$ of Higher Spin Gauge Fields. *arXiv* **2004**, arXiv:hep-th/0403059.

6. Gelfond, O.A.; Skvortsov, E.D.; Vasiliev, M.A. Higher spin conformal currents in Minkowski space. *Theor. Math. Phys.* **2008**, *154*, 294–302.
7. Bekaert, X.; Meunier, E. Higher spin interactions with scalar matter on constant curvature spacetimes: conserved current and cubic coupling generating functions. *J. High Energy Phys.* **2010**, *2010*, 116.
8. Kaparulin, D.S.; Lyakhovich, S.L.; Sharapov, A.A. Lagrange Anchor and Characteristic Symmetries of Free Massless Fields. *Symmetry Integrability Geom. Methods Appl. A* **2012**, *8*, 021.
9. Gelfond, O.A.; Vasiliev, M.A. Conserved Higher-Spin Charges in AdS_4. *Phys. Lett. B* **2016**, *754*, 187–194.
10. Landau, L.D.; Lifshitz, E.M. *The Classical Theory of Fields*; Addison-Wesley: Cambridge, UK, 1951.
11. Smirnov, P.A.; Vasiliev, M.A. Gauge Non-Invariant Higher-Spin Currents in 4*d* Minkowski Space. *Theor. Math. Phys.* **2014**, *181*, 1509–1521.
12. Fradkin, E.S.; Vasiliev, M.A. Cubic Interaction in Extended Theories of Massless Higher Spin Fields. *Nucl. Phys. B* **1987**, *291*, 141–171.
13. Fradkin, E.S.; Vasiliev, M.A. On the Gravitational Interaction of Massless Higher Spin Fields. *Phys. Lett. B* **1987**, *189*, 89–95.
14. Fronsdal, C. Massless Fields with Integer Spin. *Phys. Rev. D* **1978**, *18*, 3624.
15. Vasiliev, M.A. Free Massless Fields of Arbitrary Spin in the de Sitter space and Initial Data for a Higher Spin Superalgebra. *Fortschr. Phys.* **1987**, *35*, 741–770.
16. Metsaev, R.R. Massless mixed symmetry bosonic free fields in d-dimensional anti-de Sitter space-time. *Phys. Lett. B* **1995**, *354*, 78–84.

universe
MDPI
Article

Higher Spin Matrix Models

Mauricio Valenzuela

Facultad de Ingeniería y Tecnología, Universidad San Sebastián, General Lagos 1163, Valdivia 5110693, Chile; valenzuela.u@gmail.com

Received: 22 September 2017; Accepted: 23 October 2017; Published: 30 October 2017

Abstract: We propose a hybrid class of theories for higher spin gravity and matrix models, i.e., which handle simultaneously higher spin gravity fields and matrix models. The construction is similar to Vasiliev's higher spin gravity, but part of the equations of motion are provided by the action principle of a matrix model. In particular, we construct a higher spin (gravity) matrix model related to type IIB matrix models/string theory that have a well defined classical limit, and which is compatible with higher spin gravity in AdS space. As it has been suggested that higher spin gravity should be related to string theory in a high energy (tensionless) regime, and, therefore to M-Theory, we expect that our construction will be useful to explore concrete connections.

Keywords: higher spin gravity; string theory; matrix models

---

1. Introduction

Vasiliev's higher spin gravity (HSGR) is a gauge theory whose spectrum contains an infinite tower of massless higher spin and matter fields that enjoy infinite dimensional higher spin symmetries. In the tensionless (high-energy) limit of string theory, its spectrum is spanned by massless modes that exhibits higher spin symmetries [1] (also in the first quantized level [2]), expectedly equivalent to those of HSGR [3]. The latter arguments have led to the idea that higher spin gravity might describe a high-energy/tensionless-limit of string theory (see e.g., [4–9]), or reciprocally that string theory might appear as a broken symmetry phase of some version of higher spin gravity. This also suggests that HSGR is a manifestation of M-theory.

As for matrix models (MMs), it has been argued that M-theory should be described by a matrix model [10,11] in $0+1$ dimensions. It has been also conjectured in [12] that a certain matrix model in zero dimension should describe an effective theory of type IIB strings. Both statements suggest that matrix models might be useful for the description of non-perturbative aspects of strings and M-theory.

In this paper, we show that Vasiliev's HSGR in D spacetime dimensions can be regarded as a type of relativistic MM in D dimensions and we shall argue that certain modification of Vasiliev's equations that incorporate the action principle of matrix models will permit the construction of a new type of higher spin gravities. The latter, dubbed here *higher spin matrix models*, will be straightforwardly related to matrix models. In particular, we shall construct the type IIB higher spin matrix models that combines vanishing (higher spin gravity) curvature and constant (matter fields) covariant derivatives conditions with type IIB matrix model equations of motion [13].

2. Matrix Models and Higher Spin Gravity

There is an interesting connection between super-Yang–Mills and string theories in ten dimensions proposed in [12]. The pure interaction part of super-Yang–Mills, extended with a chemical potential term, obtained after compactification of all directions to size zero is given by the MM action,

$$S_{\mathtt{MM1}} = \left\langle \alpha\left(-\frac{1}{4}[A^I,A^J][A^I,A_J] - \frac{1}{2}\bar{\psi}\gamma^I[A_I,\psi]\right) + \beta\mathbb{1}\right\rangle, \quad I,J=0,1,...,D-1. \tag{1}$$

Here, $\langle\cdot\rangle$ is the (super)trace of the theory, which respects the correspondent cyclic properties for boson (A) and fermion (ψ) fields and α and β are some constants. When the size of the matrices A and ψ goes to infinity, they may be regarded also as Hilbert space operators and therefore they must possess a classical limit, where they become functions on a classical phase-space and their commutator product will become a Poisson bracket. The classical limit of action (1) can be identified with a string action treating the worldsheet as a phase-space and the fields A and ψ as functions on this phase-space.

Table 1. From IKKT model to type IIB string action.

IKKT Matrix Model		**String Theory**
gauge field: A^I		target space coordinate: X^I
commutator: $[\cdot,\cdot]$	– Classical limit →	Worldsheet Poisson bracket: $[\cdot,\cdot]_P$
(super)trace: $\langle\cdot\rangle$		Integral operator: $\int d^2\sigma\sqrt{-g(\sigma)}$
$S_{\mathtt{MM1}}$		$S_{\mathtt{Sch}}[Y,\psi,g]$

The conjecture of the authors [12] says that their (IKKT) matrix model is related to the type IIB Green–Schwarz action in the Schild gauge [12,14] (see Table 1),

$$S_{\mathtt{Sch}}[Y,\psi,g] = \int d^2\sigma\left(\alpha\left(\frac{1}{4\sqrt{-g(\sigma)}}[Y^I,Y^J]_P[Y_I,Y_J]_P - \frac{i}{2}\bar{\psi}\gamma^I[Y_J,\psi]_P\right) + \beta\sqrt{-g(\sigma)}\right), \quad (2)$$

where Y^I are the spacetime coordinate components of the string and ψ is an (anti-commuting) Majorana spinor. Here, the Poisson bracket is defined as

$$[X,Y]_P = \frac{\partial X}{\partial\sigma^1}\frac{\partial Y}{\partial\sigma^2} - \frac{\partial X}{\partial\sigma^2}\frac{\partial Y}{\partial\sigma^1},$$

and $\sqrt{-g(\sigma)}$ is the determinant of the induced metric on the worldsheet, which can be regarded also as an independent scalar field.

The equations of motion obtained from Label (2) are

$$[Y^J,[Y_I,Y_J]_P]_P = 0, \qquad \gamma_I[Y^I,\psi]_P = 0,$$

and from the the variation of $\sqrt{-g(\sigma)}$,

$$-\frac{\alpha}{4\sqrt{-g(\sigma)}^2}[Y^I,Y^J]_P[Y_I,Y_J]_P + \beta = 0. \quad (3)$$

Note that the current $\bar{\psi}\gamma_I\psi$ vanishes owing to anti-symmetry of the fermion product $\psi_\alpha\psi_\beta$ and the symmetry of the matrices $(C\gamma^I)^{\alpha\beta} = (C\gamma^I)^{\beta\alpha}$, where C is the conjugation matrix, and γ_I are in the Majorana representation. From Label (3), we can solve for $\sqrt{-g(\sigma)}$, and replacing back its value in action (2), it returns the Nambu–Goto form (see e.g., [15]),

$$S_{\mathtt{NG}}[Y,\psi] = \mathtt{sign}(\beta)\sqrt{2|\alpha||\beta|}\int d^2\sigma\left(\sqrt{\frac{\mathtt{sign}(\alpha\beta)}{2}[Y^I,Y^J]_P[Y_I,Y_J]_P} - i\mathtt{sign}(\alpha\beta)\frac{1}{2}\sqrt{\frac{|\alpha|}{2|\beta|}}\bar{\psi}\gamma^I[Y_I,\psi]_P\right),$$

which is equivalent to

$$S_{\mathtt{NG}}[Y,\psi] = -T\int d^2\sigma\left(\sqrt{-\frac{1}{2}[Y^I,Y^J]_P[Y_I,Y_J]_P} + 2i\bar{\psi}\gamma^I[Y_I,\psi]_P\right), \quad (4)$$

for

$$\texttt{sign}(\beta) = -1, \quad T = \sqrt{2|\alpha||\beta|}, \quad \texttt{sign}(\alpha\beta) = -1, \quad \sqrt{\frac{|\alpha|}{2|\beta|}} = 4.$$

This means,

$$\alpha = 4T, \quad \beta = -\frac{T}{8}. \tag{5}$$

Note that, in a flat background, the Poisson bracket is equivalent to the determinant of the induced metric on the workdsheet,

$$\texttt{det}(g_{IJ}\partial_\alpha Y^I \partial_\beta Y^J) = \frac{1}{2}[Y^I, Y^J]_P[Y^I, Y^J]_P, \quad \alpha, \beta = 1, 2,$$

so that Label (4) can be turned into the standard form found in many textbooks.

A variation of Label (1) more directly related to the string action (2) in the classical limit was proposed in [13],

$$S_{\mathrm{MM2}}[A, \psi, \phi] = \texttt{Str}\left(\alpha\left(-\tfrac{1}{4}\phi^{-1}[A^I, A^J][A_I, A_J] - \tfrac{1}{2}\bar{\psi}\gamma_I[A^I, \psi]\right) + \beta\phi\right), \qquad I, J = 0, 1, \ldots, D-1, \tag{6}$$

where now $\phi(\sigma)$ is the matrix associated to $\sqrt{-g}$. The equations of motion obtained from Label (6) are

$$\begin{aligned} &[Y^J, \{\phi^{-1}, [Y_I, Y_J]\}] = 0, \qquad \gamma_I[Y^I, \psi] = 0, \\ &-\tfrac{\alpha}{4}\phi^{-1}[Y^I, Y^J][Y_I, Y_J]\phi^{-1} + \beta = 0, \end{aligned} \tag{7}$$

where $\{X, Y\} := XY + YX$, is the anti-commutator. The equations of motion of the IKKT model (1) are obtained when ϕ = constant.

The matrix models (1) and (6) are in particular related to string theory, but, more generally, matrix models can be regarded as describing subspaces of non-abelian algebras (of functions in non-commutative spaces) by means of constraints obtained from an action principle. The action of a generic matrix model has the form

$$S[Y] = \langle L[Y(y)] \rangle, \tag{8}$$

where $\langle \cdot \rangle$ is the (super)trace operation, and L is a functional of functions $Y(y)$ defined on the basis of the algebra $\mathcal{Y} \ni y$. From the variation of Y, $\delta S[Y(y)] = \langle \delta Y C[Y(y)] \rangle = 0$ produces the equation of motion

$$C[Y(y)] = 0. \tag{9}$$

It is easy to show that this equation enjoy the transformation symmetry

$$\tilde{Y} = g(y) Y g^{-1}(y),$$

where $g(y)$ and its inverse $g^{-1}(y)$ are functions of the generator ys. It is enough for this purpose to assume a polynomial form of the constraints (9).

From Matrix Models to Higher Spin Gravity

Matrix models can be extended to fiber bundles locally equivalent to $\mathcal{M} \times \mathcal{U}(\mathcal{Y})$, with local sections given by a set of functions $Y(y; x)$, i.e., which at points $x = \{x^0, x^1, \ldots, x^{D-1}\} \in \mathcal{M}$ are expanded in the basis of an associative algebra $\mathcal{U}(\mathcal{Y})$ constructed from polynomial functions of the generators of a Lie (super)algebra $\mathcal{Y}$. If the representation of the algebra $\mathcal{Y}$ is given in terms of finite dimensional matrices, then $\mathcal{U}(\mathcal{Y})$ will consist of a general linear algebra or some of its associative subalgebras. For more general representations, including infinite dimensional ones, $\mathcal{U}(\mathcal{Y})$ will be equivalent to the universal enveloping algebra of $\mathcal{Y}$. We shall focus in the latter case. This is, the local sections can be expanded as

$$Y(y;x) = \sum_{0}^{\infty} \frac{1}{n!} Y^{\alpha(n)}(x) y_{\alpha(n)}, \tag{10}$$

where $y_\alpha \in \mathcal{Y}$ is an element of the Lie (super-)algebra $\mathcal{Y}$ and their symmetric products $y_{\alpha(n)} := y_{(\alpha_1} \cdots y_{\alpha_n)} \in \mathcal{U}(\mathcal{Y})$ are given in Weyl order, for definiteness, and the sum is up to infinity. Note that, considering formal Taylor expansions of the sections $Y(y;x)$ in terms of the basis generators ys, we can extend the universal enveloping algebra $\mathcal{U}(\mathcal{Y})$ to non-polynomial classes of functions/distributions (see [16,17] for their use in fractional spin gravity). If ys are finite dimensional, the expansion (10) will be truncated.

The reader familiarized with HSGR will note already a similarity of Label (10) with the fields of HSGR. Actually, their equivalence is up to determination of the algebra $\mathcal{U}(\mathcal{Y})$, which, for higher spin gravity, it is typically a Weyl algebra of multiple oscillators and their products with Clifford algebras [18–22]. As in HSGR, we shall require local invariance of the matrix model in $\mathcal{M}$-space, i.e., that the systems of constraints (9) must be constant with respect to a covariant derivative. To satisfy this requirement, we have to introduce a gauge connection, say W, and a constant curvature equation. The complete (integrable) system of constraints is given therefore by

$$\text{(kinematic constraints)} \qquad F_W = 0, \qquad D_W Y = 0, \tag{11}$$

$$\text{(rigid/MM constraints)} \qquad C(Y) = 0, \tag{12}$$

where $F_W := dW + W \wedge W$ is the curvature of the gauge connection (one-form) W and Y should be regarded now as a zero-form in the cotangent space of $\mathcal{M}$, and $D_W = d + W$ is the covariant derivative, with exterior derivative $d = dx^\mu \partial_\mu$. These equations can be split in two subsets, the "kinematic" ones (11) that involve spacetime (exterior) derivatives d, and the "rigid" ones (12) consisting of the algebraic constraints coming from the matrix model (8). Systems (11) and (12) is integrable, i.e., it is closed under repeated action of the covariant derivative since $D_W C(Y) = 0$ as consequence of $D_W Y = 0$ and $D_W D_W Y = [F_W, Y] = 0$ as consequence of $F_W = 0$, etc. Systems (11) and (12) are also gauge invariant under infinitesimal transformations,

$$\delta W = D\epsilon, \qquad \delta Y = [Y, \epsilon].$$

Solving the kinematic equations can be done absorbing the spacetime dependency in gauge-group elements g such that:

$$W = g d g^{-1}, \qquad Y = g Y_o g^{-1}, \tag{13}$$

where Y_o is a spacetime-independent gauge-algebra element. The rigid constraint (12) is left invariant so Y_o must satisfy

$$C(Y_o) = 0. \tag{14}$$

Thus, the rigid constraints encode most of the dynamics of systems (11) and (12).

As we mentioned, Equations (11) and (12) look quite similar to Vasiliev's HSGR equations (see an alternative form of Vasiliev's equations in [23]). The differences are in the details, the algebras involved and the explicit form of the constraints (12). Another important detail is the preference for the use of phase-space (deformation) quantization techniques [24,25], i.e., the use of classical functions endowed with a $\star$-product to construct the associative (non-commutative) enveloping algebras $\mathcal{U}(\mathcal{Y})$, instead of working with matrices or Hilbert-space operators (see the review [20] for further details). Though, in this section, we have not referred explicitly to the $\star$-product, and the reader may assume that the non-commutative algebras $\mathcal{Y}$ and $\mathcal{U}(\mathcal{Y})$ are constructed in terms of classical functions and a $\star$-product.

The fusion of HSGR and MMs is given by Labels (11) and (12). This means, while the constraints (12) encode the dynamics of a MM the kinetic extensions (11) will describe HSGR dynamics, i.e., the emergence of interacting generalized Lorentz connections (W) and matter fields (Y) with arbitrary spin, which will extend standard gauge gravity.

Supporting these ideas, in Ref. [26], it was noticed that the physical degrees of freedom of higher spin gravity coupled to matter fields in $2+1$ dimensions are encoded in a coordinates-free action of the rigid type (12), which is actually a matrix model equivalent to a Yang–Mills theory in even non-commutative dimensions. Actually, the equations of higher spin gravity in three dimensions are given by [27,28]:

$$F_W = 0, \quad D_W B = 0, \quad D_W S_\alpha = 0, \tag{15}$$

$$S_\alpha S^\alpha + 2i(1+B) = 0, \quad S_\alpha B + B S_\alpha = 0, \tag{16}$$

where, compared to Labels (11) and (12), the kinetic constraints are given by Labels (15) and the rigid ones by Labels (16), while the zero forms are given by $Y = \{B, S_\alpha\}$. Here α is a spinor index in three dimensions. Now, according to the authors [26], systems (15) and (16) can be reduced, using the gauge-function method (13), to the spacetime-coordinates-free equation (14) now given by the deformed oscillator algebra [29,30],

$$s_\alpha s^\alpha + 2i(1+b) = 0, \quad s_\alpha b + b s_\alpha = 0, \tag{17}$$

related to Label (16) up to spacetime dependent similarity trasformation (13), $S_\alpha = g s_a g^{-1}$, $B = g b g^{-1}$. Another form of writing Label (17) is,

$$[q,p] = i(1+\nu k), \quad \{q,k\} = \{p,k\} = 0,$$

which, for $k^2 = 1$, becomes equivalent to the Wigner deformed oscillator algebra [29]. Here, q and p are the coordinate and the conjugated momentum of a deformed oscillator and the b field has been factorized in the product of a scalar ν and the Klein operator k. It is worth mentioning that Label (17) implies $osp(1|2)$ (anti-)commutation relations with $sp(2) = \{q^2, p^2, qp + pq\}$ and supercharges q and p. This can be extended to $osp(2|2)$ treating k as an internal $u(1)$ generator and $-ikq$ and $-ikp$ as additional supercharges. The matrix model action proposed in [26] is given by

$$S_{\text{PSV}}[s,b] = \langle i s_\alpha s^\alpha b - 2b - b^2 \rangle, \tag{18}$$

from which Label (17) can be obtained variating b (even degree) and s_α (odd degree) and using the correspondent (anti-)cyclic property of the supertrace $\langle \cdot \rangle$. Thus, the equations of higher spin gravity in three dimensions are somehow a covariantization in $\mathcal{M}$ space of the Prokushkin–Segal–Vasiliev (PSV) matrix model (18). The emergence of higher spin fields on non-commutative geometries has been also observed in Refs. [31,32].

More generally, we shall refer to *higher-spin–matrix models* as the extension of matrix models and HSGR, which can be described by the system of constraints

$$\begin{aligned} F_W = 0, \quad & D_W Y = 0, \\ \delta S_{\text{MM}}&[Y] = 0, \end{aligned} \tag{19}$$

where $\delta S_{\text{MM}}[Y] = 0$, the rigid component of the equations of motion, are obtained from the variation of the action of the matrix model $S_{\text{MM}}[Y]$. We say that higher spin matrix model (19) has the same type of $S_{\text{MM}}[Y]$.

3. Deformation Quantization of Type IIB Strings

In the former section, we omitted explicit reference to the $\ast$-product. To transform our formulas into a $\ast$-product form, we shall assume that, for any two functions $F(y;x)$ and $G(y;x)$ of commutative variables x and non-commutative variables y, which can be decomposed as

$$F := f(x)g(y) = g(y)f(x), \qquad F' := f'(x)g'(y) = g'(y)f'(x),$$

where $f(x)$ and $f'(x)$ are differential forms in $\mathcal{M}$-space and $g(y)$ and $g'(y)$ are elements of the algebra $\mathcal{U}(\mathcal{Y})$, the product $F(y;x)G(y;x)$ has inserted a $\ast$-product such that

$$F(y;x)G(y;x) := (f(x)f'(x))(g(y)\ast g'(y)) = (g(y)\ast g'(y))(f(x)f'(x)).$$

Let us define now the Groenewold–Moyal $\ast$-product

$$f(y)\ast h(y) \;=\; \int \frac{d^2\xi\, d^2\zeta}{(\pi\theta)^2} \exp\left(- \tfrac{2i}{\theta}\, (\xi^1\zeta^2 - \xi^2\zeta^1)\right) f(y+\xi)\, h(y+\zeta)\,.$$

For two arbitrary functions on the world-sheet $f(y)$ and $g(y)$. The auxiliary variables ξ and ζ are also world-sheet type, and the quantization (Planck) constant θ (with units of area) is the parameter of deformation from the classical juxtaposition product $f(y)h(y) = h(y)f(y)$. For example, the $\ast$-product of world-sheet coordinates is

$$y^\alpha \ast y^\beta = y^\alpha y^\beta + i\frac{\theta}{2}\epsilon^{\alpha\beta},$$

where $y^\alpha y^\beta = y^\beta y^\alpha$ is the classical part of the product and the quantum deformation is given by the product of θ and the epsilon tensor $\epsilon^{12} = -\epsilon^{21} = 1$. The $\ast$-commutator is given by

$$[f(y), h(y)]_\ast := f(y)\ast h(y) - f(y)\ast h(y),$$

and it is simple to show that in the classical limit

$$\lim_{\theta\to 0} \frac{[X,Y]_\ast}{i\theta} = [X,Y]_P,$$

and that

$$\lim_{\theta\to 0} X\ast Y = XY = YX.$$

Now let us deform the string actions (2) and (4). This is simple since these action principles can be expressed in terms of the Poisson brackets. The deformation-quantization version of the Nambu–Goto action is obtained substituting $[X,Y]_P$ by $[X,Y]_\ast/i\theta$, so that

$$S^\theta_{\mathrm{NG}}[Y,\psi] = -T\int d^2y \left(\sqrt[\ast]{\frac{1}{2\theta^2}[Y^I,Y^J]_\ast \ast [Y_I,Y_J]_\ast} + \frac{2}{\theta}\bar\psi\ast\gamma^I[Y_I,\psi]_\ast \right),$$

where $\sqrt[\ast]{\ }$ is defined such that $\sqrt[\ast]{f}\ast\sqrt[\ast]{f} = f$. Thus, in the classical limit,

$$\lim_{\theta\to 0} S^\theta_{\mathrm{NG}}[Y,\psi] = S_{\mathrm{NG}}[Y,\psi].$$

Deforming Schild action (2) results in

$$S^\theta_{\mathrm{Sch}}[Y,\psi,\phi] = -T\int d^2y \left(\frac{1}{\theta^2}\phi^{-1}\ast[Y^I,Y^J]_\ast\ast[Y_I,Y_J]_\ast + \frac{2}{\theta}\bar\psi\ast\gamma^I[Y_I,\psi]_\ast + \frac{1}{8}\phi \right), \tag{20}$$

where we have used value (5) and ϕ^{-1} is such that $\phi{*}\phi^{-1} = 1 = \phi^{-1}{*}\phi$. In an operator representation, which can be achieved by means of a Wigner map, the equivalent of action (20) is given by Label (6). In the classical limit,

$$\lim_{\theta\to 0} S^{\theta}_{\mathtt{Sch}}[Y,\psi,\phi] = S_{\mathtt{Sch}}[Y,\psi,\phi].$$

The authors of [12,13] assume that the Poisson bracket produced in the classical limit of the respective matrix models is defined in a two-dimensional phase space (the worldsheet). This condition may be relaxed since there is no data of the phase space dimension in the matrix model. In the deformation quantization approach, we can declare the dimension of the classical phase space in the action itself

$$S^{\theta}_{\mathtt{IIB}}[Y,\psi] = -\mathcal{T}\int d^{2d}y\left(\frac{1}{\theta^2}\phi^{-1}{*}\,[Y^I,Y^J]_*{*}[Y_I,Y_J]_* + \frac{2}{\theta}\bar{\psi}{*}\gamma^I[Y_I,\psi]_* + \frac{1}{8}\phi\right), \tag{21}$$

as an example of generalization of the model [12,13]. Now, the tension $\mathcal{T}$ has been rescaled according to the choice of the dimensions of the phase space. Now, in the classical limit, what is produced is a theory of extended objects. In this case, the $*$-product will be given by

$$f(y){*}g(y) = \int \frac{d^{2d}\xi\, d^{2d}\zeta}{(\pi\theta)^{2d}} \exp\left(-\tfrac{2i}{\theta}\bar{\xi}\zeta\right) f(y+\xi)\, g(y+\zeta), \tag{22}$$

where $\bar{\xi}\zeta = \xi^{\alpha}\zeta_{\alpha}$ and we have used the symplectic matrix

$$C_{\alpha\beta} = C^{\alpha\beta} := \begin{pmatrix} 0 & \mathbb{1}_{d\times d} \\ -\mathbb{1}_{d\times d} & 0 \end{pmatrix}.$$

To raise or lower the phase space index $\bar{\xi} = \xi^{\alpha}$ according to the following conventions,

$$\xi^{\alpha} = C^{\alpha\beta}\xi_{\beta}, \qquad \xi_{\alpha} = \xi^{\beta}C_{\beta\alpha}, \quad \text{where} \quad C_{\alpha\beta} = C^{\alpha\beta}.$$

Thus, the $*$-product of two vectors in phase space yields,

$$y^{\alpha}{*}y^{\beta} = y^{\alpha}y^{\beta} + \frac{i\theta}{2}C^{\alpha\beta}.$$

The equations of motion obtained from Label (21) are

$$\begin{gathered} [Y^J,\{\phi^{-1},[Y_I,Y_J]_*\}_*]_* = 0, \qquad \gamma_I[Y^I,\psi]_* = 0, \\ \phi^{-1}{*}[Y^I,Y^J]_*{*}[Y_I,Y_J]_*{*}\phi^{-1} - \tfrac{1}{8} = 0, \end{gathered} \tag{23}$$

where $\{X,Y\}_* := X{*}Y + Y{*}X$, is the $*$–anti-commutator. The analogy to the IKKT model (1) is obtained when $\phi = 1$ and $d = 1$ in Label (22). The equations of motion are obtained from the property

$$\int d^{2d}y A{*}B = (-1)^{|A||B|}\int d^{2d}y B{*}A$$

for Grassmann parity $|A| = 0, 1$.

4. Type IIB Higher Spin Matrix Models in $D = 2, 3, 4$ Mod 8

Let us construct a specific model as an illustration of the ideas presented here. Now, making explicit reference to the star-product (22), the equations of motion are:

$$dW + W{*} \wedge W = 0, \tag{24}$$

$$dY^I + [W, Y^I]_* = 0, \qquad d\Upsilon + [W, \Upsilon]_* = 0, \qquad d\Phi + [W, \Phi]_* = 0, \tag{25}$$

$$[Y^J, \{\Phi^{-1}, [Y_I, Y_J]_*\}_*]_* = 0, \qquad \gamma_I[Y^I, \Upsilon]_* = 0, \tag{26}$$

$$\Phi^{-1}{*}[Y^I, Y^J]_*{*}[Y_I, Y_J]_*{*}\Phi^{-1} - \tfrac{1}{8} = 0, \tag{27}$$

$$I \quad \rightarrow \quad \{\mu[n] = [\mu_1\mu_2\cdots\mu_n]\,,\, \mu_1 < \mu_2 < \cdots < \mu_n, \quad \mu_k = 0, 1, ..., D-1\}. \tag{28}$$

The kinetic (11) and rigid (12) equations are now Labels (24)–(27) respectively. The rank of indices (28) will be explained below. In this specific case, the rigid equations are obtained from the variation of the action for type IIB matrix models (21). To the theory, Labels (24)–(27), we shall refer as to type IIB higher spin matrix model.

We have not declared what is the dimension of the spacetime D, neither the rank of the capital indices I, J or the algebras involved. The system above is formally integrable, but the choices of the algebras involved and of the dimension of the space-time is an engineering problem, it depends of what we want to describe. In what follows we shall solve the system (24)–(27) employing similar techniques than in reference [33].

Like in Vasiliev's HSGR, let us consider the Heisenberg algebra,

$$\mathcal{Y} = \{[y_\alpha, y_\beta]_* = i\theta C_{\alpha\beta}\,; \quad \alpha, \beta = 1, ..., 2d\},$$

whose universal enveloping algebra, i.e., the algebra of polynomials in $\mathcal{Y}$, is given by

$$\mathcal{U}(\mathcal{Y}) = \{y_{\alpha(n)}\,; n = 0, ..., \infty\}, \tag{29}$$

where the symmetric products

$$y_{\alpha(n)} := y_{(\alpha_1}{*}y_{\alpha_2}{*}\cdots{*}y_{\alpha_n)} = y_{\alpha_1}y_{\alpha_2}\cdots y_{\alpha_n}, \tag{30}$$

are by the properties of the $*$-product equivalent to the classical monomial in the right hand side. In (30) we have used standard notations for the symmetrization of the products, with the factorial normalization. Now the fields (24)–(27) can be expanded in the basis of $\mathcal{U}(\mathcal{Y})$ as follows

$$W(y; x) = \sum_{n,\alpha,} \tfrac{1}{n!} W^{\alpha(n)}(x) y_{\alpha(n)}, \tag{31}$$

$$Y^I(y; x) = \sum_{n,\alpha} \tfrac{1}{n!} Y^{I,\alpha(n)}(x) y_{\alpha(n)}, \tag{32}$$

$$\Upsilon(y; x) = \sum_{n,\alpha} \tfrac{1}{n!} \psi^{\alpha(n)}(x) y_{\alpha(n)}, \tag{33}$$

$$\Phi(y; x) = \sum_{n,\alpha} \tfrac{1}{n!} \phi^{\alpha(n)}(x) y_{\alpha(n)}, \tag{34}$$

$$n = 0, ..., \infty, \quad \alpha = 1, ..., 2d.$$

With respect to the $*$-commutator product $[\cdot,\cdot]_*$ the second order polynomials $y_{\alpha(2)}$ generate a representation of the $sp(2d)$ algebra, which for $2d = 2^{[D/2]}$ contains a representation of the anti de Sitter algebra $so(D-1, 2)$. The latter algebra can be used to make explicit the Lorentz covariance of the system (24)–(27). Indeed the variables $y_{\alpha(n)}$ transform in the spin $n/2$ adjoint representation of the Lorentz algebra in D spacetime dimensions. From this observation it is clear that **W** contains at level $n = 2$ the AdS_D Lorentz connection and for $n \neq 2$ these are higher spin gravity gauge fields, which justify the "higher spin gravity" part of the title of this paper. This is, the AdS_D space is a natural solution of the system of equations, if we put all the remaining fields to zero, for example. Now the spacetime dimension is specified to be D, according to the choice of the algebra (29).

Now we should clarify the meaning of indices (28). The rigid Equations (26) and (27) are written in a form that reminds us of the type IIB matrix model field Equations (7)–(23), but, as we shall justify, the fields involved admit more general labels, i.e., with I labeling target space Lorentz multivectors instead of just vectors. As it was shown in [33], the phase space monomials $y_{\alpha(2)}$ ($\alpha = 1, ..., 2^{[D/2]}$), in the classical level, parametrize an algebraic variety (say $\mathbf{M}$) whose coordinates are conveniently labeled by Lorentz multivectors in D dimensions. $\mathbf{M}$ admits a covariant non-commutative deformation (quantization), i.e., introducing the $\star$-product (22) in the algebra of functions in the space $\mathbf{M}$, which is Lorentz covariant. Surprisingly enough, the coordinates of the non-commutative version of $\mathbf{M}$ will satisfy the constraints (26) and (27). A posteriori, this result justifies the use of multivector labels. Thus, we can change the notation $\mathbf{Y}^I$ by $\mathbf{Y}^{[\mu_1\mu_2\cdots\mu_n]} =: \mathbf{Y}^{\mu[n]}$, $\mu_k = 0, ..., D-1$, for some values of n to be specified below. In order to observe this, let us introduce the exterior algebra of real (Majorana) Dirac matrices

$$\gamma^{\mu[n]} := \gamma^{[\mu_1} \dots \gamma^{\mu_n]}, \qquad \mu = 0, ..., D-1, \tag{35}$$

constructed from the Clifford algebra $\{\gamma_\mu, \gamma_\nu\} = 2\,\eta_{\mu\nu}$, with $\mathtt{diag}(\eta_{\mu\nu}) = (-1, 1, \cdots, 1)$. Here, n is in the set of integers $1, 2$ Mod 4 and $n \leqslant D$ for D even or $n \leqslant [D/2]$ for D odd (see [34]), for which the $\gamma^{\mu[n]}$ matrices are independent. The matrices (35) span a representation of the $sp(2^{[D/2]})$ algebra, which acts on the space of $2d = 2^{[D/2]}$-components spinors. The symmetric monomials $y_{\alpha(2)}$ also span a representation of $sp(2^{[D/2]})$, with respect to the $[\cdot,\cdot]_\star$ product, and indeed they can be given multivector labels using the bi-linear combinations provided by Label (35). Thus,

$$X^{\mu[n]} = \frac{1}{4}\bar{y}\gamma^{\mu[n]}y \tag{36}$$

are $sp(2^{[D/2]})$ generators.

To make more explicit correspondences (28) $I \leftrightarrow \mu[n]$, let us split the interval $I \in \{1, ..., \mathtt{dim}(sp(2^{[D/2]}))\}$ in subspaces

$$I \in \oplus_{n \in 1,2\,\mathtt{Mod}\,4} I_{[n]}, \qquad n = 1, 2 \;\; \mathtt{Mod}\; 4,$$

where $I_{[n]}$ are intervals of integer numbers of size $\binom{D}{n}$, in correspondence with the independent components of multivectors $X^{\mu[n]}$. To each element in the subset $I_{[n]}$, we assign a single multivector index

$$I \in I_{[n]} \quad \rightarrow \quad \mu[n] = [\mu_1\mu_2 \cdots \mu_n], \quad \mu_1 < \mu_2 < \cdots < \mu_n, \quad \mu = 0, 1, \cdots D-1.$$

For example, we have the correspondence

$$\begin{aligned} D = 2+1: \; & I_{[1]} = \{1,2,3\} \rightarrow \mu = 0,1,2,\; I_{[2]} = 4,5,6 \rightarrow \mu[2] = [01],[02],[12], \\ D = 3+1: \; & I_{[1]} = \{1, ..., 4\} \rightarrow \mu = 0,1,2,3, \\ & I_{[2]} = \{5, ..., 10\} \rightarrow \mu[2] = [01],[02],[03],[12],[13],[23], \end{aligned} \tag{37}$$

in the respective space-time dimensions. We can write now

$$\mathbf{Y}^{\mu[n]}(y; x) = \sum_{m,\alpha} \frac{1}{m!} Y^{\mu[n],\alpha(m)}(x) y_{\alpha(m)},$$

instead of Label (32), and

$$d\mathbf{Y}^{\mu[n]} + [\mathbf{W}, \mathbf{Y}^{\mu[n]}]_\star = 0,$$

instead of $d\mathbf{Y}^I + [\mathbf{W}, \mathbf{Y}^I]_* = 0$ in Label (25). Now, the meaning of the rank of indices (28) is understood, but we need to specify how we construct the invariants $X^I X_I = X^I X^J \eta_{IJ}$. Indeed, $X^I X_I$ is a $sp(2^{[D/2]})$ invariant. In multivector notations, the sp-Killing-metric looks like

$$\mathtt{sign}\,\eta_{IJ} = \mathtt{sign}\,\eta_{\mu[n]_< \nu[n]_<} = (-1)^{[n/2]}\mathtt{sign}\,\eta_{\mu_1\nu_1}\eta_{\mu_2\nu_2}\cdots\eta_{\mu_n\nu_n}\,, \tag{38}$$

where the tensor

$$\eta_{\mu[n]\,\nu[n]} = \frac{(-1)^{[n/2]}}{(n!)^2}\eta_{\mu_1\rho_1}\eta_{\mu_2\rho_2}\cdots\eta_{\mu_n\rho_n}\delta^{\rho_1\rho_2\cdots\rho_n}_{\nu_1\nu_2\cdots\nu_n}$$

is constructed from the product of the Lorentz metric tensor $\eta_{\mu\rho}$, and the indices $\mu[n]_< = [\mu_1\mu_2...\mu_n]$ are ordered, $\mu_1 < \mu_2 < ... < \mu_n$.

Now, the rigid constraints (26) and (27) look like

$$\sum_{m=1,2\ \mathtt{Mod4}}' \frac{(-1)^{[m/2]}}{m!}[\mathbf{Y}^{\mu[m]}, \{\mathbf{\Phi}^{-1}, [\mathbf{Y}_{\nu[n]}, \mathbf{Y}_{\mu[m]}]_*\}_*]_* = 0\,, \tag{39}$$

$$\sum_{m=1,2\ \mathtt{Mod4}}' \frac{(-1)^{[m/2]}}{m!}\gamma_{\mu[m]}[\mathbf{Y}^{\mu[m]}, \mathbf{\Psi}]_* = 0, \tag{40}$$

$$\sum_{m,n=1,2\ \mathtt{Mod4}}' \frac{(-1)^{[m/2]+[n/2]}}{m!n!}\mathbf{\Phi}^{-1}{}_*[\mathbf{Y}^{\mu[m]}, \mathbf{Y}^{\nu[n]}]_{*}{}_*[\mathbf{Y}_{\mu[m]}, \mathbf{Y}_{\nu[n]}]_{*}{}_*\mathbf{\Phi}^{-1} - \tfrac{1}{8} = 0, \tag{41}$$

where $\sum'$ means to count in the interval of integers n for which the $\gamma^{\mu[n]}$'s span the basis of independent matrices (35) (see comment below (35)).

4.1. Some Solutions

Let us show that

$$\mathbf{Y}^{\mu[n]} = X^{\mu[n]}, \quad \mathbf{\Psi} = 0, \qquad \mathbf{\Phi} = \mathtt{constant}, \tag{42}$$

where $X^{\mu[n]}$ is given by Label (36), solve the systems of Equations (39)–(41). It is trivial that $\mathbf{\Psi} = 0$ solves Label (40). If $\mathbf{\Phi}$ is constant, then (39) is reduced to

$$\sum_{m=1,2\ \mathtt{Mod4}}' \frac{(-1)^{[m/2]}}{m!}[\mathbf{Y}^{\mu[m]}, [\mathbf{Y}_{\nu[n]}, \mathbf{Y}_{\mu[m]}]_*]_* = 0.$$

That this equation is solved by Label (36) was shown in Ref. [33]. For that purpose, we can use the identity

$$C_{\alpha\beta}C_{\alpha'\beta'} + C_{\alpha'\beta}C_{\alpha\beta'} = \frac{1}{2^{[D/2]-1}}\sum_{m=1,2\ \mathtt{Mod}\ 4}' \frac{(-1)^{[m/2]+1}}{m!}(\gamma^{\nu[m]})_{\alpha\alpha'}(\gamma_{\nu[m]})_{\beta\beta'},$$

and that $y^\alpha y_\alpha = 0$.

Similarly to Label (39), using Label (42) and

$$y_\alpha y_\beta * y_\xi y_\zeta = y_\alpha y_\beta y_\xi y_\zeta + i\frac{\theta}{2}\left(C_{\alpha\xi}y_\beta y_\zeta + C_{\alpha\zeta}y_\beta y_\xi + C_{\beta\xi}y_\alpha y_\zeta + C_{\beta\zeta}y_\alpha y_\xi\right) - \frac{\theta^2}{4}\left(C_{\alpha\xi}C_{\beta\zeta} + C_{\alpha\zeta}C_{\beta\xi}\right),$$

we can compute the value of

$$\sum_{m,n=1,2\ \mathtt{Mod4}}' \frac{(-1)^{[m/2]+[n/2]}}{m!n!}[\mathbf{Y}^{\mu[m]}, \mathbf{Y}^{\nu[n]}]_{*}{}_*[\mathbf{Y}_{\mu[m]}, \mathbf{Y}_{\nu[n]}]_{**} = 2\theta^4 2^{2[D/2]}(1 + 2^{2[D/2]}).$$

Hence, from Label (41), we obtain

$$\mathbf{\Phi} = \pm\theta^2 2^{[D/2]+2}\sqrt{1+2^{[D/2]}}.$$

We can extend these solutions to solutions of the whole system (24)–(28) using gauge functions

$$g = \exp\left(\epsilon_{\nu[n]}(x)X^{\nu[n]}\right),$$

where $\epsilon_{\nu[n]}(x)$ are functions of the point x of the spacetime, so that

$$\mathbf{W} = g{*}dg^{-1}, \qquad \mathbf{Y}^{\mu[n]} = g{*}X^{\mu[n]}{*}g^{-1}, \quad \mathbf{\Psi} = 0, \qquad \mathbf{\Phi} = \pm\theta^2 2^{[D/2]+2}\sqrt{1+2^{[D/2]}}.$$

As we observed, we can unify multivector types of matrix models and higher spin gravity.

5. Conclusions

We have presented general class higher spin gravity models that incorporate matrix models in their own definition. The techniques are similar to standard Vasiliev's higher spin gravity, and the difference is in the internal algebras and in part of the equations of motion, which now are derived from matrix models. These theories put in closer contact higher spin gravity and string theory, as, according to e.g., [10,12,13], matrix models are candidates for non-perturbative theories of strings and M-theory. We constructed and provided solutions for a particular model related to type IIB strings, therefore called *type IIB higher spin gravity*. The matrix models constructed here extend those found in [33] in that now spacetime dimensions are added. An interesting aspect of these matrix models is that they incorporate coordinates of extended objects, as multivector coordinates of rank k are related to k-dimensional objects. For instance, part of the rigid equations of motion for (multivector) matrix model in $3+1$ dimensions (without fermions and for a constant scalar field) can be reduced to the form

$$\begin{aligned}
[\mathbf{Y}^{\nu},[\mathbf{Y}_{\mu},\mathbf{Y}_{\nu}]_*]_* - \frac{1}{2}[\mathbf{Y}^{\nu\lambda},[\mathbf{Y}_{\mu},\mathbf{Y}_{\nu\lambda}]_*]_* &= 0, \\
[\mathbf{Y}^{\nu},[\mathbf{Y}_{\mu\rho},\mathbf{Y}_{\nu}]_*]_* - \frac{1}{2}[\mathbf{Y}^{\nu\lambda},[\mathbf{Y}_{\mu\rho},\mathbf{Y}_{\nu\lambda}]_*]_* &= 0.
\end{aligned}$$

Some solutions of this system contain Plucker coordinates of planes through $3+1$ dimensions as shown in [33]. An interesting problem to address is whether the classical limit of the respective matrix models (24)–(27), or Label (24), Label (25), and Labels (39)–(41), reproduce Polyakov's string theory in $3+1$ dimensions with fine structure [35].

Note that, in this paper, we have not studied the details of the physical degrees of freedom contained in the constructed models. This can be done using similar methods such as matrix models and higher spin gravity, i.e., by means of perturbation theory around some solutions of the models—for instance, the ones provided in Section 4.1.

We would like to mention that multivector extensions of spacetime also appear in the context of E11-type string theories and supergravities [36–38] and in the study of higher spin fields dynamics in the approach [39–41]. These theories and ours suggest the existence of mutivector extended spacetimes, which may be necessary, or just convenient, for the formulation of quantum theories of gravity.

We expect that our proposal will be useful to find deeper connections between string theories, M-theory and higher spin gravity. We encourage interested readers to construct and study in detail specific models falling into the category presented here.

Acknowledgments: I would like to thank to the organizers of the *Workshop on Higher Spin Gauge Theories*, held in UMONS on April 2017, Thomas Basile, Roberto Bonezzi, Nicolas Boulanger, Andrea Campoleoni, David De Filippi and Lucas Traina, for their kind invitation and the opportunity to present the ideas here exposed. I would like to also thank Nicolas Boulanger, Patricia Ritter, Evgeny Skvortsov and Per Sundell for their insightful comments. This project was partially supported by the grants Fondecyt Regular 1140296 and Conicyt DPI 20140115.

Conflicts of Interest: The author declares no conflict of interest.

References

1. Gross, D.J. High-Energy Symmetries of String Theory. *Phys. Rev. Lett.* **1988**, *60*, 1229–1232.
2. Engquist, J.; Sundell, P.; Tamassia, L. On Singleton Composites in Non-compact WZW Models. *J. High Energy Phys.* **2007**, *2007*, 97.
3. Vasiliev, M.A. Progress in higher spin gauge theories. In Proceedings of the Quantization, Gauge Theory, and Strings, International Conference Dedicated to the Memory of Professor Efim Fradkin, Moscow, Russia, 5–10 June 2000; Volume 1 + 2, pp. 452–471.
4. Sundborg, B. Stringy gravity, interacting tensionless strings and massless higher spins. *Nucl. Phys. Proc. Suppl.* **2001**, *102*, 113–119.
5. Sezgin, E.; Sundell, P. Massless higher spins and holography. *Nucl. Phys. B* **2002**, *644*, 303–370; Erratum in **2003**, *660*, 403.
6. Bonelli, G. On the tensionless limit of bosonic strings, infinite symmetries and higher spins. *Nucl. Phys. B* **2003**, *669*, 159–172.
7. Engquist, J.; Sundell, P. Brane partons and singleton strings. *Nucl. Phys. B* **2006**, *752*, 206–279.
8. Fotopoulos, A.; Tsulaia, M. On the Tensionless Limit of String theory, Off-Shell Higher Spin Interaction Vertices and BCFW Recursion Relations. *J. High Energy Phys.* **2010**, *2010*, 86.
9. Sagnotti, A. Notes on Strings and Higher Spins. *J. Phys. A Math. Theor.* **2013**, *46*, 214006.
10. Banks, T.; Fischler, W.; Shenker, S.H.; Susskind, L. M theory as a matrix model: A Conjecture. *Phys. Rev. D* **1997**, *55*, 5112–5128.
11. Susskind, L. Another conjecture about M(atrix) theory. *arXiv* **1997**, arXiv:hep-th/9704080.
12. Ishibashi, N.; Kawai, H.; Kitazawa, Y.; Tsuchiya, A. A Large N reduced model as superstring. *Nucl. Phys. B* **1997**, *498*, 467–491.
13. Fayyazuddin, A.; Makeenko, Y.; Olesen, P.; Smith, D.J.; Zarembo, K. Towards a nonperturbative formulation of IIB superstrings by matrix models. *Nucl. Phys. B* **1997**, *499*, 159–182.
14. Schild, A. Classical Null Strings. *Phys. Rev. D* **1977**, *16*, 1722–1726.
15. Makeenko, Y. Three introductory lectures in Helsinki on matrix models of superstrings. In Proceedings of the 5th Nordic Meeting on Supersymmetric Field and String Theories, Helsinki, Finland, 10–12 March 1997.
16. Boulanger, N.; Sundell, P.; Valenzuela, M. Three-dimensional fractional-spin gravity. *J. High Energy Phys.* **2014**, *2014*, 52.
17. Boulanger, N.; Sundell, P.; Valenzuela, M. Gravitational and gauge couplings in Chern-Simons fractional spin gravity. *J. High Energy Phys.* **2016**, *2016*, 75.
18. Vasiliev, M.A. Consistent equation for interacting gauge fields of all spins in (3+1)-dimensions. *Phys. Lett. B* **1990**, *243*, 378–382.
19. Vasiliev, M.A. More on equations of motion for interacting massless fields of all spins in (3+1)-dimensions. *Phys. Lett. B* **1992**, *285*, 225–234.
20. Vasiliev, M.A. Higher spin gauge theories: Star product and AdS space. In *The Many Faces of the Superworld*; Shifman, M.A., Ed.; World Scientific: Singapore, 1999; pp. 533–610.
21. Vasiliev, M.A. Higher spin gauge theories in any dimension. *Comptes Rendus Phys.* **2004**, *5*, 1101–1109.
22. Vasiliev, M.A. Higher-Spin Theory and Space-Time Metamorphoses. *Lect. Notes Phys.* **2015**, *892*, 227–264.
23. Alkalaev, K.B.; Grigoriev, M.; Skvortsov, E.D. Uniformizing higher-spin equations. *J. Phys. A* **2015**, *48*, 015401.
24. Bayen, F.; Flato, M.; Fronsdal, C.; Lichnerowicz, A.; Sternheimer, D. Deformation Theory and Quantization. 1. Deformations of Symplectic Structures. *Ann. Phys.* **1978**, *111*, 61–110.
25. Bayen, F.; Flato, M.; Fronsdal, C.; Lichnerowicz, A.; Sternheimer, D. Deformation Theory and Quantization. 2. Physical Applications. *Ann. Phys.* **1978**, *111*, 111–151.
26. Prokushkin, S.F.; Segal, A.Y.; Vasiliev, M.A. Coordinate free action for AdS(3) higher spin matter systems. *Phys. Lett. B* **2000**, *478*, 333–342.
27. Prokushkin, S.; Vasiliev, M.A. 3-d higher spin gauge theories with matter. In Proceedings of the Theory of Elementary Particles, 31st International Symposium Ahrenshoop, Buckow, Germany, 2–6 September 1997.
28. Prokushkin, S.F.; Vasiliev, M.A. Higher spin gauge interactions for massive matter fields in 3-D AdS space-time. *Nucl. Phys. B* **1999**, *545*, 385–433.
29. Wigner, E.P. Do the Equations of Motion Determine the Quantum Mechanical Commutation Relations? *Phys. Rev.* **1950**, *77*, 711–712.

30. Yang, L.M. A Note on the Quantum Rule of the Harmonic Oscillator. *Phys. Rev.* **1951**, *84*, 788–790.
31. Sperling, M.; Steinacker, H.C. Covariant 4-dimensional fuzzy spheres, matrix models and higher spin. *J. Phys. A* **2017**, *50*, 375202.
32. Sperling, M.; Steinacker, H.C. Higher spin gauge theory on fuzzy S^4_N. *arXiv* **2017**, arXiv:1707.00885.
33. Valenzuela, M. From phase space to multivector matrix models. *arXiv* **2015**, arXiv:1501.03644.
34. Van Holten, J.W.; Van Proeyen, A. N=1 Supersymmetry Algebras in D=2, D=3, D=4 MOD-8. *J. Phys. A* **1982**, *15*, 3763–3784.
35. Polyakov, A.M. Fine Structure of Strings. *Nucl. Phys. B* **1986**, *268*, 406–412.
36. Riccioni, F.; West, P.C. E(11)-extended spacetime and gauged supergravities. *J. High Energy Phys.* **2008**, *2008*, 39.
37. West, P. E11, generalised space-time and equations of motion in four dimensions. *J. High Energy Phys.* **2012**, *2012*, 68.
38. West, P. A brief review of E theory. *Int. J. Mod. Phys. A* **2016**, *31*, 1630043.
39. Plyushchay, M.; Sorokin, D.; Tsulaia, M. GL flatness of OSp(1 | 2n) and higher spin field theory from dynamics in tensorial spaces. In Proceedings of the 5th International Workshop on Supersymmetries and Quantum Symmetries (SQS'03), Dubna, Russia, 24–29 July 2003; pp. 96–108.
40. Bandos, I.; Pasti, P.; Sorokin, D.; Tonin, M. Superfield theories in tensorial superspaces and the dynamics of higher spin fields. *J. High Energy Phys.* **2004**, *2004*, 23.
41. Bandos, I.; Bekaert, X.; de Azcarraga, J.A.; Sorokin, D.; Tsulaia, M. Dynamics of higher spin fields and tensorial space. *J. High Energy Phys.* **2005**, *2005*, 31.

universe

MDPI

Article

Infinite Spin Fields in $d = 3$ and Beyond

Yurii M. Zinoviev

Institute for High Energy Physics of National Research Center "Kurchatov Institute" Protvino, Moscow Region 142281, Russia; Yurii.Zinoviev@ihep.ru

Received: 27 July 2017; Accepted: 18 August 2017; Published: 30 August 2017

Abstract: In this paper, we consider the frame-like formulation for the so-called infinite (continuous) spin representations of the Poincare algebra. In the three-dimensional case, we give explicit Lagrangian formulation for bosonic and fermionic infinite spin fields (including the complete sets of the gauge-invariant objects and all the necessary extra fields). Moreover, we find the supertransformations for the supermultiplet containing one bosonic and one fermionic field, leaving the sum of their Lagrangians invariant. Properties of such fields and supermultiplets in four and higher dimensions are also briefly discussed.

Keywords: infinite spin fields; massive higher spins; gauge invariance; supersymmetry

---

1. Introduction

Besides the very well known finite-component massless and massive representations of the Poincare algebra, there are rather exotic so-called infinite (or continuous) spin representations (see e.g., [1,2]). In dimensions $d \geq 4$, they have an infinite number of physical degrees of freedom and so may be of some interest for the higher spins theory. Indeed, they attracted some attention recently [3–8]. One of the reasons is that, contrary to the finite-component massless fields, such representations are characterized by a dimensionful parameter (that can play the same role as the cosmological constant for the massless theories and the mass for the massive ones) and so they may provide an interesting alternative for the massless higher spin theory in the flat space. Note also that such representations can appear in the tensionless limit of the string theory.

It has been noted several times that such infinite spin representations may be considered as a limit of massive higher spin ones where spin goes to infinity and mass goes to zero while the product remains fixed. Moreover, recently, Metsaev has shown that the metric-like Lagrangian formulation for the bosonic [9] and fermionic [10] fields in AdS_d spaces with $d \geq 4$ can be constructed using exactly the same technique as was previously used for the gauge-invariant formulation of massive higher spin bosonic [11] and fermionic [12] fields.

The current paper is devoted to the frame-like formulation for such infinite spin fields. In the first (and main) section, we construct gauge-invariant Lagrangian formulation for bosonic and fermionic cases in $d = 3$. We also elaborate on the whole set of the gauge invariant objects (introducing all necessary extra fields) and rewrite our Lagrangians in the explicitly gauge-invariant form. Moreover, we managed to find supertransformations for the supermultiplet containing one bosonic and one fermionic infinite spin field that leaves the sum of their Lagrangians invariant. For this, we heavily use our previous results on the gauge-invariant formulation for massive bosonic and fermionic higher spin fields in $d = 3$ [13,14] (see also [15–17]) as well as results on the massive higher spin supermultiplets [18–20]. In the last two sections, we briefly discuss the properties of such fields and supermultiplets in $d = 4$ and $d \geq 5$ dimensions, concluding with explicit details on the forthcoming publication.

Notations and conventions We will work in the frame-like multispinor formalism (mostly the same as in [20] but we restrict ourselves to the flat Minkowski space). In this formalism, all objects are forms (3, 2, 1, 0-forms) that have totally symmetric local spinor indices. To simplify the expressions, we will use the condensed notations for the spinor indices such that, e.g.,

$$\Omega^{\alpha(2k)} = \Omega^{(\alpha_1\alpha_2\dots\alpha_{2k})}$$

Also, we will always assume that spinor indices denoted by the same letters and placed on the same level are symmetrized, e.g.,

$$\Omega^{\alpha(2k)}\zeta^{\alpha} = \Omega^{(\alpha_1\dots\alpha_{2k}}\zeta^{\alpha_{2k+1})}$$

where symmetrization uses the minimal number of terms necessary without any normalization factor. The coordinate-free description of the three-dimensional flat Minkowski space will use the background frame (one-form) $e^{\alpha(2)}$ and external derivative d

$$d \wedge d = 0$$

Basic elements of 1,2,3-form space are respectively $e^{\alpha(2)}$, $E^{\alpha(2)}$, and E where the last two are defined as the double and triple product of $e^{\alpha(2)}$:

$$e^{\alpha\alpha} \wedge e^{\beta\beta} = \varepsilon^{\alpha\beta} E^{\alpha\beta}, \qquad E^{\alpha\alpha} \wedge e^{\beta\beta} = \varepsilon^{\alpha\beta}\varepsilon^{\alpha\beta} E$$

Further on, the wedge product sign $\wedge$ will be omitted.

2. Infinite Spin Fields in $d = 3$

In this section, we develop the frame-like formalism for the massless infinite spin bosonic and fermionic fields as well as for the supermultiplet containing such fields.

2.1. Infinite Spin Boson

As we have already noted, there is a tight connection between the gauge invariant description for the massive finite spin fields and the one for the massless infinite spin fields. Recall that the general idea of such a description is to begin with the appropriate set of massless (finite component) fields and then glue them together in such a way that keeps all their gauge symmetries. This, in turn, guarantees the correct number of physical degrees of freedom. Thus, we will follow the same approach as in [13] but this time without restriction on the number of components. So, we introduce an infinite set of physical and auxiliary one-forms $\Omega^{\alpha(2k)}$, $\Phi^{\alpha(2k)}$, $1 \le k \le \infty$ as well as one-form A and zero-forms $B^{\alpha(2)}$, $\pi^{\alpha(2)}$ and φ [1]. We begin with the sum of kinetic terms for all these fields (recall that the Lagrangians are three-forms in our formalism):

$$\begin{aligned} \mathcal{L}_0 &= \sum_{k=1}^{\infty}(-1)^{k+1}[k\Omega_{\alpha(2k-1)\beta}e^{\beta}{}_{\gamma}\Omega^{\alpha(2k-1)\gamma} + \Omega_{\alpha(2k)}d\Phi^{\alpha(2k)}] \\ &\quad + EB_{\alpha(2)}B^{\alpha(2)} - B_{\alpha(2)}e^{\alpha(2)}dA - E\pi_{\alpha(2)}\pi^{\alpha(2)} + \pi_{\alpha(2)}E^{\alpha(2)}d\varphi \end{aligned} \tag{1}$$

as well as their initial gauge transformations:

$$\delta_0\Omega^{\alpha(2k)} = d\eta^{\alpha(2k)}, \qquad \delta_0\Phi^{\alpha(2k)} = d\xi^{\alpha(2k)} + e^{\alpha}{}_{\beta}\eta^{\alpha(2k-1)\beta}, \qquad \delta_0 A = d\xi \tag{2}$$

---

[1] Note that in three dimensions, such an infinite spin bosonic field (as any massive higher spin boson) has just two physical degrees of freedom, while an infinite spin fermionic field (as any massive higher spin fermion) has just one. However it is impossible to realize such representations using a finite number of components (see e.g., [6]).

Then, following a general scheme, we add to the Lagrangian a set of cross terms gluing all these components together:

$$\begin{aligned}\mathcal{L}_1 &= \sum_{k=1}^{\infty}(-1)^{k+1}[\tilde{a}_k\Omega_{\alpha(2k)\beta(2)}e^{\beta(2)}\Phi^{\alpha(2k)} + a_k\Omega_{\alpha(2k)}e_{\beta(2)}\Phi^{\alpha(2k)\beta(2)}] \\ &\quad + \tilde{a}_0\Omega_{\alpha(2)}e^{\alpha(2)}A - a_0\Phi_{\alpha\beta}E^{\beta}{}_{\gamma}B^{\alpha\gamma} + \hat{a}_0\pi_{\alpha(2)}E^{\alpha(2)}A \end{aligned} \tag{3}$$

and introduce appropriate corrections for the gauge transformations:

$$\begin{aligned}\delta_1\Omega^{\alpha(2k)} &= \frac{(k+2)}{k}a_k e_{\beta(2)}\eta^{\alpha(2k)\beta(2)} + \frac{a_{k-1}}{k(2k-1)}e^{\alpha(2)}\eta^{\alpha(2k-2)} \\ \delta_1\Phi^{\alpha(2k)} &= a_k e_{\beta(2)}\xi^{\alpha(2k)\beta(2)} + \frac{(k+1)a_{k-1}}{k(k-1)(2k-1)}e^{\alpha(2)}\xi^{\alpha(2k-2)} \\ \delta_1\Omega^{\alpha(2)} &= 3a_1 e_{\beta(2)}\eta^{\alpha(2)\beta(2)}, \qquad \delta_1\Phi^{\alpha(2)} = a_1 e_{\beta(2)}\xi^{\alpha(2)\beta(2)} + 2a_0 e^{\alpha(2)}\xi \\ \delta_1 B^{\alpha(2)} &= 2a_0\eta^{\alpha(2)}, \qquad \delta_1 A = \frac{a_0}{4}e_{\alpha(2)}\xi^{\alpha(2)}, \qquad \delta_1\varphi = -\hat{a}_0\xi \end{aligned} \tag{4}$$

Here, consistency of the gauge transformations with the Lagrangian requires:

$$\tilde{a}_k = -\frac{(k+2)}{k}a_k, \qquad \tilde{a}_0 = 2a_0$$

At last, we introduce mass-like terms for all components and appropriate corrections to the gauge transformations:

$$\mathcal{L}_2 = \sum_{k=1}^{\infty}(-1)^{k+1}b_k\Phi_{\alpha(2k-1)\beta}e^{\beta}{}_{\gamma}\Phi^{\alpha(2k-1)\gamma} + b_0\Phi_{\alpha(2)}E^{\alpha(2)}\varphi + \tilde{b}_0 E\varphi^2 \tag{5}$$

$$\delta_2\Omega^{\alpha(2k)} = \frac{b_k}{k}e^{\alpha}{}_{\beta}\xi^{\alpha(2k-1)\beta}, \qquad \delta_2\pi^{\alpha(2)} = b_0\xi^{\alpha(2)} \tag{6}$$

Now, we require that the whole Lagrangian $\mathcal{L} = \mathcal{L}_0 + \mathcal{L}_1 + \mathcal{L}_2$ will be invariant under the gauge transformations $\delta = \delta_0 + \delta_1 + \delta_2$. This produces the following general relations on the parameters:

$$(k+2)^2 b_{k+1} = k(k+1)b_k \tag{7}$$

$$\frac{2(k+2)(2k+3)}{(k+1)(2k+1)}{a_k}^2 - \frac{2(k+1)}{(k-1)}{a_{k-1}}^2 + 4b_k = 0 \tag{8}$$

as well as some relations for the lower components:

$$5{a_1}^2 - {a_0}^2 + 4b_1 = 0$$

$$\hat{a}_0^2 = 64b_1, \qquad b_0 = \frac{\hat{a}_0 a_0}{4}, \qquad \tilde{b}_0 = \frac{3{a_0}^2}{2}$$

The general solution of all these relations has two free parameters. In the massive finite spin case, it is just the mass and spin but in our case we choose a_0 and b_1 as the main parameters. Then, all other parameters can be expressed as follows:

$$b_k = \frac{4b_1}{k(k+1)^2} \tag{9}$$

$${a_k}^2 = \frac{k}{(2k+3)}[\frac{3(k+1)}{2(k+2)}{a_0}^2 - \frac{8k}{(k+1)}b_1] \tag{10}$$

Now, we are ready to analyze the solution obtained. Let us begin with the case $a_0{}^2 < 16b_1$. In general, it means that starting from some value of k, all $a_k{}^2$ become negative so that we obtain non-unitary theory. The only exceptions happen when one adjusts the values of $a_0{}^2$ and b_1 so that at some k_0 we obtain $a_{k_0} = 0$. In this case, we obtain unitary theory with the finite number of components and this case corresponds to the gauge-invariant description for the massive bosonic field with the spin $k_0 + 1$. Let us turn to the case $a_0{}^2 = 16b_1$ (this corresponds to the case $\mu_0 = 0$ in [9]). In this case, we obtain:

$$a_k{}^2 = \frac{3k}{2(k+1)(k+2)(2k+3)} a_0{}^2 \tag{11}$$

so we get a unitary theory with an infinite number of components. Note that for the case $a_0{}^2 > 16b_1$ we also obtain unitary theory but as it was shown by Metsaev [9] it corresponds to the tachyonic infinite spin field. Thus, in what follows, we will restrict ourselves to the case $a_0{}^2 = 16b_1$ only.

Naturally, all the physical properties of the solutions obtained are the same as in the metric-like formulation by Metsaev because metric-like and frame-like formalisms are equivalent and so, for the free theories, which one to use is just a matter of preference. However, for the investigation of possible interactions, the frame-like formalism may provide some advantages. In-particular, one of the nice and general features of the frame-like formalism is that for each field (physical or auxiliary) one can construct a corresponding gauge-invariant object. For the case at hand, we will follow the massive case in [17,20]. For almost all fields, corresponding gauge-invariant objects can be directly constructed from the known form for the gauge transformations given above (here, for convenience, we changed the normalization for the zero-forms $B^{\alpha(2)} \Rightarrow 2a_0 B^{\alpha(2)}$, $\pi^{\alpha(2)} \Rightarrow b_0\pi^{\alpha(2)}$):

$$\begin{aligned}
\mathcal{R}^{\alpha(2k)} &= d\Omega^{\alpha(2k)} + \frac{b_k}{k} e^\alpha{}_\beta \Phi^{\alpha(2k-1)\beta} + \frac{(k+2)}{k} a_k e_{\beta(2)} \Omega^{\alpha(2k)\beta(2)} + \frac{a_{k-1}}{k(2k-1)} e^{\alpha(2)} \Omega^{\alpha(2k-2)} \\
\mathcal{T}^{\alpha(2k)} &= d\Phi^{\alpha(2k)} + e^\alpha{}_\beta \Omega^{\alpha(2k-1)\beta} + a_k e_{\beta(2)} \Phi^{\alpha(2k)\beta(2)} + \frac{(k+1)a_{k-1}}{k(k-1)(2k-1)} e^{\alpha(2)} \Phi^{\alpha(2k-2)} \\
\mathcal{R}^{\alpha(2)} &= d\Omega^{\alpha(2)} + b_1 e^\alpha{}_\beta \Phi^{\alpha\beta} + 3a_1 e_{\beta(2)} \Omega^{\alpha(2)\beta(2)} - a_0{}^2 E^\alpha{}_\beta B^{\alpha\beta} + b_0 E^{\alpha(2)} \varphi \\
\mathcal{T}^{\alpha(2)} &= d\Phi^{\alpha(2)} + e^\alpha{}_\beta \Omega^{\alpha\beta} + a_1 e_{\beta(2)} \Phi^{\alpha(2)\beta(2)} + 2a_0 e^{\alpha(2)} A \\
\mathcal{A} &= dA - 2a_0 E_{\alpha(2)} B^{\alpha(2)} + \frac{a_0}{4} e_{\alpha(2)} \Phi^{\alpha(2)} \\
\Phi &= d\varphi - \frac{\sqrt{3}}{2} a_0{}^2 e_{\alpha(2)} \pi^{\alpha(2)} + 2\sqrt{3} a_0 A
\end{aligned} \tag{12}$$

However, to construct gauge-invariant objects for $B^{\alpha(2)}$ and $\pi^{\alpha(2)}$, one must introduce a first pair of the so-called extra fields [2] $B^{\alpha(4)}$ and $\pi^{\alpha(4)}$:

$$\begin{aligned}
\mathcal{B}^{\alpha(2)} &= dB^{\alpha(2)} - \Omega^{\alpha(2)} + b_1 e^\alpha{}_\beta \pi^{\alpha\beta} + 3a_1 e_{\beta(2)} B^{\alpha(2)\beta(2)} \\
\Pi^{\alpha(2)} &= d\pi^{\alpha(2)} + e^\alpha{}_\beta B^{\alpha\beta} - \Phi^{\alpha(2)} - \frac{1}{\sqrt{3}} e^{\alpha(2)} \varphi + a_1 e_{\beta(2)} \pi^{\alpha(2)\beta(2)}
\end{aligned} \tag{13}$$

which transform as follows:

$$\delta B^{\alpha(4)} = \eta^{\alpha(4)}, \qquad \delta\pi^{\alpha(4)} = \xi^{\alpha(4)}$$

---

2 Recall that extra fields are the fields that do not enter the free Lagrangian but are necessary for the construction of the whole set of gauge-invariant objects. Moreover, such fields play an important role in the construction of the interactions.

However, to construct gauge-invariant objects for these new fields, one must introduce the next pair of extra fields and so on. This results in the infinite chain of zero forms $B^{\alpha(2k)}$ and $\pi^{\alpha(2k)}$, $1 \leq k \leq \infty$ with the following set of gauge-invariant objects:

$$\begin{aligned}
\mathcal{B}^{\alpha(2k)} &= dB^{\alpha(2k)} - \Omega^{\alpha(2k)} + \frac{b_k}{k} e^{\alpha}{}_{\beta}\pi^{\alpha(2k-1)\beta} + \frac{(k+2)}{k} a_k e_{\beta(2)} B^{\alpha(2k)\beta(2)} \\
&+ \frac{a_{k-1}}{k(2k-1)} e^{\alpha(2)} B^{\alpha(2k-2)} \\
\Pi^{\alpha(2k)} &= d\pi^{\alpha(2k)} - \Phi^{\alpha(2k)} + e^{\alpha}{}_{\beta} B^{\alpha(2k-1)\beta} + a_k e_{\beta(2)} \pi^{\alpha(2k)\beta(2)} \\
&+ \frac{(k+1)a_{k-1}}{k(k-1)(2k-1)} e^{\alpha(2)} \pi^{\alpha(2k-2)}
\end{aligned} \tag{14}$$

Here:

$$\delta B^{\alpha(2k)} = \eta^{\alpha(2k)}, \qquad \delta\pi^{\alpha(2k)} = \xi^{\alpha(2k)}$$

Now, we have an infinite set of gauge one-forms as well as an infinite set of Stueckelberg zero-forms. As in the massive finite spin case [17,20], this allows us to rewrite the Lagrangian in the explicitly gauge-invariant form:

$$\mathcal{L} = -\frac{1}{2}\sum_{k=1}^{\infty}(-1)^{k+1}[\mathcal{R}_{\alpha(2k)}\Pi^{\alpha(2k)} + \mathcal{T}_{\alpha(2k)}\mathcal{B}^{\alpha(2k)}] + \frac{1}{2}e_{\alpha(2)}\mathcal{B}^{\alpha(2)}\Phi \tag{15}$$

By construction, each term here is separately gauge-invariant and the explicit values for all coefficients are determined by the so-called extra field decoupling conditions:

$$\frac{\delta\mathcal{L}}{\delta B^{\alpha(2k)}} = 0, \qquad \frac{\delta\mathcal{L}}{\delta\pi^{\alpha(2k)}} = 0, \qquad 2 \leq k \leq \infty$$

2.2. Fermionic Case

In this case, we will also follow the construction for the massive finite spin field [14] but this time for the infinite set of components. So, we introduce a set of one-forms $\Psi^{\alpha(2k+1)}$, $0 \leq k \leq \infty$ and a zero-form ψ^{α}. Once again, we begin with the sum of kinetic terms for all fields:

$$\frac{1}{i}\mathcal{L}_0 = \sum_{k=0}^{\infty}\frac{(-1)^{k+1}}{2}\Psi_{\alpha(2k+1)}d\Psi^{\alpha(2k+1)} + \frac{1}{2}\psi_{\alpha}E^{\alpha}{}_{\beta}d\psi^{\beta} \tag{16}$$

as well as with their initial gauge transformations:

$$\delta_0\Psi^{\alpha(2k+1)} = d\zeta^{\alpha(2k+1)} \tag{17}$$

Now we add a set of cross terms gluing them together

$$\frac{1}{i}\mathcal{L}_1 = \sum_{k=1}^{\infty}(-1)^{k+1}c_k\Psi_{\alpha(2k-1)\beta(2)}e^{\beta(2)}\Psi^{\alpha(2k-1)} + c_0\Psi_{\alpha}E^{\alpha}{}_{\beta}\psi^{\beta} \tag{18}$$

and corresponding corrections to the gauge transformations:

$$\begin{aligned}
\delta_1\Psi^{\alpha(2k+1)} &= c_{k+1}e_{\beta(2)}\zeta^{\alpha(2k+1)\beta(2)} + \frac{c_k}{k(2k+1)}e^{\alpha(2)}\zeta^{\alpha(2k-1)}, \\
\delta_1\psi^{\alpha} &= c_0\zeta^{\alpha}
\end{aligned} \tag{19}$$

At last, we add the mass-like terms for all fields and appropriate corrections to the gauge transformations:

$$\frac{1}{i}\mathcal{L}_2 = \sum_{k=0}^{\infty}(-1)^{k+1}\frac{d_k}{2}\Psi_{\alpha(2k)\beta}e^{\beta}{}_{\gamma}\Psi^{\alpha(2k)\gamma} - \frac{m_0}{2}E\psi_{\alpha}\psi^{\alpha} \tag{20}$$

$$\delta_2\Psi^{\alpha(2k+1)} = \frac{d_k}{(2k+1)}e^{\alpha}{}_{\beta}\zeta^{\alpha(2k)\beta} \tag{21}$$

Now, we require that the whole Lagrangian $\mathcal{L} = \mathcal{L}_0 + \mathcal{L}_1 + \mathcal{L}_2$ will be invariant under the gauge transformations $\delta = \delta_0 + \delta_1 + \delta_2$. This produces a number of general relations on the parameters

$$(2k+5)d_{k+1} = (2k+3)d_k \tag{22}$$

$$\frac{(k+2)(2k+1)}{(k+1)(2k+3)}c_{k+1}{}^2 - c_k{}^2 + \frac{d_k{}^2}{(2k+1)} = 0 \tag{23}$$

as well as

$$\frac{8}{3}c_1{}^2 - c_0{}^2 + 4d_0{}^2 = 0, \qquad d_0 = \frac{m_0}{3}$$

As in the bosonic case, the general solution for all these relations has two free parameters and we choose c_0 and m_0 this time. Then, all other coefficients can be expressed as follows:

$$d_k = \frac{m_0}{(2k+3)} \tag{24}$$

$$c_k{}^2 = \frac{(2k+1)}{4(k+1)}c_0{}^2 - \frac{k}{2(2k+1)}m_0{}^2 \tag{25}$$

The properties of this solution appear to be the same as in the bosonic case. Namely, for the case $m_0{}^2 > 2c_0{}^2$, in general, we obtain non-unitary theory. The only exceptions appear if one adjusts these parameters so that at some k_0 we get $c_{k_0} = 0$. In this case, we obtain unitary theory with a finite number of components which corresponds to the gauge-invariant description for a massive fermionic field with spin $k_0 + 3/2$. For the $m_0{}^2 = 2c_0{}^2$ (this corresponds to $\mu_0 = 0$ in [10]), we obtain

$$c_k{}^2 = \frac{c_0{}^2}{4(k+1)(2k+1)} \tag{26}$$

that corresponds to the unitary massless infinite spin field while for the $m_0{}^2 < c_0{}^2$, we again obtain tachyonic infinite spin case. As in the bosonic case, in what follows, we will restrict ourselves to the case $m_0{}^2 = 2c_0{}^2$ only.

Now, we proceed with the construction of the full set of gauge-invariant objects. For all one-forms, the construction is pretty straightforward (again, for convenience, we changed the normalization for the zero-form $\psi^{\alpha} \Rightarrow c_0\psi^{\alpha}$):

$$\begin{aligned}\mathcal{F}^{\alpha(2k+1)} &= d\Psi^{\alpha(2k+1)} + \frac{d_k}{(2k+1)}e^{\alpha}{}_{\beta}\Psi^{\alpha(2k)\beta} + c_{k+1}e_{\beta(2)}\Psi^{\alpha(2k+1)\beta(2)} \\ &\quad + \frac{c_k}{k(2k+1)}e^{\alpha(2)}\Psi^{\alpha(2k-1)} \\ \mathcal{F}^{\alpha} &= D\Psi^{\alpha} + d_0e^{\alpha}{}_{\beta}\Psi^{\beta} + c_1e_{\beta(2)}\Psi^{\alpha\beta(2)} - c_0{}^2E^{\alpha}{}_{\beta}\psi^{\beta}\end{aligned} \tag{27}$$

However, to construct a gauge-invariant object for the zero-form, one must introduce a first extra field:

$$\mathcal{C}^{\alpha} = d\psi^{\alpha} - \Psi^{\alpha} + d_0e^{\alpha}{}_{\beta}\psi^{\beta} + c_1e_{\beta(2)}\psi^{\alpha\beta(2)}, \qquad \delta\psi^{\alpha(3)} = \zeta^{\alpha(3)} \tag{28}$$

Then, to construct a gauge-invariant object for this field, one must introduce the second one and so on. This results in the infinite set of zero-forms with the corresponding gauge-invariant objects:

$$\begin{aligned} \mathcal{C}^{\alpha(2k+1)} &= d\psi^{\alpha(2k+1)} - \Psi^{\alpha(2k+1)} + \frac{d_k}{(2k+1)} e^{\alpha}{}_{\beta}\psi^{\alpha(2k)\beta} + c_{k+1}e_{\beta(2)}\psi^{\alpha(2k+1)\beta(2)} \\ &+ \frac{c_k}{k(2k+1)} e^{\alpha(2)}\psi^{\alpha(2k-1)} \end{aligned} \tag{29}$$

where

$$\delta\psi^{\alpha(2k+1)} = \zeta^{\alpha(2k+1)}$$

Now, we have an infinite set of one-form and zero-form fields and their gauge-invariant two and one forms. This allows us to rewrite the Lagrangian in the explicitly gauge-invariant form:

$$\mathcal{L} = -\frac{i}{2}\sum_{k=0}^{\infty}(-1)^{k+1}\mathcal{F}_{\alpha(2k+1)}\mathcal{C}^{\alpha(2k+1)} \tag{30}$$

As in the bosonic case, each term is separately gauge-invariant while the specific values of all coefficients are determined by the extra field decoupling condition:

$$\frac{\delta\mathcal{L}}{\delta\psi^{\alpha(2k+1)}} = 0, \qquad 1 \le k \le \infty$$

2.3. *Infinite Spin Supermultiplet*

It is interesting (see e.g., [1]) that, similarly to the usual massless and massive fields, such massless infinite spin fields can also form supermultiplets. In $d = 3$, the minimal supermultiplets contain just one bosonic and one fermionic field. Due to the tight relation with gauge-invariant formulation for the massive higher spin fields and supermultiplets, here we will heavily use the results of our recent paper [20]. The main difference (besides the infinite set of components) is the essentially different expressions for the coefficients a_k and c_k.

The general strategy will be to find the explicit form of the supertransformations for all fields such that all gauge-invariant two and one forms transform covariantly and to check the invariance of the Lagrangian. Let us begin with the bosonic fields. For the general case $k \ge 2$, we will use the following ansatz:

$$\begin{aligned} \delta\Omega^{\alpha(2k)} &= i\rho_k\Psi^{\alpha(2k-1)}\zeta^{\alpha} + i\sigma_k\Psi^{\alpha(2k)\beta}\zeta_{\beta} \\ \delta\Phi^{\alpha(2k)} &= i\alpha_k\Psi^{\alpha(2k-1)}\zeta^{\alpha} + i\beta_k\Psi^{\alpha(2k)\beta}\zeta_{\beta} \end{aligned} \tag{31}$$

and require that the corresponding two-form transform covariantly:

$$\begin{aligned} \delta\mathcal{R}^{\alpha(2k)} &= i\rho_k\mathcal{F}^{\alpha(2k-1)}\zeta^{\alpha} + i\sigma_k\mathcal{F}^{\alpha(2k)\beta}\zeta_{\beta} \\ \delta\mathcal{T}^{\alpha(2k)} &= i\alpha_k\mathcal{F}^{\alpha(2k-1)}\zeta^{\alpha} + i\beta_k\mathcal{F}^{\alpha(2k)\beta}\zeta_{\beta} \end{aligned} \tag{32}$$

First of all, this gives us an important relation

$$c_0{}^2 = 6a_0{}^2 \tag{33}$$

Recall that the parameters a_0 and c_0 are the main dimension-full parameters that determine the whole construction for the bosonic and fermionic fields. So this relation plays the same role as the requirement that masses of bosonic and fermionic fields in the supermultiplet must be equal. Further, we obtain explicit expressions for all parameters

$$\alpha_k{}^2 = k\hat{\alpha}^2, \qquad \beta_k{}^2 = \frac{(k+1)}{2k(2k+1)}\hat{\alpha}^2$$

$$\sigma_k{}^2 = \frac{3a_0{}^2}{4k(k+1)^2}\hat{\alpha}^2, \qquad \rho_k{}^2 = \frac{3a_0{}^2}{8k^3(k+1)(2k+1)}\hat{\alpha}^2$$

where $\hat{\alpha}$ is an arbitrary parameter that can be fixed by the normalization of the superalgebra.

For the three bosonic components that require separate consideration, we obtain:

$$\begin{aligned} \delta\Omega^{\alpha(2)} &= i\rho_1\Psi^{\alpha}\zeta^{\alpha} + i\sigma_1\Psi^{\alpha(2)\beta}\zeta_\beta - \frac{i\sqrt{3}a_0{}^2}{4}\hat{\alpha}e^{\alpha(2)}\psi^\beta\zeta_\beta \\ \delta A &= \frac{i\hat{\alpha}}{2}\Psi^\alpha\zeta_\alpha + \frac{i\sqrt{3}a_0}{2}\hat{\alpha}\psi_\alpha e^{\alpha\beta}\zeta_\beta, \qquad \delta\varphi = -i\sqrt{3}a_0\hat{\alpha}\psi^\alpha\zeta_\alpha \end{aligned} \tag{34}$$

At last, the supertransformations for the zero-forms look like:

$$\begin{aligned} \delta B^{\alpha(2k)} &= i\sigma_k\psi^{\alpha(2k)\beta}\zeta_\beta + i\rho_k\psi^{\alpha(2k-1)}\zeta^\alpha \\ \delta\pi^{\alpha(2k)} &= i\beta_k\psi^{\alpha(2k)\beta}\zeta_\beta + i\alpha_k\psi^{\alpha(2k-1)}\zeta^\alpha \end{aligned} \tag{35}$$

where all coefficients α_k, β_k, ρ_k and σ_k are the same as above.

Now, let us turn to the fermionic components. For the general case $k \geq 1$, we will consider the following ansatz:

$$\begin{aligned} \delta\Psi^{\alpha(2k+1)} &= \frac{\alpha_k}{(2k+1)}\Omega^{\alpha(2k)}\zeta^\alpha + 2(k+1)\beta_{k+1}\Omega^{\alpha(2k+1)\beta}\zeta_\beta \\ &+\gamma_k\Phi^{\alpha(2k)}\zeta^\alpha + \delta_k\Phi^{\alpha(2k+1)\beta}\zeta_\beta \end{aligned} \tag{36}$$

Then, the requirement that the corresponding two-forms transform covariantly:

$$\begin{aligned} \delta\mathcal{F}^{\alpha(2k+1)} &= \frac{\alpha_k}{(2k+1)}\mathcal{R}^{\alpha(2k)}\zeta^\alpha + 2(k+1)\beta_{k+1}\mathcal{R}^{\alpha(2k+1)\beta}\zeta_\beta \\ &+\gamma_k\mathcal{T}^{\alpha(2k)}\zeta^\alpha + \delta_k\mathcal{T}^{\alpha(2k+1)\beta}\zeta_\beta \end{aligned} \tag{37}$$

gives us the same relation on the parameters a_0 and c_0 as before and also gives:

$$\gamma_k{}^2 = \frac{3a_0{}^2}{4k(k+1)^2(2k+1)^2}\hat{\alpha}^2$$

$$\delta_k{}^2 = \frac{3a_0{}^2}{2(k+1)(k+2)(2k+3)}\hat{\alpha}^2$$

Again, there are a couple of components that need to be considered separately:

$$\begin{aligned} \delta\Psi^\alpha &= 2\beta_1\Omega^{\alpha\beta}\zeta_\beta + \delta_0\Phi^{\alpha\beta}\zeta_\beta + a_0\hat{\alpha}e_{\beta(2)}B^{\beta(2)}\zeta^\alpha + \sqrt{3}a_0\hat{\alpha}A\zeta^\alpha - \frac{\sqrt{3}a_0}{2}\hat{\alpha}\varphi e^\alpha{}_\beta\zeta^\beta \\ \delta\psi^\alpha &= \frac{2\sqrt{3}}{3}\hat{\alpha}B^{\alpha\beta}\zeta_\beta + \frac{a_0}{2}\hat{\alpha}\pi^{\alpha\beta}\zeta_\beta + \frac{\hat{\alpha}}{2}\varphi\zeta^\alpha \end{aligned} \tag{38}$$

At last, for the Stueckelberg zero-forms, we obtain:

$$\begin{aligned}\delta\psi^{\alpha(2k+1)} &= \frac{\alpha_k}{(2k+1)}B^{\alpha(2k)}\zeta^{\alpha} + 2(k+1)\beta_{k+1}B^{\alpha(2k+1)\beta}\zeta_{\beta} \\ &+\gamma_k\pi^{\alpha(2k)}\zeta^{\alpha} + \delta_k\pi^{\alpha(2k+1)\beta}\zeta_{\beta}\end{aligned} \tag{39}$$

where all parameters α_k, β_k, γ_k and δ_k are the same as before.

We have explicitly checked that the sum of the bosonic and fermionic Lagrangians is invariant under these supertransformations up to the terms proportional to the auxiliary fields $B^{\alpha(2)}$ and $\pi^{\alpha(2)}$ equations in the same way as in the case of massive higher spin supermultiplets [20].

3. Infinite Spin Fields in $d = 4$

Similarly to the three-dimensional case in $d = 4$, there exist just one bosonic and one fermionic infinite spin representation corresponding to the completely symmetric (spin-)tensors. Metric-like gauge-invariant Lagrangian formulation (valid also in $d > 4$) has been constructed recently [9,10]. Frame-like Lagrangian formulation can be straightforwardly obtained from the frame-like gauge-invariant formalism for the massive completely symmetric (spin-)tensors developed in [21]. These results will be presented elsewhere.

The complete set of the gauge-invariant objects for the massive bosonic higher spin fields in $d \geq 4$ has been constructed in [22]. It requires the following three sets of fields:

$$\Phi_{\mu}{}^{a(k),b(l)}, \quad S^{a(k),b(l)} \quad 0 \leq k \leq s-1, \quad 0 \leq l \leq k$$

$$W^{a(k),b(l)} \quad k \geq s, \quad 0 \leq l \leq s-1$$

where notation $\Phi_{\mu}{}^{a(k),b(l)}$ means that local indices correspond to the Young tableau with two rows. Thus, we have two finite sets of gauge one-forms and Stueckelberg zero-forms as well as an infinite number of gauge-invariant zero-forms. As in the three-dimensional case, one can try to consider the limit where spin goes to infinity and mass goes to zero, but in $d > 3$ it appears to be a rather involved task. As for the analogous formulation for the massive fermionic higher spin fields, to the best of our knowledge, it still remains to be elaborated.

As is quite well known, in $d = 4$, there exist two types of massive higher spin $N = 1$ supermultiplets corresponding to the integer or half-integer superspins:

$$\begin{pmatrix} & s+\frac{1}{2} & \\ s & & s' \\ & s-\frac{1}{2} & \end{pmatrix} \quad \begin{pmatrix} & s+1 & \\ s+\frac{1}{2} & & s+\frac{1}{2} \\ & s & \end{pmatrix}$$

Their explicit Lagrangian description was constructed in [23] using gauge-invariant description for massive bosonic and fermionic higher spin fields. The main idea was that the massive supermultiplet can be constructed out of the appropriately chosen set of massless supermultiplets. The decomposition of these two massive supermultiplets into the massless one is as follows:

$$\begin{pmatrix} & \Phi_{s+\frac{1}{2}} & \\ A_s & & B_s \\ & \Psi_{s-\frac{1}{2}} & \end{pmatrix} \Rightarrow \sum_{k=1}^{s} \begin{pmatrix} & \Phi_{k+\frac{1}{2}} & \\ A_k & & B_k \\ & \Psi_{k-\frac{1}{2}} & \end{pmatrix} \oplus \begin{pmatrix} \Phi_{\frac{1}{2}} \\ z \end{pmatrix}$$

$$\begin{pmatrix} & A_{s+1} & \\ \Phi_{s+\frac{1}{2}} & & \Psi_{s+\frac{1}{2}} \\ & B_s & \end{pmatrix} \Rightarrow \begin{pmatrix} A_{s+1} \\ \Psi_{s+\frac{1}{2}} \end{pmatrix} \oplus \sum_{k=1}^{s} \begin{pmatrix} & \Phi_{k+\frac{1}{2}} & \\ A_k & & B_k \\ & \Psi_{k-\frac{1}{2}} & \end{pmatrix} \oplus \begin{pmatrix} \Phi_{\frac{1}{2}} \\ z \end{pmatrix}$$

It was crucial for the whole construction that each pair of bosonic fields with equal spins must have opposite parities and one has to consider a kind of duality mixing between these fields. Moreover, such mixing arises already at the massless supermultiplets level so that even in the massless infinite spin limit these pairs do not decouple and we still have two infinite spin bosonic and two infinite spin fermionic components. It is still possible that by abandoning parity one can construct the supermultiplet containing just one bosonic and one fermionic field but it remains to be checked.

The mixing angles for the bosonic components take rather different values for the two types of supermultiplets but as can be seen from their explicit expressions in [23], in the infinite spin limit, they all become equal. At the same time, the main structural difference between them—the presence of the left most multiplet $(A_{s+1}, \Phi_{s+1/2})$—in the infinite spin limit disappears, so both types of massive supermultiplets produce the same result (up to some field re-definitions).

4. Infinite Spin Fields in $d \geq 5$

Contrary to the three- and four-dimensional cases in $d \geq 5$, there exists an infinite number of such infinite spin representations. Let us briefly reiterate how their classification arises [1]. For the massless fields, we have ${p_\mu}^2 = 0$ and by the Lorentz transformations one can always bring this vector to the canonical form $p_\mu = (1, 0, \ldots, 0, 1)$. This leads to the so-called little group (i.e., group of transformations leaving this vector intact) that, besides the group $SO(d-2)$, contains pseudo translations T_i, $i = 1, 2, \ldots, d-2$ that are specific combinations of spatial rotations and Lorentz boosts. Usual finite helicity massless representations correspond to the case where all $T_i = 0$ while to construct infinite spin representations one can follow the same root as for the Poincare group itself. Namely, one can consider eigen vectors for these pseudo translations $T_i|\xi_i >= \xi_i|\xi_i >$, ${\xi_i}^2$ being invariant. By using $SO(d-2)$ transformations, one can always bring such a vector to the form $(1, 0, \ldots, 0)$ and this, in turn, leads to the so-called short little group $SO(d-3)$, leaving this vector intact. Thus, infinite spin representations are determined by the corresponding representations of this short little group.

Now, it is clear that for the $d = 3$ and $d = 4$, this short little group is trivial; that is why we have just one bosonic and one fermionic representation while in $d \geq 5$ there exists an infinite number of them. For example, in $d = 5$ and $d = 6$, such representations can be labeled by the parameter l taking integer $l = 0, 1, 2, \ldots$ or half integer $l = \frac{1}{2}, \frac{3}{2}, \ldots$ values for the bosonic and fermionic cases correspondingly. Lagrangian formulation for such representations can be obtained from the frame-like gauge-invariant formulation for the massive mixed symmetry bosonic and fermionic fields corresponding to the Young tableau $Y(k, l)$ with two rows developed in [24–26]. Namely, one has to consider a limit where mass goes to zero, k goes to infinity while l remains fixed. This construction will be presented in the forthcoming publication, so here let us just illustrate how the spectrum of such representations appears (by the spectrum, we mean a collection of usual massless fields that we have to combine to obtain an infinite spin one).

The completely symmetric case considered before corresponds to the $l = 0$ and has the following spectrum (dot stands for the scalar field):

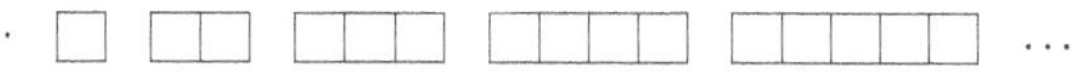

For the first non-trivial case $l = 1$, we will have two infinite chains of components:

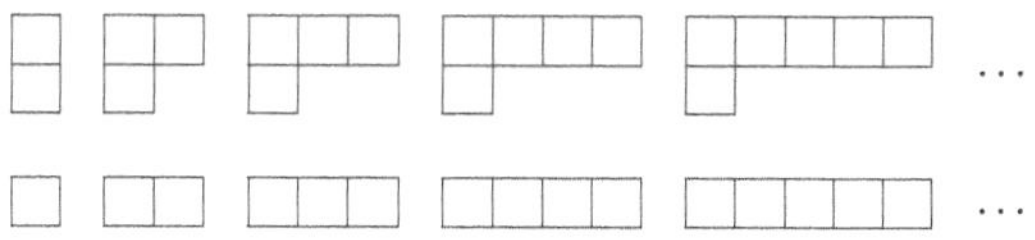

The first line begins with the anti-symmetric second rank tensor, then it contains a hook and the whole set of long hooks, while in the second line we again have completely symmetric tensors starting with the vector field this time.

Let us give here one more concrete example for $l = 3$:

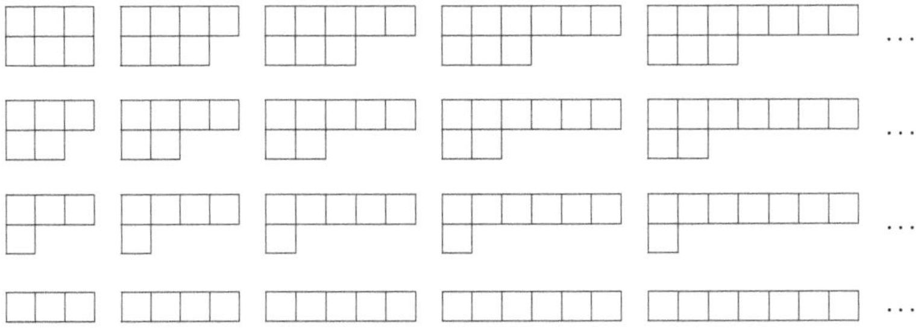

Hopefully, the general pattern is clear now. In general, in the upper left corner, we have a rectangular diagram with length l. Moving to the right, we add one box to the first row, while moving down we cut one box from the second row until we end again with the completely symmetric tensors in the bottom line.

5. Conclusions

Thus, we have seen that the same frame-like gauge-invariant formalism that has been developed for the description of massive higher spin fields can be successfully applied to the massless infinite spin case as well providing an explicit realization for the general idea that massless infinite spin representations can be obtained as an appropriate limit from the massive ones. As we have already noted, the presence of the dimensionful parameter gives hope that it may be possible to consider interactions for such fields directly in the flat space without any need to go to the anti de Sitter space. A close relationship between the frame-like gauge-invariant description for the massive higher spin fields and massless infinite spin fields means that we can try to use the same technique for the construction of possible interactions, as in the massive case. At the same time, it means that we must be ready to face the same technical difficulties as we have seen in our attempts to work with the massive high spin fields.

Acknowledgments: The author is grateful to the Joseph L. Buchbinder and Timofey V. Snegirev for collaboration. The author is also grateful to the organizers of the "Workshop on higher spin gauge theories", 26–28 April 2017, UMONS, Mons, Belgium for the kind hospitality during the workshop.

Conflicts of Interest: The author declares no conflict of interest.

References

1. Brink, L.; Khan, A.M.; Ramond, P.; Xiong, X. Continuous Spin Representations of the Poincare and Super-Poincare Groups. *J. Math. Phys.* **2002**, *43*, 6279–6295.
2. Bekaert, X.; Boulanger, N. The unitary representations of the Poincare group in any spacetime dimension. *arXiv* **2006**, arXiv:hep-th/0611263. [arXiv:hep-th/0611263].
3. Bekaert, X.; Mourad, J. The continuous spin limit of higher spin field equations. *J. High Energy Phys.* **2006**, *2006*, 115.
4. Bengtsson, A.K.H. BRST Theory for Continuous Spin. *J. High Energy Phys.* **2013**, *2013*, 108.
5. Schuster, P.; Toro, N. Continuous-spin particle field theory with helicity correspondence. *Phys. Rev. D* **2015**, *91*, 025023.
6. Schuster, P.; Toro, N. A New Class of Particle in 2 + 1 Dimensions. *Phys. Lett. B* **2015**, *743*, 224–227
7. Rivelles, V.O. Gauge Theory Formulations for Continuous and Higher Spin Fields. *Phys. Rev. D* **2015**, *91*, 125035.

8. Bekaert, X.; Najafizadeh, M.; Setare, M.R. A gauge field theory of fermionic Continuous-Spin Particles. *Phys. Lett. B* **2016**, *760*, 320–323.
9. Metsaev, R. Continuous spin gauge field in (A)dS space. *Phys. Lett. B* **2017**, *767*, 458–464.
10. Metsaev, R. Fermionic continuous spin gauge field in (A)dS space. *arXiv* **2017**, arXiv:1703.05780. [arXiv:1703.05780]
11. Zinoviev, Y.M. On Massive High Spin Particles in (A)dS. *arXiv* **2001**, arXiv:hep-th/0108192. [arXiv:hep-th/0108192]
12. Metsaev, R.R. Gauge invariant formulation of massive totally symmetric fermionic fields in (A)dS space. *Phys. Lett. B* **2006**, *643*, 205–212.
13. Buchbinder, I.L.; Snegirev, T.V.; Zinoviev, Y.M. Gauge invariant Lagrangian formulation of massive higher spin fields in (A)dS_3 space. *Phys. Lett. B* **2012**, *716*, 243–248.
14. Buchbinder, I.L.; Snegirev, T.V.; Zinoviev, Y.M. Frame-like gauge invariant Lagrangian formulation of massive fermionic higher spin fields in AdS_3 space. *Phys. Lett. B* **2014**, *738*, 258–262.
15. Boulanger, N.; Ponomarev, D.; Sezgin, E.; Sundell, P. New Unfolded Higher Spin Systems in AdS_3. *Class. Quantum Gravity* **2015**, *32*, 155002.
16. Zinoviev, Y.M. Massive higher spins in d = 3 unfolded. *J. Phys. A Math. Theor.* **2016**, *49*, 095401.
17. Zinoviev, Y.M. Towards the Fradkin-Vasiliev formalism in three dimensions. *Nucl. Phys. B* **2016**, *910*, 550–567.
18. Buchbinder, I.L.; Snegirev, T.V.; Zinoviev, Y.M. Lagrangian formulation of the massive higher spin supermultiplets in three dimensional space-time. *J. High Energy Phys.* **2015**, *2015*, 148.
19. Buchbinder, I.L.; Snegirev, T.V.; Zinoviev, Y.M. Unfolded equations for massive higher spin supermultiplets in AdS_3. *J. High Energy Phys.* **2016**, *2016*, 075.
20. Buchbinder, I.L.; Snegirev, T.V.; Zinoviev, Y.M. Lagrangian description of massive higher spin supermultiplets in AdS_3 space. *arXiv* **2017**, arXiv:1705.06163. [arXiv:1705.06163]
21. Zinoviev, Y.M. Frame-like gauge invariant formulation for massive high spin particles. *Nucl. Phys. B* **2009**, *808*, 185–204.
22. Ponomarev, D.S.; Vasiliev, M.A. Frame-Like Action and Unfolded Formulation for Massive Higher-Spin Fields. *Nucl. Phys. B* **2010**, *839*, 466–498.
23. Zinoviev, Y.M. Massive N = 1 supermultiplets with arbitrary superspins. *Nucl. Phys. B* **2007**, *785*, 98–114.
24. Zinoviev, Y.M. Towards frame-like gauge invariant formulation for massive mixed symmetry bosonic fields. *Nucl. Phys. B* **2009**, *812*, 46–63.
25. Zinoviev, Y.M. Frame-like gauge invariant formulation for mixed symmetry fermionic fields. *Nucl. Phys. B* **2009**, *821*, 21–47.
26. Zinoviev, Y.M. Towards frame-like gauge invariant formulation for massive mixed symmetry bosonic fields. II. General Young tableau with two rows. *Nucl. Phys. B* **2010**, *826*, 490–510.

MDPI
St. Alban-Anlage 66
4052 Basel
Switzerland
Tel. +41 61 683 77 34
Fax +41 61 302 89 18
www.mdpi.com

Universe Editorial Office
E-mail: universe@mdpi.com
www.mdpi.com/journal/universe

t.com/pod-product-compliance
LC

26
2B/5348

www.ingramcon
Lightning Sourc
Chambersburg P
CBHW0517032
41597CB00